Handbook
of
Biomedical
Instrumentation

Second Edition

Handbook
of
Biomedical
Instrumentation

Second Edition

R S Khandpur

Director General
Pushpa Gujral Science City, Kapurthala, Punjab

Formerly,
Director
Centre for Electronics Design and Technology of India (CEDTI)
Mohali (Chandigarh), Punjab

Director General
CEDTI, New Delhi

Head,
Medical Electronics Instruments Division
Central Scientific Instruments Organization
Chandigarh

Tata McGraw-Hill Publishing Company Limited
NEW DELHI

McGraw-Hill Offices

New Delhi New York St Louis San Francisco Auckland Bogotá Caracas
Kuala Lumpur Lisbon London Madrid Mexico City Milan Montreal
San Juan Santiago Singapore Sydney Tokyo Toronto

Information contained in this work has been obtained by Tata McGraw-Hill Publishing Company Limited, from sources believed to be reliable. However, neither Tata McGraw-Hill nor its authors guarantee the accuracy or completeness of any information published herein, and neither Tata McGraw-Hill nor its authors shall be responsible for any errors, omissions, or damages arising out of use of this information. This work is published with the understanding that Tata McGraw-Hill and its authors are supplying information but are not attempting to render engineering or other professional services. If such services are required, the assistance of an appropriate professional should be sought.

The **McGraw·Hill** Companies

Preface to the Second Edition

I am very happy to present before you the second, revised and enlarged edition of my book *Handbook of Biomedical Instrumentation*. Its revision and updation have become necessary not only because of the technological changes that have taken place in the last decade, but also because of the immense popularity of the book among professionals in the field of biomedical instrumentation, as also students and teachers in the academic institutes. I feel honoured to have assisted the teaching community in starting numerous courses on biomedical instrumentation in the engineering colleges and polytechnics across the country, which became easier, in most of the cases, due to the first edition of the book.

In the second edition, the existing material has been thoroughly revised taking into consideration the developments in technology and introduction of new and improved methods of medical diagnosis and treatment. Seven new chapters have been added including topics such as nuclear medical imaging systems covering gamma camera, PET camera and SPECT camera. The technology of lithotripsy has matured and it is not only being used for destruction of kidney stones and bile stones but also for therapeutic purposes. Description of anaesthesia machine and ventilators has been included to complete the coverage of operating room equipment. Clinical laboratory instrumentation and automated drug delivery systems are other important new chapters. A chapter on X-ray and digital radiography covers the much needed information on this vital equipment universally used in the medical facilities.

The penetration of microcontrollers and PCs in medical instrumentation has resulted in the integration of automation and built-in intelligence in medical instruments to a great extent. This has resulted in replacement of long-established recording techniques and display systems. The advantages of the PC architecture in terms of its high storage capacity of data and large screen displays have been fully exploited in clinical and research applications of biomedical instruments. Therefore, wherever it was felt necessary, reference to the use of PCs as an integral part of the medical instruments has been made in this edition.

In order to understand linkages between the life sciences and engineering techniques, it is necessary for engineers to have a fair understanding about the anatomy and physiology of the human body. A brief description of the important physiological systems, namely cardiovascular system, respiratory system and nervous system is provided in the first chapter. Special physiological systems are also described in other chapters, wherever it was felt necessary.

The new edition has been divided into three parts. Part one deals with measuring, recording and monitoring systems. Part two covers modern imaging systems whereas Part three is devoted to theraputic equipment.

The references have been thoroughly revised to include new research material from research journals from the world over. Their inclusion in the appropriate places in the text establishes the necessary link between the current status of technology vis-à-vis the field of research being persued.

When I wrote the first edition my children were young. They have now grown up, are married and have children of their own. They have been urging me to update this book. While I acknowledge their pursuation to this initiative, my heartfelt gratitude goes to my wife Mrs. Ramesh Khandpur who had to spend considerable time alone, watching TV, while I was working in my study. Often, my grand-children—Harsheen and Aashna—who are tiny-tots, would trot into my study to cheer me up with their pranks which made my task both pleasant and interesting. My thanks to all my readers who have been sending in their suggestions which have mostly been incorporated in this edition.

It is hoped that the book will enjoy the same acceptance among its readers and would prove helpful to professionals and students working in the field of biomedical instrumentation.

Chandigarh
January 31, 2003   **R S K**HANDPUR

Preface to the First Edition

During the last two decades, there has been a tremendous increase in the use of electronic equipment in the medical field for clinical and research purposes. However, it is difficult to find a book which describes the physiological basis as well as the engineering principles underlying the working of a wide variety of medical instruments. The present volume has been written to fill this gap.

The book has been designed to cater to a wide variety of readers. The users of medical instruments would find the text useful, as they would be able to appreciate the principle of operation, and the basic building blocks of the instruments they work on everyday. An attempt has been made to present the highly technical details of the instruments with descriptive and lucid explanations of the necessary information. It thus provides a useful reference for medical or paramedical persons whose knowledge of instrumentation is limited.

The field of biomedical engineering is fast developing and new departments are being established in universities, technical colleges, medical institutes and hospitals all over the world. In addition to graduate engineers involved in developing biomedical instrumentation techniques, the book will find readership in the increasing number of students taking courses in physiological measurements in technical colleges.

With the widespread use and requirements of medical electronic instruments, it is essential to have knowledgeable service and maintenance engineers. Besides having a basic knowledge of the principles of operation, it is important for them to know the details of commercial instruments from different manufacturers. A concise description of typical instruments from leading manufacturers is provided wherever deemed necessary for elucidation of the subject matter.

The book has been divided into four parts. The first part deals with recording and monitoring instruments. This part has 11 chapters.

The first chapter begins with the explanation that the human body is a source of numerous signals, highly significant for diagnosis and therapy. These signals are picked up from the surface of the body or from within. This requires electrodes of different sizes, shapes and types. Also, there are some parameters like temperature, blood flow, blood pressure, respiratory functions etc., which are to be routinely monitored. These parameters, which are basically non-electrical in nature, are converted into corresponding electrical signals by various transducers. Electrodes and transducers constitute the first building blocks of most of the diagnostic medical instruments and are, therefore, described in the first part of this book.

After picking up the signals of interest from the body, they are processed and presented in a form most convenient for interpretation. Display is generally on a picture tube for quick and visual observation or a record on graph paper. Such records facilitate a detailed study by specialists at a later convenient time. Display and recording systems, and the most commonly used biomedical recorders are covered in the subsequent three chapters.

Next is a presentation of the various types of patient monitors. The systems aid the nurses and the medical personnel to quickly gather information about the vital physiological parameters of the patient before, during and after operation, and in the intensive care ward where the patient's condition is kept under constant surveillance.

Apart from the description of conventional equipment for monitoring heart rate, blood pressure, respiration rate and temperature, a separate chapter has been included on arrhythmia monitoring instruments. This class of instruments constantly scan ECG rhythm patterns and issue alarms to events that may be premonitory or life-threatening. The chapter also includes a description of ambulatory monitoring instruments.

Foetal monitoring instrumentation is another area where considerable progress has been reported in the last few years. Instruments for foetal heart rate monitoring based on the Doppler shift have become more reliable because of better signal processing circuitry and the use of microprocessors. Intelligence is now incorporated in the cardiotocographs to provide data processing for making correlation studies of the foetal heart rate and labour activity.

Wireless telemetry permits examination of the physiological data of subjects in normal conditions and in natural surroundings without discomfort or obstruction to the person or animal under investigation. Telemetric surveillance is the most convenient method for assessing the condition of the patient during transportation within the hospital for making stress studies before discharge from the cardiac wards. The chapter on biomedical telemetry explains the techniques and instrumentation for monitoring physiological data by telemetry in a variety of situations. It also includes transmission of biomedical signals over the telephone lines for their study and analysis at a distant place.

An extensive use of computers and microprocessors is now being made in medical instruments designed to perform routine clinical measurements, particularly in those situations where data computing and processing could be considered as part of the measurement and diagnostic procedure. The use of microprocessors in various instruments and systems has been explained not only at various places in the text, but a full chapter gives a comprehensive view of computer and microcomputer applications in the medical field.

With the increasing use of monitoring and therapeutic instruments, the patient has been included as a part of an electrical circuit and thus exposed to the possibility of providing a pathway to the potentially fatal leakage currents. Such a situation particularly arises when he carries indwelling catheters. A full chapter on patient safety describes various situations requiring attention to avoid the occurrence of avoidable accidents. Precautions to be taken while designing electromedical equipment from the point of view of patient safety is also discussed.

The next part details the various measurement and analysis techniques in medicine and comprises seven chapters. The first two chapters concern the measurement of blood flow and volume.

Blood flow is one of the most important physiological parameters and is also one of the most difficult to measure. This has given rise to a variety of techniques in an effort to meet the requirements of an ideal flow metering system. Both invasive as well as non-invasive techniques have been developed. The ultrasonic Doppler technique has proved to be particularly useful in blood flow measurement. A detailed description of the modern methods of blood flow measurement including those making use of the laser Doppler technique has been given in Chapter 12. A separate chapter on cardiac output measurement details out the present state of art in this important area.

Pulmonary function testing equipment act as the additional means in automated clinical procedures and analysis techniques for carrying out a complete study of the lung function from the respiratory process. Besides the conventional pneumotachometry, several new techniques like the ultrasound spirometer and microprocessor based analysers are under development. The measurement of gases is also important for respiratory studies. Chapter 14 gives a detailed description of various instruments and systems for assessing pulmonary function.

The measurement of gases like oxygen and carbon dioxide in the blood, along with blood pH form important test parameters for studying the acid-base balance in the body. Blood gas analysers have greatly developed in the last few years. The modern microprocessor controlled instruments include automatic sample dilution and data processing. A separate chapter on blood gas analysers gives details of modern instruments and their use in clinical practice. Oximeters are covered in Chapter 16, which describes various techniques of assessing the oxygen saturation level of blood both by invasive and non-invasive techniques. A chapter on blood cell counters touches upon electronic methods of blood cell counting and microprocessor based system for making calculations important in haematology.

The third part contains four chapters on medical imaging systems. The last decade saw an unprecedented progress in this area and resulted in the evolution and development of ultrasonic, computerised tomography and NMR scanners. Ultrasound has proved a useful imaging modality because of its non-invasive character and ability to distinguish interfaces between soft tissues. Ultrasonic imaging systems are now applied to obtain images of almost the entire range of internal organs in the abdomen. The chapter on ultrasound covers extensive information on this technology and includes the physics of ultrasound, pulse echo systems including M-mode echocardiography and a variety of scanning systems and techniques. CT scanners are considered as the most significant development since the discovery of X-rays. In spite of their inherent high cost, several thousands of these are now installed in hospitals around the world. Keeping in view the impact on medical diagnostics, a detailed description of the various scanning techniques in CT systems has been given in Chapter 19. The chapter also carries information on the basic X-ray machine and image intensifiers. Thermography—the science of visualizing and interpreting the skin temperature pattern—is another technique, which stands alongside X-ray, ultrasonic and clinical examination as an aid to medical diagnostics. Keeping in view its usefulness and recognizing the non-availability of information on this topic in most of the medical electronic instrumentation books, a separate chapter has been included in this text.

The last part with six chapters is devoted to therapeutic instruments.

Two types of instruments are commonly employed to meet cardiac emergencies. These are the pacemakers and the defibrillators. The technology of implantable pacemakers has considerably developed in the past few years, resulting in the availability of pacemakers with life long guarantee of their activity. This has become possible due to improvements in power sources, low drain current circuits and better encapsulation techniques. The availability of programmable pacemakers has further helped to individualise the pacemaker treatment. Similarly, microprocessor based defibrillators have appeared in the market to give the possibility of more efficiently delivering the defibrillating discharge by appropriately adjusting the output on the basis of patient-electrode impedance. These two topics are covered in two separate chapters.

The use of high frequency in electro-surgical procedures is well established. There has not been very many changes in the basic design except for the availability of solid state versions with better safety provisions for the patients and operators. Application of lasers for bloodless surgery and for coagulation of fine structures in the small and sensitive organs of the body is now routinely practiced in many centres in the world. Separate chapters cover the high frequency electro-surgical machines and laser applications in medicine respectively.

The maintenance of renal function in acute and chronic renal failure through dialysis is a routinely practiced technique. Haemodialysis machines for use in hospitals contain a variety of monitoring and control facilities, and some of these functions have also been computerised. There have also been attempts to bring out a wearable artificial kidney so that patients suffering from this disease could enjoy a near normal life during their stay away from the dialysis centre. The chapter on haemodialysis machines includes a description of the well established machines with an indication of the efforts on the development of portable systems.

Physiotherapy instruments like the short-wave diathermy machine, microwave diathermy machine and ultrasonic therapy units have acquired an established role in the hospitals. Similarly, the technique of electro-diagnosis and electrotherapy are now routinely employed in the physiotherapy departments. An extension of this technique has been the development of small stimulators for a variety of applications like pain relief, control of micturition, epilepsy, etc. The information on these techniques is usually not available in the books on the subject. The inclusion of a full chapter on these techniques fulfils this gap.

A large number of references have been included at the end. This is to help the more interested readers to conveniently look for extra material on the subject of their interest.

I am thankful to the Director, Central Scientific Instruments Organization, Chandigarh for kind permission to publish this book. I am also grateful to various manufacturers of medical electronic instruments who supplied valuable information on the products along with some interesting photographs.

Finally, I am extremely grateful to my wife Ramesh Khandpur who helped me in correcting and comparing the typed script. I also acknowledge the assistance provided to me in this work by my children Vimal, Gurdial and Popila. All of them bore the brunt of uncalled for neglect over a long period during the preparation of the manuscript.

R S KHANDPUR

Contents

Part One

Measuring, Recording and Monitoring Instruments

Part Two

Modern Imaging Systems

Part Three

Therapeutic Equipment

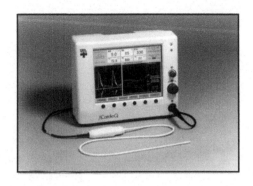

⇒ PART ONE : MEASURING, RECORDING AND MONITORING INSTRUMENTS

Fundamentals of Medical Instrumentation

During the last quarter of the century, there has been a tremendous increase in the use of electrical and electronic equipment in the medical field for clinical and research purposes. In a medical instrumentation system, the main function is to measure or determine the presence of some physical quantity that may be useful for diagnostic purposes. Therefore, many types of instrumentation systems are used in hospitals and physician's clinics.

▶ 1.1 ANATOMY AND PHYSIOLOGY

A knowledge of the structure of the living body and its function is essential for understanding the functioning of most of the medical instruments. The science of structure of the body is known as **"Anatomy"** and that of its function, **"Physiology"**.

Anatomy is classified according to the following basis:

Gross anatomy deals with the study of the structure of the organs as seen by the naked eye on dissection. It describes the shape, size, components and appearance of the organ under study.

Topographical anatomy deals with the position of the organs in relation to each other, as they are seen in sections through the body in different planes.

Microscopic anatomy (Histology) is the study of the minute structure of the organs by means of microscopy.

Cytology is a special field of histology in which the structure, function and development of the cells are studied.

Similarly, **physiology,** which relates to the normal function of the organs of the body, can be classified in different ways. For example:

Cell physiology is the study of the functions of the cells.

Pathophysiology relates to the pathological (study or symptoms of disease) functions of the organs.

In addition, classification into various sub-areas dealing with different organs can be made. For example:

Circulatory physiology is the study of blood circulation relating to functioning of the heart.

Respiratory physiology deals with the functioning of breathing organs.

▶ 1.2 PHYSIOLOGICAL SYSTEMS OF THE BODY

Human body is a complex engineering marvel, which contains various types of systems such as electrical, mechanical, hydraulic, pneumatic, chemical and thermal etc. These systems communicate internally with each other and also with an external environment. By means of a multi-level control system and communications network, the individual systems enable the human body to perform useful tasks, sustain life and reproduce itself.

Although, the coverage of detailed information on the physiological systems is outside the scope of this book, nevertheless a brief description of the major sub-systems of the body is given below to illustrate the engineering aspects of the human body.

1.2.1 The Cardiovascular System

The cardiovascular system is a complex closed hydraulic system, which performs the essential service of transportation of oxygen, carbon dioxide, numerous chemical compounds and the blood cells. Structurally, the heart is divided into right and left parts. Each part has two chambers called atrium and ventricle. The heart has four valves (Fig. 1.1):

- The Tricuspid valve or right atrio-ventricular valve—between right atrium and ventricle. It consists of three flaps or cusps. It prevents backward flow of blood from right ventricle to right atrium.
- Bicuspid Mitral or left atrio-ventricular valve—between left atrium and left ventricle. The valve has two flaps or cusps. It prevents backward flow of blood from left ventricle to atrium.
- Pulmonary valve—at the right ventricle. It consists of three half moon shaped cusps. This does not allow blood to come back to the right ventricle.
- Aortic valve—between left ventricle and aorta. Its construction is like pulmonary valve. This valve prevents the return of blood back to the left ventricle from aorta.

The heart wall consists of three layers: (i) The *pericardium,* which is the outer layer of the heart. It keeps the outer surface moist and prevents friction as the heart beats. (ii) The *myocardium* is the middle layer of the heart. It is the main muscle of the heart, which is made up of short cylindrical fibres. This muscle is automatic in action, contracting and relaxing rythmically throughout life. (iii) The *endocardium* is the inner layer of the heart. It provides smooth lining for the blood to flow.

The blood is carried to the various parts of the body through blood vessels, which are hollow tubes. There are three types of blood vessels. (i) *Arteries*—are thick walled and they carry the oxygenated blood away from the heart. (ii) *Veins*—are thin walled and carry de-oxygenated blood

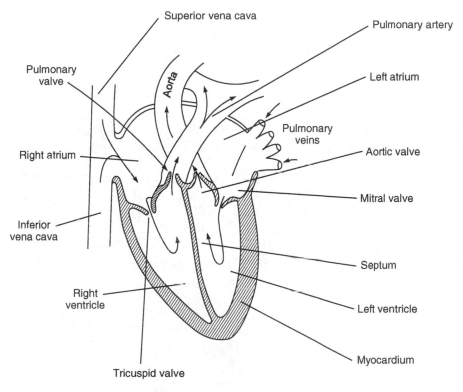

> **Fig. 1.1** *Structure of the heart*

towards the heart. (iii) *Capillaries*—are the smallest and the last level of blood vessels. They are so small that the blood cells, which make blood, actually flow one at a time through them. There are estimated to be over 800,000 km of capillaries in human being, which include all the arteries and veins, which carry blood.

From an engineering point of view, the heart which drives the blood through the blood vessels of the circulatory system (Fig. 1.2) consists of four chamber muscular pump that beats about 72 times per minute (on an average for a normal adult), sending blood through every part of the body. The pump acts as two synchronized but functionally isolated two stage pumps. The first stage of each pump (the atrium) collects blood from the hydraulic system and pumps it into the second stage (the ventricle). In this process, the heart pumps the blood through the *pulmonary circulation* to the lungs and through the *systemic circulation* to the other parts of the body.

In the pulmonary circulation, the venous (de-oxygenated) blood flows from the right ventricle, through the pulmonary artery, to the lungs, where it is oxygenated and gives off carbon dioxide. The arterial (oxygenated) blood then flows through the pulmonary veins to the left atrium.

In systemic circulation, the blood is forced through blood vessels, which are somewhat elastic. The blood flows from the left atrium to the left ventricle and is pumped through the aorta and its branches, the arteries, out into the body. Through the arterioles (small arteries), the blood is

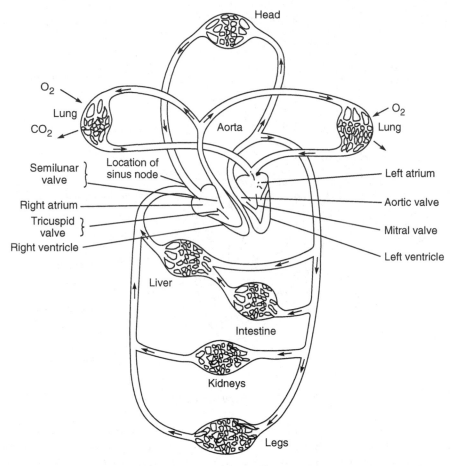

➤ **Fig.** 1.2   *The Circulatory system*

distributed to the capillaries in the tissues, where it gives up its oxygen and chemical compounds, takes up carbon dioxide and products of combustion.

The blood returns to the heart along different routes from different parts of the body. It usually passes from the venous side of the capillaries directly via the venous system to either the superior vena cava or the inferior vena cava, both of which empty into the right atrium. The heart itself is supplied by two small but highly important arteries, the coronary arteries. They branch from the aorta just above the heart. If they are blocked by coronary thrombosis, myocardial infarction follows, often leading to a fatal situation.

The heart rate is partly controlled by autonomic nervous system and partly by harmone action. These control the heart pump's speed, efficiency and the fluid flow pattern through the system.

The circulatory system is the transport system of the body by which food, oxygen, water and other essentials are transported to the tissue cells and their waste products are transported away. This happens through a diffusion process in which nourishment from the blood cell diffuses through the capillary wall into interstitial fluid. Similarly, carbon dioxide and some waste products from the interstitial fluid diffuses through the capillary wall into the blood cell.

The condition of the cardiovascular system is examined by haemodynamic measurements and by recording the electrical activity of the heart muscle (electrocardiography) and listening to the heart sounds (phonocardiography). For assessing the performance of the heart as a pump, measurement of the cardiac output (amount of blood pumped by the heart per unit time), blood pressure, blood flow rate and blood volume are made at various locations throughout the circulatory system.

1.2.2 The Respiratory System

The respiratory system in the human body (Fig. 1.3) is a pneumatic system in which an air pump (diaphragm) alternately creates negative and positive pressures in a sealed chamber (thoracic cavity) and causes air to be sucked into and forced out of a pair of elastic bags (lungs). The lungs are connected to the outside environment through a passage way comprising nasal cavities,

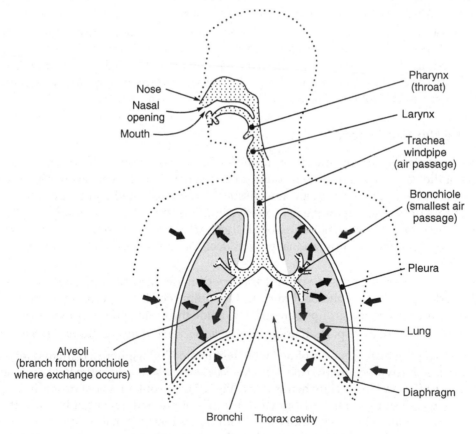

➤ Fig. 1.3 *The Respiratory system*

pharynx, larynx, trachea, bronchi and bronchioles. The passage way bifurcates to carry air into each of the lungs wherein it again subdivides several times to carry air into and out of each of the many tiny air spaces (alveoli) within the lungs. In the tiny air spaces of the lungs is a membrane interface with the hydraulic system of the body through which certain gases can defuse. Oxygen is taken into the blood from the incoming air and carbon dioxide is transferred from the blood to the air under the control of the pneumatic pump. Thus, the blood circulation forms the link in the supply of oxygen to the tissues and in the removal of gaseous waste products of metabolism. The movement of gases between blood and the alveolar air is basically due to constant molecular movement or diffusion from points of higher pressure to points of lower pressure.

An automatic respiratory control centre in the brain maintains heart pump operation at a speed that is adequate to supply oxygen and take away carbon dioxide as required by the system. In each minute, under normal conditions, about 250 ml of oxygen are taken up and 250 ml of CO_2 are given out by the body and these are the amounts of the two gases, which enter and leave the blood in the lungs. Similar exchanges occur in reverse in the tissues where oxygen is given up and CO_2 is removed. The exact amount of CO_2 expired depends upon the metabolism, the acid-base balance and the pattern of respiration. The exchange of gases takes place in the alveoli and can be achieved by the normal 15-20 breaths/min, each one involving about 500 ml of air.

The respiratory system variables which are important for assessing the proper functioning of the system are respiratory rate, respiratory air flow, respiratory volume and concentration of CO_2 in the expired air. The system also requires measurements to be made of certain volumes and capacities such as the tidal volume, vital capacity, residual volume, inspiratory reserve volume and expiratory reserve volume. The details of these are given in Chapter 13.

1.2.3 The Nervous System

The nervous system is the control and communication network for the body which coordinates the functions of the various organs. Rapid communication between the various parts, the effective, integrated activity of different organs and tissues and coordinated contraction of muscle are almost entirely dependent upon the nervous system. It is thus, the most highly developed and complex system in the body. The centre of all these activities is the brain (central information processor) with memory, computational power, decision making capability and a host of input output channels.

The nervous system consists of a central and a peripheral part. The central nervous system is (Fig. 1.4) made up of the encephalon (brain) and the spinal cord. The peripheral nervous system comprises all the nerves and groups of neurons outside the brain and the spinal cord.

The brain consists of three parts, namely, the *cerebrum*, *cerebellum* and the *brain stem.*

Cerebrum: The cerebrum consists of two well demarcated hemispheres, right and left and each hemisphere is sub-divided into two lobes: *frontal lobe and temporal lobe* in the left hemisphere and *parietal and occipital lobes* in the right hemisphere (Fig. 1.5). The outer layer of the brain is called the cerebral cortex. All sensory inputs from various parts of the body eventually reach the cortex, where certain regions relate specifically to certain modalities of sensory information. Various areas are responsible for hearing, sight, touch and control of the voluntary muscles of the body.

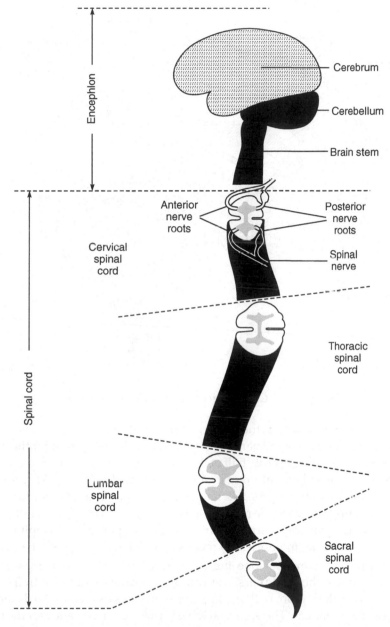

> **Fig.** 1.4 *Central nervous system, human brain and spinal cord*

The cerebral cortex is also the centre of intellectual functions. The frontal lobes are essential for intelligence, constructive imagination and thought. Here, large quantities of information can be stored temporarily and correlated, thus making a basis for higher mental functions.

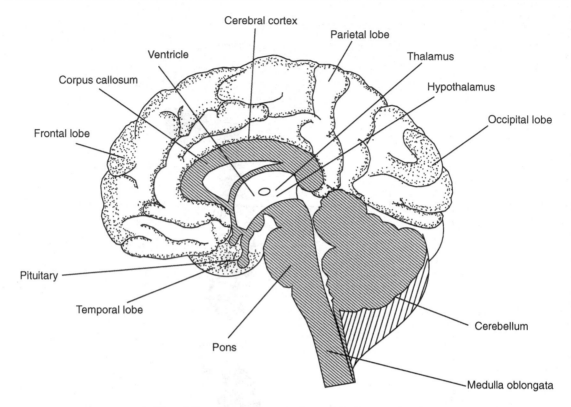

> **Fig. 1.5**   *Cut-away section of the human brain*

Each point in the motor centre in the cerebral cortex (Fig. 1.6) corresponds to a certain body movement. In the anterior part of the parietal lobe lies the terminal station for the nerve pathways conducting sensation from the opposite half of the body. The sensory centre contains counterparts of the various areas of the body in different locations of the cortex. The sensory inputs come from the legs, the torso, arms, hands, fingers, face and throat etc. The amount of surface allotted to each part of the body is in proportion to the number of sensory nerves it contains rather than its actual physical size. The visual pathways terminate in the posterior part of the occipital lobe. The rest of the occipital lobes store visual memories, by means of which we interpret what we see.

On the upper side of the temporal lobe, the acoustic pathways terminate making it as a hearing centre. This is located just above the ears. Neurons responding to different frequencies of sound input are spread across the region, with the higher frequencies located towards the front and low frequencies to the rear of the ear. The temporal lobes are also of importance for the storage process in the long-term memory.

Cerebellum: The cerebellum acts as a physiological microcomputer which intercepts various sensory and motor nerves to smooth out the muscle motions which could be otherwise jerky. It also consists of two hemispheres which regulate the coordination of muscular movements elicited by the cerebrum. The cerebellum also enables a person to maintain his balance.

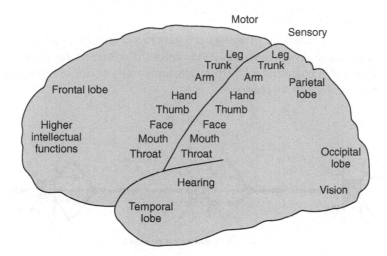

> ➤ **Fig.** 1.6   *Sites of some activity centres in the cerebral cortex*

Brain Stem: The brain stem connects the spinal cord to the centre of the brain just below the cerebral cortex. The essential parts of the brain stem are (i) *Medulla oblongata* which is the lowest section of the brain stem and contains centres for regulating the work performed by the heart, the vasomotor centres, which control blood distribution and respiratory centre which controls the ventilation of the lungs. (ii) the *pons* located just above the medulla and protruding somewhat in front of the brain stem. (iii) *midbrain* which lies in the upper part of the brain stem (iv) the *diencephalon* is located above and slightly forward of the mid brain. It has one part, the *thalamus*, which acts as a relay station for sensory pathways to the cortical sensory centre of the cerebrum. In the lower part of the diencephalon is the *hypothalamus* which has several vital centres for temperature regulation, metabolism and fluid regulation. They include the centres for appetite, thirst, sleep and sexual drive. The hypothalamus is important for subjective feelings and emotions.

Spinal Cord: The spinal cord is a downward continuation of the medulla oblongata in the brain to the level of first lumbar vertebra. It consists of a cylinder of nerve tissue about the thickness of the little finger and has a length of about 38 to 45 cms. The cord consists of white matter on the surface and gray matter inside. The white matter contain fibres running between the cord and brain only. The cord containing motor and sensory fibres is responsible for the link between the brain and the body and reflex action. In the H-shaped gray matter of the spinal cord are located the neurons that control many reflexes such as the knee reflex and the bladder-emptying reflex. The reflex action is a result of the stimulation of the motor cells by stimuli brought in by sensory nerves from the tissues.

The central nervous system consists of billions of specialized cells about half of which, called neurons, are functionally active as signal transmitters while the other half (supporting cells), maintain and nourish the neurons. The fundamental property of the neurons is the ability to transmit electrical signals, called nerve impulses, in response to changes in their environment, i.e. stimuli. The central nervous system controls the voluntary muscles of the body and is responsible for all movements and sensations.

The basic functional unit of the nervous system is the neuron. A typical neuron consists of a nucleated cell body and has several processes or branches (Fig. 1.7). The size and distribution of these branches vary greatly at different sites and in cells with different functions, but the two main kinds are: the *axone* and the *dendrite*. The dendrites normally conduct impulses toward the cell body and the axons conduct away from it.

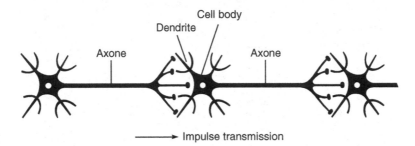

> **Fig. 1.7** *Structure of the neuron and the phenomenon of impulse transmission*

The neurons form an extremely complex network, which connects all parts of the body. While the size of the central body of the nerve cell is the same as that of other cells of the body, the overall size of the neuron structure varies from a millimetre or so in the spinal cord to over a metre in length. For example, the axones of the foot muscle originate in the lower part of the spinal cord, where the associated nerve cells are located.

The nervous system is the body's principal regulatory system and pathological processes in it often lead to serious functional disturbances. The symptoms vary greatly depending upon the part of the nervous system affected by the pathological changes. The measurements on the nervous system include recording of electroencephalogram (EEG) and muscle's electrical action potentials, electromyogram (EMG), measurement of conduction velocity in motor nerves, and recording of the peripheral nerves' action potential, electroneurogram (ENG).

1.2.4 Other Systems

There are some other important functional systems in the body, such as digestive system, excretory system, reproductive system and the biochemical system which perform vital functions required to carry out the various body functions. The coverage of all these and other systems is outside the scope of this book.

1.3 SOURCES OF BIOMEDICAL SIGNALS

Biomedical signals are those signals (phenomenon that conveys information) which are used primarily for extracting information on a biological system under investigation. The process of extracting information could be as simple as feeling the pulse of a person on the wrist or as complex as analyzing the structure of internal soft tissues by an ultrasound scanner. Biomedical signals originate from a variety of sources (Fig. 1.8) such as:

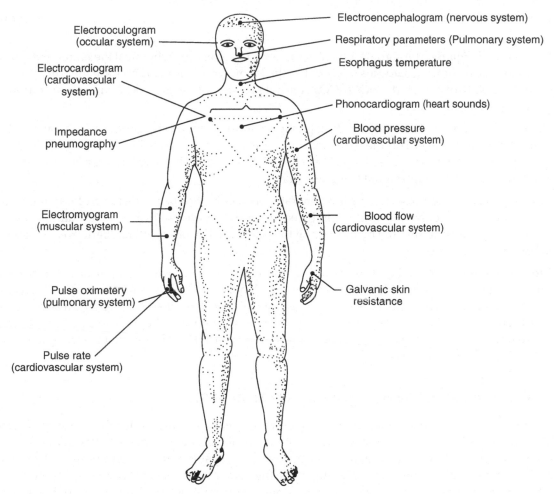

➤ **Fig.** 1.8   *Sources of biomedical signals*

Bioelectric Signals: These are unique to the biomedical systems. They are generated by nerve cells and muscle cells. Their basic source is the cell membrane potential which under certain conditions may be excited to generate an action potential. The electric field generated by the action of many cells constitutes the bio-electric signal. The most common examples of bioelectric signals are the ECG (electrocardiographic) and EEG (electroencephalographic) signals.

Bioacoustic Signals: The measurement of acoustic signals created by many biomedical pheno-mena provides information about the underlying phenomena. The examples of such signals are: flow of blood in the heart, through the heart's valves and flow of air through the upper and lower airways and in the lungs which generate typical acoustic signal.

Biomechanical Signals: These signals originate from some mechanical function of the biological system. They include all types of motion and displacement signals, pressure and flow signals etc.

The movement of the chest wall in accordance with the respiratory activity is an example of this type of signal.

Biochemical Signals: The signals which are obtained as a result of chemical measurements from the living tissue or from samples analyzed in the laboratory. The examples are measurement of partial pressure of carbon-dioxide (pCO_2), partial pressure of oxygen (pO_2) and concentration of various ions in the blood.

Biomagnetic Signals: Extremely weak magnetic fields are produced by various organs such as the brain, heart and lungs. The measurement of these signals provides information which is not available in other types of bio-signals such bio-electric signals. A typical example is that of magneto-encephalograph signal from the brain.

Bio-optical Signals: These signals are generated as result of optical functions of the biological systems, occurring either naturally or induced by the measurement process. For example, blood oxygenation may be estimated by measuring the transmitted/back scattered light from a tissue at different wavelengths.

Bio-impedance Signals: The impedance of the tissue is a source of important information concerning its composition, blood distribution and blood volume etc. The measurement of galvanic skin resistance is a typical example of this type of signal. The bio-impedance signal is also obtained by injecting sinusoidal current in the tissue and measuring the voltage drop generated by the tissue impedance. The measurement of respiration rate based on bio-impedance technique is an example of this type of signals.

▶ 1.4   BASIC MEDICAL INSTRUMENTATION SYSTEM

The primary purpose of medical instrumentation is to measure or determine the presence of some physical quantity that may some way assist the medical personnel to make better diagnosis and treatment. Accordingly, many types of instrumentation systems are presently used in hospitals and other medical facilities. The majority of the instruments are electrical or electronic systems, although mechanical systems such as ventilators or spirometers are also employed. Because of the predominantly large number of electronic systems used in medical practice, the concepts explained hereafter are mostly related to electronic medical instruments.

Certain characteristic features, which are common to most instrumentation systems, are also applicable to medical instrumentation systems. In the broadest sense, any medical instrument (Fig. 1.9) would comprise of the following four basic functional components:

Measurand: The physical quantity or condition that the instrumentation system measures is called the *measurand*. The source for the measurand is the human body which generates a variety of signals. The measurand may be on the surface of the body (electrocardiogram potential) or it may be blood pressure in the chambers of the heart.

Transducer/Sensor: A transducer is a device that converts one form of energy to another. Because of the familiar advantages of electric and electronic methods of measurement, it is the usual practice to convert into electrical quantities all non-electrical phenomenon associated with the measurand with the help of a transducer. For example: a piezo-electric crystal converts mechanical vibrations

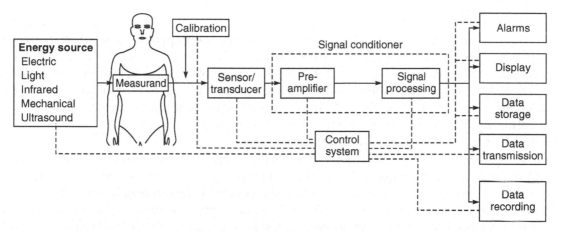

> **Fig.**1.9   *General block diagram of a medical instrumentation system*

into an electrical signal and therefore, is a transducer. The primary function of the transducer is to provide a usable output in response to the measurand which may be a specific physical quantity, property or condition. In practice, two or more transducers may be used simultaneously to make measurements of a number of physiological parameters.

Another term 'sensor' is also used in medical instrumentation systems. Basically, a sensor converts a physical measurand to an electrical signal. The sensor should be minimally invasive and interface with the living system with minimum extraction of energy.

Signal Conditioner: Converts the output of the transducer into an electrical quantity suitable for operation of the display or recording system. Signal conditioners may vary in complexity from a simple resistance network or impedance matching device to multi-stage amplifiers and other complex electronic circuitry. Signal conditioning usually include functions such as amplification, filtering (analog or digital) analog-to-digital and digital-to-analog conversion or signal transmission circuitry. They help in increasing the sensitivity of instruments by amplification of the original signal or its transduced form.

Display System: Provides a visible representation of the quantity as a displacement on a scale, or on the chart of a recorder, or on the screen of a cathode ray tube or in numerical form. Although, most of the displays are in the visual form, other forms of displays such as audible signals from alarm or foetal Doppler ultrasonic signals are also used. In addition of the above, the processed signal after signal conditioning may be passed on to:

Alarm System—with upper and lower adjustable thresholds to indicate when the measurand goes beyond preset limits.

Data Storage—to maintain the data for future reference. It may be a hard copy on a paper or on magnetic or semiconductor memories.

Data Transmission—using standard interface connections so that information obtained may be carried to other parts of an integrated system or to transmit it from one location to another.

In most of the medical instrumentation systems, some form of *calibration* is necessary at regular intervals during their operation. The calibration signal is usually applied to the sensor input or as early in the signal conditioning chain as possible.

In many measurements in the medical field, some form of stimulus or energy is given to the patient and the effect it has on the patient is measured. The stimulus may be visual in the form of flash of light or audio tone or direct electrical stimulation of some part of the nervous system. A typical example is that of recording of the evoked response with EEG machine when visual/audible stimulus is given to the subject under test.

In some situations, it is required to have automatic control of the transducer, stimulus or signal conditioning part of the system. This is achieved by using a feedback loop in which part of the output from the signal conditioning or display device is fed back to the input stage. Control and feedback may be automatic or manual. Almost all measuring and recording equipment is now controlled by microprocessors as this makes it possible to design equipment that requires minimal user intervention, calibration and set up procedure.

Measurements on the human body can be made at several levels on the functional systems and sub-systems. For example; it is easiest to make measurements on the human body as a whole due to accessible environment. Examples of measurement made on the human body are recording of electrocardiogram and measurement of temperature. The next level of measurements can be made on the major functional systems of the body such as the cardiovascular system, the pulmonary system and so on. Many of the major systems communicate with each other as well as with external environment. The functional systems can be further sub-divided into sub-systems and organs and still smaller units up to the cellular and molecular level. Measurements in the medical field are made at all these levels with specially designed instruments with appropriate degree of sophistication.

Measurements in the medical field can be classified into two types: *in vivo* and *in vitro*. *In vivo* measurement is made on or within the living organism itself, such as measurement of pressure in the chambers of the heart. On the other hand, *in vitro* measurement is performed outside the body. For example; the measurement of blood glucose level in a sample of blood drawn from the patient represent *in vitro* measurement.

1.5 PERFORMANCE REQUIREMENTS OF MEDICAL INSTRUMENTATION SYSTEMS

Information obtained from a sensor/transducer is often in terms of current intensity, voltage level, frequency or signal phase relative to a standard. Voltage measurements are the easiest to make, as the signal from the transducer can be directly applied to an amplifier having a high input impedance. However, most of the transducers produce signal in terms of current, which can be conveniently converted into voltage by using operational amplifiers with appropriate feedback.

To make an accurate measurement of voltage, it is necessary to arrange that the input impedance of the measuring device must be large compared with the output impedance of the signal source. This is to minimize the error that would occur, if an appreciable fraction of the signal source were dropped across the source impedance. Conversely, accurate measurement of current source signals necessitates that the source output impedance be large compared with the receiver input impedance. Ideally, a receiver that exhibits a zero input impedance would not cause any perturbation of the current source. Therefore, high-impedance current sources are more easily handled than low-impedance current sources.

In general, the frequency response of the system should be compatible with the operating range of the signal being measured. To process the signal waveform without distortion, the bandpass of the system must encompass all of the frequency components of the signal that contribute significantly to signal strength. The range can be determined quantitatively by obtaining a Fourier analysis of the signal. The bandpass of an electronic instrument is usually defined as the range between the upper and lower half-power frequencies.

The electrical signals are invariably accompanied by components that are unrelated to the phenomenon being studied. Spurious signal components, which may occur at any frequency within the band pass of the system are known as noise. The instruments are designed in such a way that the noise is minimised to facilitate accurate and sensitive measurement. For extraction of information from noisy signals, it is essential to enhance signal-to-noise ratio, for which several techniques have been put in practice. The simplest method is that of bandwidth reduction, although many sophisticated methods have been developed to achieve noise reduction from the noisy bio-medical signals.

The recent progress of digital technology in terms of both hardware and software, makes more efficient and flexible digital rather than analog processing. Digital techniques have several advantages. Their performance is powerful as they are able to easily implement even complex algorithms. Their performance is not effected by unpredictable variable such as component aging and temperature which can normally degrade the performance of analog devices. Moreover, design parameters can be more easily changed because they involve software rather than hardware modifications.

The results of a measurement in medical instruments are usually displayed either on analog meters or digital displays. Digital displays present the values of the measured quantities in numerical form. Instruments with such a facility are directly readable and slight changes in the parameter being measured are easily discernible in such displays, as compared to their analog counterparts. Because of their higher resolution, accuracy and ruggedness, they are preferred for display over conventional analog moving coil indicating meters. Different types of devices are available for display in numerical form.

Light emitting diodes (LED) are used in small sized seven-segment displays. These semiconductor diodes are made of gallium arsenide phosphide and are directly compatible with 5 V supplies typically encountered in digital circuitry. LEDs are very rugged and can withstand large variations in temperature. LEDs are available in deep-red, green and yellow colours.

Liquid crystal displays (LCD) are currently preferred devises for displays as they require very low current for their operation. LCDs with large screen sizes and full colour display capabilities are available commercially and are finding extensive and preferable applications in laptop computers and many portable medical instruments.

Since computers are used increasingly to control the equipment and to implement the man-machine interface, there is a growing appearance of high resolution colour graphic screens to display the course of vital signs relating to physiological variables, laboratory values, machine settings or the results of image processing methods such as magnetic resonance tomography. The analog and digital displays have been largely replaced by video display units, which present information not only as a list of numbers but as elegant character and graphic displays and sometimes as a 3 dimensional colour display. Visual display units (VDU) are usually monochrome as the CRTs in these units are coated with either white or green phosphors. Coloured video display units are employed in such applications as patient monitoring system and colour Doppler echocardiography.

A keyboard is the most common device connected into almost all form of data acquisition, processing and controlling functions in medical instruments. A keyboard can be as simple as a numeric pad with function keys, as in a calculator or complete alphanumeric and type writer keyboard with associated group of control keys suitable for computer data entry equipment. Most available keyboards have single contact switches, which are followed by an encoder to convert the key closures into ASCII (American Standard Code for Information Interchange) code for interfacing with the microprocessor.

▓▶ 1.6   INTELLIGENT MEDICAL INSTRUMENTATION SYSTEMS

Intelligent technology is pervading every area of modern society, from satellite communications to washing machines. The medical instrumentation field is no exception from this reality. In this case, the goal of intelligent devices is to assure high quality of life by providing optimal health care delivery in home care, emergency situations, diagnosis, surgical procedures and hospitalization. Medicine is now equipped with more and more signals and images taken from the human body, complex models of physiological systems and armaments of therapeutic procedures and devices. Careful observation of this process shows a congestion of the decision-making activities of medical personnel. To solve this problem, some method of integrating all patient information into a concise and interpretative form is necessary. The availability of high performance microprocessors, microcontrollers and personal computers has given powerful tools in the hands of medical professionals which offers them intelligent and efficient monitoring and management of the patients.

1.6.1   Use of Microprocessors in Medical Instruments

The application of microprocessors in medical instrumentation, has matured following a series of stages. In the first stage, the microprocessors simply replaced conventional hard wired electronic systems that were used for processing data. This resulted in more reliable and faster data. This was followed soon by use of the microprocessor to control logic sequences required in instrumentation. Thus, the microprocessor replaced programming devices as well as manual programming, making possible digital control of all of the functions of the medical instruments. With the availability of more powerful microprocessors and large data storage capacity, it has become possible to optimize the measurement conditions.

Extensive use has been made of microprocessors in medical instruments designed to perform routine clinical measurements, particularly in those situations where data computing and processing could be considered as a part of measurement and diagnostic procedure. The incorporation of microprocessors into instruments enables to have a certain amount of intelligence or decision-making capability. The decision-making capability increases the degree of automation of the instrument and reduces the complexity of the man-machine interface. Life support systems have been designed with numerous safety back-up features and real-time self-diagnostics and self-repair facilities. The reliability of many transducers has been improved and many measurements can now be made non-invasively because of the added computational ability of microprocessors. The computational capability makes possible features such as automatic calibration, operator guidance, trend displays, alarm priority and automated record keeping. Use of microprocessors in various instruments and systems has been explained at various places in the text.

Microprocessors have been used to replace the complicated instructional procedures that are now required in several medical instruments. Microprocessor based instrumentation is enabling to incorporate the ability to make intelligent judgement and provide diagnostic signals in case of potential errors, provide warnings or preferably make appropriate corrections. Already, the microprocessors are assisting in instruction-based servicing of equipment. This is possible by incorporating monitoring circuits that will provide valuable diagnostic information on potential instrumentation failure modes and guide the operator in their correction. The instrument diagnostic microprocessor programs would sense such a potential failure of the unit and the operator is informed to remove and service the defective part while the measurement work proceeds uninterrupted.

1.6.2   The Microprocessor

The microprocessor, in essence, consists of basic circuit elements such as transistors, resistors and diodes which when combined form the basic logical elements, namely AND, OR and INVERTERS. In principle, the complete operation of the microprocessor could be described by a combination of these devices. More complex circuit elements such as flip-flops, counters, registers and arithmetic logic unit are formed from these gates and go to make the complete microprocessor. Microprocessor is a single integrated circuit with 40 or even 64 or even higher connection pins.

Microprocessors are usually classified depending upon their word length. The word length of a microprocessor defines the basic resolution and memory addressing capability. For example, an 8-bit microprocessor will perform all calculations on binary numbers with 8 digits. 8 binary digits give a decimal number between 0-255.

The microprocessor's most powerful asset is its enormous speed of operation. This is possible because the microprocessor can store all the necessary instructions and data, until required in memory. *Memory* which is usually external serves as a place to store instructions that direct the activities of the Central Processing Unit (CPU) and data that are processed by the CPU (Fig. 1.10). It is arranged in two forms:

(a) *Read Only Memory* (ROM) to hold the program of instructions in binary digital form. The contents of this memory cannot be altered by the functioning of the microprocessor system.

(b) *Random Access Memory* (RAM) to hold results and variable data, for making calculations, remembering trends and assembling information for display devices.

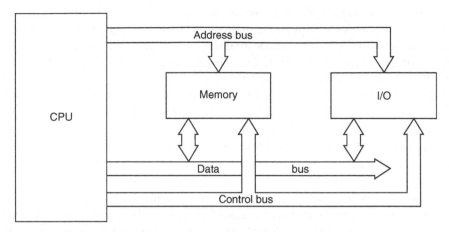

> **Fig.** 1.10   *Building blocks of a microcomputer*

Many of the address locations in a typical system are storage locations in the memory. When a memory location is addressed, the memory may store the information that resides on the data bus. This is called MEMORY WRITE. Addressing stored information to be placed on the data bus for use by the CPU constitutes a MEMORY READ.

The microprocessor can rapidly access any data stored in memory, but often memory is not large enough to store the entire data bank for particular application. This problem can be solved using external storage equipment, such as floppy disk or hard disk system. A microprocessor also requires input/output ports, through which it can communicate its results with the outside words, like a display or peripheral device or provide control signals that may direct another system.

As microprocessor systems are based on the binary numbering system, it is necessary to use multiple connections generally 8,16 or 32 between each of the integrated circuits (chips). These interconnections are usually referred to as buses. There are three buses in a microprocessor system.

Data Bus: A bidirectional path on which data can flow between the CPU and memory or input/output. It carries the actual data being manipulated.

Address Bus: A unidirectional group of lines that identify a particular memory location or input/output device.

Control Bus: It carries all the control and timing signals. It is a unidirectional set of signals that indicate the type of activity in current process. The types of activities could be memory read, memory write, input/output read, input/ output write and interrupt acknowledge.

The operation of the microprocessor and synchronisation of various activities under its control is maintained by a crystal controlled clock or oscillator, which is usually at a fixed frequency, generally greater than 5 MHz.

The heart of a microprocessor based system is the central processing unit (CPU). It requests instructions prepared by the programmer, calls for data and makes decisions related to the instructions. Based on the data, the processor determines appropriate actions to be performed by other parts of the system. Since there are many peripherals associated with the given system, the microprocessor must be capable of selecting a particular device. It identifies each device by means of a unique address code. A typical microprocessor has 16 binary address lines providing 65, 536 addressing codes. Data to and from the processor is carried across a bi-directional 8 or 16 bit wide data bus. Many processors also provide a serial data path. Several microprocessors use a multiplexed address/data bus on which both address and data are transmitted on the same signal paths. In this case, the first portion of the bus cycle transmits the address while data transfer takes place later in the cycle. This architecture is popular for microprocessors with an 8-bit data bus.

Another important link in the system is the set of input-output (I/O) interfaces. These interfaces include all the information channels between the system and the real world. There are digital ports through which programs and control commands may be loaded and from which digital data may be transmitted to peripherals such as keyboard, printers and floppy drive etc.

The assembly language of a microprocessor enables to extract the greatest run-time performance because it provides for direct manipulation of the architecture of the processor. However, it is also the most difficult language for writing programs, so it falls far from the optimal language line.

The C language which is used to develop modern versions of the Unix operating system provides a significant improvement over assembly language for implementing most applications, it is the language of choice for real time programming. It is an excellent compromise between a low level assembly language and a high level language. C is standardized and structured. C programs are based on functions that can be evolved independently of one another and put together to implement an application. These functions are to software just as black boxes are to hardware. C programs are transportable. By design, a program developed in C on one type of processor can be relatively easily transported to another.

1.6.3 The Microcontrollers

A *microcontroller*, contains a CPU, clock circuitry, ROM, RAM and I/O circuitry on a single integrated circuit package. The microcontroller is therefore, a self-contained device, which does not require a host of associated support chips for its operation as conventional microprocessors do. It offers several advantages over conventional multichip systems. There is a cost and space advantage as extra chip costs and printed circuit board and connectors required to support multichip systems are eliminated. The other advantages include cheaper maintenance, decreased hardware design effort and decreased board density, which is relevant in portable medical equipment.

Microcontrollers have traditionally been characterised by low cost high volume products requiring a relatively simple and cheap computer controller. The design optimization parameters require careful consideration of architectural tradeoffs, memory design factors, instruction size, memory addressing techniques and other design constraints with respect to area and performance. Microcontrollers functionality, however, has been tremendously increased in the recent years. Today, one gets microcontrollers, which are stand alone for applications in data acquisition system and control. They have analog-to-digital converters on chip, which enable them direct use in instrumentation. Another type of microcontroller has on-chip communication controller, which is designed for applications requiring local intelligence at remote nodes and communication capability among these distributed nodes. Advanced versions of the microcontrollers in 16-bit

configuration have been introduced for high performance requirements particularly in applications where good arithmetical capabilities are required.

1.6.4   Interfacing Analog Signals to Microprocessors

It is well-known that we live in an analog world. Virtually, all information we need to acquire from
the human body and eventually analyze is in the analog form i.e. the signals consist of many
waveforms that continuously vary as a function of time. Examples include electrocardiograph,
pressure signals and pulse waveform.

For interfacing analog signals to microprocessors/microcomputers, use is made of some kind
of data acquisition system. The function of this system is to acquire and digitize data, often from
hostile clinical environments, without any degradation in the resolution or accuracy of the signal.
Since software costs generally far exceed the hardware costs, the analog/digital interface structure
must permit software effective transfers of data and command and status signals to avail of the
full capability of the microprocessor.

The analog interface system, in general, handles signals in the form of voltages. The physical
parameters such as temperature, flow, pressure, etc. are converted to voltages by means of
transducers. The choice and selection of appropriate transducers is very important, since the data
can only be as accurate as the transducer.

Figure 1.11 shows a block diagram of a universal interface circuit for connecting analog signals
to microprocessors. It basically comprises a multiplexer, instrumentation (buffer) amplifier, a
sample-and-hold circuit, analog-to-digital converter (ADC), tristate drivers and control logic. These
components operate under the control of interface logic that automatically maintains the correct
order of events.

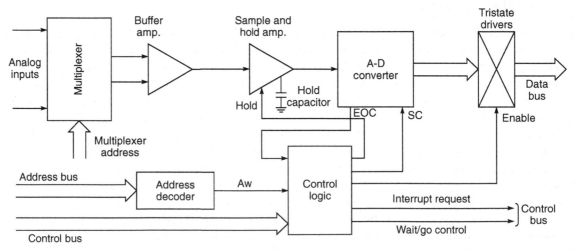

> **Fig. 1.11**   *Interfacing analog signals to microcomputers*

Multiplexer: The function of the multiplexer is to select under address control, an analog input
channel and connected to the buffer amplifier. The number of channels is usually 8 or 16.

Depending on its input configuration, the multiplexer will handle either single ended or differential signals.

The address logic of most multiplexers can perform both random and sequential channel selection. For real time systems, the random mode permits the multiplexer to select any channel when the program responds to a peripheral service request. Sequential channel selection, as the name imploys involves addressing each channel in order.

Buffer Amplifier: The buffer amplifier conditions the selected input signal to a suitable level for application to the A/D converter. Driven by the multiplexer, the buffer amplifier, which is usually an instrumentation amplifier, provides impedance buffering, signal gain and common mode rejection. It has a high input impedance, 100 Mohms or more to reduce the effects of any signal distortion caused by the multiplexer. The high input impedance also minimizes errors due to the finite on-resistance of the multiplexer channel switches.

To improve system sensitivity, the amplifier boosts the input signal. If it is required to have analog signals of differing ranges, connected to the multiplexer input, then a programmable gain amplifier would be preferable where the gain would be set in accordance with the multiplexer selection address. The use of programmable gain amplifiers removes the necessity to standardize on the analog input ranges.

Sample and Hold Circuit: The A/D converter requires a finite time for the conversion process, during which time the analog signal will still be hanging according to its frequency components. It is therefore necessary to sample the amplitude of the input signal, and hold this value on the input to the A/D converter during the conversion process. The sample and hold circuit freezes its output on receipt of a command from the control circuit, thereby providing an essentially constant voltage to the A/D converter throughout the conversion cycle.

The sample hold is essentially important in systems having resolution of 12-bits or greater, or in applications in which real time inputs are changing rapidly during a conversion of the sampled value. On the other hand, a sample hold may not be required in applications where input variation is low compared to the conversion time.

A/D Converter: The A/D converter carries out the process of the analog-to-digital conversion. It is a member of the family of action/status devices which have two control lines— the start conversion or action input line and the end of conversion or status output line.

An A/D converter is a single chip integrated circuit having a single input connection for the analog signal and multiple pins for digital output. It may have 8,12,16 or even more output pins, each representing an output bit. The higher the number of bits, the higher the precision of conversion. Each step represents a change in the analog signal: 8-bits gives 256 steps, 12-bits provides 4096 steps and we get 32768 steps with 16 output bits.

The key parameters in A/D converters are:

- *Resolution* of the A/D converter is a measure of the number of discrete digital code that it can handle and is expressed as number of bits (binary). For example, for an 8-bit converter, the resolution is 1 part in 256.
- *Accuracy* is expressed as either a percentage of full scale or alternatively in bits of resolution. For example, a converter may be termed 12-bit accurate if its error is 1 part in 4096. The

sources of error contributing to the inaccuracy of a converter or linearly, gain, error and off-set error.

- *Integral non-linearity* is a measure of the deviation of the transfer function from a straight line.
- *Off-set error* is a measure of the difference of the analog value from the ideal at a code of all zeroes.
- *Gain* error represents the difference in slope of the transfer function from the ideal.
- *Speed* of an A/D converter is generally expressed as its conversion time, i.e. the time elapsed between application of a convert command and the availability of data at its outputs. The speed of D/A converter is measured by its settling time for a full scale digital input change.

Each of the above parameters is temperature-dependent and they are usually defined at 25° C.

Tri-state Drivers: The tri-state drivers provide the necessary isolation of the A/D converter output data from the microprocessor data bus and are available as 8-line units. Thus, for the 10 or 12 bit converters, two drivers would be required which would be enabled by two different read addresses derived from the address decoder.

Some A/D converters have in-built tri-state drivers. However, because of their limited drive capability, they can be used only on lightly loaded buses. For heavily loaded systems, as in microcomputers, the built-in drivers are permanently enabled and separate tri-state drivers employed for the data bus isolation.

Control Logic: The control logic provides the necessary interface between the microprocessor system but the elements of the acquisition unit in providing the necessary timing control. It is to ensure that the correct analog signal is selected, sampled at the correct time, initiate the A/D conversion process (start-conversion = SC) and signals to the microprocessors on completion of conversion (End of conversion = EOC).

Output Interface: Digital output signals often have to be converted into analog form so that they can be used and acted upon by external circuits, e.g., oscilloscope, chart recorder, etc. Therefore, digital-to-analog (D/A) converters are used for converting a signal in a digital format into an analog form. The output of the D/A converter is either current or voltage when presented with a binary signal at the input.

The input coding for the D/A converter is similar to the output coding of the A/D converter, while full-scale outputs are jumper-selectable for 0 to ± 1, ± 5 and ± 10 V. D/A converters generally deliver the standard 4 to 20 mA output and loading can range from 50 W to 4 kW. The important parameters which govern the choice of an A/D converter or D/A converter are resolution, measurement frequency, input characteristics, offset error, noise, microprocessor compatibility and linearity, etc.

1.6.5 PC Based Medical Instruments

An area of intense commercial activity in the field of computers is due to the popularity of the so called personal computers (PC) or home computers. The low cost and increasing power of the personal computers are making them popular in the medical field. Also, software for personal computers is largely commercially available and the users can purchase and use it conveniently. Personal computers are now well established and widely accepted in the medical field for data collection, manipulation and processing and are emerging as complete workstations for a variety of applications. A personal computer becomes a workstation with the simple installation of one or more "instruments-on-a-board" in its accessory slots, and with the loading of the driver software that comes with each board. The concept has proven to be ideal instrument, providing a low cost yet highly versatile computing platform for the measurement, capture, analysis, display and storage of data derived from a variety of sources.

Fig. 1.12 illustrates the typical configuration of a PC based workstation. It is obvious that the system is highly flexible and can accommodate a variety of inputs, which can be connected to a PC for analysis, graphics and control. Basic elements in the system include sensors or transducers that convert physical phenomena into a measurable signal, a data acquisition system (a plug-in instrument/acquisition board), an acquisition/analysis software package or programme and computing platform. The system works totally under the control of software. It may operate from either the PC's floppy and/or hard disk drive. Permanent loading or unloading of driver files can be accomplished easily. However, for complex applications, some programming in one or several of the higher level programming languages such as 'C language' may be needed. Data received from the measurements can be stored in a file or output to a printer, plotter or other device via one of the ports on the computer.

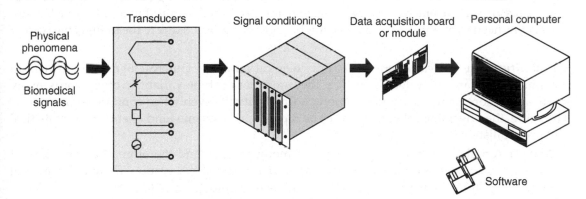

> **Fig. 1.12** *Typical configuration of PC based medical instrument*

PC based medical instruments are gaining in popularity for several reasons including price, programmability and performance specifications offered. Software development, rather than hardware development, increasingly dominates new product design cycles. Therefore, one of the most common reasons why system designers are increasingly choosing PC and PC architecture is for its rich and cost effective software tool set. This includes operating systems, device drivers, libraries, languages and debugging tools. Several examples of PC based medical instruments can be found at various places in the book.

▐▐▐▶ 1.7   GENERAL CONSTRAINTS IN DESIGN OF MEDICAL INSTRUMENTATION SYSTEMS

Medical equipment are primarily used for making measurements of physiological parameters of the human body and also in some cases a stimulus or some kind of energy is applied to the human body for diagnosis and treatment. Some of the important factors, which determine the design of a medical measuring instrument, are:

- Measurement Range: Generally the measurement ranges are quite low compared with non-medical parameters. Most signals are in the microvolt range.
- Frequency Range: Most of the bio-medical signals are in the audio frequency range or below and that many signals contain dc and very low frequency components.

These general characteristics of physiological signals limit the practical choices available to designers of medical instruments. Besides, there are some additional constraints, which need to be considered while designing a measurement system for medical applications. Some of these are:

Inaccessibility of the Signal Source: One of the major problems in making measurements from a living system is the difficulty in gaining access to the source of the physiological variable being measured. For example; measurement of intracranial pressure in the brain requires the placement of a sensor in the brain, which is quite a difficult task. Besides, the physical size of many sensors may put a constraint for its use on the area of interest. Evidently, such inaccessible physiological variables must be measured indirectly. The typical example of making indirect measurement of blood pressure on the brachial artery is that of using cuff based Korotoff method. In such cases, corrections need to be applied to data that might have been affected due to the indirect measuring process.

Variability of Physiological Parameters: Physiological variables of interest for measurement from the human body are rarely deterministic as they are generally time-variant. In other words, many medical measurements vary widely among normal patients even when conditions are similar. Therefore, the physiological variable must be represented by some kind of empirical, statistical and probabilistic distribution function.

Many internal anatomical variations exist among patients and therefore, the variability of physiological parameters from one patient to another is a normal observation. Therefore, statistical methods are employed in order to establish relationships among variables.

Interference among Physiological Systems: Many feedback loops exist among physiological systems and many of the interrelationships amongst them contribute to this inherent variability of physiological signals. In other words, stimulation of one part of a given system generally affects all other parts of that system in some way. Also, unlike many complex non-medical systems, a biological system is of such a nature that it is not possible to turn it off and remove parts of it during measurement procedure to avoid interference from undesirable physiological signals.

Transducer Interface Problems: All measurement systems are affected in some way by the presence of the measuring transducer. The problem gets compounded while making measurement on the living system where the physical presence of the transducer may change the reading significantly. Also, the presence of a transducer in one system can affect responses in other systems.

Adequate care needs to be taken while designing a measuring system to ensure that the loading effect of the transducer is minimal on the source of the measured variable.

High Possibility of Artifacts: The term artifact refers to an undesirable signal that is extraneous to the physiological variable under measurement. The examples of artifacts are: 50 Hz electrical interference, cross talk and noise generated within the measuring instrument. A major source of artifacts in medical instruments is due to the movement of the subject. Many of the transducers are sensitive to the movement and therefore, the movement of the subject result in generating spurious signals, which may even be large enough to obscure the signal of interest. This type of situation puts a heavy demand on the signal conditioning part of the measurement system.

Safe Levels of Applied Energy: Nearly all biomedical measurements require some form of energy to be applied to the living tissue or some energy gets applied as an incidental consequence of transducer operation. For example, ultrasonic imaging techniques depend upon externally applied ultrasound energy to the human body. Safe levels of the various types of energy on the human subjects are difficult to establish. However, designers of medical instruments depend upon a large number of studies carried out by numerous researchers, which establish the threshold of adverse affects by the applied energy.

Patient Safety Considerations: Medical instruments have to be physically connected to the patient in some way or the other. In case it happens to be an electric or electronic equipment, the possibility of an electric shock hazard is very strong unless adequate measures have been taken in the design of the equipment. In addition, the equipment is used by non-technical medical and paramedical staff and their safety needs also to be ensured. Various organizations at national and international level have laid down specific guidelines to provide for the safety and effectiveness of the medical devices intended for use on human subjects.

Reliability Aspects: In case of life saving equipment like defibrillators, their failure to operate or provide desired output can become a potential life threat for the patient. Therefore, equipment must be reliable, simple to operate and capable of withstanding physical abuse due to transportation within the hospital or in the ambulances and exposure to corrosive chemicals.

Human Factor Considerations: As a result of the increasing complexity of medical devices and systems, the demand on physicians and paramedical staff using the equipment have continued to grow. The equipment requires a high amount of information exchange between itself and the user in order to monitor and control the technical functions of the system. Further more, medical staff generally have only little experience in working with complex technical system. There is a risk that the medical staff is not able to master the equipment adequately for every task. This inadequacy can increase the probability of error and reduce the quality and reliability of a clinical procedure. As a result, the desired or intended performance of the whole system may not be achieved due to deficiencies in man-machine interaction. The user interface design issues therefore assume more and more importance in case of medical equipment.

Government Regulations: During the initial stages of introduction of technology and a range of diagnostic and therapeutic devices in the medical field, there was almost no government control on their design, testing and sales. Situation is rapidly changing and government regulations are being introduced to ensure that the equipment perform their intended function and are safe to operate and function. Designers of medical instruments should therefore be fully conversant with all such regulations on a particular product or system issued by national and international agencies.

It is thus obvious that there are many factors that impose constraints on the design of medical instruments. In addition to these, there are general considerations, which need to be considered into the initial design and development of a medical instrument. These factors are:

Signal Considerations: Type of sensor, sensitivity, range, input impedance, frequency response, accuracy, linearity, reliability, differential or absolute input.

Environmental Considerations: Signal-to-noise ratio, stability with respect to temperature, pressure, humidity, acceleration, shock, vibration, radiation etc.

Medical Considerations: Invasive or non-invasive technique, patient discomfort, radiation and heat dissipation, electrical safety, material toxicity etc.

Economic Considerations: Initial cost, cost and availability of consumables and compatibility with existing equipment.

Obviously, a project for a commercial medical instrument is quite complex which must take into consideration several factors before it is launched for design and development. In addition, the association of the engineering design team with motivated medical professionals is essential for the success of the project. This association is useful not only during the development process, but also for the clinical trials of the product so developed.

1.8   REGULATION OF MEDICAL DEVICES

The medical instrumentation industry in general and hospitals in particular are required to be most regulated industries. This is because when measurements are made on human beings and by the human beings, the equipment should not only be safe to operate but must give intended performance so that the patients could be properly diagnosed and treated. Adequate measures need to be evolved so that the users of medical equipment are not subject to legal, moral and ethical issues in their practice since they deal with the health of the people which could at times be as vital as the question of life and death. To minimize such type of problems, various countries have introduced a large number of codes, standards and regulations for different types of equipment and facilities. It is therefore, essential that engineers understand their significance and be aware of the issues that are brought about by technological and economic realities.

Regulations: A regulation is an organization's way of specifying that some particular standard must be adhered to. These are rules normally promulgated by the government.

Standards: A standard is a multi-party agreement for establishment of an arbitrary criterion for reference. Alternatively, a standard is a prescribed set of rules, conditions or requirements concerned with the definition of terms, classification of components, delineation of procedures, specifications of materials, performance, design or operations, measurement of quality and quality in describing materials, products, systems, services or practice. Standards exist that address systems (protection of the electrical power distribution system from faults), individuals (measures to reduce potential electric shock hazards) and protection of the environment (disposal of medical waste).

Codes: A system of principles or regulations or a systematized body of law or an accumulation of a system of regulations and standards. The most familiar code in USA is the National Electric Code issued by National Fire Protection Association (NFPA). In India, it is the National Electric Code issued by the Bureau of Indian Standards. In general, a code is a compilation of standards relating to a particular area of concern. For example; a state government health codes contain standards relating to providing health care to the state population.

Specifications: Documents used to control the procurement of equipment by laying down the performance and other associated criteria. These documents usually cover design criteria, system performance, materials and technical data.

Standards, codes and regulations may or may not have legal implications depending upon whether the promulgating organization is government or private.

1.8.1 Types of Standards

There are in general three type of standards for medical devices:

Voluntary Standards: Developed through a consensus process where manufacturers, users, consumers and government agencies participate. They carry no inherent power of enforcement but provide a reference point of mutual understanding.

Mandatory Standards: Required to be followed under law. They are incumbent on those to whom the standard is addressed and enforceable by the authority having jurisdiction.

Proprietary Standards: Developed either by a manufacturer for its own internal use or by a trade association for use by its members. They can be adopted as voluntary or mandatory standards with the consensus/approval of the concerned agencies.

1.8.2 Regulatory Requirements

In 1976, the US Congress approved Medical Device Amendments to the Federal Food, Drug and Cosmetic Act which empowered the Food and Drug Administration (FDA) to regulate nearly every facet of the manufacture and sale of medical and diagnostic devices. The term "Medical Device" in this law means "any item promoted for a medical purpose that does not rely on chemical action to achieve its intended effect". Further amendments to the Act have been made with the primary purpose to ensure the safety and efficacy of new medical devices prior to marketing of the devices. This is accomplished by classifying the devices into three classes based on the principle that devices that pose greater potential hazards should be subject to more regulatory requirements.

Class–I

General Controls: A device for which the controls authorized by law are sufficient to provide reasonable assurance of the safety and effectiveness of the device. Manufacturers are required to perform registration, pre-marketing notification, record keeping, labeling, reporting of adverse experiences and good manufacturing practices. Obviously, these controls apply to all three classes.

Class–II

Performance Standards: Apply to devices for which general controls alone do not provide reasonable assurance of safety and efficacy, and for which existing information is sufficient to establish a performance standard that provides this assurance. However, until performance standards are developed by regulation, only general controls apply.

Class–III

Pre-market Approval: Apply to devices which are used to support or sustain human life or to prevent impairment of human health, devices implanted in the body and devices which present a potentially un-reasonable risk of illness or injury. These are highly regulated devices and require manufacturers to prove their safety and effectiveness prior to their market release.

It may be of interest to note that software which is being increasingly used in medical devices has become an area of utmost importance because several serious accidents have been traced to software bugs and problems. In view of this, there is an increased requirement for maintaining traceability of devices to the ultimate customer, post marketing surveillance for life-sustaining and life-supporting implants, and hospital reporting system for adverse incidents.

New medical devices can be introduced in the market by two path ways. For devices that perform a new function or operate on a new principle, the FDA requires premarket approval. For a device that duplicates the function of a device already in the market and if the device is substantially equivalent to the existing device, the approval is granted. Such type of regulatory requirements exist in some other countries also.

1.8.3   Standards Related Agencies

Most countries of the world have their own internal agencies to set and enforce standards. For example, in India, the agency responsible for laying down standards for various products and services is Bureau of Indian Standards (BIS). However, in the present world of international cooperation and trade, it has become necessary to adopt uniform standards which could be applicable across national boundaries. There are two organisations at the international level which are active in this area.

International Electro-technical Commission (IEC): Deals with all matters relating to standards for electrical and electronic items. Membership in the IEC is held by a national committee for each nation. One of the notable standards developed under IEC is 60601–1, Safety of Medical Electrical Equipment, Part–I: General Requirements for Safety (1988) and its Amendment (1991) and the document 60601-1-1, Safety Requirements for Medical Electrical Systems.

International Organization for Standardization (ISO): This organization oversees aspects of device standards other than those related to electro-technology. The purpose of the ISO is to facilitate international exchange of goods and services and to develop mutual cooperation in intellectual, scientific, technological and economic ability.

In addition, many agencies promulgate regulations and standards in the areas of electrical safety, fire safety, technology management, occupational safety, radiology, nuclear medicine, clinical laboratories, bio-safety, infection control, anaesthesia equipment, power distribution and medical gas systems. There are thousands of applicable standards, clinical practice guidelines, laws and regulations. In addition, voluntary standards are issued by a large number of organizations and mandatory standards by numerous government agencies. Biomedical Engineers are therefore, advised to consult the relevant international/national standards for effective discharge of their professional duties.

Bioelectric Signals and Electrodes

▶ 2.1 ORIGIN OF BIOELECTRIC SIGNALS

The association of electricity with medical science dates back to the 18th century when Galvani demonstrated that most of the physiological processes were accompanied with electrical changes. This discovery formed the basis of the explanation of the action of living tissues in terms of bioelectric potentials. It is now well established that the human body, which is composed of living tissues, can be considered as a power station generating multiple electrical signals with two internal sources, namely muscles and nerves. Normal muscular contraction is associated with the migration of ions which generates potential differences measurable with suitably placed electrodes. For example, the heart and the brain produce characteristic patterns of voltage variations which when recorded and analyzed are useful in both clinical practice and research. Potential differences are also generated by the electrochemical changes accompanied with the conduction of signals along the nerves to or from the brain. These signals are of the order of a few microvolts and give rise to a complicated pattern of electrical activity when recorded. The fact that the activity of the living tissues is due to the potential changes in them suggested the use of external electricity for the diagnosis of certain diseases affecting muscles and nerves, for the augmentation or replacement of a deficient natural activity or for the restoration of a palsied muscle.

Bioelectric potentials are generated at a cellular level and the source of these potentials is ionic in nature. A cell consists of an ionic conductor separated from the outside environment by a semipermeable membrane which acts as a selective ionic filter to the ions. This means that some ions can pass through the membrane freely where as others cannot do so. All living matter is composed of cells of different types. Human cells may vary from 1 micron to 100 microns in diameter, from 1 mm to 1 m in length, and have a typical membrane thickness of 0.01 micron (Peter Strong, 1973).

Surrounding the cells of the body are body fluids, which are ionic and which provide a conducting medium for electric potentials. The principal ions involved with the phenomena of producing cell potentials are sodium (Na^+), potassium (K^+) and chloride (Cl^-). The membrane of excitable cells readily permits the entry of K^+ and Cl^- but impedes the flow of Na^+ even though there may be a very high concentration gradient of sodium across the cell membrane. This results in the concentration of the sodium ion more on the outside of the cell membrane than on the inside. Since sodium is a positive ion, in its resting state, a cell has a negative charge along the inner surface of its membrane and a positive charge along the outer portion. The unequal charge distribution is a result of certain electrochemical reactions and processes occurring within the living cell and the potential measured is called the resting potential. The cell in such a condition is said to be polarized. A decrease in this resting membrane potential difference is called depolarization.

The distribution of positively charged ions on the outer surface and negatively charged ions inside the cell membrane results in the difference of potential across it and the cell becomes, in effect, a tiny biological battery. Experiments have shown that the internal resting potential within a cell is approximately −90 mV with reference to the outside of the cell. When the cell is excited or stimulated, the outer side of the cell membrane becomes momentarily negative with respect to the interior. This process is called depolarization and the cell potential changes to approximately +20 mV. Repolarization then takes place a short time later when the cell regains its normal state in which the inside of the membrane is again negative with respect to the outside. Repolarization is necessary in order to re-establish the resting potential. This discharging and recharging of the cell produces the voltage waveforms which can be recorded by suitable methods using microelectrodes. A typical cell potential waveform so recorded is shown in Fig. 2.1.

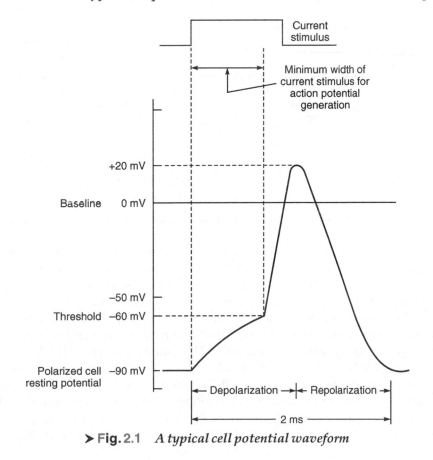

> **Fig. 2.1** *A typical cell potential waveform*

The wave of excitation while propagating in the muscle causes its contraction. The contraction wave always follows the excitation wave because of its lower velocity. This phenomenon is found with the skeletal muscles, the heart muscle and the smooth muscles. In its turn, every contraction (movement) of a muscle results in the production of an electric voltage. This voltage occurs in the muscle in such a way that the moving muscle section is always negative with respect to its surroundings. These voltages are called action potentials because they are generated by the action of the muscles. After complete contraction, repolarization takes place resulting in the relaxation of the muscle and its returning to the original state. Figure 2.2 shows electrical activity associated with one contraction in a muscle.

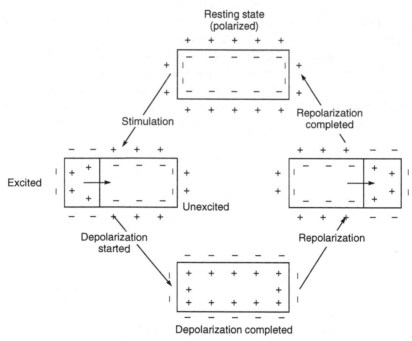

> **Fig. 2.2**　*Electrical activity associated with one contraction in a muscle*

The currents involved in bioelectricity are unlike the currents involved in electronics. Bioelectric currents are due to positive and negative ion movement within a conductive fluid. The ions possess finite mass and encounter resistance to movement within the fluid for they have limited speeds. The cell action potential, therefore, shows a finite rise time and fall time. It may be noted that a cell may be caused to depolarize and then repolarize by subjecting the cell membrane to an ionic current. However, unless a stimulus above a certain minimum value is applied, the cell will not be depolarized and no action potential is generated. This value is known as the stimulus threshold. After a cell is stimulated, a finite period of time is required for the cell to return to its pre-stimulus state. This is because the energy associated with the action potential is developed from metabolic processes within the cell which take time for completion. This period is known as refractory period.

The bioelectric signals of clinical interest, which are often recorded, are produced by the coordinated activity of large groups of cells. In this type of synchronized excitation of many cells, the charges tend to migrate through the body fluids towards the still unexcited cell areas. Such charge migration constitutes an electric current and hence sets up potential differences between various portions of the body, including its outer surface. Such potential differences can be conveniently picked up by placing conducting plates (electrodes) at any two points on the surface of the body and measured with the help of a sensitive instrument. These potentials are highly significant for diagnosis and therapy. The primary characteristics of typical bioelectric signals are given in Table 2.1.

• Table 2.1 *Bioelectric Signals*

Parameter	Primary signal characteristics	Type of Electrode
Electrocardiography (ECG)	Frequency range: 0.05 to 120 Hz Signal amplitude: 0.1 to 5 μV Typical signal: 1 μV	Skin electrodes
Electroencephalo-graphy (EEG)	Frequency range: 0.1 to 100 Hz Signal amplitude: 2 to 200 μV Typical signal: 50 μV	Scalp electrodes
Electromyography (EMG)	Frequency range: 5 to 2000 Hz Signal amplitude: 0.1 to 5 μV	Needle electrodes
Electroretinography (ERG)	Frequency range: dc to 20 Hz Signal amplitude: 0.5 μV to 1 μV Typical signal: 0.5 μV	Contact electrodes
Electro-oculography (EOG)	Frequency range: dc to 100 Hz Signal amplitude: 10 to 3500 μV Typical signal: 0.5 μV	Contact electrodes

2.1.1 Electrocardiogram (ECG)

The recording of the electrical activity associated with the functioning of the heart is known as electrocardiogram. ECG is a quasi-periodical, rhythmically repeating signal synchronized by the function of the heart, which acts as a generator of bioelectric events. This generated signal can be described by means of a simple electric dipole (pole consisting of a positive and negative pair of charge). The dipole generates a field vector, changing nearly periodically in time and space and its effects are measured on the surface. The waveforms thus recorded have been standardized in terms of amplitude and phase relationships and any deviation from this would reflect the presence of an abnormality. Therefore, it is important to understand the electrical activity and the associated mechanical sequences performed by the heart in providing the driving force for the circulation of blood.

The heart has its own system for generating and conducting action potentials through a complex change of ionic concentration across the cell membrane. Located in the top right atrium near the entry of the vena cava, are a group of cells known as the sino-atrial node (SA node) that initiate the heart activity and act as the primary pace maker of the heart (Fig. 2.3). The SA node is 25 to 30 mm in length and 2 to 5 mm thick. It generates impulses at the normal rate of the heart,

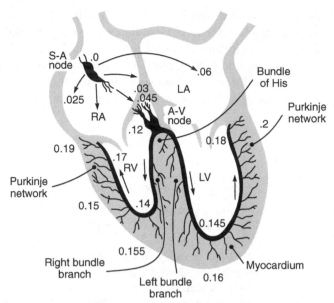

> **Fig.** 2.3   *The position of the sino-atrial node in the heart from where the impulse re-sponsible for the electrical activity of the heart originates. The arrow shows the path of the impulse.*
>
> *Note: The numbers like 0.18, 0.145, 0.15, 0.2 ... etc. indicate the time taken for the impulse to travel from the S-A node to various parts of the heart*

about 72 beats per minute at rest. Because the body acts as a purely resistive medium, the potential field generated by the SA node extends to the other parts of the heart. The wave propagates through the right and left atria at a velocity of about 1 m/s. About 0.1 s are required for the excitation of the atria to be completed. The action potential contracts the atrial muscle and the impulse spreads through the atrial wall about 0.04s to the AV (atrio-ventricular) node. This node is located in the lower part of the wall between the two atria.

The AV node delays the spread of excitation for about 0.12 s, due to the presence of a fibrous barrier of non-excitable cells that effectively prevent its propagation from continuing beyond the limits of the atria. Then, a special conduction system, known as the bundle of His (pronounced as hiss) carries the action potential to the ventricles. The atria and ventricles are thus functionally linked only by the AV node and the conduction system. The AV node delay ensures that the atria complete their contraction before there is any ventricular contraction. The impulse leaves the AV node via the bundle of His. The fibres in this bundle, known as Purkinje fibres, after a short distance split into two branches to initiate action potentials simultaneously in the two ventricles.

Conduction velocity in the Purkinje fibres is about 1.5 to 2.5 m/s. Since the direction of the impulse propagating in the bundle of His is from the apex of the heart, ventricular contraction begins at the apex and proceeds upward through the ventricular walls. This results in the contraction of the ventricles producing a squeezing action which forces the blood out of the ventricles into the arterial system. Figure 2.3 shows the time for action potential to propagate to various areas of the heart.

The normal wave pattern of the electrocardiogram is shown in Fig. 2.4. The *PR* and *PQ* interval, measured from the beginning of the *P* wave to the onset of the *R* or *Q* wave respectively, marks the time which an impulse leaving the SA node takes to reach the ventricles. The *PR* interval normally lies between 0.12 to 0.2 s. The *QRS* interval, which represents the time taken by the heart impulse to travel first through the interventricular system and then through the free walls of the ventricles, normally varies from 0.05 to 0.10s.

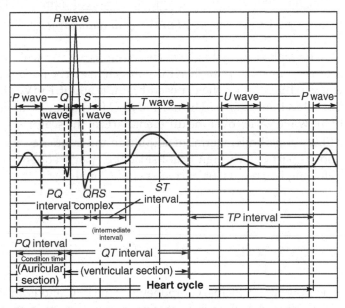

> **Fig. 2.4** *Normal wave pattern of an ECG waveform recorded in the standard lead position*

The *T* wave represents repolarization of both ventricles. The *QT* interval, therefore, is the period for one complete ventricular contraction (systole). Ventricular diastole, starting from the end of the *T* wave extends to the beginning of the next *Q* wave. Typical amplitude of *QRS* is 1 mV for a normal human heart, when recorded in lead 1 position.

2.1.2 Electroencephalogram (EEG)

The brain generates rhythmical potentials which originate in the individual neurons of the brain. These potentials get summated as millions of cell discharge synchronously and appear as a surface waveform, the recording of which is known as the electroencephalogram (Fig. 2.5).

The neurons, like the other cells of the body, are electrically polarized at rest. The interior of the neuron is at a potential of about –70 mV relative to the exterior. When a neuron is exposed to a stimulus above a certain threshold, a nerve impulse, seen as a change in membrane potential, is generated which spreads in the cell resulting in the depolarization of the cell. Shortly afterwards, repolarization occurs.

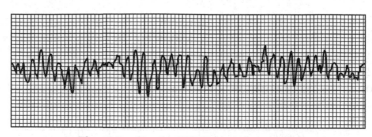

> **Fig. 2.5**   *Typical EEG signal waveform*

The EEG signal can be picked up with electrodes either from the scalp or directly from the cerebral cortex. The peak-to-peak amplitude of the waves that can be picked up from the scalp is normally 100 μV or less while that on the exposed brain, is about 1 mV. The frequency varies greatly with different behavioural states. The normal EEG frequency content ranges from 0.5 to 50 Hz. The nature of the wave varies over the different parts of the scalp.

The variations in EEG signals both in terms of amplitude and frequency are of diagnostic value. Frequency information is particularly significant since the basic frequency of the EEG range is classified into the following five bands for purposes of EEG analysis:

Delta (δ)	0.5–4 Hz
Theta (θ)	4–8 Hz
Alpha (α)	8–13 Hz
Beta (β)	13–22 Hz
Gamma (γ)	22–30 Hz

The alpha rhythm is one of the principal components of the EEG and is an indicator of the state of 'alertness' of the brain. It serves as an indicator of the depth of anaesthesia in the operating room. The frequency of the EEG seems to be affected by the mental activity of a person. The wide variation among individuals and the lack of repeatability in a given person from one occasion to another makes the analysis a difficult proposition. However, certain characteristic EEG waveforms can be conveniently related to gross abnormalities like epileptic seizures and sleep disorders.

Besides the importance of the frequency content of the EEG pattern, phase relationships between similar EEG patterns from different parts of the brain are also being studied with great interest in order to obtain additional knowledge regarding the functioning of the brain. Another important measurement is the recording of 'evoked response', which indicates the disturbance in the EEG pattern resulting from external stimuli. The stimuli could be a flash of light or a click of sound. Since the responses to the stimuli are repeatable, the evoked response can be distinguished from the rest of the EEG activity by averaging techniques to obtain useful information about the functioning of particular parts of the brain.

2.1.3   Electromyogram (EMG)

The contraction of the skeletal muscle results in the generation of action potentials in the individual muscle fibres, a record of which is known as electromyogram. The activity is similar to that observed in the cardiac muscle, but in the skeletal muscle, repolarization takes place much more rapidly, the action potential lasting only a few milliseconds. Since most EMG measurements are made to obtain an indication of the amount of activity of a given muscle, or a group of muscles, rather than of an individual muscle fibre, the EMG pattern is usually a summation of the individual action potentials from the fibres constituting the muscle or muscles being studied. The electrical activity of the underlying muscle mass can be observed by means of surface electrodes on the skin. However, it is usually preferred to record the action potentials from individual motor units for better diagnostic information using needle electrodes.

In voluntary contraction of the skeletal muscle, the muscle potentials range from $50\,\mu V$ to $5\,mV$ and the duration from 2 to 15 ms. The values vary with the anatomic position of the muscle and the size and location of the electrode. In a relaxed muscle, there are normally no action potentials. A typical EMG signal is shown in Fig. 2.6.

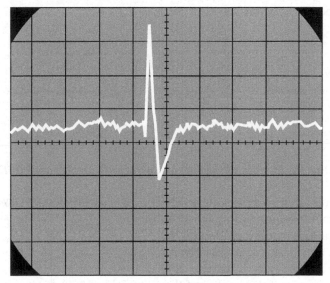

➤ **Fig. 2.6**   *Waveshape of a typical EMG signal*

➤ 2.2   RECORDING ELECTRODES

Bioelectric events have to be picked up from the surface of the body before they can be put into the amplifier for subsequent record or display. This is done by using electrodes. Electrodes make a transfer from the ionic conduction in the tissue to the electronic conduction which is necessary for making measurements. Electrodes are also required when physiological parameters are measured by the impedance method and when irritable tissues are to be stimulated in electrotherapy. Two types of electrodes are used in practice-surface electrodes and the deep-seated electrodes. The surface electrodes pick up the potential difference from the tissue surface when placed over it without damaging the live tissue, whereas the deep-seated electrodes indicate the electric potential difference arising inside the live tissue or cell. The same classification can be applied to electrodes used for the stimulation of muscles.

Electrodes play an important part in the satisfactory recording of bioelectric signals and their choice requires careful consideration. They should be comfortable for the patients to wear over long periods and should not produce any artefacts. Another desirable factor is the convenience of application of the electrodes.

2.2.1 Electrode-Tissue Interface

The most commonly used electrodes in patient monitoring and related studies are surface electrodes. The notable examples are when they are used for recording ECG, EEG and respiratory activity by impedance pneumography. In order to avoid movement artefacts and to obtain a clearly established contact (low contact impedance) an electrolyte or electrode paste is usually employed as an interface between the electrode and the surface of the source of the event. Figure 2.7 (a, b) represent the electrode-tissue interface.

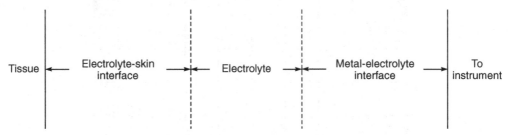

> **Fig. 2.7(a)**   *Electrode-tissue interface for surface electrodes used with electrode jelly*

The characteristic of a surface electrode composed of a metal electrode and attached to the surface of the body through an electrolyte (electrode jelly) are dependent upon the conditions at the metal-electrolyte interface, the electrolyte-skin interface and the quality of the electrolyte.

Metal-Electrolyte Interface: At the metal-electrolyte transition, there is a tendency for each electrode to discharge ions into the solution and for ions in the electrolyte to combine with each electrode. The net result is the creation of a charge gradient (difference of potential) at each electrode, the spatial arrangement of which is called the electrical double layer (Fig. 2.7(c)). The double layer is known to be present in the region immediately adjacent to the electrode and can be represented, in its simplest form, as two parallel sheets of charge of opposite sign separated by a thin film of dielectric. Therefore, the metal-

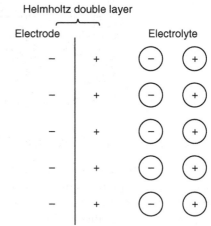

> **Fig. 2.7(b)**   *Electrode tissue interface circuit involves transfer of electrons from the metal phase to an ionic carrier in the electrolyte, a charge double layer (capacitance) forms at the interface*

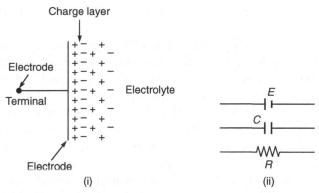

> **Fig. 2.7(c)** *(i) Charge distribution at electrode-electrolyte interface*
> *(ii) Three components representing the interface*

electrolyte interface appears to consist of a voltage source in series with a parallel combination of a capacitance and reaction resistance. The voltage developed is called the half-cell potential.

To a first-order approximation, the half-cell potential is equal to the electrode potential of the metal, if the electrodes were used in a chemical measuring application. All electrode potentials are measured with respect to a reference electrode, usually that of hydrogen absorbed on platinum black. This is an inconvenient electrode to make and, therefore, other alternative electrodes which may have fairly stable and repeatable potential (e.g. calomel electrode) are employed. Electrode potentials of some of the commonly used metals in the electrochemical series with respect to Hydrogen are given in Table 2.2.

- Table 2.2 *Electrode Potentials of some Metals with Respect to Hydrogen*

Metal	Ionic symbol	Electrode potential
Aluminium	Al^{+++}	−1.66 V
Iron	Fe^{++}	−0.44 V
Lead	Pb^{++}	−0.12 V
Hydrogen	H+	0
Copper	C^{++}	+0.34 V
Silver	Ag+	+0.80 V
Platinum	Pt+	+1.2 V
Gold	Au+	+1.69 V

This table shows that the electrode potentials are appreciable when dissimilar metals are used. They also exist, though of smaller magnitude, even if electrodes of similar materials are employed. The lowest potential has been observed to be in the silver-silver chloride electrodes. The values of the capacitance and the resistance depend upon many factors which include the current density, temperature, type and concentration of the electrolyte and the type of metal used.

The difference in half-cell potentials that exists between two electrodes is also called 'offset potential'. The differential amplifiers used to measure potentials between two electrodes are generally designed to cancel the electrode offset potential so that only the signals of interest are recorded. The electrode offset potential produced between electrodes may be unstable and unpredictable. The long-term change in this potential appears as baseline drift and short-term changes as noise on the recorded trace. If electrodes are used with ac-coupled amplifiers, the long term drift may be partially rejected by the low frequency characteristics of the amplifier. But it will depend upon the rate of change of electrode offset potential in relation to the ac-coupling time constant in the amplifier. For example, if the electrode offset potential drift rate is 1 mV/s, satisfactory results can only be obtained if the low frequency response of the amplifier is 1 Hz.

Also, the absolute value of the electrode offset potential is rarely significant except when it may exceed the maximum dc differential offset of the amplifier. In such a case, the trace may go out of the monitor screen or the pen in a recording instrument shifts to the extreme end of the chart paper, and then it will not be possible to bring them back. Silver-silver chloride electrodes have been found to give almost noise free characteristics. They are also found to be acceptable from the point of view of long-term drift. Electrodes made of stainless steel are generally not acceptable for high sensitivity physiological recordings. This is because stainless steel electrodes in contact with a saline electrolyte produce a potential difference of 10 mV between the electrodes, whereas this value is 2.5 mV for silver-silver chloride electrodes. Some representative values of potential between electrodes in electrolytes are given in Table 2.3. Staewen (1982) discusses various aspects concerning dc offset voltage standard for pregelled ECG disposable electrodes.

• Table 2.3   *Potential between Electrolytes in Electrodes (Courtesy: Geddes and Baker 1975)*

Electrode metal	Electrolyte	Potential difference between electrodes
Stainless steel	Saline	10 mV
Silver	Saline	94 mV
Silver-silver chloride	Saline	2.5 mV
Silver-silver chloride(11 mm disc)	ECG paste	0.47 mV
Silver-silver chloride (sponge)	ECG paste	0.2 mV

Warburg (1899) in his pioneering studies discovered that a single electrode/electrolyte interface can be represented by a series capacitance C and resistance R as shown in Fig. 2.8(a). However, C and R are unlike real capacitors and resistors because their values are frequency and current-density dependent. Often, these components are called the polarization capacitance and resistance. Warburg found that, for low current density, the reactance X of C ($1/2\pi fC$) equals R; both varied almost inversely as the square root of frequency, i.e. $R = X = k/\sqrt{f}$, where k is a constant. The consequence of this relationship is that the phase angle θ is constant at $\pi/4$ for all frequencies. However, only a limited number of studies have tested the accuracy of the Warburg model (Ragheb and Geddes, 1990).

It has been observed that the Warburg series RC equivalent does not adequately represent the behaviour of an electrode/electrolyte interface as this equivalent does not truly account for the very low-frequency behaviour of the interface. It is well known that such an interface can pass direct current. Therefore, a resistance R_f placed in parallel with the Warburg equivalent is more

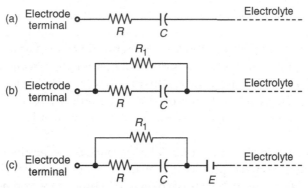

> ➤ **Fig.** 2.8 (a) *Warburg equivalent for an electrode-electrolyte interface*
> (b) *Addition of the faradic leakage resistance R_f to account for the direct current properties*
> (c) *Half-cell potential E of the electrode-electrolyte interface*

appropriate. Figure 2.8(b) shows this equivalent circuit in which R_f represents the Faradic leakage resistance. The value of R_f is high in the low-frequency region and is dependent on current density, increasing with a increase in current density.

To complete the equivalent circuit of an electrode/electrolyte interface, it is necessary to add the half-cell potential E. This is the potential developed at the electrode/electrolyte interface. The value of E depends on the species of metal and the type of electrolyte, its concentration and temperature. Figure 2.8(c) illustrates the complete equivalent circuit of a single electrode/electrolyte interface.

Electrolyte-Skin Interface: An approximation of the electrolyte-skin interface can be had by assuming that the skin acts as a diaphragm arranged between two solutions (electrolyte and body fluids) of different concentrations containing the same ions, which is bound to give potential differences. The simplest equivalent representation could then be described as a voltage source in series with a parallel combination of a capacitance and resistance. The capacitance represents the charge developed at the phase boundary whereas the resistance depends upon the conditions associated with ion-migration along the phase boundaries and inside the diaphragm.

The above discussion shows that there is a possibility of the presence of voltages of non-physiological origin. These voltages are called contact potentials.

The electrical equivalent circuit of the surface electrode suggests that the voltage presented to the measuring instrument from the electrode consists of two main components. One is the contact potential and the other is the biological signal of interest. The contact potential depends upon several factors and may produce an interference signal which exceeds several times the useful signal. The contact potential is found to be a function of the type of skin, skin preparation and composition of the electrolyte.

When bioelectric events are recorded, interference signals are produced by the potential differences of metal-electrolyte and the electrolyte-skin interface. Normally, these potential differences are connected in opposition during the recording procedure, and in the case of a truly reversible and uniform electrode pair, their difference would be nil. However, in practice, a difference of potential—may be extremely small—is found to exist between electrodes produced even under conditions of utmost care during manufacture. Also, some of the elements in the equivalent circuit are time-dependent and are bound to show slow variations with time.

The main reason for this rate of change is due to a relative displacement affecting chiefly the potential of the metal-electrolyte transition. Other factors responsible for variations of potential difference with time can possibly be temperature variations, relative displacement of the components in the system and changes in the electrolyte concentration, etc. (Odman and Oberg, 1982).

If ac signals are to be recorded, the potential difference between the two electrodes will not interfere with the useful signals, provided that the contact potential difference between the electrodes is constant. However, if the rate of change with time of the contact potential falls within the frequency spectrum of the signal under test, an error will be produced. The problem of difference of contact potentials becomes serious in case dc signals such as EOG are to be recorded. Any variation in the contact potential would greatly alter the character of the signal to be recorded which may itself be of extremely low amplitude—of the order of a few microvolts.

Based on the above mentioned considerations, it is possible to construct the circuit in which a pair of electrodes is placed in electrolytic contact with a subject. The electrodes are used to measure a bioelectric event and are connected to a differential amplifier. Three potentials are found to exist in this circuit (Fig. 2.9), one is due to the bioelectric event (E_b) and the other two are non-physiologic

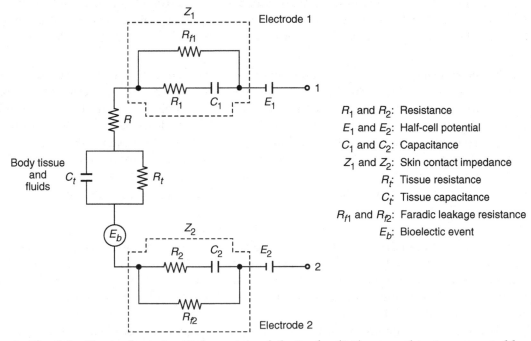

R_1 and R_2: Resistance
E_1 and E_2: Half-cell potential
C_1 and C_2: Capacitance
Z_1 and Z_2: Skin contact impedance
R_t: Tissue resistance
C_t: Tissue capacitance
R_{f1} and R_{f2}: Faradic leakage resistance
E_b: Bioelectic event

➤ **Fig. 2.9** *Equivalent circuit for a pair of electrodes (1,2) on a subject represented by $R\,R_t, C_t$. Embedded in the subject is a bioelectric generator E_b (after Tacker and Geddes, 1996)*

and represent the half-cell potentials (E_1 and E_2) of the electrodes. Z_1 and Z_2 are the skin contact impedances of these electrodes and R is the tissue resistance or resistance of the bioelectric generator. This circuit shows that the impedance of the electrodes would be high in the low-frequency region and it would decrease with increasing frequency. It is further clear that in the measurement of a bioelectric signal, it is essential to minimize potential drops across the electrode impedance. This is achieved by making the skin-contact impedance as low as possible and making the input impedance of the measuring device as high as possible.

2.2.2   Polarization

If a low voltage is applied to two electrodes placed in a solution, the electrical double layers are disturbed. Depending on the metals constituting the electrodes, a steady flow of current may or may not take place. In some metal/liquid interfaces, the electrical double layer gets temporarily disturbed by the externally applied voltage, and therefore, a very small current flows after the first surge, thus indicating a high resistance. This type of electrode will not permit the measurement of steady or slowly varying potentials in the tissues. They are said to, be polarized or nonreversible. Thus, the phenomenon of polarization affects the electro-chemical double layer on the electrode surface and manifests itself in changing the value of the impedance and voltage source representing the transition layer. Parsons (1964) stated that electrodes in which no net transfer of charge takes place across the metal-electrolyte interface can be termed as perfectly polarized. Those in which unhindered exchange of charge is possible are called non-polarizable or reversible electrodes. The ionic double layer in metals of these electrodes is such that they allow considerable current to flow when a small voltage is applied, thus offering a low resistance.

Although polarizable electrodes are becoming less common, they are still in use. They usually employ stainless steel and are used for resting ECGs or other situations where there is small likelihood that the electrodes would be exposed to a large pulse of energy (such as a defibrillation discharge) in which case they would retain a residual charge, become polarized, and will no longer transmit the relatively small bioelectric signals, thus becoming useless.

Non-polarizing electrodes on the other hand, are designed to rapidly dissipate any charge imbalance induced by powerful electrical discharges such as a defibrillation procedure. Rapid depolarization enables the immediate reappearance of bioelectric signals on the monitor after defibrillation. For this reason, non-polarizing electrodes have become the electrodes of choice for monitoring in the intensive care units and stress testing procedures. Historically, these electrodes employ a conducting metal with a silver/silver-chloride (Ag/AgCl) surface in contact with the conducting gel.

The choice of metals for electrodes is not determined only by their susceptibility to polarization, but other factors such as mechanical properties, skin irritation or skin staining, etc. have also to be taken into consideration. A detailed comprehensive review of electrodes for measurement of bioelectronic events is given by Geddes and Baker (1975).

The LeeTec Corporation, USA has devised a tin-backed electrode, Tracets MP-3000 (Fig. 2.10) which is non-polarizable and performs electrically as well as or better than similar electrodes employing silver/silver-chloride (Montecalvo and Rolf, 1990). U.S. Patent 4,674,512 describes the construction of this non-polarizing ECG electrode, which employs no silver. This represents a

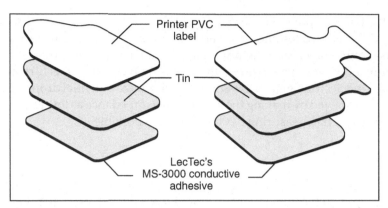

> ➤ **Fig. 2.10** *Resting ECG electrode Lec Tec MP-3000-a multipurpose monitoring and diagnostic non-polarizable electrode*

new era for electrocardiology where silver is no longer a critical electrode component for quality performance.

2.2.3 Skin Contact Impedance

The bioelectrical events are usually recorded by means of metallic electrodes placed on the surface of the body. The electrical activity generated by various muscles and nerves within the body is conducted to the electrode sites through the body tissues, reaches the electrodes through the skin-electrode transition and is then conducted by direct wire connection to the input circuit of the recording machine. The impedance at the electrode-skin junction comes in the overall circuitry of the recording machine and, therefore, has significant effect on the final record. Skin electrode impedance is known as the contact impedance and is of a value much greater than the electrical impedance of the body tissue as measured beneath the skin. The outer horny layer of the skin is responsible for the bulk of the skin contact impedance and, therefore, a careful skin preparation is essential in order to obtain best results.

Measurement of Skin Contact Impedance: A convenient method to measure the contact impedance at any individual electrode is shown in Fig. 2.11. This method has been suggested by Miller (1969). The three electrodes, A, B and C, have contact impedance respectively of Z_a, Z_b and Z_c. An oscillator provides a constant current in the frequency range of 0.1–100 Hz through the 47 kΩ series resistor. By suitably positioning the switch, a sensitive oscilloscope can be used to monitor either the voltage dropped across the 1 kΩ resistor or the voltage dropped across Z_b. The voltage drop across Z_b can be neglected since the input impedance of the oscilloscope used with an input probe is usually high. From the voltage dropped across the 1 kΩ resistor it is possible to calculate the circuit current and thus to obtain a value for Z_c. Using this technique, the skin contact impedance of the following types of electrodes were measured by Hill and Khandpur (1969).

- Plastic cup self-adhesive electrodes (Boter *et al*, 1966)
- Metal plate limb electrodes used with conducting jelly
- Metal plate electrodes used with conducting plastic (Jenkner, 1967)

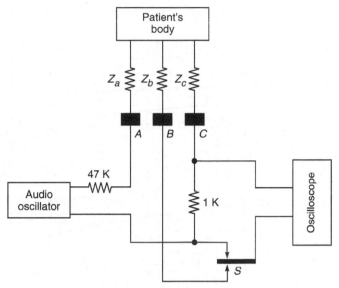

> ➤ **Fig. 2.11** *Arrangement for measurement electrode skin-contact impedance for surface electrodes*

- Dry multi-point limb electrodes (Lewes, 1966)
- Dry multi-point suction chest electrodes
- Self-adhesive multi-point chest electrodes used with conducting jelly
- Self-adhesive gauze electrodes
- Self-adhesive dry multi-point chest electrodes (Lewes and Hill, 1967)

Representative plots of contact impedance versus frequency are shown in Fig. 2.12.

Usually the contact impedance in respect of surface electrodes used for recording of ECG is measured (Grimnes, 1983) at 10–20 Hz because most of the energy content of the ECG is concentrated below 30 Hz. Geddes and Baker (1968) used a synchronous rectifier with a phase-sensitive detector to continuously measure the resistive and reactive components of the impedance.

2.2.4 Motion Artefacts

Motion artefact is a problem in biopotential measurements. The problem is greatest in cardiac stress laboratories where the exercise ECG is recorded. The problem is also serious in coronary care units where patients are monitored for relatively long periods. Motion of the subject under measurement creates artefacts which may even mask the desired signal or cause an abrupt shift in the baseline. These artefacts may result in a display being unreadable, a recording instrument exceeding its range, a computer yielding incorrect output or a false alarm being triggered by the monitoring device. Tam and Webster (1977) concluded that the skin-electrolytic paste interface is the major source of motion artefact. When a metal electrode contacts an electrolytic paste, a half-cell potential is generated at the electrode-paste interface. Kahn (1965) demonstrated that when polarizable metal-plate electrodes are used, the electrode-paste interface can be a source of motion

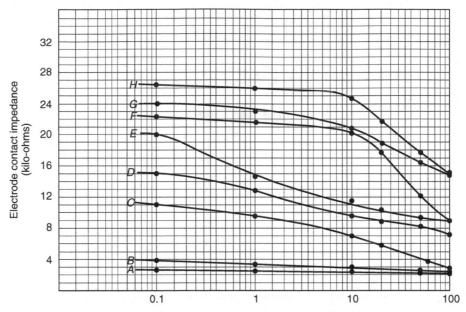

> ➤ **Fig.** 2.12   *Electrode skin-contact impedance versus signal frequency for different types of electrodes (after Hill and Khandpur, 1969)*

artefact. When the paste is agitated, the half-cell potential varies because of the altered metallic ion gradient at the interface. He recorded a 1 mV offset potential change from a silver-silver chloride electrode exposed to a flowing stream of saline solution, as contrasted to 30 mV change for some silver electrodes.

Motion artefact is reduced to a negligible magnitude by skin abrasion. However, when the skin is abraided, it is more susceptible to irritants. The possible sources for skin irritation include the electrode, the paste and the adhesive. When large currents flow through metallic electrodes, migration of some ions into the skin can cause irritation. However, silver-silver chloride electrodes do not cause much problem since silver chloride is almost insoluble in a chloride containing solution. Therefore, when these electrodes are used, the skin irritation is mostly caused by the paste and or the adhesive. Most commercial pastes produce about the same irritation when used on unprepared skin. They cause itching due to restricted perspiration, and reddening of the skin directly under the electrodes appears in 2–4 days. Thakor and Webster (1985) studied the sources of artefacts, means of reducing them using skin preparations, the electrode designs and their placement on the chest for long-term ambulatory ECG.

▥➤ 2.3 SILVER-SILVER CHLORIDE ELECTRODES

One of the important desirable characteristics of the electrodes designed to pick up signals from biological objects is that they should not polarize. This means that the electrode potential must not vary considerably even when current is passed through them. Electrodes made of silver-silver chloride have been found to yield acceptable standards of performance. By properly preparing and selecting the electrodes, pairs have been produced with potential differences between them of only fractions of a millivolt (Feder, 1963). Standing voltage of not more than 0.1 mV with a drift over 30 min. of about 0.5 mV was achieved in properly selected silver-silver chloride electrodes by Venables and Sayer (1963). Silver-silver chloride electrodes are also nontoxic and are preferred over other electrodes like zinc-zinc sulphate, which also produce low offset potential characteristics, but are highly toxic to exposed tissues. Silver-silver chloride electrodes meet the demands of medical practice with their highly reproducible parameters and superior properties with regard to long-term stability.

Production of Silver-Silver Chloride Electrodes: Silver-silver chloride electrodes are normally prepared by electrolysis. Two silver discs are suspended in a saline solution. The positive pole of a dc supply is connected to the disc to be chlorided and the negative pole goes to the other disc. A current at the rate of $1 \, mA/cm^2$ of surface area is passed through the electrode for several minutes. A layer of silver chloride is thus deposited on the surface of the anode. The chemical changes that take place at the anode and cathode respectively are:

$$NaCl \qquad Na^+ + Cl^-$$
$$Cl^- + Ag^+ \quad \rightarrow \quad AgCl$$

The positively charged sodium ions generate hydrogen when they reach the cathode surface.

$$2Na^+ + 2H_2O + 2 \text{ electrons} \rightarrow 2NaOH + H_2$$

To prepare silver-silver chloride electrodes of good quality, only pure silver should be used and the saline solution should be made from analar grade sodium chloride. Before chloriding, silver must be cleaned—preferably by the electrolytic method.

Geddes *et al.* (1969) investigated the effect of the chloride deposit on the impedance-frequency characteristics of the silver-silver chloride electrodes. They demonstrated that the impedance was different for different layers of chloride and that there is an optimum chloriding, which gives the lowest impedance. They concluded that the lowest electrode-electrolyte impedance in the frequency range of 10 Hz to 10 kHz was found to occur with a chloride deposit ranging between 100 and $500 \, mAs/cm^2$ of electrode area. To achieve this deposit by manipulation of current and time, the minimum constant chloriding current density should be $5 \, mA/cm^2$ of electrode area.

Higher values may be used with a corresponding reduction in time to achieve the 100–500 mAs/cm^2 chloride deposit. With this chloride deposit, the electrode electrolyte impedance was found to be resistive. The use of a chloride deposit in excess of this range did not alter the resistive nature of the electrode-electrolyte impedance although it increased its magnitude. Cole and Kishimoto (1962), however found that the chloride deposit for achieving the lowest impedance is $2000 \, mAs/cm^2$. Geddes (1972) confirmed that an optimal coating of silver chloride applied to a silver electrode minimizes the electrical impedance. This is supported by Getzel and Webster (1976) who concluded that silver chloride may be applied to cleaned silver electrodes in the amount of $1050–1350 \, mA \, s/cm^2$ in order to reduce the impedance of the electrodes. However, to further reduce the impedance of the electrodes, they should be coated with at least $2000 \, mAs/cm^2$ of silver chloride followed by immersion in a photographic developer for 3 minutes. A second layer of silver chloride, however, did not result in any further reduction in impedance. Grubbs and Worley (1983) obtained a lower and more stable impedance electrode by placing a heavier initial chloride coat on an etched silver electrode, and then electrolytically removing a portion of that coat.

▐▷ 2.4   ELECTRODES FOR ECG

2.4.1   Limb Electrodes

The most common type of electrodes routinely used for recording ECG are rectangular or circular surface electrodes (Fig. 2.13). The material used is german silver, nickel silver or nickel plated steel. They are applied to the surface of the body with electrode jelly. The typical value of the contact impedance of these electrodes, which are of normal size, is nearly 2 to 5 kΩ when measured at 10 Hz. The electrodes are held in position by elastic straps. They are also called limb electrodes as they are most suitable for application on the four limbs of the body. The size of the limb electrodes is usually 3 × 5 cm and they are generally made of german silver, an alloy of zinc, copper and nickel. They are reusuable and last several years.

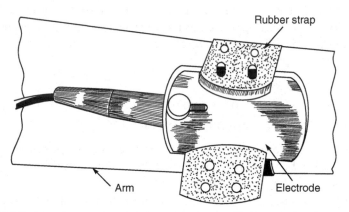

> **Fig. 2.13**   *ECG plate electrode. The electrode is usually fastened to the arm or leg with a perforated rubber strap which keeps it in position during ECG recording*

Limb electrodes are generally preferred for use during surgery because the patient's limbs are relatively immobile. Moreover, chest electrodes cannot be used as they would interfere with the surgery.

Limb electrodes are not suitable for use in long-term patient monitoring because the long flowing leads are inconvenient to the patient. Also, the electromyographic voltages generated by the activity of the limb muscles makes them unsuitable for use when monitoring conscious and semi-conscious patients.

Suction-cup electrode is commonly used to record the unipolar chest leads. It has a high contact impedance as only the rim of the electrode is in contact with the skin. The electrode is popular for its practicality, being easily attachable to fleshy parts of the body. Electrode jelly forms the vacuum seal.

2.4.2   Floating Electrodes

Limb electrodes generally suffer from what is known as motion artefacts caused due to the relative motion at the interface between the metal electrode and the adjacent layer of electrode jelly, Kahn

(1965) and Boter et al (1966). The interface can be stabilized by the use of floating electrodes in which the metal electrode does not make direct contact with the skin. The electrode (Fig. 2.14) consists of a light-weight metalled screen or plate held away from the subject by a flat washer which is connected to the skin. Floating electrodes can be recharged, i.e. the jelly in the electrodes can be replenished if desired.

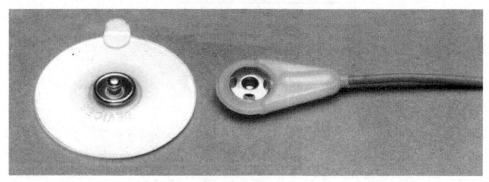

> **Fig. 2.14** *Light weight floating electrode with press stud for long-term monitoring of ECG*

Patten *et al* (1966) have described spray-on chest electrodes where a conducting spot is developed on the skin by spraying a film of conducting adhesive (mixture of Duco cement, silver powder and acetone). Connection with the instrument is established with silver-plated copper wires fixed in the conducting adhesive. The type of electrodes are extremely light-weight and do not make use of electrode jelly. This makes them ideal for use in monitoring the ECG of exercising subjects and aeroplane pilots as they give rise to minimal motion artefacts. The contact impedance shown by these electrodes is of the order of 50 kΩ.

Completely flexible ECG electrodes for the long-term monitoring of ECG during space flight are reported by Sandler *et al* (1973). These electrodes were made of silver-impregnated silastic rubber and were found to be comfortable to wear. They were also evaluated for use during exercise or prolonged monitoring as may be necessary in an intensive care or coronary care unit.

2.4.3 Pregelled Disposable Electrodes

Electrodes which are employed in stress testing or long term monitoring, present additional problems because of the severe stresses, perspiration and major body movement encountered in such studies. Both design considerations and application techniques of electrodes used in electrocardiography are necessary to prevent random noise on the baseline, baseline wandering and skin contact over extended periods causing a loss of signal. To overcome problems due to prolonged application, special disposable electrodes have been developed. Figure 2.15(a) illustrates the principle of a pregelled electrode while Fig. 2.15(b) shows a cross-section of such an electrode. The main design feature of these electrodes which helps in reducing the possibility of artefacts, drift and baseline wandering is the provision of a high-absorbancy buffer layer with isotonic electrolyte. This layer absorbs the effects of movement of the electrode in relationship to

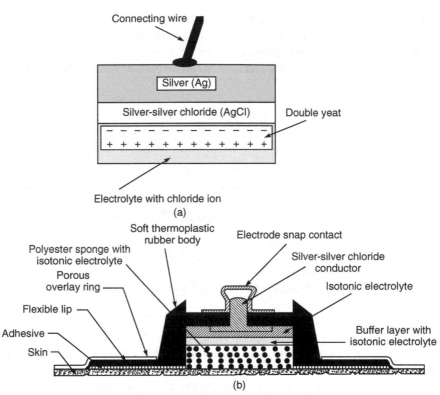

> **Fig.** 2.15   (a) *Principle of pre-gelled ECG electrode made of silver-silver chloride. The*
>   *electrode has electrolyte layers that are made of a firm gel which has*
>   *adhesive properties. The firm gel minimizes the disturbance of the*
>   *charge double layer*
>   (b) *Cross-section of a typical pre-gelled electrode*

the skin, and attempts to maintain the polarization associated with the half-cell potential constant. Since perspiration is the most common cause of electrode displacement, the use of an additional porous overlay disc resists perspiration and ensures secure placement of the electrode on the skin even under stress conditions. Figure 2.16 show a typical pregelled electrode.

Various manufacturers offer common features in pregelled electrode construction. The lead wire's female connector "snaps" on, allowing a convenient snap-on pull-off connection with a 360 rotation providing mechanical and electrical connection. The plastic eyelet or sensor has a diameter of 0.5–1.5 cm and is electroplated with silver up to a thickness of 10 μm. The surface of the Ag layer is partially converted to AgCl.

The tape is made from one of the adhesive coated occulusive foams made from a plastic, such as polyethylene, or a porous backing, such as non-woven cloth. Tapes used for first aid dressings are suitable. The electrode diameters range from 4–6 cm.

Some advantages of these electrodes as compared to plates and welch bulbs are : as there is no risk of infection which is possible with reusable electrodes, their smaller size makes them less prone to detachment and also less time is required per ECG procedure.

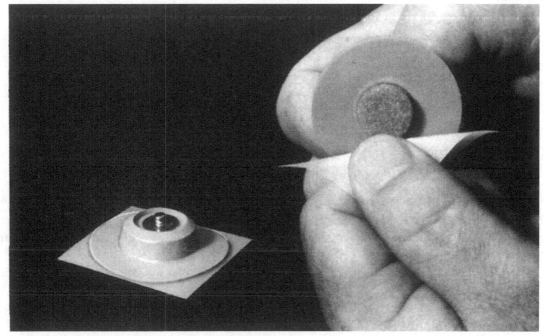

> **Fig. 2.16** *Disposable pre-gelled ECG electrode. A porous tape overlaying ring placed over the electrode resists perspiration and ensures positive placement under stress conditions (Courtesy: Mar Avionics, USA)*

Skin-electrode adhesion is an important performance criterion. Partial electrode lift would cause a gel dry out and intermittent contact artefacts, while premature electrode fall off would increase monitoring costs. The adhesives used to secure electrodes to the skin are usually pressure-sensitive adhesives which implies that a force must be applied to achieve adhesion. The adhesive should have good bonding capability; internal strength so that upon removal, objectionable "stringy" residue will not remain on the skin; good temperature stability; immunity to oxidation from air pollutants, saline and other common solutions; resistance to water, isopropyl alcohol, saline and other common solutions used in hospitals and low potential for skin irritation.

Trimby (1976) describes an alternative design to reduce motion artefacts generated by mechanical shocks. Ideally, the thin layer (critical layer) of electrolyte just under the metal piece should be cushioned from all mechanical shocks. Figure 2.17 shows the construction of fluid column electrodes (HP 14245A, HP 14248A) in which the critical layer is protected by a semi-rigid collar. These and several other commercially available electrodes are surrounded on the first few tenths of a millimetre by a plastic collar. However, even with this mechanical stabilization, a strong force in the vertical direction will be transmitted up the electrolyte column and will disturb the interface area. Motion artefact will still be seen in the trace. The rigid collar serves to minimize interference from minor pulls on the lead wire.

Pregelled ECG surface electrodes are manufactured with 0.3–1.5 g of electrolyte paste (gel, cream, or jelly) in contact with the sensor which forms a conducting bridge with the skin. A high value of $[Cl^-]$ common in pastes renders the electrode more non-polarizable and decreases the

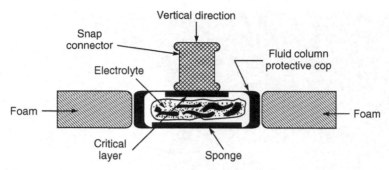

> **Fig. 2.17**   *Construction of a fluid column electrode*

electrolyte skin impedance. However, it must not be high enough to cause skin irritation. Ideally, gels should possess the following properties:

- Stay moist for the intended shelf life and during use. This is controlled by including a humectant in the gel.
- Prevent micro-organism and mould growth. Generally, the gels contain a bactericide/fungicide and may be disinfected using gamma radiation.
- Provide low electrolyte skin impedance by having ionic salts and surfactants.
- Cause minimum skin irritation, for which gels should have a pH range of 3.5–9.

Electrode-skin contact impedance with respect to pregelled disposable ECG electrodes was measured by Klingler *et al* (1979 b) using a 10 Hz ac current source. It was found that the skin on the average contributed a resistance of 564 Ω to the equivalent electrode impedance if mildly abraided, but contributed 54.7 kΩ if applied to clean, dry skin. They further found that over 90% of electrodes on abraided skin will have less than 5 kΩ impedance imbalance. The curve for clean, dry skin shows a very significant imbalance. In fact 20% of the electrodes will exceed 15 kΩ imbalance.

ECG electrodes are used in conjunction with cardiac monitors or ECG recorders which invariably have dc-input bias currents. Klingler *et al* (1979 b) studied the effects of these small dc currents on the offset potentials of disposable ECG electrodes. They found that after periods ranging from a few minutes to several days, the electrodes tested (four brands of silver-silver chloride and two brands of stainless steel) exhibited offset potentials exceeding 200 mV after subjection to dc bias currents over 200 nA. All silver-silver chloride electrodes were able to withstand bias currents of 200 nA, with minimal changes in offset for periods up to seven days. On the other hand, the stainless steel electrodes exhibited large offset potentials within minutes after subjection to bias currents of only 100 nA. Based on this study, a 200 nA limit on the dc-input bias current for cardiac monitors is suggested.

The modern ECG monitors are generally provided with inputs protected against defibrillator overloads. The high defibrillating currents are harmlessly bypassed through neon or diode breakdown circuits. However, this unidirectional current passes through and tends to polarize the electrodes. Usually, the standards on ECG monitors require that the trace be readable within 5 s after three or fewer defibrillator discharges. This implies that the electrode polarization voltage must return to below 300 mV within a few seconds after application of the defibrillating voltages.

Schoenberg *et al* (1979) developed a standard test method for evaluating defibrillation recovery characteristics of disposable ECG electrodes.

ECG pregelled electrodes can be characterized electrically by tests developed by the Association for the Advancement of Medical Instrumentation (AAMI), USA to establish a reasonable safety and efficacy level in clinical use of electrodes. In abridged form the standards are as follows:

1. *Direct-current offset voltage:* A pair of electrodes connected gel-to-gel, after 1 min stabilization period must exhibit offset voltage no greater than 100 mV.

2. *Combined offset instabiliy and internal noise:* A pair of electrodes connected gel-to-gel shall after 1 min. stablization generate a voltage no greater than 150 μV in the passband.

3. *Alternating-current impedance:* The average value of 10 Hz for at least 12 electrode pairs connected at a level of impressed current not exceeding 100 mA peak to peak shall not exceed 2 kΩ. None of the individual pair impedances shall exceed 3 kΩ.

4. *Defibrillation overload recovery:* The absolute value of polarization potential of a pair of electrodes connected gel-to-gel shall not exceed 100 mV, 5s after each of four capacitor discharges. The capacitor should be 10 μF charged to 200 V and discharged through the electrode pair with 100 Ω in series.

5. *Bias current tolerance:* The observed dc voltage offset change across an electrode pair connected gel-to-gel shall not exceed 100 mV when subjected to a continuous 200 μA dc current over the period recommended by the manufacturer for the clinical use of the electrodes. In no case shall this period be less than 8 hours.

2.4.4 Pasteless Electrodes

ECG monitoring electrodes, in a majority of the cases, are metal plates applied to the skin after cleaning and application of a coupling-electrolyte in the form of an electrode paste or jelly. Such preliminary preparation can be sometimes irritating and time consuming. Also, it is often not done satisfactorily, resulting in problems like poor quality signals and baseline drift, etc. Another disadvantage of using electrode jelly is that during long-term monitoring there is likely to be patient-skin reactions as the electrode-skin interface dries out in a matter of a few hours. The electrodes need to be periodically removed for jelly replenishments, thus causing further discomfort due to repetitive skin preparation. In addition, bacterial and fungal growth can take place under electrodes worn over long periods. Also, in conductive electrodes, shifts in electrode position at the electrode-skin interface appear as baseline drift, particularly when the subject moves. Therefore, any attempt of using a dry electrode that may dispense with the practice of skin preparation would look attractive.

Capacitive Electrodes: A metal plate electrode in direct contact with the skin though makes a very high resistive contact and has a considerable capacitive contact too with the skin (Stevens, 1963). By using a very high input impedance amplifier, it is possible to record a signal through the tissue-electrode capacitance. Lopez and Richardson (1969) describe the construction of electrodes which can be capacitively coupled to the subject. The electrode consists of an aluminium plate which is anodized on the surface to be placed in contact with the skin. The ohmic resistance of the anodized electrode is about 1 to 30 GΩ (1000–30,000 MΩ). Two such electrodes are applied to the subject without any preparation of the skin and the output of the source followers is connected to a conventional electrocardiograph. Wolfson and Neuman (1969) also designed a capacitively coupling electrode and used a high input impedance amplifier having a MOSFET in the input stage arranged in a source-follower configuration. The capacitances encountered in such type of electrodes range from about 5000 to 20,000 pf/cm^2 of the electrode area (Geddes, 1972).

Conrad (1990) illustrates the construction of a capacitive electrode which is formed from conductive silicon wafer, oxygen diffused into one surface produces a silicon dioxide layer, which serves as the dielectric. A high performace FET operational amplifier in unity gain configuration acts as an impedance changer to permit use with systems designed for paste type electrodes. The insulated, capacitively coupled electrode is used on unprepared skin, which acts as one plate of the capacitor, the substrate acts as the other plate.

Luca *et al* (1979) designed an electrode and amplifier as an integrated unit, so that the assembly could be used in the front end of the commonly used biomedical recorders. The arrangement (Fig. 2.18) basically comprises a metal shell which performs a dual function as a housing for the electrode and as the ground contact. The shell is made of highly pure titanium metal measuring $30 \times 15 \times 7$ mm. Two FETs are cemented with epoxy glue in the middle of the shell, their centres spaced 10 mm apart. The recording surfaces are formed by the cases of the two FETs. The cans have a diameter of 4.5 mm and are made of stainless steel. The rectangular border of the shell acts as the ground contact and the remainder of the shell forms a shield against interfering radiation. The source leads of the two FETs are connected to the differential inputs of an instrumentation amplifier. The amplifier (Analog Devices 521) has a high ac input impedance (> 100 MΩ).

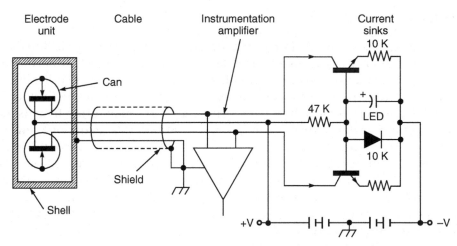

> **Fig. 2.18** *Schematic diagram of integrated electrode and amplifier arrangement for pasteless operation (after Luca et al, 1979; reproduced by permission of Med. and Biol. Eng. and Comp.)*

Among the drawbacks of the dry or pasteless electrodes are high electrode-skin resistance and the strong sensitivity to motion. Whereas a high electrode-skin resistance can be overcome by sufficiently high values of the amplifier input resistance, no universal remedy for eliminating motion artefacts from dry bioelectrodes has so far been suggested. Skin preparation certainly minimises motion artefacts (Burbank and Webster, 1978), but this time consuming manipulation loses the practicability of dry electrodes for routine clinical use and is also less acceptable to the patient.

An important characteristic observed with dry surface electrodes is that with the passage of time, the resistance of a dry electrode placed on dry human skin decreases exponentially. Geddes *et al* (1973) report that measurements made by them on silver disc electrodes (1.7 cm diameter) showed that for 10 determinations, the average initial resistance was about 1.36 MΩ. After about 20 min. the resistance of the electrode-subject circuit had dropped to about one-tenth of its initial value. The fall in impedance is attributed to the presence of small amounts of perspiration accumulated under the electrode.

Air-Jet ECG Electrodes: Wohnhas (1991) describes a novel air-jet electrode which employs Bernoulli technology to achieve constant and secure electrode contact resulting in quality tracings while minimizing artefact and maximizing baseline stability.

Air-jet electrodes (Fig. 2.19) are Ag-AgCl electrodes encased within a contoured medical silicon cup bounded by a skin-engaging rim. The contact area (pill) is anchored to a layer of synthetic, sintered carbon by a titanium screw. The miniature silver venturi air-jet bisects the sintered layer of synthetic carbon.

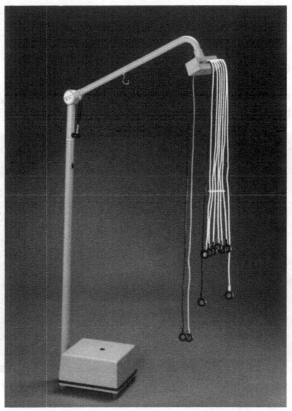

➤ **Fig. 2.19**   *Air-jet electrodes (Courtesy: M/s Medi-Globe, USA)*

Ambient air is drawn to a small compressor and passed to the electrodes at a constant pressure of 3.8 psi. As the air passes through the venturi air-jets, the electrodes produce a constant vacuum of 2.0 psi at the point of contact. The constant vacuum allows the electrodes to remain securely attached to the body. The fact that the air-jet functions with a positive flow of air ensures that no hair, skin particles or liquids clog the system. The electrode system eliminates gels, tapes and glue residues and generally reduces the need to sand or shave the patient's skin.

The combination of two special features—silver/silver chloride pills and sintered carbon base-plates—optimize the effectiveness of the air-jet electrodes. The electrode pills possess the advantage of low contact resistance. The micro-pressed carbon base-plate located behind the electrode pill serves to accurately transmit the electronic signal acquired from the electrode pill to the venturi air-jet and on to the stainless steel leadwire, and to securely position the air-jets within the electrode housing.

The venturi air-jet electrodes are integral components of the system comprising the ECG electrodes and patient cable in one device and is compatible with any ECG recorder on the market. The system uses a compressor to pump ambient air into a distribution box. From the distribution box, ten PVC hoses carry air and the stainless steel leadwires to the air jet electrodes. The electrode system is effective for both resting and exercise ECG applications. Heavy skin moisture is not a problem for the air jet electrodes—all excess liquids are literally blown out of the system via the air-jets. Also, the electrodes clean themselves when sprayed with a suitable disinfectant and submerged into water or alcohol while the system is running.

▐▐▶ 2.5   ELECTRODES FOR EEG

Among the most commonly used electrodes for EEG (electroencephalogram) recording are the chlorided silver discs (Fig. 2.20) having approximately 6–8 mm diameters. Contact with the scalp

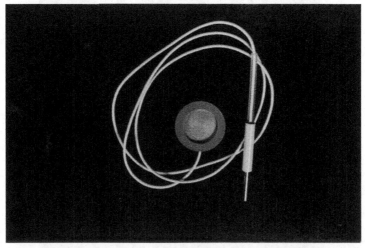

➤ **Fig. 2.20**   *EEG electrode which can be applied to the surface of the skin by an adhesive tape (Courtesy: In Vivo Metrics, USA)*

is made via an electrolytic paste through a washer of soft felt. They have ac resistance varying from 3–20 kΩ. Small needle electrodes are sometimes used for carrying out special EEG studies when they are inserted subcutaneously. Silver ball or pellet electrodes covered with a small cloth pad are useful when electrical activity is to be recorded from the exposed cortex, but they have high dc resistances.

Hector (1968) describes a pad electrode (Fig. 2.21(a)) which is made from a silver rod belled out at the end and padded with a sponge, or a similar material, contained in gauze. It is screwed into an insulated mount and held in place on the head with a rubber cap. To hold three such electrodes, an adjustable tripod mount is employed. Another type of EEG electrode consists of multiple fine chlorided silver wires (Fig. 2.21(b)) fixed in a rigid plastic cup. The plastic cup is fixed to the scalp with an adhesive. It is filled with jelly through a hole in the top. In this electrode, contact with the tissue is made via an electrolyte bridge so that jelly in contact with the electrode metal is not disturbed by scalp movement. To avoid metal junctions which may get corroded with electrolyte, the silver wires are used as the output lead. The large surface area and excess of silver chloride favour stability.

2.6 ELECTRODES FOR EMG

Electrodes for electromyographic work are usually of the needle type (Fig. 2.22(a)). Needle electrodes are used in clinical electromyography, neurography and other electrophysiological investigations of the muscle tissues underneath the skin and in the deeper tissues. The material of the needle electrode is generally stainless steel. In spite of the fact that stainless steel is unfavourable electrode material from the point of view of noise, it is preferred in EMG work due to its mechanical solidity and low price. Needle electrodes are designed to be fully autoclavable and in any case they should be thoroughly sterilized before use.

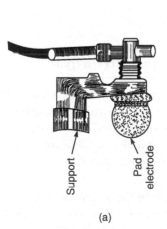

Support   Pad electrode

(a)

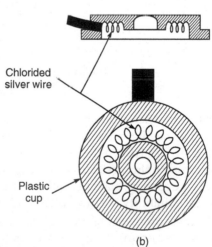

Chlorided silver wire

Plastic cup

(b)

➤ **Fig. 2.21**   (a) *Pad electrode for recording EEG (after Hector, 1968)*
　　　　　　　(b) *EEG electrode consisting of chlorided silver wire in plastic cup. Jelly is inserted from the hole kept for the purpose*

Needle electrodes come in various forms. The monopolar needle electrode usually consists of a Teflon coated stainless steel wire which is bare only at the tip. It is found that after the needle has been used a number of times, the Teflon coating will recede, increasing the tip area. The needle must be discarded when this occurs. Bipolar (double coaxial) needle electrodes contain two insulated wires within a metal cannula. The two wires are bared at the tip and provide the contacts to the patient. The cannula acts as the ground. Bipolar electrodes are electrically symmetrical and have no polarity sense.

A concentric (coaxial) core needle electrode contains both the active and reference electrode within the same structure. It consists of an insulated wire contained within a hypodermic needle (Fig. 2.22(b)). The inner wire is exposed at the tip and this forms one electrode. The concentric needle is very convenient to use and has very stable electrical characteristics. Care should be taken to maintain the surface electrode in good condition in order to avoid artefacts. Concentric needle electrodes are made by moulding a fine platinum wire into a hypodermic needle having an outside diameter less than 0.6 mm. One end of the needle is bevelled to expose the end of the wire and to provide easy penetration when the needle is inserted. The surface area of the exposed tip of the wire may be less than 0.0005 mm^2.

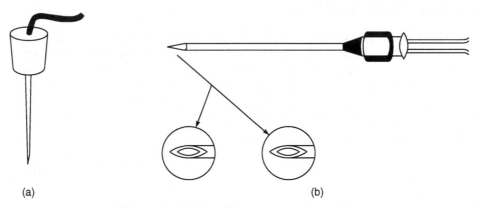

(a)   (b)

> **Fig. 2.22** (a) *Needle type EMG electrode*
>   (b) *Hypodermic needle type EMG electrode*

Multi-element needle electrodes are used to pick up the signals from individual fibres of muscle tissue. Special needles are available using 25-micron diameter electrode surfaces and having up to 14 pick up surfaces down the side of one needle. From the point of view of construction, needle electrodes are the simplest. However, edging of the needle point to the suitable angle, providing a proper plastic coating, making them resistant against thermal and chemical stresses and ensuring histological suitability is a difficult manufacturing process.

For the measurement of potentials from a specific part of the brain, longer needles are actually inserted into the brain. The needles are precisely located by means of a map or atlas of the brain. Generally, a special instrument called a stereotaxic instrument is used to hold the subject's head and guide the placement of electrodes. Often, these electrodes are implanted to permit repeated measurements over an extended period of time.

The ground electrode for EMG studies usually consists of a conducting strip which is inserted into a saline soaked strap and wrapped around the patient's limb. The ground electrode is usually positioned over bony structures rather than over large muscle masses, in the vicinity of the recording and stimulating electrodes, and where possible, equidistant from them. Surface electrodes are employed for recording gross electrical activity from a particular group of underlying muscles in nerve-conduction velocity measurements. A single surface electrode may also be used as the reference (indifferent) electrode with monopolar needle electrodes. Surface electrodes can be easily and quickly attached and are generally comfortable to wear over long periods. Surface electrodes usually consist of square or circular metal (chlorided silver) plates with leadoff wires attached. They are held in place by straps or adhesive tapes. To reduce electrical resistance between the skin and the electrode, the use of saline soaked felt pads or a small amount of electrode gel between the electrode surface and the skin is recommended. Disposable, adhesive type electrodes are also used for EMG work.

Bhullar *et al* (1990) describe the design and construction of a selective non-invasive surface electrode to study myoelectric signals. The electrode recording surfaces are two concentric steel rings. A third ring attached to the casing of the electrode is the earth contact. The rings are separated from each other by Teflon, the insulating material. Figure 2.23 shows a schematic diagram of the electrode configuration. The small surface area of the electrode plates, the small physical size and the concentric arrangement produce the effect of recording signals mainly from fibres near to the axis of the electrode and thereby make the electrode much more selective. The concentric ring instead of the normal passive electrode configuration also obviates the problem of electrode alignment relative to the direction of the muscle fibres. The results of tests undertaken with these electrodes showed that it was able to pick up individual motor unit action potentials at moderate force levels.

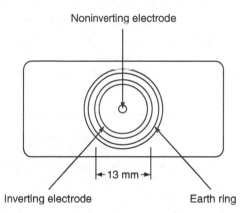

➤ **Fig. 2.23** *Schematic diagram of the electrode configuration to study myoelectric signals (after Bhullar et al., 1990)*

Crenner *et al* (1989) constructed a special electrode which allows recording of electrical signals from their muscular layers, specifically to collect electromyographic signals of the gastrointestinal tract. The active electrode is surrounded by a ring which avoids the recording of interfering signals.

▨➤ 2.7 ELECTRICAL CONDUCTIVITY OF ELECTRODE JELLIES AND CREAMS

Conducting creams and jellies have for long been used to facilitate a more intimate contact between the subject's skin and the recording electrodes. The outer horny layer of the skin is responsible for the bulk of the skin contact impedance, and for this reason careful skin preparation is essential in order to obtain the best results. The recording site should first be cleaned with an ether-meth mixture. In addition to having good electrical conductivity, the electrode jelly must have a particular chloride ion concentration (about 1%) close to the physiological chloride concentration. This is primarily important for long-term monitoring because it should not produce a harmful diffusion between the jelly and the body. It is to be particularly ensured that the jelly chosen is of a bland nature and does not contain soap or phenol which can produce a marked irritation of the skin after a few hours.

The electrical conductivity of different makes of electrode cream can be measured (Hill and Khandpur, 1969) by means of the Schering ac bridge circuit. The cream is placed in a perspex conductivity cell of known dimensions and the resistive component of the cell impedance is measured at 10 Hz, the conductivity being calculated from the cell dimensions.

The contact impedance of the skin depends upon the type of electrolyte used and the time (Trimby, 1976). Figure 2.24 shows the effect of these parameters. A low concentration sodium chloride electrolyte has 0.5% NaCI and a high concentration electrolyte has a concentration in the range of 5 to 10% NaCI. The impedance is found to fall rapidly to 40% between 7 to 30 min. Stabilization occurs at about 30 to 45 min. An interesting observation from this figure is that while pre-rubbing the skin will lower the initial impedance value, the final value after using a high concentration electrolyte becomes nearly the same.

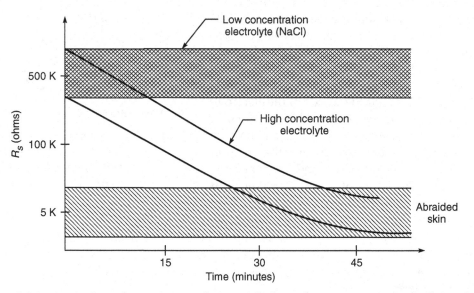

> **Fig. 2.24**   *Variation of contact impedance with electrolyte concentration and time (re-drawn after Trimby, 1976; Courtesy: Hewlett Packard, USA)*

Electrode jelly can be replaced in certain cases by using a conducting plastic as an interface between the electrode and the surface of the body. Jenkner (1967) used silastic S-2086 by Dow Corning with EEG electrodes and showed that contact resistance was almost the same as with a conventional electrode which make use of electrode jelly.

ⅢⅢ▶ 2.8 MICROELECTRODES

To study the electrical activity of individual cells, microelectrodes are employed. This type of electrode is small enough with respect to the size of the cell in which it is inserted so that penetration by the electrode does not damage the cell. The size of an intracellular microelectrode is dictated by the size of the cell and the ability of its enveloping membrane to tolerate penetration by the microelectrode tip. Single-living cells are rarely larger than 0.5 mm (500 microns) and are usually less than one-tenth of this size. Typical microelectrodes have tip dimensions ranging from 0.5 to 5 microns. The tips of these electrodes have to be sufficiently strong to be introduced through layers of tissues without breaking.

Two types of microelectrodes are generally used: metallic (Fig. 2.25(a)) and glass microcapillaries (Fig. 2.25(b)). Metallic electrodes are formed from a fine needle of a suitable metal drawn to a fine tip. On the other hand, glass electrodes are drawn from Pyrex glass of special grade. These microcapillaries are usually filled with an electrolyte. The metal microelectrodes are used in direct contact with the biological tissue and, therefore, have a lower resistance. However, they polarize with smaller amplifier input currents. Hence, they tend to develop unstable electrode offset potentials and are therefore not preferred for steady state potential measurements. On the other hand, in case of glass microelectrodes, improved stability can be obtained by properly choosing the metal and the electrolyte so that the small current passing through their junction may not be able to modify the electrical properties of the electrodes. Also, the glass microelectrode has a substantial current carrying capacity because of the large surface contact area between the metal and the electrolyte.

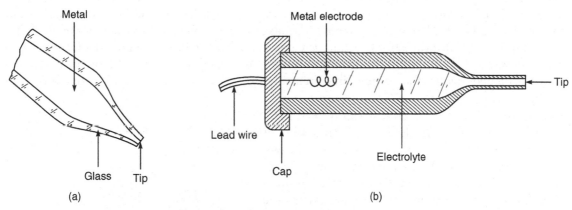

(a)   (b)

> **Fig. 2.25** (a) *Microelectrodes—metal microelectrodes*
> (b) *Microelectrodes—micropipette or micro capillaries electrode*

The microelectrodes have a very high impedance as compared to conventional electrodes used for recording ECG, EEG, etc. The high impedance of a metal microelectrode is due to the characteristics of the small area metal-electrolyte interface. Similarly, a micropipet tip is filled with an electrolyte which substitutes an electrolytic conductor of small cross-sectional area, which gives a micropipet its high resistance. Because of high impedance of microelectrodes, amplifiers with extremely high input impedances are required to avoid loading the circuit and to minimize the effects of small changes in interface impedance.

2.8.1 Glass Microcapillary Electrodes

Several methods exist for producing microelectrodes of wide variety and shapes. For drawing electrodes of uniform and accurate diameter, it is essential to maintain constant timing, temperature, strength and direction of pull. These factors are difficult to control when the electrodes are drawn manually. The mechanical method employs gravitational force for extension and the electrodes which are drawn in one or more stages can readily produce capillary tubes between 3–30 μm diameter, but great difficulty is encountered in producing electrodes of less than 1 μm. The most commonly used method for making small tip micropipet consists of the circumferential application of heat to a small area of glass tubing which is placed under some initial tension. When the glass softens, the tension is increased very rapidly and the heat is turned off. Proper timing, controlled adjustment of the amount of heat as well as the initial and final tensions and cooling result in the production of microcapillaries with controlled dimensions.

2.8.2 Metal Microelectrodes

Metal electrodes with very fine tips used for recording from single cells have the advantage over glass micropipetes of being relatively robust. Steel microelectrodes can be made from ordinary darning needles but preferably they should be of good stainless steel wire. They can be easily made up to 10 μm diameter but great care has to be taken for diameters as small as 1 μm. These very small tips are not very satisfactory as they are extremely brittle and have very high input impedance. Hubel (1957) described a method to make tungsten microelectrodes with a tip diameter of 0.4 μm. He used electropointing technique which consists in etching a metal rod while the metal rod is slowly withdrawn from the etching solution, thus forming a tapered tip on the end of the rod. The etched metal is then dipped into an insulating solution for placing insulation on all but the tip.

Figure 2.26 shows the cross-section of a metal microelectrode. In this electrode, a thin film of precious metal is bonded to the outside of a drawn glass microelectrode. This arrangement offers lower impedance than the microcapillary electrode, infinite shelf life and reproduciable performance, with ease of cleaning and maintenance. The metal—electrolyte interface is between the metal film and the electrolyte of the cell.

Skrzypek and Keller (1975) illustrated a new method of manufacturing tungsten microelectrode permitting close control of microelectrode parameters. In this technique, the tips are dc electroetched to diameters below 500° A and completely covered

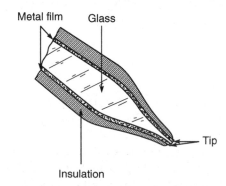

➤ Fig. 2.26 *Microelectrode—thin metal film coated micropipette*

with polymethyl methacrylate. An electron beam from a scanning electron microscope is then used to expose a precise area on the tip for later removal by chemical methods. Recording results with these electrodes suggested good desirable recording characteristics, i.e. ability to isolate and hold single cells.

Tungsten is preferred for constructing micro-electrodes due to its mechanical strength and its apparent inertness. Although tungsten itself is reactive, a surface layer of tungsten oxide will, in most situations, protect the metal against corrosion. The electrical properties of tungsten microelectrodes made with a taper of the tip of about 1:10 and insulated with lacquer leaving a tip length of about 10–100 μm were studied by Zeuthen (1978). The resting potential in saline was found to be –0.3 V relative to a silver-silver chloride reference electrode for input currents less than 10^{12} A. The small signal impedance was ideally that of a capacitor 0.4 pF/μm^2 between 10 and 1000 Hz. Imperfect insulation at the tip caused this impedance to be increasingly resistive. The electrochemical properties of tungsten showed that it behaves as an inert metal within the potentials where it is usually used in biological experiments.

Jobling *et al* (1981) constructed an active microelectrode array using IC fabrication technology to record extracellular potentials in nervous tissues. The array substantially reduces the noise caused by electrostatic pick-up with a good long-term stability.

Physiological Transducers

3.1 INTRODUCTION

Transducers are devices which convert one form of energy into another. Because of the familiar advantages of electric and electronic methods of measurement, it is the usual practice to convert all non-electric phenomenon associated with the physiological events into electric quantities. Numerous methods have since been developed for this purpose and basic principles of physics have extensively been employed. Variation in electric circuit parameters like resistance, capacitance and inductance in accordance with the events to be measured, is the simplest of such methods. Peizo-electric and photoelectric transducers are also very common. Chemical events are detected by measurement of current flow through the electrolyte or by the potential changes developed across the membrane electrodes. A number of factors decide the choice of a particular transducer to be used for the study of a specific phenomenon. These factors include:

- The magnitude of quantity to be measured.
- The order of accuracy required.
- The static or dynamic character of the process to be studied.
- The site of application on the patient's body, both for short-term and long-term monitoring.
- Economic considerations.

3.2 CLASSIFICATION OF TRANSDUCERS

Many physical, chemical and optical properties and principles can be applied to construct transducers for applications in the medical field. The transducers can be classified in many ways, such as:

(i) By the process used to convert the signal energy into an electrical signal. For this, transducers can be categorized as:

Active Transducers—a transducer that converts one form of energy directly into another. For example: photovoltaic cell in which light energy is converted into electrical energy.

Passive Transducers—a transducer that requires energy to be put into it in order to translate changes due to the measurand. They utilize the principle of controlling a dc excitation voltage or an ac carrier signal. For example: a variable resistance placed in a Wheatstone bridge in which the voltage at the output of the circuit reflects the physical variable. Here, the actual transducer is a passive circuit element but needs to be powered by an ac or dc excitation signal.

(ii) By the physical or chemical principles used. For example: variable resistance devices, Hall effect devices and optical fibre transducers.

(iii) By application for measuring a specific physiological variable. For example: flow transducers, pressure transducers, temperature transducers, etc.

Biomedical parameters which are commonly encountered are listed alongwith their characteristics and corresponding transducers in Tables 3.1 to 3.3.

ⅢⅢ▶ 3.3 PERFORMANCE CHARACTERISTICS OF TRANSDUCERS

A transducer is normally placed at the input of a measurement system, and therefore, its characteristics play an important role in determining the performance of the system. The characteristics of a transducer can be categorized as follows:

3.3.1 Static Characteristics

Accuracy: This term describes the algebraic difference between the indicated value and the true or theoretical value of the measurand. In practice, accuracy is usually expressed as a percentage of full scale output or percent of reading or ± number of digits for digital readouts.

Precision: It refers to the degree of repeatability of a measurement. Precision should not be confused with accuracy. For example: an uncompensated offset voltage in an operational amplifier may give very reproducible results (high precision), but they may not be accurate.

Resolution: The resolution of a transducer indicates the smallest measureable input increment.

Sensitivity: It describes transfer ratio of output to input.

Drift: It indicates a change of baseline (output when input is zero) or of sensitivity with time, temperature etc. It may be noted that the sensitivity of the device does not change the calibration curve if shifted up or down.

Linearity: It shows closeness of a transducer's calibration curve to a specified straight line with in a given percentage of full scale output. Basically, it reflects that the output is in some way proportional to the input.

Threshold: The threshold of the transducer is the smallest change in the measurand that will result in a measureable change in the transducer output. It sets a lower limit on the measurement capability of a transducer.

Noise: This is an unwanted signal at the output due either to internal source or to interference.

• Table 3.1   *Signals from Cardiovascular System*

Parameter	Primary signal characteristics	Transducer required
Blood pressure (arterial, direct)	Frequency range: dc to 200 Hz (dc to 40 Hz usually adequate) Pressure range: 20 to 300 mm Hg; slightly negative in left ventricle	Unbonded wire strain gauge pressure transducer Bonded semiconductor strain gauge transducer Capacitance type pressure transducer Differential transformer
Blood pressure (arterial, indirect)	Frequency range: dc to 5 Hz Pressure range: 20 to 300 mm Hg Microphone: 10 to 100 Hz, signal voltage depends upon the type of microphone used	Low frequency microphone for picking up Korotkoff sounds
Blood pressure (venous, direct)	Frequency range: dc to 40 Hz Pressure range: −5 to +20 mm Hg	Strain gauge pressure transducer with higher sensitivity
Peripheral arterial blood pressure pulse wave	Frequency range: 0.1 to 50 Hz usually adequate Pulse trace similar to arterial blood pressure waveform	Finger or ear lobe pick-up (light source and photocell) and piezo-electric pick-up or by impedance method
Blood flow (aortic or venous)	Rate: 0 to 300 ml/s Frequency range: 0 to 100 Hz	Tracer methods Electromagnetic flowmeter Ultasonic Doppler flowmeter
Cardiac output (blood flow)	Frequency range: 0 to 60 Hz (0 to 5 Hz usually adequate) Blood flow measured as litres per minute (4 to 25 litre/min)	Dye dilution methods Integration of aortic blood flow function Thermal dilution method Electromagnetic flowmeter and Integrator.
Heart rate	Rate: 25 to 300 beats per min. Normal human heart rate at rest 60 to 90 beats/min. Normal foetal heart rate: 110 to 175 beats/min.	Derived from ECG or arterial blood pressure waveform or photo-electric plethysmograph From foetal phonocardiogram ultra-sonic method, scalp electrodes for foetal ECG
Phonocardiogram (heart sounds)	Frequency range: 20 to 2000 Hz Signal voltage depends upon the type of microphone used	Crystal or moving coil microphone
Ballistocardiogram (BCG)	Frequency range: dc to 40 Hz	Infinite period platform with strain gauge accelerometer
Impedance cardiogram (Rheocardiography)	Frequency range: dc to 50 Hz Impedance range: 10 to 500 Ω Typical frquency employed for measuring impedance: 20 kHz to 50 kHz	Surface or needle electrodes
Oximetry	Frequency range: 0 to 60 Hz; 0 to 5 Hz usually adequate	Photoelectric pulse pick-up Photoelectric flow-through cuvette

• Table 3.2 *Signals from Respiratory System*

Parameter	Primary signal characteristics	Transducer required
Respiration rate	Normal range: 0 to 50 breaths/min	Thermistor Impedance pneumography electrodes Microswitch method CO_2 detectors Strain gauge transducer Doppler shift transducer
Pneumotachogram (respiratory flow rate)	Frequency range: dc to 50 Hz	Fleisch pneumotachograph BMR spirometers
Impedance pneumograh	Frequency range: dc to 30 Hz (from demodulated carrier) Typical carrier frequency used: 20 kHz to 50 kHz	Surface or needle electrodes
Tidal volume (Volume/breath)	Frequency range: dc to 5 Hz Typical value for adult human: 600 ml/breath	Direct from spirometer Integrated from pneumotachogram
Minute ventilation (vol/min)	6 to 8 l/min	Integrated from pneumotachogram
Gases in expired air	CO_2–normal range: 0 to 10% End tidal CO_2 (human) 4 to 6% N_2O – Range 0 to 100% Halothane–Range 0 to 3%	Infrared sensors Mass spectrometery
Diffusion of inspired gases (nitrogen washout technique)	Normal range of nitrogen concentration differential: 0 to 10%	Discharge tube nitrogen analyzer
Pulmonary diffusing capacity (using carbon monoxide)	Normal range(human): 16 to 35 ml CO/mmHg/min	Infrared absorption by CO
Dissolved Gases and pH		
pH	Signal range: 0 to ± 700 mV	Glass and calomel electrodes
Partial pressure of dissolved CO_2(P_{CO_2})	Range: 1 to 1000 mm Hg Normal signal range: 0 to ± 150 mV	Direct recording CO_2 electrode
Partial pressure of dissolved oxygen (P_{O_2})	Normal measurement range: 0 to 800 mm Hg Hyperbaric P_{O_2} range: 800 to 3000 mm Hg	Polarographic electrodes
Na^+, Cl^- Cations	Signal range: 60 mV/decade	Specific ion-sensitive electrodes

Hysteresis: It describes change of output with the same value of input but with a different history of input variation. For example: hysteresis is observed when the input/output characteristics for a transducer are different for increasing inputs than for decreasing inputs. It results when some of the energy applied for increasing inputs is not recovered when the input decreases.

Span: It indicates total operating range of the transducer.

• Table 3.3   *Physical Quantities*

Parameter	Primary signal characteristics	Transducer required
Temperature	Frequency range: dc to 1 Hz	Thermocouples Electrical resistance thermometer Thermistor Silicon diode
Galvanic skin resistance (GSR)	Resistance range: 1 k to 500 kW	Surface eletrodes similar to that of ECG electrodes
Plethysmogram (volume measurement)	Frequency range: dc to 30 Hz	Qualitative Plethysmography by: -photoelectric method -impedance method -piezo-electric method
Tocogram (uterine contr- ations measurement during labour)	Frequency range: dc to 5 Hz	Strain gauge transducers
Isometric force, dimensional change		Strain gauge (resistance or semiconductor

Saturation: In a transducer the output is generally proportional to the input. Sometimes, if the input continues to increase positively or negatively, a point is reached where the transducer will no longer increase its output for increased input, giving rise to a non-linear relationship. The region in which the output does not change with increase in input is called the saturation region.

Conformance: Conformance indicates closeness of a calibration curve to a specified curve for an inherently non-linear transducer.

3.3.2   Dynamic Characteristics

Only a few signals of interest in the field of medical instrumentation, such as body temperature, are of constant or slowly varying type. Most of the signals are function of time and therefore, it is the time varying property of biomedical signals that is required to consider the dynamic characteristics of the measurement system. Obviously, when a measurement system is subjected to varying inputs, the input-output relation becomes quite different form that in the static case. In general, the dynamic response of the system can be expressed in the form of a differential equation.

For any dynamic system, the order of the differential equation that describes the system is called the order of the system. Most medical instrumentation systems can be classified into zero-, first-, second-, and higher-order systems.

A **zero-order system** has an ideal dynamic performance, because the output is proportional to the input for all frequencies and there is no amplitude or phase distortion. A linear potentiometer used as a displacement trasduser is a good example of a zero-order transducer.

The **first-order** transducer or instrument is characterized by a linear differential equation. The temperature transducers are typical examples of first order measuring devices since they can be characterized by a single parameter, i.e. time constant 'T'. The differential equation for the first-order system is given by

$$y + T \, dy/dx = x(t)$$

where x is the input, y the output and $x(t)$, the time function of the input.

A transducer or an instrument is of **second-order** if a second-order differential equation is required to describe the dynamic response. A typical example a second-order system is the spring—mass system of the measurement of force. In this system, the two parameters that characterize it are the natural frequency f_n (or angular frequency $w_n = 2\pi f_n$), and the damping ratio 'z'. The second-order differential equation for this system is given by

$$(1/w^2_n) \, d^2y/dx^2 + (2z/w_n) \, dy/dx + y = x(t)$$

where w_n is in radians per second and 'z' is a non-dimensional quantity. It may be noted that in this system, mass, spring and viscous-damping element oppose the applied input force and the output is the resulting displacement of the movable mass attached to the spring.

In the second-order systems, the natural frequency is an index of speed of response whereas the damping ratio is a measure of the system stability. An under-damped system exhibits oscillatory output in response to a transient input whereas an over-damped system would show sluggish response, thereby taking considerable time to reach the steady-state value. Therefore, such systems are required to be critically damped for a stable output.

Another important parameter to describe the dynamic performance of a transducer is the **response time**. It characterizes the response of a transducer to a step change in the input (measurand) and includes rise time, decay time and time constant.

3.3.3   Other Characteristics

There are many other characteristics which determine the performance and choice of a transducer for a particular application in medical instrumentation systems. Some of these characteristics are:

- Input and output impedance
- Overlead range
- Recovery time after overload
- Excitation voltage
- Shelf life
- Reliability
- Size and weight

⬛⬛⬛▶ 3.4   DISPLACEMENT, POSITION AND MOTION TRANSDUCERS

These transducers are useful in measuring the size, shape and position of the organs and tissues of the body. Specifically, the following measurements are made:

Position: Spatial location of a body or point with respect to a reference point.

Displacement: Vector representing a change in position of a body or point with respect to a reference point. Displacement may be linear or angular.

Motion: Change in position with respect to a reference system.

Displacement transducers can be used in both direct and indirect systems of measurement. For example: direct measurements of displacement could be used to determine the change in diameter of the blood vessels and the changes in volume and shape of the cardiac chambers. Indirect measurements of displacement are used to quantify the movement of liquids through the heart valves. For example: detection of movement of the heart indirectly through movement of a microphone diaphragm. Displacement measurements are of great interest because they form the basis of many transducers for measuring pressure, force, acceleration and temperature, etc. The following types of transducers are generally used for displacement, position and motion measurements.

3.4.1 Potentiometric Transducers

One of the simplest transducers for measuring displacement is a variable resistor (potentiometer) similar to the volume control on an audio electronic device. The resistance between two terminals on this device is related to the linear or angular displacement of a sliding tap along a resistance element (Fig. 3.1).

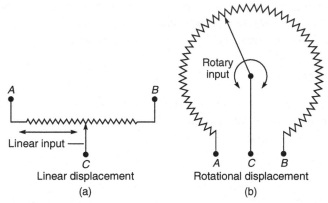

> **Fig. 3.1** *Principle of a potentiometric displacement transducer for the measurement of (a) linear displacement (b) angular displacement*

When the fixed terminals of the potentiometer are connected to the power supply, either ac or dc, output voltage at the wiper varies with the displacement of the object. Single turn, multi-turn (2 to 10 turns), linear, logarthmic and sectored potentiometers are available for different applications.

3.4.2 Variable Capacitance

When the distance between a pair of metallic plates forming an electrical capacitance is altered, there is a change in the capacitance according to the relation:

$$C = 0.0885 \, k \cdot A / d$$

where   C = capacitance in micro-microfarads

d = distance between the plates in cm

A = area of each identical plate in cm^2

k = dielectric constant of the medium separating the two plates

Each of the quantities in this equation can be varied to form a displacement transducer. By moving one of the plates with respect to the other, the capacitance will vary inversely with respect to the plate separation. This will give a hyperbolic displacement—capacitance characteristic. However, if the plate separation is kept constant and the plates are displaced laterally with respect to one another so that the area of overlap changes, a displacement can be produced—capacitance characteristic that can be linear, depending upon the shape of the actual plates.

The inverse relationship between C and d means that the sensitivity increases as the plate separation decreases. Hence, it is desirable to design a capacitor transducer with a small plate separation and a large area. However, in the case of a displacement transducer, it is the change in capacitance that is proportional to the displacement, the sensitivity will be independent of plate area but increases as d approaches zero.

For measurement of displacement, the capacitor is made a part of an LC oscillator wherein the resulting frequency change can be converted into an equivalent output voltage. However, lack of repeatability and the difficulty in properly positioning the transducer makes the measurement a difficult task.

Capacitance transducers are very sensitive displacement transducers. Therefore, they are required to be thermally insulated and the connecting cables have to be made as small as possible in length to avoid involvement of the cable capacitance in the measuring circuit. They have the disadvantage that a special high frequency energizing system is needed for their operation.

3.4.3   Variable Inductance

Changes in the inductance can be used to measure displacement by varying any of the three coil parameters given in the following equation

$$L = n^2 G \mu$$

Where L = inductance of the coil

n = number of turns in a coil

μ = permeability of the medium

G = geometric form factor

The inductance can be changed either by varying its physical dimensions or by changing the permeability of its magnetic core. The core having a permeability higher than air can be made to move through the coil in relation to the displacement. The changes in the inductance can be measured using an ac signal which would then correspond to the displacement.

Another transducer based on inductance is the variable reluctance transducer. In this transducer, the core remains stationary but the air gap in the magnetic path of the core is varied to change the effective permeability. It may however be noted that the inductance of the coil in these cases is usually not related linearly to the displacement of the core or the size of the air gap, especially if large displacements are to be measured.

3.4.4   Linear Variable Differential Transformer (LVDT)

An extremely useful phenomenon frequently utilized in designing displacement pick up units is based upon variations in coupling between transformer windings, when the magnetic core of the transformer is displaced with respect to the position of these two windings. These transducers can be conveniently used for the measurement of physiological pressures. Generally, a differential transformer (Fig. 3.2) designed on this principle is employed for this purpose. The central coil is the energizing or primary coil connected to a sinewave oscillator. The two other coils (secondary coils) are so connected that their outputs are equal in magnitude but opposite in phase.

With the ferromagnetic core symmetrically placed between the coils, and the two secondary coils connected in series, the induced output voltage across them would be zero.

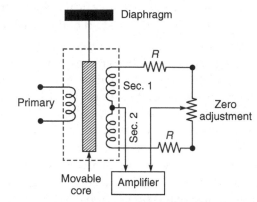

> **Fig.** 3.2   *Principle of LVDT pressure trans-ducer. The diagram shows the diffrential transformer and bridge circuit for detection of diffrential signals*

When the core is moved, the voltage induced in one of the secondary windings exceeds that induced in the other. A net voltage output then results from the two secondaries. The phase of the output will reverse if the core moves past the central position. A simple bridge circuit can be employed to detect the differential signals thus produced. The signal can be further processed to directly display a calibrated output in terms of mm of displacement. Since there is always some capacitive coupling between windings of differential transducers, it produces a quadrature component of induced voltage in the secondary windings. Because of the presence of this component, it is usually not possible to reduce the output voltage to zero unless the phase of the backing voltage is also altered.

Differential transformer displacement transducers generally work in conjunction with carrier amplifiers. Typical operating excitation of these transducers is 6 V at 2.5 kHz. Since the output voltage of the LVDT is proportional to the excitation voltage, the sensitivity is usually defined for a 1 Volt excitation. Commercial devices typically have a sensitivity of 0.5 to 2.0 mV per 0.001 cm displacement for a 1 Volt input. Full scale displacements of 0.001 to 25 cm with a linearity of ± 0.25 % are available.

3.4.5   Linear or Angular Encoders

With the increasing use of digital technology in biomedical instrumentation, it is becoming desirable to have transducers which can give the output directly in the digital format. Transducers are now available for the measurement of linear or angular displacements, which provide output in the digital form. These transducers are basically encoded disks or rulers with digital patterns photographically etched on glass plates. These patterns are decoded using a light source and an array of photodetectors (photo-diodes or photo-transistors). A digital signal that indicates the position of the encoding disk is obtained, which thereby represents the displacement being measured. Figure 3.3 shows typical patterns on the digital encoders.

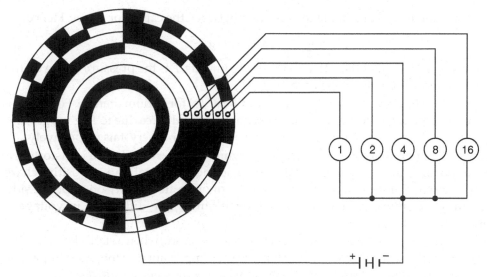

> **➤ Fig. 3.3** *Spatial encoder using a binary counting system*

The encoder consists of a cylinderical disk with the coding patterns arranged in concentric rings, having a defined number of segments on each ring. The number of segments on the concentric rings decrease, in a binary count (32-16-8-4-2), from a total of 32 (16 conductive and 16 non-conductive) on the outside ring to two on the inside ring. Each angular position of the disk would have a different combination of segments which will indicate the position of the shaft on which the disk is mounted. Since the outer ring of the disk encoder has 32 distinct areas, the resolution would be $1/32 \times 360° = 11$–$1/4°$ (1 digit). The resolution can be improved by increasing the number of segments on each ring, thereby decreasing the angle corresponding to each segment.

3.4.6 Piezo-electric Transducers

The piezo-electric effect is a property of natural crystalline substances to develop electric potential along a crystallographic axis in response to the movement of charge as a result of mechanical deformation. Thus, piezo-electricity is pressure electricity. The reverse effect is also present. When an electrical field is applied, the crystal changes shape. On application of pressure, the charge Q developed along a particular axis is given by

$$Q = dF \text{ coulomb}$$

where d is the piezo-electric constant (expressed in Coulombs/Newton, i.e. C/N) and F is the applied force. The change in voltage can be found by assuming that the system acts like a parallel—plate capacitor where the voltage E_o across the capacitor is charge Q divided by capacitance C. Therefore

$$E_o = \frac{Q}{C} = \frac{dF}{C}$$

The capacitance between two parallel plates of area 'a' separated by distance 'x' is given by

$$C = \epsilon \frac{a}{x}$$

where ϵ is the dielectric constant of the insulator between the capacitor plates. Hence,

$$E_o = \frac{d}{\epsilon} \cdot \frac{F}{a} \cdot x = g \cdot P \cdot x$$

where $d/\epsilon = g$ is defined as the voltage sensitivity in volts, P is the pressure acting on the crystal per unit area and x is the thickness of the crystal.

Typical values for d are 2.3 C/N for quartz and 140 C/N for barium titanate. For a piezo-electric transducer of 1 cm^2 area and 1 mm thickness with an applied force due to 10 g weight, the output voltage is 0.23 mV and 14 mV for the quartz and barium titanate crystals respectively.

The priniciple of operation is that when an asymmetrical crystal lattice is distorted, a charge re-orientation takes place, causing a relative displacement of negative and positive charges. The displaced internal charges induce surface charges of opposite polarity on opposite sides of the crystal. Surface charge can be determined by measuring the difference in voltage between electrodes attached to the surfaces.

Piezo-electric materials have a high resistance, and therefore, when a static deflection is applied, the charge leaks through the leakage resistor. It is thus important that the input impedance of the voltage measuring device must be higher than that of the piezo-electric sensor.

In order to describe the behaviour of piezo-electric elements, equivalent circuit techniques have been utilized. The analogies for electro-mechanical system which have been used for piezo-electric structures are:

Electrical Units	Analogous Mechanical Units
Voltage	Force
Current	Velocity
Charge	Displacement
Capacitance	Compliance
Inductance	Mass
Impedance	Mechanical impedance

Figure 3.4 shows the two basic equivalent circuits. If the constants in the two circuits are suitably related, the circuits are equivalent at all frequencies. The mechanical terminals represent the face or point of mechanical energy transfer to or from the piezo-electric element. The inductance symbol represents the effective vibrating mass of the element. The transformer symbol represents an ideal electro-mechanical transformer—a device that transforms voltage to force and vice versa, and current to velocity and vice versa. The transformation ratio N:1 in the circuit A is the ratio of voltage input to force output of the ideal transformer and also the ratio of velocity input to current output. The transformation ratio 1:N in circuit B has similar significance. The capacitance symbols on the electrical side represent electrical capacitances. The capacitance symbol on the mechanical side represent mechanical compliances.

A large number of crytsallographic substances are known but the most extensively used piezo-electric crystals are quartz, tourmaline, ammonium dihydrogen phosphate, Rochelle salt, lithium sulphate, lead zirconate and barium titanate.

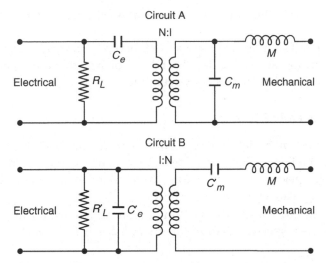

> **Fig. 3.4** *Equivalent circuits for piezo-electric elements*

Quartz is the most stable natural crystal with high mechanical and thermal stability and has volume resistivity higher than 10^4 ohm-cm and small internal electric loss. Rochelle salt has low mechanical strength, high volume resistivity of 10^{12} ohm–cm and is affected by humidity. Ammonium dihydrogen phosphate has high thermal stability, volume resistivity of 10^{12} ohm– cm and high sensitivity. Barium titanate ceramic is a ferroelectric crystal and has small voltage output. The voltage output versus strain input characteristic is linear only on a very small range and is affected by humidity. They have a range of measurement from 1N to 2000N, linearity within ± 1%, are simple in structure, stiff in nature, robust in use and preferred for dynamic measurements.

The piezo-electric transducers manufactured by M/S Brush Clevite, UK are sold under the trade name PZT which indicates a Lead Zirconate–Lead Titanate composition. They are available in many types for specific applications such as:

- PZT-2 is used for high frequency ultrasonic transducer applications and delay lines.
- PZT- 4 is used for high power acoustic radiating transducers for use in deep submersion type applications and as the active element in electrical power generating systems.
- PZT-5A has high sensitivity, high time stability and resistivity at elevated temperatures. They are mostly used in hydrophones and instrument applications.
- PZT-5H has higher sensitivity and higher diaelectric constant than PZT-5A. However, it has a lower temperature stability and limited working temperature range.
- PZT-8 is a high power material which has much lower diaelectric losses under high electric drive.

Piezo-electric materials are also available as polymeric films such as polyvinylidene fluoride (PVDF). These films are very thin, light weight and flexible and can be cut easily for adaptation to uneven services. These films are not suitable for resonance applications due to the low mechanical quality factor but are extensively used in acoustical broad band applications for microphones and loudspeakers.

Piezo-electric transducers find numerous applications in the medical instrumentation field. They are used in ultrasonic scanners for imaging and blood flow measurements. They are used in the detection of Korotkoff sounds in non-invasive blood pressure measurements and in external and internal phonocardiography. The details of their construction and associated requirements and characteristics are given in the respective chapters.

3.4.7 Other Displacement Sensors

The position and motion can be detected using optical transducers. Both amplitude and position of the transmitted and reflected light can be used to measure displacement. An optical fibre can also be used to detect displacement by measuring the transmitted light intensity or phase difference between the measuring beam and a reference beam.

Similarly, ultrasonic, microwave, X-rays, and nuclear radiations can be used to sense position. These are covered in the respective chapters at other places in the book.

▸ 3.5   PRESSURE TRANSDUCERS

Pressure is a very valuable parameter in the medical field and therefore many devices have been developed to effect its transduction to measurable electrical signals. The basic principle behind all these pressure transducers is that the pressure to be measured is applied to a flexible diaphragm which gets deformed by the action of the pressure exerted on it. This motion of the diaphragm is then measured in terms of an electrical signal. In its simplest form, a diaphragm is a thin flat plate of circular shape, attached firmly by its edge to the wall of a containing vessel. Typical diaphragm materials are stainless steel, phosphor bronze and beryllium copper.

Absolute pressure is pressure referred to a vacuum. Gauge pressure is pressure referred to atmospheric pressure. The commonly used units for pressure are defined at 0°C as

P = 1 mm Hg = 1 torr = 12.9 mm blood = 13.1 mm saline = 13.6 mm H_2O = 133.0 dyn/cm^2 = 1330 bar = 133.31 Pa = 0.133 kPa (Pa = Pascal)

For faithful reproduction of the pressure contours, the transducing system as a whole must have a uniform frequency response at least up to the 20th harmonic of the fundamental frequency of the signal. For blood pressure recording (which is at a rate of say 72 bpm or 1.2 Hz), the system should have a linear frequency response at least up to 30 Hz.

The most commonly employed pressure transducers which make use of the diaphragm are of the following types:

- *Capacitance manometer*—in which the diaphragm forms one plate of a capacitor.
- *Differential transformer*—where the diaphragm is attached to the core of a differential transformer.
- *Strain gauge*—where the strain gauge bridge is attached to the diaphragm.

Displacement transducers can be conveniently converted into pressure transducers by attaching a diaphragm to the moving member of the transducer such that the pressure is applied to the diaphragm. The following pressure transducers are commercially available.

3.5.1 LVDT Pressure Transducer

LVDT pressure transducer consists of three parts: a plastic dome with two female Luer-Lok fittings, a stainless steel diaphragm and core assembly and a plastic body containing the LVDT coil assembly. The transducers are available commercially in a complete array of pressure ranges with corresponding sensitivities, volume displacements and frequency response characteristics. LVDT pressure transducers are available in two basic diaphragm and core assemblies. The first for venous and general purpose clinical measurements has a standard-size diaphragm with internal fluid volume between the dome and diaphragm of less than 0.5 cc. The second design with higher response characteristics for arterial pressure contours has a reduced diaphragm area and internal volume of approximately 0.1 cc. The gauge can be sterilized by ethylene oxide gas or cold liquid methods. The LVDT gauges offer a linearity better than ±1%.

Arterial pressure transducers consist of very small differential transformers having tiny cores. A typical commercially available transducer has an outside diameter of 3.2 mm and a length of 9 mm, the movement of the movable core is ±0.5 mm and gives the full range of electrical output as 125 mV/mm (62.5 mV in either direction). Baker *et al* (1960) designed a miniature differential transformer pressure transducer which could be mounted directly on a dog's heart. The size of the transducer was 12.5 × 6.25 mm and gave an average sensitivity of 38 mV per mmHg per volt applied to the input coils. The zero drift has been measured at less than 0.2 mmHg/°C.

3.5.2 Strain Gauge Pressure Transducers

Nearly all commercially available pressure monitoring systems use the strain gauge type pressure transducers for intra-arterial and intravenous pressure measurements. The transducer is based upon the changes in resistance of a wire produced due to small mechanical displacements. A linear relation exists between the deformation and electric resistance of a suitably selected gauge (wire, foil) over a specified range.

The figure of merit which describes the overall behaviour of the wire under stress is determined from the "Gauge Factor" which is defined as

$$g = \frac{\Delta R/L}{\Delta L/L}$$

where ΔR = Incremental change in resistance due to stress

R = Resistance of an unstretched wire

ΔL = Incremental change in length

L = Unstretched length of wire

Accordingly, the gauge factor gives information on the expected resistance change or output signal at maximum permissible elongation. The gauge factor determines to a large extent the sensitivity of the wire when it is made into a practical strain gauge. The gauge factor varies with the material. So, it is advisable to select a material with a high gauge factor. But the wire used should be selected for the minimum temperature coefficient of resistance.

Table 3.4 shows the gauge factor for different materials along with their temperature coefficient of resistance.

• Table 3.4   *Gauge Factors for Different Materials*

Material	Gauge factor	Temperature coefficient of resistance ($\Omega/\Omega/°C$)
Constantan	2.0	2×10^{-6}
Platinum	6.1	3×10^{-3}
Nickel	12.1	6×10^{-3}
Silicon	120	6×10^{-3}

From sensitivity considerations, semiconductor silicon seems to be the obvious choice but it has been seen that it is highly temperature dependent. Techniques have, however, been developed to partially compensate for this temperature effect by the use of thermistors and combinations of suitably chosen p-type and n-type gauges. Silicon strain gauges can be made to have either positive or negative gauge factors by selectively doping the material. In this way, for a strain which is only compressive, both positive and negative resistance changes can be produced. Double sensitivity can be achieved by using two gauges of each type arranged in the form of a bridge circuit and can be made to respond to strains of the same sign in each of the four arms.

Unbonded Strain Gauges: Most of the pressure transducers for the direct measurement of blood pressure are of the unbonded wire strain gauge type. The arrangement consists in mounting strain wires of two frames which may move with respect to each other.

The outer frame is fixed and the inner frame which is connected to the diaphragm upon which the pressure acts, is movable. A pressure P applied in the direction shown (Fig. 3.5) stretches wires B and C and relaxes wires A and D. These wires form a four arm active bridge. The moving frame is mounted on springs which brings it to the central reference position when no pressure is applied to the diaphragm. It has been observed that even after employing utmost care during manufacture, it is not possible to produce a zero output signal of the transducer at zero pressure. This may be due to the inhomogeneous character of the wire and the inevitable dimensional inaccuracies of machining and assembly. Extra resistors are, therefore, placed in the gauge housing

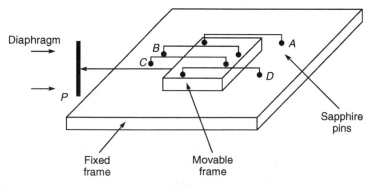

➤ **Fig. 3.5**   *Schematic diagram of an unbonded strain gauge pressure transducer*

(Fig. 3.6) to adjust the same electrically. Zeroing of the bridge can be accomplished by a resistor R_x connected in series thereto. Similarly, a series connected resistor R_t in the other arm of the bridge serves to compensate for zero point drift, caused by temperature changes.

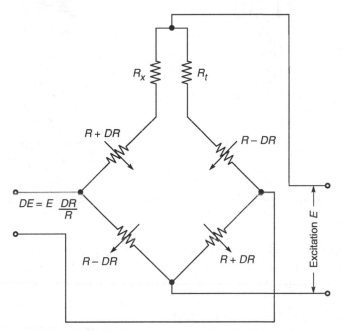

> **Fig.** 3.6   *Resistors mounted in the pressure gauge housing to enable adjustments of electrical zero at zero pressure conditions*

The unbonded strain gauge transducers are preferred when low pressure measurements are to be made since hysteresis errors are much less than would be the case if wire gauges were bound to the diaphragm. Unbonded strain gauge transducers can be made sufficiently small, which are even suitable for mounting at the tip of a cardiac catheter.

Bonded Strain Gauges: The bonded strain gauge consists of strain-sensitive gauges which are firmly bonded with an adhesive to the membrane or diaphragm whose movement is to be recorded. In practice, it is made by taking a length of very thin wire (for example, 0.025 mm dia) or foil which is formed into a grid pattern (Fig. 3.7 a,b) and bonded to a backing material. The backing material commonly used is paper, bakelized paper or and similar material. For pressure measurement purposes, a strain gauge constructed as above, is attached to a diaphragm. The deflection of the diaphragm under pressure causes a corresponding strain in the wire gauge. Since the deflection is proportional to pressure, a direct pressure resistance relation results.

By using a pair of strain gauges and mounting them one above the other, the changes in the resistances of the two gauges arising from the changes in the ambient temperature can be cancelled. Also, one strain gauge would increase while the other would decrease in resistance when the pressure is applied to them.

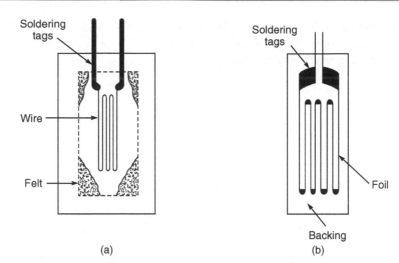

(a)   (b)

> **Fig. 3.7(a)** *Wire strain gauge a bonded type strain guage pressure tranducer*
>   **(b)** *Etched foil strain gauge*

Silicon Bonded Strain Gauges: In recent years, there has been an increasing tendency to use bonded gauges made from a silicon semiconductor instead of from bonded wire or foil strain gauges. This is because of its higher gauge factor resulting in a greater sensitivity and potential for miniaturization.

Figure 3.7(c) shows an arrangement in which the positive-doped (*p*-doped) silicon elements of a Wheatstone bridge are diffused directly on to a base of negative-doped (*n*-doped) silicon. Although, semiconductor strain gauges are very sensitive to variations in temperature, the inclusion of eight elements to form all four resistive arms of a bridge eliminates this problem by exposing all of the elements to the same temperature.

In the bonded silicon semiconductor strain gauge, the conventional wire or foil element is replaced by a single chip of silicon, processed to a finished size less than 0.0125 mm thick and 0.25 mm wide. The strip is then mounted on a substrate of epoxy resin and glass fibre with nickel strips for the electrical connections. Gold wires are used for interconnection between silicon and nickel. The complete assembly is coated with epoxy resin to protect it from

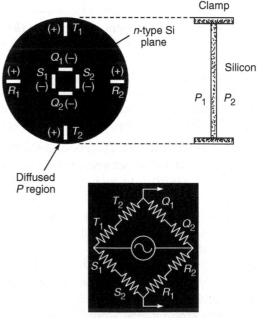

> **Fig. 3.7(c)** *Diffused p-type strain gauge*

environmental conditions. Lead wire resistance and capacitance also change with temperature. Therefore, lead wires from the strain gauge to the Wheatstone bridge must also be temperature compensated. Compensation for temperature variation in the leads can be provided by using the three lead method. In this method, two of the leads are in adjacent legs of the bridge which cancels their resistance changes and does not disturb the bridge balance. The third lead is in series with the power supply and is, therefore, independent of bridge balance.

The bridge power supply is regulated and temperature compensated. The piezo-resistive elements are the four arms of a Wheatstone bridge and consist of four p-doped (boron) regions, diffused into the edged chip of the n-type silicon. The transducer has a range of 0–1 atmosphere. It has hysteresis error of ±0.1% full scale and a temperature coefficient of ±2–3 mV/°C. It gives an output of ±0.75 V per 100 mmHg. The high output permits the use of this transducer with practically any dc recorder.

⇒ 3.6 TRANSDUCERS FOR BODY TEMPERATURE MEASUREMENT

The most popular method of measuring temperature is by using a mercury-in-glass thermometer. However, they are slow, difficult to read and susceptible to contamination. Also, reliable accuracy cannot be attained by these thermometers, especially over the wide range which is now found to be necessary. In many of the circumstances of lowered body temperature, continuous or frequent sampling of temperature is desirable, as in the operating theatre, post-operative recovery room and intensive care unit, and during forced diuresis, massive blood transfusion, and accidental hypothermia. The continuous reading facility of electronic thermometers obviously lends itself to such applications. Electronic thermometers are convenient, reliable and generally more accurate in practice than mercury-in-glass thermometers for medical applications. They mostly use probes incorporating a thermistor or thermocouple sensor which have rapid response characteristics. The probes are generally reusable and their covers are disposable.

Small thermistor probes may be used for oesophageal, rectal, cutaneous, subcutaneous, intramuscular and intravenous measurements and in cardiac catheters. Thermocouples are normally used for measurement of surface skin temperature, but rectal thermocouple probes are also available. Resistance thermometers are usually used for rectal and body temperature measurement. The resistance thermometer and thermistor measure absolute temperature, whereas thermocouples generally measure relative temperature.

3.6.1 Thermocouples

When two wires of different materials are joined together at either end, forming two junctions which are maintained at different temperatures, a thermo-electromotive force (emf) is generated causing a current to flow around the circuit. This arrangement is called a thermocouple. The junction at the higher temperature is termed the hot or measuring junction and that at the lower temperature the cold or reference junction. The cold junction is usually kept at 0°C. Over a limited range of temperature, the thermal emf and hence the current produced is proportional to the temperature difference existing between the junctions. Therefore, we have a basis of temperature measurement, since by inserting one junction in or on the surface of the medium whose temperature is to be measured and keeping the other at a lower and constant temperature (usually $0°C$), a measurable emf is produced proportional to the temperature difference between the two junctions. The reference junction is normally held at $0°C$ inside a vacuum flask containing melting ice.

The amount of voltage change per degree of temperature change of the junction varies with the kinds of metals making up the junction. The voltage sensitivities of thermocouples made of various metals are given in Table 3.5.

• Table 3.5   *Thermal emf for Various Types of Thermocouples*

Type	Thermocouple	Useful range °C	Sensitivity at 20°C (mV/°C)
T	Copper-constantan	−150 to + 350	45
J	Iron-constantan	−150 to + 1000	52
K	Chromel-alumel	−200 to + 1200	40
S	Platinum-platinum (90%) rhodium (10%)	0 to + 1500	6.4

Figure 3.8 illustrates the emf output versus temperature in degrees for each of the commonly used thermocouples. Two important facts emerge from this graph: (i) the sensitivity (slope) of each curve is different, and (ii) none of the curves has a perfectly linear rate of change of emf output per degree F. This shows that each type of thermocouple has a unique, non-linear response to temperature. Since most recording or display devices are linear, there is a need to linearize the output from the thermocouples so that the displayed output gets correlated with actual temperature. The sensitivity of a thermocouple does not depend upon the size of the junction or the wires forming it as the contact potentials developed are related to the difference in the work function of the two metals.

For medical applications, a copper-constantan combination is usually preferred. With the reference junction at $0°C$ and the other at $37.5°C$, the output from this thermocouple is 1.5 mV. Two types of measuring instruments can be used with thermocouples to measure potential differences of this order. In one, moving coil movements are used as millivoltmeters to measure the thermocouple emf. They are directly calibrated in temperature units. Usually in clinical thermocouple instruments, reflecting galvanometers or light spot galvanometers are preferred to measure and display temperature. If the thermocouple voltages are small (less than 1 mv), they can be readily measured with a precision dc potentiometer having a Weston cadmium-mercury cell as a reference. They can also be read directly on a digital voltmeter or by using a chopper stabilized dc amplifier followed by a panel meter of the analog or digital type.

Much experimentation with thermocouple circuits has led to the formulation of the following empirical laws which are fundamental to the accurate measurement of temperature by thermoelectric means: (i) the algebraic sum of thermoelectric emf's generated in any given circuit containing any number of dissimilar homogeneous metals is a function only of the temperatures of the junctions, and (ii) if all but one of the junctions in such a circuit are maintained at some

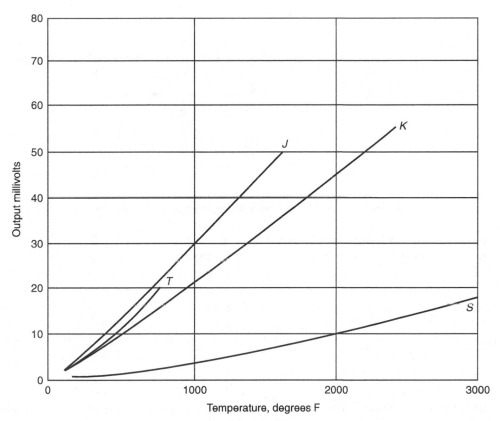

> **Fig.** 3.8 *EMF versus temperature characteristics of major types of thermocouples (Courtesy: Gould Inc., USA)*

reference temperature, the emf generated depends only upon the temperature of that junction and can be used as a measure of temperature. It is thus evident that the junction temperature can be determined if the reference junction is at a different, but known temperature. It is also feasible in the use of oven-controlled and electrically simulated reference junctions.

The temperature of the measuring junction can be determined from the thermo emf only if the absolute temperature of the reference junction is known. This can be done by either of the following methods:

1. By measuring the reference temperature with a standard direct reading thermometer.
2. By containing the reference junction in a bath of well defined temperature, e.g. a carefully constructed and used ice bath which can give an accuracy of 0.05°C with a reproducibility of 0.001°C.

A simpler method is to use a reference-temperature compensator which generates an emf that will exactly compensate for variations in the reference-junction temperature. Figure 3.9 illustrates a thermally sensitive bridge which is designed to generate an emf that varies with the enclosure temperature (normally ambient) in such a way that variations in the cold junction are nullified.

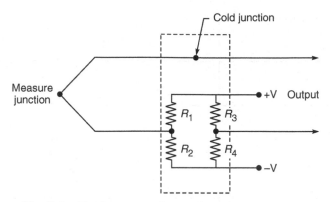

> **Fig. 3.9** *Bridge type reference junction compensator*

In this circuit, R_2 is a temperature-sensitive component that is thermally bonded to the cold junction thermocouple. The resistance-temperature curve of R_2 matches the emf temperature characteristic of the thermocoupled material. The voltage change across R_2 is equal and opposite to the cold junction thermal volt-age over a limited ambient temperature range. This system introduces errors if the enclosure temperature undergoes wide variations. How-ever, for moderate fluctuations (±5°C), it enables an effective reference temperature stability of ±0.2°C to be achieved. This type of reference temperature simulation is well adapted to zero suppression of a large temperature. By specifying the reference temperature of or near the temperature to be recorded, small temperature changes can be recorded.

Precise, easily calibrated, integrated circuit (LM 135 series from National Semiconductors) temperature sensors can be conveniently used in cold junction compensation circuits. They are suitable in compensation circuits over a –55° to + 150°C temperature range.

The small size and very fast response of thermocouples make them suitable for intracellular transient temperature measurements and for measuring temperatures from within the body at sites like the oesophagus, rectum, etc. They can be inserted into catheters and hypodermic needles. Special needles are commercially available that enable a thermocouple to be subcutaneously implanted, the needle being withdrawn to leave the thermocouple in place. Mekjavic *et al* (1984) have illustrated the construction and evaluation of a thermocouple probe for measuring oesophageal temperature.

Cain and Welch (1974a) developed thin film Cu-Ni micro-thermocouple probes for dynamic as well as static temperature measurements in biological tissue. These probes which use a quartz substrate exhibit response times of less than a millisecond with thermal properties similar to tissues. Their thermoelectric emf of 21 μmV/°C is linearly dependent on temperature over the range normally encountered in biological measurements. Probe tip diameters as small as 10 and 30 μm have been fabricated. These small sizes enable the measurement of temperature profiles of the retina of the eye. These probes have been experimentally inserted through the sclera of the eye and kept in contact with the biological tissue for periods of 4 to 6 hours. Their linearity, thermal emf and response time make it possible to record the response to laser irradiation for temperature rises from 1 to 45°C and from 1 ms to 10s (Cain and Welch, 1974b).

3.6.2   Electrical Resistance Thermometer

The temperature dependence of resistance of certain metals makes it convenient to construct a temperature transducer. Although most of the metals can be used, the choice, however, depends upon the linearity and sensitivity of the temperature resistance characteristics. The resistance R_t of a metallic conductor at any temperature t is given by:

$$R_t = R_0(1 + \alpha t)$$

where R_0 = resistance at 0°C and

α = temperature coefficient of resistivity

If R_0 and α are known, a measurement of R_t shall directly give the value of temperature. The increase in resistance is linear over the range 0–100°C for commonly used materials in resistance thermometry.

Normally platinum or nickel are used for resistance thermometry since they can be readily obtained in a pure form and are comparatively stable. Thermometers constructed from a coil of these metals have been used for the measurement of skin, rectal and oesophageal temperature. The coefficient of resistivity for platinum is 0.004 Ω/Ω °C. The measurement of resistance is generally carried out by using a Wheatstone bridge in which all leads must be of constantan to keep their own temperature resistance changes minimal. The comparator resistances must also be temperature stable. The bridge can be operated from either direct or alternating current, but to neglect any electrochemical changes or polarization in the circuit, an ac bridge is recommended.

Figure 3.10 shows the simple bridge circuit employed in resistance thermometry. A and B are fixed resistances. C is a variable resistance made from constantan which has a very low temperature coefficient of resistance. The measuring coil and its connecting leads are placed in one arm of the bridge circuit with a dummy pair of leads connected in the opposite arm. In this manner, changes in resistance of the coil leads with ambient temperature are cancelled out by the corresponding changes in the dummy or compensating leads.

The simple Wheatstone bridge circuit is basically a non-linear device when it is operated away from its null-balance point; therefore, it is important to understand the degree of non-linearity to be

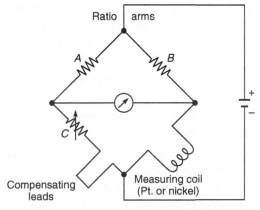

➤ Fig. 3.10   *Circuit arrangement of a metal resistance thermometer*

encountered. Platinum resistance sensors are linear within ±10% between −200 and +600°C. To achieve better linearization, Foster (1974) describes a circuit which makes use of a nonlinear amplifier. Positive feedback is added around the input amplifier, making its effective gain increase as the sensor resistance increases with temperature. The linearization achieved is typically accurate within ±0.5°C in the worst case over the specified range. The analog linearization method has the advantage that there are no discontinuities in the displayed temperature as the temperature changes. This is sometimes a problem when piece-wise linear compensation networks or read only memories are used for linearization.

Resistance thermometers are particularly suitable for remote reading and can be made of inert materials like platinum. They are very stable and show almost no hysteresis with large temperature excursions. However, the size of the coils used presently is such that in the medical field, the resistance thermometer probes are best suited only for rectal probe use. They are bulky and are not suitable for mounting in needles for the measurement of tissue temperature.

3.6.3 THERMISTORS

Thermistors are the oxides of certain metals like manganese, cobalt and nickel which have large negative temperature coefficient of resistance, i.e. resistance of the thermistor shows a fall with increase in temperature. The general resistance-temperature relation for a thermistor is given by:

$$R = Ae^{B/T}$$

where　R = resistance of the thermistor in Ω

　　　　T = absolute temperatue

　　　　A and B are constants

Thermistors when used for measuring temperature have many advantages over thermocouples and resistance thermometers. They are summarized as follows:

- They show a considerably high sensitivity—about 4% change in resistance per degree centigrade.
- Since the thermistors themselves are of high resistance, the resistances of the connecting leads are of small influence and therefore, no compensating leads are necessary.
- The time constant can be made quite small by easily reducing the mass of the thermistor. Hence the measurements can be taken rather quickly.
- They can be had in a variety of shapes making them suitable for all types of applications.
- They are small in size—about the size of the head of a pin and can be mounted on a catheter or on hypodermic needles.
- Since the thermistors are available in a large range of resistance values, it makes it much easier to match them in the circuits.

The large change in resistance with temperature means that a comparatively simple bridge circuit is sufficient. The thermistors have inherently non-linear resistance-temperature characteristics. But by the proper selection of the values of the bridge resistances (bridge in itself is a non-linear arrangement), it is possible to get nearly linear calibration of the indicating instrument over a limited range.

There is drawback in using thermistors when multichannel temperature indicators are made. This is due to the fact that it is not possible to make a batch of them with uniform characteristics. The normal tolerance on resistance is ±20% but matching down to ±1% is available at higher cost. Recalibration of the instrument is thus required whenever a particular probe has to be replaced.

For linearization of thermistors over a limited temperature range (Sapoff, 1982), two approaches are usually employed. Figure 3.11(a) and (b) show these techniques, which are:

(i) If the thermistor is supplied with a constant current and the voltage across the thermistor is used to indicate the temperature, linearization can be obtained by shunting the thermistor with a selected resistor R_p. The objective is to make the point of inflection of the parallel combination coincide with the mid-scale temperature (Fig. 3.11(a)).

(ii) When the current through the thermistor, for a fixed applied difference, is used to indicate the temperature, the series arrangement is employed (Fig. 3.11(b)).

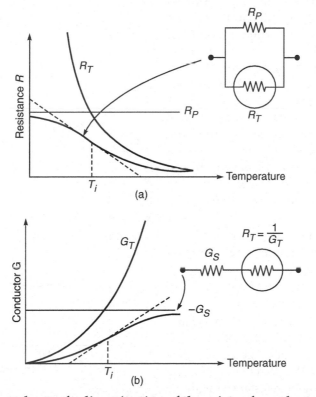

> **Fig. 3.11(a)** *Two schemes for linearization of thermistor for a short range and their characteristics. (a) Parallel combination of a thermistor and a resistor*
> **(b)** *Series combination of as thermistor and a resistor*

By using this technique, the maximum deviation from linearity observed is 0.03°C for every 10°C. It may, however, be noted that improved linearity achieved results in a decrease in the effective temperature coefficient of the combination. For example, the temperature coefficient is 3.3%°C for a thermistor, when optimally linearized with a series resistance. More complex circuit arrangements are needed to achieve better linearization over wider ranges of temperature.

Cronwell (1965) has presented a review of the techniques used for linearizing the thermistors. Thermistors which provide a linear change in resistance with a change in temperature have been given the name 'Linistors'.

An operational amplifier circuit designed by Stockert and Nave (1974) provides a linear relation between the output voltage and temperature from 10 to 50°C using a non-linear thermistor as the temperature transducer. The paper discusses the method for calculating the circuit values. Allen (1978) suggests the use of a microprocessor as a 'look-up' table of thermistor resistance-temperature values, using linear interpolation between the values to determine the temperature.

Thermistors are made in a wide variety of forms suitable for use in medical applications. They are available as wafers required for applying to the skin surface, rods which can be used for rectal, oral, or similar insertions, and tiny beads so small that they can be mounted at the tip of a hypodermic needle for insertion into tissues. These tiny thermistors have very small thermal 'time constants' when they are properly mounted. 'Time constant', the standard measure of thermistor probe response time, is the time required for a probe to read 63% of a newly impressed temperature. Generally, the time constants are obtained by transferring the probe from a well stirred water bath at 20°C to a like bath at 43°C. Approximately, five 'time constants' are required for a probe to read 99% of the total change.

Thermistors with positive thermo-resistive coefficients are called Posistors (PTC). They exhibit a remarkably high variation in resistance with increase in temperature. The posistors are made from barium titanate ceramic.

PTCs are characterized by an extremely large resistance change in a small temperature span. Figure 3.12 shows a generalized resistance temperature plot for a PTC (Krelner, 1977). The temperature at which the resistance begins to increase rapidly is referred to as the switching temperature. This point can be changed from below 0°C to above 120°C. Thus, PTCs have a nearly constant resistance at temperatures below the switching temperature but show a rapid increase in resistance at temperatures above the switch temperature.

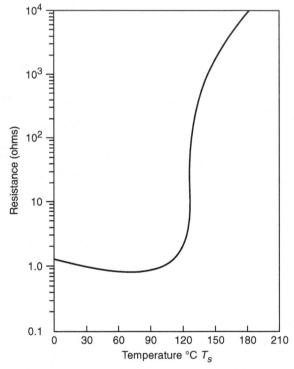

> **Fig. 3.12** *A PTC thermistor switches abruptly when heated to the switching temperature, resistance changes by four decades (Courtesy: Keystone Carbon Co., USA)*

Thermistor probes for measurement of body temperature consist of thermistor beads sealed into the tip of a glass tube. The bead is protected by this glass housing and the rapid response is also maintained. Probes suitable for rectal and oesophageal applications usually contain a thermistor bead mounted inside a stainless steel sheath. Figure 3.13 shows a variety of probes suitable for different applications in the medical field.

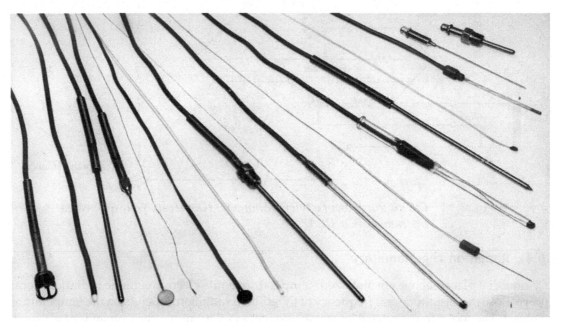

> **Fig.** 3.13 *Different shapes of thermistor probes (Courtesy: Yellow Springs Instruments Co., USA)*

Probes should be sterilized by using a chemical solution such as 70% alcohol or Dakin's solution (sodium hypochlorite in neutral buffer). The cleaning agents Zephiran and Haemo-sol are likewise suitable. Probes should not be boiled or actoclaved except where it is specifically mentioned.

The instruments which give direct reading of the temperature at the thermistor position are known by the name telethermometers because of their ability to use leads which are hundreds of feet long without a significant decrease in accuracy. The continuous signal is also suitable for recording without amplification. Figure 3.14 is a typical circuit diagram of a telethermometer. The instrument works on a 1.5 V dry cell which has an operating life of 1000 h. The addition of a second thermistor in the opposite arm of the bridge can make the circuit twice as sensitive and permits the use of a lower sensitivity meter.

For specific applications like pyrogenic studies, it is often necessary to have a multichannel system, which should automatically scan 3, 7 or 11 probes in sequence with 20 s, 1 min. or 5 min. readings per probe and record the measured readings on a recorder. For this, a motorized rotary switch can be employed. The recorder output in these instruments is 0 to 100 mV with zero output at the highest temperature.

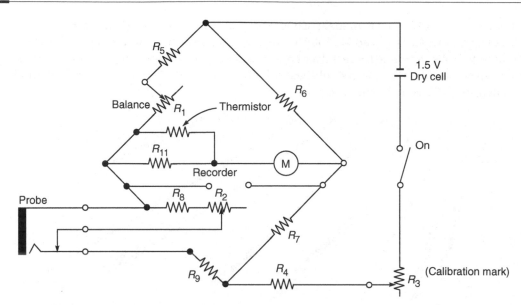

> ➤ **Fig.** 3.14   *Circuit diagram of a telethermometer (Courtesy: Yellow Springs Instruments Co. USA)*

3.6.4   Radiation Thermometery

Any material placed above absolute zero temperature emits electromagnetic radiation from its surface. Both the amplitude and frequency of the emitted radiation depends on the temperature of the object. The cooler the object, the lower the frequency of the emitted electromagnetic waves, and lesses the power emitted. The temperature of the object can be determined from the power emitted. Infrared thermometers measure (Ring, 1988) the magnitude of infrared power (flux) in a broad spectral range, typically from 4 to 14 micrometers. They make no contact with the object measured. The detectors used for measuring the emitted infrared radiation are thermopiles (series connected thermocouple pairs), pyroelectric sensors, Golay cells, photoconductive cells and photovoltaic cells. All these devices can be mounted in specially designed housing to measure surface temperature without direct contact with the body. Table 3.6 summarizes salient specifications for various types of temperature measuring devices.

Hand-held infrared scanners are now available for monitoring the pattern of skin temperature changes, particularly for tympanic membrane temperature measurements. This measurement is based on the development of a new type of sensor called "Pyroelectric Sensor".

A pyroelectric sensor develops an electric charge that is a function of its temperature gradient. The sensor contains a crystalline flake which is preprocessed to orient its polarized crystals. Temperature variation from infrared light striking the crystal changes the crystalline orientation, resulting in development of an electric charge. The charge creates a current which can be accurately measured and related to the temperature of the tympanic membrane.

Infrared thermometers have significant advantages over both glass and thermistor thermometers used orally, rectally or axillarily. They eliminate reliance on conduction and instead measure the

• Table 3.6   *Comparison of Electrical Temperature-Sensing Techniques*

	RTD	*Thermistor*	*Thermocouple*	*Radiation*
Accuracy	0.01° to 0.1°F	0.01° to 1°F	1° to 10°F	0.2°F
Stability	Less than 0.1% drift in 5 years	0.2°F drift/year	1° drift/year	Same as thermocouple
Sensitivity	0.1 to 10 ohms/°F	50 to 500 ohms/F°	50 to 500 μ volts/°F	Same as thermo-couple
Features	Greatest accuracy over wide spans; greatest stability	Greatest sensitivity	Greatest economy; highest range	Fastest response; no contaminations; easiest to use

body's natural radiation. They use an ideal measurement site—the tympanic membrane of the ear which is a function of the core body temperature. It is a dry, non-mucous membrane site that minimizes risks of cross-contamination. The disadvantage of infrared thermometers is their high cost as compared to other types of thermometers.

3.6.5   Silicon Diode

The voltage drop across a forward biased silicon diode is known to vary at the rate of 2 mV/°C. This suggests that a silicon diode can be used as a temperature sensor. Griffths and Hill (1969) describe the technique and circuit diagram for the measurement of temperature using a silicon diode.

The circuit (Fig. 3.15) is designed to monitor body temperature in the range of 34–40°C with an accuracy of 2.5%. The temperature-sensing diode D_1 is connected to the non-inverting input of an operational amplifier. The gain of the amplifier is 500. With D_1 mounted in a water bath at 37°C, R_1 is adjusted to give an amplifier output voltage of zero. Using a centre zero meter, the scale can be calibrated to ±3°C.

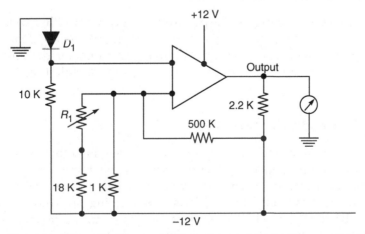

➤ **Fig. 3.15**   *Use of a silicon diode as a temperature sensor (redrawn after Griffths and Hill, 1969)*

The disadvantage of using a diode as a temperature sensor is the requirement of a stable calibration source. Soderquist and Simmons (1979) explain the use of a matched transistor pair which has predictable differential base-emitter voltage relationship which can be exploited as a temperature sensor. It is advantageous to have a diode or transistor sensor fabricated on a chip with interfacing electronics by integrated circuit (IC) technology. Several integrated temperature sensors have been developed and some of these are available commercially. AD 599 is an integrated circuit temperature transducer manufactured by Analog Devices. The transducer produces an output proportional to absolute temperature. For supply voltages between +4 and +30 V, it acts as a high impedance, constant-current regulator passing 1 μA/K. The transducer uses a fundamental property of the silicon transistor from which it is made to realise its temperature proportional characteristics.

3.6.6   Chemical Thermometery

Liquid crystals which are commonly used for digital displays are composed of materials which have liquid like mechanical properties but possess the optical properties of single crystals. These crystals demonstrate remarkable changes in their optical properties when temperature is varied. The change in the colour of the crystal related to temperature is used to measure the surface temperature. A liquid crystal film encapsulated in a plastic housing can be attached to the body surface. Some systems use flexible liquid crystal plates so that better contact over the body surface can be achieved. These are single use thermometers and there is less likelihood of cross-contamination.

The clinical use of liquid crystal thermometry has not become popular because of its low thermal resolution (± 0.5°C) and slow response time (> 60 sec.). In addition, contact thermometry might modify the temperature of the measured skin due to undue pressure over the skin surface.

▷ 3.7   PHOTOELECTRIC TRANSDUCERS

Photoelectric transducers are based on the principle of conversion of light energy into electrical energy. This is done by causing the radiation to fall on a photosensitive element and measuring the electrical current so generated with a sensitive galvanometer directly or after suitable amplification. There are two types of photoelectric cells—photovoltaic cells and photomissive cells.

3.7.1   Photovoltaic or Barrier Layer Cells

Photovoltaic or barrier layer cells usually consist of a semiconducting substance, which is generally selenium deposited on a metal base which may be iron and which acts as one of the electrodes. The semiconducting substance is covered with a thin layer of silver or gold deposited by cathodic deposition in vacuum. This layer acts as a collecting electrode. Figure 3.16 shows the construction of the barrier layer cell. When radiant energy falls upon the semiconductor surface, it excites the electrons at the silver-selenium interface. The electrons are thus released and collected at the collector electrode.

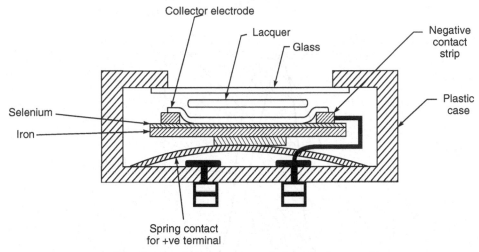

> **Fig.** 3.16 *Construction of a barrier layer cell*

The cell is enclosed in a housing of insulating material and covered with a sheet of glass. The two electrodes are connected to two terminals which connect the cell with other parts of the electrical circuit.

Photovoltaic cells are very robust in construction, need no external electrical supply and produce a photocurrent sometimes stronger than other photosensitive elements. Typical photocurrents produced by these cells are as high as 120 μA/lumen. At constant temperature, the current set up in the cell usually shows a linear relationship with the incident light intensity. Selenium photocells have very low internal resistance, and therefore, it is difficult to amplify the current they produce by dc amplifiers. The currents are usually measured directly by connecting the terminals of the cell to a very sensitive galvanometer.

Selenium cells are sensitive to almost the entire range of wavelengths of the spectrum. However, their sensitivity is greater within the visible spectrum and highest in the zones near to the yellow wavelengths. Figure 3.17 shows spectral response of the selenium photocell and the human eye.

Selenium cells have a high temperature coefficient and therefore, it is very necessary to allow the instrument to warm up before the readings are commenced. They also show fatigue effects. When illuminated, the photocurrent rises to a value several percent above the equilibrium value and then falls off gradually. When connected in the optical path of the light rays, care should be taken to block all external light and to see that only the light from the source reaches the cell.

Selenium cells are not suitable for operations in instruments where the levels of illumination change rapidly, because they fail to respond immediately to those changes. They are thus not suitable where mechanical choppers are used to interrupt light 15–60 times a second.

3.7.2 Photoemissive Cells

Photoemissive cells are of three types: (a) high vacuum photocells, (b) gas-filled photocells and (c) photomultiplier tubes. All of these types differ from selenium cells in that they require an external

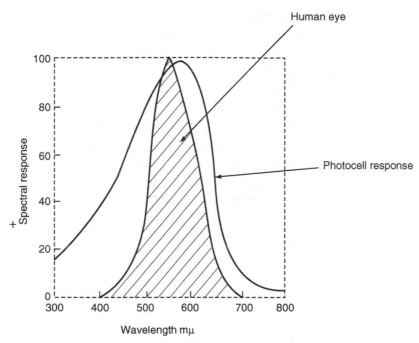

> **Fig. 3.17**   *Spectral response of a human eye and a selenium photocell*

power supply to provide a sufficient potential difference between the electrodes to facilitate the flow of electrons generated at the photosensitive cathode surface. Also, amplifier circuits are invariably employed for the amplification of this current.

High Vacuum Photoemissive Cells: The vacuum photocell consists of two electrodes—the cathode having a photosensitive layer of metallic cesium deposited on a base of silver oxide and the anode which is either an axially centered wire or a rectangular wire that frames the cathode. The construction of the anode is such that no shadow falls on the cathode. The two electrodes are sealed within an evacuated glass envelope.

When a beam of light falls on the surface of the cathode, electrons are released from it, which are drawn towards the anode which is maintained at a certain positive potential. This gives rise to a photocurrent which can be measured in the external circuit. The spectral response of a photoemissive tube depends on the nature of the substance coating the cathode, and can be varied by using different metals or by variation in the method of preparation of the cathode surface. Cesium-silver oxide cells are sensitive to the near infrared wavelengths. Similarly, potassium-silver oxide and cesium-antimony cells have maximum sensitivity in the visible and ultraviolet regions. The spectral response also depends partly on the transparency to different wavelengths of the medium to be traversed by the light before reaching the cathode. For example, the sensitively of the cell in the ultraviolet region is limited by the transparency of the wall of the envelope. For this region, the use of quartz material can be avoided by using a fluorescent material, like sodium salicylate, which when applied to the outside of the photocell, transforms the ultraviolet into visible radiations.

Figure 3.18 shows the current-voltage characteristics of a vacuum photoemissive tube at different levels of light flux. They show that as the voltage increases, a point is reached where all the photoelectrons are swept to the anode as soon as they are released which results in a saturation photocurrent. It is not desirable to apply very high voltages, as they would result in a excessive dark current without any gain in response.

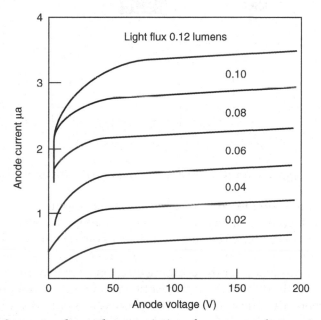

> **Fig.** 3.18   *Current voltage characteristics of a vacuum photoemissive tube at different levels of light flux*

Figure 3.19 shows a typical circuit configuration usually employed with photoemissive tubes. Large values of phototube load resistor are employed to increase the sensitivity up to the practical limit. Load resistances as high as 10,000 MΩ have been used. This, however, almost puts a limit, as further increase of sensitivity induces difficulties in the form of noise, non-linearity and slow response. At these high values of load resistors, it is very essential to shield the circuit from moisture and electrostatic effects. Therefore, special type of electrometer tubes, carefully shielded and with a grid cap input are employed in the first stage of the amplifier.

Gas-filled Photoemissive Cells: This type of cell contains small quantities of an inert gas like argon, whose molecules can be ionized when the electrons present in the cell posses sufficient energy. The presence of small quantities of this gas prevent the phenomenon of saturation current, when higher potential differences are applied between the cathode and anode. Due to repeated collisions of electrons in the gas-filled tubes, the photoelectric current produced is greater even at low potentials.

Photomultiplier Tubes: Photomultiplier tubes are used as detectors when it is required to detect very weak light intensities. The tube consists of a photosensitive cathode and has multiple cascade

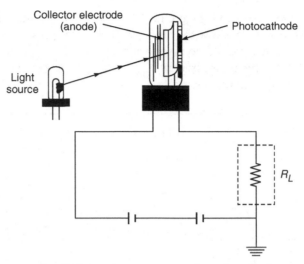

> **Fig. 3.19**   *Typical circuit configuration employed with photoemissive tubes*

stages of electron amplification in order to achieve a large amplification of the primary photo-current within the envelope of the phototube itself. The electrons generated at the photocathode are attracted by the first electrode, called dynode, which gives out secondary electrons. There may be 9–16 dynodes (Fig. 3.20). The dynode consists of a plate of material coated with a substance having a small force of attraction for the escaping electrons. Each impinging electron dislodges secondary electrons from the dynode. Under the influence of positive potential, these electrons are accelerated to the second dynode and so on. This process is repeated at the successive dynode, which are operated at voltages increasing in steps of 50–100 V. These electrons are finally collected at the collector electrode.

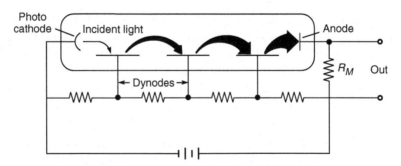

> **Fig. 3.20**   *Principle of photomultiplier tube*

The sensitivity of the photomultiplier tube can be varied by regulating the voltage of the first amplifying stage. Because of the relatively small potential difference between the two electrodes, the response is linear. The output of the photomultiplier tube is limited to an anode current of a few milliamperes. Consequently, only low intensity radiant energy can be measured without causing any appreciable heating effect on the electrodes surface. They can measure light intensities about

10^7 times weaker than those measurable with an ordinary phototube. For this reason, they should be carefully shielded from stray light. The tube is fairly fast in response to the extent that they are used in scintillation counters, where light pulses as brief as 10^{-9} s duration are encountered. A direct current power supply is required to operate a photomultiplier, the stability of which must be at least one order of magnitude better than the desired precision of measurement; for example, to attain precision of 1%, fluctuation of the stabilized voltage must not exceed 0.1%.

Fatigue and saturation can occur at high illumination levels. The devices are sensitive to electromagnetic interference and they are also more costly than other photoelectric sensors. Photomultipliers are not uniformly sensitive over the whole spectrum and in practice, manufacturers incorporate units best suited for the frequency range for which the instrument is designed. In the case of spectrophotometers, the photomultipliers normally supplied cover the range of 185 to 650 nm. For measurements at long wavelengths, special red sensitive tubes are offered. They cover a spectral range from 185 to 850 nm but are noticeably less sensitive at wavelengths below 450 nm than the standard photomultipliers.

Photomultiplier tubes may be damaged if excessive current is drawn from the final anode. Since accidental overload may easily occur in a laboratory and tubes are expensive to replace, it is advisable to adopt some means of protection from overloads.

3.7.3 Silicon Diode Detectors

The photomultiplier which is large and expensive and requires a source of stablized high voltage can be replaced by a silicon diode detector (photodiode, e.g. H.P.5082–4220). This diode is useable within a spectral range of 0.4–1.0 μm, in a number of instruments (spectrophotometers, flame photometers). The photodiode can be powered from a low voltage source.

These detectors when integrated with an operational amplifier have performance characteristics which compare with those of a photomultiplier over a similar wavelength range. Figure 3.21 shows the spectral response of silicon diode detectors. The devices being solid state are mechanically robust and consume much less power. Dark current output and noise levels are such that they can be used over a much greater dynamic range.

3.7.4 Diode Arrays

Diode arrays are assemblies of individual detector elements in linear or matrix form, which in a spectrophotometer can be mounted so that the complete spectrum is focussed on to an array of appropriate size. The arrangement does not require any wavelength selection mechanism and the output is instantaneously available. However, resolution in diode arrays is limited by the physical size of individual detector elements, which at present is about 2 nm.

The diode array photodetectors used in the Hewlett Packard spectrophotomer Model 8450A consists of two silicon integrated circuits, each containing 211 photosensitive diodes and 211 storage capacitors. The photodiode array is a PMOS (p-channel metal-oxide semiconductor) integrated circuit that is over 1.25 cm long. Each photosensitive diode in the array is 0.05 by 0.50 mm and has a spectral response that extends well beyond the 200–800 nm range.

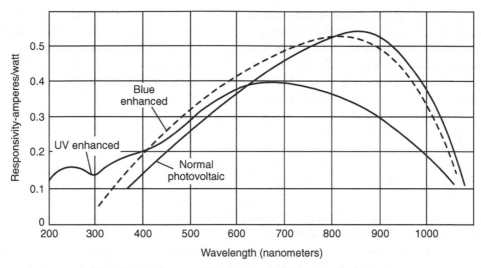

> **Fig.** 3.21   *Spectral response of silicon diode detectors*

A functional block diagram of the diode array chip is shown in Fig. 3.22. In parallel with each of the 211 photodiodes is a 10pF storage capacitor. These photo diode-capacitor pairs are sequentially connected to a common output signal line through individual MOSFET switches. When a FET switch is closed, the preamplifier connected to this signal line forces a potential of −5V on to the capacitor-diode pair. When the FET switch is opened again, the photocurrent causes the capacitor to discharge towards zero potential. Serial read-out of the diode array is accomplished by means of a digital shift register designed into the photodiode array chip.

The diode arrays typically exhibit a leakage current less than 0.1 pA. This error increases exponentially with temperature, but because the initial leakage value is so low, there is no need to cool the array at high ambient temperatures.

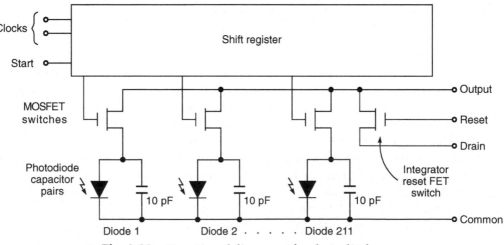

> **Fig.** 3.22   *Functional diagram of a photodiode array*

▓▓▓▶ 3.8 OPTICAL FIBRE SENSORS

The development of optical fibres has given rise to a number of transducers which find applications in the medical field. The ability of these fibres to transmit light over great distances with low power loss and the interaction of light with a measured system provide the basis of these sensing devices. These sensors are electrically passive and consequently immune to electromagnetic disturbances. They are geometrically flexible and corrosion resistant. They can be miniaturized and are most suitable for telemetry applications.

A great number of optical fibre sensors have been developed for biomedical applications (Martin et al, 1989). The potential of optical fibres in the sensing of chemical species has led to the development of a number of optical fibre chemical sensors. They have also been devised for the measurement of physical parameters such as temperature, pressure and displacement. However, the most direct use of optical fibres in medicine is for gaining access to otherwise inaccessible regions whether for imaging these areas, as in endoscopy or as the delivery system for light in laser surgery. However, fibre-optic sensors (Walt, 1992) which are predominantly used for physiological measurements are included in the following coverage.

The optical transducers are based on glass or plastic fibres, about 100 to 250 μm in diameter, as found in fibre-optic communication systems. The initial optical fibre had poor transmission characteristics, but within a decade, fibre losses were reduced from 1000 db/km in 1966 to below 1 db/km in 1976. Such improvements in the manufacture and theoretical understanding of light transmission in optical beam have lead to their wide spread use in a variety of applications, including that of clinical medicine.

3.8.1 Advantages of Optical Fibre Sensors

- Optical fibre sensors are non-electrical and hence are free from electrical interference usually associated with electronically based sensors.
- They are immune from cross-talk.
- There is a high degree of mechanical flexibility associated with the fibre optic and this combined with its reduced size, allows access to otherwise inaccessible areas of the body.
- They are suitable for telemetry applications as bulk of the instrumentation can be at a reasonable distance from the patient.
- These sensors do not involve any electrical connection to the patient body, thereby ensuring patient safety.
- More than one chemical species can be measured with a single sensor by employing more than one probe detection wavelength offering substantial economic advantage.
- These devices are intrinsically safe, involving low optical power—generally a few milliwatts.
- The sensors are capable of observing a sample in its dynamic environment, no matter how distant, difficult to reach or hostile the environment.
- The cost is low enough to make the sensors disposable for many applications.

3.8.2   Types of Optical Fibre Sensors

Optical fibre sensors are generally classified into three types as follows:

- Photometric sensors
- Physical sensors
- Chemical sensors

Photometric Sensors: Several types of measurements can be made by using the optical fibre as a device for highly localized observation of the spectral intensity in the blood or tissue. Light emanating from a fibre end will be scattered or fluoresced back into the fibre, allowing measurement of the returning light as an indication of the optical absorption or fluorescence of the volume at the fibre tip. The variations in the returning light are sensed using a photodetector. Such sensors monitor variations either in the amplitude or frequency of the reflected light.

Amplitude Measurements: The most widely used photometric sensor in the amplitude measurement category is the oximeter. This device measures the oxygen saturation of blood based on the fact that haemoglobin and oxyhaemoglobin have different absorption spectra. The details of this type of sensor are given in Chapter 10. The use of fibre-optic catheters allows oxygen saturation to be monitored intra-arterially.

Blood flow measurement based on dye densitometry is closely related to oximetry. A dye, commonly indocyanine green is injected into the blood and its concentration monitored by its absorption at an appropriate wavelength. The time variation of dye concentration can then be used to calculate cardiac output by dilution techniques. The details of these devices are given in Chapter 12.

Monitoring the amplitude of the reflected or transmitted light at specific wavelengths can provide useful information concerning the metabolic state of the tissue under investigation. The technique is non-invasive and fibre-optics play an important role as the technique enables very small areas of tissue to be examined so that metabolism at a localized level can be followed. The method is based on fluorometry and depends upon the direct observation of tissue and blood luminescence using fibre-optic light guide to connect the instrument to the tissue.

Frequency Measurements: The second category of photometric sensors using fibre-optic light guide is based on frequency changes in the signal. The most common example is that of laser Doppler velocimetry. In this method, light from a laser, normally helium/neon, is sent via a fibre onto the skin surface. The moving red blood cells scatter the light and produce a Doppler frequency shift because of their movement. When the light, shifted and unshifted in frequency is mixed, a spectrum of beat frequencies is obtained. Using a number of different processing techniques on the beat frequency spectrum, the information on the blood flow can be obtained. This technique is given in detail in Chapter 11.

Physical Sensors: Two of the most important physical parameters that can be advantageously measured using fibre optics are temperature and pressure. These sensors are based on the attachment of an optical transducer at the end of an optical fibre.

Temperature Sensors: The production of localized and controlled hyperthermia (elevated temperatures in the range of 42–45°C or higher) for cancer treatment by electromagnetic energy, either in the radio frequency or microwave frequency range, poses a difficult temperature measurement problem. Traditional temperature sensors, such as thermistors or thermocouples, have metallic components and connecting wires which perturb the incident electromagnetic (EM) fields and may even cause localized heating spots and the temperature readings may be erratic due to interference. This problem is overcome by using temperature sensors based on fibre-optics. These devices utilise externally induced changes in the transmission characteristics of the optical fibres and offer typical advantages of optical fibres such as flexibility, small dimensions and immunity from EM interference.

One of the simplest types of temperature sensors consists of a layer of liquid crystal at the end of optical fibres, giving a variation in light scattering with temperature at a particular wavelength. Figure 3.23 shows ray-path configuration of a temperature sensor which utilizes a silica-core silicon-clad fibre, with an unclad terminal portion immersed in a liquid which replaces the clad. A temperature rise causes a reduction in the refractive index of the liquid clad fibre section. Therefore, the light travelling from the silicon-clad fibre to the liquid-clad fibre undergoes an attenuation which decreases by increasing temperature. The light from an 860 nm light-emitting diode (LED) is coupled into the fibre. The light reflected backwards is sent along the same fibres and the light amplitude modulation induced by the thermo-sensitive clading applied on the distal end of the fibre is detected and processed. Scheggi *et al* (1984) constructed a miniature temperature probe for medical use with a 0.8 mm external diameter and 0.5 mm internal diameter. The sensitivity achieved was ±0.1°C in the temperature interval 20–50°C.

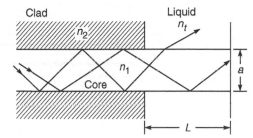

➤ Fig. 3.23 *Principle of a temperature sensor based on variation of refractive index with temperature. The diagram shows optical ray path configuration*

Another type of temperature sensor is based upon the temperature dependence of the band edge absorption of infrared light in GaAs (gallium arsenide) crystal, proposed and developed by Christensen (1977). The variation of band-gap energy with temperature (thermal wavelength shift) provides a measureable variation in the transmission efficiency of infrared light through the crystal.

In the temperature measuring system (Fig. 3.24(a)) based on this principle, light is emitted by an LED, transmitted to and from the crystal via optical fibres and measured by a photodetector. No metal parts are used in the temperature probe design, resulting in transparency of the probe to elecromagnetic fields. Single sensor probe with an outer diameter of 0.6 mm and a four point temperature sensor probe (Fig. 3.24(b)) of 1.2 mm-diameter based on this technique are commercially available.

Fluoroptic temperature sensors (Culshaw, 1982) are other useful devices which can be used for tissue temperature measurement. They contain a rare earth phosphor which is illuminated by a white light along a short length of large core optical fibre. The light excites the phosphor which emits a number of lines. By using filters, two of these lines at 540 and 630 nm are selected, and the ratio of their intensities is a single valued function of the temperature of the phosphor. By

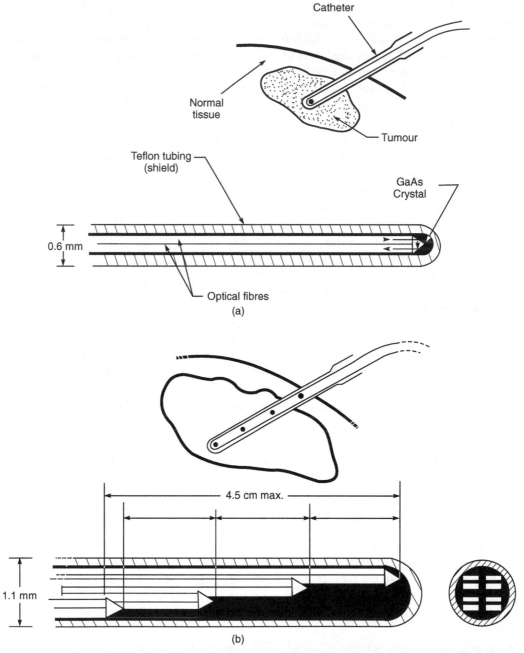

> **Fig.** 3.24   *Temperature sensor based on change of wavelength of infrared light in gallium arsenide crystal (a) single point sensor (b) multipoint sensor (Courtesy: M/s Chlinitherm, USA)*

measuring this ratio, an exact measure of the temperature may be made. The measurement is independent of the output light intensity. Resolution of 0.1°C over the range –50 to +250°C is reported with this technique.

Pressure Sensors: Measurement of intracranial and intracardiac pressure are both important and can be performed using fibre optic sensors. For intracranial pressure measurement, the device is based on a pressure balancing system. Here static pressure is to be monitored and a sensor based on the deflection of a cantilever mirror attached to a membrane has been demonstrated. Deflection of the membrane causes the light emitted frame centre optical fibre to be reflected differentially towards either of the two light-collecting fibres located on each side of the control fibre. The ratio of the light collected by two different fibres is sensed and suitable feedback air pressure is applied to the interior of the probe through a pneumatic connecting tube, balancing the membrane to its null position and providing a readout of the balancing pressure.

A similar sensor based on the deflection of a mirror has been developed for monitoring intravascular pressure. For intravascular use, dynamic pressure measurement is needed and hence the sensors should not only be small but also have good frequency response in order to follow the pressure waveforms faithfully.

Chemical Sensors: The development of optical fibre sensors for chemical species has attracted much interest. The ability of these fibres to transmit light over great distances with low power loss and the interaction of light with a measurand provide the basis of these sensing devices.

The basic concept of a chemical sensor based on optical fibres (Sertz, 1984) is illustrated in Fig. 3.25. Light from a suitable source is applied to the fibre and is directed to a region where the light interacts with the measurement system or with a chemical transducer. The interaction results in a modulation of optical intensity and the modulated light is collected by the same or another optical fibre and measured by photo-detection system.

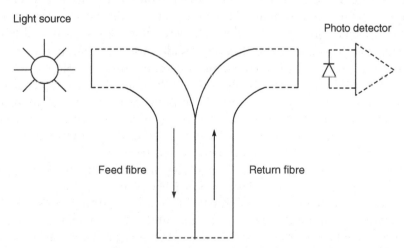

> **Fig.** 3.25   *An optical fibre chemical sensor (after Narayana Swamy and Sevilla, 1988)*

The optical sensing of chemical species is based on the interaction of these entities with light. When light strikes a substance, a variety of interaction may occur between the photons of the electromagnetic radiation and the atoms and molecules of the substance. These interactions involve an exchange of energy and may lead to absorption, transmission, emission, scattering or reflection of light. The quantized nature of this energy transfer produces information about the composition of the system and forms the basis of the spectrosopic method of chemical analysis (Narayana Swamy and Sevilla, 1988).

Two types of optical fibre sensors have been developed for measuring chemcial species:

- *Spectroscopic Sensors:* This type of sensor detects the analyte species directly through their characteristic spectral properties. In these sensors, the optical fibre functions only as a light guide, conveying light from the source to the sampling area and from the sample to the detector. Here, the light interacts with the species being sensed.

- *Chemical Sensors:* In the chemical sensors, a chemical transduction system is interfaced to the optical fibre at its end. In operation, interaction with analyte leads to a change in optical properties of the reagent phase, which is probed and detected through the fibre optic. The optical property measured can be absorbance, reflectance or luminescence. These sensors have a great specificity as a consequence of the incorporation of the chemical transduction system.

ⅢⅢ▷ 3.9   BIOSENSORS

Biosensors combine the exquisite selectivity of biology with the processing power of modern microelectronics and optoelectronics to offer powerful new analytical tools with major applications in medicine, environmental studies, food and processing industries.

All chemical sensors consist of a sensing layer that interacts with particular chemical substances, and a transducer element that monitors these interactions. Biosensors are chemical sensors in which the sensing layer is composed of biological macro-molecules, such as antibodies or enzymes.

Today, the term biosensor is used to describe a wide variety of analytical devices based on the union between biological and physico-chemical components. The biological component can consist of enzymes, antibodies, whole cells or tissue slices and is used to recognize and interact with a specific analyte. The physico-chemical component, often referred to as the transducer, converts this interaction into a signal, which can be amplified and which has a direct relationship with the concentration of the analyte. The transducer may use potentiometric, amperometric, optical, magnetic, colorimetric or conductance change properties.

Biosensors offer the the specificity and sensitivity of biological-based assays packaged into convenient devices for an in situ use by lay personnel. For example, in the medical field biosensors allow clinical analyses to be performed at the bedside, in critical care units and doctors' offices rather than in centralized laboratories. The subsequent results of the test can then be acted upon immediately thus avoiding the delays associated with having to send samples to and having to wait for results from centralized laboratories

The critical areas of biosensor construction are the means of coupling the biological component to the transducer and the subsequent amplification system. Most of the early biosensors immobilized enzymes on selective electrodes, such as the Clark O_2 electode, which measured one of the reaction products (e.g. O_2) of the enzyme-analyte interaction. The most successful biosensor to-date is the home blood glucose monitor for use by people suffering from diabetes. The biosensor in this instrument relies upon enzymes that recognise and catalyze reactions of glucose with the generation of redox-active species that are detected electrochemically.

Figure 3.26 shows the construction of this type of sensor. If the immobilized enzyme is soluble glucose oxidase between the two membranes, it becomes a glucose sensor. It works on the principle that in the presence of glucose, oxygen is consumed, providing a change in the signal from a conventional oxygen electrode.

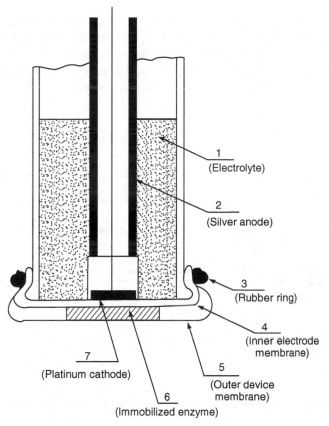

> **Fig. 3.26** *Constructional details of an enzyme utilizing sensor with oxygen electrode as the underlying analytical tool*

The chemical reaction of glucose with oxygen is catalyzed in the presence of glucose oxidase. This causes a decrease in the partial pressure of oxygen (pO_2), and the production of hydrogen peroxide by the oxidation of glucose to gluconic acid as per the following reaction:

$$\text{Glucose} + O_2 \xrightarrow{\text{Glucose oxidase}} \text{Gluconic acid} + H_2O_2$$

The changes in all of these chemical components can be measured in order to determine the concentration of glucose.

For constructing the sensor, glucose oxidase entrapped in a polyacrylamide gel was used. In general, the response time of such types of bioelectrodes as slow and subsequent work has concentrated on closer coupling of the biological component to the transducer. The present technology in biosensors disposes of the coupling agent by direct immobilization of the enzyme onto an electrode surface, making the bio-recognition component an integral part of the electrode transducer. The major disadvantage of enzymatic glucose sensors is the instability of the immobilized enzyme. Therefore, most glucose sensors operate effectively only for short periods of time.

A number of alternative approaches have been investigated to develop a glucose sensor. An important principle that can be used for this purpose depends upon the fluorescence-based, reversible competitive affinity sensor. The sensing element consists of a 3 mm hollow dialysis tube remotely connected to a fluorimeter via a single optical fibre (Fig. 3.27). It contains a carbohydrate receptor, Canavalin A, immobilized on its inner surface and a fluorescein-labelled indicator as a competing agent.

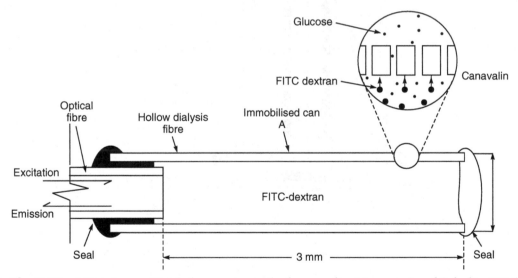

> **Fig. 3.27**　*Miniature optical glucose sensor (redrawn after Mansouri and Schultz, 1984)*

The analyte glucose in the external medium diffuses through the dialysis membrane and competes for binding sites on a substrate (Canavalin A), with FITC-dextran. The sensor is arranged so that the substrate is fixed in a position out of the optical path of the fibre end. It is bound to the inner wall of a glucose-permeable hollow fibre fastened to the end of an optical fibre. The hollow fibre acts as the container and is impermeable to the large molecules of the fluorescent indicator. Increasing glucose concentration displaces labelled FITC-dextran from the Canavalin A, causing it to be free to diffuse into the illuminated solution volume. The optical field that extends from the fibre sees only the unbound indicator. At equilibrium, the level of free fluorescein in the hollow fibre is measured as fluorescence intensity and is correlated to the external glucose concentration.

Biosensors are the most appropriate technology in areas where traditional laboratory analyses, with their associated cost and time requirement, are not a suitable soluation. One such area in which sampling and laboratory analysis are clearly approriate is the detection of hazardous gas escapes in the workplace. The protection of workers in the chemical industry is ideally achieved with personal monitors based on biosensors technology that can be worn on the clothing and which will give an immediate audible warning of gas escape. For this purpose, biosensors for detection of hydrogen cyanide have been developed.

Besides the medical field, biosensors have tremendous applications in the food and beverage industries. Although several biosensors have been developed over the past few years and there are already numerous working biosensors, various problems still need to be resolved. Most complex problems awaiting solution are their limited lifetime, which restrict their commercial viability, necessitating improvements in their stability.

▸ 3.10 SMART SENSORS

Although an accepted industry definition for a smart sensor does not currently exist, it is generally agreed that they have tight coupling between sensing and computing elements. Their characteristics, therefore, include: temperature compensation, calibration, amplification, some level of decision-making capability, self-diagnostic and testing capability and the ability to communicate interactively with external digital circuitry. Currently available smart sensors are actually hybrid assemblies of semiconductor sensors plus other semiconductor devices. In some cases, the coupling between the sensor and computing element is at the chip level on a single piece of silicon in what is referred to as an integrated smart sensor. In other cases, the term is applied at the system level. The important role of smart sensors are:

Signal Conditioning: The smart sensor serves to convert from a time-dependent analog variable to a digital output. Functions such as linearization, temperature compensation and signal processing are included in the package.

Tightening Feedback Loops: Communication delays can cause problems for systems that rely on feedback or that must react/adapt to their environment. By reducing the distance between sensor and processor, smart sensors bring about significant advantages to these types of applications.

Monitoring/Diagnosis: Smart sensors that incorporate pattern recognition and statistical techniques can be used to provide data reduction, change detection and compliation of information for monitoring and diagnostic purposes, specially in the health sector.

Smart sensors divert much of the signal processing workload away from the general purpose computers. They offer a reduction in overall package size and improved reliability, both of which are critical for in situ and sample return applications. Achieving a smart sensor depends on integrating the technical resources necessary to design the sensor and the circuitry, developing a manufacturable process and choosing the right technology.

A typical example of a smart sensor is a pressure sensor (MPX5050D) with integrated amplification, calibration and temperature compensation introduced by Motorola (Frank, 1993). The sensor typically uses piezioresistive effect in silicon and employs bipolar integrated circuit processing techniques to manufacture the sensor.

By laser trimming thin film resistors on the pressure sensor, the device achieves a zero-pressure offset-voltage of 0.5 V nominal and full-scale output voltage of 4.5 V, when connected to a 5.0 V supply. Therefore, the output dynamic range due to an input pressure swing of 0–375 mmHg is 4.0 V. The performance of the device compares favourably to products that are manufactured with direct components.

CHAPTER

4

Recording Systems

⑉➤ 4.1 BASIC RECORDING SYSTEM

Recorders provide a permanent visual trace or record of an applied electrical signal. There are many types of recorders utilizing a variety of techniques for writing purposes. The most elementary electronic recording system is shown in Fig. 4.1. It consists of three important components. Firstly, the electrode or the transducer. The electrode picks up the bioelectrical potentials whereas the transducer converts the physiological signal to be measured into a usable electrical output. The signal conditioner converts the output of the electrode/transducer into an electrical quantity suitable for operating the writing system. The writing system provides a visible graphic representation of the quantity of the physiological variable of interest.

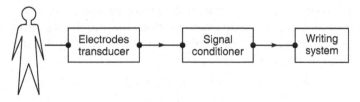

➤ **Fig. 4.1** *A basic electronic recording system*

In medical recorders, the signal conditioners usually consist of a preamplifier and the main amplifier. Both these types of amplifiers must satisfy specific operating requirements such as input impedance, gain and frequency response characteristics for a faithful reproduction of the input signal.

To make the signal from any transducer compatible with the input signal required for the driver amplifier of the display or recording system, it is usual to arrange to normalize the electrical signals produced by each transducer. This is done in the signal conditioner which adjusts its output to a common signal level, say one volt. The necessary adjustments of gain and frequency response are provided by the signal conditioners. By this means, it is possible to interchange the signal conditioners to record any one of the physical or bioelectric events on the same writing channel.

The writing systems which are available in many forms constitute the key portion of the recording instrument. The commonly used writing systems are the galvanometer type pen recorder, the inkjet recorder and the potentiometric recorder. While the electrodes and transducers have been described in Chapters 2 and 3, the writing methods and signal conditioners are illustrated in this chapter.

ⅢⅢ▶ 4.2   GENERAL CONSIDERATIONS FOR SIGNAL CONDITIONERS

Information obtained from the electrodes/transducers is often in terms of current intensity, voltage level, frequency or signal phase relative to a standard. In addition to handling specific outputs from these devices, signal conditioners used in biomedical instruments perform a variety of general purpose conditioning functions to improve the quality, flexibility and reliability of the measurement system. Important functions performed by signal conditioners, before the signal is given to a display or recording device, are illustrated below:

Signal Amplification: The signals available from the transducers are often very small in magnitude. Amplifiers boost the level of the input signal to match the requirements of the recording/display system or to match the range of the analog-to-digital convertor, thus increasing the resolution and sensitivity of the measurement.

Bioelectric measurements are basically low-level measurements, which involve amplifying and recording of signals often at microvolt levels. The problem of electrical noise makes these measurements a difficult proposition and calls for both expert technical knowledge and skillful handling of the signal in the circuit design. Noise can produce errors in measurements and completely obscure useful data. It is a special problem in applications where low-level signals are recorded at high off-ground voltages, or transmitted over distance or obtained in electromagnetic noise environments. Using signal conditioners located closer to the signal source, or transducer, improves the signal-to-noise ratio of the measurement by boosting the signal level before it is affected by the environmental noise.

Frequency Response: Modern biomedical instruments are designed to handle data with bandwidths from dc up to several hundred cycles per second. Electrical or mechanical filters cannot separate useful signals from the noise when their bandwidths overlap. Instruments and recording systems that work satisfactorily for steady state or low frequency data are generally inadequate to meet this requirement. On the other hand, recording systems that will faithfully reproduce such data are inherently more susceptible to external noise and, therefore, they must be designed to eliminate the possibility of signal contamination from noise.

The bioelectric signals often contain components of extremely low frequency. For a faithful reproduction of the signal, the amplifiers must have excellent frequency response in the sub-audio frequency range. This response should be down to less than one hertz which is a very frequent requirement.

In all RC-coupled amplifiers, low frequency response is limited by the reactance of the interstage coupling capacitors. To achieve the low frequency response required for medical applications, the amplifier must have large values of coupling capacitance. The disadvantage of using large capacitors is that they can cause blocking of the amplifier in cases of high-level input, arising due to switching transients or other high-level inputs. Because of the long time constant introduced by these large coupling capacitors, several seconds may elapse before the capacitors have discharged back to the normal levels. The amplifier, therefore, becomes momentarily unreceptive following each occurrence of overdriving signals. This type of problem does not exist in direct coupled amplifiers simply because there are no coupling capacitors.

Although the direct coupled amplifier gives an excellent frequency response at low frequencies, it tends to drift. The drift is a slow change of output having no relation with the input signal applied to the amplifier. Since the frequency response of the RC-coupled amplifier does not extend all the way down to dc, it does not drift. In medical amplifiers, the advantages of both types of coupling can be obtained in one amplifier. Typically, all stages except one are direct coupled. The one RC coupled stage prevents the drifting of the output. Suitable measures are taken to prevent blocking of the amplifier due to overdrive by quickly discharging the coupling capacitor automatically after occurrence of the overdriving input.

It is not desirable to have the frequency response of the amplifiers much above the highest signal frequency of interest. Excessive bandwidth allows passage of noise voltages that tend to obscure the bioelectric signal. Another reason for limiting the response of an amplifier to the signal bandwidth is to minimize the tendency towards oscillation due to stray feedback. High frequency response, therefore, is deliberately limited as a means of noise reduction.

Filtering: A filter is a circuit which amplifies some of the frequencies applied to its input and attenuates others. There are four common types: high-pass, which only amplifies frequencies above a certain value; low-pass, which only amplifies frequencies below a certain value; band-pass, which only amplifies frequencies within a certain band; and band stop, which amplifies all frequencies except those in a certain band.

Filters may be designed using many different methods. These include passive filters which use only passive components, such as resistors, capacitors and inductors. Active filters use amplifiers in addition to passive components in order to obtain better performance, which is difficult with passive filters. Operational amplifiers are frequently used as the gain blocks in active filters. Digital filters use analog-to-digital converters to convert a signal to digital form and then use high-speed digital computing techniques for filteration.

Additionally, signal conditioners can include filters to reject unwanted noise within a certain frequency range. Almost all measuring and recording applications are subject to some degree of 50 Hz noise picked up from power lines or machinery. Therefore, most signal conditioners include low-pass filters designed specifically to provide maximum rejection of 50 Hz noise. Such filters are called "notch" filters.

Filters can be classified as digital and analog filters. They differ by the nature of the input and the output signals. An analog filter processes analog inputs and generates analog outputs. A digital filter processes and generates digital data. The processing techniques followed are also different. Analog filters are based on the mathematical operator of differentiation and digital filters require no more than addition, multiplication and delay operators.

Digital filters have several advantages over analog filters. They are relatively insensitive to temperature, ageing, voltage drift and external interference as compared to analog filters. Their response is completely reproducible and predictable, and software simulation can exactly reflect product performance.

Isolation: Improper grounding of the system is one of the most common causes of measurement problems and noise. Signal conditioners with isolation can prevent these problems. Such devices pass the signal from its source to the measurement device without a physical or galvanic connection by using transformer, optical or capacitive coupling techniques. Besides breaking ground loops, isolation blocks high voltage surges and rejects high common mode voltages.

Excitation: Signal conditioners are sometimes also required to generate excitation for some transducers. Strain gauges, thermistors, for example, require external voltage or current excitation. Signal conditioning part of the measurement system usually provides the excitation signal. Strain gauges are resistance devices in a Wheatstone bridge configuration, which require bridge completion circuitry and an excitation source.

Linearizaion: Another common signal conditioning function is linearization. Many transducers such as thermocouples and thermistors have a non-linear response to changes in the phenomenon being measured. Signal conditioners include either hardware-based or software-based linearization routines for this purpose.

Signal conditioners, therefore, perform an extremely useful function in a measuring and recording system as they determine the range, accuracy and resolution of the system.

ⅠⅠⅠⅠ▶ 4.3 PREAMPLIFIERS

Modern multi-channel biomedical instruments and recorders are usually modularly designed to meet both existing and anticipated requirements. Numerous configurations provide for every measurement need, with or without interchangeable plug-in preamplifiers, which provides a choice of signal conditioners for a large selection of analog measurements. Conventional preamplifiers offer a wide range of input sensitivities to cover virtually all signal sources. Calibrated zero suppression to expand desired portions of an input signal, and selectable low-pass filtering facilities to reject noise or unwanted signal components are available on these amplifiers.

For biophysical measurements, the amplifiers employed include: (i) ac/dc universal amplifier with special features such as capacity neutralization, current injection, low leakage current and low dc drift suitable for intracellular measurements through high resistance fluid-filled electrodes or to make extracellular recordings through metal microelectrodes for EMG, EEG, EOG, etc. (ii) an ECG amplifier with full 12 lead selection and patient isolation (iii) a transducer amplifier suited for bridge measurements on strain gauges, strain gauge based blood pressure transducers, force transducers, resistance temperature devices and direct low level dc input signals and (iv) a dc amplifier used in conjunction with standard thermistor probes for the accurate measurement of temperature within the range of medical applications.

Various types of amplifiers which are generally used are as follows:

Differential amplifier is one which will reject any common mode signal that appears simultaneously at both amplifier input terminals and amplifies only the voltage difference that appears across its input terminals. Most of the amplifiers used for measuring bioelectric signals are of the differential type.

Ac coupled amplifiers have a limited frequency response and are, therefore, used only for special medical applications such as electrocardiograph machine. For electrocardiograms, an ac amplifier with a sensitivity, giving 0.5 mV/cm, and frequency response up to 1 kHz and an input impedance of 2 to 5 MΩ is used. For such applications as retinography, EEG and EMG, more sensitive ac amplifiers are required, giving a chart sensitivity of say 50 μV/cm with a high input impedance of over 10 MΩ.

Carrier amplifiers are used with transducers which require an external source of excitation. They are characterized by high gain, negligible drift, extremely low noise and the ability to operate with resistive, inductive or capacitive type transducers. They essentially contain a carrier oscillator, a bridge balance and calibration circuit, a high gain ac amplifier, a phase-sensitive detector and a dc output amplifier.

DC amplifiers are generally of the negative feedback type and are used for medium gain applications down to about 1 mV signal levels for full scale. They are not practical for very low level applications because of dc drift and poor common-mode rejection capabilities. They are usually employed as pen drive amplifiers in direct writing recorders.

Chopper input dc amplifiers are preferred for low level inputs to instrumentation systems because of their high sensitivity, negligible drift and excellent common mode rejection capability. Their high frequency response is limited to about one half of the input chopper frequency.

Chopper stabilized dc amplifiers are used for low level but preferably wideband applications such as oscilloscopes, tape recorders and light beam oscilloscope recorders. These are complex amplifiers having three amplifiers incorporated in the module. This includes an ac amplifier for signals above about 20 Hz, a dc chopper input amplifier for signals from about 20 Hz down to dc plus a wideband feedback stabilized dc amplifier.

DC bridge amplifiers are employed with resistive transducers which require an external source of excitation. Essentially, the amplifier comprises of a stable dc excitation source, a bridge balance and calibration unit, a high gain differential dc amplifier and a dc output amplifier. They can be used as conventional dc high gain amplifiers and offer operating simplicity and high frequency response. These amplifiers are necessary for transducers used to measure temperature and blood pressure. The sensitivity in these cases may be 50 μV/cm with an input impedance of 50 kΩ.

4.3.1 Differential Amplifier

Medical amplifiers designed for use in the input stage (preamplifiers) are mostly of the differential type. These type have three input terminals out of which one is arranged at the reference potential and the other two are live terminals. The differential amplifier is employed when it is necessary to measure the voltage difference between two points, both of them varying in amplitude at different rates and in different patterns. Heart-generated voltages picked up by means of electrodes on the arms and legs, and brain-generated voltages picked up by the electrodes on the scalp are typical examples of signals whose measurement requires the use of differential amplifiers.

The differential amplifier is an excellent device for use in the recording systems. Its excellence lies in its ability to reject common-mode interference signals which are invariably picked up by electrodes from the body along with the useful bioelectric signals. Also, as a direct coupled amplifier, it has good stability and versatility. High stability is achieved because it can be insensitive to temperature changes which is often the source of excessive drift in other configurations. It is versatile in that it may be adapted for a good many applications, e.g. applications requiring floating inputs and outputs or for applications where grounded inputs and/or outputs are desirable.

The working of a differential amplifier can be explained with the help of Fig. 4.2 where the two transistors with their respective collector resistances (R_1 and R_2) form a bridge circuit. If the two resistors and the characteristics of the two transistors are identical, the bridge is perfectly balanced and the potential difference across the output terminals is zero.

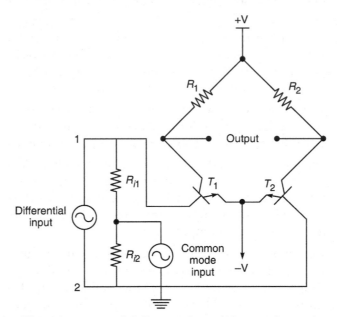

> **Fig.** 4.2 *Typical differential amplifier configuration*

Let us now apply a signal at the input terminals 1 and 2 of this circuit. The signal is to be such that at each input terminal, it is equal in amplitude but opposite in phase with reference to the ground. This signal is known as the differential mode signal. Because of this signal, if the collector current of T_1 increases, the collector current of T_2 will decrease by the same amount, and the collector voltage of T_1 will decrease while that of T_2 will increase. This results in a difference voltage between the two output terminals that is proportional to the gain of the transistors.

On the other hand, if the signal applied to each input terminal is equal in amplitude and is in the same phase (called the common-mode input signal), the change in current flow through both transistors will be identical, the bridge will remain balanced, and the voltage between the output terminals will remain zero. Thus, the circuit provides high gain for differential mode signals and no output at all for common-mode signals. Resistances Ri_1 and Ri_2 are current limiting resistances for common-mode signals.

The ability of the amplifier to reject these common voltages on its two input leads is known as common-mode rejection and is specified as the ratio of common-mode input to differential input to elicit the same response. It is abbreviated as CMRR (Common-mode rejection ratio). CMRR is an important specification referred to the differential amplifier and is normally expressed as decibels. CMRR of the preamplifiers should be as high as possible so that only the wanted signals find a way through the amplifier and all unwanted signals get rejected in the preamplifier stage. A high rejection ratio is usually achieved by the use of a matched pair of transistors in the input stage of the preamplifier and a large 'tail' resistance in the long-tailed pair to provide maximum negative feedback for inphase signals. The technique of long-tailing (a technique used to current drive an active device) improves the CMRR in differential amplifiers without upsetting the gain for the desired signal. Very high CMRR can be achieved with the use of an active long-tail.

In order to be able to minimize the effects of changes occurring in the electrode impedances, it is necessary to employ a preamplifier having a high input impedance. It has been found that a low value of input impedance gives rise to a considerable distortion of the recordings (Hill and Khandpur, 1969).

Figure 4.3 is an equivalent circuit for the input of an ECG amplifier, in which:

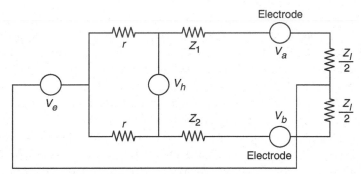

> **Fig.** 4.3 *Equivalent circuit for the input of an ECG amplifier*

V_h represents the voltage signal generated by the heart.

V_e represents unwanted inphase signal picked up from the mains wiring and other sources.

Z_I is the total input impedance of the preamplifier.

Z_1 and Z_2 are the skin contact impedances of the electrodes.

The resistance r represents tissue and blood resistance which is negligibly low as compared with other impedances.

If the amplifier is perfectly balanced by equal inphase voltages, V_a and V_b, at the electrodes would give rise to a zero output signal. However, the voltages V_a and V_b depend, in practice, on the values of Z_1 and Z_2. It can be shown that the electrical interference signal V_e will give rise to the same output signal as would a desired signal, from the patient, of amplitude

$$\frac{Z_2 - Z_1}{Z_I / 2} \cdot V_e$$

Hence, the discrimination factor between desired and undesired signals is given by

$$\frac{Z_I / 2}{Z_2 - Z_1}$$

Assuming a common mode rejection ratio of 1000:1 and a difference of electrode skin contact impedance as 1 kΩ ($Z_2 - Z_1 = 1$ kΩ), then

$$\text{CMRR} = \frac{Z_I}{2(Z_2 - Z_1)}$$

$$1000 = \frac{Z_I}{2 \times 1000}$$

$$Z_I = 2\,\text{M}\Omega$$

If $Z_2 - Z_1 = 5$ kΩ.

then, $Z_I = 10$ MΩ

This shows that a high input impedance is very necessary in order to obtain a high CMRR. Also, the electrode skin resistance should be low and as nearly equal as possible. Winter and Webster (1983) reviewed several design approaches for reducing common-mode voltage interference in biopotential amplifiers.

The design of a good differential amplifier essentially implies the use of closely matched components which has been best achieved in the integrated circuit form. High gain integrated dc amplifiers, with differential input connections and a provision for external feedback have been given the name operational amplifiers because of their ability to perform mathematical operations. These amplifiers are applied for the construction of ac or dc amplifiers, active filters, phase inverters, multi-cvibrators and comparators, etc. by suitable feedback arrangement, and therefore find a large number of applications in the medical field.

Figure 4.4 shows a single op-amp in a differential configuration. The common mode rejection for most op-amps is typically between 60 dB and 90 dB. This may not be sufficient to reject common mode noise generally encountered in biomedical measurements. Also, the input impedance is not very high to handle signals from high impedance sources. One method to increase the input

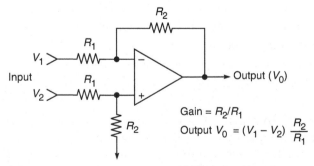

➤ **Fig. 4.4** *A single op-amp in a differential configuration*

impedance of the op-amp is to use field effect transistors (FET) in the input differential stage. A more common approach is to use an instrumentation amplifier in the preamplifier stage.

4.3.2 Instrumentation Amplifier

The differential amplifier is well suited for most of the applications in biomedical measurements. However, it has the following limitations:

- The amplifier has a limited input impedance and therefore, draws some current from the signal source and loads them to some extent.
- The CMRR of the amplifier may not exceed 60 dB in most of the cases, which is usually inadequate in modern biomedical instrumentation systems.

These limitations have been overcome with the availability of an improved version of the differential amplifier, whose configuration is shown in Fig. 4.5. An instrumentation amplifier is a precision differential voltage gain device that is optimized for operation in an environment hostile to precision measurement. It basically consists of three op-amps and seven resistors. Basically, connecting a buffered amplifier to a basic differential amplifier makes an instrumentation amplifier.

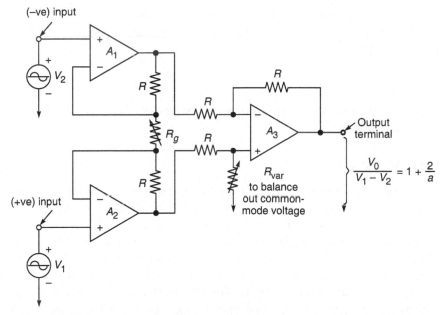

> **Fig. 4.5** *Schematic diagram of an instrumentation amplifier*

In the figure shown above, op-amp A_3 and its four equal resistors R form a differential amplifier with a gain of 1. Only A_3 resistors have to be matched. The variable resistance R_{var} is varied to balance out any common-mode voltage. Another resistor R_g, is used to set the gain using the formula

$$\frac{V_0}{V_1 - V_2} = 1 + \frac{2}{a}$$

Where $a = R_g/R$

V_1 is applied to the +ve input terminal and V_2 to the –ve input terminal. V_0 is proportional to the difference between the two input voltages.

The important characteristics of the instrumentation amplifier are:

- Voltage gain from differential input $(V_1 - V_2)$ to single ended output, is set by one resistor.
- The input resistance of both inputs is very high and does not change as the gain is varied.
- V_0 does not depend on common-mode voltage, but only on their difference.

If the inputs are prone to high voltage spikes or fast swings, which the op-amps cannot cope with, they may be protected using back-to-back connected diodes at their inputs. However, this reduces the input impedance value substantially and also limits the bandwidth.

The instrumentation amplifier offers the following advantages for its applications in the biomedical field:

- Extremely high input impedance
- Low bias and offset currents
- Less performance deterioration if source impedance changes
- Possibility of independent reference levels for source and amplifier
- Very high CMRR
- High slew rate
- Low power consumption

Good quality instrumentation amplifiers have become available in single IC form such as µA725, ICL7605, LH0036, etc.

4.3.3 Carrier Amplifiers

To obtain zero frequency response of the dc amplifier and the inherent stability of the capacitance coupled amplifier, a carrier type of amplifier is generally used. The carrier amplifier consists of an oscillator and a capacitance coupled amplifier. The oscillator is used to energize the transducer with an alternating carrier voltage. The transducers, which require ac excitation, are those whose impedance is not purely resistive. Example can be of a capacitance based pressure transducer whose impedance is mainly capacitative with a small resistive component. The frequency of the excitation voltage is usually around 2.5 kHz. The transducer shall change the amplitude of the carrier voltage in relation to the changes in the physiological variable being measured. The output of the transducer therefore, would be an amplitude modulated (AM) signal (Fig. 4.6). The modulated ac signal can then be fed to a multi-stage capacitance coupled amplifier. The first stage produces amplification of the AM signal. The second stage is so constructed that it can respond only to signal frequency of the carrier. It can be further amplified in the following stage. After amplification, the signal is demodulated in a phase-sensitive demodulator circuit. This helps to extract amplified signal voltage after the filter circuit. The voltage produced by the demodulator can then be applied to the driver stage of the writing system.

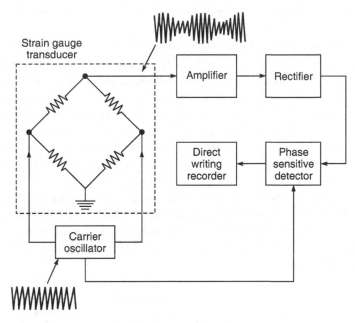

> ➤ **Fig.** 4.6   *Block diagram of carrier amplifier*

Carrier amplifiers can be used with a resistance strain gauge transducer such as a semi-conductor strain gauge. When used with pressure gauges, a calibration control is provided on the carrier amplifier. This enables direct measurements of the blood pressure from the calibrated graphic recorder.

Lock-in amplifier is a useful version of the carrier technique designed for the measurement of low-level signals buried in noise. This type of amplifier, by having an extremely narrow-width output band in which the signal is carried, reduces wideband noise and increases the signal-to-noise ratio. Thus, the difference between carrier amplifier and lock-in amplifier is that the former is a general purpose instrument amplifier while the latter is designed to measure signals in a noisy background.

In principle, the lock-in amplifier works by synchronizing on a single frequency, called the reference frequency. This frequency is made to contain the signal of interest. The signal is modulated by the reference frequency in such a way that all the desired data is at the single reference frequency whereas the inevitable noise, being broadband, is at all frequencies. This permits the signal to be recovered from its noisy background.

4.3.4   Chopper Amplifier

The chopper amplifier is a useful device in the field of medical electronics as it gives another solution to the problem of achieving adequate low frequency response while avoiding the drift problem inherent in direct coupled amplifiers. This type of amplifier makes use of a chopping device, which converts a slowly varying direct current to an alternating form with amplitude proportional to the input direct current and with phase dependent on the polarity of the original signal. The alternating voltage is then amplified by a conventional ac amplifier whose output is rectified back to get an amplified direct current. A chopper amplifier is an excellent device for signals of narrow bandwidth and reduces the drift problem.

Figure 4.7 shows a simplified block diagram of a single-ended chopper stabilized amplifier. The amplifier achieves its ultra low dc offset voltage and bias current by chopping the low frequency components of the input signal, amplifying this chopped signal in an ac amplifier (A_1) and then demodulating the output of the ac amplifier. The low frequency components are derived from the input signal by passing it through the low-pass filter, consisting of R_2, C_2 and R_2. The chopping signal is generated by the oscillator. The filtered output is then further amplified in a second stage of dc amplification (A_2). High frequency signals, which are filtered out at the input of the chopper channel, are coupled directly into the second stage amplifier. The result of this technique is to reduce the dc offsets and drift of the second amplifier by a factor equal to the gain of the chopper channel. The ac amplifier introduces no offsets. Minor offsets and bias currents exist due to imperfect chopping, but these are extremely small. The amplifier modules contain the chopper channel, including switches and switch-driving oscillator built on the module; only the dc power is supplied externally.

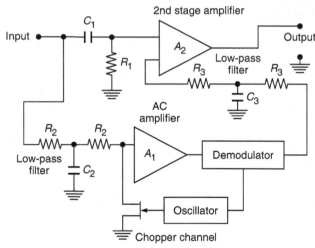

> **Fig.** 4.7 *Simplified block diagram of a single ended chopper-stabilized operational amplifier*

Due to the extremely low dc offset and dc drift associated with the chopper-stabilized amplifier, the signal resolution is limited only by the noise present in the circuit. Thus, it is desirable to design the feedback networks and external wiring to minimize the total circuit noise. When the full bandwidth of the amplifier is not required, it is advisable that a feedback capacitor be used to limit the overall bandwidth and eliminate as much high frequency noise as possible. Shielding of feedback components is desirable in chopper amplifiers. It is particularly necessary in electrically noisy environments. Use of shielded wire for summing junction leads is also recommended. Typical voltage drift in chopper-stabilised amplifiers is $0.1\,\mu V/^0C$ and current drift as $0.5\,pA/^\circ C$.

The great strength of the chopper-stabilized amplifier is its insensitivity to component changes due to ageing, temperature change, power supply variation or other environmental factors. Thus, it is usually the best choice where both offset voltage and bias current must be small over long periods of time or when there are significant environmental changes. Both bias current and offset voltage can be externally nulled. Chopper amplifiers are available in both single-ended as well as differential input configurations. Chopper amplifiers find applications in the medical field in amplification of small dc signals of a few microvolts. Such order of amplitudes are obtainable from transducers such as strain gauge pressure transducers, temperature sensors such as thermistors and strain gauge myographs, when they are used as arms of a dc Wheatstone bridge. A chopper amplifier is also suitable for use with a thermocouple.

4.3.5   Isolation Amplifiers

Isolation amplifiers are commonly used for providing protection against leakage currents. They break the ohmic continuity of electric signals between the input and output of the amplifier. The isolation includes different supply voltage sources and different grounds on each side of the isolation barrier. Three methods are used in the design of isolation amplifiers: (i) transformer isolation (ii) optical isolation (iii) capacitive isolation.

The transformer approach is shown in Fig. 4.8. It uses either a frequency-modulated or a pulse-width-modulated carrier signal with small signal bandwidths up to 30 kHz to carry the signal. It uses an internal dc–to-dc converter comprising of a 20 kHz oscillator, transformer, rectifier and filter to supply isolated power.

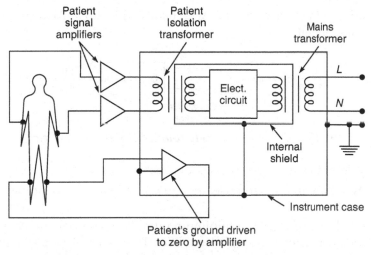

> **Fig. 4.8**   *Isolation amplifier (transformer type)*

Isolation could also be achieved by optical means in which the patient is electrically connected with neither the hospital line nor the ground line. A separate battery operated circuit supplies power to the patient circuit and the signal of interest is converted into light by a light source (LED).

This light falls on a phototransistor on the output side, which converts the light signal again into an electrical signal (Fig. 4.9), having its original frequency, amplitude and linearity. No modulator/demodulator is needed because the signal is transmitted optically all the way.

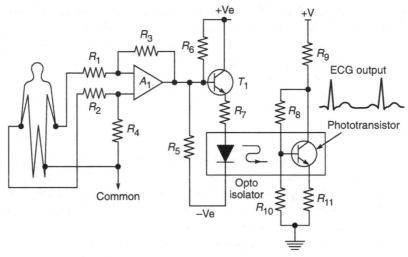

➤ **Fig. 4.9**   *Optically isolated isolation amplifier*

The capacitive method (Fig. 4.10) uses digital encoding of the input voltage and frequency modulation to send the signal across a differential capacitive barrier. Separate power supply is needed on both sides of the barrier. Signals with bandwidths up to 70 kHz can be conveniently handled in this arrangement.

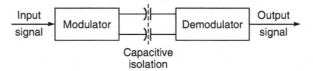

➤ **Fig. 4.10**   *Capacitively coupled isolation amplifier*

The relative merits of the three types of isolation techniques are:

- All three types are in common use, though the transformer isolation amplifier is more popular.
- Opto-coupled amplifier uses a minimum number of components and is cost effective, followed by the transformer coupled amplifier. The capacitor coupled amplifier is the most expensive.
- Opto-isolated amplifiers offer the lowest isolation voltage (800 V continuous) between input and output; transformer coupled 1200 V and capacitance coupled 2200 V.
- Isolation resistance levels are of the order of 10^{10}, 10^{12} and 10^{12} ohms for transformer coupled, opto-coupled and capacitance coupled amplifiers respectively.

- Gain stability and linearity are best for capacitance coupled versions—0.005%, and on par for the transformer and opto-coupled amplifier—0.02%.

Electrical isolation is the most commonly used technique to ensure patient protection against electrical hazards. Instruments such as electrocardiographs, pressure monitors, pressure transducers, pacemakers and others have been designed to electrically separate the portion of the circuit to which the patient is connected from the portion of the circuit connected to the ac power line and ground.

4.4 SOURCES OF NOISE IN LOW LEVEL MEASUREMENTS

4.4.1 Electrostatic and Electromagnetic Coupling to ac Signals

The distributed capacitance between the signal conductors and from the signal conductors to the ground provides a low impedance ac path, resulting in signal contamination from external sources like power lines and transformers. Similarly, the alternating magnetic flux from the adjacent power line wires induces a voltage in the signal loop which is proportional to the rate of change of the disturbing current, the magnitude of the disturbing current and the areas enclosed by the signal loop. It is inversely proportional to the distance from the disturbing wire to the signal circuit. Unequal distances of the two signal carrying conductors from the disturbing current wire result in unequal mutual inductances, which cause the magnetic field to produce a noise voltage across the amplifier input terminals.

Low-level signals are sensitive to external contamination especially in the case of high source impedance. Referring to Fig. 4.11, it is obvious that the currents generated by various noise signals will flow through the signal source impedance Z and result in an unwanted addition to the

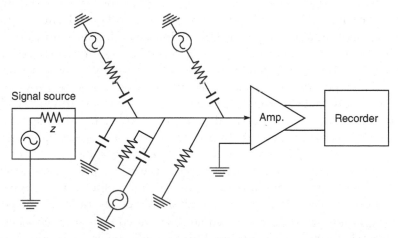

➤ Fig. 4.11 *Currents produced by various forms of noise flow through the signal source impedance and become an unwanted addition to the useful signal. The noise amplitude is directly proportional to signal source impedance (Courtesy: Gould Inc. USA)*

bioelectric or transducer signal. This may include electromagnetic noise pick-up, electrostatic pick-up and the unwanted current generated by a ground loop between two separate grounds on the same signal circuit. The magnitude of these unwanted signals will be directly proportional to the signal source impedance as shown by the relationship:

Amplifier input signal $= E + I Z$

where E = normal signal amplitude

$\quad\quad\; Z$ = impedance of signal source

$\quad\quad\; I$ = current generated by noise

It is obvious that as the signal source impedance approaches zero, so will the noise input to the amplifier. In fact, low-source impedance effectively shunts out the noise.

To prevent noise pick-up from electrostatic fields, low-level signal conductors are surrounded by an effective shield. This is usually a woven metal braid around the signal pair, which is placed under an outside layer of insulation. A more effective shielding is provided by a special type of signal cable, which has lapped foil shields, plus a low resistance drain wire instead of the conventional braided wire shield.

The easiest and generally the best way to protect a signal cable against external electromagnetic disturbances is to twist the circuit conductors closely together to electrically cancel the effect of an external magnetic field. The shorter the lay of the twist, the greater the noise rejection. Thus, electromagnetic coupling is reduced by shielding, wire twisting and proper grounding which provide a balanced signal pair with satisfactory noise rejection characteristics.

4.4.2 Proper Grounding (Common Impedance Coupling)

Placing more than one ground on a signal circuit produces a ground loop which may generate so much noise that it may completely obscure the useful signal. The term 'grounding' means a low-impedance metallic connection to a properly designed ground grid, located in the earth. Stable grounding is necessary to attain effective shielding of low level circuits to provide a stable reference for making voltage measurements and to establish a solid base for the rejection of unwanted common-mode signals. There are generally two grounding systems—a system ground and a signal ground. All low-level measurements and recording systems should be provided with a stable system ground to assure that electronic enclosures and chassis operating in an electromagnetic environment are maintained at zero potential. In most instances, the third copper conductor in all electrical circuits, which is firmly tied to both electric power ground—the building ground and the water system, will provide a satisfactory system ground. In the signal ground, on the other hand, it is necessary to ensure a low noise signal reference to the ground. This ground should be a low-impedance path to wet earth to minimize the introduction of spurious voltages into the signal circuitry. It is important to note that a signal circuit should be grounded at one point only.

Two separate grounds are seldom at the same absolute voltage. If we connect more than one ground to the same signal circuit, an unwanted current will flow in the ground loop thus created. This current combines itself with the useful signal (Fig. 4.12). Also, there is a second ground loop through the signal cable-shield from the signal source to the amplifier. The current in the shield is coupled to the signal pair through the distributed capacitance in the signal cable. This current

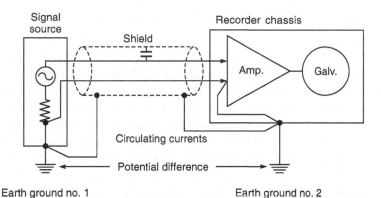

> **Fig. 4.12** *Ground loop created by more than one ground on a signal circuit. The potential difference between earth ground no. 1 and earth ground no. 2 causes current to circulate in the signal cable shield and also in the lower signal conductor, producing two separate ground loops*

then flows through the output impedance of the signal source and back to the ground, thus adding a second source of noise to the useful signal. Either one of these ground loops generates a noise signal that is larger than a typical millivolt useful signal. Ground loops are eliminated by the floating lower input terminal of the amplifier. The amplifier enclosure is still solidly grounded to earth-round No. 2 but this will not create a ground loop, since the amplifier enclosure is insulated from the signal circuit. The ground-loop through the signal cable is removed by grounding the shield only at the signal source which is the proper configuration for minimum noise pick-up (Fig. 4.13).

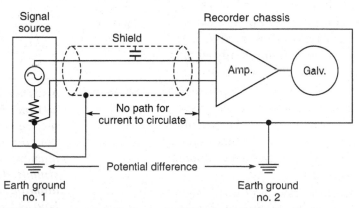

> **Fig. 4.13** *Eliminating multiple grounds. The ground loop in the lower signal lead has been broken by removing the jumper wire to earth ground no. 2. The ground loop in the cable shield has been broken by removing its connection to earth ground no. 2. (Courtesy: Gould Inc., USA)*

ⅢⅢ▶ 4.5   BIOMEDICAL SIGNAL ANALYSIS TECHNIQUES

4.5.1   Fourier Transform

In biomedical instrumentation, sensors/transducers pick up signals from biologic tissue with the objective of finding out their health and well being. Signal processing employs sophisticated mathematical analysis tools and algorithm to extract information burried in these signals received from various sensors and transducers. Signal processing algorithms attempt to capture signal features and components that are of diagnostic value. Since most signals of bio-origin are time varying, there is a need to capture transient phenomena when studying the behaviour of bio-signals.

There are two different presentations of the same experimental data, known as domains. These are the time-domain in which the data are recorded as a series of measurements at successive time intervals and the frequency domain in which the data are represented by the amplitude of its sine and cosine components at different frequencies. For example; for recording and display purposes, the biomedical signals are represented in the time domain, i.e. the signal is represented by means of its value (y-axis) on the time axis (x-axis). In the frequency domain any biomedical signal may be described as consisting of sine waves and having different amplitudes and phases (y-axis) as a function of the frequency (x-axis). The transformation between the two representations is given by the Fourier Transform (FT).

The basic motivation for developing frequency analysis tools is to provide a mathematical and pictorial representation for the frequency components that are contained in any given signal. The term *spectrum* is used when referring to the frequency content of a signal. The process of obtaining the spectrum of a given signal using the basic mathematical tools is known as *frequency* or *spectral analysis*. Most biomedical signals of practical interest can be decomposed into a sum of sinusoidal signal components. For the class of periodic signals, such a decomposition is called a *Fourier series*. For the class of finite energy signals, the decomposition is called the *Fourier transform*.

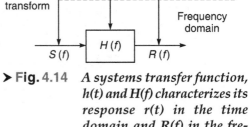

Referring to Fig. 4.14, a system's response to a varying input signal $s(t)$, with a frequency spectrum $S(f)$, can essentially be described interchangeably, by the response $r(t)$ in the time domain (as a time history) or $R(f)$ in the frequency domain (as a frequency spectrum).

Key system characteristics operate on the signal to produce the response to the stimulus. In the

> **Fig. 4.14**   *A systems transfer function, h(t) and H(f) characterizes its response r(t) in the time domain and R(f) in the frequency domain, to receive stimuli, s(t) and S(f). The two domains are related by the Fourier transforms*

frequency domain, the operation can be expressed as a simple product. The ratio of response to stimulus is called the *transfer function*, $H(f)$

$$H(f) = R(f)/S(f)$$

The time domain and frequency domains are closely connected via the Fourier transform.

4.5.2 Fast Fourier Transform (FFT)

Using the Fourier Transform can become tedious and time consuming even when using computers and especially when a large number of points have to be considered. The situation has considerably been eased with the introduction of the Fast Fourier Transform (FFT) algorithm, which expands the signal into sine and cosine functions. The frequency spectrum is computed for each discrete segment of the signal. The output of the FFT algorithm is a set of coefficients, two for each frequency component in the signal's spectrum. One coefficient (A) is multiplied by the cosine, or amplitude portion of the component. The other (B) is multiplied by the sine, or phase portion, of the component. Each component in the FFT series can then be represented as:

$A \cos(\omega t) + i B \sin(\phi)$

Where ω = Angular frequency of the component

$\quad\quad A$ = An FFT coefficient

$\quad\quad B$ = An FFT coefficient

$\quad\quad \phi$ = The phase angle of the component

$\quad\quad i$ = The imaginary number, $\sqrt{-1}$

$\quad\quad t$ = time

The number of frequency components and pairs of FFT coefficients necessary to represent a given waveform is a function of the highest frequency to be resolved and the sample rate used.

Figure 4.15 illustrates the decomposition of a typical bioelectric signal (EEG waveform) into its basic frequency components and then displays them as a frequency spectrum. The diagram shows frequency components along with the amplitude present in each frequency. Once the frequency spectrum of a given segment of the signal has been calculated, a number of techniques are available to display the information.

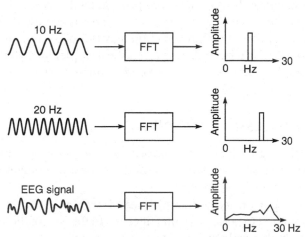

> **Fig. 4.15** *EEG signal decomposed into basic frequency components*

The FFT method assumes the signal to be stationary and is therefore insensitive to its varying features. However, most biomedical signals are non-stationary and have highly complex time frequency characteristics. The stationary condition for the non-stationary signal can be satisfied by dividing the signals into blocks of short segments, in which the signal segment can be assumed to be stationary. This method, is called the ***short-time Fourier transform (STFT)***. However, the problem with this method is the length of the desired segment. Choosing a short analysis window may cause poor frequency resolution. On the other hand, a long analysis window may improve the frequency resolution but compromises the assumption of stationarity within the window.

4.5.3   Wavelet Transform

An emerging method to analyze non-stationary biomedical signals is the ***wavelet transform***. The wavelet method acts as a mathematical microscope in which we can observe different parts of the signal by just adjusting the focus. In practice, it is not necessary for the wavelet transform to have continuous frequency (scale) parameters to allow fast numerical implementations; the scale can be varied as we move along the sequence. Therefore, the wavelet transform has very good time resolution at high frequencies and good frequency resolution at low frequencies. In biomedical engineering, wavelet transform have been widely used in many research areas including spatial filtering, edge detection, feature extraction, data compression, pattern recognition, speech recognition, image compression and texture analysis. For example, Sahambi *et al* (2000) has developed an algorithm for beat-by-beat *QT* interval variability using wavelet approach.

Wavelets are a relatively new signal processing method. A wavelet transform is almost always implemented as a bank of filters that decompose a signal into multiple signal bands. It separates and retains the signal features in one or a few of these bands. Thus, one of the biggest advantages of using the wavelet transform is that signal features can be easily extracted. In many cases, a wavelet transform outperforms the conventional FFT when it comes to feature extraction and noise reduction.

Another method of signal analysis is that of ***adaptive filter*** that can continuously adjust itself to perform optimally under the changing circumstances. This is achieved by correcting the signal according to the specific application. The correction may be enhancement or some reshaping, for which a correction algorithm is required. This can be best implemented digitally. Most adaptive filters, therefore, are implemented by means of computers or special digital signal processing chips.

▶ 4.6   SIGNAL PROCESSING TECHNIQUES

When we pass a signal through a device that performs an operation, as in filtering, we say we have processed the signal. The type of operation performed may be linear or non-linear. Such operations are usually referred to as ***signal processing.***

The operations can be performed with a physical device or software. For example a digital computer can be programmed to perform digital filtering. In the case of digital hardware operations (logic circuits), we have a physical device that performs a specified operation. In contrast, in the digital processing of signals on a digital computer, the operations performed on a signal consist a of number of mathematical operations as specified by the software.

Most of the signals encountered in biomedical instrumentation are analog in nature, i.e. the signals are functions of a continuous variable such as time or space. Such signals may be processed by analog systems such as filters or frequency analyzers or frequency multipliers. Until about two decades ago, most signal processing was performed using specialized analog processors. As digital systems became available and digital processing algorithms were developed, digital processing became more popular. Initially, digital processing was performed on general purpose microprocessors. However, for more sophisticated signal analysis, these devices were quite slow and not found suitable for real-time applications. Specialized designs of microprocessors have resulted in the development of digital signal processors, which although perform a fairly limited number of functions, but do so at very high speeds.

Digital signal processor (popularly known as DSP) requires an interface (Fig. 4.16) between the analog signal and the digital processor, which is commonly provided by an analog-to-digital converter. Once the signal is digitized, the DSP can be programmed to perform the desired operations on the input signal. The programming facility provides the flexibility to change the signal processing operations through a change in the software, whereas hardwired machines are difficult to configure. Hence programmable signal processors are common in practice. On the other hand when the signal processing operations are well defined, as in some applications, a hardwired implementation of the operations can be optimized so that it results in cheaper and faster signal processors. In cases when the digital output from a processor is to be given to a user in the analog form, a DA converter is required.

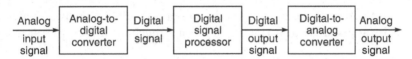

> **Fig. 4.16**   *Basic elements of a digital signal processor (DSP) system*

DSPs are available as single chip devices and are commercially available. The most widely used DSP family is the TMS320 from Texas Instruments. Another range of processors available from Motorola is the DSP56001. For the sake of comparison of speed, the 16-bit Motorola 68,000 microprocessor can handle 2,70,000 multiplications per second while the DSP56001 is capable of 10,000,000 multiplications per second, thus giving an increase in speed of 37 times. Because of the flexibility to reconfigure the DSP operations, they are used in most of the modern biomedical instruments for signal processing applications like transformation to the frequency domain, averaging and a variety of filtering techniques.

▶ 4.7   THE MAIN AMPLIFIER AND DRIVER STAGE

Normal practice in the design of recorders is to have a preamplifier followed by the main amplifier which in turn is connected to the driving stage. The main amplifier thus forms an intermediate stage between the preamplifier and the output stage. The gain and frequency response of the main amplifier is adjusted to give adequate pen deflection and frequency response.

The pen motor is driven by a dc driver stage feeding a four transistor output stage operating the galvanometer. A bridge arrangement is preferred because of the low power efficiency of conventional push-pull power amplifiers. Referring to Fig. 4.17, the current through T_1 and T_4 increases as that through T_2 and T_3 decreases. Thus when T_1 and T_4 approximate to short circuits,

T_2 and T_3 are nearly cut-off and almost all the circuit current passes through the galvanometer coil. T_1 and T_2 function as emitter loads. T_1 and T_4 operate as amplifiers having T_1 and T_3 as collector loads. Resistors R_1 to R_4 secure the correct biasing of T_2 and T_4.

The output stages of the amplifier are required to produce more output for a given input signal level as the signal frequency increases. This is achieved in the output amplifiers by the use of a frequency selective network in the negative feedback line of the driver and output stages so that the gain of the amplifier increases with frequency.

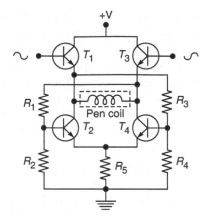

➤ **Fig.** 4.17   *Bridge output stage for driving galvanometer coil in direct writing recorders*

Also, the signal fidelity can be maintained in a linear recording system only when it has a linear phase shift characteristic and constant gain over the band of frequencies in the signal of interest. The pen motors need to be protected from excessive transient energy being applied during switch on or switch off. A delay circuit holds the pen motor gain low for a short time after switch on. When the power supply is switched off, the output goes rapidly to zero, reducing pen motor gain in advance of power fall. Pen over-deflection is limited both electrically by adjustable output swing limits and mechanically by pen stops.

ⅢⅢ➤ 4.8   WRITING SYSTEMS

The final and most important stage in any recorder is the writing system. For a faithful reproduction of the input signal, three basic conditions must be satisfied by the individual parts of the system. These requirements are linearity over the required range of signal amplitudes and an adequate passband for the frequencies involved without producing any phase shift between the input and recorded signal. Recorders are selected according to the frequency response of the data, accuracy requirements, the type of chart record that is desired and the number of data channels that must be recorded on a single piece of chart paper. According to frequency response, recorders fall into the following groups:

Potentiometric recorders usually provide a frequency response of 1 Hz at 25 cm peak-to-peak or up to 6 Hz at reduced amplitude. Chart paper is inexpensive and the writing method is generally capillary ink or fibre-tip pen.

Direct-writing ac recorders provide a frequency response up to 60 Hz at 40 mm peak-to-peak or up to 200 Hz at reduced amplitudes. The most common type of direct recorder is the stylus type which directly writes on the paper moving beneath it. The stylus can be made to write by several methods, but the most commonly employed is a heated stylus writing on specially prepared heat sensitive paper. An ink system can also be used but difficulty is experienced in obtaining a uniform flow with the additional problem of clogging of ink in the pen if the recorder is kept unused over long periods.

Ink-jet recorder gives a frequency response up to 1000 Hz. It employs inexpensive plain paper as the writing system makes use of a jet of ink.

Electrostatic recorder employs an electrostatic writing process and works for frequencies up to 10 kHz. The accuracy of the peak-to-peak amplitude is 0.1% at 1000 Hz, 0.2% at 1500 Hz, 2% at 5000 Hz, 20% at 15 kHz. It is independent of signal amplitude and the number of channels.

Thermal array recorder uses a specially designed linear thermal array head and thus dispenses with pens, pen motor or linkages. Frequency response is independent of trace amplitude and the number of channels. A 20 Hz sine wave is defined by 10 adjacent segments. The accuracy of peak-to-peak amplitude is 0.2% up to 32 Hz, 2% at 100 Hz and 20% at 320 Hz. A single pulse of duration greater than 625 μs will be recorded with full amplitude. It uses a plain, heat sensitive paper.

Ultra-violet recorders with mirror galvanometer arrangement and an ultra-violet light source gives-frequency response up to 2000 Hz. They make use of special UV sensitive paper, which requires careful storage. For higher frequency records, UV recorders using a fibre-optic cathode ray tube are used. Their response goes up to several MHz. These recorders are no longer in use.

Cathode ray oscilloscopes are widely used for the display of waveforms encountered in the medical field. These waveforms can be recorded from the CRO screen by running a photographic film through a recording camera fixed in front of the screen. Recorders are either of the single channel type or of the type which record several channels simultaneously. Each channel usually carries its own independent recording system to avoid interference from other channels.

ⅢⅢ▶ 4.9 DIRECT WRITING RECORDERS

In the most commonly used direct writing recorders, a galvanometer activates the writing arm called the pen or the stylus. The mechanism is a modified form of the D'Arsonval meter movement. This arrangement owes its popularity to its versatality combined with reasonable ruggedness, accuracy and simplicity.

A coil of thin wire, wound on a rectangular aluminium frame is mounted in the air space between the poles of a permanent magnet (Fig. 4.18). Hardened-steel pivots attached to the coil frame fit into jewelled bearings so that the coil rotates with a minimum of friction. Most often, the pivot and jewel is being replaced by a taut band system. A light-weight pen is attached to the coil. Springs attached to the frame return the pen and coil always to a fixed reference point.

When current flows through the coil, a magnetic field is developed which interacts with the magnetic field of the permanent magnet. It causes the coil to change its angular position as in an electric motor. The direction of rotation depends upon the direction of flow of current in the coil. The magnitude of pen deflection is proportional to the current flowing through the coil. The writing stylus can have an ink tip or it can have a tip that is the contact for an electro-sensitive, pressure sensitive or heat sensitive paper. If a writing arm of fixed length is used, the ordinate will be curved. In order to convert the curvilinear motion of the writing tip into a rectilinear motion, various correcting mechanisms have been devised to change the effective length of the writing arm as it moves across the recording chart.

Taut band instruments are preferred over pivot and jewel type instruments because they have the advantages of increased electrical sensitivity, elimination of friction, better repeatability and increased service life.

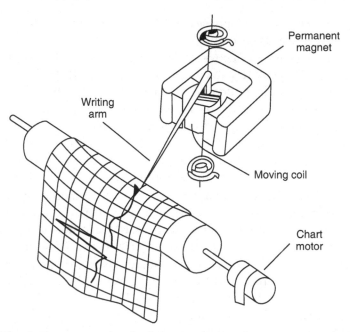

> ➤ **Fig.** 4.18　*Principle of a direct writing galvanometric recorder*

Of the several writing methods available, the ink recording method is widely used in slow speed as well as in high speed recorders. The writing pen depends upon the capillary action of the ink for its performance. The pen tip may be of stainless steel or tungsten carbide or a small glass nozzle. The ink reservoir is usually placed slightly above the plane of writing to facilitate the flow of ink. The ink used must flow out from the pen tip in an even manner so that the trace is continuous and no gaps are produced.

Writing quality of a recorder is important to ensure clear non-smudging traces of uniform width in both steady state and dynamic recording. The trace should be thin, yet well defined and uninterrupted, to allow best resolution of measurements. These problems spurred the development of the pressurised ink system. This system uses high pressure, high viscosity ink to overcome inertial effects within the pen tube and assure a continuous flow of ink even at high pen velocities. Writing fluid, in these systems, is supplied by a central ink reservoir that produces pressures of 15 to 20 psi and forces viscous ink into the microscopic pores in the chart paper surface. Any ink above the surface is sheared off by the pen tip leaving a permanent trace that is dry and uniform at all writing speeds. Typically, two ounces of viscous ink will provide up to 500 h of recording without refilling. This system is used in the Brush Mark 200 series direct writing recorders.

Most of the portable recorders use the heated stylus writing system wherein recording paper moves over a steel knife edge kept at right angles to the paper motion and the stylus moves along the knife edge. The hot stylus burns off the white cellulose covering of the heat sensitive paper, exposing the black under-surface of the paper thus forming the trace.

Paper Drive: The usual paper drive is by a synchronous motor and a gear box. The speed of the paper through the recorder is determined by the gear ratio. If it is desired to change the speed of the paper, one or more gears must be changed. Certain instruments are of the fixed speed variety, i.e. there is no provision for changing the rate at which the paper moves under the pen. Many applications, for example, ECG machines, have a single speed.

A constant speed is the basic requirement of the paper drive because the recorded events are time correlated. The frequency components of the recorded waveform can be determined if it is known how far the paper moved past the pen position as the record was being taken. Some types of medical recorders incorporate arrangements for several chart speeds. In such cases, the gear box has a single fixed reduction ratio and the speed variation over the total range is achieved by digital electronic means. This technique gives a wide ranging crystal controlled accuracy with exact speed ratios, without the high power consumption and relatively large steps of the stepping system, and the added reliability of fewer moving parts compared to a mechanical gear change system. Most of the recorders contain an additional timing mechanism that prints a series of small dots on the edge of the paper as it moves through the recorder. This "time marker" produces one mark each second or at some other convenient time interval.

4.9.1 Performance Characteristics of Galvanometers Used in Direct Writing Recorders

The galvanometers used in the recorders generally resemble the corresponding types in the indicating instruments except that they have a lighter arm carrying a pen in place of the pointer. The pen rests lightly on a chart which moves at a uniform speed in a direction perpendicular to that of the deflection of the pen. Owing to the friction of the pen on the chart and due to the necessarily greater weight of the moving system, the design of the galvanometer must be somewhat modified if it is to be used for recording purposes.

In the direct writing recorders, there are several forces which act upon the moving system. The three basic forces are: (i) the deflecting force, (ii) the controlling force, (iii) the damping force.

The deflecting force results from the current which flows in the coil and is supplied to it from the driving amplifier. This force causes the moving system of the recorder to move from its zero position. A controlling force applied to it will limit the otherwise indefinite movement of the pen motor and ensures that the same magnitude of the deflection is always obtained for a given value of the quantity to be recorded. The damping force is necessary in order to bring the movement system to rest in its deflected position quickly. In the absence of damping, owing to the inertia of the moving parts, the pen would oscillate about the final deflecting position for some time before coming to rest. The function of damping is to absorb energy from the oscillating system and to bring it promptly in its equilibrium position.

Damping Control and Frequency Response of Pen Motors: Galvanometers are characterized by five important parameters: frequency response, sensitivity (current, voltage and wattage), phase angle, damping and power dissipation. Each one of these parameters is not only important individually, but is also dependent upon the others. Thus, the galvanometer selected to obtain the desired results under certain specified conditions must therefore be a compromise of all these characteristics (Mercier, 1973).

Step Function Response: A pen galvanometer is unable to instantaneously follow the deflection arising when a step change in current is applied to it as would occur when a calibration signal is applied. The response of the moving coil lags behind the driving signal due to the mechanical inertia of the movement. With a good direct writing recorder, the rise time is of the order of 3 ms to attain a 1 cm deflection when the 1 mV calibration button is pressed. The effect of inertia also causes the coil to tend to overshoot its final deflection. The amount of overshoot depends upon the value of the damping factor. This is taken as unity when the galvanometer is 'critically' damped. Under these conditions, the coil will deflect smoothly to take up its final position in the shortest possible time without an overshoot. If the damping factor is substantially less than the critical value, the writing arm will overshoot its final deflected position and execute a damped simple harmonic frequency of almost the natural frequency of the coil and arm assembly.

Figure 4.19 illustrates the step function response for a galvanometer with nominal 65% damping. To standardize the curve, response is shown versus free period of 1/natural frequency so that the difference in galvanometer sensitivities can be eliminated. The deflections are shown as ratios of the direct current deflections. The manner in which a galvanometer follows a sudden current rise depends upon the damping ratio.

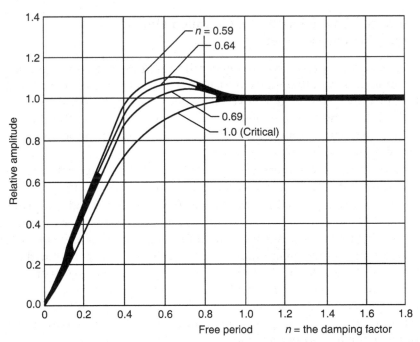

> **Fig. 4.19** *The effect of applying a step input current to a galvanometer with varying degrees of damping. The ordinate is the ratio of the deflection at a particular time to the steady-state deflection. The abcissa is the ratio of time to the periodic time of the galvanometer natural frequency.*

The degree of damping depends upon the resistance in the coil circuit. The value of the resistance required to be placed in parallel with a given galvanometer for correct damping is specified by the manufacturers. However, when such a galvanometer is to be used in a recording system and it is driven by the power amplifier, it is necessary to readjust the damping resistance so that the output impedance of the amplifier is taken care of.

Phase Angle: Figure 4.20 shows the theoretical phase angle between a sinusoidal current applied to the galvanometer of any undamped natural frequency. The frequency of the current is shown as the ratio of this frequency to the undamped natural frequency of the galvanometer. Phase angle is also dependent upon the amount of damping. For any frequency ratio and amount of damping, the phase angle is shown directly.

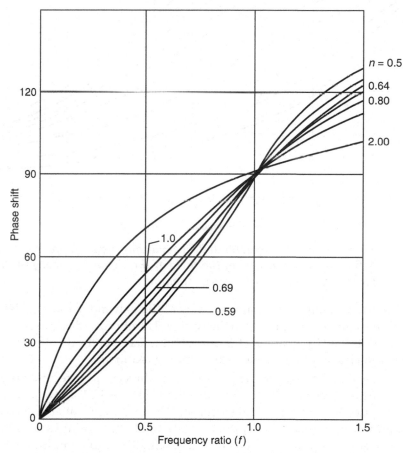

➤ **Fig.** 4.20 *The phase shift between the sinusoidal driving signal and the resultant coil motion as a function of the signal frequency to the natural frequency*

Frequency Response and Sensitivity: Galvanometers when damped should give a frequency response flat within ±5% from dc to 60% of their undamped natural frequency. The natural frequency is calculated by multiplying the required frequency response by 1.6. Under certain conditions, it may be desirable to utilize a galvanometer with higher natural frequency than that calculated as outlined above, but it may be noted, a lower sensitivity will be the result of the accompanying higher frequency response.

Figure 4.21 represents the theoretical deflection of a galvanometer to a sinusoidal current, constant in magnitude but variable in frequency. To make the curves applicable to a galvanometer of any undamped frequency and sensitivity, the frequency of the current is shown as the ratio of

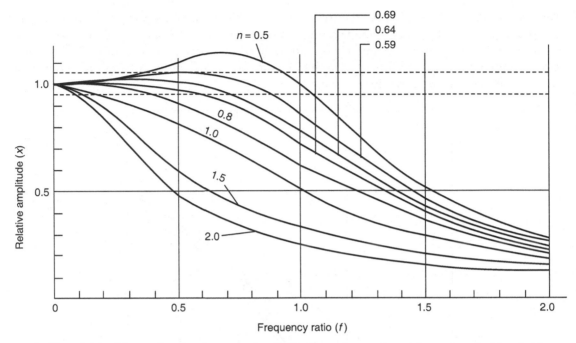

> **Fig.** 4.21   *The ratio of galvanometer deflection at any signal frequency to the static deflection as a function of the ratio of the signal frequency to the natural frequency for various degrees of damping*

this frequency to the undamped natural frequency of the galvanometer. The deflection is also dependent upon the amount of damping in the system. The effective damping of the moving coil assembly can be controlled by arranging the output impedance of the deflection amplifier to fall with frequency in such a way as to provide the damping required.

The maximum band of flat frequency response for any direct writing recorder is always limited by some specific peak-to-peak amplitude on the chart. Figure 4.22 shows a typical frequency versus amplitude curve for a direct writing recorder. The recorder is capable of writing full scale amplitude (50 mm chart width) up to approximately 40 Hz, 50% of full scale amplitude up to 60 Hz, 20% of full scale amplitude at 100 Hz and about 10% of full amplitude at 200 Hz. This characteristic curve is typical of all direct recorders and indicates that increasing the pen motor power would not help very much. Therefore, for higher frequency response, it becomes imperative to go to some form of light beam recorder where the mass of the moving element is much smaller.

4.9.2 Linearity Considerations in Direct Writing Recorders

Ideally, a linear writing system should produce a pen deflection that is directly proportional to the input current given to the pen motor. This implies that equal increments of input would cause equal changes of trace amplitude in any region of the chart.

In many industrial and general purpose recorders, the pen generates curvilinear traces. The tip of the pen in such cases traces out an arc on the recording chart with the deflection of the

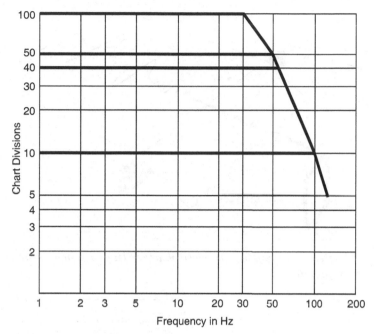

➤ Fig. 4.22 *Frequency versus amplitude curve for a direct writing recorder*

galvanometer coil. The system is simple, reliable and cheap but may not be acceptable due to the following disadvantages: (i) special chart paper is required, ruled to match the pen radius, and (ii) curvilinear traces are often difficult to correlate to their mathematical shapes. It makes the analysis somewhat more difficult.

The curvilinear arc at the pen tip can be converted into a rectilinear trace with more easily interpretable rectangular coordinates. Two methods are in common use. In one, an electrically heated stylus moves on an arc across a special heat-sensitive paper as the paper is drawn over a straight fixed knife (Fig. 4.23) edge. A different portion of the stylus end contacts the paper for each angle of displacement. The arrangement gives a straight line across the width of the paper with the paper stationary. However, the method is subject to a degree of geometric non-linearity, which may be around 30% for a coil rotation of 18 degrees from the chart centre line to the chart edge.

The second method (Fig. 4.24) makes use of a rotary to linear linkage, which eliminates geometric errors inherent with conventional knife edge systems. When the pen motor armature rotates, it drives the pen support arm. The coupling provided by a flexible metal band causes the pen shaft at the upper end of the pen support arm to rotate in the opposite direction since the lower end of the flexible band is firmly attached to a stationary ring on the top of the pen motor housing. As the pen support arm swings in an arc about the axis of the pen motor armature, the pen tip travels in a straight line and produces a deflection that is directly proportional to the angular movement of the pen motor armature. The angular displacement of the pen motor coil is translated into a straight line motion at the pen tip by the equal and opposite curvilinear path of the pen support arm. This is achieved to an accuracy of 0.2% in recorders, which employ this type of arrangement.

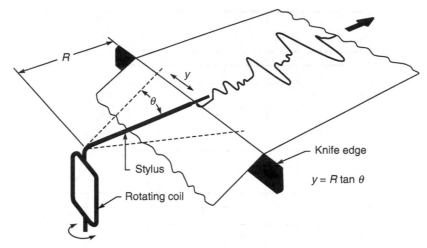

> **Fig.** 4.23 *A heated stylus system for rectilinear writing on a heat-sensitive paper over a knife edge*

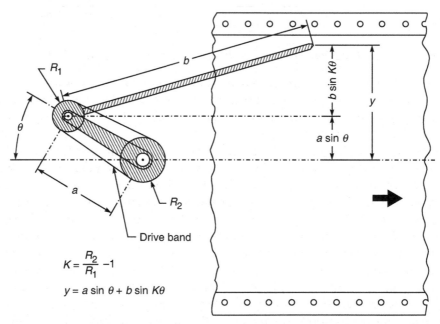

> **Fig.** 4.24 *Parallel motion mechanism for rectilinear writing. The mechanism keeps the pen tip nearly perpendicular to the direction of chart motion*

A high degree of linearity, accuracy and rapid response is achieved by using closed loop pen position feedback systems, such as the one employed in Brush Recorder Mark 200. This is the result of four separate but mutually dependent devices. These are:

(i) A position feedback pen motor that develops more than 250 G's at the pen tip and maintains position accuracy within 0.25%.

(ii) A precise non-contact pen position sensing transducer with infinite resolution called the Metrisite.

(iii) A rectilinear transfer linkage which translates angular pen motor movement into rectilinear pen travel without introducing hysteresis.

(iv) A pressurized fluid writing system that ensures uniform traces at all writing speeds.

Figure 4.25 illustrates the principle of closed-loop pen position feedback, which automatically eliminates virtually all sources of error between the incoming signal and pen position. Actual pen position on the chart is continuously sensed by the contact-free Metrisite transducer located inside the pen-motor. Output from the Metrisite is fed to a summing junction on the pen drive amplifier where it is compared to the input signal. The difference between the Metrisite output and the incoming signal produces an "error" voltage, which is amplified and instantaneously corrects the

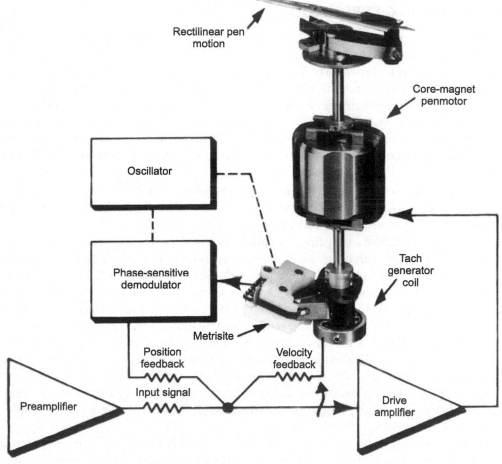

> **Fig. 4.25** *Servoloop feedback pen motor system for fast and accurate recordings (Courtesy Gould Inc., USA)*

pen position to correspond with the incoming signal. An important advantage of the closed-loop pen position feedback is that it can produce position stiffness at the pen tip, which is 10 to 100 times greater than the conventional spring restored recorders. Errors due to friction are virtually eliminated and the pen motor heating is reduced. Absolute linearity is maintained across the entire chart since the pen motor has no torsion spring, never works against mechanical restraint and uses operating power only for changing pen position. Thus, recording accuracy at the pen tip is limited only by the output accuracy of the Metrisite transducer and the ability of the pen motor to apply the necessary corrective forces.

Much depends on the character of the pen position transducer. In a rapidly moving system such as a recorder, any tendency to wear will reduce accuracy because of dirt, or other contaminants would quickly destroy reliability. A good pen position sensor is essential in order to combine fast dynamic response and good recording accuracy in a single instrument.

4.10   THE INK JET RECORDER

Certain biomedical applications like phonocardiography and electromyography require recording of signals of which the frequency spectrum extends much above the high frequency response of the conventional pen recorders. The high frequency response of the stylus direct recorders is limited by the large moment of inertia of the moving parts and friction of the writing arm over paper. Moreover, stylus systems can usually only record amplitudes up to 30 mm without colliding with each other and causing damage. Overlapping of the adjacent tracings is thus not possible. This means that with multi-channel recorders, the stylus recording system requires very wide paper for tracing large amplitudes since the maximum deflection must be recorded without overlapping.

An elegant method of increasing the frequency response extending to several hundred cycles and combining the advantages of the direct recorder input signal with those of the photographic recorder is embodied in the jet recording system.

The technique consists of a very fine jet of ink, which replaces the light beam of a photographic recorder or the stylus of a direct recorder. The jet of ink is produced when the ink is expelled from a nozzle of an extremely fine bore at high pressure. The ink is squirted over the chart moving beneath the jet. The method is usable up to frequencies of about 1000 Hz. The arrangement (Fig. 4.26) consists of a glass capillary tube placed between the poles of an electromagnet. The coils of this electromagnet are connected to the output amplifier and are driven by the amplified signal. Attached to this capillary is a very small cylindrical permanent magnet. The variations of current corresponding to the signal variation in the electromagnet coil produce a varying magnetic field, which interacts with the field of the permanent magnet. This interaction results in the deflection of the capillary tube. The capillary tube is shaped in the form of a nozzle at one end and ink is supplied to this tube at a high pressure from the other end. The ink coming out of the nozzle strikes the paper and the signal waveform is traced.

Reliable functioning of the jet system demands that the ink be properly filtered to ensure that even small particles are kept back so that they do not block the nozzle. The high pressure necessary for jet recording is produced by a pump and is adjustable between 20–50 atmospheres.

The inherent high frequency response of the jet recording system is due to the significant reduction in the inert mass of the moving assembly. In fact, this mass is so low that it compares with that of the photographic system used in earlier days. With very little mass of the moving parts, the jet recording mechanism needs only very low driving power despite the high inherent frequency. Therefore, it is possible to record deflections of ±30 mm measured from the base line. From this it is seen that amplitudes of more than 60 mm can be traced with the jet recorder. Overlapping of adjacent tracings causes no difficulty because liquid jets cannot penetrate each other as is the case with light beams. Theoretically, the points of impact of the individual jets do not lie on the same time ordinate of the recording paper. In practice, however, the shifting is not noticed since it is less than the evaluation accuracy. Since the jet

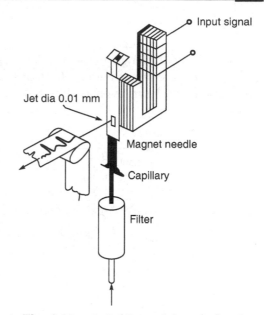

> **Fig.** 4.26 *An inkJet writing mechanism*

recorder can write on the recording paper without friction, linear tracing is ensured even with very small amplitudes.

Recording of large amplitudes means that there would be considerable deviation of linearity at the two extreme ends on the recording paper. This is because of the tangent error. Suitable techniques are adopted to compensate for this error at larger deflections. Compensation is achieved by employing a linear magnetic field rather than the usual radial field. The effective magnetic field is then proportional to the cosine of the angle of deflection and thus reduces the sensitivity at larger deflections.

As with photographic recorders, the recording mechanism of the jet recorder is liquid damped. This is to obtain balanced tracing over the whole frequency range. The damping medium is conducted in a tube sealed at one end. Its viscosity is so high that it does not run out even if the recording mechanism is turned upside down.

The jet recorder makes use of normal untreated paper, which is much cheaper than the heat sensitive paper used in the heated-stylus paper recorders. Therefore, it provides an economic recording method, particularly for multi-channel applications.

⭢ 4.11 POTENTIOMETRIC RECORDER

For the recording of low frequency phenomena, strip chart recorders based on the potentiometric null-balance principle are generally used. The operating principle of a potentiometer recorder is shown in Fig. 4.27. A slide wire *AB* is supplied with a constant current from a battery S. The slide wire is constructed from a length of resistance wire of high stability and uniform cross section such that the resistance per unit length is constant. The unknown dc voltage is fed between the moving contact *C* and one end *A* of the slide wire. The moving contact is adjusted so that the current flowing through the detector is zero. At that moment the unknown input voltage is proportional to the length of the wire *AC*. In practice, the slide wire is calibrated in terms of span voltage, the typical span being 100, 10 or 1 mV. The moving contact of the slide wire is made to carry a pen, which writes on a calibrated chart moving underneath it.

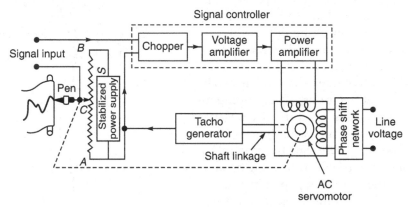

> **Fig. 4.27**   *Schematic diagram of a self-balancing potentiometric recorder*

For obtaining the null-balance, a self-balancing type potentiometer is generally used. The balancing of the input unknown voltage against the reference voltage is achieved by using a servo-system. The potential difference between the sliding contact *C* and the input dc voltage is given to a chopper type dc amplifier in place of a galvanometer. The chopper is driven at the mains frequency and converts this voltage difference into a square wave signal. This is amplified by the servo-amplifier and then applied to the control winding of a servo motor. The servo motor is a two-phase motor whose second winding is supplied with a 50 Hz mains supply that works as a reference phase winding. The motor is mechanically coupled to the sliding contact. The motor turns to move the pen and simultaneously varies the voltage of the sliding contact such that the potential difference between the input voltage and reference voltage is zero. The circuit operates in such a manner that the motor moves in one direction if the voltage across the ac is greater and in the opposite direction if it is less than the input voltage. The servo motor is shaft-coupled to a techno-generator, which provides the necessary damping to the servo motor. It slows down as it approaches balance position and thus minimizes the overshoot.

The servo motor generally used to drive the pen in the self-balancing potentiometric recorders is the ac two-phase induction motor. The motor has two separate stator windings, which are physically perpendicular to each other. The out-of-phase alternating currents in the two stator windings produce a rotating magnetic field. This rotating magnetic field induces a voltage in the rotor and the resulting current in the armature produces an interacting field, which makes the rotor to turn in the same direction as the rotating magnetic field. To produce a rotating magnetic field, the ac voltage applied to one stator winding should be 90° out-of-phase with the voltage applied to the other winding. It can be done either in the power amplifier, which supplies the control winding, or in a phase-shift network for the line winding. For no input signal, obviously, the rotor does not turn. When a voltage is applied to the input, the rotor would turn slowly and its speed would increase with the magnitude of the input voltage. Power amplifiers are required to supply the necessary alternating current to the control winding of the servo motor. They are essentially class B push-pull amplifiers.

The chart is driven by a constant speed motor to provide a time axis. Therefore, the input signal is plotted against time. The recorders of this type are called T-Y recorders. If the chart is made to move according to another variable, then the pen would move under the control of the second variable in the X direction. Such type of recorders are called X-Y recorders. The chart drive can also be provided by a stepper motor, which is controlled by a reference frequency from a stable oscillator. Choice of speeds is achieved by a programmable frequency divider and applied to the motor. The chart drive accuracy is thus independent of the power line frequency. Using a stepper motor instead of a synchronous motor eliminates the need for a mechanical transmission to provide different chart speeds and substantially increases reliability.

Wire wound or conductive plastic potentiometers are used as position feedback elements in most servo recorders. However, the electromechanical contact between slider and slidewire limits the reliability of such recorders and requires regular maintenance. A non-contacting ultrasonic pen position transducer permits high reliability and maintenance-free operation of a recording potentiometer. The ultrasonic transducer measures distance ratios as the ratio of the propagation times of an ultrasonic pulse travelling from a transmitter past two sensors on a magnetostrictive delay line. Figure 4.28 shows the basic construction of the position feedback ultrasonic transducer. Ultrasonic pulses are generated and detected using coils through which the magnetostrictive line passes. Three coils (N_1, N_0, N_2) are used. Coil N_0 is attached to the recorder pen and is used to generate magnetostrictive pulses in the line. When an electrical pulse is applied to N_0, its magnetic field produces a magnetostrictive pulse in the line. This stress impulse propagates at ultrasonic speed along the line. This impulse reaches fixed coils N_1 and N_2 after times T_1 and T_2 respectively and pulses are induced in the coils as shown in the diagram.

The propagation times T_1 and T_2 correspond to the distances D_1 and D_2 divided by the speed of propagation V of the ultrasonic pulse in the line. The ratio of propagation times equals the ratio of corresponding distances derived as follows:

$$X = \frac{(T_1 - T_2)}{(T_1 + T_2)} = \frac{2x}{D_1 + D_2}$$

where $T_1 = D_1/V, T_2 = D_2/V$

x = deviation of N_0 from centre point between N_1 and N_2

X = computed time (and distance) ratio

Stegenga (1980) explains the circuit used with an ultrasonic feedback transducer used in recording potentiometers. The ultrasonic transducer has been found to be a practicable position feedback element.

Potentiometric recorders can be conveniently used to record a number of slowly varying physiological signals. In particular, it is possible to produce a record of the patient's condition over a period of 24 h. An electrically driven switch connects the input signals in turn to the recorder, which would then print a dot on the calibrated chart. The position of the dot on the chart gives the value of the parameter at that moment.

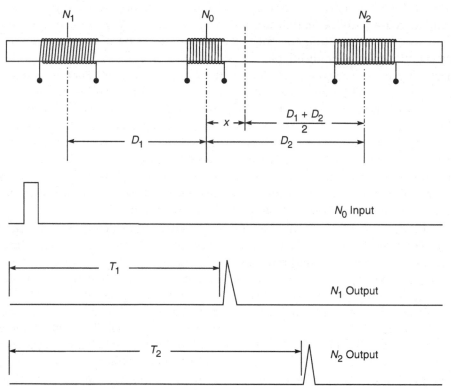

> **Fig. 4.28** *Position feedback ultrasonic transducer for potentiometric recorders*

▮▮▮▶ 4.12 DIGITAL RECORDERS

Digital recorders use a linear fixed array of small recording elements under which the paper moves. This is in contrast with the conventional recorders that use a moving pen or stylus. The stylus in these recorder is a large number of fixed stylii, each one of which corresponds to one amplitude of signal to be recorded. Signals are thus reproduced as discrete values at discrete times. Analog as well as digital signals can be processed. In the analog case, sampling and digitization are part of the recording process. The accuracy of array recorders is determined by the act of sampling and digitization.

As all processing occurs in digital form and recording is a matter of generating the correct addresses of writing points to be activated, there is no problem in writing figures alongwith signal tracings. Alphanumeric information to be added to the chart can be supplied via the keyboard. A display unit that visualizes the traces stored in memory before they are recorded on the chart can be provided, thus, giving an opportunity for saving the paper. Time compression or expansion is possible as data can be sampled at a high rate (expansion), stored in memory and then recorded as they are released at a reduced rate. A bandwidth higher than that determined by the writing frequency can be achieved. In this way, fast transients that cannot be handled online can be accurately reproduced. A large number of input channels can be made available through a multi-channel data acquisition system (Collier, 1991).

The obvious advantage of a linear array recorder is the absence of moving parts—nothing moves but the paper. Inconveniences resulting from mechanical moving parts are not present and there is no problem of overshoot. The two commonly used techniques in linear array recording technology are: thermal and electrostatic.

4.12.1 Thermal Array Recorder

The thermal array writing technique helps to record analog traces, grid lines, trace identification and alphanumerics on plain thermal sensitive paper. The array is composed of 512 thermal stylii spaced at four per millimetre in a linear array. The array is 128 mm long by 0.25 mm wide.

Array writing is a three-step process. First, the individual drivers for the stylii to be heated are selected and latched. Next, all stylii to be driven are heated simultaneously by an applied voltage. Finally, the paper is moved an increment equal to or less than the width of one element. This process is repeated up to 200 times per second. This writing technique makes it possible to write any combination of dots anywhere on the record. If during the writing period, the analog input signal changes value, a line segment is printed connecting the lowest and highest amplitudes of the signal during the next writing period. Thus, a sine wave of 20 Hz is actually made up of 10 discrete line segments. A 20 Hz square wave, on the other hand, appears as an exact reproduction due to its ability to lay down the transient line all at once.

Thermal writing occurs when specially treated, thermally sensitive paper is heated to 90°–110°C. The paper, which is initially white or off-white, turns a dark colour (usually blue or black) where heated. The dark colour is caused by a chemical reaction resulting from heating the paper. The functioning of the thermal array writing system is basically digital. A raster line (a series of 1's and 0's, one digit corresponding to each of the 512 thermal stylii, a 1 indicating on, a 0 off) is fed to the driver circuitry, which activates the individual stylii to be heated. Each raster line fed to the drive circuitry is a summation of raster lines from several sources. The grid generator raster line makes up the dots forming the grid lines. The raster line from the annotation section allows the printing of preamp settings and chart speed. Each analog plug-in channel accepts one input and converts it to a raster line. The analog signal is sampled by a discrete analog-to digital converter 1600 times per second. Each sample generates a 512 bit raster line with all zeroes except for a single 1, representing the position on the thermal array corresponding to the value of the analog signal. The plug-in makes at least eight conversions between prints, combining each new raster line with the old raster line and filling between the highest and lowest values with ONES. After eight or more conversions, the raster line from this plug-in is combined with all other raster lines and fed to the thermal array drive circuitry.

The heart of the system is the unique thermal array head (Fig. 4.29) and attendant drive circuitry. The head is made up of three components; the heater bar, the heat exchanger and the drive circuitry. The heater bar is basically a thick film microcircuit. It comprises 5 mil conductors on 10 mil centres connected to a common bus by thick film resistors. The resistors form a linear array of 512 active heating elements spaced at 4 per mm. Utilization of a high heat conductivity substrate allows quick conduction of excess heat to the exchanger attached to the back of the substrate.

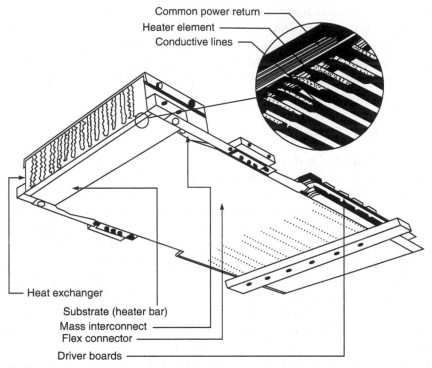

Common power return
Heater element
Conductive lines

Heat exchanger
Substrate (heater bar)
Mass interconnect
Flex connector
Driver boards

➤ **Fig. 4.29** *Principle of a thermal array recorder (Courtesy: Gould Inc., USA)*

The heat exchanger is cooled using forced air. The efficiency of the heat exchanger allows the head to have a 25% duty cycle. Even when the recorder is printing all black at the highest chart speed, there is no problem of overheating the head. The head drive circuitry is contained on two boards, which are half flex, half rigid. The interconnect pattern is etched directly on the flex portion of the board which is then connected directly to the heater bar. Thus, there is only one mass interconnect point improving reliability and providing ease of head replacement.

The heater bar may be easily removed from the heat sink and disconnected from the drive circuitry. A replacement can be just as easily installed and connected without special tools or fixtures making the head field replaceable. The thermal array writing system has no pens to wear out, no ink to fuss with and no pen motors or linkages to cause problems. The only moving part is the paper drive mechanism. Hence, shortcomings inherent to mechanical linkages such as overshoot, hysteresis and poor reliability are avoided. For the same reason, a fast transient response time of 625 μs full scale is possible. With no linkages in the way, all traces can go full scale and overlap in any desired relationship. Because grids are selected and printed with the traces, low cost plain thermal paper can be used.

With the advent of array recorders, frequency response needs to be redefined. In moving pen recorders, frequency response is defined for a sine wave at some specified amplitude. A sine wave is used because it is most easily reproduced by a writing system with moving parts with associated inertia. With array recorders, there is no inertia and in fact they can reproduce square waves at a higher frequency than sine waves.

In defining frequency response for array recorders, three parameters must be used. Out of these parameters which one is more important depends on the specific application. The parameters are "peak capture", "bandwidth" and "waveform response". Peak capture is defined as the shortest duration pulse which can be represented at true value by the recorder. The array recorder has a peak capture rating of 625 µs, that is any pulse lasting longer than 625 µs will be recorded at full value. Bandwidth is defined as the maximum sine wave frequency which can be recorded with a specified accuracy. From the classical sampling theory, an envelope ripple can occur before the usual-3 dB point is reached, and the envelope ripple is regarded as highly objectionable. So, we must speak of bandwidth as the maximum sine wave frequency without envelope ripple more than ±0.5 stylus. In this case only a black strip whose width equals the signal amplitude will be printed because even at maximum paper speed the line segments are contiguous. No waveform or frequency analysis will be possible but envelope (and transient peak) response is still very useful. The bandwidth of a recorder is 100 Hz with an accuracy of 2%. Waveform response is defined as that frequency at which a sine wave contains enough line segments to be aesthetically pleasing. It is specified as a frequency at which each cycle is made up of some minimum number of line segments (usually 10 or 20). As described earlier, each waveform is made up of line segments. Typically, a 20 Hz signal would have 10 line segments per cycle. Brimbal and Robillard (1990) describe thermal array recorders for high speed applications, whereas Gaskill (1991) explains the relationship between recorder resolution and print head density.

All medical instruments that require oscillographic recorders are presently preferring thermal array printers. They aim to create and store arbitrary backgrounds, print multi-channels with full scale full overlapping of waveforms, store and print annotated channels and create graphics, all at the same time. Embedded software provides freedom to create output anyway the user wishes to see it and change the way it looks, instantly.

4.12.2   Video Printers

Hardly two decades ago, capturing medical images was limited to X-ray photographic prints taken by a camera mounted on the front of a display monitor. The introduction of video technology to medicine has revolutionized medical imaging with both video tape and video stills being used to record almost every medical procedure. The early users of electronic medical imaging were obstetricians and gynaecologists, who used the technology to generate ultrasound images of foetus. Video technology was introduced in the early 1980s with the availability of low cost black and white thermal video printers. The printers displaced the film cameras by providing quality images at a lower cost.

Black and white printers were designed specifically for use with medical equipment, such as ultrasound systems and cardiac cath labs. They can be easily interfaced to medical devices with a standard video signal and produce hard copy images in 4 to 25 seconds. Each heat element on the thermal head of a black and white printer can apply 256 different steps of heat that produce 256 different densities for each spot (pixel), resulting in a 256 gradation for each pixel.

The next major step came in the late 1980s with the development of the colour video printer. They were particularly used in recording colour flow Doppler ultrasound and surgical endoscopy. Video printers use a dye transfer sublimation thermal printing method in which thousands of heating elements come into contact with yellow, magenta and cyan pigments on chemical coated plastic film. The amount of heat emitted by each heating element controls the amount of dye transferred to the paper. Most video printers support 256 shades of colour per primary colour, which translates to over 16 million (256^3) colours. The result is a photo realistic, full 24-bit colour image.

First generation printers offered single–frame memory only. This meant that one could capture images and print them, but the printer remained useless during print time. Modern printers come with multi-frame memory in which the printers continue to capture images while other images are being printed. This enable the doctor to view several images and print only those which are clinically most relevant.

4.12.3　Electrostatic Recorder

Electrostatic recorders are high frequency analog recorders which employ a high resolution electrostatic device to produce records on a wide, low cost paper at chart speeds up to 250 mm per second. By eliminating moving writing parts, the electrostatic writing process disposes of the characteristic moving pen problems like: inertia effects such as overshoot or low-frequency response limits, linkage effects such as non-linearity, hysteresis, and the inability to overlap traces, and preprinted grids that move with paper movement and expand or shrink with changing humidity conditions.

The Gould ES 1000 (Fig. 4.30) electrostatic writing system is composed of three elements: the imaging head, the toning head, and the vacuum knife. The imaging head is composed of a linear array of 1000 wire elements, spaced 4 per mm, for a total length of 250 mm. On each side of the array are 32 copper bars called shoes. As the paper moves over the image head, a negative voltage is applied to selected wire elements and a positive voltage is applied to the closest shoes. This places a negative point charge on the paper at the point where the wire element was. The paper then passes the toner head and positively charged ink particles adhere to where the paper had negative charge. A vacuum knife finally removes all excess toner and particles, making the image with charged particles. Exposure to air causes the adhesive-coated particles to permanently bond to the paper and the record emerges from the machine completely dry.

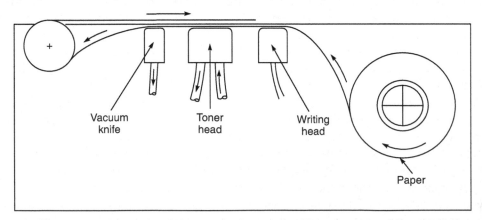

Vacuum
knife

Toner
head

Writing
head

Paper

> **Fig.** 4.30　*Principle of electrostatic recorder (Courtesy: Gould Inc., USA)*

The paper advances in increments less than or equal to the width of a single element, or 0.25 mm. The imaging elements are energized at a constant rate of 1000 times/s. If during the incremental period the analog signal changes value, then a line segment is printed connecting the lowest and highest amplitudes during that period. Thus, a sine wave of 100 Hz is actually made up of 10 discrete line segments. A 100 Hz square wave, on the other hand, appears as an exact reproduction due to the ability of the system to lay down the transient line all at once. The electrostatic writing system allows the production of any combination of dots anywhere on the chart. Plain electrostatic paper is used, with grid lines printed as selected, along with trace identification and all desired alphanumerics.

Dot spacing along the amplitude axis is 0.25 mm. This allows 0.1% resolution. Along the time axis, the top chart speed of 250 mm/s produces dot spacing of 0.25 mm. Because the printing rate is constant, this resolution increases to 40 dots/mm at a chart speed of 25 mm/s, for example. The chart is driven at continuous selected speed by a servo-controlled dc motor. Six speed choices are provided from 5 to 250 mm/s. Speeds can be controlled either manually or remotely by a 3-bit word signal.

In defining frequency response for the electrostatic recorders, three measures similar to thermal array recorders must be used. The measures are "peak capture", "bandwidth" and "waveform response".

Peak capture rating of the electrostatic recorder is 40 microsecond. The bandwidth of the recording system is 5000 Hz with an accuracy of 2% or 15 kHz with an accuracy of 20%. However, in this case, only a black strip whose width equals the signal amplitude will be printed. No waveform or frequency analysis will be possible. Each waveform is made up of line segments and a 100 Hz signal would have 10 line segments per cycle.

▓▶ 4.13   INSTRUMENTATION TAPE RECORDERS

Magnetic tape recording techniques and equipment have found extensive use in the hospital set-up. The fact that the signal is always available in electrical form, makes it possible to record the whole of an experimental procedure on a tape and then play it back for display on the CRT at a later time. The use of computers in the medical field has further broadened the field of magnetic tape recording. The information fed into the computers is coded and stored on magnetic tape, thus forming the memory banks for the digital computers.

Magnetic tape recording offers some useful features over other methods of recording. It permits the recording of signals, with suitable techniques, from dc up to several MHz. As the recordings of the tape can be erased any number of times, the tape becomes re-usable, thus offering economy in the recording process. The ability to alter the time base of the recorded events on the tape is something which no other recording medium provides. The events can be played back either faster or slower than they actually occurred. This permits the use of miniature tape recorders for ambulatory monitoring. Since the tape can be played back any number of times, it permits extracting every bit of useful information from the recording. It is also possible to have a very wide dynamic range of recording which may be in excess of 50 decibels. This permits an accurate and linear recording from full-scale signal level down to its 1/3%.

The most familiar method of recording signals on the magnetic tape, is the direct recording process. The electrical signal to be recorded is amplified by the recording amplifier and it is then fed into the recording head where corresponding magnetic fields are produced. The varying magnetic field is transferred to the tape in the form of magnetic patterns in accordance with the signal variations with the tape moving past the head. On replay, the recording process is repeated in the reverse order. The magnetized tape is pulled past and the magnetic field induces a current in the playback head coil. This is then amplified before being passed to a loud speaker, a recorder or any other display device.

4.13.1 Tape Recording with Frequency Modulation

The inherent limitation of the direct recording process on the magnetic tape is its inability to record signals of very low frequencies. The frequency range of signals which we come across in the medical field varies from dc to several hundred Hz. There is a method to handle these signals by employing a carrier frequency which is frequency-modulated by the signal to be recorded. On replay, the signal has to be demodulated and passed through a low-pass filter for removing the carrier and other unwanted frequencies. The circuit details of an FM tape recording system for biomedical application are given by Smith *et al.* (1979).

An important application of the FM recording technique is its flexibility in using "frequency-deviation multiplexing" where a number of carrier frequencies are separately modulated by different input signals. The resulting signals can then mix linearly and the composite signal can be recorded using the direct recording process. This method offers the facility of recording many channels of signal information on one track of the tape and utilizing the wide bandwidth and the linearity of the direct recording technique.

FM recording technique requires that the tape should move across the heads at a precisely uniform speed, because any speed variations introduced into the tape will cause an unwanted modulation of the carrier frequency. This will result in noise in the output signal and would limit the accuracy and dynamic range of the FM system. The variation of speed of tape movement would cause wow and flutter and the problem is particularly acute in the frequency multiplexed system.

4.13.2 Tape Recording with Digital Technique

Another method of recording information on the magnetic tape is the digital recording process. This process has been growing rapidly in importance as a result of the widespread application of digital computers. In this method, a sampling technique is used to measure a varying signal. The sampled readings are then converted into a code consisting of a group of binary digits. Recording is accomplished by magnetizing the tape to saturation in either of its two possible directions at discrete points along its length.

The advantages of the digital recording process are its inherent capability of extremely high orders of accuracy, its insensitivity to tape speed variations and a simple recording and reproducing electronic circuitry.

The major problem of digital recording process is its sensitivity to tape drop-out errors. Since all information is contained in the presence or absence of pulses, on replay we cannot tolerate the loss of pulses or the generation of spurious pulses caused by tape imperfections. This puts a limitation on the practical minimum duration for pulses, thus affecting the pulse packing density. The process also needs the data to be digitized at the source or special digital transducers must be employed.

CHAPTER

5

Biomedical Recorders

▥▶ 5.1 ELECTROCARDIOGRAPH

The electrocardiograph (ECG) is an instrument, which records the electrical activity of the heart. Electrical signals from the heart characteristically precede the normal mechanical function and monitoring of these signals has great clinical significance. ECG provides valuable information about a wide range of cardiac disorders such as the presence of an inactive part (infarction) or an enlargement (cardiac hypertrophy) of the heart muscle. Electrocardiographs are used in catheterization laboratories, coronary care units and for routine diagnostic applications in cardiology.

Although the electric field generated by the heart can be best characterized by vector quantities, it is generally convenient to directly measure only scalar quantities, i.e. a voltage difference of mV order between the given points of the body. The diagnostically useful frequency range is usually accepted as 0.05 to 150 Hz (Golden *et al* 1973). The amplifier and writing part should faithfully reproduce signals in this range. A good low frequency response is essential to ensure stability of the baseline. High frequency response is a compromise of several factors like isolation between a useful ECG signal from other signals of biological origin (myographic potentials) and limitations of the direct writing pen recorders due to mass, inertia and friction. The interference of non-biological origin can be handled by using modern differential amplifiers, which are capable of providing excellent rejection capabilities. CMRR of the order of 100–120 dB with 5 kΩ unbalance in the leads is a desirable feature of ECG machines. In addition to this, under specially adverse circumstances, it becomes necessary to include a notch filter tuned to 50 Hz to reject hum due to power mains. The instability of the baseline, originating from the changes of the contact impedance, demands the application of the automatic baseline stabilizing circuit. A minimum of two paper speeds is necessary (25 and 50 mm per sec) for ECG recording.

5.1.1 Block Diagram Description of an Electrocardiograph

Figure 5.1 shows the block diagram of an electrocardiograph machine. The potentials picked up by the patient electrodes are taken to the lead selector switch. In the lead selector, the electrodes are selected two by two according to the lead program. By means of capacitive coupling, the signal is connected symmetrically to the long-tail pair differential preamplifier. The preamplifier is usually a three or four stage differential amplifier having a sufficiently large negative current feedback, from the end stage to the first stage, which gives a stabilizing effect. The amplified output signal is picked up single-ended and is given to the power amplifier. The power amplifier is generally of the push-pull differentical type. The base of one input transistor of this amplifier is driven by the preamplified unsymmetrical signal. The base of the other transistor is driven by the feedback signal resulting from the pen position and connected via frequency selective network. The output of the power amplifier is single-ended and is fed to the pen motor, which deflects the writing arm on the paper. A direct writing recorder is usually adequate since the ECG signal of interest has limited bandwidth. Frequency selective network is an $R–C$ network, which provides necessary damping of the pen motor and is preset by the manufacturer. The auxiliary circuits provide a 1 mV calibration signal and automatic blocking of the amplifier during a change in the position of the lead switch. It may include a speed control circuit for the chart drive motor.

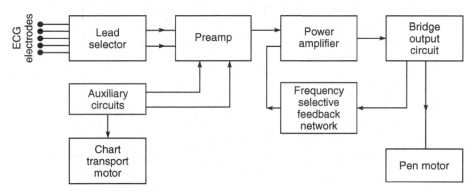

> ➤ **Fig.** 5.1 *Block diagram of an ECG machine*

A 'stand by' mode of operation is generally provided on the electrocardiograph. In this mode, the stylus moves in response to input signals, but the paper is stationary. This mode allows the operator to adjust the gain and baseline position controls without wasting paper.

Electrocardiograms are almost invariably recorded on graph paper with horizontal and vertical lines at 1 mm intervals with a thicker line at 5 mm intervals. Time measurements and heart rate measurements are made horizontally on the electrocardiogram. For routine work, the paper recording speed is 25 mm/s. Amplitude measurements are made vertically in millivolts. The sensitivity of an electrocardiograph is typically set at 10 mm/mV.

Isolated Preamplifier: It had been traditional for all electrocardiographs to have the right leg (RL) electrode connected to the chassis, and from there to the ground. This provided a ready path for any ground seeking current through the patient and presented an electrical hazard. As the microshock hazard became better understood, particularly when intracardiac catheters are employed, the necessity of isolating the patient from the ground was stressed. The American Heart Association guidelines state that the leakage current should not be greater than 10 microamperes when measured from the patient's leads to the ground or through the main instrument grounding wire with the ground open or intact. For this, patient leads would have to be isolated from the ground for all line operated units.

Figure 5.2 shows a block diagram of an isolation preamplifier used in modern electro-cardiographs. Difference signals obtained from the right arm (RA), left arm (LA) and right leg (RL) is given to a low-pass filter. Filtering is required on the input leads to reduce interference caused by electrosurgery and radio frequency emissions and sometimes from the 50 kHz current used for respiration detection. The filter usually has a cut off frequency higher than 10 kHz. A multistage filter is needed to achieve a suitable reduction in high frequency signal.

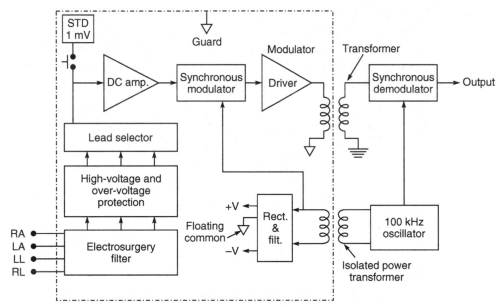

> **Fig. 5.2** *Block diagram of an isolation preamplifier (transformed-coupled) commonly used in modern ECG machines*

The filter circuit is followed by high voltage and over voltage protection circuits so that the amplifier can withstand large voltages during defibrillation. However, the price of this protection is a relatively high amplifier noise level arising from the high series resistance in each lead.

The lead selector switch is used to derive the required lead configurations and give it to a dc-coupled amplifier. A dc level of 1 mV is obtained by dividing down the power supply, which can be given to this amplifier through a push button for calibration of the amplifier. Isolation of the patient circuit is obtained using a low capacitance transformer whose primary winding is driven from a 100 kHz oscillator. The transformer secondary is used to obtain an isolated power supply of ±6 V for operating the devices in the isolated portion of the circuit and to drive the synchronous modulator at 100 kHz, which linearly modulates an ECG signal given to it. The oscillator frequency of 100 kHz is chosen as a compromise so that reasonable size transformers (higher the frequency the smaller the transformer) could be used and that the switching time is not too fast, so that inexpensive transistors and logic circuitry can be utilized. A square wave is utilized to minimize the power requirements of the driven transistors. A synchronous demodulator is chosen to give low noise performance utilizing switching FET's.

Isolation of the patient preamplifier can also be obtained using an optical isolator. The high common-mode rejection of the amplifier is obtained by proper shielding. The effective capacitance from the input leads to the earth is made negligible. The preamplifier circuitry should is preferably be shielded in a separate case.

To minimize the common-mode signal between the body of the patient and the floating ground, a right leg drive circuit (Fig. 5.3) is used. The common-mode signals after amplification in a preamplifier are inverted and fed back to the right leg electrode, reducing the common mode voltage on the input with respect to the floating ground. Winter and Webster (1983) examined optimal design parameters for a driven-right-leg circuit.

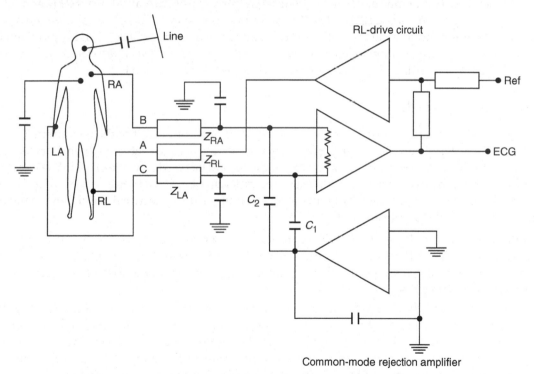

> **Fig. 5.3** *Improvement in CMRR using right leg drive (Courtesy: Hewlett Packard, USA)*

The presence of stray capacitance at the input of the preamplifier causes common-mode currents to flow in LA and RA, resulting in a voltage drop at the electrode resistors. An imbalance of the stray capacitance or the electrode resistors causes a difference signal. This difference signal can be almost eliminated, in that the common-mode currents of stray capacitances are not allowed to flow through the electrode resistors but are neutralized by currents delivered to stray capacitances from the common-mode rejection amplifier. In other words, the potentials at A, B and C are equalised through an in-phase component of the common-mode voltage, which the amplifier delivers via C_1 and C_2 to LA and RA. As a result, the potentials at A, B and C are kept equal, independent of the imbalance in the electrode resistors and stray capacitance.

The modern ECG machines with their completely shielded patient cable and lead wires and their high common-mode rejection, are sufficiently resistant to mains interference. However, there could be locations where such interference cannot be eliminated by reapplying the electrodes or moving the cable, instrument or patient. To overcome this problem, some ECG machines have an additional filter to sharply attenuate a narrow band centred at 50 Hz. The attenuation provided could be up to 40 dB. In this way, the trace is cleaned up by the substantial reduction of line frequency interference.

Isolation amplifiers are available in the modular form. One such amplifier is Model 274 from Analog Devices. This amplifier has the patient safety current as 1.2 μA at 115 V ac 60 Hz and offers a noise of 5 μV pp. It has a CMRR of 115 dB, differential input impedance of 10^{12} Ω paralleled with 3 pF and common mode impedance as 10^{11} Ω and a shunt capacitance of 20 pF. It is optimized for signal frequencies in the range of 0.05 to 100 Hz. Metting van Rijn *et al* (1990) detail out methods for high-quality recording of bioelectric events with special reference to ECG.

5.1.2   The ECG Leads

Two electrodes placed over different areas of the heart and connected to the galvanometer will pick up the electrical currents resulting from the potential difference between them. For example, if under one electrode a wave of 1 mV and under the second electrode a wave of 0.2 mV occur at the same time, then the two electrodes will record the difference between them, i.e. a wave of 0.8 mV. The resulting tracing of voltage difference at any two sites due to electrical activity of the heart is called a "LEAD" (Figs 5.4 (a)-(d)).

Bipolar Leads: In bipolar leads, ECG is recorded by using two electrodes such that the final trace corresponds to the difference of electrical potentials existing between them. They are called standard leads and have been universally adopted. They are sometimes also referred to as Einthoven leads (Fig. 5.4(a)).

In standard lead I, the electrodes are placed on the right and the left arm (RA and LA). In lead II, the electrodes are placed on the right arm and the left leg and in lead Ill, they are placed on the left arm and the left leg. In all lead connections, the difference of potential measured between two electrodes is always with reference to a third point on the body. This reference point is conventionally taken as the "right leg". The records are, therefore, made by using three electrodes at a time, the right leg connection being always present.

In defining the bipolar leads, Einthoven postulated that at any given instant of the cardiac cycle, the electrical axis of the heart can be represented as a two dimensional vector. The ECG measured from any of the three basic limb leads is a time-variant single-dimensional component of the vector. He proposed that the electric field of the heart could be represented diagrammatically as a triangle, with the heart ideally located at the centre. The triangle, known as the *"Einthoven triangle"*, is shown in Fig. 5.5. The sides of the triangle represent the lines along which the three projections of the ECG vector are measured. It was shown that the instantaneous voltage measured

Bipolar Limb Leads

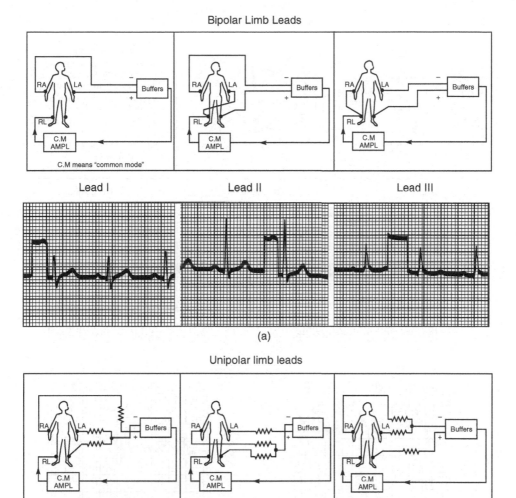

(a)

Unipolar limb leads

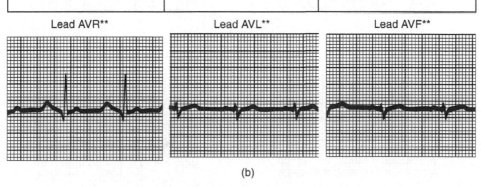

(b)

➤ **Fig. 5.4** *Types of lead connections with typical ECG waveforms (a) bipolar limb leads (b) unipolar limb leads (Courtesy: Hewlett Packard, USA)*

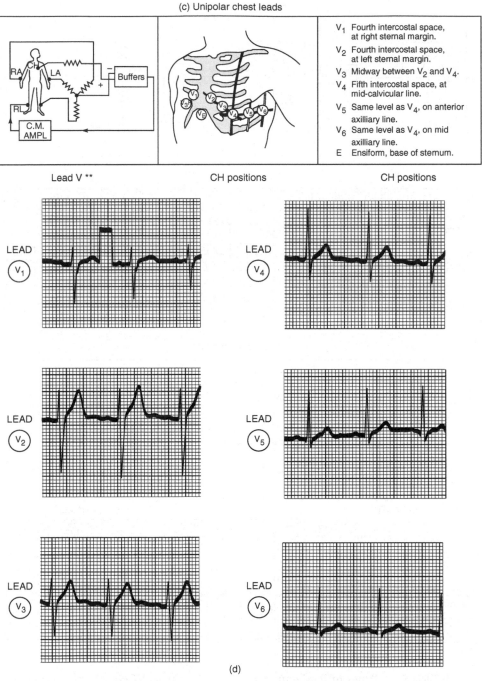

(c) Unipolar chest leads

V_1 Fourth intercostal space, at right sternal margin.

V_2 Fourth intercostal space, at left sternal margin.

V_3 Midway between V_2 and V_4.

V_4 Fifth intercostal space, at mid-calvicular line.

V_5 Same level as V_4, on anterior axilliary line.

V_6 Same level as V_4, on mid axilliary line.

E Ensiform, base of sternum.

Lead V **   CH positions   CH positions

(d)

> **Fig.** 5.4 *Types of lead connections with typical ECG waveforms (c) position of the chest lead in unipolar precordial lead recording (d) C leads (Courtesy: Hewlett Packard, USA)*

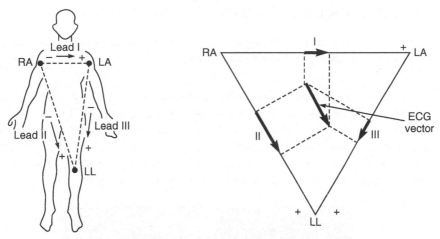

> **Fig. 5.5** *The Einthoven triangle for defining ECG leads*

from any one of the three limb lead positions is approximately equal to the algebraic sum of the other two or that the vector sum of the projections on all three lines is equal to zero.

In all the bipolar lead positions, *QRS* of a normal heart is such that the *R* wave is positive and is greatest in lead II.

Unipolar Leads (V Leads): The standard leads record the difference in electrical potential between two points on the body produced by the heart's action. Quite often, this voltage will show smaller changes than either of the potentials and so better sensitivity an be obtained if the potential of a single electrode is recorded. Moreover, if the electrode is placed on the chest close to the heart, higher potentials can be detected than normally available at the limbs. This lead to the development of unipolar leads introduced by Wilson in 1894. In this arrangement, the electrocardiogram is recorded between a single exploratory electrode and the central terminal, which has a potential corresponding to the centre of the body. In practice, the reference electrode or central terminal is obtained by a combination of several electrodes tied together at one point. Two types of unipolar leads are employed which are: (i) limb leads, and (ii) precordial leads.

(i) *Limb leads* In unipolar limb leads (Fig. 5.4(b)), two of the limb leads are tied together and recorded with respect to the third limb. In the lead identified as AVR, the right arm is recorded with respect to a reference established by joining the left arm and left leg electrodes. In the AVL lead, the left arm is recorded with respect to the common junction of the right arm and left leg. In the AVF lead, the left leg is recorded with respect to the two arm electrodes tied together.

They are also called augmented leads or 'averaging leads'. The resistances inserted between the electrodes-machine connections are known as 'averaging resistances'.

(ii) *Precordial leads* The second type of unipolar lead is a precordial lead. It employs an exploring electrode to record the potential of the heart action on the chest at six different positions. These leads are designated by the capital letter 'V' followed by a subscript numeral, which represents the position of the electrode on the pericardium. The positions of the chest leads are shown in Fig. 5.4(c).

5.1.3   Effects of Artefacts on ECG Recordings

Abnormal patterns of ECG may be due to pathological states or on occasion they may be due to artefacts. To diagnose the presence of undesirable artefacts on the ECG trace, a few recordings are illustrated below:

Interference from the Power Line: Power line interference is easily recognizable since the interfering voltage in the ECG would have a frequency of 50 Hz (Fig. 5.6(a)). This interference may be due to the stray effect of the alternating current on the patient or because of alternating current fields due to loops in the patient cable. Other causes of interference are loose contacts on the patient cable as well as dirty electrodes. When the machine or the patient is not properly grounded, power line interference may even completely obscure the ECG waveform.

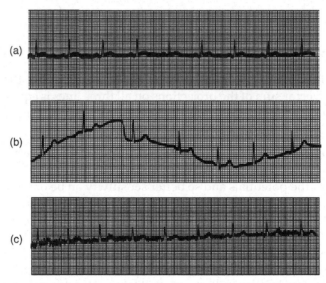

> **Fig.** 5.6   (a) *ECG recording with regular spreading of the curve with super*
>   *imposed 50 Hz power line interference signals*
>   (b) *Recording with irregular trembling of the ECG trace without*
>   *wandering of the base line but otherwise normal ECG trace*
>   (c) *ECG trace without wandering of the base line*

The most common cause of 50 Hz interference is the disconnected electrode resulting in a very strong disturbing signal. It is often strong enough to damage the stylus of an unprotected direct writing recorder, and therefore needs quick action.

Sometimes static charges on the synthetic uniform of the operator may result in a random noise on the trace. This noise is very difficult to remove except in those machines which have very high CMRR. The noise can be reduced by partially shielding the patient by means of the bed springs. Connection of the springs to the instrument case helps to compensate for a poor CMRR (Spooner, 1977).

Electromagnetic interference from the power lines also results in poor quality tracings. Electrical equipment such as air-conditioners, elevators and X-ray units draw heavy power-line current, which induce 50 Hz signals in the input circuits of ECG machines. Due to unbalanced linkages, common mode rejection circuits almost prove ineffective against them. A practical solution to minimize this problem is physical separation between the interference causing sources and the patient. Levkov *et al* (1984) developed a method of digital 50 Hz interference elimination by computing the interference amplitudes and subtracting these data from the original signal, thereby greatly reducing the requirements of amplifiers, shielding, earthing, electrode quality and application procedures.

Electrical power systems also induce extremely rapid pulses or spikes on the trace, as a result of switching action. Use of a transient suppressor in the mains lead of the machines helps to solve this problem.

Shifting of the Baseline: A wandering baseline (Fig. 5.6(b)) but otherwise normal ECG trace is usually due to the movement of the patient or electrodes. The baseline shift can be eliminated by ensuring that the patient lies relaxed and the electrodes are properly attached. Baseline wander is usually observed immediately after application of the electrodes. It is due to a relatively slow establishment of electrochemical equilibrium at the electrode-skin interface. This can be minimized by selecting the proper electrode material, which will reach equilibrium quickly with a good electrode jelly.

Muscle Tremor: Irregular trembling of the ECG trace (Fig. 5.6(c)), without wandering of the baseline occurs when the patient is not relaxed or is cold. It is generally found in the case of older patients. Muscle tremor signals are especially bothersome on limb leads when a patient moves or the muscles are stretched. Therefore, for long-term monitoring, the electrodes are applied on the chest and not on the limbs. For normal routine ECG recordings, the patient must be advised to get warm and to relax so that muscle tremor from shivering or tension is eliminated.

The most critical component of the ECG recorder is the patient cable. The conventional PVC insulation gets degraded and becomes rigid and breakable because of the arification of the softener. Some manufacturers supply a patient-cable made of silicon-rubber, which provides better elasticity over long periods.

5.1.4 Microprocessor Based ECG Machines

Microprocessor technology has been employed in the electrocardiographs to attain certain desirable features like removal of artefacts, baseline wander, etc. using software techniques. Automatic centering of the tracing is another feature which can be similarly achieved.

Microprocessor-based ECG machines can perform self-testing each time they are switched on. These machines are programmed to check lead continuity and polarity and also indicate lead fall-off or reversal. Use of digital filters considerably improves signal quality during recording and problems like baseline drift and excessive mains hum are thus automatically corrected. For this, the programs are stored in EPROM to obtain good quality tracings. Minimising baseline drift without distorting the signal helps in monitoring the ECG of exercising or ambulatory subjects.

The frequency components of ECG signals are low enough for the microprocessor to perform reliable data acquisition, processing and display. For example, the highest frequency components in the waveform are in the *QRS* complex which typically last 60 to 100 ms. In this case, a sampling rate of 200 samples per second, which will yield 12–20 points in the *QRS* complex, is considered adequate (Hsue and Graham, 1976). Since the microprocessor instruction times are of the order of 5 μs, this permits approximately 1000 instructions between samples, which is sufficient for data acquisition, processing and storage. In these cardiographs, the ECG is converted to digital form for waveform preprocessing and then reconverted into analog form for display or telephonic transmission to the central computer. Using microprocessors, the need for the technician to switch from one lead to another during the recording process is eliminated Microprocessors have also been used for simultaneous acquisition of multiple leads. Stored programs in the ROM direct the operator about entering of patient data and display error codes. An important advantage of microprocessor-based ECG machines is that it offers the potential for reducing the complexity of analysis algorithms by preprocessing the data. Chapter 7 details out methods of ECG analysis for automatic determination of various kinds of arrhythmias.

5.1.5 Multi-channel ECG Machine

Most of the electrocardiographs used for clinical purposes are single channel machines, i.e. the machine contains one amplifier channel and one recording system. Such machines usually carry a multi-position switch, by means of which the desired lead connection can be selected. Only one lead at a time can be recorded with such type of instruments.

Multi-channel ECG machines are also available. They carry several amplifier channels and a corresponding number of recording pens. This facilitates recording of several ECG leads simultaneously and thus considerably reduces the time required to complete a set of recordings. Another advantage of multi-channel recording is that the waveforms are recorded simultaneously and they can be shown in their proper time relationship with respect to each other.

Modern multi-channel ECG machines use microprocessor to capture the heart signals from a standard 12-lead configuration, sequencing the lead selector to capture four groups of three lead signals and switching groups every few seconds. Figure 5.7 shows the block diagram of a three-channel microprocessor based ECG machine. The operating program controlling the lead selection and other operations is stored in a ROM. The ECG signals selected by the microprocessor are amplified, filtered and sent to a three-channel multiplexer. The multiplexed analog signals are then given to an analog-to-digital converter. For a 10 μV resolution referred to as the input (Fostik *et al*, 1980), it is necessary to use a 10-bit A-D converter. Ten bits provide resolution of one part in 1024 ($2^{10} = 1024$), which for a 10 mV peak-to-peak input range equals 10 μV. A suitable 10-bit A-D converter is Analog Devices 7570. The maximum conversion rate of the device for 10-bit words is 25000 per second. The sampling rate is usually 200–1000 samples/s. The microprocessor stores the digitized signals in a RAM. The contents of the RAM are sent to a digital-to-analog converter for reconstructing the analog signals (Shackil, 1981). The analog signals are demultiplexed and passed to the video display or chart recorder.

We have seen that older versions of ECG machines recorded one lead at a time and then evolved to three simultaneous leads. This necessitated the use of switching circuits to generate the various

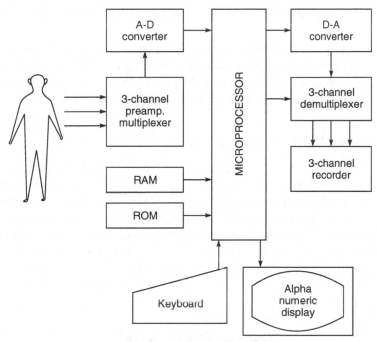

➤ **Fig. 5.7** *Block diagram of a microprocessor based three channel ECG machine*

12 leads. This is now eliminated in modern digital ECG machines by the use of an individual single-ended amplifier for each electrode on the body. Each potential signal is then digitally converted and all the ECG leads can be calculated mathematically in software. This would necessitate a 9- amplifier system. The machines make use of a 16-bit A–D converter, all within a small amplifier lead stage. The digital signals are optically isolated and sent via a high speed serial link to the main ECG instrument. Here the 32-bit CPU and DSP (Digital Signal Processor) chip perform all the calculation and hard copy report is generated on a standard A-4 size paper. These machines are therefore called page writers.

The recording system also operates digitally. The machine output is produced by an X-Y drive mechanism which uses drive wheels to move the ECG chart paper in the paper axis direction while moving a carriage mounted pen in the carriage axis direction. Each direction of movement is caused by identical low inertia dc servo motors with attached encoders to provide position feed back for the respective servo chips (integrated circuit). Drive from the servo motors is provided by toothed belts. The microprocessor controls the plotting process by sending plot commands to the X- and Y-axis servo chips. These commands are in the form of a pulse train. The servo chips control the motors and keep track of the position error between the current and the desired position of the motor. Each of the motor drives is activated in the direction which reduces this position error. Motor position is fed back to the respective servo chip by a shaft encoder mounted on the motor. This position feedback decreases the servo chips position error as the motor approaches the desired position. The responsible servo chip communicates to the microprocessor should a servo error overflow occur, i.e. of a servo position has been lost. This is an irrecoverable error, which causes the microprocessor to shut down the whole recorder assembly and send back an error message.

Modern ECG machines also incorporate embedded software for an automatic interpretation of the ECG. These programs are quite sophisticated involving waveform recognition, calculation of amplitude, duration, area and shape of every *P* wave, *QRS* complex, *T* wave and *ST* segment in every lead and rhythm analysis. The basis of these analysis are covered in Chapter 7.

⁛⁛▶ 5.2   VECTORCARDIOGRAPH (VCG)

Vectorcardiography is the technique of analyzing the electrical activity of the heart by obtaining ECG's along three axes at right angles to one another and displaying any two of these ECGs as a vector display on an X-Y oscilloscope. The display is known as a vectorcardiogram (VCG). In contrast, the electrocardiogram which displays the electrical potential in any one single axis, the vectorcardiogram displays the same electrical events simultaneously in two perpendicular axes. This gives a vectorial representation of the distribution of electrical potentials generated by the heart, and produces loop type patterns (Fig. 5.8) on the CRT screen. Usually a photograph is taken of each cardiac cycle. From such pictures, the magnitude and orientation of the *P, Q, R, S* and *T* vector loops are determined.

Vectorcardiogram [QRS T]

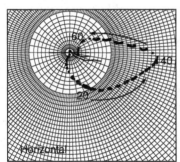 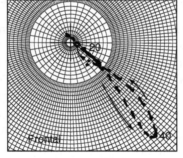 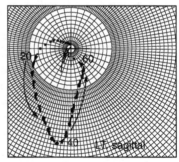

▶ Fig. 5.8   *Typical normal loop patterns recorded in three planes on a direct writing vectorcardiograph*

VCG illustrates the phase differences between the voltages and also the various leads from which it is derived. The major information that it provides is the direction of depolarization and repolarization of the atria and the ventricles. Each vectorcardiogram exhibits three loops, showing the vector orientation of the *P* wave, the *QRS* axis and the *T* wave. Because of the high amplitude associated with *QRS,* loops from the *QRS* complex predominate. An increase in horizontal and vertical deflection sensitivities is normally required to adequately display the loops resulting from the *P* wave and *T* wave. Bourne (1974) describes circuit details of an automated vector ECG recording system.

The VCG has been demonstrated to be superior to the standard 12-lead scalar electrocardiogram in the recognition of undetected atrial and ventricular hypertrophy, sensitivity in identification of myocardial infarction and capability for diagnosis of multiple infarctions in the presence of fascicular and bundle branch blocks (Benchimol and Desser, 1975).

5.3 PHONOCARDIOGRAPH (PCG)

The phonocardiograph is an instrument used for recording the sounds connected with the pumping action of the heart. These sounds provide an indication of the heart rate and its rhythmicity. They also give useful information regarding effectiveness of blood pumping and valve action.

Heart sounds are diagnostically useful. Sounds produced by healthy hearts are remarkably identical and abnormal sounds always corelate to specific physical abnormalities. From the beginning till today, the principal instrument used for the clinical detection of heart sounds is the acoustical stethoscope. An improvement over the acoustal stethoscope, which usually has low fidelity, is the electronic stethoscope consisting of a microphone, an amplifier and a head set. Electronic stethoscopes can detect heart sounds which are too low in intensity or too high in frequency to be heard in a purely acoustal instrument. The phonocardiographs provide a recording of the waveforms of the heart sounds. These waveforms are diagnostically more important and revealing than the sounds themselves.

5.3.1 Origin of Heart Sounds

The sounds are produced by the mechanical events that occur during the heart cycle. These sounds can be from the movement of the heart wall, closure of walls and turbulence and leakage of blood flow. A typical recording of these sounds is illustrated in Fig. 5.9. The first sound, which corresponds to the *R* wave of the ECG, is longer in duration, lower in frequency, and greater in intensity than the second sound. The sound is produced principally by closure of the valves between the upper and lower chambers of the heart, i.e. it occurs at the termination of the atrial

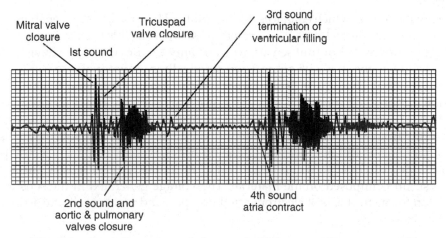

> **Fig. 5.9** *Basic heart sounds in a typical phonocardiogram recording*

contraction and at the onset of the ventricular contraction. The closure of the mitral and tricuspid valve contributes largely to the first sound. The frequencies of these sounds are generally in the range of 30 to 100 Hz and the duration is between 50 to 100 ms. The second sound is higher in pitch than the first, with frequencies above 100 Hz and the duration between 25 to 50 ms. This sound is produced by the slight back flow of blood into the heart before the valves close and then by the closure of the valves in the arteries leading out of the ventricles. This means that it occurs at the closure of aortic and the pulmonic valves.

The heart also produces third and fourth sounds but they are much lower in intensity and are normally inaudible. The third sound is produced by the inflow of blood to the ventricles and the fourth sound is produced by the contraction of the atria. These sounds are called diastolic sounds and are generally inaudible in the normal adult but are commonly heard among children.

5.3.2 Microphones for Phonocardiography

Two types of microphones are commonly in use for recording phonocardiograms. They are the contact microphone and the air coupled microphone. They are further categorized into crystal type or dynamic type based on their principle of operation.

The *crystal microphone* contains a wafer of piezo-electric material, which generates potentials when subjected to mechanical stresses due to heart sounds. They are smaller in size and more sensitive than the dynamic microphone.

The *dynamic type* microphone consists of a moving coil having a fixed magnetic core inside it. The coil moves with the heart sounds and produces a voltage because of its interaction with the magnetic flux.

The phonocardiogram depends extensively on the technical design of the microphone, since it does not transform the acoustic oscillations into electrical potential uniformly for all frequencies. Therefore, the heart sound recordings made with a microphone are valid only for that particular type of microphone. As a consequence, microphones of various types cannot, as a rule, be interchanged.

A new acoustic sensor, which enhances the audibility of heart sounds and enables recording of quantitative acoustic spectral data is described by Kassal *et al*, 1994. This device is a polymer-based adherent differential-output sensor, which is only 1.0 mm thick. The device is compliant and can be applied to the skin with gel and two-sided adhesive material, and can conform to the contour of the patient's body. The device can be used for phonocardiography, lung sounds and the detection of Korotkoff sounds. The device is not a microphone and does not detect acoustic pressure, rather it actually discriminates against it. Instead, the sensor detects the motion of the skin that results from acoustic energy incident upon it from within the soft tissue. Its principle sensing components is PVDF (poly-vinylidene fluoride), which is a piezo-electric polymer. It produces charges of equal magnitude and opposite polarity on opposite surfaces when a mechanical strain is imposed on the material. The voltage generated in the sensor due to the flexing motion forms the basis of electronic stethoscopes, and real time digital acoustic spectral analysis of heart sounds.

5.3.3 Amplifiers for Phonocardiography

The amplifier used for a phonocardiograph has wide bandwidth with a frequency range of about 20 to 2000 Hz. Filters permit selection of suitable frequency bands, so that particular heart sound frequencies can be recorded. In general, the high frequency components of cardiovascular sound have a much smaller intensity than the low frequency components and that much information of medical interest is contained in the relatively high frequency part of this spectrum (Bekkering and Vollenhoven, 1967). Therefore, high-pass filters are used to separate the louder low frequency components from the soft and interesting high frequency murmurs. Experiments have shown that the choice of different filters does not have to be very critical but in general, sets of four or five high-pass filters with different cut-off frequencies and slopes are used in the commercially available instruments. PCG amplifiers usually have gain compensation circuits to increase the amplification of high frequency signals, which are usually of low intensity. The frequencies at the higher end of the range are of particular significance in research applications.

The appropriate filter characteristics may be selected to attenuate the unwanted frequencies at filter slopes of 12 dB/octave or 24 dB/octave. This is based on the fact that cardiac vibrations follow the inverse square law which is 12 dB/octave, i.e. as the frequency of the sound is increased, the intensity decreases approximately 12 dB/octave over a portion of the sound spectrum. The 12 dB/octave approximation is valid from 50–200 Hz and the 24 dB/octave from 200–800 Hz. Below 50 Hz, 6 dB/octave is found to be the best choice. A filter with a 12 dB/octave slope causes the intensity of the unwanted sounds to decrease to 0.25 times the original, when the frequency of the sound is one half its original value. Figure 5.10 shows heart sound amplifier characteristics.

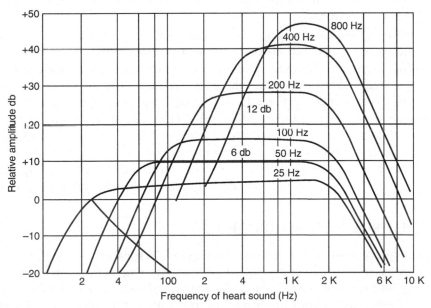

➤ **Fig. 5.10** *Characteristics of amplifiers with commonly employed filters in phonocardiography (Courtesy: Hewlett Packard, USA)*

5.3.4   Writing Methods for Phonocardiography

In order to obtain a faithful reproduction of all the frequency components, the phonocardiogram requires a recording system capable of responding to 2000 Hz. Light beam galvanometers have no trouble in meeting this demand. But these galvanometers are expensive and require more power from the amplifiers when used for recording high frequencies, of the order specified above. Direct writing recorders with an upper frequency response of about 150 Hz cannot be used to write frequencies that lie beyond their working range. This drawback is overcome by using the "Envelope Detection" technique. The technique consists in using an artificial frequency, say 100 Hz, in the heart sound amplifier. This is employed to oscillate the stylus so that the high frequency sounds are modulated by 100 Hz. The shape of the sound wave is thus retained and reproduced with a response of 100 Hz. The heart sounds can now be recorded on a direct writing recorder. This procedure, however, can only record heart sound intensity picked up every 10 ms, and events taking place within this time interval of 10 ms are not recorded.

Recording of phonocardiograms by direct recording methods has been made possible by the ink jet recorder. Also, digital recorders such as the electrostatic recorder or the thermal recorder are also suitable for phonographic recordings.

Many phonocardiograms also have a provision for recording the patient's electrocardiogram on the same chart. Simultaneous recording of the phonocardiogram and the electrocardiogram displays both the sounds and the electrical activity of the heart in their proper time relationship. This facilitates the work of the clinician to correlate sounds with the phase of the heart cycle during which they occur. From such time correlations it is easier to identify, specifically, the nature of the defect. The phonocardiogram is more informative than an electrocardiogram for monitoring heart valve actions.

While recording a phonocardiogram, the microphone picks up not only the sounds and murmurs at the body surface, but also all extraneous noises in the immediate vicinity of the patient. Therefore, common utilities like fans, air-conditioners and other noise producing gadgets working nearby create vibrations within the same frequency range as the heart sounds and murmurs and will result in artefacts on the recording. This fact emphasizes the importance of having a relatively quiet area for phonocardiography. The walls and ceiling of the room should be acoustic-tiled and the recording system placed on some cushioning material, to minimize external and internal noise. It is advisable to secure the microphone to the patient's chest with a strap because if it is hand held, its output may vary in accordance with the pressure applied. It is only with practice that one learns to apply a constant pressure to the microphone and the patient.

▶ 5.4   ELECTROENCEPHALOGRAPH (EEG)

Electroencephalograph is an instrument for recording the electrical activity of the brain, by suitably placing surface electrodes on the scalp. EEG, describing the general function of the brain activity, is the superimposed wave of neuron potentials operating in a non-synchronized manner in the physical sense. Its stochastic nature originates just from this, and the prominent signal groups can be empirically connected to diagnostic conclusions.

Monitoring the electroencephalogram has proven to be an effective method of diagnosing many neurological illnesses and diseases, such as epilepsy, tumour, cerebrovascular lesions, ischemia and problems associated with trauma. It is also effectively used in the operating room to facilitate anaesthetics and to establish the integrity of the anaesthetized patient's nervous system. This has become possible with the advent of small, computer-based EEG analyzers. Consequently, routine EEG monitoring in the operating room and intensive care units is becoming popular.

Several types of electrodes may be used to record EEG. These include: Peel and Stick electrodes, Silver plated cup electrodes and Needle electrodes.

EEG electrodes are smaller in size than ECG electrodes. They may be applied separately to the scalp or may be mounted in special bands, which can be placed on the patient's head. In either case, electrode jelly or paste is used to improve the electrical contact. If the electrodes are intended to be used under the skin of the scalp, needle electrodes are used. They offer the advantage of reducing movement artefacts. EEG electrodes give high skin contact impedance as compared to ECG electrodes. Good electrode impedance should be generally below 5 kilohms. Impedance between a pair of electrodes must also be balanced or the difference between them should be less than 2 kilohms. EEG preamplifiers are generally designed to have a very high value of input impedance to take care of high electrode impedance.

EEG may be recorded by picking up the voltage difference between an active electrode on the scalp with respect to a reference electrode on the ear lobe or any other part of the body. This type of recording is called 'monopolar' recording. However, 'bipolar' recording is more popular wherein the voltage difference between two scalp electrodes is recorded. Such recordings are done with multi-channel electroencephalographs.

EEG signals picked up by the surface electrodes are usually small as compared with the ECG signals. They may be several hundred microvolts, but 50 microvolts peak-to-peak is the most typical. The brain waves, unlike the electrical activity of the heart, do not represent the same pattern over and over again. Therefore, brain recordings are made over a much longer interval of time in order to be able to detect any kind of abnormalities.

5.4.1 Block diagram description of Electroencephalograph

The basic block diagram of an EEG machine with both analog and digital components is shown in Fig. 5.11.

Montages: A pattern of electrodes on the head and the channels they are connected to is called a montage. Montages are always symmetrical. The reference electrode is generally placed on a non-active site such as the forehead or earlobe. EEG electrodes are arranged on the scalp according to a standard known as the 10/20 system, adopted by the American EEG Society (Barlow *et al*, 1974). Traditionally, there are 21 electrode locations in the 10/20 system. This system involves placement of electrodes at distances of 10% and 20% of measured coronal, sagittal and circumferential arcs between landmarks on the cranium (Fig. 5.12). Electrodes are identified according to their position on the head: **Fp** for frontal-polar, **F** for frontal, **C** for central, **P** for parietal, **T** for temporal and **O** for occipital. Odd numbers refer to electrodes on the left side of the head and even numbers represent those on the right while **Z** denotes midline electrodes. One electrode is labelled isoground and

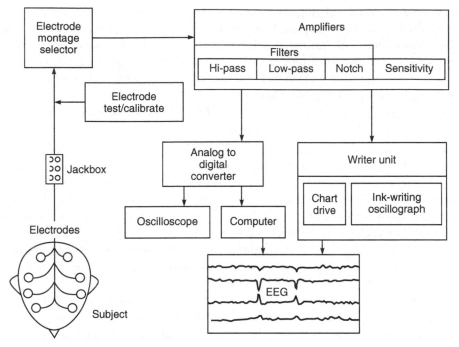

> **Fig. 5.11** *Schematic diagram of an EEG machine (after Isley et al, 1998)*

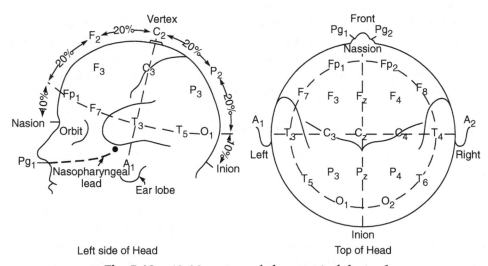

Left side of Head   Top of Head

> **Fig. 5.12** *10-20 system of placement of electrodes*

placed at a relatively neutral site on the head, usually the midline forehead. A new montage convention has recently been introduced in which electrodes are spaced at 5% distances along the cranium. These electrodes are called closely spaced electrodes and have their own naming convention (Fig. 5.13).

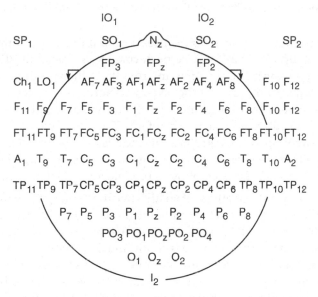

> ➤ **Fig. 5.13** *Pictorial representation of closely spaced electrodes*

Electrode Montage Selector: EEG signals are transmitted from the electrodes to the head box, which is labelled according to the 10–20 system, and then to the montage selector. The montage selector on analog EEG machine is a large panel containing switches that allow the user to select which electrode pair will have signals subtracted from each other to create an array of channels of output called a montage. Each channel is created in the form of the input from one electrode minus the input from a second electrode.

Montages are either bipolar (made by the subtraction of signals from adjacent electrode pairs) or referential (made by subtracting the potential of a common reference electrode from each electrode on the head). In order to minimize noise, a separate reference is often chosen for each side of the head e.g. the ipsilateral ear. Bipolar and referential montages contain the same basic information that is transformable into either format by simple substration as long as all the electrodes, including reference, are included in both montages and linked to one common reference. Many modern digital EEG machines record information referentially, allowing easy conversion to several different bipolar montages. The advantage of recording EEG in several montages is that each montage displays different spatial characteristics of the same data.

Preamplifier: Every channel has an individual, multistage, ac coupled, very sensitive amplifier with differential input and adjustable gain in a wide range. Its frequency response can be selected by single-stage passive filters. A calibrating signal is used for controlling and documenting the sensitivity of the amplifier channels. This supplies a voltage step of adequate amplitude to the input of the channels. A typical value of the calibration signal is 50 uV/cm.

The preamplifier used in electroencephalographs must have high gain and low noise characteristics because the EEG potentials are small in amplitude. In addition, the amplifier must have very high common-mode rejection to minimise stray interference signals from power lines and other electrical equipments. The amplifier must be free from drift so as to prevent the slow movement of the recording pen from its centre position as a result of changes in temperature, etc.

EEG amplifiers must have high gain in the presence of unbalanced source resistances and dc skin potentials at least up to 100 mV. Noise performance is crucial in EEG work because skin electrodes couple brain waves of only a few microvolts to the amplifier. Each individual EEG signal should be preferably amplified at the bedside. Therefore, a specially designed connector box, which can be mounted near the patient, is generally employed with EEG machines. This ensures the avoidance of cable or switching artefacts. The use of electrode amplifiers at the site also eliminates undesirable cross-talk effects of the individual electrode potentials. The connector box also carries a circuit arrangement for measuring the skin contact impedance of electrodes with ac. Thus, poor electrode-to-skin contacts above a predetermined level can be easily spotted out.

Sensitivity Control: The overall sensitivity of an EEG machine is the gain of the amplifier multiplied by the sensitivity of the writer. Thus, if the writer sensitivity is 1 cm/V, the amplifier must have an overall gain of 20,000 for a 50 μV signal. The various stages are capacitor coupled. An EEG machine has two types of gain controls. One is continuously variable and it is used to equalize the sensitivities of all channels. The other control operates in steps and is meant to increase or reduce the sensitivity of a channel by known amounts. This control is usually calibrated in decibels. The gain of amplifiers is normally set so that signals of about 200 μV deflect the pens over their full linear range. Artefacts, several times greater than this, can cause excessive deflections of the pen by charging the coupling capacitors to large voltages. This will make the system unusable over a period depending upon the value of the coupling capacitors. To overcome this problem, most modern EEG machines have de-blocking circuits similar to those used in ECG machines.

Filters: Just like in an ECG when recorded by surface electrodes, an EEG may also contain muscle artefacts due to contraction of the scalp and neck muscles, which overlie the brain and skull. The artefacts are large and sharp, in contrast to the ECG, causing great difficulty in both clinical and automated EEG interpretation. The most effective way to eliminate muscle artefact is to advise the subject to relax, but it is not always successful. These artefacts are generally removed using low-pass filters. This filter on an EEG machine has several selectable positions, which are usually labelled in terms of a time constant. A typical set of time constant values for the low-frequency control are 0.03, 0.1, 0.3 and 1.0 s. These time constants correspond to 3 dB points at frequencies of 5.3, 1.6, 0.53 and 0.16 Hz.

The upper cut-off frequency can be controlled by the high frequency filter. Several values can be selected, typical of them being 15, 30, 70 and 300 Hz.

Some EEG machines have a notch filter sharply tuned at 50 Hz so as to eliminate mains frequency interference. These however have the undesirable property of 'ringing' i.e. they produce a damped oscillatory response to a square wave calibration waveform or a muscle potential. The use of notch filters should preferably be restricted to exceptional circumstances when all other methods of eliminating interference have been found to be ineffective.

The high frequency response of an EEG machine will be the resultant of the response of the amplifier and the writing part. However, the figure mentioned on the high frequency filter control of most EEG machines generally refers to the amplifier. The typical frequency range of standard EEG machines is from 0.1 Hz to 70 Hz, though newer machines allow the detection and filtering of frequencies up to several hundred Hertz. This may be of importance in some intracranial recordings.

Noise: EEG amplifiers are selected for minimum noise level, which is expressed in terms of an equivalent input voltage. Two microvolts is often stated as the acceptable figure for EEG recording. Noise contains components at all frequencies and because of this, the recorded noise increases with the bandwidth of the system. It is therefore important to restrict the bandwidth to that required for faithful reproduction of the signal. Noise level should be specified as peak-to-peak value as it is seen on the record rather than rms value, which could be misleading.

Writing Part: The writing part of an EEG machine is usually of the ink type direct writing recorder. The best types of pen motors used in EEG machines have a frequency response of about 90 Hz. Most of the machines have a response lower than this, and some of them have it even as low as 45 Hz. The ink jet recording system, which gives a response up to 1000 Hz, is useful for some special applications.

Paper Drive: This is provided by a synchronous motor. An accurate and stable paper drive mechanism is necessary and it is normal practice to have several paper speeds available for selection. Speeds of 15, 30 and 60 mm/s are essential. Some machines also provide speed values outside this range. A time scale is usually registered on the record by one or two time marker pens, which make a mark once per second. Timing pulses are preferably generated independently of the paper drive mechanism in order to avoid difference in timing marks due to changes in paper speed.

Channels: An electroencephalogram is recorded simultaneously from an array of many electrodes. The record can be made from bipolar or monopolar leads. The electrodes are connected to separate amplifiers and writing systems. Commercial EEG machines have up to 32 channels, although 8 or 16 channels are more common.

Microprocessors are now employed in most of the commercially available EEG machines. These machines permit customer programmable montage selection; for example, up to eight electrode combinations can be selected with a keyboard switch. In fact, any desired combination of electrodes can be selected with push buttons and can be memorized. These machines also include a video monitor screen to display the selected pattern (montage) as well as the position of scalp sites with electrode-to-skin contact. Individual channel control settings for gain and filter positions can be displayed on the video monitor for immediate review. Therefore, a setting can be changed by a simple push button operation while looking at the display.

Modern EEG machines are mostly PC based, with a pentium processor, 16-MB RAM, atleast a 2 GB hard disk, cache memory and a 4 GB DAT tape drive. The system can store up to 40 hours of EEG. The EEG is displayed on a 43 cm colour monitor with a resolution of 1280×1024 pixels. The user interface is through an ASCII keyboard and the output is available in the hard copy form through a laser printer.

5.4.2 Recording of Evoked Potentials

If an external stimulus is applied to the a sensory area of brain, it responds by producing an electrical potential known as the 'evoked potential'. The most frequently used evoked potentials for clinical testing include brainstem auditory evoked responses, visual evoked responses and somatosensory evoked potentials.

Evoked potential, recorded at the surface of the brain, is the integrated response of the action of many cells. The amplitude of the evoked potential is of the order of 10 microvolts. The evoked potentials are generally superimposed with electroencephalograms. Therefore, it is necessary to remove the EEG by an averaging technique while making evoked potential measurements. Since the background EEG and other unwanted signals often appear irregular, or do not synchronize to stimuli, they are markedly reduced by averaging across multiple trials. In general, averaging reduces noise proportional to the square route of the number of trials. Most of the improvements in signal-to-noise ratio occur within 40 to 100 trials.

Since many evoked potential components are of short duration, about 2 ms to 1 sec., rapid sampling rates are needed to digitally record such low level potentials. Usually, the sampling rate is 1000/second. The amplitude of evoked potentials are normally measured on a vertical scale with sample points measured as bits on a logarithmic resolution scale. Resolution of voltage is usually sufficient with a 8-bit recording, although 10-to 16-bit A/D systems are becoming available.

5.4.3   Computerized Analysis of EEG

Assessment of the frequency and amplitude of the EEG is crucial for rapid and accurate interpretation. This involves the need for constant analysis of the EEG signal by a skilled technician and the acquisition of volumes of recording paper. Therefore, modern machines make use of computerized EEG signal processing to extract and present the frequency and amplitude information in simple, visually enhanced formats that are directly useful to the clinician (Isley et al, 1998).

Frequency Analysis: It takes the raw EEG waves, mathematically analyzes them and breaks them into their component frequencies. The most popular method of doing this is called the Fast-Fourier Transform or FFT.

Fast-Fourier Transformation of the digitized EEG waveform is a mathematical transformation of a complex waveform (having varying frequency and amplitude content) into simpler, more uniform waveforms (such as different sine waves of varying amplitudes). In this method, the EEG signal is converted into a simplified waveform called a spectrum. It is then separated into frequency bands at intervals of 0.5 Hz over a range of 1 to 32 Hz. The re-distribution of electrical activity in the brain among certain frequency bands or the predominance of one band over the others correlates with specific physiologic and pathologic conditions (Fig. 5.14(a)). The spectral analysis transforms the analog EEG signal recorded on the time axis into a signal displayed on the frequency axis.

Amplitude Analysis: Changes in the EEG amplitude can indicate clinical changes. The amplitude changes result in changes in the power of the resulting frequency spectrum. As the amplitude increases, so does the power. The most common number reflecting EEG amplitude is the total power of the EEG spectrum. Due to the microvolt amplitude of the EEG, power is either in nanowatts or picowatts. The *power spectrum* is calculated by squaring the amplitudes of the individual frequency components. The powers of the individual frequency bands are also commonly used and expressed as an absolute or a percentage of the total power. For example, 25% Alpha would indicate that 25% of the total power is derived from the amplitudes of the Alpha waves.

Several different *display formats* have been developed for visually enhancing the computer-processed information. These are

Compressed Spectral Array (CSA): In this format, a series of computer-smoothed spectral arrays are stacked vertically, usually at two second intervals, with the most recent EEG event at the bottom and the oldest at the top. Peaks appear at frequencies, which contain more power or make larger contributions to the total power spectrum. Since the origin of the plots shifts vertically with time, this produces a pseudo three-dimensional graph (Fig. 5.14(b)). With this method, it is easy to pick up changes in frequency and amplitude of each sample over a longer period of time as it compresses a large amount of data into a compact, easy to read trend.

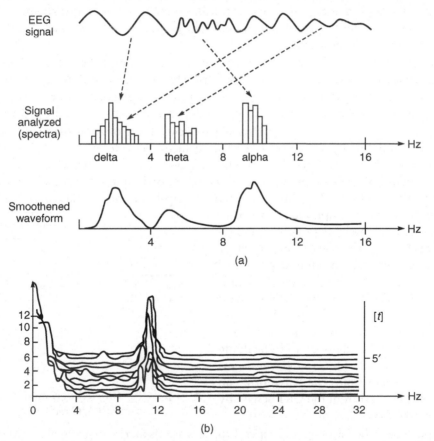

(a)

(b)

➤ **Fig. 5.14** (a) *Typical ECG waveform broken down into frequency components*
   (b) *Mathematical and display techniques used to generate the compressed spectral array format*

Dot-density modulated Spectral Array (DSA): It is another method for displaying the power spectra. This format displays a power spectrum as a line of variable intensities and/or densities with successive epochs again stacked vertically as in the CSA plots. Areas of greatest density represent frequencies, which make the greatest contribution to the EEG power spectrum. An advantage of the DSA format is that no data is hidden by the peaks as in the CSA display. DSA displays could be in the form of gray or colour-scaled densities.

5.5 ELECTROMYOGRAPH (EMG)

Electromyograph is an instrument used for recording the electrical activity of the muscles to determine whether the muscle is contracting or not; or for displaying on the CRO and loudspeaker the action potentials spontaneously present in a muscle or those induced by voluntary contractions as a means of detecting the nature and location of motor unit lesions; or for recording the electrical activity evoked in a muscle by the stimulation of its nerve. The instrument is useful for making a study of several aspects of neuromuscular function, neuromuscular condition, extent of nerve lesion, reflex responses, etc.

EMG measurements are also important for the myoelectric control of prosthetic devices (artificial limbs). This use involves picking up EMG signals from the muscles at the terminated nerve endings of the remaining limb and using the signals to activate a mechanical arm. This is the most demanding requirement from an EMG since on it depends the working of the prosthetic device.

EMG is usually recorded by using surface electrodes or more often by using needle electrodes, which are inserted directly into the muscle. The surface electrodes may be disposable, adhesive types or the ones which can be used repeatedly. A ground electrode is necessary for providing a common reference for measurement. These electrodes pick up the potentials produced by the contracting muscle fibres. The signal can then be amplified and displayed on the screen of a cathode ray tube. It is also applied to an audio-amplifier connected to a loudspeaker. A trained EMG interpreter can diagnose various muscular disorders by listening to the sounds produced when the muscle potentials are fed to the loud-speaker. The block diagram (Fig. 5.15) shows a typical set-up for EMG recordings. The oscilloscope displays EMG waveforms. The tape recorder is included in the system to facilitate playback and study of the EMG sound waveforms at a later convenient time. The waveform can also be photographed from the CRT screen by using a synchronized camera.

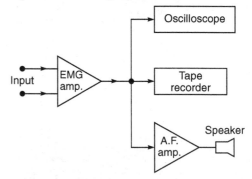

➤ **Fig. 5.15** *Block diagram of a typical set-up for EMG recording*

The amplitude of the EMG signals depends upon various factors, e.g. the type and placement of electrodes used and the degree of muscular exertions. The needle electrode in contact with a single muscle fibre will pick up spike type voltages whereas a surface electrode picks up many overlapping spikes and therefore produces an average voltage effect. A typical EMG signal ranges from 0.1 to 0.5 mV. They may contain frequency components extending up to 10 kHz. Such high froquency signals cannot be recorded on the conventional pen recorders and therefore, they are usually displayed on the CRT screen.

Modern EMG machines are PC based (Fig. 5.16) available both in console as well as laptop models. They provide full colour waveform display, automatic cursors for marking and making

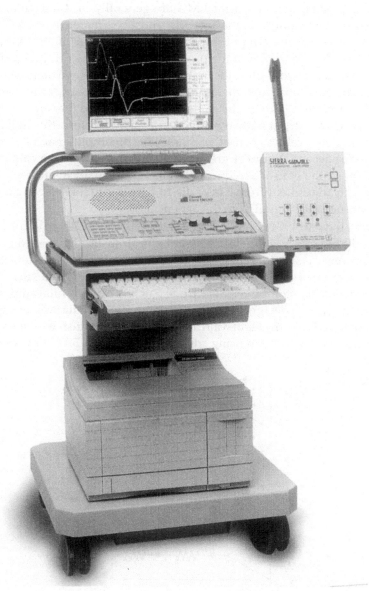

> **Fig. 5.16** *PC based digital EMG recording and reviewing system for 2 to 4 channels. It includes a pentium processor, hard disk, DAT-tape or optical disk storage, laser printer, high resolution colour monitor (Courtesy: M/s Cadwell, USA)*

measurements and a keyboard for access to convenient and important test controls. The system usually incorporates facilities for recording of the EMG and evoked potentials. The stimulators are software controlled. For report generation in the hard copy form, popular laser printers can be used.

Preamplifier: The preamplifiers used for EMG are generally of differential type with a good bandwidth. The input impedance of the amplifier must be greater than 2×50 MΩ. Present day electronic devices easily provide input impedances of the order of 10^{12} ohms in parallel with 5 picofarads. It is preferable to mount the preamplifiers very near the subject using very small electrode leads, in order to avoid the undesirable effects of stray capacitance between connecting cables and the earth. Also, any movement of the cable from the output of the electrode will not generate significant noise signals in the cable, which feeds into the subsequent amplifier. The preamplifier provides an output with low impedance and, therefore, the high frequencies do not get attenuated even if long cables are used to connect the preamplifier and the rest of the machine. The common-mode rejection should be greater than 90 dB up to 5 kHz. A calibrating square wave signal of 100 μV (peak-to-peak) at a frequency of 100 Hz is usually available. The main amplifier has controls for gain adjustment from 5 mV/div to 10 mV/div for selecting the sensitivity most appropriate to the incoming signal from the patient.

Basmajian and Hudson (1974) suggested the use of a preamplifier to amplify the EMG signals picked up by needle electrodes at the electrode site before transmitting them along wires. The effect of electrical interference is substantially reduced and the microphonic artefacts generated in the wires due to movement of the subject are virtually eliminated. When surface EMGs are to be measured, it is convenient to combine the electrode pair and a differential amplifier within a single module. Johnson *et al* (1977) designed a miniature amplifier circuit fully encapsulated in epoxy resin with two small silver electrodes of 6 mm diameter, exposed flush with the base of the module. The electrode is attached to the skin using adhesive tape. Figure 5.17 shows a circuit diagram of

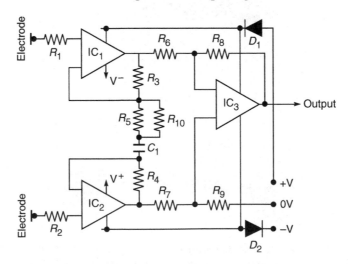

> **Fig. 5.17** *Pre-amplifier circuit for an EMG machine (redrawn after Johnson et al, 1977; by permission of Med. and Bio. Eng. and Comp.)*

the preamplifier. The amplifier design provides for a flat frequency response between 10 Hz and 1 kHz, with a CMRR of 100 dB at the mains frequency. The noise level was found to be 2 mV rms and the input impedance greater than 10 MΩ.

The two ICs in the input stage act as voltage followers, which present the desired high input impedance to the electrodes. They are coupled via C_1 and R_5 to provide a high differential signal gain. Capacitor C_1 determines the low frequency performance of the circuit. It also eliminates the effects, at the output, of any dc offsets due to IC_1, and IC_2 or any imbalance in electrode potentials. The second stage IC_3 provides further differential signal gain, while rejecting common-mode signals. The overall gain of the amplifier is 1000.

Input impedance of the amplifier must be higher by several orders than the electrode impedance. Also, selection of the electrode type without the knowledge of the amplifier's input resistance results in distorted records and considerable error. The larger the surface of the electrode, the less input resistance is allowable. For example, a needle electrode with a surface of 15,000 μm^2 may need an amplifier with input impedance of 5 MΩ, while a needle electrode with a surface of 500 μm^2 will ensure a record with acceptable distortion by means of an amplifier with minimum input impedance of 100 MΩ.

The capacitance present parallel with the input resistance of the amplifier reduces the frequency response of the amplifier as well as lowers the common-mode rejection at higher frequencies. Owing to these, the electrode cable, the extension cable and the input stage of the amplifier require careful designing. Generally, shielded cables are used which reduce the disturbing signals but at the same time, the parasite capacitance will increase. By careful design, a capacitance value of 50 pF or less can be achieved for the input stage. McRobbie (1990) illustrate a rapid recovery EMG preamplifier without AC coupling capacitors.

To ensure patient safety, the subject should be electrically isolated from any electrical connection to the power line or ground. This isolation is achieved either through the use of optical isolators or through the use of isolation transformers.

Low Frequency and High Frequency Filters: These are used to select the passband of the incoming signal and to modify the progressive reduction in voltage output which occurs at either end of the frequency spectrum roll-off. The low frequency 3 dB point may be selected over the range of 0.016 to 32 Hz while the high frequency 6 dB point can be selected over the range 16 Hz to 32 kHz. Thus, the passband may be varied over a very wide range but is normally made as narrow as possible, subject to the requirements of the particular application in order to restrict displayed noise.

Signal Delay and Trigger Unit: Sometimes, it is necessary to examine the signals from individual fibres of muscle tissue. For this purpose, special needles are available with a 25 micron diameter electrode surface and up to 14 pick-up surfaces down the side of one needle. These 14 points are scanned sequentially to determine which point is acquiring the largest signal. This point is then considered as the reference and its signal is used to trigger the sweep. Signals from the remaining 13 points are then scanned sequentially and recorded with respect to the reference signal. To examine these signals, it is necessary to trigger the sweep from the signal and to delay the signal so that the whole of its leading edge is displayed. The delay is achieved by passing the digitized signal through the shift register or random access memory into the recirculate mode to obtain a non-fade display of a transient phenomenon.

EMG machines have a provision for the selection of sweep speeds from 0.05–500 ms per division on the CRT.

Integrator: The integrator is used for quantifying the activity of a muscle. Lippold (1952) established that a linear relationship exists between the integrated EMG and the tension produced by a muscle. The integrator operates by rectifying an incoming EMG signal, i.e. by converting all negative potentials to identical positive deflections so that the EMG pattern consists of positive deflections only. The area under the rectified potentials is accumulated using a low-pass filter so that the module output, at any time, represents the total area summed from a selected starting time. The integrator indicates the EMG activity either as a variable frequency saw tooth waveform or as a steady deflection. In the former case, the output curve is a measure of the total electrical activity per second, recorded from a muscle during voluntary contraction within the analysis time. The slope of this curve, measured as the number of resets per second, can be used to detect changes in the number of motor units firing over a period of time. The steady deflection or mean voltage mode is used in plotting mean voltage v. isometric tension curves of muscle interference patterns during voluntary contraction, to show changes in muscle activity due to neuro-muscular disease such as muscular dystrophy, poliomyelitis, etc. Different time constants determine the amount of smoothing applied to the output signal. When rapid changes in the signal have to be followed, the shortest time constant provides maximum smoothing of the signal and the most easily read mean value.

Stimulators: The stimulators incorporated in the EMG machines are used for providing a single or double pulse or a train of pulses. Stimulus amplitude, duration, repetition and delay are all adjustable and facilities are provided for external triggering. The output is either of the constant voltage type or of the constant current type. The constant voltage type stimulator provides square wave pulses with amplitudes in the range of 0–500 V, a pulse duration of 0.1–3 ms and frequency between 0-100 Hz. Output of the constant current generator can be adjusted between 0 to 100 mA.

Usually, the electromyographic changes in an advanced diseased state are readily recognized on an oscilloscope display and by the sound from a loudspeaker. However, since the loss of muscle fibres, and therefore, the action potential changes are relatively small in early or mild disease states, changes in the EMG signals may be obscured by the usual variability of action potentials. Quantitative analysis is thus necessary to determine when the waveforms have changed beyond the normal range. Quantities measured for such analysis include zero-crossing rate, peak rate, negative wave duration and wave rise time. These time-domain techniques are somewhat different from the classical frequency spectrum and correlation function methods, but are much simpler to implement with electronic techniques. Fusfeld (1978) details out circuits used to implement quantitative analysis of the electromyogram.

5.6 OTHER BIOMEDICAL RECORDERS

5.6.1 Apexcardiograph

An apexcardiograph records the chest-wall movements over the apex of the heart. These movements are in the form of vibrations having a frequency range of 0.1 to about 20 Hz. The transducer required for recording these movements is similar to that employed for a phonocarddiagraph (PCG) but which has a frequency response much below the audio range. It can be an air-coupled microphone or a contact microphone. The apexcardiograph has limited applications. It is, however, useful in the diagnosis of the enlargement of the heart chambers and some type of valvular disorders.

5.6.2 Ballistocardiograph (BCG)

A ballistocardiograph is a machine that records the movement imparted to the body with each beat of the heart cycle. These movements occur during the ventricular contraction of the heart muscle when the blood is ejected with sufficient force.

In BCG, the patient is made to lie on a table top which is spring suspended or otherwise mounted to respond to very slight movements along the head axis. Sensing devices are mounted on the table to convert these movements into corresponding electrical signals. The sensors usually are piezo-electric crystals, resistive elements or permanent magnets, moving with respect to fixed coils. In all such cases, the output of the sensor is amplified and fed to an oscilloscope or to a chart recorder. BCG has so far been used mainly for research purpose only. It is rarely used in routine clinical applications.

5.6.3 Electro-oculograph (EOG)

Electro-oculography is the recording of the bio-potentials generated by the movement of the eye ball. The EOG potentials are picked up by small surface electrodes placed on the skin near the eye. One pair of electrodes is placed above and below the eye to pick up voltages corresponding to vertical movements of the eye ball. Another pair of electrodes is positioned to the left and right of the eye to measure horizontal movement. The recording pen is centred on the recording paper, corresponding to the voltage changes accompanying it. EOG has applications mostly for research and is not widely used for clinical purposes.

5.6.4 Electroretinograph (ERG)

It is found that an electrical potential exists between the cornea and the back of the eye. This potential changes when the eye is illuminated. The process of recording the change in potential when light falls on the eye is called electroretinography. ERG potentials can be recorded with a pair of electrodes. One of the electrodes is mounted on a contact lens and is in direct contact with the cornea. The other electrode is placed on the skin adjacent to the outer corner of the eye. A reference electrode may be placed on the forehead. A general purpose direct writing recorder may be used for recording electroretinograms.

The magnitude of the ERG voltage depends upon the intensity and duration of the light falling on the eye. It may be typically about 500 μV.

ⅢⅢ▷ 5.7 BIOFEEDBACK INSTRUMENTATION

Feedback is a common engineering term and refers to its function to control a process. When this concept is applied to biological processes within the body, it is known as biofeedback. Here again, biofeedback is a means for gaining control of the body processes to create a specially required psychological state so as to increase relaxation, relieve pain and develop healthier and more comfortable life patterns. The technique involves the measurement of a variable produced by the body process and compares it with a reference value. Based on the difference between the measured and reference value, action is taken to bring the variable to the reference value.

It may be noted that biofeedback is not a treatment. Rather, biofeedback training is an educational process for learning specialized mind/body skills. Through practice, one learns to recognize physiological responses and to control them rather than having them control us. The objective of biofeedback training is to gain self-regulatory skills which help to adjust the activity in various systems to optimal levels.

Many different physiological processes have been evaluated for possible control by biofeedback methods. However, the following four neural functions are commonly employed:

- Emotions or Electrodermal Activity (Galvanic skin response measurements)
- Muscle tension or EMG (Electromyograph measurements)
- Temperature/sympathetic pattern (Thermistor readings)
- Pulse (Heart rate monitoring)

Electrodermal activity is measured in two ways: BSR (basal skin response) and GSR (galvanic skin response) is a measure of the average activity of the sweat glands and is a measure of the phasic activity (the high and low points) of these glands. BSR gives the baseline value of the skin resistance where as GSR is due to the activity of the sweat glands. The GSR is measured most conveniently at the palms of the hand, where the body has the highest concentration of sweat glands. The measurement is made using a dc current source. Silver-silver electrodes are used to measure and record the BSR and GSR. Figure 5.18 shows the arrangement for measuring these parameters. The BSR output is connected to an RC network with a time constant of 3 to 5 seconds which enables the measurement of GSR as a change of the skin resistance.

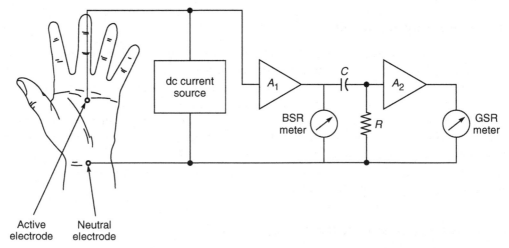

> **Fig. 5.18** *Block diagram for measurement and record of Basal Skin Resistance (BSR) and Galvanic Skin Response (GSR)*

Biofeedback instrumentation for the measurement of EMG, temperature and pulse/heart rate is not different from other instruments used for the measurement of physiological variables. Transducers and amplifiers are employed to measure the variable that is to be controlled by the feedback process. The magnitude of the measured variable or changes in the magnitude are converted into a suitable visual or auditory stimulus that is presented to the subject. Based on the stimulus, the subject learns to control the abnormal conditions. Reports have appeared in literature regarding applications of biofeedback to control migraine headaches, to slow down heart rate, etc. Biofeedback techniques have been greatly refined and computerized biofeedback training and psychological computer-assisted guidance programs in the privacy of one's home are now a reality.

Patient Monitoring Systems

▪▪▶ 6.1 SYSTEM CONCEPTS

The objective of patient monitoring is to have a quantitative assessment of the important physiological variables of the patients during critical periods of their biological functions. For diagnostic and research purposes, it is necessary to know their actual value or trend of change. Patient monitoring systems are used for measuring continuously or at regular intervals, automatically, the values of the patient's important physiological parameters. There are several categories of patients who may need continuous monitoring or intensive care. Critically ill patients recovering from surgery, heart attack or serious illness, are often placed in special units, generally known as *intensive care units*, where their vital signs can be watched constantly by the use of electronic instruments. The long-term objective of patient monitoring is generally to decrease mortality and morbidity by: (i) organizing and displaying information in a form meaningful for improved patient care, (ii) correlating multiple parameters for clear demonstration of clinical problems, (iii) processing the data to set alarms on the development of abnormal conditions, (iv) providing information, based on automated data, regarding therapy and (v) ensuring better care with fewer staff members.

During a surgical operation, the patient is deprived of several natural reaction mechanisms, which normally restore abnormalities in his physical condition or alert other people. Indications or alarms that cannot be given by the patient himself can be presented by patient monitoring equipment. Besides this, in special cases, it is not uncommon for surgical procedures to last for several hours. During these lengthy operative procedures, it is difficult for the anaesthesiologist and the surgeon to maintain intimate contact with the patient's vital signs and at the same time attend to anaesthesia, surgery, fluid therapy and many other details that are required under such circumstances. Also, when a patient is connected to a life-support apparatus, e.g. heart-lung machine or ventilator, correct functioning of these has to be monitored as well. A patient monitoring system thus better informs the surgeon and the anaesthesiologist of the patient's condition. With patient monitoring systems, the risk that surgery involves has been considerably reduced since it is possible to detect the complications before they prove dangerous as suitable remedial measures can be taken well in time.

The choice of proper parameters, which have a high information content, is an important issue in patient monitoring. It is, however, generally agreed that monitoring of the following biological functions is often needed. Electrocardiogram (ECG), heart rate (instantaneous or average), pulse rate, blood pressure (indirect arterial blood pressure, direct arterial blood pressure or venous blood pressure), body temperature and respiratory rate. In addition to these primary parameters, electroencephalogram (EEG), oxygen tension (pO_2) and respiratory volume also become part of monitoring in special cases. In addition to these, equipment such as defibrillators and cardiac pacemakers are routinely needed in the intensive care wards.

The general requirements for patient monitoring equipment have not changed much over the past few decades. However, today's equipment monitors more parameters and processes more information. Trends in monitoring include software control, arrhythmia monitoring, haemodynamics monitoring, monitoring during transportation of the patient and increased user friendliness. With more than 10 parameters to be monitored and scores of calculations to be made, the requirement for an easy-to-use user interface has assumed great significance.

Monitoring is generally carried out at the bedside, central station and bedside with a central display. The choice amongst these is dependent upon medical requirements, available space and cost considerations.

6.2 CARDIAC MONITOR

The most important physiological parameters monitored in the intensive care unit are the heart rate and the morphology or shape of the electrical waveform produced by the heart. This is done to observe the presence of arrhythmias or to detect changes in the heart rate that might be indicative of a serious condition. Thus, a cardiac monitor is specifically useful for monitoring patients with cardiac problems and the special areas in the hospitals where they are generally used are known as cardiac care units or coronary care units (CCU). These instruments are also called 'Cardioscopes' and comprise of:

- Disposable type pregelled electrodes to pick up the ECG signal.
- Amplifier and a cathode ray tube (CRT) for the amplification and display the ECG which enable direct observation of the ECG waveform.
- A heart rate meter to indicate average heart rate with audible beep or flashing light or both with each beat.
- An alarm system to produce signal in the event of abnormalities occurring in the heart rate.

The cardioscope is basically similar to the conventional oscilloscope used for the display of waveforms in electronic laboratories. They have the usual circuit blocks like vertical and horizontal amplifiers, the time base and the EHT (extra high tension) for the cathode ray tube. However, they differ in two important aspects as compared to the conventional instrument. These are slower sweep speeds and a long persistence screen. The slow sweep is an outcome of the low frequency character of the ECG signal. The slow sweep speed necessitates the use of a long persistence screen so as to enable a convenient observation of the waveform. Without a long persistence screen, one can only see a moving dot of light instead of a continuous trace. Typically for a 13-cm screen, total sweep time is usually kept as 2.5 or 5 s. In this way, one can observe at least four heart beats in a single sweep period.

Most of the present day cardioscopes are designed to be used at the bedside. Some of them are even portable and can work on storage batteries. A large screen with about 50 cm screen size instruments are usually mounted in one corner of the operating room at a height at which it is possible to conveniently observe the waveforms being displayed. Small cardioscopes using 3″ diameter cathode ray tubes are mounted on anaesthesia trollies. These are called "Anaesthesia monitors". These monitors are use by the anaesthetist for continuous monitoring of the ECG of anaesthetized patients.

Low frequency waveforms, which are available from various physiological parameters do not show up well on a conventional oscilloscope CRT screen because the scan is so low that most of the wave-track is dark during the scan. This is particularly true in large screen displays. Non-fade monitors using digital memories have been developed to overcome the problem of the fading of slow scanning CRT displays. By this technique, it is possible to generate a rolling waveform display that produces an effect comparable to a pen write-out. The display is thus continuous, bright and flicker-free on a normal non-storage CRT. If required, the image can be held indefinitely, selectively erased, allowed to roll across the screen or made to simulate a normal non-memory CRT according to the sequence of operations in the read-out circuit. These type of displays are especially suitable for patient monitoring applications because the stored information can be readily transmitted to a remote viewing station or a chart recorder without losing track of incoming signals.

With conventional oscilloscopes, there is an unavoidable break in the chain of information between the end of sweep and the beginning of the next, because of sweep circuit retrace and hold-off requirements. Furthermore, this lost display usually extends while the oscilloscope waits for the next trigger. If the train of events being observed is sine wave, the loss of information during the hold-off period may be of no consequence. But if, for example, we are looking for arrhythmias in a tape recorded ECG display with a conventional oscilloscope, a complete, QRS complex could easily be missed. By using non-fading type roll-mode display, waveforms can be paraded across the CRT screen in a continuous stream with no disruptions of any kind. The viewing window depends upon the sweep rate. Figure 6.1 shows the same waveform train displayed by a conventional oscilloscope and by the memory monitor in the roll-mode. Moreover, the ability to 'freeze' the display in the roll mode at any point makes it easy to capture important data, which can be outputted to a chart recorder.

Figure 6.2 shows the basic system for incorporating digital storage in the oscilloscopic displays. Essentially, the system carries out high-speed real-time sampling of an incoming analog waveform, followed by the digital measurement of each successive sample and the subsequent storage of the stream of data. Once the data is stored in the digital form, it can be recalled for conversion back to analog form or for other processing operations. This 'replay' process can be continuous and its speed can be chosen to provide a non-fading, flicker-free trace on the CRT, irrespective of the speed of the original recording, or to provide a low-speed output to drive a conventional chart recorder.

Any digital method of waveform recording will have an analog–digital converter, which feeds data corresponding to the input signal into a digital store in a controlled 'write' cycle. The data is retrieved via a similar controlled 'read' cycle and is reformed via a digital/analog converter for display. As the regenerated signal is based on a finite number of measurements of the input signal, it is inevitably degraded as compared to the original. Two important factors governing the final

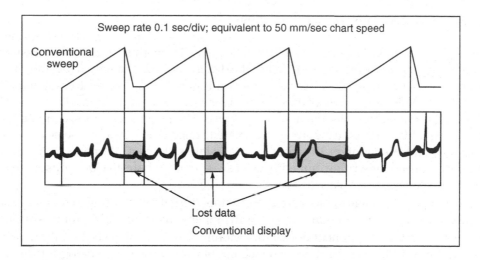

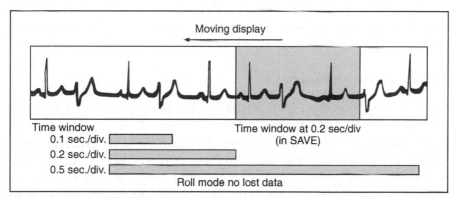

> **Fig. 6.1** *This figure illustrates, how the roll mode in digital storage can capture signals without any loss of data. The time window is selectable (Courtesy: Tektronix, USA)*

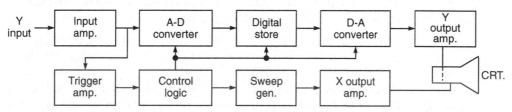

> **Fig. 6.2** *Block schematic of an oscilloscope display system incorporating digital storage*

resolution are the sample rate and word length. The former must be high enough to provide sufficient resolution on the time axis, while the latter depends on the number of bits provided by the analog/digital converter or store which determines the number of levels between zero and full scale on the vertical axis (Y-axis). In actual operation, the selected trigger signal initiates a scan and writes in sequence into each address of the 1024 word length store. The writing sequence depends upon the instructions from the 10-stage write address counter, which is controlled by the time base speed control. The write-cycle control logic is designed to update or refresh the store whenever a trigger pulse is received and hence the display follows changes in the input waveform as they occur. A separate read address counter continuously scans all addresses at a fixed rate and drives the Y-deflection system of the CRT via a digital/analog converter. At the same time, the time base ramp generator is initiated, with the required speed, by the address counter at the start of the scan. The address input to the store is alternated, if necessary, by a data selector between the write and read address counters. Read out generally takes place on alternate clock pulses which are otherwise unused for writing; but with a fast sweep rate, where all clock pulses are required while writing, read out takes place between successive writing scans.

Two basic types of storage devices are used to store digital information in memory monitors: shift registers and random access memories. Both of them are equally good for this application.

The other important component of memory monitors is the analog-to-digital converter. Any book on digital techniques would describe techniques for converting analog information into digital form. Out of the methods available for A to D conversion, the counter or dual ramp methods are very effective for slow conversion rates. For higher conversion rates, the successive approximation, tracking, parallel or flash techniques are preferred.

6.2.1 Selection of System Parameters

In a digital memory display system, operating parameters should be carefully selected to maintain signal accuracy, bandwidth and fidelity. The following are important parameters:

Sampling Rate: Signals with high bandwidth require sampling to be carried out at a high rate to faithfully obtain all its features. A high sampling rate, however, necessitates a large memory to store all the data. Therefore, a compromise must be worked out. For ECG, AHA (American Heart Association) recommends a sampling rate of 500 samples per second for computer analysis. However, for routine monitoring purposes, such a high sampling rate is not needed. A sampling rate, which is three times the highest frequency component to be displayed, has been found to be generally adequate. For monitoring purposes the bandwidth is usually limited to 50 Hz. Therefore, a sampling rate of 150 or above will be satisfactory. Most of the single channel memory display systems have a sampling rate of 250 samples/s and multi-channel displays work on 180 samples/s.

Word Length: The ECG signals, before they are coded in the digital form, are amplified in a preamplifier and brought to a level of 0–1 V. The accuracy of conversion of analog signals depends on the number of bits used in the conversion. The greater the number of bits per word, the greater will be the resolution, as a greater number of levels will be available to accurately define the value of the sampled signal. In practice, however, the ultimate resolution of a given design is limited by the noise in the various analog and switching circuits and by the linearity and monotonicity of the converter. Usually, an 8-bit word is chosen, which defines 28 or 256 different words or levels that can be resolved.

Memory Capacity: The contents of each channel memory are displayed with each sweep across the CRT screen. Therefore, the stored signal must roll on the screen with a time interval, which must be convenient for viewing. For example, a display of 5 s of data (on a 13 cm CRT screen), with he spot moving at a rate of 25 mm/s and at a sampling rate of 200 samples/s, would require a 1000 word memory. Increasing either the sampling rate or the display time interval will increase the memory size proportionately.

6.2.2 Cardiac Monitor Using Digital Memory

Modern ECG monitors not only include the non-fade display facility but also display heart rate along with the ECG trace. Figure 6.3 shows a block diagram of a single channel cardioscope with digital memory. The ECG signal is sensed differentially by the RA (right arm) and LA (left arm) electrodes and is amplified by an isolation ECG amplifier. The patient circuit is isolated by using a transformer and by modulating the 102 kHz carrier signal with amplified and filtered ECG. The modulated carrier is demodulated, amplified and applied to an analog-to-digital converter. The converter samples the waveform at a rate of 250 samples/s, converts each sample to an 8-bit parallel word, and enters the word into the recirculating memory where it replaces the oldest word stored.

The recirculating memory, usually employed, consists of eight 1024-stage shift registers operating at a clock rate of 250 kHz. The output of each shift register is fed back to its input, so that the contents of the shift registers recirculate continuously. Hence, a waveform acquired at a rate of 250 samples/s is available at a rate of 250,000 samples/s. The samples are reconverted into an analog signal for presentation on the CRT display. The sweep time is so arranged that it matches the time to read out 1024 samples. The most recent four seconds of the original ECG waveform are traced in four milliseconds. The relatively fast repetition rate of the stored information causes the displayed waveform to appear bright with no fading.

At any time, the freeze control can be activated to obtain a fixed display on the CRT screen. In this position, feeding of new signals into the memory is discontinued and all values in the memory then return to the same positions after each recirculation through the memory. The display in the frozen position helps to observe a particular event more conveniently. For the display of heart rate on the CRT screen, the original ECG waveform is shaped and supplied to the tachometer circuit to compute and display the heart rate.

Adjustable heart rate alarm limits are indicated visually and in the audible form. Operation of these limits is carried out by using two comparators to sense when the heart rate goes beyond set limits.

Both types of CRTs, viz. electrostatic deflection type or electromagnetic deflection type, are used to construct non-fade display monitors. However, it is more convenient to use tubes with electromagnat deflection because of their small size. The deflection coils of CRTs having electromagnetic deflection have to be specially designed to give a frequency response of about 50 kHz in the vertical section and a low frequency response in the horizontal section for slow sweep operations.

As compared to the directed beam display, it will be preferable to use the raster scan deflection technique because of cost, brightness and power considerations, which make this technique highly desirable for electromagnetic deflection type tubes. However, the normal scanning rates would result in a display of waveforms with widely spaced dots. The separation of these dots proves unacceptable to physicians and nurses who are used to studying smooth waveforms. This problem

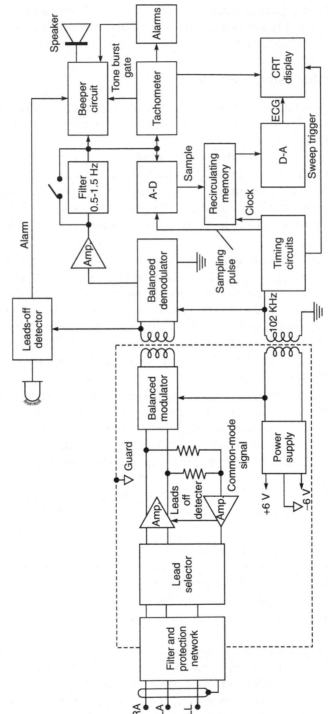

▶ **Fig. 6.3** *Block diagram of cardioscope using digital memory (redrawn after Grobstein and Gatzke, 1977; by permission of Hewlett Packard, USA)*

can be solved by using a high raster frequency (70 kHz instead of the usual 15 kHz) combined with beam width shaping on individual lines.

There is at present an intense competition between the long-established cathode ray tube (CRT) and the newly emerging flat panel technologies for the display of alphanumeric and graphic information. Flat panels offer advantages of wide-angle visibility and brightness, resistance to physical shock and immunity to electromagnetic disturbances. Compared to CRTs, flat panels are thin, save space and have less weight and bulk. They employ line-addressed, dot matrics imaging, rather than the complex, pixel-by-pixel raster pattern of CRTs. They are in production in formats ranging from 278×128 pixels to 1024×864 pixels, with screen diagonals up to 19 inches. Flat panels have found particular favour in medical instruments for high information rate applications, such as patient monitoring systems and ventilators.

The three most common flat panel technologies are: liquid crystal displays, plasma displays and electro-luminescent displays.

It will be of interest to diagramatically illustrate some of the special features of cardiac monitors, particularly in respect of the frequency response, input circuitry and other special features, which distinguish them from electrocardiographs.

Frequency Response of Cardioscopes: Some monitors have two selectable frequency response modes, namely Monitor and Diagnostic. In the 'Monitor' mode or 'Filter-in' mode, both the low and high frequency components of an electrocardiogram are attenuated. It is used to reduce baseline wander and high frequency noise. The Monitor mode bandwidth is generally 0.4 to 50 Hz (3 dB points). In the 'Diagnostic' mode, the instrument offers expanded bandwidth capability of 0.05 to 100 Hz. Some instruments include a 50 Hz notch filter to improve the common mode rejection ratio and this factor should be kept in mind while checking the frequency response of the instrument.

Input Circuit: Figure 6.4 shows a complete input circuit used in present day cardiac monitors. There are three prominent circuit blocks: (i) low-pass filter circuit, to suppress RF interference,

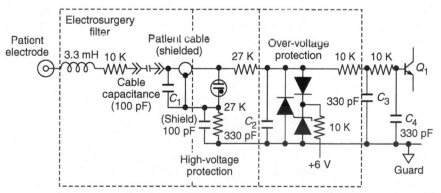

> **Fig. 6.4** *Details of the input network in a cardiac monitor. The features include low pass filter to suppress RF interference and voltage clamps to prevent defibrillator pulses from damaging the sensitive input amplifier (redrawn after Grobstein and Gatzke, 1977)*

(ii) high voltage protection circuit, similar to electrocardiographs, to provide voltage clamp in the presence of defibrillator pulses, and (iii) over voltage protection circuit.

Electrosurgery Interference: RF interference occurs due to any one of two possible modes. The first and usually the most severe is due to conduction, i.e. the RF energy is actually carried via the patient into the monitor. The second is radiation, by which the RF energy is transmitted through the air and is induced into the circuits of the monitoring instrument, its leads and cables.

Electrosurgery machines generate RF signals within a range of 0.4 to 5 MHz with peak-to-peak amplitudes of 100 to 1000 V, pulse modulated at rates from 1.5 to 25 kHz for coagulating or 120 Hz for cutting. Cardiac monitors are often used in operation theatres, where the RF is applied through a pointed scalpel at the point of incision and the return path for the current is through a wide area electrode on the opposite side of the patient's body. The ECG signal, on the other hand, is of the order of 1 mV with frequency components below 100 Hz. The ECG input amplifier, the ECG electrode-skin interface and the scalpel-tissue interface are the main sites where rectification occurs. Moreover, the common mode RF signal gets converted into a normal mode signal by an imbalance in the capacitance between input to ground at the differential amplifier input.

In order to reduce interference caused by electrosurgery and radiofrequency emissions, it is very essential to use filters at the input of leads of the monitoring instruments. Grobstein and Gatzke (1977) give the design criteria for constructing such a filter. To reduce the interference to acceptable levels, the ratio of amplifier sensitivities to the ECG signal (1 mV at less than 100 Hz) and to the electrosurgery machine (100 V at 1 MHz or so), should be at least 10^5. If we use a filter having a cut off frequency above 10 kHz, a three pole filter is needed to achieve this level of reduction. Five poles are used in the monitors, three provided by the RC filters within the instrument and two provided by the electrosurgery filters and the conductor-to-shield capacitance of the cable (usually 1000 pF). The equivalent network for the ECG machine input cable is as shown in Fig. 6.5. The imbalance in capacitance from each input of the differential amplifier to the ground also injects a differential RF signal that can be rectified and can produce interference. The imbalance necessary to provide a 100 µV signal is surprisingly very small, of the order of 0.001 pF at 1 MHz. To minimize this problem, a guard is used, which besides shielding the input circuits from electromagnetic radiation also helps in equalizing the capacitance from input to ground of each amplifier input.

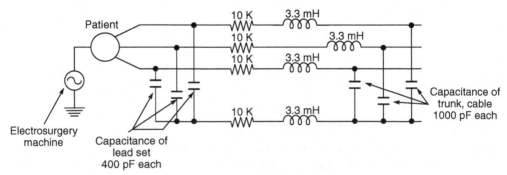

> **Fig. 6.5** *Equivalent network for the ECG machine input cable (after Grobstein and Gatzke, 1977)*

The following precautions should be taken to achieve good ECG display in the presence of electrosurgery interference (Benders, 1976):

- The electrosurgery return plate should be directly under the surgical site, as far as possible.
- The ECG electrodes should be placed at the maximum possible distance away from the surgical site.
- The electrodes should be equidistant from the surgical site.
- All monitoring electrodes must be placed either on the frontal surface or on the posterior surface.
- Only shielded ECG patient cables and electrode leads must be used.

Leads Off Detector: The "leads off' detector circuit usually works on the principle that loss of body contact of either the RA or LA electrode causes a rather high impedance change at the electrode/ body contact surface, consequently causing a loss of bias at the appropriate amplifier input. This sudden change makes the amplifier to saturate, producing maximum amplitude waveform. This waveform is rectified and applied to a comparator that switches on an alarm circuit (leads off) when the waveform exceeds a certain amplitude.

Quick Recovery Circuit: In order to avoid problems due to dc drift and supply voltage variations, ECG amplifiers are usually constructed with at least one stage of ac coupling. The circuit employs a high-value series capacitor in between two amplifying stages, a high value is used to maintain a good low frequency response. The presence of this capacitor, however, creates the problem of a very long recovery time after an over voltage appears at the input of the amplifier due to conditions like leads off, excessive patient movement, defibrillator operation, etc. This problem is overcome by using a quick recovery circuit. This circuit basically provides a fast discharge of the coupling capacitor once it has been excessively charged. Figure 6.6 shows the working principle of this circuit, in which the over range voltage for A_2 is determined by the voltage dividers (R_1, R_2, R_3, R_4, respectively)—the voltage may be plus or minus. In either case, the plus or minus over range input to the differential amplifier A_3 causes the output to go positive. This turns 'on' the FET Q, which rapidly discharges C, causing the output baseline to return to zero level.

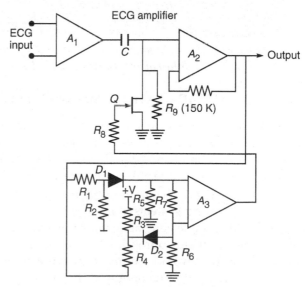

> **Fig. 6.6** *Principle of operation of quick recovery circuit*

Cardiac monitors are also available which display more than one channel of information such as ECG and delayed ECG with digital heart rate display and alarm facility. The display is for 4 second. The two channels can also be cascaded to have a

8 second display (Fig. 6.7). They use time-multiplexing of complete sweeps to avoid the excessive band-width introduced by electronic switching or the chopper technique. Each channel, in these instruments, can be independently controlled for updating or freezing the information.

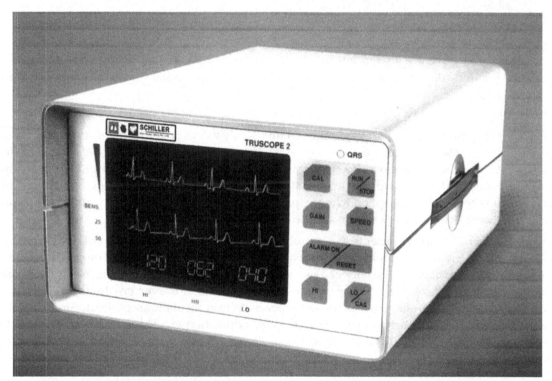

> **Fig. 6.7** *Two channel digital storage ECG waveform display system, with digital display of heart rate (Courtesty: M/s Schiller Health Care India Pvt. Ltd.)*

6.3   BEDSIDE PATIENT MONITORING SYSTEMS

Bedside monitors are available in a variety of configurations from different manufacturers. They are designed to monitor different parameters but the common feature amongst all is the facility to continuously monitor and provide non-fade display of ECG waveform and heart rate. Some instruments also include pulse, pressure, temperature and respiration rate monitoring facilities.

The advent of microcomputers has marked the beginning of a fundamentally new direction in patient monitoring systems. Such systems are intended to replace the traditional monitoring devices with a single general purpose unit capable of recognizing the nature of the signal source and processing them appropriately. The hardware responsible for physiological signal analysis, information display and user interaction is actually a set of firmware modules implemented in terms of a microcomputer program. The firmware gives the system its functional personality and the usual switches, knobs, dials and meters can be replaced by a touch-sensitive character display.

A typical example of a microprocessor-based bedside patient monitoring instrument is shown in Fig. 6.8. The system is designed to display an electrocardiogram, heart rate with high and low

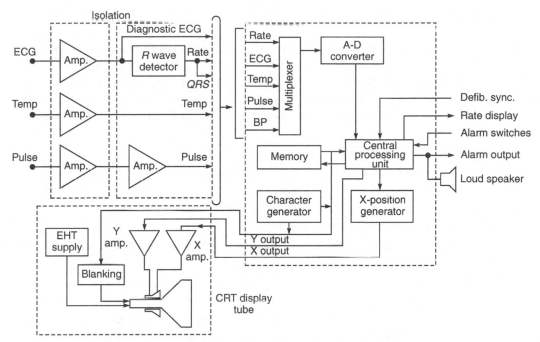

> **Fig. 6.8**   *Block diagram of the bedside patient monitor (Courtesy: Albury Instruments, U.K.)*

alarms, pulse rate, dynamic pressure or other waveforms received from external preamplifiers. It also gives immediate and historical data on the patient for trend information on heart rate, temperature, and systolic and diastolic blood pressures for periods up to eight hours. The system basically consists of three circuit blocks: **Preamplifier section, Logic boards** and **Display part**. The preamplifiers incorporate patient isolation circuits based on optical couplers. The ECG waveform has facilities for lead-off detection, 'pacer' detection and quick recovery circuit for overload signals. Various amplified signals are carried to a multiplexer and then to an analog-to-digital converter, included in the logic board. The central processing unit along with memory gives X and Y output for the CRT display. The character generator output is mixed with the Y output for numeric display on the CRT. The alarm settings, selection switches for different parameters and the defibrillator synchronization system communicate with the CPU. The alarm signals are also initiated under its control. The memory comprises 5 K bytes ROM and 3.25 K bytes RAM with 256 samples of ECG. Eight seconds delayed ECG is available for recording purposes.

Several important trends in the design and function of bedside monitors have emerged in the past few years. More bedside units are now software based, a feature that facilitates changes and updates in function by the simple replacement of computer memory chips. Wider use of on-board microprocessors also permits bedside monitors to perform increasingly sophisticated signal-processing tasks. Advances in monitoring the haemodynamic parameters are particularly noteworthy. New smart algorithms help to carry out automatic calculations of indices of cardiovascular functions and artefact removal tasks. The trend in ECG monitoring is towards display and analysis of data from multiple leads. Several manufacturers now include arrhythmia monitoring, including the monitoring of the ST segment of the ECG, as a standard feature in bedside monitors.

While increasingly sophisticated monitoring capabilities have been added to bedside monitors, many monitors today are much easier to use than their predecessors were. Improvements in software and features such as touch screen make today's bedside monitors a user-friendly equipment.

Patient monitors are also known as vital sign monitors as they are primarily designed to measure and display vital physiological parameters. Figure 6.9 shows the system block diagram of a patient monitoring system. It basically consists of the modular parts for measurement of the following:

- ECG and respiration measuring electronics
- Blood pressure (non-invasive) measuring electronics, pump and tubing
- Blood pressure (invasive)
- Temperature measuring electronics
- Pulse probe and SpO_2 (pulse oximetry)
- Microprocessor board including analog signal multiplexer, A–D converter and real time clock
- Video control board to convert the CPU commands into video signal
- Video display module
- Transformer and power supply board to generate necessary voltages
- Mother board including signal buses and analog input signal buffers
- Keyboard

The whole system works under the control of the processor 80C32, which works at 16 MHz. The processor boards for each of the above parameters are described in the subsequent sections. Figure 6.10 shows a typical vital signs monitor.

▷ 6.4   CENTRAL MONITORS

With central monitoring, the measured values are displayed and recorded at a central station. Usually, the signal conditioners are mounted at the bedside and the display and alarms, etc. are located in a central station.

The central station monitoring equipment may incorporate a multi-microprocessor architecture to display a flexible mixture of smooth waveforms, alphanumerics and graphics on a single cathode ray tube. This presents all the information at a glance and thus assists the hospital staff in several ways. First, it generates audible and visual alarms if preset vital sign limits are exceeded. It is important that the central station announces these emergencies without generating too many false alarms, arising due to patient movements, etc. Secondly, it displays the patient's vital sign data. By watching this data, the attending staff can detect problems before they reach the alarm stage. Trend plots of vital signs aid in guiding the patient's therapy. Thirdly, it provides a recording of the ECG and sometimes of other parameters, especially of the few seconds just before an alarm, which shows what kind of irregularity led to the alarm.

Central stations are primarily designed for coronary care patients to display ECG waveforms and heart-rate information for eight patients. The display shows (Fig. 6.11) four seconds of real

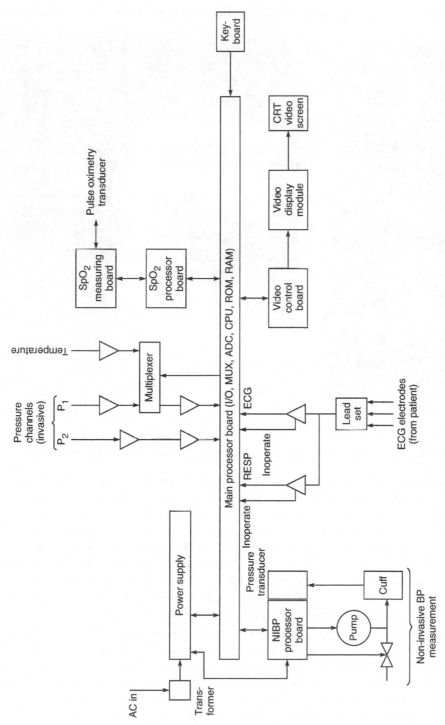

▶ **Fig. 6.9** *Block diagram of bed side monitor*

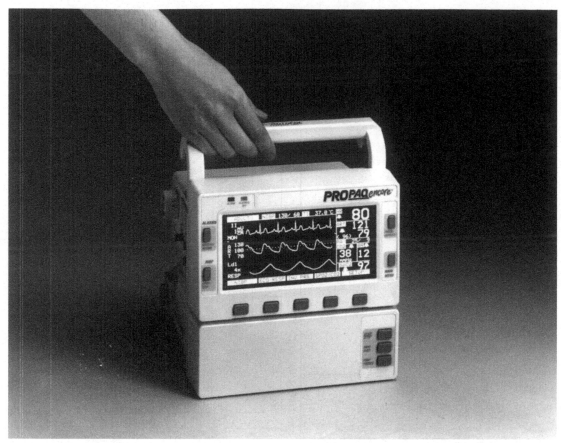

➤ **Fig. 6.10**   *Vital Signs Monitor (Courtesy: Protocol Systems Inc., USA)*

time ECG waveform and alarm information. A long trend (for either 9 or 24 h) or short trends (90 min) may be selected for display for observation and/or documentation. The information for the central monitor is collected from the bedside. Each bedside cable contains as many as fifteen analog signals representing physiological parameters, which may include several blood pressures, ECG, heart rate, respiration, end-tidal CO_2 and temperature. Status information such as alarm signals is also carried by the same cable. The 80 or so incoming physiological values are then sampled and digitized at appropriate rates by a 10-bit analog-to-digital converter. ECG waveforms are sampled every two milliseconds to maintain the 0–100 Hz bandwidth. Slowly varying variables such as temperature are sampled every four seconds.

The display part has two subsections-raster type display for waveforms and a conventional 300×260 picture-element bit map for alphanumerics and graphics. To make the waveforms look smooth, a 1200 line vertical raster is used. The display subsection also manages a 16 K-word memory, which is used as temporary storage for waveform and hard copy data. ECG waveform data for each patient is continually stored temporarily to provide a delayed (typically 8 s) waveform output for recording. The delay is used to capture a snapshot of any ECG abnormality

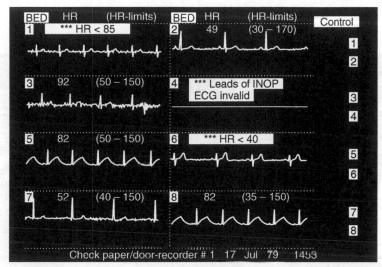

> **Fig. 6.11** *Typical displays on the patient information centre. The display depicts information on the heart rate, alarm limits, display of ECG waveforms from four to eight patients (Courtesy: Hewlett Packard, USA)*

leading to an alarm. If the recorder happens to be busy, the display processor stores it until a recorder becomes available.

The trend memory can hold patient data for 24 h. It contains 6 K words of CMOS RAM, which requires very little power in the standby mode and, therefore, can be connected to a battery back-up power supply. The patient data is thus held for at least 24 h after a power failure.

There has been an increasing realization with regard to assuring minimum breakdown of patient monitoring equipment. Microcomputer based self-test provision in the system helps the equipment to automatically test, analyze and diagnose itself for failures. When some parts fail, others take over their functions to minimize the impact of the failure on the system operation.

Alarm notifications are typically audible and/or visual. Audible alarms can be distinguished by varying the pitch, volume, duration and sequencing of the tones. Visual alarms can be indicated by varying the colour of the display on the monitor screen. Alarm conditions can also be captured in hard copy documentation via automatic generation of a recording at the time of the event (Slye, 1995). Traditional limit alarms need to be set up by the nursing staff, usually process each input independently and have unacceptably frequent false alarms. Dodd (1993) explains the use of neural network techniques to minimize the false alarm indications frequently encountered in a patient monitoring environment.

The increasing use of workstations in central monitoring installations have made it possible to monitor a large number of patients on a single monitor. Patient and equipment status are indicated by simple, colour graphic symbols. The ease of networking the monitors on the existing computer network wiring has resulted in flexible systems. The patient's bedside monitors can be viewed and operated from the central station. Due to the large storage capacity of the workstations, all the waveforms and numerics can be automatically captured, stored and retained to allow access to data for up to three days after the patient is discharged. Figure 6.12 shows the view of a modern central patient monitoring station.

➤ **Fig. 6.12**   *PC based central station (Courtesy: M/s Protocol Systems Inc. USA)*

▥▶ 6.5   MEASUREMENT OF HEART RATE

Heart rate is derived by the amplification of the ECG signal and by measuring either the average or instantaneous time intervals between two successive R peaks. Techniques used to calculate heart rate include:

- *Average calculation*   This is the oldest and most popular technique. An average rate (beats/min) is calculated by counting the number of pulses in a given time. The average method of calculation does not show changes in the time between beats and thus does not represent the true picture of the heart's response to exercise, stress and environment.

- *Beat-to-beat calculation*   This is done by measuring the time (T), in seconds, between two consecutive pulses, and converting this time into beats/min., using the formula beats/min. = $60/T$. This technique accurately represents the true picture of the heart rate.

- *Combination of beat-to-beat calculation with averaging*   This is based on a four or six beats average. The advantage of this technique over the averaging techniques is its similarity with the beat-to-beat monitoring system.

The normal heart rate measuring range is 0–250 beats/min. Limb or chest ECG electrodes are used as sensors.

6.5.1 Average Heart Rate Meters

The heart rate meters, which are a part of the patient monitoring systems, are usually of the average reading type. They work on the basis of converting each R wave of the ECG into a pulse of fixed amplitude and duration and then determining the average current from these pulses. They incorporate specially designed frequency to a voltage converter circuit to display the average heart rate in terms of beats per minute.

6.5.2 Instantaneous Heart Rate Meters

Instantaneous heart rate facilitates detection of arrhythmias and permits the timely observation of incipient cardiac emergencies. Calculation of heart rate from a patient's ECG is based upon the reliable detection of the QRS complex (Thakor, *et al* 1983). Most of the instruments are, however, quite sensitive to the muscle noise (artefact) generated by patient movement. This noise often causes a false high rate that may exceed the high rate alarm. A method to reduce false alarm is by using a QRS matched filter, as suggested by Hanna (1980). This filter is a fifteen sample finite-impulse-response-filter whose impulse response shape approximates the shape of a normal QRS complex. The filter, therefore, would have maximum absolute output when similarly shaped waveforms are input. The output from other parts of the ECG waveform, like a T wave, will produce reduced output.

Figure 6.13 is a block diagram of the scheme. The ECG is sampled every 2 ms. Fast transition and high amplitude components are attenuated by a slew rate limiter which reduces the amplitude of pacemaker artefacts and the probability of counting these artefacts as beats. Two adjacent 2 ms samples are averaged and the result is a train of 4 ms samples. In order to remove unnecessary high frequency components of the signal, a 30 Hz, infinite-impulse-response, Butterworth filter is employed. This produces 8 ms samples in the process. Any dc offset with the signal is removed by a 1.25 Hz high-pass filter. The clamped and filtered ECG waveform is finally passed through a

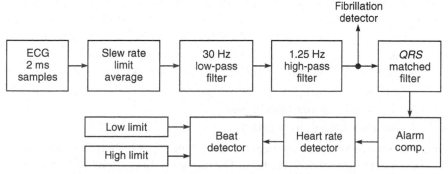

> **Fig. 6.13** *Block diagram of the cardiotachometer based on matched QRS filter (redrawn after Hanna, 1980; by permission of Hewlett Packard, USA)*

QRS matched filter. The beat detector recognizes *QRS* complexes in the processed ECG waveform value that has occurred since the last heart beat. If this value exceeds a threshold value, a heart beat is counted. The beat interval averaged over several beats is used to calculate the heart rate for display, alarm limit comparison, trending and recorder annotation. The threshold in this arrangement gets automatically adjusted depending upon the value of the *QRS* wave amplitude and the interval between the *QRS* complexes. Following each beat, an inhibitory period of 200 ms is introduced during which no heart beat is detected. This reduces the possibility of the T wave from getting counted. The inhibitory period is also kept varied as an inverse function of the high rate limit, with lower high rate limits giving longer inhibitory periods.

Based on the power spectra estimation of the *QRS* complex, Thakor *et al* (1984 b) have suggested that a bandpass filter with a centre frequency of 17 Hz and a *Q* of five, yields the best signal to noise ratio. Such a simple filter should be useful in the design of heart rate meters, arrhythmia monitors and implantable pacemakers.

The subject of reliable detection of *R*-wave continues to be of great interest for the researchers. Besides the hardware approach, a number of software based approaches have been reported in literature. Since the ultimate aim of detecting the R-wave is to automate the interpretation of ECG and detect arrhythmias, they are best covered in the succeeding chapter.

⑈▶ 6.6 MEASUREMENT OF PULSE RATE

Each time the heart muscle contracts, blood is ejected from the ventricles and a pulse of pressure is transmitted through the circulatory system. This pressure pulse when travelling through the vessels, causes vessel-wall displacement, which is measurable at various points of the peripheral circulatory system. The pulse can be felt by placing the finger tip over the radial artery in the wrist or some other location where an artery seems just below the skin. The timing and wave shape of the pressure pulse are diagnostically important as they provide valuable information.

The pulse pressure and waveform are indicators for blood pressure and flow. Instruments used to detect the arterial pulse and pulse pressure waveforms in the extremities are called plethysmographs. Most plethysmograph techniques respond to a change in the volume of blood as a measure of blood pressure.

The pulse gives a measure of pulse wave velocity and can be recorded and compared with the ECG signal (Fig. 6.14). The pulse wave travels at 5 to 15 m/s, depending on the size and rigidity of the arterial walls. The larger and more rigid the artery walls, the greater the velocity. The velocity is 10–15 times faster than blood flow, and is relatively independent of it.

The methods used for the detection of volume (pulse) changes due to blood flow are:

- Electrical impedance changes
- Strain gauge or microphone (mechanical)
- Optical changes (changes in density)

An electric impedance method measures the impedance change between two electrodes caused by the change in blood volume between them. The change in impedance (0.1 ohm) may be small as compared to the total impedance (several hundred ohms). The impedance is measured by applying an alternating current between electrodes attached to the body. An alternating signal (10–100 kHz) is used (rather than dc) in order to prevent polarization of the electrodes.

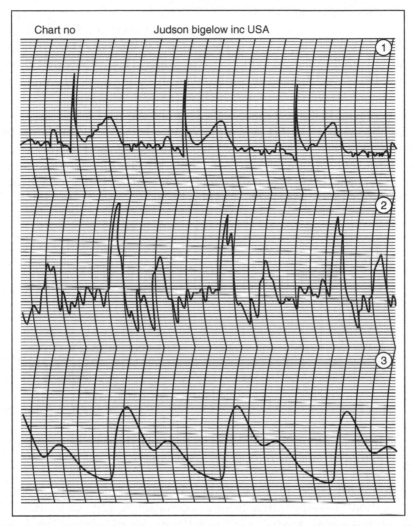

Chart no   Judson bigelow inc USA

> **Fig. 6.14** *Pulse pick up, showing time relationship with electrocardiogram (i) ECG (ii) crystal microphone pulse pick-up (iii) photoelectric pulse pick-up*

The mechanical method involves the use of a strain gauge connected to a rubber-band placed around a limb or finger. Expansion in the band due to change in blood volume causes a change in resistance of the strain gauge. In another technique, a sensitive crystal microphone is placed on the skin's surface to pick up the pulsation.

The most commonly used method to measure pulsatile blood volume changes is by the *photoelectric method*. Two methods are common: Reflectance method and transmittance method.

In the *transmittance method* (Fig. 6.15(a)) a light-emitting diode (LED) and photoresistor are mounted in an enclosure that fits over the tip of the patient's finger. Light is transmitted through the finger tip of the subject's finger and the resistance of the photoresistor is determined by the amount of light reaching it. With each contraction of the heart, blood is forced to the extremities and the amount of blood in the finger increases. It alters the optical density with the result that the light transmission through the finger reduces and the resistance of the photoresistor increases accordingly. The photoresistor is connected as part of a voltage divider circuit and produces a voltage that varies with the amount of blood in the finger. This voltage that closely follows the pressure pulse and its waveshape can be displayed on an oscilloscope or recorded on a strip-chart recorder.

The arrangement used in the *reflectance method* of photoelectric plethysmography is shown in Fig. 6.15(b). The photoresistor, in this case, is placed adjacent to the exciter lamp. Part of the light rays emitted by the LED is reflected and scattered from the skin and the tissues and falls on the photoresistor. The quantity of light reflected is determined by the blood saturation of the capillaries and, therefore, the voltage drop across the photoresistor, connected as a voltage divider, will vary in proportion to the volume changes of the blood vessels.

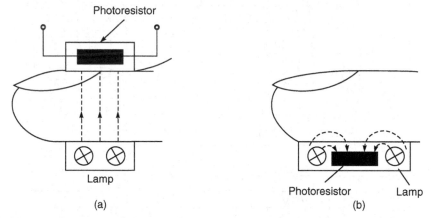

> **Fig. 6.15** *Arrangement of photoresistor and lamp in a finger probe for pulse pick-up:*
> *(a) transmission method, (b) reflectance method*

The LED phototransistor-photoplethysmograph transducer (Lee et al, 1975) consists of a Ga-As infrared emitting diode and a phototransistor in a compact package measuring $6.25 \times 4.5 \times 4.75$ mm. The peak spectral emission of the LED is at $0.94\,\mu m$ with a 0.707 peak bandwidth of $0.04\,\mu m$. The phototransistor is sensitive to radiation between 0.4 and $1.1\,\mu m$ (Fig. 6.16).

For pulse rate measurement, a photoelectric transducer suitable for use on the finger or ear lobe is used. The signal from the photocell is amplified and filtered (0.5 to 5 Hz passband) and the time interval between two successive pulses is measured. The measuring range is 0–250 bpm. Careful placement and application of the device is essential in order to prevent movement artefacts due to mechanical distortion of the skin.

Figure 6.17 shows the block diagram for processing the plethysmographic signal detected from a photoelectric transducer. The circuit consists of two parts, a LED oscillator and driver, which produce 300 Hz, 50 μs infrared light pulses to the finger probe attached to the patient, and a phototransistor that picks up the attenuated light. The electrical signal obtained from the phototransistor is amplified and its peak value is sampled and filtered. An automatic gain control

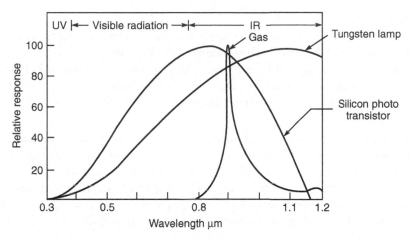

➤ **Fig. 6.16** *Relative spectral response for silicon phototransistor and the radiant spectral distribution of a tungsten lamp and a gallium-aresenide lamp (after Lee et al. 1975; reproducd by permission of IEEE Trans. Biomed. Eng.)*

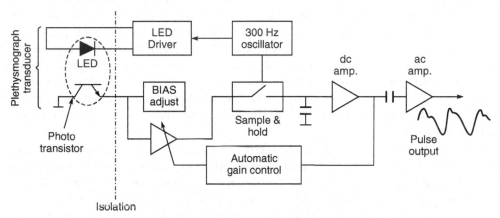

➤ **Fig. 6.17** *Block diagram for processing plythysmographic signal*

circuit adjusts the amplifier gain to yield a constant average pulse height at the output. The ac component with a frequency in the heart rate range (0.8–5 Hz), is further amplified to output the plethysmographic pulse rate form. This signal is transmitted across the isolation barrier, demodulated, low-pass filtered and transmitted to the analog multiplexer resident on the CPU board.

A *Piezo-electric* crystal can also be used to detect the pulse wave at certain places of the peripheral system where considerable displacement of the tissue layer above the artery is involved. The arrangement consists of a piezo-electric crystal clamped in a hermetically sealed capsule subject to displacement stresses. The displacement can be transmitted to the crystal through a soft rubber diaphragm. The crystal can be connected to an ECG recorder for recording the pressure pulse waveform.

There is another variation of the finger plethysmograph in which an air-coupled piezo-electric transducer is employed. As the volume of blood in the finger varies during the cardiac cycle, slight changes occur in the size of the finger. These changes can be transmitted as pressure variations in the air column inside the plastic tubing. A piezo-electric transducer at the end of the tube converts the pressure changes to a corresponding electrical signal. This signal can then be amplified and displayed. Similarly, a semiconductor strain gauge can be used to detect the displacement of the vessel wall due to a pulse wave.

Monitoring the peripheral pulse is more useful and dependable than monitoring the heart rate derived from ECG in the case of a heart block because it can immediately indicate the cessation of blood circulation in the limb terminals. Moreover, a photoelectric pick-up transducer is much easier to apply than the three ECG electrodes. The amplitude of the plethysmographic signal obtained is also quite large as compared to the ECG signal and therefore, gives better signal-to-noise ratio. However, the technique is severely subject to motion artefacts.

⫸ 6.7   BLOOD PRESSURE MEASUREMENT

Blood pressure is the most often measured and the most intensively studied parameter in medical and physiological practice. The determination of only its maximum and minimum levels during each cardiac cycle supplemented by information about other physiological parameters is an invaluable diagnostic aid to assess the vascular condition and certain other aspects of cardiac performance. Pressure measurements are a vital indication in the successful treatment and management of critically ill patients in an intensive cardiac care or of patients undergoing cardiac catheterization. The tremendous research and development for an automatic blood pressure monitor has resulted in several methods but only very few have been commercialized due to certain practical difficulties.

Blood is pumped by the left side of the heart into the aorta, which supplies it to the arterial circuit. Due to the load resistance of the arterioles and precapillaries, it loses most of its pressure and returns to the heart at a low pressure via highly distensible veins. The right side of the heart pumps it to the pulmonary circuit, which operates at a lower pressure. The heart supplies blood to both circuits as simultaneous intermittent flow pulses of variable rate and volume. The maximum pressure reached during cardiac ejection is called systolic pressure and the minimum pressure occurring at the end of a ventricular relaxation is termed as diastolic pressure. The mean arterial pressure over one cardiac cycle is approximated by adding one-third of the pulse pressure (difference between systolic and diastolic values) to the diastolic pressure. All blood pressure measurements are made with reference to the atmospheric pressure.

Typical haemodynamic pressure values are shown in Fig. 6.18. The nominal values in the basic circulatory system are as follows:

Arterial system	30–300	mmHg
Venous system	5–15	mmHg
Pulmonary system	6–25	mmHg

The most frequently monitored pressures, which have clinical usefulness in medium and long-term patient monitoring, are the arterial pressure and the venous pressure. There are two basic methods for measuring blood pressure—direct and indirect.

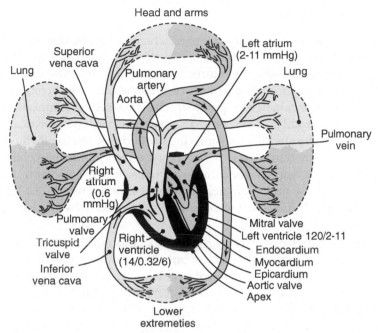

> **Fig.** 6.18 *Typical haemodynamic pressure values present in the basic circulatory system (Courtesy: Hewlett Packard, USA)*

The indirect methods consist of simple equipment and cause very little discomfort to the subject but they are intermittent and less informative. They are based on the adjustment of a known external pressure equal to the vascular pressure so that the vessel just collapses. On the other hand, the direct methods provide continuous and much more reliable information about the absolute vascular pressure from probes or transducers inserted directly into the blood stream. But the additional information is obtained at the cost of increased disturbance to the patient and complexity of the equipment.

Blood pressure readings vary with subjects and, among other variables, with the location of the transducer. If manometric blood pressure readings are not taken at heart level, they should be compensated to correspond to the readings at heart level. For example, if a mercury-manometer reading is taken at h mm below heart level, the reading is high due to the weight of a column of blood h mm high (this weight is ρ gh). The compensation factor is simply the ratio of densities:

For Mercury, $\rho = 13.6 \, \text{g/cm}^3$

For blood, $\rho = 1.055 \, \text{g/cm}^3$

$$\text{Ratio} = \frac{13.6}{1.055} = 12.9$$

The equivalent reading at heart level is thus:

$$\text{mmHg reading} = \frac{(\text{mm above or below heart level})}{12.9}$$

If the manometer is above heart level, add the correction; if below, subtract the correction.

6.7.1 Direct Methods of Monitoring Blood Pressure

The direct method of pressure measurement is used when the highest degree of absolute accuracy, dynamic response and continuous monitoring is required. The method is also used to measure the pressure in deep regions inaccessible by indirect means. For direct measurement, a catheter or a needle type probe is inserted through a vein or artery to the area of interest. Two types of probes can be used. One type is the catheter tip probe in which the sensor is mounted on the tip of the probe and the pressures exerted on it are converted to the proportional electrical signals. The other is the fluid-filled catheter type, which transmits the pressure exerted on its fluid-filled column to an external transducer. This transducer converts the exerted pressure to electrical signals. The electrical signals can then be amplified and displayed or recorded. Catheter tip probes provide the maximum dynamic response and avoid acceleration artefacts whereas the fluid-filled catheter type systems require careful adjustment of the catheter dimensions to obtain an optimum dynamic response.

Measurement of blood pressure by the direct method, though an invasive technique, gives not only the systolic, diastolic and mean pressures, but also a visualization of the pulse contour and such information as stroke volume, duration of systole, ejection time and other variables. Once an arterial catheter is in place, it is also convenient for drawing blood samples to determine the cardiac output (by dye dilution curve method), blood gases and other chemistries. Problems of catheter insertion have largely been eliminated and complications have been minimized. This has been due to the development of a simple percutaneous cannulation technique; a continuous flush system that causes minimal signal distortion and simple, stable electronics which the paramedical staff can easily operate.

A typical set-up of a fluid-filled system for measuring blood pressure shown in Fig. 6.19. Before inserting the catheters into the blood vessel it is important that the fluid-filled system should be thoroughly flushed. In practice a steady flow of sterile saline is passed through the catheter to prevent blood clotting in it. As air bubbles dampen the frequency response of the system, it should be ensured that the system is free from them.

Figure 6.20 shows a simplified circuit diagram commonly used for processing the electrical signals received from the pressure transducer for the measurement of arterial pressure. The transducer is excited with a 5 V dc excitation. The electrical signals corresponding to the arterial pressure are amplified in an operational amplifier or a carrier amplifier. The modern preamplifier for processing pressure signals are of the isolated type and therefore comprise of floating and grounded circuits similar to ECG amplifiers. The excitation for the transducer comes from an amplitude controlled bridge oscillator through an isolating transformer, which provides an interconnection between the floating and grounded circuits. An additional secondary winding in the transformer is used to obtain isolated power supply for the floating circuits. The input stage is a differential circuit, which amplifies pressure change, which is sensed in the patient connected circuit. The gain of the amplifier can be adjusted depending upon the sensitivity of the transducer. After RF filtering, the signal is transformer-coupled to a synchronized demodulator for removing the carrier frequency from the pressure signal.

For the measurement of systolic pressure, a conventional peak reading type voltmeter is used. When a positive going pressure pulse appears at A, diode D_3 conducts and charges C_3 to the peak

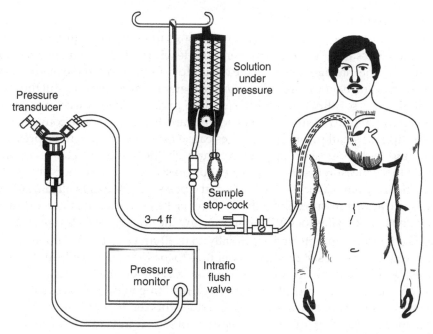

> **Fig. 6.19**   *Typical set up of a pressure measuring system by direct method*

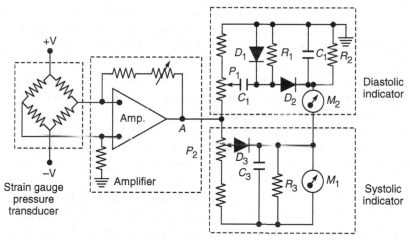

> **Fig. 6.20**   *Circuit diagram for measurement of systolic and diastolic blood pressure*

value of the input signal, which corresponds to the systolic value. Time constant R_3C_3 is chosen in such a way that it gives a steady output to the indicating meter.

The value of diastolic pressure is derived in an indirect way. A clamping circuit consisting of C_1 and D_1 is used to develop a voltage equal to the peak-to-peak value of the pulse pressure. This voltage appears across R_1. Diode D_2 would then conduct and charge capacitor C_2 to the peak value of the pulse signal. The diastolic pressure is indicated by a second meter M_2 which shows the difference between the peak systolic minus the peak-to-peak pulse pressure signal. The mean arterial pressure can also be read by using a smoothing circuit when required.

Central venous pressure (CVP) measurements made with needle cannulation techniques prove extremely useful in the management of acute circulatory failure and in the maintenance of blood volume in difficult fluid balance problems. Simple water manometers are still the most common measuring device in use, although highly sensitive pressure transducers are preferred when accurate measurements are required. However, the transducers cannot be conveniently mounted at the catheter tip and small positional changes cause large errors in venous pressure. Infusing intravenous fluids while measuring pressure through the same catheter is another problem encountered in these measurements. Central venous pressure is usually measured from a catheter located in the superior vena cava. The CVP reflects the pressure of the right atrium and is sometimes referred to as right atrial pressure. The catheter can even be located in the right atrium. Major peripheral veins used as entry sites for CVP monitoring are the brachial, subclavian and jugular veins.

Catheters used for CVP monitoring are usually 25 to 30 cm long. Long catheters, is they remain in place over extended periods of time are susceptible to the formation of fibrin sheaths along their outer surfaces. Besides this, air can be aspirated into a catheter that is situated in an area of low pressure (as compared to the atmospheric pressure), resulting in thrombo-embolic complications. A continuous infusion of heparin solution will reduce this tendency. Also, it should be ensured that there is no possibility of air intake. Development of the Swan-Ganz catheter- a balloon tipped, flexible catheter that can be flow-directed from a peripheral vein into the pulmonary artery, has made routine clinical monitoring of pulmonary artery pressure possible (Swan and Ganz, 1970). Information about pulmonary artery wedge pressure or end diastolic pressure in the pulmonary artery gives a good indication of the left atrial pressure. This is a very valuable parameter in predicting and treating left ventricular failure in myocardial infarction in patients undergoing cardiac surgery.

Clinical experience has demonstrated the difficulty in maintaining a high-quality arterial pulse waveform during direct measurements of blood pressure. Minute leaks in the stopcocks permit a small quantity of blood to enter the catheter where it clots. Even with a highly leak proof system, clots still form at the catheter tip due to the small volume of blood which may enter as a result of gauge volume displacement (0.04 mm^3 per 100 mmHg) and any volume displacement of minute entrapped air bubbles. This type of clotting at the catheter tip can be avoided by using the continuous flush system. Pressure transducers presently available incorporate a continuous flush arrangement. The source of fluid for the flushing system (Fig. 6.21) is a plastic bag (600 ml), which is filled with normal saline and kept at a pressure of 300 mmHg. The high pressure fluid then flows through a Millipore filter (0.22μ) which is essential to prevent clogging of the fine bore resistance element and which also serves to filter any bacteria found in the solution. Continuous flush is achieved by using a large resistive element to convert the pressure source to a flow source. With a 0.05 mm diameter glass tubing 1 cm long, flow across the element with 300 mmHg pressure is about 3 ml/h. It is found that large flow rates can cause significant error when using the small diameter catheter. Flow rates of 3 ml/h for adults and 0.5 ml/h for children have been found to be adequate. To initially fill the transducer and catheter, a fast flush feature is needed. This is done by using a rubber valve in the system which when operated permits a fast flush, fills the transducer, and purges the air bubbles from the flush system.

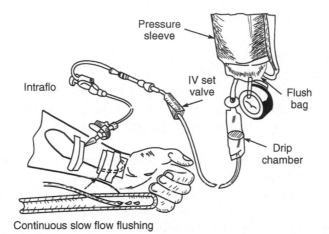

Continuous slow flow flushing

> **Fig.** 6.21 *Continuous slow-flow flushing arrangement for pressure measurements (Courtesy: Hewlett Packard, USA)*

Venous pressure measurement can be made by using a strain gauge transducer and a similar electronic signal processing circuitry. The transducers should be of higher sensitivity to give more accurate results at lower pressures. Since the blood pressure is always referred to as the atmospheric pressure at the height of the heart, a correction must be applied while making venous pressure measurements to compensate for the difference of level between the heart and the site of measurement. A correction of 7.8 mmHg is applied for every 10 cm. The site of measurement is below the height of the heart.

Frequency Response and Damping Adjustment of the Fluid-filled Catheters The frequency components of a normal pressure pulse consist of a zero frequency (dc) component, a fundamental component at the heart rate and harmonics of the fundamental rate. To record a pressure pulse without any distortion, the measuring system should be capable of recording all the frequency components with equal amplification and phase shift. The range of uniform frequency response starting from dc is determined from an estimation of the highest heart rate expected and the number of harmonics to be taken into account. Various authors have suggested that blood pressure pulses contain from 6 to 20 significant harmonics, but generally it is accepted that frequencies up to 10th harmonics produce significant components in the pressure pulse. If we assume the heart rate to be 90 beats per minute, the same shall be the rate of the arterial blood pressure waves. This means that the upper frequency response should be at least 15 Hz for a heart rate of 90 per minute.

An important factor which requires special consideration in relation to a fluid column pressure measuring system is the natural frequency or the resonant frequency of the system. The measuring system can respond accurately only for frequencies well below the natural frequency. A simplified equation which defines the natural frequency of the system is given by

$$F = \frac{D}{4} \sqrt{\frac{1}{\pi L} \times \frac{\Delta P}{\Delta V}}$$

(i)

where D = diameter of the fluid column

L = length of the fluid column

ΔP = pressure change

ΔV = volume change for a given ΔP

This equation assumes that the mass of the moving element of the transducer is very small as compared to the mass of the fluid. It also ignores the specific gravity of the fluid. ΔP is the change in pressure corresponding to a total volume change of ΔV by the application of the assumed pressure.

Fluid column systems usually have a very low natural frequency to be commensurate with the requirements, due to their large inertia and compliance. Therefore, to accurately record the pressure pulses, some form of compensation is necessary to improve the frequency response of the system. This compensation is called damping. In most pressure measuring systems, damping is provided by the viscous resistance of the liquid in the catheter and is given by

$$D = \left[\frac{4_\eta}{(\gamma_c)^3}\right]\left[\frac{l_c}{\pi E_\rho}\right]^{1/2} \qquad \text{(ii)}$$

where D = damping coefficient

ρ = liquid density in g/cm^3

η = viscosity of the liquid in poises

γ_c = radius of the catheter bore in cm

l_c = length of the catheter bore in cm

E = volume elasticity of the sensing element in dynes/cm^5

This equation shows that the damping increases inversely as the cube of the catheter diameter decreases. Equations (i) and (ii) become contradictory because natural frequency decreases directly with the diameter of the fluid column while the damping ratio increases. Therefore, a compromise must be reached to obtain a maximum flat frequency response.

Figure 6.22 shows the frequency response curve of a catheter manometer system if filled by the usual method with cold water. To obtain a uniform flat response, the system has to be suitably damped. As in the case of recorders, for optimum frequency response, damping coefficient should be set at 0.7. The damping can be varied by introducing an adjustable constriction into the flow line. Two methods are common. A series damper makes use of a capillary introduced between the catheter and the manometer. With series damping, it is possible to effect a considerable decrease in the height of the resonance peak. However, the resonance peak is shifted to a lower value, which makes the frequency response worse than without damping. The other method is by using a parallel damper consisting of a variable needle resistance parallel to the manometer and in series with a distensible plastic tube connected to a syringe. The parallel damper is able to flatten the resonance peak and a flat response curve is achieved almost up to the original peak.

Pressure waveform distortion due to transmission in fluid-filled catheters can be compensated by electronic means. The compensator is based on the fact that fluid-filled catheters can be characterized as second order systems and the distortions introduced by catheters are typical

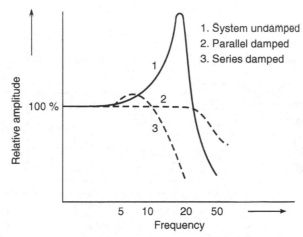

> **Fig. 6.22** *Frequency response of a fluid-filled catheter system*

of the output of such systems. It is obvious that a compensator whose transfer function is the exact inverse of such a second order system would facilitate recovery of the original waveform shape.

In an attempt to evaluate the pressure distortion due to clinical catheter-manometer systems, it becomes necessary to establish their linearity. A system is considered linear if its transfer function coefficients are independent of pressure and time in the applicable zone of pressure and frequency. It may be noted that linearity is to be individually established for different catheters having differing cross-sectional structures.

Special Considerations for the Design of Pressure Transducers for Medical Applications: Physiological pressure transducers are usually linked directly to the patient's heart and hence they must ensure complete safety of the patient. It is for this reason that the construction of the transducers should be such that they provide patient-safe isolation. Figure 6.23 shows the construction of one such transducer from M/s American Optical, USA. This transducer has three modes of isolation: (i) External isolation of the case with a plastic sheath, which provides protection from extraneous voltages. (ii) Standard internal isolation of the sensing (bridge) elements from the inside of the transducer case and from the frame. (iii) Additional internal isolation of the frame from the case and the diaphragm, in case of wire breakage.

Thus, isolation of the patient/fluid column from electrical excitation voltage is assured, even in the event of failure of the standard internal isolation. Typically, transducers have maximum leakage of 2 microamperes at 120 V ac 60 Hz.

Pressure transducers are commonly used in intensive care units, cardiac catheterization and other situations in which there is a possibility of a defibrillator being used on the patient. If the transducer breaks down with the application of a defibrillating shock, the current to earth through the transducer would be a few miliamperes, assuming that the catheter impedance is 1 MΩ. This is probably not significant as far as the patient is concerned, in view of the very high current flow in between the defibrillator paddles. However, there is a possibility of irreversible damage, which may be done to the transducer. Transducers, therefore, should be able to withstand high voltages

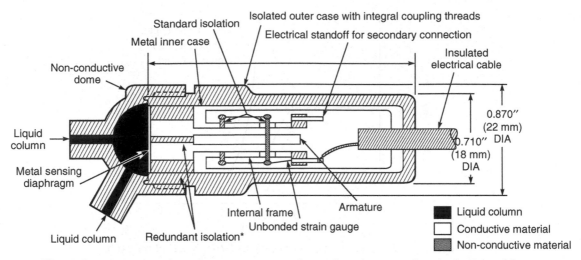

> **Fig. 6.23** *Cross-section of a pressure transducer showing complete isolation of the patient (Courtesy: American Optical, USA)*

arising from electrocautery or defibrillator procedures. The minimum breakdown voltage for this purpose should be 10,000 V dc.

Physiological pressure transducers need to be sterilized before every use. This is a costly and time consuming exercise and results in transducer wear and tear due to repeated cleanings and sterilization. This problem has been overcome by designing a presterilized disposable dome with a built-in membrane that provides a sterile barrier between the patient and transducer.

Pressure transducers are sensitive devices. They are adversely affected by steam autoclaving and Gamma irradiation methods and therefore these methods should not be used. Ethylene oxide gas sterilization followed by aeration is a preferred method. Chemical sterilization is also acceptable provided the procedure is based on a thorough knowledge of the agent and the process. Ultrasonic cleaning is damaging to the transducer and should not be used. Sharp and hard objects should not be used particularly on or near the diaphragm. Gross soilage and particulate material and residue should be removed by soaking the transducer only for a few minutes in an acceptable cleaning solution or disinfectant. For best operational performance, the transducers should not be exposed to a temperature higher than 65°C (maximum for short duration) or 50°C for extended periods of time. The transducer should be thoroughly cleaned as soon as practicable after each application and maintained in a clean condition for optimum long-term service.

One of the commonest type of abuses of physiological transducers is applying a pressure in excess of the working pressure either by standing on the pressure line or by flushing out the transducer with the output closed. A weight of 2 kg applied to a 1 ml disposable syringe is known to produce a pressure of about 10,000 mmHg. The calibration of the transducers should not substantially change (more than 1%) after the maximum rated pressure is applied. The diaphragm rupture pressure values are different for pressure transducers of different makes, and are usually in the value range of 3,000 to 10,000 mmHg.

Catheter Tip Pressure Transducers: Fluid-filled catheter and external pressure transducer arrangements for the measurement of intravascular pressures have limited dynamic response.

The problem has been solved to some extent, by the use of miniature catheter tip pressure transducers. One such transducer which makes use of a semiconductor strain gauge pressure sensor is described by Miller and Baker (1973). The transducer is 12 mm long and has a diameter of 1.65 mm. It is mounted at the tip of a No. 5 French Teflon catheter, 1.5 m long. The other end of the catheter carries an electrical connector for connecting the transducer to a source of excitation and a signal conditioner. A hole in the connector provides an opening to the atmosphere for the rear of the diaphragm. The active portion of the transducer consists of a silicon-rubber diaphragm with an effective area of 0.75 mm^2. Pressure applied to the diaphragm is transmitted by a linkage to two silicon strain gauges, which form a half bridge, while the other half is constituted by the electrical circuit.

The transducer gives a high output (0.1 V per 300 mmHg) for 3.5 V excitation and shows an excellent thermal stability of ±0.15 mmHg/°C over an ambient temperature range of 25–40°C. The transducer has a working range of ±300 mmHg with a volume displacement of 2×10^{-3} mm^3 per 100 mmHg. The dynamic response of the transducer permits high fidelity recording of pressure transients, anywhere in the vascular system. With a natural resonant frequency of 15 kHz, it is ideally suitable for studies requiring accurate analyses of pressures and pressure waveforms, particularly the derivatives of waveforms. The transducer has an epoxy resin insulation over the strain gauge element within the tip and gives a leakage current of less than 0.5 µA for an applied voltage of 150 V. Since the transducer system is made to work on 5.4 V excitation, leakage current from this source will be of the order of 0.0216 µA, which is much less than any ac level considered dangerous for catheters placed directly within the heart chambers.

Nichols and Walker (1974) used a Miller PC-350 catheter tip transducer, a commercial version of the above described transducer (manufactured by Miller Instruments Inc., Houston, Texas, USA) in experimental and clinical work and found it working satisfactorily as compared to fluid-filled systems (Fig. 6.24). Derived variables such as the maximum rate of change of the left ventricular pressures (*dp/dt*) could be accurately recorded, but were found to be consistently lower

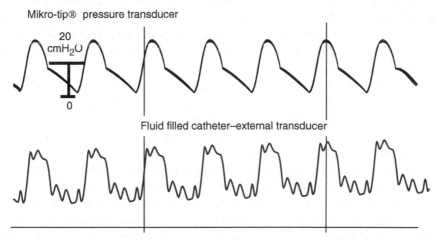

➤ **Fig. 6.24** *A high fidelity pressure tracing from mikro-tip pressure sensor in the pulmonary artery and a simultaneous tracing with artefact from a fluid-filled catheter (Courtesy: Miller Instruments Inc. USA)*

than those measured with fluid-filled systems. Catheter tip transducers with a fluid-velocity sensor can be used for making high fidelity pressure and velocity measurements simultaneously, at one or more locations in the heart (Fig. 6.25).

Matsumoto *et al* (1978) point out that the catheter tip pressure transducers do not remain stable over variations in temperature and with long-term clinical use. Safety problems may also be experienced due to the direct connection between external electronic devices and the heart. To overcome these problems, they proposed the application of fibre optics to the catheter tip pressure transducer for the measurement of intracardiac pressures. The pressure is detected by the photo-electric transducer element whose output voltage is proportional to the applied pressure. Comparison of waveforms and frequency analysis of waveforms reveal that the fibre optic catheter has characteristics by no means inferior to the tip transducer type catheter.

6.7.2   Indirect Methods of Blood Pressure Measurement

The classical method of making an indirect measurement of blood pressure is by the use of a cuff over the limb containing the artery. This technique was introduced by Riva-Rocci for the determination of systolic and diastolic pressures. Initially, the pressure in the cuff is raised to a level well above the systolic pressure so that the flow of blood is completely terminated. Pressure in the cuff is then released at a particular rate. When it reaches a level, which is below the systolic pressure, a brief flow occurs. If the cuff pressure is allowed to fall further, just below the diastolic pressure value, the flow becomes normal and uninterrupted.

The problem here finally reduces to determining the exact instant at which the artery just opens and when it is fully opened. The method given by Korotkoff and based on the sounds produced by flow changes is the one normally used in the conventional sphygmomanometers. The sounds first appear (Fig. 6.26) when the cuff pressure falls to just below the systolic pressure. They are produced by the brief turbulent flow terminated by a sharp collapse of the vessel and persist as the cuff pressure continues to fall. The sounds disappear or change in character at just below diastolic pressure when the flow is no longer interrupted. These sounds are picked up by using a microphone placed over an artery distal to the cuff. The sphygmomanometric technique is an ausculatory method; it depends upon the operator recognizing the occurrence and disappearance of the Korotkoff sounds with variations in cuff pressure.

A number of automated blood pressure measuring instruments have been designed which make use of the Riva-Rocci method. They operate in a manner analogous to that employed by a human operator, but differ in the method of detecting the pulsations of blood flow at the systolic and diastolic levels. Frequency bands that best discriminate the Korotkoff sounds at systole and diastole from the sounds immediately preceding these events must be defined for achieving a high degree of reliability in the automatic electronic blood pressure instruments. Golden *et al* (1974) carried out a special analysis of seven Korotkoff sounds centred about the systolic and diastolic ausculatory events and found that a maximum increase in amplitude at the systolic transition occurred in the 18–26 Hz band. Similarly, a maximum decrease in spectral energy of diastolic Korotkoff sounds, at ausculatory cessation, was observed within a 40–60 Hz passband.

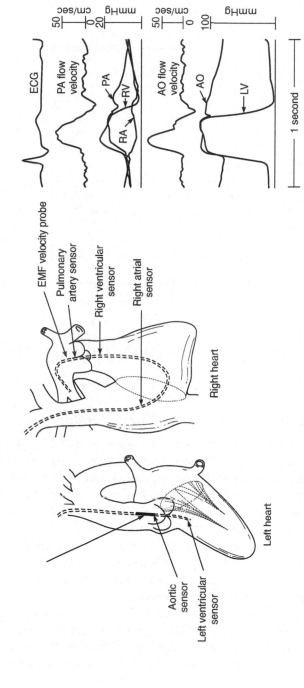

▶ **Fig. 6.25** *Simultaneous recording of pulmonary artery and aortic flow velocity signals with high fidelity pressure waveforms using mikro-tip transducers (Courtesy: Miller Instruments Inc. USA)*

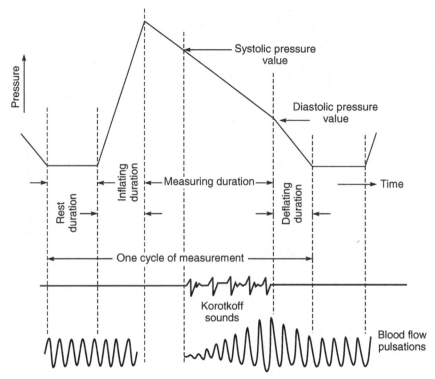

> **Fig. 6.26** *Principle of blood pressure measurement based on Korotkoff sounds*

6.7.2.1 *Automatic Blood Pressure Measuring Apparatus using Korotkoff's Method*

The method consists in putting a cuff around the upper part of the patient's arm and applying a microphone over the brachial artery. The compressed air required for inflating the cuff is provided by a pumping system incorporated in the apparatus. Usually the inflating is done to a preset pressure level, well beyond the systolic value at the rate of approximately 30 mmHg/s. The pressure in the cuff is then decreased at a relatively slow pace at the rate of 3–5 mmHg/s. The cuff is to be applied in such a way that the veins are not occluded.

While air is allowed to leak from the cuff, the Korotkoff sounds are picked up by a special piezo-electric microphone. The corresponding electrical signals are fed to a preamplifier. The amplified signals are then passed on to a bandpass filter having a bandwidth of 25 to 125 Hz. With this passband, a good signal-to-noise ratio is achieved when recording Korotkoff sounds from the brachial artery beneath the lower edge of the cuff. The system is so designed that the appearance of the first Korotkoff sound switches in the systolic manometer and locks the reading on the indicating meter. In a similar way, the diastolic value is fixed by the last Korotkoff sound. The cuff is completely deflated, automatically, after an interval of 2–5 s after the determination of the diastolic value.

Instruments operating on this principle are subject to serious errors, particularly in restless patients, unless steps are taken to ensure protection against artefacts. One method of doing this is to design the control system in such a way that when pressure is registered, the first sound must be followed by a second one within the preset interval. If this is not the case, the recorded value is automatically cancelled and the measurement starts again with the subsequent sounds. The measuring accuracy of such type of instruments is not very high and the error is usually of the same order (±5 mmHg) as is obtained in clinical sphygmomanometers.

A complete cycle of measurement consists of cuff pumping, controlled deflation, picking up and evaluation of the Korotkoff sounds, fixing of the systolic and diastolic pressure and then a complete deflation of the cuff. The cycle is initiated by a time delay and the operation is controlled by a command pulse. Manual operation is, however, always possible.

A number of shortcomings limit the application of the Riva-Rocci method. Probably the most serious among them is that the measurement is not continuous. Even for a particular single measurement, a number of heart cycles intervene between the determination of the systolic and diastolic pressures. Moreover, large errors are common since the pressure applied to the exterior vessel wall is not necessarily identical to that in the cuff, but is attenuated by the intervening tissue and an exact state of flow cannot be precisely determined. This problem is so severe that in most of the measurements made with instruments based on this principle, the diastolic pressure value is less reliable than the systolic. The diastolic pressure can be determined with greater accuracy and more reliability if the microphone output is amplified and fed to a chart recorder. The recorder can be calibrated in terms of pressure by feeding simulated signals corresponding to 60, 120, 180, 240 and 300 mmHg.

The automatic built-in pump system of inflating and deflating the cuff must be provided with safety devices so that the patient does not experience any discomfort in the case of system failure. Provision should be made for immediate switching off of the pump when the pressure in the cuff reaches the preset maximum over-systolic value and in no case should the pressure in the cuff be allowed to exceed 300 mmHg. An additional arrangement must switch off the pump at any pressure after 20 s from starting and deflate the cuff at a constant rate. These devices shall ensure that the pressure in the cuff will not reach too high a value and that no pressure is kept longer than approximately 20 s.

6.7.2.2 *The Rheographic Method*

A fully automatic apparatus for measuring systolic and diastolic blood pressures has been developed using the ordinary Riva-Rocci cuff and the principle of rheographic detection of an arterial pulse. Here, the change in impedance at two points under the occluding cuff forms the basis of detection of the diastolic pressure.

In this method, a set of three electrodes (Fig. 6.27), which are attached to the cuff, are placed in contact with the skin. A good contact is essential to reduce the skin electrode contact impedance. Electrode **B** which acts as a common electrode is positioned slightly distal from the mid-line of the cuff. Electrodes **A** and **C** are placed at a certain distance from the electrode **B**, one distally and the other proximally. A high frequency current source operating at 100 kHz is connected to the electrodes **A** and **C**. When we measure the impedance between any two electrodes before pressurizing the cuffs, it shows modulation in accordance with the blood flow pulsations in the artery. Therefore, arterial pulses can be detected by the demodulation and amplification of this modulation.

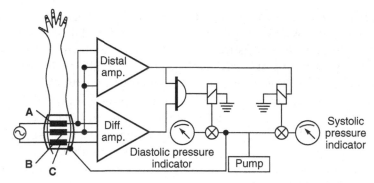

> **Fig. 6.27**   *Rheographic method of indirect blood pressure measurement*

When the cuff is inflated above the systolic value, no pulse is detected by the electrode **A**. The pulse appears when the cuff pressure is just below the systolic level. The appearance of the first distal arterial pulse results in an electrical signal, which operates a valve to fix a manometer pointer on the systolic value. As long as the pressure in the cuff is between the systolic and diastolic values, differential signal exists between the electrodes **A** and **C**. This is because the blood flow is impeded underneath the occluding cuff and the pulse appearing at the electrode **A** is time delayed from the pulse appearing at **C**. When the cuff pressure reaches diastolic pressure, the arterial blood flow is no longer impeded and the differential signal disappears. A command signal is then initiated and the diastolic pressure is indicated on the manometer.

In the rheographic method of measuring blood pressure, the cuff need not be precisely positioned as in the case with the Korotkoff microphone, which is to be fixed exactly above an artery. Also the readings are not affected by ambient sounds.

Similar to the rheographic method of measuring arterial blood pressure non-invasively, is the photo-electric plethysmograph, which detects cardiac volume pulse and transduces it to an electrical signal. The rate of pulsatile blood-volume change in a peripheral site, such as the earlobe or finger tip is used for indicating the magnitude of the arterial systolic blood pressure after calibration against a standard blood pressure measuring means. The correlation between the rate of the peripheral pulsatile blood-volume change and the arterial systolic blood pressure is more close if these quantities are measured during a portion of each cardiac cycle, i.e. period of diastole or systole and for just one instant during this period.

6.7.2.3 *Differential Auscultatory Technique*

The "differential auscultatory technique" is a non-invasive method for accurately measuring blood pressure. A special cuff-mounted sensor consisting of a pair of pressure sensitive elements, isolates the signal created each time the artery is forced open.

Figure 6.28 illustrates how high frequency pulses are created each time, the intra-arterial pressure exceeds the cuff pressure. As long as the cuff pressure exceeds the pressure in the artery, the artery is held closed, and no pulse is generated. However, as soon as the intra-arterial pressure rises to a value, which momentarily exceeds the cuff pressure, the artery "snaps" open; and a pulse is created. Once the artery is open, blood flows through it giving rise to the low frequency pressure

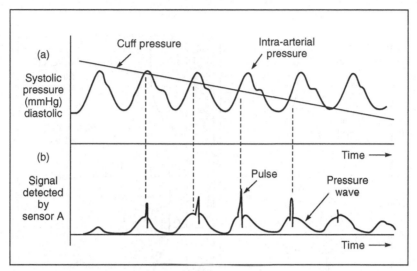

> **Fig.** 6.28 (a)   *Diagram showing the relationship between cuff pressure and*
> *intra-arterial pressure*
> (b)   *Signal created by the relative pressure changes*

wave signal, which lasts until the arterial pressure again drops below the cuff pressure. This process is repeated until the cuff pressure drops to a value below the diastolic.

Figure 6.29 is a cut away view of an arm with a cuff partially occluding the brachial artery. Each time the artery opens, the signal shown in Fig. 6.30 is created. Note that this signal consists of a slowly rising, low frequency component (in the frequency range of 0.5–5 Hz) with a fast "pulse" (frequencies approximately 10–80 Hz) superimposed on it. This signal is denoted by the arrows marked A in Fig. 6.29 transmitted from the artery to both the sensor and the air bag in the cuff.

Due to the air bag characteristics, the high frequency component is highly attenuated, leaving only the low-frequency signal, as shown in Fig. 6.30(b). Therefore, only the low frequency signal is transmitted to the side of the sensor facing the air bag, as denoted by the arrows marked B in Fig. 6.29. Since most artefact signals (unwanted signals due to motion, etc.) fall in a frequency range below 10 Hz, they are also transmitted to both sides of the sensor.

The systolic pressure is determined as the pressure at which the first opening of the artery occurs, as shown by the first pulse in Fig. 6.30(c) , because this pulse is created the first time the artery is forced open by intra-arterial pressure. Similarly, diastolic value is determined as the pressure at which the differential signal essentially disappears, because this corresponds to the last time the artery is forced open. The differential sensor subtracts the side "B" signal from the side "A" signal, thereby cancelling out the pressure wave component and the motion artefact signals, and the higher frequency Korotkoff signals are isolated.

6.7.2.4   *Oscillometric Measurement Method*

The automated oscillometric method of non-invasive blood pressure measurement has distinct advantages over the auscultatory method. Since sound is not used to measure blood pressure in

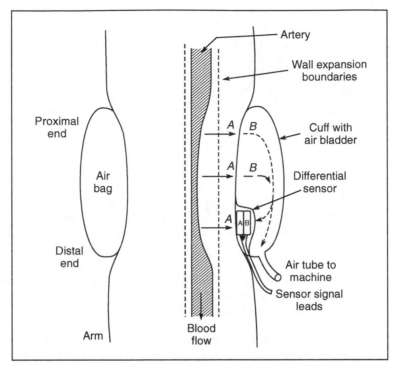

> **Fig. 6.29**   *Cut-away view showing signal detection*

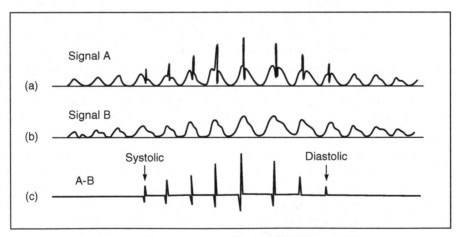

> **Fig. 6.30**   (a) *Signal generated by artery as air is bled from cuff*
>
> (b) *Signal 'A' after filtering by air bag (equivalent of oscillometric signal created by expansion of arterial walls; frequencies are in 0.5 – 5 Hz range)*
>
> (c) *Differential signal due to opening and closing of artery (frequency range 10-80 Hz)*

the oscillometric technique, high environmental noise levels such as those found in a busy clinical or emergency room do not hamper the measurement. In addition, because this technique does not require a microphone or transducer in the cuff, placement of the cuff is not as critical as it is with the auscultatory or Doppler methods. The oscillometric method works without a significant loss in accuracy even when the cuff is placed over a light shirt sleeve. The appropriate size cuff can be used on the forearm, thigh, or calf as well as in the traditional location of the upper arm. A disadvantage of the oscillometric method, as well as the auscultatory method, is that excessive movement or vibration during the measurement can cause inaccurate readings or failure to obtain any reading at all.

The oscillometric technique operates on the principle that as an occluding cuff deflates from a level above the systolic pressure, the artery walls begin to vibrate or oscillate as the blood flows turbulently through the partially occluded artery and these vibrations will be sensed in the transducer system monitoring cuff pressure. As the pressure in the cuff further decrease, the oscillations increase to a maximum amplitude and then decrease until the cuff fully deflates and blood flow returns to normal.

The cuff pressure at the point of maximum oscillations usually corresponds to the mean arterial pressure. The point above the mean pressure at which the oscillations begin to rapidly increase in amplitude correlates with the diastolic pressure (Fig. 6.31). These correlations have been derived and proven empirically but are not yet well explained by any physiologic theory. The actual determination of blood pressure by an oscillometric device is performed by a proprietary algorithm developed by the manufacturer of the device.

The oscillometric method is based on oscillometric pulses (pressure pulses) generated in the cuff during inflation or deflation. Blood pressure values are usually determined by the application of mathematical criteria to the locus or envelope formed by plotting a certain characteristic, called the oscillometric pulse index, of the oscillometric pulses against the baseline cuff pressure (Fig. 6.32). The baseline-to-peak amplitude, peak-to-peak amplitude, or a quantity based on the partial or full time-integral of the oscillometric pulse can be used as the oscillometric pulse index. The baseline cuff pressure at which the envelope peaks (maximum height) is generally regarded as the MAP (mean arterial pressure). Height-based and slope-based criteria have been used to determine systolic and diastolic pressures.

An envelope that has been normalized with respect to the peak index can also be used for the determination of the oscillometric blood pressure. The ECG-gating technique has been used to assist in the identification of oscillometric pulse signals. Measurement sites for oscillometric blood pressure measurement include the upper arm, forearm, wrist, finger and thigh.

Most of the patient monitoring systems are based on the oscillometric measuring principle. Figure 6.33 shows the major functional parts of a NIBP (non-invasive blood pressure) measuring system.

An air pump is used to automatically inflate the patient cuff. The pump is of a membrane type and is enclosed in a foam rubber filled casing to attenuate noise. The pneumatic unit includes damping chambers to (i) prevent a rapid increase of pressure caused by the pump, (ii) slow down the pressure change in the measurement of infant and (iii) smooth down rapid pressure pulses caused by the bleed valve. A safety valve prevents accidental cuff over-pressurization and operates nominally at 330 mmHg. A bleed valve is incorporated to release the cuff pressure. The opening of

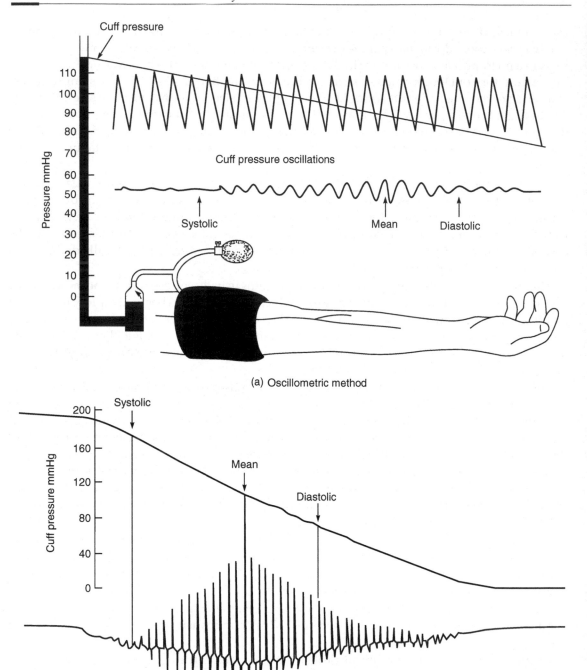

(a) Oscillometric method

(b) Oscillations in cuff pressure

➤ **Fig. 6.31**　*Illusration of oscillometric method of blood pressure measurement*

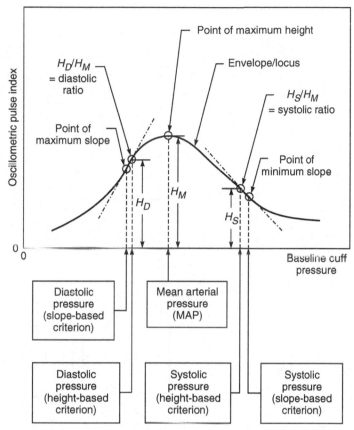

> **Fig. 6.32** *Criteria for oscillometric blood pressure determination*

the valve is pulse–width controlled between 100% (valve fully open) and 0% (valve fully closed). The driving signal frequency is 40 Hz. For quickly deflating the cuff, an exhaust valve is provided. The solenoid (magnetic) valves and the air pump are controlled by an open collector darlington driver circuit. A watch dog timer prevents a prolonged inflation. The operation of the pump and the operation of the valves are all under microprocessor control.

A piezo-resistive pressure transducer is connected to the cuff. It measures the absolute pressure of the blood pressure cuff and the pressure fluctuations caused by the arterial wall movement. The pressure transducer is excited by a 4 mA constant current source. The output of the pressure transducer is a differential signal, which is amplified, in a differential amplifier with a gain of 30. This is followed by zero-control and gain control circuits. A dc channel is used to measure the static or non-oscillating pressure of the blood pressure cuff. The ac component of the pressure data is amplified to allow the processor to analyze the small cuff pressure fluctuations, which are used as a basis for blood pressure determination. This is followed by a second order high-pass filter to effectively block out the dc component of the pressure signal. The next stage is a low-pass filter, which blocks the offset voltages.

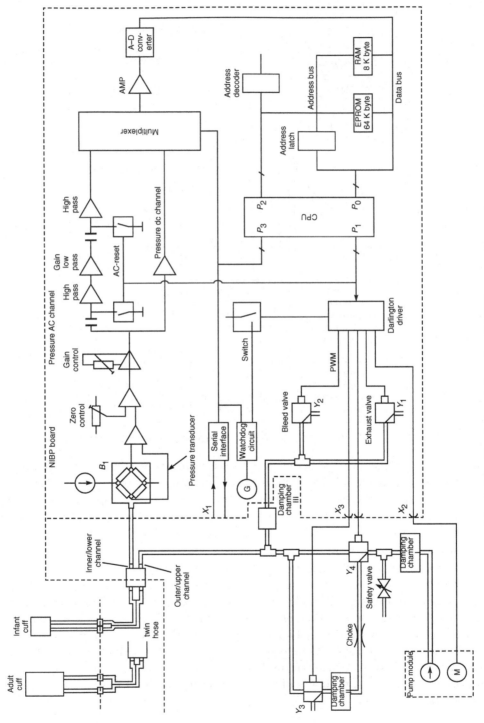

The analog multiplexer circuit is used to select either the dc or ac channel to the A-D convertor, whose output goes to the processor circuit. The 8051 microcontroller alongwith its associated memory circuits allow the control of the whole measuring procedure.

6.7.2.5   *Ultrasonic Doppler Shift Method*

Automatic blood pressure monitors have also been designed based on the ultrasonic detection of arterial wall motion. The control logic incorporated in the instrument analyzes the wall motion signals to detect the systolic and diastolic pressures and displays the corresponding values.

As explained in Chapter 11, the observed Doppler frequency can be expressed as:

$$\Delta f = \frac{2V_t}{\lambda_c} \tag{i}$$

where Δf = Doppler frequency (Hz)
   V_t = velocity of the object (m/s)
   λ_c = carrier wavelength (m)

For blood pressure measurement, the brachial artery is the object from where the ultrasound gets reflected. Arterial movement produces the Doppler frequency shift.

$$\lambda_c = \frac{V_c}{f_c} \tag{ii}$$

where λ_c = wavelength (in metres) of the carrier frequency in the medium
   V_c = velocity of the carrier frequency in the medium (1480 m/s in water)
   f_c = carrier frequency in the medium (2 MHz)

Substituting these values in equation (ii),

$$\lambda_c = \frac{1480}{2 \times 10^6} = 0.74 \times 10^{-3}\, m$$

The Doppler frequency is expressed as

$$\Delta f = \frac{2V_t}{0.74 \times 10^{-3}} = 2.7 \times 10^3\, V_t\,(Hz) \tag{iii}$$

Thus, Δf varies directly with the target velocity, i.e. the motion of the brachial artery.

To measure blood pressure, the Doppler frequency shift due to the snapping action of the artery must be known. The arterial movement with the opening and closing of the artery is 5×10^{-3} m, assuming that the snapping occurs in 0.1 s (Δt), the arterial wall velocity is

$$V_t = \frac{\Delta d}{\Delta t} = \frac{5 \times 10^{-3}}{0.1} = 50 \times 10^{-3}\, m/s$$

By substituting values for V_t, and λ_c into the Doppler equation (iii), the artery motion Doppler frequency is

$$2.7 \times 10^3\, V_t = 2.7 \times 10^3 \times 50 \times 10^{-3} = 135\, Hz$$

Instruments making use of ultrasonic Doppler-shift principle for the measurement of blood flow are based on the detection of the frequency shift ascribed to back scattering from moving blood particles. On the other hand, the blood pressure instrument filters out these higher frequency reflections and senses the lower frequency refractions originating from the movement of the relatively slow moving arterial wall.

In principle, the instrument consists of four major subsystems (Fig. 6.34). The power supply block converts incoming ac line voltage to several filtered and regulated dc voltages required for the pneumatic subsystem in order to inflate the occlusive cuff around the patient's arm.

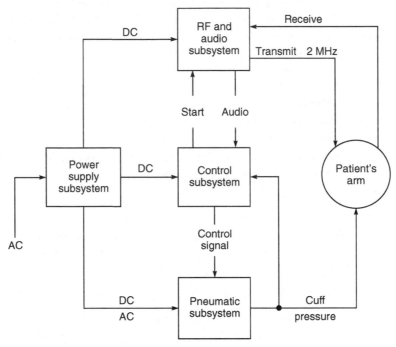

> **Fig.** 6.34 *Major subsystems in ultrasonic blood pressure monitor (Courtesy: Roche Medical Electronics Division, USA)*

At the same time, control subsystem signals gate-on the transmitter in the RF and audio subsystem, thereby generating a 2 MHz carrier, which is given to the transducer located in the cuff. The transducer converts the RF energy into ultrasonic vibrations, which pass into the patient's arm. The cuff pressure is monitored by the control subsystem and when the pressure reaches the preset level, further cuff inflation stops. At this time, audio circuits in the RF and audio subsystems are enabled by control subsystem signals, and the audio signals representative of any Doppler frequency shift are thus able to enter the control subsystem logic.

The control subsystem signals the pneumatic subsystem to bleed off the cuff pressure at a rate determined by the preset bleed rate. As air bleeds from the cuff, the frequency of the returned RF is not appreciably different from the transmitted frequency as long as the brachial artery remains occluded. Till then, there are no audio signals entering the control subsystem.

At the systolic pressure, the occluded artery snaps open and the arterial blood flow starts. This artery motion results in a Doppler shift in the returning ultrasonic vibrations. The converted audio frequency signal is recognized as tentative systolic by the control subsystem logic. Four valid artery returns must be recognized in order to register the tentative systole and for it to become fixed as true systole. This reduces the possibility of artefacts from recording a false systole reading. As a further check, the audio returns are examined for width and rate of occurrence to prevent artefacts from being accepted as true artery returns. Pulses more than 125 milliseconds wide or occurring more frequently than every 250 milliseconds are rejected by the control subsystem logic. This sets the upper limit (240 bpm) on the patient's heart rate to which this instrument can function. The lower limit of response is 24 bpm. At diastole, cuff pressure equals or slightly exceeds the arterial wall pressure. As a result, wall snapping ceases and the RF and audio subsystem no longer receive the Doppler shifted returns. The reading is registered, and the cuff pressure is allowed to deflate rapidly to atmospheric pressure. The readings are held fixed until a new measurement cycle is initiated.

An occlusive cuff is placed on the arm (Fig. 6.35) in the usual manner, with an ultrasonic transducer on the arm over the brachial artery. The cuff is inflated first to above systolic pressure and then deflated at a specified rate. A low energy ultrasonic beam (less than 50 mw/cm^2) at a frequency of 2 MHz is transmitted into the arm. The portion of the ultrasound that is reflected by the arterial wall shifts in frequency when the wall of the artery moves. Above systolic, the vessel remains closed due to the pressure of the occluding cuff, and the monitor signals are not received. As the cuff pressure falls to the point where it is just overcome by the brachial artery pressure, the artery wall snaps open. This opening wall movement, corresponding to the occurrence of the first Korotkoff sound, produces a Doppler-shift which is interpreted by logic in the instrument as systolic and displayed accordingly. With each subsequent pulse wave, a similar frequency shift is

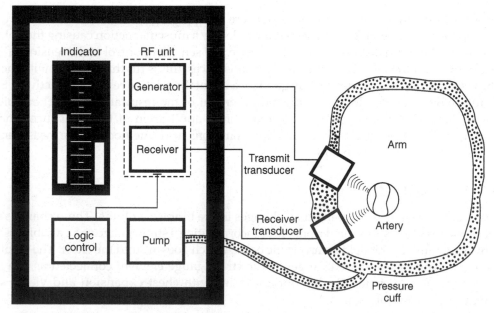

> **Fig. 6.35** *Measurement of blood pressure using ultrasonic Doppler-shift principle*

produced until at the diastolic pressure the artery is no longer occluded. Its rapid motion suddenly disappears and the Doppler-shift becomes relatively small. The instrument notes the sudden diminution in the amplitude of the Doppler shift and cuff pressure at this point is displayed as diastolic pressure. Special electronic circuits used in the instrument help to discriminate against extraneous motion artefacts. A coupling medium is essential between the transducer and the patients' skin for the efficient transmission of ultrasonic energy.

Unlike the Korotkoff method, the instruments based on the ultrasonic Doppler-shift principle often provide reliable blood pressure measurements in severe hypotensive states, at unfavourable sites such as the popliteal artery, in neonates where no other indirect method of measurement is feasible, in patients too obese for successful auscultation, under unfavourable conditions such as high ambient noise, and in many species of laboratory research animals.

6.8   MEASUREMENT OF TEMPERATURE

The transducer normally used for temperature measurement in a patient monitoring system is a thermistor. Changes in resistance of the thermistor with changes in temperature are measured in a bridge circuit and indicated on a calibrated meter. The measuring range is 30–42°C.

In a patient monitoring system, provision for two channel temperature measurements are usually made. Similar to ECG monitoring, the output circuits are isolated through opto-couplers. Provision for inoperate conditions are also included in such type of monitoring equipment.

6.9   MEASUREMENT OF RESPIRATION RATE

The primary functions of the respiratory system are to supply oxygen and remove carbon dioxide from the tissues. The action of breathing is controlled by a muscular action causing the volume of the lung to increase and decrease to effect a precise and sensitive control of the tension of carbon dioxide in the arterial blood. Under normal circumstances, this is rhythmic action with the result that the respiration rate provides a fairly good idea about the relative respiratory activity. Several techniques have been developed for the measurement of the respiration rate. The choice of a particular method depends mostly upon the ease of application of the transducer and their acceptance by the subject under test. Some of the commonly used methods for the measurement of respiration rate are explained below.

6.9.1   Displacement Method

The respiratory cycle is accompanied by changes in the thoracic volume. These changes can be sensed by means of a displacement transducer incorporating a strain gauge or a variable resistance element. The transducer is held by an elastic band, which goes around the chest. The respiratory movements result in resistance changes of the strain gauge element connected as one arm of a Wheatstone bridge circuit. Bridge output varies with chest expansion and yields signals corresponding to respiratory activity.

Changes in the chest circumference can also be detected by a rubber tube filled with mercury. The tube is fastened firmly around the chest. With the expansion of the chest during an inspiratory phase, the rubber tube increases in length and thus the resistance of the mercury from one end of this tube to the other changes. Resistance changes can be measured by sending a constant current through it and by measuring the changes in voltage developed with the respiratory cycle.

6.9.2 Thermistor Method

Since air is warmed during its passage through the lungs and the respiratory tract, there is a detectable difference of temperature between inspired and expired air. This difference of temperature can be best sensed by using a thermistor placed in front of the nostrils by means of a suitable holding device. In case the difference in temperature of the outside air and that of the expired air is small, the thermistor can even be initially heated to an appropriate temperature and the variation of its resistance in synchronism with the respiration rate, as a result of the cooling effect of the air stream, can be detected. This can be achieved with thermistor dissipations of about 5 to 25 mW. Excessive thermistor heating may cause discomfort to the subject. The thermistor is placed as part of a voltage dividing circuit or in a bridge circuit whose unbalance signal can be amplified to obtain the respiratory activity. The method is simple and works well except in the case of some patients who object to having anything attached to their nose or face. This method is found to satisfy the majority of clinical needs including for operative and post-operative subjects.

Occasionally, unconscious patients display a tendency for the uncontrolled tongue to block the breathing system. Under such systems, we are often faced with the situation that not a single millilitre of air is inhaled but the patient's thorax is carrying out large, even though frustral breathing motions. In this condition, the impedance pneumograph and switch methods will show the correct state. Putting the thermistor in a tracheal cannula is not simple. There it is very soon covered with excretions. In the case of suffocated patients with no spontaneous respiration motions, those few millilitres that pass through the cannula are sufficient to drive the breath rate meter. This is a drawback in the technique of using thermistors for the detection of respiration rate.

6.9.3 Impedance Pneumography

This is an indirect technique for the measurement of respiration rate. Using externally applied electrodes on the thorax, the impedance pneumograph measures rate through the relationship between respiratory depth and thoracic impedance change. The technique avoids encumbering the subject with masks, tubes, flowmeters or spirometers, does not impede respiration and has minimal effect on the psychological state of the subject. Impedance method for measuring respiration rate consists in passing a high frequency current through the appropriately placed electrodes on the surface of the body (Fig. 6.36) and detecting the modulated signal. The signal is modulated by changes in the body impedance, accompanying the respiratory cycle. The electrode used for impedance pneumograph are of the self-adhesive type. Contact with the skin is made through the electrode cream layer for minimizing motion artefacts. The electrodes, when the skin is properly prepared, offer an impedance of 150 to 200 Ω. The change in impedance corresponding to each respiratory cycle is of the order of 1% of the base impedance.

The two electrode impedance pneumograph is convenient for use with quiet subjects. Movement artefacts are produced due to changes in the electrode contact impedance, in case the subject is

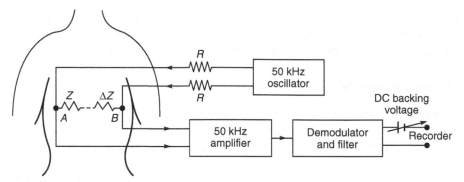

> **Fig. 6.36**   *Principle of impedance pneumograph (two electrode method)*

moving. These artefacts can be significantly reduced by using a four electrode impedance pneumograph. In this case, the output from the oscillator is applied to the two outer electrodes. By doing so, the main oscillator current does not flow through the contact impedance of the measuring electrodes. This system is useful for monitoring restless subjects such as babies.

To avoid the stimulation of sensory receptors, nerves and muscle, currents higher in frequency than 5 kHz must be used for the measurement of physiological events by impedance. Frequencies lower than 5 kHz are particularly hazardous since ventricular fibrillation may be produced with substantial current flow. The use of higher frequencies not only provides the protection sought in the avoidance of tissue stimulation, but also provides the safe use of currents of magnitude, which could be lethal if the frequencies were lower.

Electrical impedance changes associated with physiological activity have been studied extensively. Some of the physiological quantities which have been measured and recorded by the impedance method include respiration, blood flow, stroke volume, autonomic nervous system activity, muscle contraction, eye movement, endocrine activity and activity of the brain cells.

The impedance-based method of measuring respiration rate is commonly employed in patient monitoring systems. The electrodes used for this purpose are the same as those used for ECG measurement. The dynamic measuring range of the amplifier is 0.1 to 3.0 Ω with a frequency response of 0.2 to 3.0 Hz corresponding to respiratory rate of 12 to 180 per minute. The amplifier operates within an impedance window established by the static impedance level (approx. 3 k ohms) and its output produces a respiratory waveform from which respiratory rate is derived.

6.9.4 CO_2 Method of Respiration Rate Measurement

Respiration rate can also be derived by continuously monitoring the CO_2 contained in the subject's alveolar air. Measurement of CO_2 in expired air is otherwise useful in several ways; for example, for originally setting up the respirator and in making adjustments to it afterwards, supervising patients suffering from respiratory paralysis, and other cases where there is respiratory involvement.

The measurement is based on the absorption property of infrared rays by certain gases. Suitable filters are required to determine the concentration of specific gases (like CO_2, CO, and NO_2) constituting the expired air. Rare gases and diatomic gases do not absorb infrared rays.

When infrared rays are passed through the expired air containing a certain amount of CO_2, some of the radiations are absorbed by it. There is a proportional loss of heat energy associated with the rays. The detector changes the loss in heating effect of the rays into an electrical signal. This signal is used to obtain the average respiration rate.

Figure 6.37 shows the arrangement for the detection of CO_2 in expired air. Two beams of equal intensity of infrared radiations from the hot-wire spirals fall on one half of each of the condenser microphone assembly. The detector has two identical portions separated by a thin, flexible metal diaphragm. The detector is filled with a sample of pure CO_2. Because of the absorption of CO_2 in the analysis cell, the beam falling on the test side of the detector is weaker than that falling on the reference side. The gas in the reference side would, therefore, be heated more than that on the analysis side. As a result, the diaphragm is pushed slightly to the analysis side of the detector. The diaphragm forms one plate of a capacitor. The infrared beams are chopped at 25 Hz and the alternating signal which appears across the detector is amplified, shaped and suitably integrated to give the respiration rate.

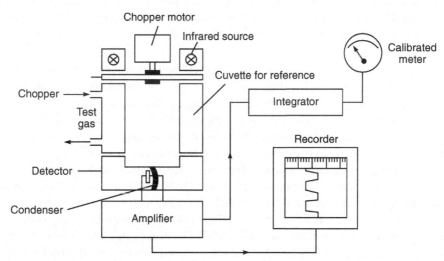

➤ **Fig. 6.37** *Schematic diagram for detection of CO_2 in the expired air for continuous monitoring of respiration rate*

6.9.5 Apnoea Detectors

Apnoea is the cessation of breathing which may precede the arrest of the heart and circulation in several clinical situations such as head injury, drug overdose, anaesthetic complications and obstructive respiratory diseases. Apnoea may also occur in premature babies during the first weeks of life because of their immature nervous system. If apnoea persists for a prolonged period, brain function can be severely damaged. Therefore, apnoeic patients require close and constant observation of their respiratory activity. Apnoea monitors are particularly useful for monitoring the respiratory activity of premature infants.

Several contactless methods are available for monitoring the respiration of infants. The most successful apnoea monitors to-date have been the mattress monitors. These instruments rely for their operation on the fact that the process of breathing redistributes an infant's weight and this is detected by some form of a pressure sensitive pad or mattress on which the infant is nursed. The mattress, in its simplest form, is a multi-compartment air bed, and in this case the weight redistribution forces air to flow from one compartment to another. The air flow, is detected by the cooling effect it produces on a heated thermistor bead. Though the technique is simple, the main disadvantage with the air mattress is the short-term sensitivity variation and the double peaking effect when inspiration or expiration produce separate cooling of the thermistor.

Alternatively, a capacitance type pressure sensor in the form of a thin square pad is usually placed under or slightly above the infant's head. Respiratory movements produce regular pressure changes on the pad and these alter the capacitance between the electrode plates incorporated in the pad. This capacitance change is measured by applying a 200 kHz signal across the electrodes and by detecting the current flow with a phase-sensitive amplifier. Two types of electrodes can be used: (i) 70 mm plates, 350 mm apart in a plastic tube which is placed alongside the body; (ii) 250 mm long, 60 mm diameter cylinders placed one on either side of the body. This system is much too sensitive to people moving nearby and thus an electrically screened incubator is essential for the infant.

Impedance pneumography is another practical method to monitor the breathing of the patient. The technique also enables the simultaneous monitoring of the heart rate and respiration. The heart rate is known to drop during apnoea. Monitoring the heart rate and respiration thus gives an extra measure of security. Electrodes measure the effort to breath and not the actual ventilation (Kulkarni, 1991). Impedance pneumography has certain inherent disadvantages. One is that the placement of the electrodes is very critical and the other is cardiovascular artefact. This results from the detection of movement between the electrodes because of the cardiovascular system, rather than due to respiration. Apnoea monitors need to be designed to reject this artefact. Silvola (1989) describes a new non-invasive piezo-electric transducer for the recording of respiration, heart rate and body movements using the PVDF (polyvinylidenefluoride) polymer film. The transducer consists of an area of about 1000 cm^2 PVDF film (length 40–50 cm, width 20–30 cm) with a thickness of 40 µm. The PVDF elements are placed directly on the bed mattress without being fixed on the skin. The recordings can be performed when the subject is lying on their back, stomach or on their side.

Apnoea monitors are generally designed to give audio-visual signals under apnoeic conditions when no respiration occurs within a selectable period of 10, 20 or 30 s. The apnoea monitors are basically motion detectors and are thus subject to other motion artefacts also which could give false readings. The instruments must, therefore, provide means of elimination of these error sources. Figure 6.38 shows a block diagram of an apnoea monitor. It basically consists of an input amplifier circuit, motion and respiration channels, a motion/respiration discrimination circuit, and an alarm circuit. The input circuit consists of a high input impedance amplifier which couples the input signal from the sensor pad to the logic circuits. The sensor may be a strain gauge transducer embedded in the mattress. The output of the amplifier is adjusted to zero volts with offset adjustment provided in the amplifier. The amplified signal goes to motion and respiration channels connected in parallel. The motion channel discriminates between motion and respiration

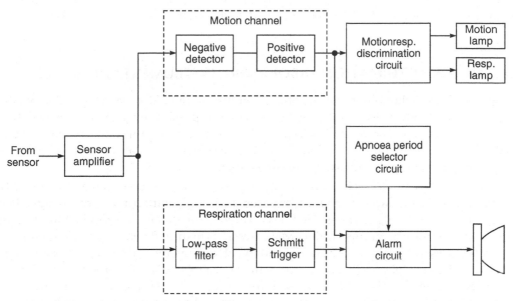

> ➤ **Fig. 6.38** *Block diagram of apnoea monitor (Courtesy: B-D Electrodyne, USA)*

as a function of frequency. In the case of motion signals, high level signals above a fixed threshold are detected from the sensor. In the respiration channel, a low-pass filter is incorporated. Low frequency signals below 1.5 Hz (respiration) cause the output of the Schmidt trigger circuit to pulse at the respiration rate. Higher frequency signals, above 1.5 Hz (motion), cause the output of the trigger to go positive. Absence of the signal (apnoea) causes the output of the Schmidt trigger to go negative.

The outputs of the motion and the respiration signals are combined in a comparator circuit, which compares the polarities of the motion and respiration channel signals to indicate respiration. The presence of respiration is indicated by a flashing lamp. The output of the discrimination detector also goes to an apnoea period selector circuit, a low frequency alarm oscillator and driver, a tone oscillator and audio amplifier connected to a speaker. Audible alarm is given at a frequency of 800–1000 Hz, which is pulsed at 2 Hz.

An alternative method of detecting apnoea is based on electromagnetic induction. It consists in passing a high frequency alternating current through a transmitting coil and creating an alternating magnetic field. The transmitting coil is placed at some distance from the infant. The receiving coil is applied on to the abdomen of the infant. The alternating magnetic field induces an emf in the receiving coil. The movement of the abdominal wall with the infant's respiration results in inducing an amplitude-modulated signal in the receiving coil. If this amplitude-modulated signal is demodulated, the modulation frequency corresponding to the respiration frequency can be recovered.

Another contactless method for monitoring the breathing activity of premature babies is by the use of microwave energy. It operates by directing low intensity microwave energy (10 GHz) at the individual to be monitored and detecting this energy, after its reflection from the moving surface. The difference between the transmitted and received microwave frequencies (Doppler-shift) provide a signal voltage, which can be amplified, and used as an indicator of the continuance of respiratory activity.

6.10 CATHETERIZATION LABORATORY INSTRUMENTATION

An important field where continuous patient monitoring can aid in carrying out sophisticated procedures which otherwise would have been either impossible or difficult to carry out is the cardiac catheterization laboratory. These procedures result in huge amounts of data which need to be acquired processed, analyzed, correlated and stored. The calculations involved can be done more accurately and rapidly by using computer-based systems.

A cardiac catheterization laboratory is a place for carrying out specialized catheterization procedures, providing surgical facilities for pacemaker implantation and carrying out advanced research in biomedical engineering. Facilities in catheterization laboratories include systems for recording of electrocardiograms, intra-cardiac pressures in the left ventricle (LV) and pulmonary artery, atrial pressure, right atria pressure and dye densitometer and then processing to get the pulmonary artery systolic/end diastolic pressure, LV systolic/diastolic pressures cardiac output (CO) and cardiac index (CI), stroke volume, intra-cardiac potentials, heart sounds, oximetry, etc. Number of channels available in each case depends upon the procedure to be carried out. Usually two channels for pressure monitoring are considered to be sufficient. Various signals from the patient are recorded on a multi-channel graphic recorder and simultaneously displayed on a large screen monitor mounted at a height to facilitate convenient viewing from a distance. A multi-channel analog FM tape recorder is added to the system for the storage of physiological data. The data can be replayed at a later stage for future studies and detailed analysis.

The computerized system provides the physician with an immediate pressure waveform analysis to allow a step-by-step assessment of the progress of catheterization. It helps to make rapid decisions regarding the course and duration of the procedure. The heart of the system is the online analysis of pressure waveform for calculations of systolic pressure, diastolic pressure, mean pressure, pressure derivative (dp/dt), etc. in the chambers of the heart, arteries or the aorta. Additional calculations like systolic ejection period, diastolic filling period, stroke volume, cardiac output and valve areas, etc. are also calculated. Figure 6.39 shows a system configuration employed in a typical catheterization lab.

The information is obtained online during the catheterization procedure from a selected ECG lead, up to three pressure transducers to measure oxygen saturation and a dye dilution sensor for measuring the cardiac output. A summary of the analyzed pressure gradients, valve areas and cardiac output is displayed on the video monitor.

6.10.1 Pressure Measurements

To make pressure measurements, the operator identifies the pressure sites and starts pressure sampling. Five or eleven successive beats of the pressure signal are sampled. Although the analysis is done for three or nine beats, five or eleven beats are sampled to allow for the possibility of an incomplete first beat and for the part of the last waveform necessary to define, among other things, LV end diastolic pressure. Nine beat analysis is used for automatically rejecting PVCs and other

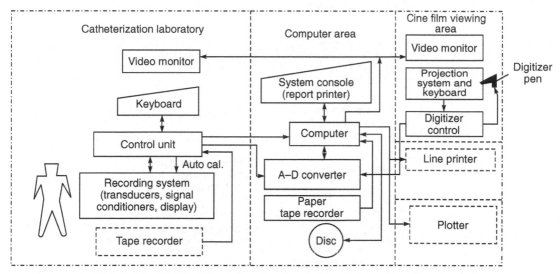

> **Fig. 6.39** *Typical instrumentation layout in a catheterization lab*

arrhythmias and for averaging out respiratory variations. Three beat analysis increases the speed of analysis and is generally used where the rhythm is regular. This concept of 'statistically selected beats' is employed to minimize the effect of signal distortion by arrhythmias and the occasional premature ventricular contractions elicited by the catheter position. Figure 6.40 shows a flow chart for pressure signal processing in a catheterization laboratory.

The computer smooths noise fluctuation from the raw pressure data by use of a low-pass Fourier filter, which also permits accurate dp/dt calculation. The cut-off frequencies are preset to 17 Hz for ventricular pressure and 6 Hz for non-ventricular pressures. The control unit of the computer generates a *QRS* timing pulse from the electrocardiogram and this is used by the computer for reference timing of the pressure waveforms.

The search for peak systolic pressure begins at 50 ms after the *QRS* pulse (Fig. 6.41). To reject the possibility of detecting an early erroneous peak in the presence of catheter fling, the search is continued for 100 ms after a peak is found. Any large value found within this 100 ms is accepted as the new peak. The maximum dp/dt is found for each beat by searching forward from the *QRS* onset +50 ms to peak systolic pressure.

Another important parameter calculated is V_{max}, which is the isovolumetric index of the cardiac function or contractility (Mirsky *et al*, 1974). This is obtained by extrapolating the downslope of the $dp/kP\, dt$ versus P curve back to zero pressure. In this expression, P is the total left ventricular pressure, and K a stiffness constant, with a value of 30.

Beginning diastolic pressure is detected in the interval between the maximum peak systolic pressure and a point 100 ms before the next *QRS* onset pulse. Maximum (P_{max}) and minimum (P_{min}) pressure are measured in that interval and the computer searches for the first point on the downstroke where the pressure curve slope has a magnitude less than $\dfrac{(P_{max} - P_{min})}{2}$ mmHg/s. If

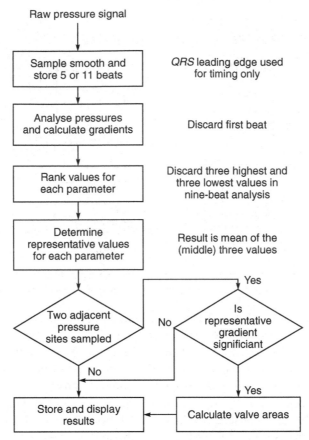

> **Fig. 6.40** *Flow chart for pressure signal processing in cath. Lab (Courtesy: Hewlett Packard, USA)*

this slope is not found before the minimum pressure point is reached, beginning diastolic pressure is taken as the minimum pressure.

End diastolic pressure is located between the *QRS* onset-20 ms and the minimum *dp/dt* on the next pressure complex. Starting from the maximum *dp/dt* point, a backward search is made for end diastole, which is the first point encountered whose slope is less than $(P_{max} - P_{min})$ mmHg/s. Pattern recognition of direct and indirect (pulmonary artery wedge) arterial pressure involves the timing of pressure oscillations with the electrocardiogram and the setting of maximum local values for waves in the appropriate halves of the *R-R* interval.

Among the valvular parameters measured and/or calculated are outflow gradients and inflow gradients. Gradients can be determined from either simultaneous or non-simultaneous pressures on each side of the valve. Pressures that are not sampled simultaneously are matched beat-by-beat. The pressure sites used for valve gradients are those closest to the valve.

Systolic ejection period begins at the pressure on the ventricular pressure waveform equal to the diastolic arterial pressure. If the two do not occur simultaneously, the computer shifts the arterial

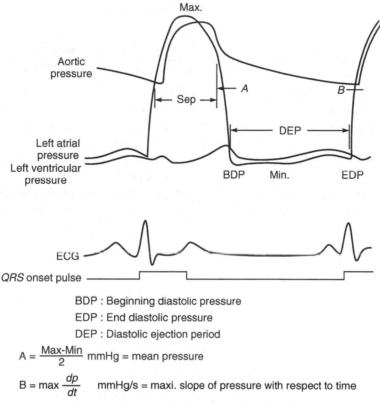

BDP : Beginning diastolic pressure
EDP : End diastolic pressure
DEP : Diastolic ejection period

$$A = \frac{Max-Min}{2} \text{ mmHg} = \text{mean pressure}$$

$$B = \max \frac{dp}{dt} \quad \text{mmHg/s} = \text{maxi. slope of pressure with respect to time}$$

➤ Fig. 6.41 *Determination of ventricular pressure measurement points, systolic ejection period and diastolic filling period (Courtesy: Hewlett Packard, USA)*

pressure waveform backward relative to the ventricular pressure waveform by the required time delay. The computer then searches for a second crossing point between both pressures. The search starts at a point on the arterial pressure waveform corresponding to the beginning of the ventricular systole. The crossing point marks the end of the systolic ejection period.

The search for the diastolic filling period starts at the point detected as beginning diastolic pressure on the ventricular waveform and proceeds backward until a crossing between the ventricular and atrial or shifted pulmonary wedge pressure is found. The search for the end of diastolic pressure on the ventricular beats moves forward until an intersection is found. If no crossing is found, then the end of the diastolic filling period is set at end diastolic pressure.

Some other useful calculations made during catheterization procedures are as follow:

$$\text{Valve area (cm}^2\text{)} = \frac{\text{Valve flow (ml/s)}}{C\sqrt{\text{Valve gradient (mmHg)}}}$$

Where $C = 40$ for mitral valve and 44.5 for aortic, pulmonic and tricuspid valves.

The formula is applicable only to stenotic valves (Grolin and Grolin, 1957). All valve areas greater than 3.5 cm^2 are not reported.

$$\text{Valve flow (mitral and tricuspid valves)} = \frac{\text{Cardiac output (ml/min)}}{\text{Diastolic filling period (s/min)}}$$

$$\text{Valve flow (aortic, pulmonic valves)} = \frac{\text{Cardiac output (ml/min)}}{\text{Systolic ejection period (s/min)}}$$

Computerized measurement of blood flow can be made either by the dye dilution or Fick method. Oxygen saturation readings of the blood sample withdrawn through a cuvette oximeter are accomplished by sampling directly the output of the red and infrared cell of the instrument four times per second. Logarithmic conversion in the computer permits values to read as percent saturation of the blood with oxygen. This instrument is used for measuring cardiac output by the indicator dilution method. Dye injected into the circulation is detected by continuously sampling from a downstream sampling site. The clinical findings are entered into the selective retrieval file and preparation of the final report and patient assessment are essentially complete at the conclusion of each procedure.

Arrhythmia and Ambulatory Monitoring Instruments

Ⅲ▷ 7.1 CARDIAC ARRHYTHMIAS

Any disturbance in the heart's normal rhythmic contraction is called an arrhythmia. Patients undergoing a seemingly uneventful recovery from myocardial infarction may develop cardiac arrest as a direct and immediate result of ventricular fibrillation. It is possible to treat and reverse many of these dangerous episodes if they could be detected early and an advance warning of their onset could be made available. The necessity for early detection of these harbingers of catastrophic arrhythmias led to the establishment of coronary care units in hospitals in the early 1960s for the intensive monitoring and treatment of patients with acute myocardial infarction. The attempt in these units was to effectively carry out resuscitation techniques such as cardiac massage and transthoracic defibrillation. During the last decade, attention was focussed on early detection and treatment of arrhythmias and the emphasis of treatment switched from resuscitation to aggressive prophylactic therapy.

Detailed and extensive examination of ECG records has shown that abnormalities in the functioning of the heart invariably manifest themselves in ECG waveform. Diagnostic statements as observed from the ECG records are classified into two classes: (i) *morphological statements*—primarily based on ECG waveshapes that attempt to describe the state of the working muscle masses and (ii) *rhythm statements*—concerned with the site and rate of the cardiac pacemaker and the propagation of impulses through the conduction system.

Sometimes, irritation occurs in the ventricles, the self-triggering impulse does not arrive through the AV node and thus travels a different and slower path in spreading over the ventricles. The *QRS* then becomes widened, and is classified as ventricular ectopic beat. Thus, an ectopic beat is a beat, which starts in an abnormal location in the heart and is often premature, therefore also called premature ventricular contraction (PVC), i.e. it occurs sooner than the next expected beat. Ventricular ectopic beats result in an abnormal depolarization sequence, the ECG displays an abnormal *QRS* morphology, often with a pronounced increase in width and change in amplitude. Figure 7.1 shows waveforms corresponding to some common types of arrhythmias.

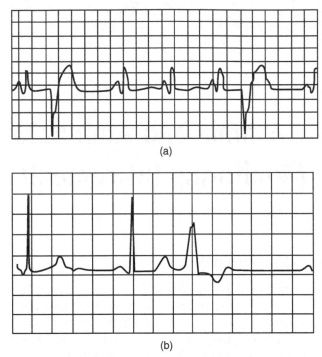

(a)

(b)

> **Fig.** 7.1   (a)   *ECG waveform shows ventricular ectopic beat on second pulse*
>   (b)   *Widened QRS-classified as a ventricular ectopic beat*

7.2   ARRHYTHMIA MONITOR

An arrhythmia monitor is basically a sophisticated alarm system. It is not an ECG interpretation system. It constantly scans ECG rhythm patterns and issues alarms to events that may be premonitory or life threatening. Arrhythmia monitors are available in various degrees of sophistication, but all are ventricular oriented, detecting most of the significant ventricular arrhythmias including ventricular premature beats which comprise the majority of such events. While the complex computerised systems are useful for multi-patient set-ups and can help detect arrhythmias of a wide variety at graded alarm levels, the comparatively simpler instruments mostly look for widened *QRS* waves and heart timing for premature beats. With the availability of low cost PCs, desk-top arrhythmia monitors are now available at lower cost, these provide comparative monitoring facilities which were earlier derived from large computerized systems.

The arrhythmias, which the instruments are designed to detect, are premature *QRS* complexes, widened *QRS* complexes and runs of widened complexes. Because each patient's ECG may differ, the instruments generally base their determination of abnormal or ectopic beats upon a reference obtained from the patient himself. Therefore, any arrhythmia monitoring instrument will operate in the following sequence:

- Stores a normal *QRS* for reference, particularly *QRS* width and *R–R* interval. An external ECG recorder is automatically activated during the store normal mode so that the reference heart beats may be visually examined and determined as to whether they are truly representative.
- Initiates an alarm automatically, when ectopic beats are detected—either the ventricular prematured or widened varieties.
- Gives alarm light signals whenever the prematured or widened ectopic beats exist up to the rate of 6/min or 12/min.
- Detects and triggers alarm when artifacts are present at the source, e.g. muscle tremor due to patient movement, base line shift and improperly connected electrodes.

Figure 7.2 shows the major signal processing tasks performed by most automated arrhythmia monitoring and analysis systems. The function of each of these blocks is as follows:

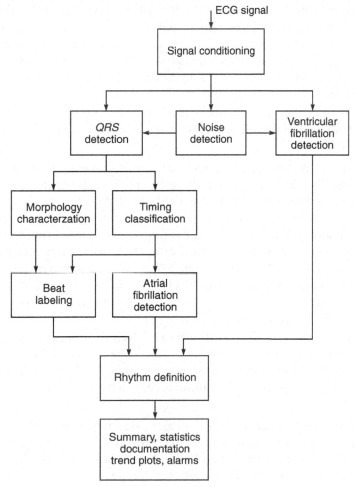

> **Fig. 7.2** *Block diagram of basic arrhythmia monitoring system*

Signal Conditioning: Single or multiple ECG leads may be used for arrhythmia monitoring. Although the first generation arrhythmia systems were used on a single lead, the present day systems analyze two or more leads simultaneously. ECG signal is amplified, filtered (0.05–100 Hz for diagnostic purposes, 1–40 Hz for monitoring purposes) and digitized using an 8-or 12-bit analog-to-digital convertor with a typical sampling rate of 250 Hz.

Noise Detection: Inspite of analog or digital filtering performed on the ECG signal, some unwanted noise and artifact still remain. Baseline wander, motion artifact and muscle noise all have some energy that overlaps the ECG signal spectrum. Using specialized signal processing techniques, unwanted noise and artifact are minimized.

For example, electrode motion artifact is the most troublesome for arrhythmia analyzers, since it contains considerable power in the ECG signal band. This can be detected by measuring the electrode-skin contact impedance. Motion of the electrode with respect to the skin causes changes in the contact impedance generating electrical artifact. The impedance signal can thus provide an independent measure of electrode motion, which can be helpful to prevent false alarm.

The ECG waveform is processed by two digital filters: a detection filter and a classification filter. The detection filter removes low frequency noise (baseline wander) and muscle artifact. *P* waves and *T* waves are diminished. This filter helps avoid an erroneous detection of tall *T* waves as beats. Even though the shape of the *QRS* is distorted, the output from the detection filter is used only for beat detection.

The classification filter removes signal irregularities, and preserves the important features of the *QRS*. So, the resulting ECG output can be used for feature measurements and beat classification (Hewlett Packard, 1999a).

QRS Detection: Arrhythmia monitors require reliable *R* wave detectors as a prerequisite for subsequent analysis. The steep, large amplitude variation of the *QRS* complex is the obvious characteristic to use and this is the function of the *R* wave detector. Any subsequent analysis is entirely dependent upon the output from the *R* wave detector and, therefore, it is important that this unit should function reliably. Most analog devices use various filtering methods to extract the *QRS* complex by attenuating *P* and *T* waves and artifacts. Since the maximum of the *QRS* energy spectrum is in the vicinity of 10 Hz (Clynes *et al.*, 1970), the filter is designed to have a bandwidth of about 15 Hz with a centre frequency of 10–12 Hz. By using a bandpass filter rather than a low-pass filter, the amplitude of low frequency noise as well as the low frequency components of the ECG will be reduced without affecting the *QRS*.

QRS detection is now almost universally performed digitally in a two-step process. The ECG is first preprocessed to enhance the *QRS* complex while suppressing noise, artifact and non-*QRS* portions of the ECG. The output of the preprocessor stage is subjected to a decision rule that confirms detection of *QRS* if the processor output exceeds a threshold. The threshold may be fixed or adaptive.

Morphology Characterization: This is based on analyzing the shape of the *QRS* complexes and separating beats into groups or clusters of similar morphology. Most algorithms for real time arrhythmia analysis maintain no more than 10–20 clusters at a time, in order to limit the amount of computation needed to assign a *QRS* complex to a cluster.

Timing Classification: It involves categorization of the *QRS* complexes as on time, premature or late. The observed *R–R* interval is compared to an estimate of the expected *R-R* interval. An *R-R* interval will be declared premature if it is less than 85% of the predicted interval. Similarly, an *R–R* interval is long if it is greater than 110% of the predicted value.

Beat Labelling: A physiologic label is assigned to each *QRS* complex. The possible beat labels that can be attached by a beat classification module include the following: normal, supraventricular premature beat, PVC, etc. This is the most complex form of the algorithm and is rarely disclosed by the manufacturers.

Rhythm Labelling: This is the final stage in arrhythmia analysis. It is based on defined sequences of *QRS* complexes. The analysis systems are heavily oriented towards detecting ventricular arrhythmias, particularly single PVCs. Special detectors are employed to identify rhythms such as atrial fibrillation or ventricular fibrillation.

Atrial Fibrillation Detection: It is based on detecting abnormal rhythms from the timing sequence of *QRS* complexes.

Ventricular Fibrillation: This is usually detected by frequency domain analysis. The system is characterized as a narrow-band, low frequency signal with energy concentrated in a band around 5–6 Hz. It can be distinguished from noise (16–18 Hz) by appropriately designing bandpass filters.

Summary Statistics: These characterize the cardiac rhythm over long time periods. These statistics may be presented in the form of a table or graphically. Trend plots of heart rate and abnormal beats are particularly useful to the clinician.

Alarms: These are necessary to bring to the attention of the nursing staff the serious arrhythmias suffered by the patient with appropriate alarms.

Advanced software techniques and sophisticated algorithms are constantly being developed for the automated analysis of arrhythmias. Recent efforts are focusing on the possible application of artificial intelligence methodology into the design of such systems particularly to ensure optimal alarm strategies.

▶ 7.3 QRS DETECTION TECHNIQUES

There are several methods and computer programs in existence for the automatic detection of *QRS* complexes. These include the use of digital filters, non-linear transformations, decision processors and template matching. Generally, two or more of these techniques are used in combination in a *QRS* detector algorithm.

A popular approach in the detection of arrhythmias is based on template matching. A model of the normal *QRS* complex, called a template, is derived from the ECG complex of a patient under normal circumstances. This template is stored and compared with the subsequent incoming real-time ECG to look for a possible match, using a mathematical criterion. A close enough match to the template represents a detected *QRS* complex. If a waveform does not match the available template but is a suspected abnormal *QRS* complex, it is treated as a separate template, and future suspected *QRS* complexes are compared with it. Obviously, the system requires considerable memory for storing the templates. Alternatively, algorithms have been developed based on digital filters to separate out normal and abnormal *QRS* complexes.

7.3.1   ST/AR Arrhythmia Algorithm

The ST/AR (ST and Arrhythmia) algorithm from Hewlett Packard is a multi-lead ECG algorithm designed for both arrhythmias and ST segment monitoring. The algorithm processes the ECG signals for both paced and non-paced patients and performs several actions on the incoming ECG waveform, including filtering the signal, detecting and classifying the *QRS*, generating heart rate, identifying ectopic events and rhythms and generating alarms, when necessary.

In order to reliably detect the *QRS*, the detection threshold is kept as 0.15 mV to prevent the detection of *T* waves or baseline noise as *QRS* complexes during a complete heart block or asystole. For optimal performance and to prevent false alarms, the lead selected for monitoring should have adequate amplitude. Figure 7.3 shows the arrangement for generating the *QRS* detection signal using multiple leads. The contribution from each ECG lead to the *QRS* detection signal is propotional to its measured quality based on the waveform amplitude, and the amount of muscle and baseline noise. The weighting factors are updated atleast every 200 ms to allow for quick adaptation to signal quality changes (Hewlett Packard, 1999(a)).

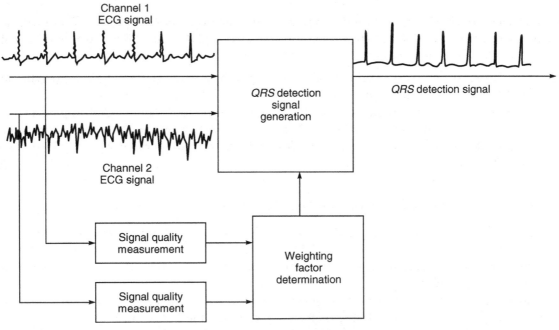

> **Fig. 7.3**   *Generating the QRS detection signal in ST/AR system*

The *QRS* detector checks the *QRS* detection signal for the presence of the peak of an *R* wave. Search begins after an absolute refractory period from the previously identified *QRS* complex. The value used for the refractory period is 192 ms. This helps to prevent a *T* wave from being identified as an *R* wave.

After a *QRS* complex is identified, a search is made on each lead independently in the area prior to the *R* wave to determine if there is an associated *P* wave. This area is 200 ms wide and ends 120 ms before the *R* wave peak (Fig. 7.4). To be accepted as a *P* wave, it must be atleast 1/32 of the *R* wave height and the *P–R* interval must be close to the average *P–R* interval. *P* wave detection is used to differentiate between a Sinus Rhythm and a Supraventricular (SV) Rhythm.

After detection of the *QRS*, a number of features which represent beat characteristics and which can be used to discriminate between different type of beats are measured. The features measured are: height, width, area and timing (*R–R* interval).

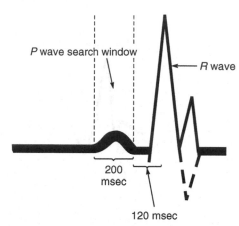

> ➤ **Fig. 7.4** *P-wave detection technique*

Once the *QRS* is detected and measured, the beat is labelled. Labelling means that the algorithm assigns the complex one of the following labels: normal (**N**), supraventricular premature (**S**), ventricular ectopic (**V**), paced (**P**), questionable (**?**) and learning (**L**). If the signal quality is not good, the algorithm assigns the label "inoperative (**I**)" and "artifact (**A**)".

Beat labelling is based on the use of template families to represent recurring morphologies. For each patient, up to 16 different active template families can be created for each individual lead. To keep the template family information current, they are dynamically created and replaced as the patient's beat shape changes. When a beat is detected, it is matched against the stored waveform templates for that patient. This process involves overlaying the beat on the template and using a mathematical procedure to measure the differences between the two shapes. Figure 7.5 illustrates the technique of template matching.

A separate detector continuously examines the ECG signal for ventricular fibrillation. If a flutter or sinusoidal waveform persists for more than 4 seconds in any ECG channel, then the monitor alarms for ventricular fibrillation.

Normally, the heart rate is computed by averaging the most recent 12 *R–R* intervals. This average gives a stable estimate of the heart rate even when the rhythm is irregular. Alarms are activated by the alarm generator wherein higher priority alarms such as asystole take precedence and supersede lower priority alarms, such as low heart rate.

Any computerized arrhythmia algorithm would not make 100% accurate analysis of all patients. Data bases are available for the testing of arrhythmia detection algorithms. These data bases consist of records of patient ECG waveform, together with a set of annotation files in which each beat has been labelled by an expert cardiologist. One such data base is from the American Heart Association (AHA), distributed by the Emergency Care Research Institute (ECRI), USA which contains data from 80 patients, two of whom are paced. The performance results from the ST/AR arrhythmia algorithm (Hewlett Packard 1999b) are shown to be giving very high accuracy and reliability (above 95%), in both single lead and multi-lead arrangements.

Friesen *et al* (1990) compared noise sensitivity of nine types of *QRS* detection algorithms. He established that an algorithm using a digital filter had the best performance for the composite noise corrupted data.

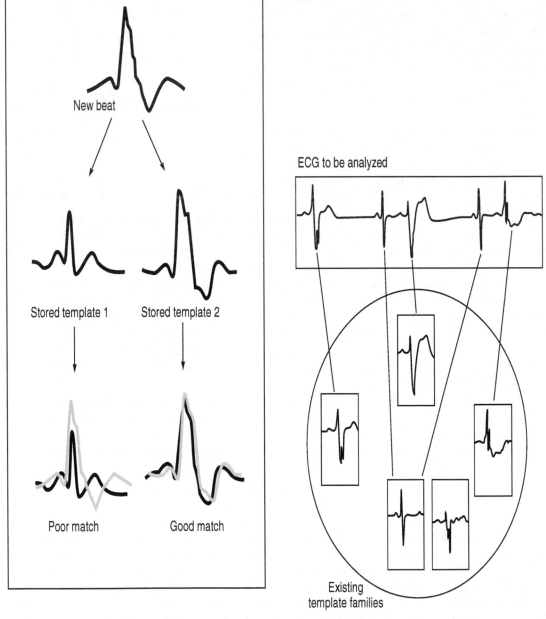

> **Fig. 7.5**   *Template matching technique for detection of arrhythmias*

7.3.2 Data Compression and Processing of the ECG Signal by AZTEC

To handle the relatively high data rate of ECG signals for real-time analysis of rhythm, it is necessary to use some technique of data compression so that less memory is required to store the waveforms and the overall processing time is less. Cox and Nolle (1968) suggested AZTEC (Amplitude-Zone-Time-Epoch-Coding)—a pre-processing program for real time ECG rhythm analysis at an approxi-mate data reduction rate of 10 : 1.

Basically, AZTEC regards the ECG signal as a series of straight line segments, each segment representing a sequence of consecutive points with approximately the same amplitude. Therefore, AZTEC's first step is to convert the ECG signal into such segments, some of short and others of long duration. For example, *QRS* complex is converted into a series of short line segments as the signal amplitude changes rapidly. A series of lines, each containing four samples (with 500 samples/s) or less, is considered to be adequately represented by a constant rate of voltage change or slope as long as the voltage difference between adjacent lines does not change sign. The slope is terminated by a line longer than the four samples or a change in signs. The slope duration and the voltage between the lines bounding the slope are then stored as the next pair of data words. Thus AZTEC also examines sequences of short segments moving in the same direction of amplitude and replaces them with slope segments, thereby further compressing the data. Figure 7.6 illustrates an ECG signal and its resulting AZTEC representation, which is a caricature of the original ECG signal. It is composed of sequences of straight line segments and sloping line segments. Further, straight line segments are stored in the computer as voltage and time duration whereas the sloping line segments are stored as voltage variation (+ or –) and time duration. AZTEC also contains digital logic, which is able to detect base line wander. Also, the sum of absolute values of all slopes is used for noise detection, but sudden bursts of muscle artifact or an excursion from baseline is distinguishable from an ECG wave only if the condition persists.

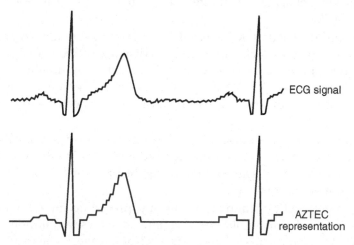

➤ **Fig.** 7.6   *ECG signal and its resulting AZTEC representation (after Cox and Nolle, 1968; by permission of IEEE Trans. Biomed. Eng.)*

Accurate description and delineation of ECG waveform by a complex series of scans through the AZTEC data is the next step. The module for processing this function identifies QRS complex and P and T waves. It systematically determines each complex's beginning and end, and performs the measurement of four QRS parameters (height, duration, offset and R–R interval). This module functions as follows:

- The module runs once every second, depending upon the quality of the signal. If the signal is 'chaotic' (a signal of poor quality that cannot be processed) or 'noisy' (a signal of intermediate quality), no processing is carried out.

- The memory is then scanned to find the QRS complex until the operation is temporarily suspended to give the artifact time to terminate in the case of a poor quality signal.

- Starting with the last QRS found and processed, R wave is looked for after a refractory period of 120 or 200 ms. The scan looks for a slope greater than 1/3rd the height of the R wave of the patient's 'normal' QRS. If no such slope is found within the interval, the scan is repeated with the slope criterion relaxed to 1/6th of the height. If no QRS is detected in a distance of two times the normal 'R–R' interval, less 1/4th the previous R–R interval, then it is a case of 'missed beat'.

- After finding the 'R' wave, the QRS complex is determined by the following criteria: (a) it must be V-shaped and must not contain (wiggles) lines and slopes < 3/8th the largest slope and have a duration longer than 32 ms, (b) the other slope of the complex must be at least 3/8th the amplitude of the largest slope found. A scan is then performed on each side of the complex until a `wiggle' with a duration greater than 32 ms is found. These two points define the limits of QRS.

- After establishing the QRS and its extent, the QRS performs the four measurements required to compare it with the patient's 'normal' QRS. These measurements refer to QRS height, duration, offset and R–R interval. The basis for making these measurements is shown in Fig. 7.7.

- In order to confirm that the QRS detected as per the above mentioned procedure is the right one, the processor goes back and scans for a distance of 300 ms following this assumed QRS. It is quite probable that an abnormally big sized P wave might have been mistaken as a R wave. If more than one R wave possibilities are observed, the one with the highest amplitude is taken as the true R wave and the process repeated to determine the complex limits.

- A T wave following a QRS is also searched. A wave is considered as a T wave if (i) it occurs within a 200 ms interval, (ii) its peak occurred within 1/3rd the R–R interval +80 ms or within 240 ms whichever is greater, following the preceding R wave, and (iii) its height is half the height of a normal R wave.

- P waves are detected by scanning for 300 ms, preceding the QRS, and measuring the minimum and maximum amplitudes encountered. If these amplitudes are consistent with the patient's 'normal' P wave, the wave is taken as P wave. P waves are used to determine the sinus rhythm—(more than 50% of the beats are preceded by P waves.)

The four measurements of QRS and a fifth measurement called 'distance D' based on height, duration and offset to those of the patient's 'normal' QRS are compared. Beats are classified on the

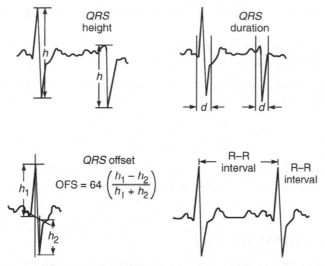

> ➤ **Fig.** 7.7 *Detection criterion for QRS is based upon QRS height, QRS duration, QRS offset, QRS polarity and R-to-R interval (reproduced with permission of Hewlett Packard, USA)*

basis of their deviation from the normal and may be definitely normal, probably normal, probably PVC or definitely PVC. Only those beats are used for updating the 'stored' normal which are definitely normal. A table is prepared on the value of D calculated on the basis of 'normal' and current values. The table provides 120 possibilities, which are isolated and the result in terms of probability of PVCs is presented.

7.3.3 Detection of Ventricular Fibrillation

Most of the arrhythmia monitoring systems do not always generate an alarm during ventricular fibrillation, as a *QRS* complex with known and established characteristics is not found to be present. Ventricular fibrillation is characterized by a sinusoidal ECG waveform resulting from uncoordinated heart muscle activity. Therefore, the detector must be capable of using the sinusoidal characteristics of the waveform and generate alarm under such conditions. The block diagram of the detector is shown in Fig. 6.13.

The detection of ventricular fibrillation is based upon two criteria: the frequency of the sinusoid and the filter leakage fraction. The average physiologically occurring frequency filter leakage fraction is computed as follows.

The average period of the ECG waveform for one second is first determined by the following equation:

$$T = 2\pi \sum_{1s} \frac{V_n}{|V_n - V_{n-1}|}$$

where T = the number of sample points in an average period

   V_n = the value of the *n*th sample

The period (T) is then used to adjust a notch filter, the notch to be positioned at the average frequency corresponding to the average period determined above. This filtering is achieved by adding two samples over the average half period.

$$(\text{Filter output})_n = V_n + V_{n-T/2}$$

The unfiltered signal and filter output are summed over a one second period to calculate the 'filter leakage fraction' and is given by:

$$\text{Filter leakage fraction} = \sum_{1\,s} \frac{|V_n + V_{n-T/2}|}{|V_n| + |V_{n-T/2}|}$$

For an ideal sinusoidal waveform, this fraction will be zero. As the ventricular fibrillation waveform is not an ideal sinusoid, a higher leakage fraction is used as the threshold for detection of this condition.

▷ 7.4 EXERCISE STRESS TESTING

Stress test or exercise electrocardiography is used when the diagnosis of coronary arterial disease is suspected or to determine the physical performance characteristics of a patient. The test involves the recording of the electrocardiogram during dynamic or occasionally isometric exercise. The diagnostic value of exercise testing primarily concerns either depression or elevation of the ST segment present in myocardial ischemia.

Dynamic exercise is performed by the patient who walks on a treadmill on which the speed and elevation can be adjusted, manually or automatically, to suit a variety of graded exercise protocols. Alternatively, the patient may be asked to pedal an electrically braked bicycle ergometer. Both the treadmill and ergometer can be used as stand-alone devices for testing physical fitness. Advanced ergometers and treadmills can store and display activity data, transfer it to a PC and download patient data from the PC.

7.4.1 Treadmill Test

There are two basic kinds of exercise protocols used in treadmill tests:

- *The Balke-Ware Protocol:* It uses a constant speed of 3.3 miles/hour (5.3 km/hour), with progressive increments in the load every 2 minutes. This is achieved by increasing the grade or incline of the motor-driven treadmill. The exercising subject therefore walks "up-hill" and his own body weight serves as the load.
- *The Bruce Protocol:* It uses simultaneous increments in both speed and treadmill grade at intervals of 3 minutes.

Both the protocols are satisfactory for most clinical purposes. However, they may have to be modified to suit the condition of the individual being subjected to exercise testing.

7.4.2 Bicycle Test

In this test, the speed is usually kept constant and incremental resistance is applied either mechanically (friction belt or brake pads) or electrically. The work load can be precisely measured in terms of kilograms per metre per minute. In bicycle ergometry, physiological measurements are more easily obtained with the subject comfortably seated and the torso relatively immobile.

The stress test laboratory is equipped with an exercise device, a PC-based ECG display and recording system, a blood pressure measuring instrument (sphygmomanometer) and a defibrillator. Sophisticated algorithms (digital filters and programs for the morphologic recognition) have been developed in order to obtain stabilized artifact-free ECG tracings. The system displays in real time three user selectable leads, a reference ECG complex (template) and the current mean beat on which the system detects stress-induced changes. Automatic measurements of ST level and slope, current and maximum heart rate are displayed in real time and can be printed as alphanumeric data or as trends. The usefulness of the exercise stress test lies in assessing the extent and severity of the cardiovascular disease and it remains a useful tool in providing the important prognostic information.

The exercise test has become an established tool for the diagnosis of coronary artery disease due to changes observed in the ECG induced by exercise. The principal ECG abnormalities that are recognised as manifestations of ischemia are ST segment depression and ST segment elevation. The current standard of determining the ST segment measurement is by measuring the voltage difference between the value at a point 60–80 ms after the J point and the isoelectric baseline. The isoelectric baseline is either between the *P* and *Q* waves (the *P-R* interval) or in front of the *P*-wave (the *T-P* interval) (Fig. 7.8). Investigations have revealed that the slope of the ST segment significantly influences the abnormal ST shift. Figure 7.9 shows normal and abnormal shifts observed in the ST segment. The algorithm used for ST segment monitoring is based on beat detection and classification information provided by multi-lead arrhythmia algorithm and performs several additional actions on the ECG waveforms including filtering the signals, measuring ST values and generating ST alarms.

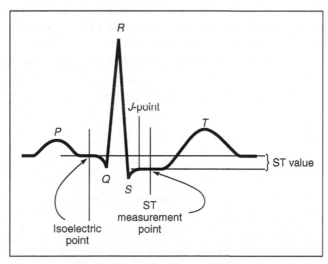

➤ **Fig. 7.8**   *ST segment measurement*

The ST segment is a signal of low amplitude and low frequency content. Therefore, a sampling rate of 250 samples/s is adequate. To ensure that the ST segment can be measured accurately, the

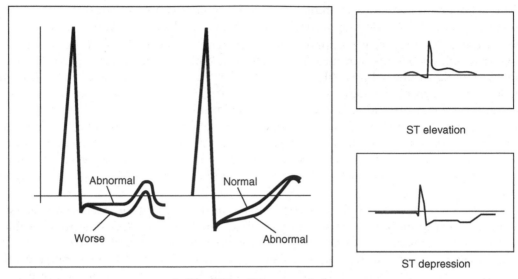

➤ Fig. 7.9   *ST segment slope*

incoming ECG signals must have a low-end bandwidth of 0.05 Hz. This is to ensure that no signal distortion is introduced in the ST segment. A special ST filter with a higher low-end bandwidth of 0.67 Hz is used to further remove unwanted baseline noise. Since ST segment values do not change very rapidly, every single beat need not be measured. Instead, reliable ST measurement can extracted using signal averaging techniques. Generally, all *QRS* complexes detected by the arrhythmia algorithm within a discrete 15 second period are saved and used for ST segment analysis. Statistical analysis techniques are used and the largest deviation in the value of the ST segment from the reference is selected and displayed (Hewlett Packard, 1997). Alfonso *et al* (1996) compare various algorithms employed for processing stress ECG signals.

It may be observed that the exercise test when applied in an appropriate manner and interpreted by an experienced cardiologist is a vary useful tool in the functional assessment of normal and abnormal cardiovascular physiology, particularly its capability to provide prognostic information.

ⅢⅢ➤ 7.5   AMBULATORY MONITORING INSTRUMENTS

The traditional medical examination involves a number of chemical, physical and electro-physiological measurements. These measurements are of very short duration and comprise no more than a physiological snapshot of the patient's condition. However, when one wants to perform functional tests on a patient, which are expected to have some relationship to his behaviour in normal life, the measurements have to be made over a long period. Ambulatory monitoring concerns itself with the extension of such measurements into the time domain on unrestricted ambulatory (mobile) patients during everyday stress and activity as well as during periods of sleep. Therefore, the precise objective of ambulatory monitoring is to record one or more physiological variables continuously or repeatedly, without interference with the spontaneous activities of the subject by the restraints of conventional laboratory instrumentation and without influencing the variable being measured. Ambulatory monitoring is not only an invaluable aid to the physician in the differential diagnosis of many unexplained symptoms like dizziness, syncope and palpitation but it also provides accurate data for the evaluation of drug therapy, stress testing, artificial pacemakers, status of myocardial infarction and several other problems in research programmes. The technique is so well established now that it is predicted that within the next decade, ambulatory monitoring departments will become a common feature in the hospital service, accepted as a matter of course just like the X-ray or pathology department. Ambulatory monitoring of ECG is called '**Holter Cardiography**', after Dr Norman Holter who introduced this concept in 1962.

Currently, two types of systems are available for ambulatory ECG monitoring:

(i) Tape-based or solid-state systems that provide continuous recording of the complete 24–72 hours of ECG waveforms. All of the data is available to the system and screening expert for retrieval, review and editing.

(ii) Event, time-activated or patient-demand recorders that are not conventional, continuous tape-recorders, but systems that can provide limited monitoring for a specific type of cardiac event, such as ST segment analysis. These recorders can capture data both and during preceding the symptoms when a patient activates the unit by pressing a button. Event recorders document and record abnormal beats but provide no analysis. In order to document the abnormal rhythm that precedes the symptomatology, a memory loop can document the abnormal event prior to activation of the system at the time of the symptoms (Handelsman, 1990). An excellent review of ambulatory cardiac event recorders is provided by Benz (1999).

7.5.1 Data Recording

The core of the modern ambulatory monitoring system is a multi-channel sub-miniature tape recorder running normally at a speed of 2 mm/s. At this speed, a C-120 entertainment cassette will run for 24 h. The recorders are designed to be fitted with a variety of plug-in circuit boards adapted for different signals or transducers. The main areas of interest in ambulatory monitoring are centred on the cardiovascular system, with particular reference to the control of cardiac rhythm disturbances, along with treatment of hypertension and the diagnosis and treatment of ischaemic heart disease. Another area of interest is EEG recording, with particular reference to epilepsy. Ambulatory monitoring of respiration has also considerably developed in the last few years into a rapidly growing area of clinical research, which is beginning to make a substantial contribution to respiratory medicine. The main non-clinical application of ambulatory monitoring is for studies in work physiology and environmental health and it continues to make considerable impact on these areas.

During replay, the tape is run at 120 mm/s (60 times the recording speed) to achieve rapid manual or automatic scanning of ambulatory records. The tape recorders used for this purpose have some special features as compared to the usually available entertainment tape recorders. A block diagram of a two channel recorder is shown in Fig. 7.10. The tape recorder offers single-sided 24 h recording facility, with three recording channels—two for ECG and one for timing, for the precise correlation of recorded events to patient activity. As in conventional circuits, the ECG

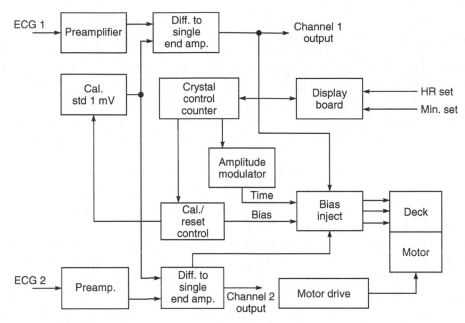

> **Fig. 7.10** *Block diagram of the recording unit for ambulatory monitoring (Courtesy: American Optical, USA)*

channels feature high input impedance, differential inputs consisting of transient protection mode switching and differential amplifier stages, followed by a single ended amplifier, with provision for fast transient recovery and reset.

Calibration is automatic, as it switches the ECG channels to receive the 1 mV pulse output of a calibration circuit. The timing channel provides duty cycle coding for odd versus even minutes as well as half hour transitions. Timing channel information is amplitude modulated at low frequency so that the recorder user may mark significant episodes without disturbance to either the ECG or timing information. Time coding is a function of a crystal controlled counter, which controls biasing to the three recording channels. Other facilities incorporated are a cassette interlock that prevents use of battery power without a cassette in place and the automatic shut down on low battery.

Conventional tape and solid-state systems commonly use data compression. This process requires digitization at least at 250 samples/s. Digitization through tape or solid-state systems results in the loss of the original ECG data by compression since digitization only records certain points along the waveform. However, for practical purposes, the loss in data may not be highly significant so long as the device retains the necessary components for accurate reproduction of the various waveforms necessary for clinical evaluation.

7.5.2 Data Replay and Analysis

One difficult problem, which is inevitably faced when examining long-term recordings, is the almost overwhelming quantity of data which becomes available. Even a single channel 24 hour

ECG recording will contain in excess of 100,000 beats. Early replay and analysis equipment relied on visual inspection of the replayed signals in accelerated time. This can become exceedingly tedious and is subject to error. For example, many events of clinical significance are quite transitory and may occupy less than one minute of real time, which means that they last perhaps one or two seconds on playback and in the course of replaying, a tape which runs for about half an hour, can easily be missed. More complex equipment now has the ability to work largely automatically and is able to recognise abnormalities and to write these out on a conventional pen recorder for subsequent examination. The displays in the form of *R–R* interval histograms, have also been found to give important diagnostic information. The analyser part in the automatic scanning of ambulatory records look for four arrhythmic conditions. These are bradycardia, tachycardia, dropped beat and premature beat. A threshold control is associated with each of these and when the appropriate threshold is exceeded, an alarm condition is generated.

A typical example of a modern two channel ECG analysis system incorporating precise tape control and data processing is shown in Fig. 7.11. The most prominent operational feature of this system is its dual microprocessors: control and timing and acquisition and display CPUs.

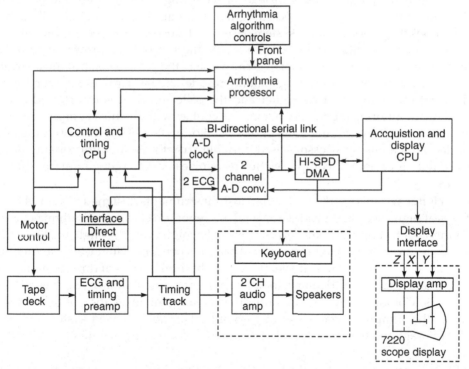

> **Fig. 7.11** *Functional diagram of data replay and analysis system for ECG (Courtesy: American Optical, USA)*

The control and timing CPU has overall system control responsibility. It also handles individual functions such as keyboard and direct writer interface, tape deck control, timing data processing, and arrhythmia count totalizing via a high speed interrupt system. The control and timing program is primarily resident in the 8 K PROM with spill-over occupying additional RAM. The RAM resident program portion is loaded during system initialization at start-up from the data acquisition and display PROM via a bidirectional serial communications link. The acquisition and display CPU is responsible for the scan-mode A-D conversion and storage of 72 s worth of ECG data for each of the two channels. Digital data storage is in RAM in circular buffer fashion for ready access to facilitate display generation. This CPU is also responsible for accessing the stored data in order to produce a variety of system scan mode CRT displays, as well as ECG strip and trend-mode report write-outs.

The arrhythmia processor provides arrhythmia detection sounds, ectopic count tabulation and a trend report write-out. After adjustment of detection sensitivities, the arrhythmia detector output can be set to either sound distinctive tones for different types of detected abnormalities thereby alerting the operator to their occurrence or to automatically stop the scanner and display the detected abnormality.

A convenient way of managing enormous ECG data would be to perform real-time rhythm analysis and discard normal signals instead of recording the total information on magnetic tape—the continuous 24-hour electrocardiogram of an ambulatory patient. Tompkins (1978 &1980) describes the basics of such a system based on the use of a microcomputer. It is designed to detect an abnormal rhythm and store in its RAM the 16 s of ECG prior to the time it sensed the abnormal event. By using a manual override switch, the patient can put the instrument into alarm mode to capture data during symptomatic episodes. In the alarm mode, a communication is established with a remote host computer and the collected data is transferred before it can continue in the monitoring mode once again. The contact with the host computer is established by the patient himself by dialling a telephone. Thakor et al (1984 a) discussed the design and implementation of a microprocessor-based portable arrhythmia monitor for ambulatory ECG, which does not store normal complexes but recognizes and triggers an alarm on significant arrhythmias.

Modern Holter systems are PC-based diagnostic instruments designed to scan 24 hour analog Holter tapes at high speed and produce an analysis report on cardiac arrhythmia event activity. The arrhythmia detection and analysis is performed at high speed (as high as 240× real time) on C-60 cassettes recorded on the Holter recorder. Analysis is performed for mean heart rate, minimum and maximum heart rates, premature ventricular arrhythmias, runs of three or more ventricular ectopics, premature supraventricular ectopic beats, supraventricular tachycardia and ST segment measurements. Colour coded beat identification greatly enhances the ability to discriminate and validate the computer analysis of abnormal beats. The PC-based system offers multiple patient data storage and retrieval capability. Holter reports can be generated on a high speed laser printer.

In addition to recording and analysing ECG, blood pressure is another parameter which has been studied extensively through ambulatory monitoring. The method consists of inserting a small diameter plastic cannula into the radial or brachial artery, and a connecting tube about 75 cm long, from the cannula to the transducer, which is worn in a chest harness by the patient. The recordings are made on a miniature tape recorder. The unit uses semiconductor gauge pressure transducer and the perfusion is maintained by a miniature peristaltic pump. The frequency response of the recorder and replay unit reduce the bandwidth to 10 Hz. This is, however, satisfactory for accurate pressure measurements in peripheral arteries (Pickering and Stott, 1980). The system has been used for recording ambulatory blood pressure for periods up to 72 hour, with satisfactory results.

Figure 7.12 shows a non-invasive blood pressure measuring system for ambulatory subjects. This allows the patient to pursue normal daily activities while blood pressure measurements are automatically computed and recorded throughout the ambulatory period. Repetitive blood pressure measurement is achieved through pre-selection of time intervals for an automatic patient cuff inflation cycle. The pre-selected time intervals may be over-ridden by means of manual switching to activate, restart or terminate the cuff inflation cycle. The apparatus utilizes a standard pneumatic cuff with a piezo-electric microphone transducer held in place with an adhesive disc for reliable detection of Korotkoff sounds. Detected Korotkoff sounds are exhibited as systolic and diastolic readings on a digital liquid crystal display with a resolution of 1 mmHg. Accuracy of

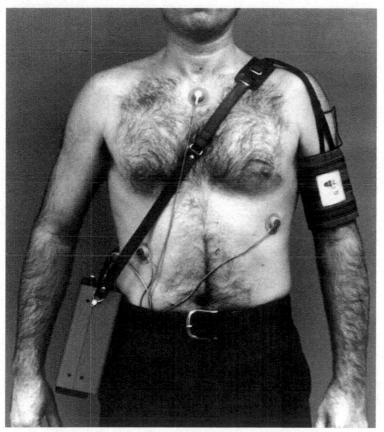

➤ **Fig. 7.12** *Attachment of electrodes for ECG and cuff for arterial blood pressure measurement of ambulatory subjects (Courtesy: Del Mar Avionics, USA)*

blood pressure measurement is achieved through the use of a closed loop design for electro-pneumatic bleed-down of the patients cuff and repetitive electronic sensing of Korotkoff sounds. Sometimes, the sounds are gated by ECG *R* wave signals, which are also simultaneously recorded. This reduces artifact since Korotkoff sounds are only accepted within a limited time period following the *R* wave. The instrument is powered by a rechargeable battery which will operate for 24 h or for 192 blood pressure readings with intervals of 7.5 min between readings. The instrument ensures patient safety as it automatically releases cuff pressure if inflation and deflation cycles exceed a 2 min period. Additionally, the ECG signal controls cuff-bleed-down at a rate of 3 mm per heart beat to ensure accurate readings of systolic and diastolic pressure.

Foetal Monitoring Instruments

During the last decade, several techniques, devices and instruments have become available which provide reasonably reliable information and data instantaneously about the foetus during its intrauterine life or at the time of delivery. Nevertheless, the development of foetal instrumentation has been a difficult task because of the complex nature of the problems involved. The foetus while in the uterus is mechanically shielded from the outside world so that it can safely develop there. This means that only a limited amount of information can be obtained directly about the foetal condition. The only information which is readily available about the foetus and which can be picked up from the maternal abdominal wall are the electrical potential of the foetal heart activity and the foetal heart sound signals. These signals when picked up are mixed up with the corresponding maternal signals. The isolation and interpretation of the mixed up foetal signals requires expert handling. Consequently, the obstetrician is faced with the problem of having very few parameters available on which to base a diagnosis of foetal well-being or distress. In most cases, the condition of the foetus is assessed by studying the blood flow in the foetal heart and its heart rate. The foetal heart rate (FHR) yields important information about the status of the foetus, and therefore, has become a widely studied parameter in maternity cases.

Foetal heart rate monitoring in the labour ward has generally been carried out on an intermittent basis. It has been traditional to listen to the foetal heart sounds at intervals of up to every 15 minutes. This is done by using the Pinard stethoscope. Nevertheless, this technique does not provide a complete picture of the foetal heart beat during the majority of contractions, specially if they are strong, nor is it feasible to listen continuously. Thus, perhaps the most valuable information of the foetal heart rate occurring during a contraction goes unrecorded (Day *et al.*, 1968). Moreover, the subjective element of human variability in counting compromises the data further and the absence of permanent foetal heart rate records renders comparison of different patients unsatisfactory.

▶ 8.1 CARDIOTOCOGRAPH

An assessment of the condition of the foetus can be made during labour from the foetal heart action. Simultaneously, recording beat-to-beat foetal heart rate and uterine activity provides basic information for assessing the compensatory potential of the foetal circulatory system. The instrument which carries out a continuous and simultaneous recording of the instantaneous foetal heart rate and labour activity is called cardiotocograph. In addition to detecting long-term bradycardia or tachycardia, this instrument helps in the evaluation of foetal heart rate response of the undisturbed circulatory system and response stimulated by uterine contractions. In the undisturbed, healthy foetus, oscillation of the FHR is normal whereas, absence of FHR oscillation is considered a sign of potential foetal distress (Gentner and Winkler, 1973). Uterine contraction may or may not cause a response in the FHR. To determine the prognostic significance of a response, the shape and time relationship of the change in FHR, with respect to the contraction is usually studied.

Cardiotocographs are designed to measure and record foetal heart rate on a beat-to-beat basis rather than on an average basis. Normally, an accuracy of measurement may be 2–3% for classification of responses. Sensitivity of 20 bpm/cm of recording chart allows adequate reading of the recorded FHR. Labour activity and FHR traces are usually recorded simultaneously on the same time scale. Chart speed of 1–2 cm/min is adequate to provide sufficient resolution of the stimulus-response relationship. In addition, foetal monitoring instrumentation should allow the user a choice of various clinically accepted monitoring methods, be simple to operate and result in minimum patient annoyance within the constraints of high quality data presentation.

The following methods are commonly employed in most of the cardiotocographic monitoring during labour:

Method	Foetal Heart Rate	Uterine Contraction
Indirect (external)	1. Abdominal foetal electrocardiogram 2. Foetal phonocardiogram 3. Ultrasound techniques (narrow beam and wide-angle transducer)	Tocodynamometry (using tocotonometer to sense changes in uterine tension transmitted to the abdominal skin surface)
Direct (internal)	Foetal ECG with scalp electrode (spiral, clip or suction electrode attached to the presenting part of the foetus)	Intrauterine pressure measurement (using a fluid-filled intracervical catheter with strain gauge transducer)

▶ 8.2   METHODS OF MONITORING FOETAL HEART RATE

8.2.1   Abdominal Foetal Electrocardiogram (AFECG)

Foetal electrocardiogram is recorded by suitably placing the electrodes on the mother's abdomen and recording the combined maternal and foetal ECG. Figure 8.1 shows a typical recording of foetal ECG picked up from the abdomen. The maximum amplitude of FECG (*R* wave) recorded during pregnancy is about 100 to 300 μV. This magnitude is much smaller than in the typical adult ECG which is about 1 mV in the standard lead connection. The amplitude is still lower in some stages of pregnancy and may not be even properly detected. Low signal amplitude places very stringent requirements on the recording of the FECG if the signal-to-noise ratio (SNR) is to be kept high. Hence, the usual precautions of obtaining good ECG records are more carefully observed. They include low electrode skin contact impedance, proper electrode material with low depolariza-

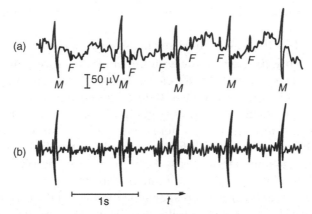

> **Fig. 8.1** *Abdominal recording of foetal electrocardiogram. Bandwidth (a) 0.2 – 200 Hz*
> *(b) 15-40 Hz*

tion effects and placement of the electrodes at appropriate positions. The signals must be properly shielded, the equipment properly grounded and the patient electrically isolated from the equipment. Van Bemmel *et al* (1971) suggest that the best place for abdominal electrodes is when one electrode is near the umbilicus and the other above the symphysis.

The foetal heart rate is computed from the foetal ECG by appropriately shaping the foetal *QRS* wave. The foetus heart rate is approximately twice that of the normal adult ranging approximately from 110 to 180 bpm. The main problem in processing the foetal heart signals is the poor SNR. There are periods, particularly during birth, during which instantaneous computation of FHR is not possible because of excessive noise. Therefore, specific signal properties are made use of to improve SNR.

The major sources of noise in the foetal ECG signal recorded from the maternal abdomen are (i) amplifier input noise, (ii) maternal muscle noise (EMG), (iii) fluctuations in electrode polarization potential, and (iv) maternal ECG. For practical purposes, the first three of these sources can be considered as random whereas the maternal ECG is a periodic noise source. The frequency spectrum of each noise source partially overlaps that of the foetal ECG and therefore, filtering alone is not sufficient to achieve adequate noise reduction. Mains frequency noise pick-up, which is normally a problem in physiologic recordings, is usually eliminated by the use of a notch filter.

A method for dealing with this type of noise problem in foetal ECG, first suggested by Hon and Lee (1963) and later refined by several investigators (Van Bernmel *et al.* 1968), utilize the technique of signal averaging to improve the ratio of signal to random noise. The maternal ECG component is effectively removed from each lead recording by creating an average maternal waveform at each of its occurrences in the recording. The maternal component can then either be subtracted directly from the foetal ECG recording before averaging the foetal waveforms or corresponding portions of the separated maternal component can be averaged in parallel with the selected foetal waveforms and the resulting residual maternal signal subtracted from the foetal waveform average at the end of the process. The latter technique is more efficient as fewer calculations are required. A particularly difficult problem encountered when applying signal averaging to the foetal ECG is the selection of a signal to trigger the averaging process. The signal must bear the same precise temporal relationship to each waveform if the average is to be coherent. Unlike the signal averaging often employed in stimulus-response work, there is no external stimulus to trigger the start of the averaging procedure. Thus, the ECG signal itself must be used to provide that trigger.

Figure 8.2 shows a block diagram of the abdominal FECG processing circuit for computing foetal heart rate. After proper placement of the electrodes, the signals are amplified in a preamplifier which provides a very high input impedance (100 MΩ) and a high sensitivity and good common mode rejection ratio (up to 120 dB). The input stage should preferably be kept isolated so that any earth leakage currents that may develop under fault conditions comply with the safety requirements. The preamplifier is a low-noise differential amplifier that has a wide dynamic range. A sizable common-mode signal manages to pass through the input amplifier, a circumstance to be expected whenever electrodes spaced a few centimetres apart are attached to the human body in a hospital environment. Power line hum is responsible for most of the common-mode interfering signal. This is suppressed by a notch filter following the input amplifier.

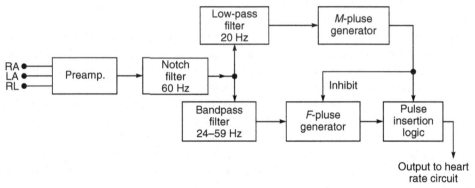

> **Fig. 8.2** *Block diagram of the abdominal foetal electrocardiogram processing circuit (after Courtin et al. 1977)*

The signal path then splits into two channels: the maternal ECG channel or *M* channel and the foetal or *F* channel. Since the frequency spectrum of the foetal ECG differs somewhat from the maternal ECG, some initial signal separation is achieved by using the appropriate bandpass filtering in each channel. Polarity recognition circuits in each channel accommodate signals of either polarity. After filtering, the *M* signal is assured of being the largest signal component in the *M* channel, so it can be detected on the basis of peak amplitude. It is used to generate a blanking pulse for use in the *F* channel and in the pulse-insertion logic circuits. The *F* channel has a 30 ms pulse generator that is triggered by the foetal ECG. It is inhibited, however, by the blanking pulse from the *M* channel, so it will not generate a pulse in response to the maternal ECG signal feeding through to the *F* channel.

Foetal ECG signal detected via electrodes placed on the mother's abdomen is complex and requires attenuation of maternal signals for obtaining FHR. Also, due to the overlapping of the foetal ECG with the maternal ECG, about 20% to 50% of the expected pulses may be missing.

Therefore, the pulse train generated in the *F* channel is fed to logic circuits. These determine the rate at which the *F* channel pulses occur and if the timing indicates that there should be an *F* pulse at a time when one is blanked or missing, a pulse is inserted into the F channel output pulse stream. However, the logic circuits will not insert two pulses in a row, so there is no danger that the instrument will continue to output normal pulses when no foetal ECG is present. The logic circuits also keep track of the maternal heart rate. If the M and F channels have exactly the same rate, they inhibit the F channel output during the maternal P wave. This precaution is taken because otherwise it could be possible that when no foetal ECG is detected, the F channel would respond to the maternal P wave and generate a train of pseudo F pulses.

The substitution logic requires a delay time to establish a missing foetal trigger pulse. On the one hand, this delay has to be longer than the maximum permissible change in heart period (14 bpm change from 50–64 bpm = 262 ms) and on the other hand, it has to be shorter than the shortest period duration (216 bpm = 285.7 ms). It is thus kept as 270 ms. The range of FHR measurement is limited to 40–240 bpm because of the substitution logic. Thereafter, the output of logic circuits go to standard heart rate computing circuits.

Clinical trials have shown that the AECG technique is usually effective in the most cases except in those rare cases where the amniotic fluid fails to provide adequate electrical coupling from foetus to mother. However, during labour, the uterine and abdominal wall electromyogram signals tend to obliterate the FECG signal, making FHR counting quite difficult. At present, the abdominal FECG, therefore, does not seem to offer a practical reliable means of FHR monitoring during labour and delivery. Microprocessor-based signal averager for analysis of foetal ECG in the presence of noise has been reported by Wickham (1982).

8.2.2 Foetal Phonocardiogram

Foetal heart sounds can be picked up from the maternal abdomen by a sensitive microphone. The heart sounds in the form of mechanical vibrations have to pass through tissue structure and the signals picked up are rather weak because of distance effects and the small size of foetal heart valves. Moreover, the heart sounds are greatly disturbed by maternal movements and external noise. To pick up the heart sounds, it is essential that the transducer be properly placed and its impedance carefully matched. A crystal microphone is used for picking up phono signals. The phono transducer signals are amplified by a low noise preamplifier and fed to a bandpass filter which rejects all frequencies outside the 70 to 110 Hz range. The preamplifier is incorporated in the transducer housing to minimize interference signals being picked up. Much of the random noise is eliminated during this process and the record on paper after this stage is called foetal phonocardiograph.

From the normal foetal heart action, generally two sounds (Fig. 8.3) are produced corresponding to the contraction and relaxation of the heart muscles. These two bursts of heart sounds are mixed up with unwanted signals which may succeed in passing through the filters. Such situations are quite complicated and require elaborate electronic circuitry to get one pulse for each foetal heart beat. This is achieved by using the repetitive properties of the FHR considering the highest FHR to be 210 and the lowest FHR as 50/min. A certain interval between the two heart beats, when computed, is stored into a memory for comparison with the following interval. The latter is only accepted if it does not differ more than a certain number of beats (±7 bpm) from the stored interval. By this method, a reasonably reliable heart rate measurement can be made.

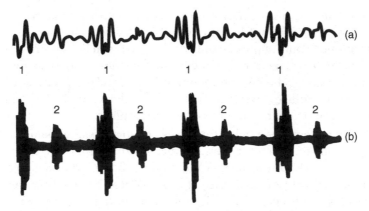

> **Fig. 8.3**   *Foetal phonocardiogram: (a) unfiltered heart sounds, (b) filtered heart sounds*

To ensure this, a detected heart sound triggers a one-shot multi-vibrator that inhibits succeeding heart sounds from reaching the following circuits for the duration of the one-shot. The circuit must be able to operate a 4-to-1 range (50 to 210 beats/min. or 1.2 to 0.285 s/period). This necessitates designing in the ability to adjust the one-shot ontime to the heart rate. If the time between two triggers is less than 400 ms, the duration of the blanking pulse produced is 273 ms. If it is more than 400 ms, then the blanking pulse is extended to 346 ms.

Figure 8.4 shows a block diagram of the arrangement used for obtaining a variable pulse duration to inhibit triggering by the second heart sound. After peak detection, the processed pulses operate a one-shot circuit which gives a fixed pulse width of 230 ms. The output of one-shot (2) triggers a variable pulse width multi-vibrator (3) which adds and gives either 43 or 116 ms time depending on the heart rate. The pulse width at the output will be either 230 + 43 = 273 ms or 230 + 116 = 346 ms.

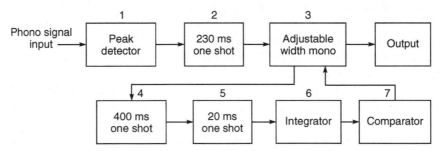

> **Fig. 8.4**   *Block diagram of the circuit arrangement used for obtaining variable pulse duration to inhibit triggering by second heart sound*

To detect the heart frequency, the 400 ms one-shot (4) is used. If the period duration is greater than 400 ms, the one-shot will deliver a pulse. The negative slope of this pulse is used to trigger the 20 ms one-shot (5). These 20 ms pulses are integrated by the integrator (6) and the output of this integrator is compared with a fixed voltage -V. If the output of the integrator is more negative than -V, the output of comparator (7) will become positive. Consequently, the reference level of the one-shot (3) is shifted to more a positive level and pulse width of this one-shot increases from 43 ms to 116 ms, giving the total pulse width at the output between 273 ms to 346 ms. The integrator (6) is used to delay the change in the time constant and to make sure that a change of on-time takes place only if several (3 to 4) heart beats with the longer period duration (below 150 bpm) are present.

No output pulse will occur, if the period between two pulses is less than 400 ms. The 20 ms pulses are, therefore, not generated and the integrator discharges slowly from the negative output voltage to a positive output voltage. If the output of the integrator (6) is less negative than $-V$, the output of comparator (7) will become negative. Now the reference level of the one-shot changes in such a way that the time varies from 116 to 43 ms, resulting in a pulse width of 273 ms.

Phonocardiography provides a basically cleaner signal than does ultrasound, thus allowing a greater chance of detecting a smooth baseline FHR. Unfortunately, phonocardiography is more susceptible to artefacts introduced from ambient noise, patient movement or other intra-abdominal sounds. Thus, even with phonocardiography, the baseline FHR may have an apparent increase in variability that may not be real. Although phonocardiography has some advantages, it has almost become obsolete for clinical monitoring because of its tendency to pick up too much background noise, signal loss during uterine contractions and the general difficulty of obtaining a good signal.

8.2.3   FHR Measurement from Ultrasound Doppler Foetal Signal

An important clinical instrument for obstetric applications which makes use of the Doppler shift principle is the foetus blood flow detector. The technique is extended to derive an integrated rate of the foetus heart from blood flow signals and to display it on a suitable display system. In obstetric applications, the site of investigation varies from 5 to 20 cm below the surface of the abdomen (Fielder, 1968). This depends upon the patient and the stage of pregnancy. For obstetric studies, ultrasonic frequency of about 2 to 2.5 MHz is usually employed, whereas in the study of blood flow in arteries and superficial blood vessels frequencies around 5–10 MHz are preferred. The level of ultrasonic energy transmitted into the body is generally kept between 10–15 mW/cm^2. Assuming a maximum of 50% conversion efficiency, this would mean that the transducer should be powered with an electrical energy below 30 mW/cm^2.

The Doppler-shift based ultrasound foetal blood flow detectors use hand-held probes which may be either pencil-shaped or flat and contain two piezo-electric crystals. The probe is coupled to the patient's skin by means of an acoustic gel. This is done to exclude any air from the interface. The presence of air severely attenuates the ultrasound, the problem being more acute during early pregnancy. The transmitting crystal emits ultrasound (2 – 2.5 MHz) and the back-scattered ultrasound is detected by the receiving crystal. The back-scattered ultrasound frequency would be unchanged if the reflecting object is stationary. If the reflecting object is moving, as would be the foetal heart blood vessels, then the back-scattered frequency is higher as the blood cell is approaching the probe, and lower if it is moving away from the probe. The magnitude of the frequency shift (Δf) varies according to the following formula:

$$\Delta f = (2 f_o u \cos \theta)/c$$

where, f_o is the transmitted frequency, u is the blood velocity, $\cos \theta$ is the cosine of the angle of the sound beam and the object's direction and c is the velocity of the sound wave in the tissue.

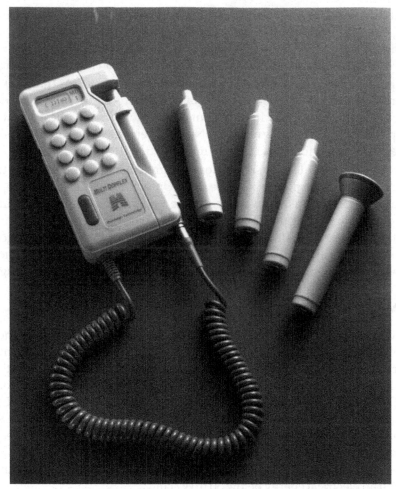

➤ **Fig. 8.5(a)**   *Ultrasonic blood flow detector (Courtesy: M/s Huntleigh Health Care, USA)*

Representing the frequency shift as sound is the simplest method of processing the Doppler signal and displaying the same. Battery operated portable instruments are commercially available. Figure 8.5(a) shows a typical instrument of this type which also displays the foetal heart rate and Fig. 8.5(b) shows it in use.

Blood flow detectors based on ultrasonic Doppler shift can detect foetal pulse as early as the tenth week of pregnancy and in nearly all cases by the twelfth week. At about 20 weeks it is possible to detect multiple pregnancies especially if two instruments are used together and the pulse rates compared. Intrauterine death of the foetus can also be diagnosed. Later after about 25 weeks of pregnancy, a distinctive sound from the placenta helps to determine its location and facilitates diagnosis of placenta praevia. Blood flow through the umbilical cord can also be heard at this stage. Such types of instruments are also useful during labour to assess the condition of the foetus.

Ultrasonic Doppler foetal heart signal is easy to obtain but it is difficult to process and to get consistent trigger pulses required for instantaneous beat-to-beat rate measurement. This is mainly because the signal usually has a lot of fading and the level, spectra and envelope waveform change rapidly. We can hear the signal through a loudspeaker with a scarce chance of failing to recognize any beat, but a simple electronic circuit may fail to trigger from this signal. Still, the ultrasound Doppler shift method is more practical and easy to use during labour. It is currently the most reliable method for detecting the FHR pattern that is interpretable.

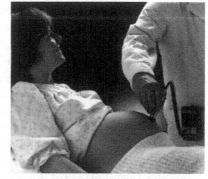

➤ **Fig. 8.5(b)** *Ultrasonic foetal heart beat detector in use*

Signal processing for FHR determination can be based either on detecting the foetal heart valve motion or on detecting the heart wall motion. The heart valve motion detection technique is based on the distinct ultrasound frequency shift produced by the fast opening and closing of the heart valves. The technique, however, requires that the ultrasound beam must be directed against the relatively small heart valves involving a longer search period and frequent repositioning of the transducer. Therefore, it is not preferred for continuous monitoring applications. Movements of the foetal heart wall are slower as compared to valve movements and, therefore, produce a smaller frequency shift. This signal is less precise than the heart valve signal and tends to produce more jitter on the FHR trace. However, since it is much easier to obtain these signals and transducer repositioning is necessary less often, they are better suited for continuous monitoring. In order to reduce jitter on the trace, the usual practice is to incorporate a signal smoothing circuit with an averaging time constant over a window of approximately three heart periods. This will nevertheless result in lesser beat-to-beat variability details than those obtained with scalp electrodes.

Improved artefact rejection is usually accomplished in the Doppler mode of operation by some form of short-term averaging. The averaged measure may lack the variability of the beat-to-beat, but it does provide adequate detail of the baseline trend of the FHR. Tuck (1981) suggests a two second average rate from only those intervals which have been determined to be valid based on the criterion that they should be within ±10% of the most previous valid interval. The interval testing algorithm suggested by him does not accept intervals greater than 1000 ms (corresponding to rates < 60 bpm) and less than 250 ms (rates > 240 bpm), nor interval changes greater than ±10% of the last valid accepted interval. However, should this inacceptance continue to occur, then a new valid interval is recognized after three successive intervals fall within ±10% of each other and so the process continues. This averaging approach offers an improvement over the continuous averaging techniques by allowing an optimized trade-off between FHR measurement error and lack of response in tracking the true FHR. This approach can be easily implemented in a microprocessor-based instrument.

Two types of ultrasonic transducers for FHR measurement are in common use. They are the narrow beam and the wide-angle beam types. The narrow beam transducer uses a single ultrasound transmitter/receiver piezo-electric crystal pair. The maximum ultrasound intensity is generally kept below 25 mW/cm^2. The typical transducer diameter is 25 mm. The narrow beam transducer is very sensitive and produces a good trigger signal for instantaneous heart rate determination. However, it takes time to detect a good signal and, therefore, frequent transducer repositioning is necessary.

The broad beam transducers are available in many configurations. The transducers comprise a number of piezo-electric crystals mounted in such a way as to be able to detect foetal heart movements over a wider area. In one arrangement, the ultrasonic transducer is arranged in the shape of a clover-leaf so that it provides a large area of ultrasonic illumination which allows the monitoring considerable lateral and descending foetal motion before requiring repositioning. The transducer housing is flexible to permit it to follow the contour of the abdomen regardless of shape changes with contractions. The transducer has three crystals on the other side acting as transmitters whereas the crystal placed at the centre acts as a receiver. An alternative arrangement is the array transducer which has one transmitter and six peripheral ceramic receiving crystals (Fig. 8.6). The transmitting crystal emits a 40° divergent beam so that at 10 cm from the skin surface the beam covers an area of approximately 10 cm diameter. This construction ensures continuous recording of the foetal heart activity without the need to reposition the transducer which is otherwise necessitated due to normal foetal movement. The transducer has a diameter of 6 cm and can be held in place either by a simple buckle or a stretch belt.

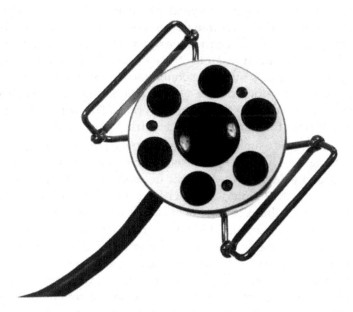

> **Fig. 8.6**   *Multireceiver transducer*

Analysis of ultrasonic Doppler signals using a speech spectrograph shows that frequency components in the range of 100–1000 Hz tended to be more distinctly related to the foetal heart cycle than components lying outside this frequency range. The bandpass filter, therefore, enhances the signal/noise ratio—the noise in this context being, for example, foetal movements at low frequencies and maternal placental blood flow at high frequencies.

With ultrasonic Doppler signals, there remains the possibility of more than one burst in each cardiac cycle being detected. In some instruments, this difficulty is overcome by the dead-time generator, which inactivates the detector for a period of 0.3 sec after an amplitude burst has been detected. This dead time is chosen on a compromise basis: it defines a maximum heart rate (200 bpm) that can be detected, while at rates which are less than half this maximum, i.e. 100 bpm, it is conceivable to 'double-count' the signal. Although the total signal processing in many instruments goes far in minimizing this frequency-doubling possibility, the effect remains a fundamental limitation of using the foetal ultrasonic Doppler signal for recording heart rate.

The principle of ultrasonic Doppler-shift based FHR measuring circuit is shown in the block diagram in Fig. 8.7. This arrangement can be used both with a wide angle beam as well as a narrow beam transducer. The transmitted signal that leaks into the receiving path serves as a local-oscillator signal for the mixing diodes in the demodulator. The output of the demodulator is dc except in the presence of a Doppler-shift frequency. The reflected signal is some 90 to 130 dB lower in amplitude than the transmitted signal. The high overall gain in the receiving channel (+110 dB) requires special measures to minimize the effects of interference. One measure used is a low noise, low distortion oscillator for the transmitter. This reduces interference caused by oscillator harmonics beating with radio and TV signals. Other measures involve filters in the transducer connected for attenuating high-intensity high frequency radiation that could drive the amplifiers into a non-linear operating region. The high frequency section of the circuits is surrounded by both magnetic and electrical shields.

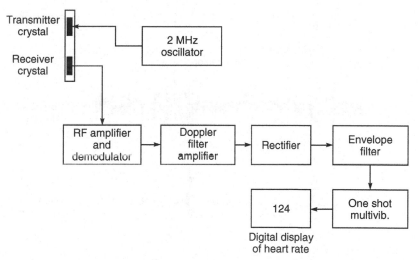

> **Fig. 8.7** *Ultrasonic Doppler-shift based FHR measuring circuit (after Courtin et al., 1977)*

Depending upon the transducer used, i.e. array or narrow beam, the filter circuits can be selected to match the Doppler-shifted frequency components. A bandpass filter centred on 265 Hz isolates the Doppler frequencies resulting from the movement of the heart walls. The array transducer used with this circuit gives a broad ultrasonic beam that does not require careful positioning to obtain a strong Doppler return from the relatively large heart walls.

The availability of the foetal heart Doppler signal from the twelfth week of gestation onwards, its usually good signal-to-noise ratio and the lack of maternal interference renders it practical for measuring each and every heart beat interval.

Lauersen *et al* (1976) describe a system which enhances ultrasonic reflections from certain distances from the transducer and reduce those from others. This allows the operator to 'range in' on the best sounding foetal heart Doppler to optimize signal clarity. The depth ranging capability is accomplished by superimposing a digital code on the transmitted 2 MHz continuous ultrasonic wave and awaiting the return of the code. As the ultrasonic propagation time through tissue is known, the expected time from various distances is known. By the use of a correlation technique the system accepts only ultrasonic reflections at the distances selected (Fig. 8.8) by the operator. The correlation technique (Tuck 1982) also enhances the selected reflection three fold (10 dB), and reduces reflections from other distances by 30 fold (30 dB). So, by selecting the depth which produces the loudest, clearest, foetal heart Doppler, all returns from this area are enhanced and other Doppler sources are largely excluded. Doppler signals contain a multiplicity of components represating atrial contractions, A-V valve closure, A-V valve opening and to some extent aortic and pulmonic valve motion. As the Doppler signal changes or becomes less distinct, greater

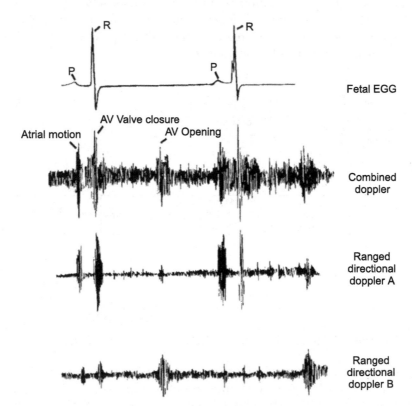

> **Fig. 8.8** *Doppler signal corresponding to different events in the cardiac cycle (after Laursen et al., 1976)*

artifactual jitter is produced and long-term variability may be obscured. Lauersen *et al* (1976) report that the Doppler from the ranging auto-correlation system may be further improved by electronically selecting the Doppler signal from the moving heart valves either going toward or away from the transducer for processing. This provides an additional 20-fold signal clarification.

Studies have shown that ultrasonic power levels generally used in foetal monitoring produce no chromosome damage and/or other apparent damage to the full term foetus, even after long exposures (Abdulla *et al* 1971). The Doppler signal is far more complex than the foetal electro-cardiogram. It presents no easy single point for counting as does the *R* wave of the FECG. Also, the signal components are inconsistent in amplitude, shape and presence or absence. Therefore, the electronic circuits must accommodate rapid gross amplitude changes, appearance and disappearance of components and must attempt to count on the same component with each heart beat. Takeuchi and Hogaki (1977) developed a special auto-correlation processor for obtaining a reliable beat-to-beat record from ultrasonic Doppler foetal heart signals. The basic concept of the development is to make a extremely quick real-time auto-correlation algorithm equivalent to the beat-to-beat heart rate meter, by adoptively controlling the algorithm itself according to the "present heart rate". Figure 8.9 shows the system block diagram. The system consists of preprocessor, correlator, post-processor and system controller.

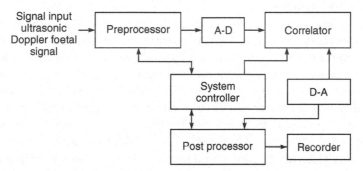

> **Fig.** 8.9 *Block diagram of the auto-correlation processor for FHR measurement (adapted after Takeuchi and Hogaki, 1977)*

The input signal comes from the ultrasound Doppler foetal signal detector. The signal goes through the bandpass filter, automatic gain control circuit (AGC) and envelope detector. The bandpass filter passes the heart valve signal and the higher frequency part of the heart wall signal. The envelope signal is again passed through another set of bandpass filter and AGC circuit. This removes dc and lower frequency components and only the useful components are put for correlation after A-D conversion. The correlator has a data storage of 4 bit×256 word with a sampling rate of 200 samples/s and a 16 bit × 256 word in correlation storage. The system has special external logic control facility to its operating mode which can be used to optimize the system in real time according to FHR, i.e. to control the data length effective for correlation computation just near to one beat-to-beat interval. Hence, the correlation output responds actually to beat-to-beat changes in the heart rate. The effective data length is controlled by scaler, by changing the refreshing rate of correlation storage. The correlation output is read out in cyclic mode. A simple clocked peak detector measures the location of the first fundamental peak location. The reciprocal value of the peak distance (i.e. heart rate) is obtained as analog voltage and put to the recorder.

Most foetal monitors check for potential false FHR with a credence check. The credence check detects beat-to-beat interval changes which exceed the likely physiologic rate of change of the FHR. In some instruments, the recorder pen may be held for a few seconds until several rhythmic heart beats are found and, if none are found, it may lift off the paper until rhythmic beats are found. In other systems the recorder pen may immediately lift. Because random noise or low signals may appear rhythmic for short periods of time, they tend to produce erratic lift and splatter. Experience has shown that at least 90% of patients who are calm and have normal presentation can be monitored throughout labour by a Dopper FHR system. When patients are too restless, too obese or have an unusual presentation, foetal scalp electrocardiography is preferred over the ultrasound Doppler method for detecting foetal heart signals.

Signal response is not always as good after the rupture of the membranes as before this stage. On many patients, however, ultrasound monitoring can be continued after the rupture of the membranes with extremely satisfactory results. However, once the membranes have ruptured and dilatation of the cervix is 1.5–2 cms or more and if satisfactory and accurate performance with the external ultrasound system is not obtained, the ECG facility should then be used and monitoring continued with scalp electrodes.

The percentage of successful results of monitoring with the ultrasound has been found to be very much dependent upon the initial correct positioning of the transducer and the adjustment of the transducer position from time to time throughout labour as and when the need arises. The correct site is not necessarily that where the best sound is obtained, and it is necessary to try various positions on the abdomen to obtain the most reliable ratemeter operation. If the umbilical cord comes between the transducer and the foetal heart, it can lead to bad counting. The same can occur if the transducer is sited over an anterior placenta due to the relatively long delay between the heart action and the actual blood flow at the placental termination of the cord.

While monitoring FHR with the ultrasound method, the ratemeter and chart recorder may give inaccurate results. Many of these are due to operating conditions like the lack of a coupling medium between the skin and the transducer, the transducer being too loosely held in position, accidental moving of the transducer by the patient, patient sitting up or turning on the side and movement of the foetus in the uterus. Inaccuracy due to these factors can be eliminated by the careful placement of the transducer and by checking the response on the audio output.

8.2.4　FHR Measurement with Direct FECG

When the membranes are not ruptured, the foetal heart activity (FECG) can be recorded by using abdominal or trans-abdominal electrodes. Usually, a hooked electrode of nichrome wire 0.5 mm in diameter can be placed subcutaneously in the foetal buttock through a puncture of the maternal abdomen. This minimizes interferences from maternal ECG.

However, after the rupture of the membranes, it is possible to attach an electrode directly to the foetal scalp. Such electrodes are called scalp electrodes. The use of scalp electrodes gives an excellent record of the electrocardiographic signals (Fig. 8.10) as the signal-to-noise ratio is much higher than in the abdominal recordings. *QRS* amplitudes of several hundred (50 to 300) microvolts

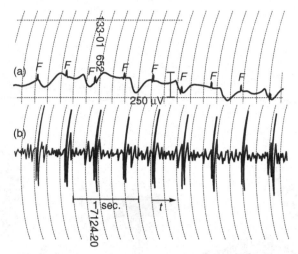

> **Fig. 8.10** *Intrauterine recording with scalp electrodes of foetal ECG during delivery with filter response (a) 12 Hz – 200 Hz, (b) 15-40 Hz*

are easily obtainable. Maternal ECG signals are rarely present in the direct lead connections. For this reason, foetal monitoring with scalp electrodes is the most reliable during the birth process. While this method has the advantage of producing very clean foetal heart frequency curves, its use is limited to only during delivery periods.

Three electrodes (Fig. 8.11) are required for the direct detection of the ECG signal. One of the electrodes, called the scalp electrode, is attached to the presenting part of the foetus. The other signal electrode is fixed contacting the vaginal wall. The indifferent electrode is strapped to the maternal leg. Conducting jelly is used with the leg electrode to ensure good electrical contact with the skin. An applicator is provided for attaching the scalp electrode. Three types of electrodes have been used for direct FECG studies, clip, spiral and suction (Fig. 8.12). The clip is an adaptation of a surgical skin clip applied to the foetal scalp. It causes little damage to the foetal skin or vaginal tissue. The spiral electrode consists of a double small metallic spiral wire and can be 'screwed' into the upper layers of the skin of the presenting part. This electrode provides larger amplitude FECG signals than the clip electrode. The suction electrode is held on the scalp by suction. It is not preferred because it can get dislodged. Spiral electrodes are easy to apply and the problem of their dislodgement is nearly absent. The foetal ECG signal measured in this way has similar significance to that of the vertical leads in conventional electrocardiography, although the amplitude of the signal is attenuated to that which would be evident post-natally. In a breech delivery, where the foetal electrode is in fact attached to the foetal buttock, the foetal ECG is comparable with the conventional aVF lead of electrocardiography, while in the normal vertex delivery, the foetal ECG is inverted with respect to the conventional aVF. This is of practical importance, since although the ECG is a biphasic signal, a knowledge of the dominant polarity of the *QRS* complex is useful in achieving consistent triggering of the ratemeter. Because the FECG has severe baseline wandering, the display is often filtered to keep the FECG in view on the display. This Filtering distorts the *ST* segment and the *T* wave, rendering it difficult to judge *ST-T* changes. These changes are late changes and are almost always preceded by the more common signs of foetal stress. Most physicians

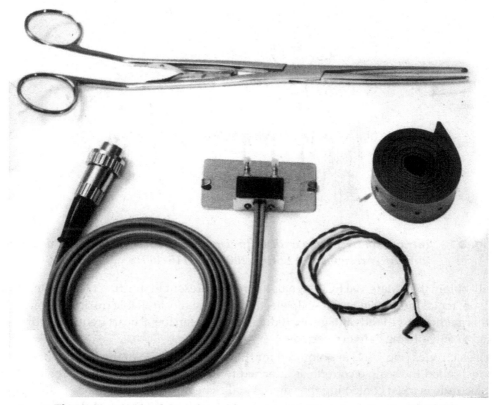

> **Fig. 8.11**   *Scalp electrodes and accessories (Courtesy: Sonicaid, U.K.)*

feel that when the FECG is necessary, the minimal risk of using direct monitoring techniques is well worth the benefits obtained from the monitoring information.

The circuit arrangement for processing FECG signals is standard except for having capabilities for handling low level signals. The input signal is passed through an electrically isolated amplifier and then filtered by a mains frequency 'notch filter' to eliminate power line interference. A bandpass filter suppresses the maternal ECG signal, if any. Together with the notch filter, the bandpass filter has a lower cut-off frequency of 30 Hz and an upper cut-off at 45 Hz. The available signal from a scalp electrode is 20 μV to 3 mV. There is one disadvantage of direct ECG monitoring. Internal monitoring may carry potential maternal risks of infection and perforation of the uterus, as well as scalp injury and infection for the foetus.

▏▶ 8.3   MONITORING LABOUR ACTIVITY

During labour, the uterus muscle starts contractions of increasing intensity in a bid to expel out the child. The intrauterine pressure can reach values of 150 mmHg or more during the expulsion period. However, a normal patient in spontaneous active labour will demonstrate uterine con-

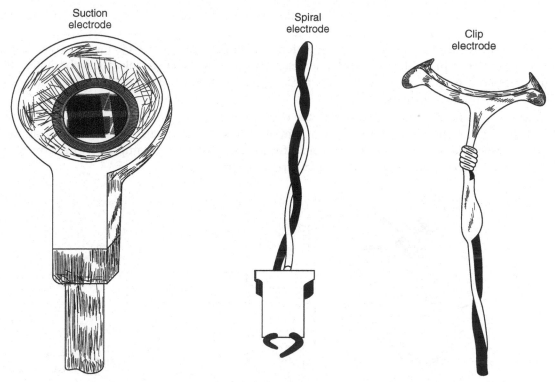

➤ Fig. 8.12   *Three types of electrodes for studying direct FECG*

tractions occurring at intervals of three to five minutes, with a duration of 30 to 70 s and peak intensity of 50 to 75 mmHg. Each uterine contraction diminishes placental perfusion and acts as a transient stress to the foetus, which may be damaged by excessive contractility or by prolonged duration of labour. Some patients will spontaneously exhibit much lower uterine activity, in terms of intensity and frequency of contractions than others but will still show progressive cervical dilatation and an otherwise normal progress of labour.

The labour activity can be recorded either in terms of the intra-uterine pressure measured directly by means of a catheter or a relative indication of the labour intensity measured through an external transducer. A plot of the tension of the uterine wall is obtained by means of a spring loaded displacement transducer. The transducer performs a quasi-isometric measurement of the tension of the uterus. The transducer carries a protruding tip which is pressed to the mother's abdomen with a light force to ensure an effective coupling. The protruding surface of the transducer is displaced as the tension in the uterus increases. This movement is converted into an electrical signal by a strain gauge in the transducer housing. The abdominal transducer provides a reliable indication of the occurrence frequency, duration and relative intensity of the contraction.

The toco-transducers are location sensitive. They should be placed over the fundus where there is maximum motion with the contractions. The toco-tonometer transducer cannot be used in the same place as the foetal heart rate detector, thus the patient must have two transducers on her abdomen.

To sense uterine contractions externally, it is necessary to press into the uterus through the abdominal wall. Resistance to pressure is measured either by the motion of a spring or the force needed to prevent a button from moving. External strain gauges are used to measure and record the bending of a spring. In some instruments, a crystal which changes electrical characteristics with applied pressure is used to measure force against a plunger. This method is automatic and provides pertinent information.

Figure 8.13 shows a block diagram of the circuit which measures labour activity externally. The transducer output is amplified in an ac amplifier. The low frequency labour activity signal is obtained from the synchronous detector and is further amplified by a dc amplifier. The activity can be either displayed on a meter or on a direct writing chart recorder.

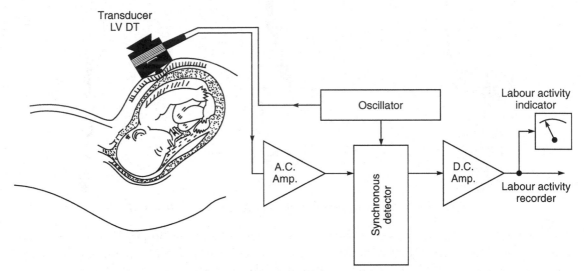

➤ **Fig. 8.13** *Block diagram of of labour activity monitor (external method)*

The labour-activity transducers are pressure transducers that drive circuits for obtaining an electrical indication of pressure by conventional means. The pressure channel on the recorder is provided with a positioning control. This is done because the baseline is affected by the static pressure on the transducer that results from the tension on the belt holding the transducer in place. The control permits the operator to position the baseline on the zero-level line of the recording chart.

In external toco-tonometry, movement of the foetus may be superimposed on the labour activity curve. Stress imposed on the foetal circulatory system by the uterine contractions, foetal movements or other factors are seen in the response of the foetal heart to these stimuli and are studied in the correct time relationship.

The internal method measures intra-uterine pressure (IUP) via a fluid-filled catheter. The catheter is inserted into the uterus through a guide after the rupture of the foetal membranes. After allowing free flow of amniotic fluid to ensure correct placement, the distal end of the catheter is usually attached to a pressure transducer of the type used for cardiac studies. Changes in amniotic pressure are easily transmitted to the gauge by the incompressible fluid in the catheter. The pressure transducer converts the catheter pressure into an electrical signal which can be displayed on the strip chart recorder. Strain gauges, though very accurate, tend to drift up to several mmHg/h or drift with temperature changes. Therefore, when continuous monitoring is employed, it is necessary to set zero and calibrate the transducer frequently. The peak pressure may vary according to which catheter is placed in the uterus. It is necessary to flush the catheter system to avoid any blockage and to maintain the frequency response. The major applications of IUP measurement are accurate assessment of the pressure during contractions and measurement of tonus, both impossible by indirect means.

Although the system is inherently capable of having great accuracy, catheter-obtained uterine contraction data may be distorted or inaccurate. The IUP may be accurately recorded only as long as a fluid pool is sustained around the tip of the catheter and leakage is completely controlled by the descending foetal head (Caspo, 1970). Since there is no real control of catheter placement, it may slip into an isolated pocket and receive very high pressure, especially if there is little fluid. Also, the uterus only approximates a closed fluid chamber, pressures are not necessarily transmitted equally to all segments. Open segments tend to lose fluid and thus may generate lower pressures. One study showed that IUP varies by as much as 25% at different points in the uterus. Thus, the physiological measurement does not approach the instrument in accuracy or reproducibility.

8.4 RECORDING SYSTEM

Instantaneous "beat-to-beat" rate is displayed on a calibrated linear scale or digitally displayed with a range from 50 to 210 bpm. A two-channel chart recorder is incorporated in instruments used for monitoring labour activity. One channel records FHR on a calibrated chart in beats per minute (50–210 bpm) while the other channel is used for recording uterine contractions calibrated 0-100 mmHg. The standard chart speed is usually 1 or 2 cm/min. Both the contraction transducer and the foetal heart transducer are held together in position using stretch belts or bandages. The recorder usually uses thermal writing and thus avoids the possibility of running out of ink. In one system, the stylus is a thick film resistor. To make the operation quieter, contactless position feedback is provided by a capacitive transducer on the galvanometer shaft. This contactless feedback also enhances reliability by eliminating mechanical parts that could wear out. The galvanometer, which needs a frequency response of only 3 Hz, is positioned by a servo motor through a silent step-down belt drive. Recording sensitivity is 20 bpm/cm giving a basic resolution of 1 bpm for seeing small changes in the heart rate.

The chart paper is advanced by a direct-drive stepper motor eliminating the usual gear train. Paper speed is changed simply by switching to a different motor drive frequency, rather than by shifting gears. The paper magazine is designed to make loading the chart paper an extremely easy task.

The ability to record large amounts of relatively artefact-free data, especially after the onset of labour in the form of foetal electrocardiogram (FECG), foetal heart rate (FHR), and uterine contractions has been an encouraging and a valuable step forward. However, new problems have been generated. It is indeed impossible to analyze even in a rudimentary manner, about 9000 foetal ECG complexes and various FHR and uterine contraction patterns associated with even an hour of labour, with a large percentage being, by known criteria, normal. Despite such monitors, intelligence has still to be provided by the clinical staff observing and interpreting the records. Transferring information in this way from a chart record to a clinically useful result relies on the perception of patterns and summation of activity over long periods of monitoring.

The first task of recognition of patterns is most amenable to the human observer. The second, long-term survey of several metres of chart records to deduce trends, is naturally difficult for human interpreters. Time available for the study of the records, the problems of translating a long analog recording into meaningful data about trends and recognition of ominous shifts in the foetal condition are other factors contributing to difficulties in manually processing the huge volumes of data. Advocates of some form of machine intelligence in foetal monitoring contend that all the criticism against human unreliability of interpretation can be solved by computer surveillance of the FHR and IUP patterns. Many centres, therefore, have undertaken studies for examining the feasibility of data reduction techniques with the use of a digital computer. These systems collect data from foetal monitors, analyze the patterns and produce results and interpretations. Some systems also produce contraction, dilatation and effacement time curves as well as records of treatment and obstetrical history.

Cardiotocographs using microprocessors for working out several computations on FHR and IUP data are commercially available. Such systems compute beat-to-beat foetal heart rate with a resolution of 1 bpm. The FHR variability data is plotted in a bar graph form on strip chart recorder from every 256 usable heart beats. This is computed in direct FECG mode and calculated only between contractions. Data is edited by omitting beat-to-beat changes which are greater than 20 bpm. The system stores the heart rate for variability computations if a contraction is not present and if the beat-to-beat change is not greater than 12 bpm. The uterine activity is plotted also in bar graph form every 10 minutes.

The microprocessor also computes conditions of bradycardia and gives an audio alert alarm after 35 s. The processing circuit contains the usual elements like: PROM containing the program controlling the CPU, RAM for temporary storage of data by the CPU, input/output ports for controlling flow of information into and out of the CPU, digital-to-analog converters for generating F signals for the recorder and mode logic for decoding signals determined by the type of transducers connected to the monitor.

Biomedical Telemetry and Telemedicine

▷ 9.1 **WIRELESS TELEMETRY**

Wireless telemetry permits examination of the physiological data of man or animal under normal conditions and in natural surroundings without any discomfort or obstruction to the person or animal under investigation. Factors influencing healthy and sick persons during the performance of their daily tasks may thus be easily recognized and evaluated. Wireless bio-telemetry has made possible the study of active subjects under conditions that so far prohibited measurements. It is, therefore, an indispensable technique in situations where no cable connection is feasible. (Gandikola, 2000).

Using wireless telemetry, physiological signals can be obtained from swimmers, riders, athletes, pilots or manual labourers. Telemetric surveillance is most convenient during transportation within the hospital area as well for the continuous monitoring of patients sent to other wards or clinics for check-up or therapy.

9.1.1 Modulation Systems

The modulation systems used in wireless telemetry for transmitting biomedical signals makes use of two modulators. This means that a comparatively lower frequency sub-carrier is employed in addition to the VHF, which finally transmits the signal from the transmitter. The principle of double modulation gives better interference free performance in transmission and enables the reception of low frequency biological signals. The sub-modulator can be a FM (frequency modulation) system or a PWM (Pulse Width Modulation) system, whereas the final modulator is practically always an FM system.

Frequency Modulation: In frequency modulation, intelligence is transmitted by varying the instantaneous frequency in accordance with the signal to be modulated on the wave, while keeping the amplitude of the carrier wave constant. The rate at which the instantaneous frequency varies is the modulating frequency. The magnitude to which the carrier frequency varies away from the centre frequency is called "Frequency Deviation" and is proportional to the amplitude of the modulating signal. Usually, an FM signal is produced by controlling the frequency of an oscillator by the amplitude of the modulating voltage. For example, the frequency of oscillation for most oscillators depends on a particular value of capacitance. If the modulation signal can be applied in such a way that it changes the value of capacitance, then the frequency of oscillation will change in accordance with the amplitude of the modulating signal.

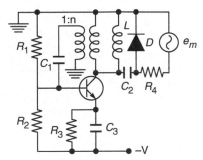

> **Fig. 9.1** *Circuit diagram of a frequency modulator using varactor diode*

Figure 9.1 shows a tuned oscillator that serves as a frequency modulator. The diode used is a varactor diode operating in the reverse-biased mode and, therefore, presents a depletion layer capacitance to the tank circuit. This capacitance is a function of the reverse-biased voltage across the diode and, therefore, produces an FM wave with the modulating signal applied as shown in the circuit diagram. This type of circuit can allow frequency deviations of 2–5% of the carrier frequency without serious distortion.

Pulse Width Modulation: Pulse width modulation method has the advantage of being less perceptive to distortion and noise. Figure 9.2 shows a typical pulse width modulator. Transistors Q_1 and Q_2 form a free-running multi-vibrator. Transistors Q_3 and Q_4 provide constant current sources for charging the timing capacitors C_1 and C_2 and driving transistors Q_1 and Q_2. When Q_1 is 'off' and Q_2 is 'on', capacitor C_2 charges through R_1 to the amplitude of the modulating voltage e_m. The other side of this capacitor is connected to the base of transistor Q_2 and is at zero volt. When Q_1 turns 'on' switching the circuit to the other stage, the base voltage of Q_2 drops from approximately

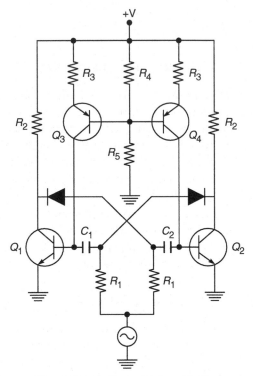

> **Fig. 9.2** *Pulse width modulator*

zero to $-e_m$. Transistor Q_2 will remain 'off' until the base voltage charges to zero volt. Since the charging current is constant at I, the time required to charge C_2 and restore the circuit to the initial stage is:

$$T_2 = \frac{C_2}{I} \cdot e_m$$

Similarly, the time that the circuit remains in the original stage is:

$$T_1 = \frac{C_1}{I} \cdot e_m$$

This shows that both portions of the astable period are directly proportional to the modulating voltage.

When a balanced differential output from an amplifier such as the ECG amplifier is applied to the input points 1 and 2, the frequency of the astable multi-vibrator would remain constant, but the width of the pulse available at the collector of transistor Q_2 shall vary in accordance with the amplitude of the input signal.

In practice, the negative edge of the square wave is varied in rhythm with the ECG signal. Therefore, only this edge contains information of interest. The ratio $P:Q$ (Fig. 9.3) represents the momentary amplitude of the ECG. The amplitude or even the frequency variation of the square wave does not have an influence on the $P:Q$ ratio and consequently on the ECG signal. The signal output from this modulator is fed to a normal speech transmitter, usually via an attenuator, to make it suitable to the input level of the transmitter.

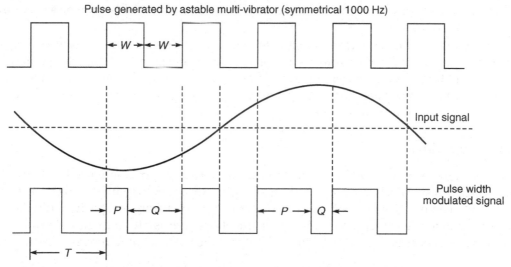

> **Fig.** 9.3 *Variation of pulse width with amplitude of the input signal*
> *W = Pulse width as generated by the multi-vibrator.*
> *P = Variable pulse width; variation in accordance with input signal.*
> *Q = Off-period, which also gets varied as the pulse width P varies with the amplitude of the input signal.*

9.1.2 Choice of Radio Carrier Frequency

In every country there are regulations governing the use of only certain frequency and bandwidth for medical telemetry. Therefore, the permission to operate a particular telemetry system needs to be obtained from the postal department of the country concerned. The radio frequencies normally used for medical telemetry purposes are of the order of 37, 102, 153, 159, 220 and 450 MHz. The transmitter is typically of 50 mW at 50 Ω, which can give a transmission range of about 1.5 km in the open flat country. The range will be much less in built-up areas. In USA, two frequency bands have been designated for short range medical telemetry work by the FCC (Federal Communications

Commission). The lower frequency band of 174–216 MHz, coincides with the VHF television broadcast band (Channels 7–13). Therefore, the output of the telemetry transmitter must be limited to avoid interference with TV sets. Operation of telemetry units in this band does not normally require any licence. In the higher frequency band of 450–470 MHz, greater transmitter power is allowed, but an FCC licence has to be obtained for operating the system.

Radiowaves can travel through most non-conducting material such as air, wood, and plaster with relative ease. However, they are hindered, blocked or reflected by most conductive material and by concrete because of the presence of reinforced steel. Therefore, transmission may be lost or be of poor quality when a patient with a telemetry transmitter moves in an environment with a concrete wall or behind a structural column. Reception may also get affected by radio frequency wave effects that may result in areas of poor reception or null spots, under some conditions of patient location and carrier frequency. Another serious problem that is sometimes present in the telemetry systems is the cross-talk or interference between telemetry channels. It can be minimized by the careful selection of transmitter frequencies, by the use of a suitable antenna system and by the equipment design.

The range of any radio system is primarily determined by transmitter output power and frequency. However, in medical telemetry systems, factors such as receiver and antenna design may make the power and frequency characteristics less significant. The use of a higher-powered transmitter than is required for adequate range is preferable as it may eliminate or reduce some noise effects due to interference from other sources.

9.1.3 Transmitter

Figure 9.4 shows a circuit diagram of the FM transmitter stage commonly used in medical tele-metry. The transistor T acts in a grounded base Colpitts R.F oscillator with L_1 and C_1 and C_2 as the tank circuit. The positive feedback to the emitter is provided from a capacitive divider in the collector circuit formed by C_1 and C_2. Inductor L_1 functions both as a tuning coil and a transmitting antenna. Trim capacitor C_2 is adjusted to precisely set the transmission frequency at the desired

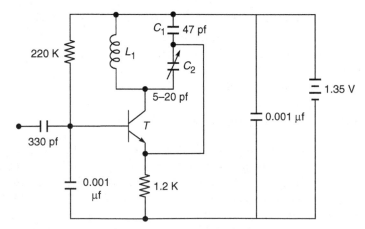

➤ **Fig. 9.4**   *Typical circuit diagram of a FM telemetry transmitter*

point. In this case, it is within the standard FM broadcast band from 88 to 108 MHz. Frequency modulation is achieved by variation in the operating point of the transistor, which in turn varies its collector capacitance, thus changing the resonant frequency of the tank circuit. The operating point is changed by the sub-carrier input. Thus, the transmitter's output consists of an RF signal, tuned in the FM broadcast band and frequency modulated by the sub-carrier oscillator (SCO), which in turn is frequency modulated by the physiological signals of interest. It is better to use a separate power source for the RF oscillator from other parts of the circuit to achieve stability and prevent interference between circuit functions (Beerwinkle and Burch, 1976).

9.1.4 The Receiver

In most cases, the receiver can be a common broadcast receiver with a sensitivity of $1\mu V$. The output of the HF unit of the receiver is fed to the sub-demodulator to extract the modulating signal. In a FM/FM system, the sub-demodulator first converts the FM signal into an AM signal. This is followed by an AM detector which demodulates the newly created AM waveform. With this arrangement, the output is linear with frequency deviation only for small frequency deviations. Other types of detectors can be used to improve the linearity.

In the PWM/FM system, a square wave is obtained at the output of the RF unit. This square wave is clipped to cut off all amplitude variations of the incoming square wave and the average value of the normalized square wave is determined. The value thus obtained is directly proportional to the area which in turn is directly proportional to the pulse duration. Since the pulse duration is directly proportional to the modulating frequency, the output signal is directly proportional to the output voltage of the demodulator. The output voltage of the demodulator is adjusted such that it can be directly fed to a chart recorder. The receiver unit also provides signal outputs where a magnetic tape recorder may be directly connected to store the demodulated signal.

Successful utilization of biological telemetry systems is usually dependent upon the user's systematic understanding of the limits of the system, both biological and electrical. The two major areas of difficulty arising in biotelemetry occur at the system interfaces. The first is the interface between the biological system and the electrical system. No amount of engineering can correct a shoddy, hasty job of instrumenting the subject. Therefore, electrodes and transducers must be put on with great care. The other major area of difficulty is the interface between the transmitter and receiver. It must be kept in mind that the range of operation should be limited to the primary service area, otherwise the fringe area reception is likely to be noisy and unacceptable. Besides this, there are problems caused by the patient's movements, by widely varying signal strength and because of interference from electrical equipment and other radio systems, which need careful equipment design and operating procedures.

9.2 SINGLE CHANNEL TELEMETRY SYSTEMS

In a majority of the situations requiring monitoring of the patients by wireless telemetry, the parameter which is most commonly studied is the electrocardiogram. It is known that the display of the ECG and cardiac rate gives sufficient information on the loading of the cardiovascular system of the active subjects. Therefore, we shall first deal with a single channel telemetry system suitable for the transmission of an electrocardiogram.

9.2.1 ECG Telemetry System

Figure 9.5 shows the block diagram of a single channel telemetry system suitable for the transmission of an electrocardiogram. There are two main parts:

- The *Telemetry Transmitter* which consists of an ECG amplifier, a sub-carrier oscillator and a UHF transmitter along with dry cell batteries.
- *Telemetry Receiver* consists of a high frequency unit and a demodulator, to which an electrocardiograph can be connected to record, a cardioscope to display and a magnetic tape recorder to store the ECG. A heart rate meter with an alarm facility can be provided to continuously monitor the beat-to-beat heart rate of the subject.

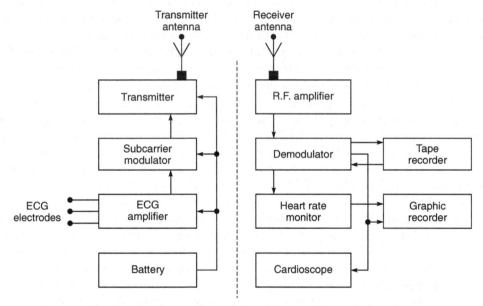

> **Fig.** 9.5   *Block diagram of a single channel telemetry system*

For distortion-free transmission of ECG, the following requirements must be met: (Kurper *et al*, 1996).

- The subject should be able to carry on with his normal activities whilst carrying the instruments without the slightest discomfort. He should be able to forget their presence after some minutes of application.
- Motion artefacts and muscle potential interference should be kept minimum.
- The battery life should be long enough so that a complete experimental procedure may be carried out.
- While monitoring paced patients for ECG through telemetry, it is necessary to reduce pacemaker pulses. The amplitude of pacemaker pulses can be as large as 80 mV compared to 1–2 mV, which is typical of the ECG. The ECG amplifiers in the transmitter are slew rate (rate of change of output) limited so that the relatively narrow pacemaker pulses are reduced in amplitude substantially.

Some ECG telemetry systems operate in the 450–470 MHz band, which is well-suited for transmission within a hospital and has the added advantage of having a large number of channels available. The circuit details of an ECG telemetry system are described below:

Transmitter: A block diagram of the transmitter is shown in Fig. 9.6. The ECG signal, picked up by three pre-gelled electrodes attached to the patient's chest, is amplified and used to frequency modulate a 1 kHz sub-carrier that in turn frequency-modulates the UHF carrier. The resulting signal is radiated by one of the electrode leads (RL), which serves as the antenna. The input circuitry is protected against large amplitude pulses that may result during defibrillation.

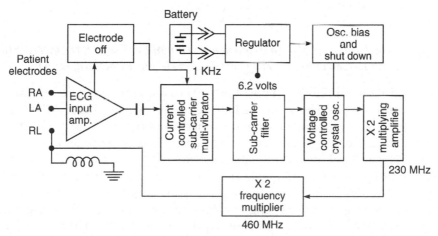

> **Fig.** 9.6 *Block diagram of ECG telemetry transmitter (redrawn after Larsen et al., 1972; by permission of Hewlett Packard, USA)*

The ECG input amplifier is ac coupled to the succeeding stages. The coupling capacitor not only eliminates dc voltage that results from the contact potentials at the patient-electrode interface, but also determines the low-frequency cut-off of the system which is usually 0.4 Hz. The sub-carrier oscillator is a current-controlled multi-vibrator which provides ±320 Hz deviation from the 1 kHz centre frequency for a full range (± 5 mV) ECG signal. The sub-carrier filter removes the square-wave harmonic and results in a sinusoid for modulating the RF carrier. In the event of one of the electrodes failing off, the frequency of the multi-vibrator shifts by about 400 Hz. This condition when sensed in the receiver turns on an 'Electrode inoperative' alarm.

The carrier is generated in a crystal-controlled oscillator operating at 115 MHz. The crystal is a fifth overtone device and is connected and operated in the series resonant mode. This is followed by two frequency doubler stages. The first stage is a class-C transistor doubler and the second is a series connected step recovery diode doubler. With the output power around 2 mW, the system has an operating range of 60 m within a hospital.

Receiver: The receiver uses an omnidirectional receiving antenna which is a quarter-wave monopole, mounted vertically over the ground plane of the receiver top cover. This arrangement works well to pick up the randomly polarized signals transmitted by moving patients.

The receiver (Fig. 9.7) comprises an RF amplifier, which provides a low noise figure, RF filtering and image-frequency rejection. In addition to this, the RF amplifier also suppresses local oscillator

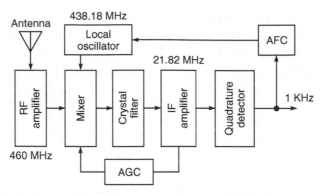

> **Fig. 9.7** *Block diagram of high frequency section of ECG telemetry receiver (adapted from Larsen et al , 1972; by permission Hewlett Packard, USA)*

radiation to –60 dBm to minimize the possibility of cross-coupling where several receivers are used in one central station. The local oscillator employs a crystal (115 MHz) similar to the one in the transmitter and × 4 multiplier and a tuned amplifier. The mixer uses the square law characteristics of a FET to avoid interference problems due to third-order intermodulation. The mixer is followed by an 8-pole crystal filter that determines the receiver selectivity. This filter with a 10 kHz bandwidth provides 60 dB of rejection for signals 13 kHz from the IF centre frequency (21.82 MHz). The IF amplifier provides the requisite gain stages and operates an AGC amplifier which reduces the mixer gain under strong signal conditions to avoid overloading at the IF stages. The IF amplifier is followed by a discriminator, a quadrature detector. The output of the discriminator is the 1 kHz sub-carrier. This output is averaged and fed back to the local oscillator for automatic frequency control. The 1 kHz sub-carrier is demodulated to convert frequency-to-voltage to recover the original ECG waveform. The ECG is passed through a low-pass filter (Fig. 9.8) having a cut-off frequency of 50 Hz and then given to a monitoring instrument. The 1 kHz sub-carrier is examined to determine whether or not a satisfactory signal is being received. This is done by establishing a window of acceptability for the sub-carrier amplitude. If the amplitude is within the window, then the received signal is considered valid. In the case of AM or FM interference, an 'inoperative' alarm lamp lights up.

Different manufacturers use different carrier frequencies in their telemetry equipment. Use of the FM television band covering 174 to 185.7 MHz (VHF TV channels 7 and 8) is quite common. However, the output is limited to a maximum of 150 μV/m at a distance of 30 m to eliminate interference with commercial television channels.

Some transmitters are also provided with special arrangements like low transmitter battery and nurse call facility. In both these situations, a fixed frequency signal is generated, which causes a deviation of the sub-carrier and when received at the receiver, actuate appropriate circuitry for visual indications.

Also, some telemetry systems include an out-of-range indication facility. This condition is caused by a patient lying on the leads or the patient getting out of range from the receiver capability. For this, the RF carrier signal from the transmitter is continuously monitored. When this signal level falls below the limit set, the alarm will turn on.

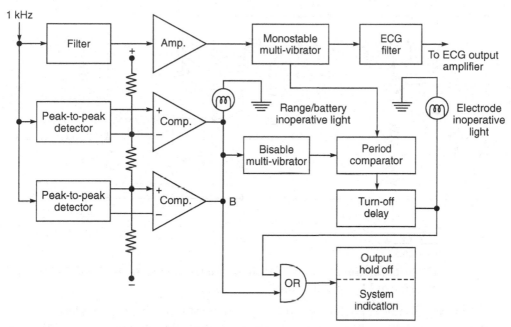

➤ Fig. 9.8 *Schematic diagram of ECG demodulation and 'inoperate' circuits in ECG telemetry receiver (after Larsen et al. 1972; by permission of Hewlett Packard, USA)*

For the satisfactory operation of a radiotelemetry system, it is important to have proper orientation between the transmitting and receiving antennas. There can be orientations in which none of the signals radiated from the transmitting antenna are picked up by the receiving antenna. It is therefore important to have some means for indicating when signal interference or signal dropout is occurring. Such a signal makes it possible to take steps to rectify this problem and informs the clinical staff that the information being received is noise and should be disregarded.

9.2.2 Temperature Telemetry System

Systems for the transmission of alternating potentials representing such parameters as ECG, EEG and EMG are relatively easy to construct. Telemetry systems which are sufficiently stable to telemeter direct current outputs from temperature, pressure or other similar transducers continuously for long periods present greater design problems. In such cases, the information is conveyed as a modulation of the mark/space ratio of a square wave. A temperature telemetry system based on this principle is illustrated in the circuit shown in Fig. 9.9.

Temperature is sensed by a thermistor having a resistance of 100 Ω (at 20°C) placed in the emitter of transistor T_1. Transistors T_1 and T_2 form a multi-vibrator circuit timed by the thermistor, $R_1 + R_2$ and C_1. R_1 is adjusted to give 1:1 mark/space ratio at midscale temperature (35–41°C). The multi-vibrator produces a square wave output at about 200 Hz. Its frequency is chosen keeping in view the available bandwidth, required response time, the physical size of the multi-vibrator, timing capacitors and the characteristics of the automatic frequency control circuit of the receiver.

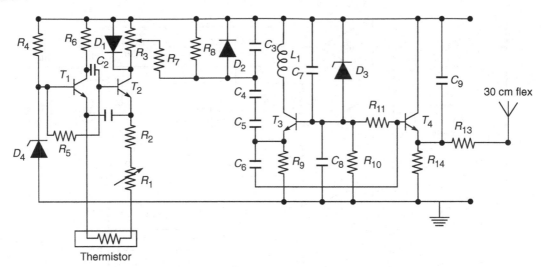

> **Fig.** 9.9   *Circuit diagram of a temperature telemetry system (after Heal, 1974; by permission of Med. & Biol. Eng.)*

This is fed to the variable capacitance diode D_2 via potentiometer R_3. D_2 is placed in the tuned circuit of a RF oscillator constituted by T_3. Transistor T_3 forms a conventional 102 MHz oscillator circuit, whose frequency is stabilized against supply voltage variations by the Zener diode D_3 between its base and the collector supply potential. T_4 is an untuned buffer stage between the oscillator and the aerial. The aerial is normally taped to the collar or harness carrying the transmitter.

On the receiver side, a vertical dipole aerial is used which feeds a FM tuner, and whose output, a 200 Hz square wave, drives the demodulator. In the demodulator, the square wave is amplified, positive dc restored and fed to a meter where it is integrated by the mechanical inertia of the meter movement. Alternatively, it is filtered with a simple RC filter to eliminate high ripple content and obtain a smooth record on a paper. A domestic FM tuner can be used for this purpose. Temperature measurements in this scheme were made with a thermistor probe having a temperature coefficient of approximately – 4% per degree centigrade. It produces a change of mark/space ratio of about 20% over a temperature range of 5°C. With a span of ±3°C, the system is found to be linear. The circuit is designed to operate on 5.4 V, 350 mAh battery which gives a continuous operation for 100 hours.

▷ 9.3   MULTI-CHANNEL WIRELESS TELEMETRY SYSTEMS

Medical measuring problems often involve the simultaneous transmission of several parameters. For this type of application, a multi-channel telemetry system is employed. Multi-channel telemetry is particularly useful in athletic training programs as it offers the possibility of simultaneously surveying several physiological parameters of the person being monitored.

With appropriate preamplifiers, the multi-channel systems permit the transmission of the following parameters simultaneously depending upon the number of channels required, ECG and heart rate, respiration rate, temperature, intravascular and intra-cardiac blood pressure.

In multi-channel telemetry, the number of sub-carriers used are the same as the number of signals to be transmitted. Each channel therefore has its own modulator. The RF unit—the same for all channels—converts the mixed frequencies into the transmission band. Similarly, the receiver unit contains the RF unit and one demodulator for each channel.

Pulse width modulation is better suited for multi-channel biotelemetry systems. Such systems are insensitive to carrier frequency shifts and have high noise immunity. FM-FM systems for similar use may have low power consumption and high baseline stability, but they are more complicated and turn out to be more expensive. They can be troubled by interference between different channels. Techniques for separation usually require expensive and complex filters and even with these, cross-talk can still be a problem. Since the FM-FM system employs a separate sub-carrier frequency for each data channel, it generally involves a high cost. Similarly, pulse-position amplitude modulation easily gets into synchronization difficulties caused by noise and thus results in a loss of the information transmitted. On the other hand, advantages of pulse-duration modulation include lower sensitivity to temperature and battery voltage changes and its adaptability to miniaturization due to availability of suitable integrated circuits.

For multi-channel radiotelemetry, various channels of information are combined into a single signal. This technique is called *multiplexing*. There are two basic methods of multiplexing. These are:

- *Frequency–division multiplexing:* The method makes use of continuous-wave sub-carrier frequencies. The signals frequency–modulate multiple subcarrier oscillators, each being at such a frequency that its modulated signal does not overlap the frequency spectra of the other modulated signals. The frequency modulated signals from all channels are added together through a summing amplifier to give a composite signal in which none of the parts overlap in frequency. This signal then modulates the RF carrier of the transmitter and is broadcast.

- *Time–division multiplexing:* In this technique, multiple signals are applied to a commutator circuit. This circuit is an electronic switch that rapidly scans the signals from different channels. An oscillator drives the commutator circuit so that it samples each signal for an instant of time, thereby giving a pulse train sequence corresponding to input signals. A frame reference signal is also provided as an additional channel to make it easy to recognize the sequence and value of the input channels.

9.3.1 Telemetry of ECG and Respiration

An FM-FM modulated radiotelemetry transmitter (Fig. 9.10) for detecting and transmitting ECG and respiration activity simultaneously on a single carrier frequency in the FM broadcast band is described by Beerwinkle and Burch (1976). Respiration is detected by the impedance pneumographic principle by using the same pair of electrodes that are used for the ECG. A 10 kHz sinusoidal constant current is injected through electrodes E_1 and E_2 attached across the subject's thoracic cavity. The carrier signal is generated by a phase shift oscillator. The varying thoracic impedance associated with respiration produces an ac voltage whose amplitude varies with a change in impedance. The amplitude varying carrier is amplified by an amplifier A_1. An amplifier filter A_3 recovers the respiration signal by using rectifiers and a double pole filter. The ECG signal,

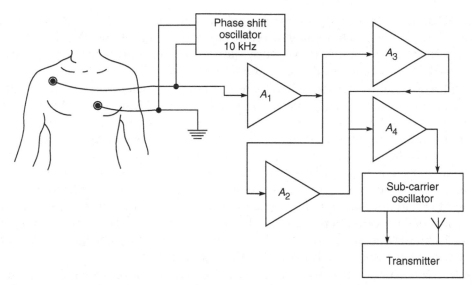

> **Fig.** 9.10   *Schematic diagram of FM-FM modulated radiotelemetry transmitter for ECG and respiration activity simultaneously (adapted from Beerwinkle and Burch, 1976; by permission of IEEE Trans. Biomed. Eng.)*

detected by electrodes E_1 and E_2 is amplified in A_1 along with the respiratory signal. It is passed through a low-pass Butterworth filter stage A_2 which passes the ECG signal but blocks respiratory signal. The amplified ECG signal is then summed up with the preprocessed respiration signal in A_4.

The output of A_4 is a composite signal which is supplied to an astable multi-vibrator which acts as a voltage-controlled sub-carrier oscillator operating at 7350 + 550 Hz. The sensitivity of the sub-carrier modulation system is 650 Hz/mV for the ECG signal and a 40 Hz/Ω change in the case of the respiration signal when the total thoracic impedance is between 600 and 800 Ω. The output of the sub-carrier oscillator is then fed to a RF oscillator for transmission. The circuit requires less than 185 μA from a 1.35 V mercury battery. Signals can be transmitted over distances up to 15 m for about four weeks before replacing the battery.

9.3.2   Obstetrical Telemetry System

There has been a great deal of interest to provide greater freedom of movement to patients during labour while the patient is continuously monitored through a wireless link. Thus, from a central location, it is possible to maintain a continuous surveillance of cardiotocogram records for several ambulatory patients. In the delivery room, telemetry reduces the encumbering instrumentation cables at the bedside. Moreover, when an emergency occurs, there is no loss of monitoring in the vital minutes needed for patient transfer.

The patient carries a small pocket-sized transmitter which is designed to pick up signals for foetal heart rate and uterine activity. The foetal heart rate is derived from foetal ECG which is obtained via a scalp electrode attached to the foetus after the mother's membranes are ruptured.

Uterine activity is measured via an intra-uterine pressure transducer. If only foetal ECG is measured, the patient herself can indicate uterine activity or foetal movement by using a hand-held push button.

The receiver (Fig. 9.11) located away from the patient, is connected to a conventional cardio-tocograph. If the patient exceeds the effective transmission range or the electrode has a poor contact, it is appropriately transmitted for corrective action.

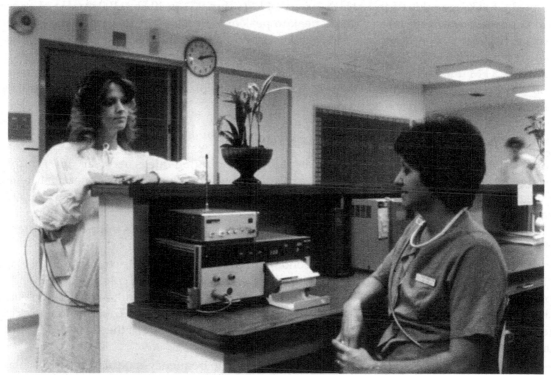

> **Fig. 9.11** *Telemetry receiving system for monitoring foetal heart rate and uterine contractions in use (Courtesy: Hewlett Packard, USA)*

The telemetry system uses FM/FM modulation, with a carrier of 450 to 470 MHz and an RF power output of 2 mW into 50Ω load measured from RL electrode to a ground plane under transmitter. The input signal range in the input for the ECG channel is 100 μV to 1 mV with a frequency band 1 to 40 Hz. The toco channel has a sensitivity of 40μV/ V/mmHg, and by using a high sensitivity transducer, it can be 5μV/ V/mmHg. The strain gauge transducer is excited with 0.25 V_{rms} for 40μV/V/mmHg at 2.4 kHz. The frequency response of this channel is 3 Hz $\pm$ 0.5 Hz.

9.3.3 Telemetry in Operating Rooms

The use of telemetry in operating rooms seems to be particularly attractive as it offers a means of achieving a high degree of patient safety from electric shock as well as elimination of the hanging inter-connecting patient leads which are necessary in direct-wired equipment. Normally, there are several parameters which are of interest while monitoring surgical patients, the most common being ECG, blood pressure, peripheral pulse and EEG.

Basically, in a four channel system, the signal encoding is based upon frequency modulation of the four sub-carriers centred at 2.2, 3.5, 5.0 and 7.5 kHz, respectively. The system is designed to give a bandwidth of dc to 100 Hz at the 3 dB point and the discriminator provides a 1.0 V dc output for a 10% shift in the sub-carrier associated with each channel. This is obtained from 2 mV peak-to-peak of ECG signal (0.05 to 100 Hz), 100 μV of EEG signal (1 to 40 Hz), 100 mmHg of arterial blood pressure (dc to 40 Hz) and 400 μV peak-to-peak of peripheral pulse (0.1 to 40 Hz). The four sub-carriers are summed and used to frequency-modulate a radio frequency carrier oscillator which is tuned to a spot frequency within the commercial FM band. The transmitted signals are tuned by a FM tuner whose output is fed into a fourth-channel discriminator which separates the sub-carriers through filtering and demodulates each using a phase-locked loop. The demodulated signals are displayed on an oscilloscope.

9.3.4 Sports Physiology Studies through Telemetry

Monitoring of pulmonary ventilation, heart rate and respiration rate is necessary for a study of energy expenditure during physical work, particularly for sports such as squash, handball, tennis, track, etc. For this purpose, the transmitter uses pulse duration modulation, i.e. each channel is sampled sequentially and a pulse is generated, the width of which is proportional to the amplitude of the corresponding signal. At the end of a frame, a synchronization gap is inserted to ensure that the receiving system locks correctly onto the signal.

Each channel is sampled 200 times a second. With each clock pulse, the counter advances one step, making the gates to open sequentially. At the opening of a particular gate, the corresponding physiological signal gets through to a comparator where it is compared with the ramp. As soon as the ramp voltage exceeds the signal voltage, the comparator changes state. Thus, the time required for the comparator to change state would depend upon the amplitude of the signal. The counter and gates serve as a multiplexer.

The pulse train at the output of the comparator is used to frequency-modulate the RF oscillator in the 88–108 MHz band. The transmitter is designed to work in a range of 100 m, which can be extended by using a whip antenna. For recording ECG, the electrodes are placed at the sternum. The pulmonary ventilation and respiration rate are derived from a mass flow transducer. At the receiving end, the system contains an FM tuner and circuitry to convert the pulse width coded signals back to analog signals and a multi-channel pen recorder to display the physiological signals. Figure 9.12 shows a three channel telemetry system for monitoring the physiological data of a sprinter.

▶ 9.4   MULTI-PATIENT TELEMETRY

The establishment of instrumented coronary care units have resulted in substantial reduction in the mortality rates of hospitalized patients. When a patient's condition has stabilized within a few days, it is necessary that he is monitored during the early stages of increased activity and exertion to determine if his heart has sufficiently recovered. This can be conveniently done by the use of

> ➤ **Fig.** 9.12 *A three channel telemetry system to monitor the physiological data of a sprinter*

telemetry which provides a sort of intermediate stage of care that smoothens the patient's transition back to a normal life. It thus permits surveillance of suspected coronaries without the unnatural constraints of confining the patient to bed. The main advantage of a multi-patient single parameter telemetry system is that patients making satisfactory recovery can vacate the hard wired instrument beds in the ICU/CCU units, which provides a positive psychological effect. The patients regain mobility after an extended period of confinement thereby improving their muscle tone and circulation. Transmitters as small as $8 \times 6.25 \times 2.25$ cm in size and weighing less than 115 g, including battery are commercially available. Data from different patients is received at the nurses central station. The station may have the facility of non-fade display of received waveforms, an ECG recorder which gets activated when the patient goes into alarm, loose lead/loss of signal alarm. The heart rate of each patient is derived and displayed simultaneously with a digital display. Multi-patient telemetry is usually done using crystal controlled circuits, which provide frequency stability to within $\pm 0.0015\%$. Codes are necessarily provided on both the transmitter as well as on the receiver units to indicate their calibrated frequencies.

The multi-patient telemetry systems, having utility mostly with cardiac patients, should have transmitters provided with defibrillator protection to 5000 V, 400 Watt sec. pulse. ECG waveforms should not be seriously affected in the presence of pacemaker pulses of 2.5 ms in width and at rates up to 150 bpm.

▸ 9.5 IMPLANTABLE TELEMETRY SYSTEMS

Implantable telemetry systems allow the measurement of multiple physiological variables over long periods of time without any attachment of wires, restraint or anaesthesia to the monitored subjects. Above all, no sensors need to be attached even to the body surface. Most of the work in implantable telemetry has been used exclusively in animal research. Single or multi-channel systems have been used successfully to monitor ECG, EEG, blood pressure, blood flow, temperature, etc. For a multi-channel operation, a time-multiplex system is used to handle from 3 to 10 channels.

The telemetry transmitters most often have to be made as small as possible so that they do not cause any disturbance to the subject under investigation. They are made with subminiature, passive and active components which are to be kept minimum in number to avoid difficulties encountered in interconnection and to minimize electrical interaction between them—which may make the circuit unstable. These difficulties are greatly reduced by using thin film hybrid circuit technology. This can permit greater complexity in design and the designer is not constrained by the need to reduce the component numbers at the expense of circuit reliability. Transmitters in thin film circuits have been made by several researchers particularly for monitoring blood flow and temperature.

Another difficult problem faced while designing implantable transmitters is that of energy supply. The weight and size of the batteries must be minimal and the operating life must be maximal. The energy consumption in the circuit can be minimized by using micropower operational amplifiers and CMOS components for multiplexing, tuning and switching operations in multi-channel telemetry systems. Besides this, the circuit design has to be such that the transmitter energy consumption should be minimized and in the case of implantable transmitters, it should be particularly possible to turn the transmitter 'on' only when required. This is usually done by having a magnetically operated switch which allows the transmitter to be turned 'off' externally during idle periods.

9.5.1 Implantable Telemetry System for Blood Pressure and Blood Flow

In animal research, it is often necessary to obtain information about the blood flow over a period of several months. This requirement is best met by the use of implantable flowmeters. Electromagnetic flowmeters are not suitable for implant purposes, since they consume a lot of power and give rise to baseline shift due to a variety of reasons. Ultrasonic Doppler shift principle is the most widely used technique for implantable blood flowmeter.

In this method, blood velocity information is converted to an electrical signal by means of two ultrasonic transducers which are mounted in a rigid cuff surrounding the vessel. One of the transducers is driven by a high frequency power source and the second receives the scattered energy with a shifted frequency. Figure 9.13 shows the block diagram of the flowmeter in which the implantable portion is shown within the dashed box. High frequency power for the flow transducer is generated by the 6 MHz oscillator. The 6 MHz AM receiver converts the incoming ultrasonic signal to an audio frequency signal by synchronous detection. Data recovery is accomplished by an internal 100 MHz, FM transmitter and an external commercial FM receiver. A

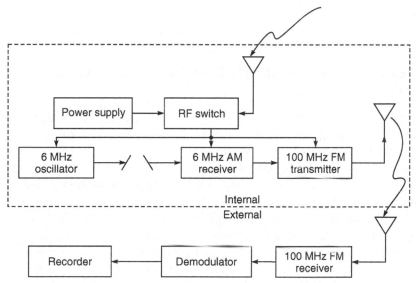

> **Fig.** 9.13 *Block diagram of an implantable blood flowmeter based on ultrasonic Doppler shift principle (after Dipietro and Meindl, 1973)*

demodulator, external to the body, converts the Doppler shift frequency to a flow estimate. Basically, the demodulator measures the zero-crossing rate of the Doppler signal which is proportional to the blood velocity.

Rader *et al* (1973) describe a miniature totally implantable FM/FM telemetry system to simultaneously measure blood pressure and blood flow. Pressure is detected by a miniature intravascular transducer by placing it directly in the blood stream. It measures 6.5 mm in diameter and is 1 mm thick. Four semiconductor strain gauges connected in a conventional four-arm bridge are bonded to the inner surface of the small pressure sensing diaphragm. The sensor produces approximately 30 mV/300 mmhg. The flow is sensed and measured by an extravascular inter-ferometric ultrasonic technique.

Barbaro and Macellari (1979) explain the construction of a radiosonde for the measurement of intracranial pressures. The main sources of error, consisting of the thermal drift of the electronic components and particularly of the pressure transducer are eliminated by simultaneously transmitting information regarding temperature, so that the data can be corrected accordingly. A strain gauge pressure transducer is implanted epidurally and is connected in a flexible manner to the body of the radiosonde. A thermistor is placed next to the pressure transducer. The sub-carrier oscillator is essentially an astable multi-vibrator operating at 4 kHz, and whose time constant is determined alternately by the two transducers. The switching of the transducers is done at 100 kHz. The sub-carrier oscillator then amplitude-modulates a radio frequency carrier of 1050 kHz, which can be conveniently received by the common commercial receivers. The tuning coil of the carrier oscillator acts as the transmitting aerial. The radiosonde is powered from outside through electromagnetic coupling and therefore contains circuits for converting the RF power into dc voltage. The total circuit is enclosed in a case of polypropylene, which is tolerated well by the human tissue. A film of silicone rubber covering the case further improves this capability. Cheng *et al* (1975) used a Pitran transducer for telemetering intracranial pressures.

ⅢⅢ▶ 9.6   TRANSMISSION OF ANALOG PHYSIOLOGICAL SIGNALS OVER TELEPHONE

Telephony provides another convenient method of sending physiological signals over telephone lines for remote processing. The method has the advantage that individual patients can be managed in remote areas. By sending ECG and other signals over the telephone lines, a patient can communicate with the doctor or specialist from his home while lying in bed. Another necessity for such a transmission is to use telephone lines for the collection of data for a central computer, from anaesthetized patients undergoing surgery in operating theatres and from conscious patients in intensive care or recovery rooms, for maintenance of records for future reference.

Telephony deals, normally, with the transmission of human speech from one place to a second distant place. Human speech consists of a large number of frequency components of different values from about 100 to 4000 Hz having different amplitude and phase relations between them. On the other hand, most of the bioelectric signals and other physiological signals like ECG, respiration, temperature and blood pressure consist of frequency components, which are much below the audio band width permitted by telephone-line systems. Some system of frequency or pulse code modulation has to be employed for the transmission of such signals. However, most of the work reported in literature is based on the use of the frequency modulation technique.

A telephone telemetry technique for transmitting and receiving medical signals is shown in Fig. 9.14. For frequency modulation, a modulator is used with a centre frequency of 1500 Hz. This frequency is modulated ±200 Hz for a 1 V peak-to-peak signal. This deviation is linear within 1% range. The demodulator consists of an audio amplifier, a carrier rejection filter and a low-pass integrator output circuit to recover the input signal. Both at the transmitting as well as receiving ends, coupling or isolation transformers are used to match the standard telephone line impedance.

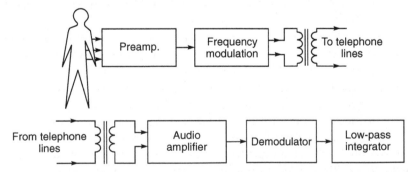

> ▶ **Fig.** 9.14   *Arrangement for transmission of analog signals over telephone lines*

This technique makes use of a wired electrical connection of the amplified signal to the telephone transmission system. This is largely due to the technical difficulties involved with carbon microphones normally used as transmitters in telephone handsets.

Acoustic coupling both at the transmitting as well as receiving ends is desirable in case the telephone transmission system is to be more widely employed. This however, necessitates the use of a superior type of carbon microphones. Transmission based on acoustic coupling has been successfully demonstrated with the development of an ECG telephone transmitter for the remote monitoring of potential cardiac patients, pacemaker studies and routine but short-term rhythm sampling. After proper amplification and filtration of the ECG signal, it is given to a voltage controlled oscillator (VCO). The centre frequency of the oscillator is set at 2000 Hz with a deviation of ±250 Hz. The output of the VCO is given to a dynamic earpiece, which provides sufficient audio signal for transmission. Coupling of the earpiece to a standard telephone handset is accomplished by clamping the two using a foam rubber gasket to ward off extraneous noise. The receiver is a standard data set, which reconstructs the waveform using zero-crossing detection technique. Even though the oscillator output is a square wave, it is satisfactory for single channel FM telephony, since the upper harmonics are attenuated by the telephone line bandpass.

Real-time trans-telephonic ECG transmitters do not store, but can transmit in real time a patient's ECG to a remote receiver. These are called 'event recorders'. They provide physicians with ECGs from patients where such information facilitates timely therapeutic decision making. ECG transmissions are typically made using continuous frequency-modulated (FM) audible tones emitted from a speaker built into the event recorder. This tone passes through the microphone on a telephone handset, over telephone lines, to the remote receiver. For example, when a patient calls a physician's office to transmit the ECG, he only needs to press a SEND button and hold the speaker on the event recorder close to the mouthpiece on the telephone handset. This type of acoustic transmission is compatible with telephone systems worldwide. The transmission normally takes place at a speed equivalent to real time, i.e. an ECG recorded over five minutes takes five minutes to transmit. Receivers are located at a physician's office, hospital or monitoring service and are used to receive and print-out the ECG.

Modern event recorders (Benz, 1999) use modems, which transmit the ECG digitally, requiring computerized receiving capability and a direct connection to the telephone system. However, this digital transmission is not necessarily compatible with all telephone systems throughout the world.

9.6.1 Multi-channel Patient Monitoring Telephone Telemetry System

Single and multi-channel systems for the transmission of electrocardiograms have been widely employed as remote diagnostic aids for cardiac patients recovering at home and for pacemaker performance follow up. There is however, an increasing need for multi-channel parameter monitoring, especially the simultaneous transmission of ECG, blood pressure, respiration and also temperature. Rezazadeh and Evans (1988), developed a remote vital signs monitor using a dial-up telephone line. A frequency modulation system using 750 Hz, 1750 Hz and 2750 Hz was employed. Figure 9.15 shows the block diagram of the transmitter.

The physiological signals after amplification to a nominal 1 volt peak-to-peak amplitude are input to an adder circuit. This adds an appropriate dc level to the signals prior to their connection to the corresponding channel of the voltage controlled oscillator (VCO). After multiplexing, the signals generated by the VCO, a low-pass filter cutting off at 3500 Hz is used to prevent out of band signals

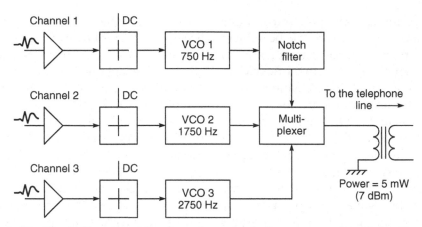

> **Fig. 9.15** *Block diagram of the three channel telephone transmitter (after Rezazadeh and Evans, 1988)*

entering the line. A 1:1 600 Ω isolating transformer is required to link the system with the telephone line. The transmitter uses the frequency division multiplexing (FDM) technique to accommodate the 3-channels of data within the 3100 Hz available bandwidths of the telephone lines.

The three physiological signals frequency modulate three different VCOs. The VCO centre frequencies are 750, 1750 and 2750 Hz. This frequency spacing ensures that the spectral components of one channel do not directly overlap signals on either of the other modulated channels. The VCO used was the Intersil 8038, which can generate three output waveforms: square, triangular and sinewave. The sinewave is preferred for use in a multi-channel system where cross-talk is to be minimized. The bandwidths occupied by each channel was calculated as 600 Hz, leaving 400 Hz as the guardband between the channels to prevent inter-channel cross-talk.

On the receiver side (Fig. 9.16), the multiplexed signals are filtered after being terminated by the isolating transformer. A second order low-pass filter with a roll-off frequency set at 3 kHz to

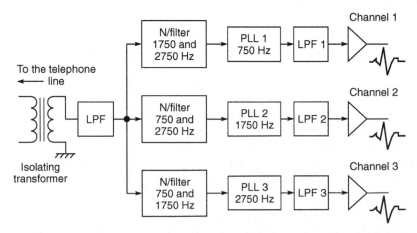

> **Fig. 9.16** *Block diagram of the three channel telephone receiver (after Rezazadeh and Evans, 1988)*

prevent high frequency line noise is used. The main component of each receiver channel is a phase-locked loop (PLL) used as a frequency demodulator. The PLLs are set to lock at frequencies of 750, 1750 and 2750 Hz for the first, second and third channels respectively. The Signetics NE 565, was selected as the PLL. After low-pass filtering the multiplexed signals, filters are used as band-reject notch filters, followed by PLL detectors in each channel. Finally, low-pass filtering of each output channel is needed to remove carrier ripple noise.

The channels were found to be identical in response to within ± 1 dB, cross-talk better than −45 dB and total harmonic distortion. The system was used over a 40 kms distance on normal telephone lines.

▧▧▷ 9.7 TELEMEDICINE

Telemedicine is the application of telecommunications and computer technology to deliver health care from one location to another. In other words, telemedicine involves the use of modern information technology to deliver timely health services to those in need by the electronic transmission of the necessary expertize and information among geographically dispersed parties, including physicians and patients, to result in improved patient care and management, resource distribution efficiency and potentially cost effectiveness (Bashshur, 1995).

Advanced information technology and improved information infrastructure the world over have made telemedicine an increasingly viable health care service delivery alternative, measured in clinical, technical and economic terms. However, most existing telemedicine programs, at present, are operating in an investigational settings. Issues such as telemedicine technology management and other barriers such as professional, legal and financial are still under debate.

From a technology stand point, the telemedicine technology includes hardware, software, medical equipment and communications link. The technology infrastructure is a telecommunication network with input and output devices at each connected location.

9.7.1 Telemedicine Applications

Although telemedicine can potentially affect all medical specialities, the greatest current applications are found in radiology, pathology, cardiology and medical education.

Teleradiology: Radiological images such as X-ray, CT or MRI images can be transferred from one location to another location for expert interpretation and consultation. The process involves image acquisition and digitization.

Telepathology: To obtain an expert opinion on the microscopic images of pathology slides and biopsy reports from specialists.

Telecardiology: Telecardiology relates to the transmission of ECG, echocardiography, colour Doppler, etc.

In addition, telemedicine is being advantageously used for:

Tele-education: Delivery of medical education programmes to the physicians and the paramedics located at smaller towns who are professionally isolated from major medical centres.

Teleconsultation: Specialist doctors can be consulted either by a patient directly or by the local medical staff through telemedicine technology. In the latter case, the patient is substituted by his/her electronic patient record (EPR) which has complete information on the physical and clinical aspects of the patient.

Depending on the level of interaction required, the telecommunication infrastructure requirement also varies: from a normal telephone, low-bit rate image transmission, real-time video transfer to video conferencing.

9.7.2   Telemedicine Concepts

Store and Forward concept involves compilation and storing of information relating to audio, video images and clips, ECG, etc. The stored information in the digital form is sent to the expert for review, interpretation and advice at his/her convenience. The expert's opinion can be transmitted back without any immediate compulsion on the consultant's time during his/her busy professional schedule.

Real Time telemedicine involves real-time exchange of information between the two centres simultaneously and communicating interactively. It may include video conferencing, interviewing and examining the patients, transmission of images of various anatomic sites, auscultation of the heart and lung sounds and a continuous review of various images.

9.7.3   Essential Parameters for Telemedicine

Telemedicine systems are based on multi-media computing, which not only support live multi-way conversations between physicians, patients and specialists but can also facilitate off-line consultations among health-care team members. It is however, advisable to create a detailed electronic patient record so that necessary information can be accessed, when desired. The following components relating to a patient are considered essential from the point of view of telemedicine:

- *Primary Patient Data:* Name, age, occupation, sex, address, telephone number, registration number, etc.
- *Patient History:* Personal and family history and diagnostic reports.
- *Clinical information:* Signs and symptoms are interpretations of data obtained from direct and indirect patient observations. Direct observations include data obtained from the senses (sight, sound, touch, smell, etc.) and through mental and physical interaction with the patient, while indirect observations include data obtained from diagnostic instruments such as temperature, pulse rate, blood pressure.
- *Investigations:* Complete analysis reports of haemotology and biochemistry tests, stool and urine examination.
- *Data and Reports:* Radiographs, MRI, CT, ultrasound and nuclear medicine images and reports; pathology slides, electrocardiogram, spirogram.

In addition, there is a need to have video-conferencing facility for online consultations.

Figure 9.17 shows the principle and various sub-systems used in a telemedicine set-up.

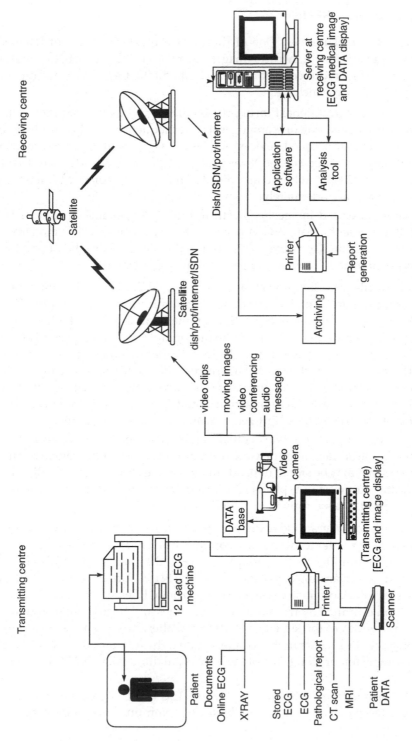

➤ **Fig. 9.17** *Block diagram of a typical telemedicine system*

9.7.4   Telemedicine Technology

Transmission of Medical Images: One of the most important aspect of telemedicine is the acquisition and transmission of medical images such as X-rays, CT, MRI, histopathology slides, etc. These images are first required to be converted into digital form. The usual types of diagnostic images used in telemedicine include:

- Images stored on traditional film or print media (e.g. X-ray film) and converted into digital format by direct imaging or scanning in a raster sequence under controlled lighting conditions. CCD (charge coupled devices) and laser-based scanners are commercially available for digitizing the film recorded X-ray images. A typical 11-by-17 inch chest film requires atleast 2000 by 2000 pixels and an optical dynamic range of at least 4000 to 1 (12-bits) to represent the image adequately.
- Computer-generated images (e.g. ultrasound, CT) available in standard video format (NTSC), computer format (SVGA), or computer-file format (TIF). In modern digital radiography systems, the X-ray image is stored in the computer in the digital format. Being a filmless system, it does not require any further digitization.

The American College of Radiologists and the National Electrical Manufacturers Association jointly developed a 13-part Digital Imaging and Communications in Medicine (DICOM) standard (ACR-NEMA, 1993) for the interconnection of medical imaging devices, particularly for radiological imaging equipment such as digital radiography, CT (Computed Tomography), MRI (Magnetic Resonance Imaging) and PACS (Picture Archiving and Communication Systems). In addition, ACR, 1994 is another standard, which deals with the transmission of radiological images from one location to another for the purposes of interpretation and consultation. The standard includes equipment guidelines for digitization of matrix images digitized in arrarys of $0.5 \, k \times 0.5$ $k \times 8$-bits and large matrix images digitized in arrarys of $2 \, k \times 2 \, k \times 12$-bits.

At the transmitting end, there is usually a requirement for the local storage of image data, particularly when the storage and forward concept is adopted in the telemedicine system. The number of images that may be stored depends upon the size of the storage facility (Hard disk) and the amount of data compression applied to the images before storage. The storage requirements for various imaging modalities are given in Table 9.1.

Transmission of Video Images: Telemedicine applications generally require video and individual still-frame images for interactive visual communication and medical diagnosis. National Television System Committee (NTSC) adopted, in 1953, the analog signal format used in USA for the broadcast and cable transmission of television. The NTSC format consists of 30 image frames/ s. Each frame is an interlaced raster scan of 512 horizontal lines sized to obtain a 4:3 aspect ratio. Vertical resolution is limited by the number of scan lines and horizontal resolution is limited by the specified bandwidth of the signal—4.2 MHz. An alternative expression of resolution is the term television lines (TVL), the number of alternating light and dark bars an observer can resolve in the vertical dimension or along 75% of the horizontal dimension. NTSC luminance resolution is 336 TVL.

At the resolution and frame rate associated with NTSC video, digital data are produced at a rate of over 100 Mbps (Mega bits per seconds). Obviously, communication and storage limitations and

• Table 9.1   **Storage Requirement for Imaging Techniques (adapted from K.G. Baxter *et al*, 1991)**

Modality	Image Size (pixels)	Dynamic Range (bits)	Average number of images per exam	Average storage requirement per exam (M Bytes)
Computed tomography	512 × 512	12	30	15.0
Magnetic resonance imaging	256 × 256	12	50	6.5
Digital subtraction angiography	1,000 × 1,000	8	20	20.0
Digital Fluorography	1,000 × 1,000	8	15	15.0
Ultrasound imaging	512 × 512	6	36	9.0
Nuclear medicine	128 × 128	8	26	0.4
Computed radiography	2,000 × 2,000	10	4	32.0
Digitized film	4,000 × 4,000	12	4	128.0

costs necessitate that some form of compression be used to reduce this data rate. The compression and decompression of the signal is usually carried out by a device known as codec (coder/decoder). This device identifies and removes both spatial and temporal redundancies and reconstructs the video in such a way that the missing data are not readily perceived. Codec normally employed in telemedicine applications operate at data rates between 336 kbps (kilo bits per second) and 1536 kbps.

Images are obtained from direct visualization by a video camera and a lens system for direct observation (e.g. a skin lesion) or an optical adapter to a conventional scope (e.g. laparoscope, microscope, otoscope) that provides magnification or remote access using fiber optics. The most commonly used digital camera is based on the use of charge coupled device (CCD). Incident light exposes an arrary of discrete light-sensitive regions called pixels, which accumulate electric charge in proportion to light intensity. The CCD sensor in a camera with a single sensor (single-chip camera) contains a microscopic array of red, green and blue filters covering the pixels, while a camera with three CCD sensors (3-chip camera) contains large prisms and filters to form separate red, green and blue images. After a specific integration time, the accumulated charges in the pixels are sequentially converted into conventional analog intensity and colour signals. Single CCD cameras offer 450 TVL (Television Lines) or higher resolution while high quality 3-chip CCD cameras offer 700 TVL or higher resolution.

Video is captured one frame at a time typically in a 640 by 480 pixel format, with an intensity scale typically consisting of eight bits for monochrome and 24 bits for colour (8-bit each for red, green and blue). Other formats exist for high-resolution cameras; a 1024 by 1024 pixel format at 8,10 or 12 bits per pixel can be achieved with a capture time, which depends in part on the time needed to integrate a sufficient number of photons from dark areas of the image.

Transmission of Digital Audio: Audio channels are usually provided for diagnostic instruments such as an electronic stethoscope or Doppler ultrasound. To reproduce heart and lung sounds accurately, an electronic stethoscope must have a uniform frequency response from 20 Hz to

2 kHz, while Doppler ultrasound requires a uniform frequency response from 100 Hz to 10 kHz. Audio used for conversation and medical diagnosis in a telemedicine system must be digitized and compressed before it can be combined with digital video and other information. Typical audio compression algorithms, operate at data rates from 16 Kbps to 64 Kbps. Medical diagnostic applications which require fidelity at higher audio frequencies will require higher data rates; 120 Kbps is sufficient to reproduce the full auditory frequency spectrum from 20 Hz to 20 kHz over a dynamic range of 90 dB which is adequate for the normal physiologic hearing range.

9.7.5   Video Conferencing

One of the essential components in a telemedicine system is the video conferencing facility, which permits real-time transmission of both audio and video information. A number of internationally recognized standards have been established by TSS (Telecommunication Standardization Sector, formerly Consultative Committee on International Telephone and Telegraph, CCITT) to assure a high degree of functional compatibility between like equipment supplied by different manufacturers and also to standardize the interface protocols which access and control the communications network.

Telemedicine is facing a number of key issues that must be addressed before it can realize its full potential. First, standards must be set, ranging from a common language to the standardization of computer architecture, so that health care providers can share data. Another key issue is licensing. Many states or countries may not allow a consultation via telemedicine unless the consulting physician is licensed in that state or country. Medical reimbursement of expenses for telemedicine is also an important factor, which would have a major influence on the growth of telemedicine.

9.7.6   Digital Communication Systems

Telemedicine primarily demands a continuous and reliable communication link for the exchange of information. There are various digital communication services available today for this purpose.

POTS: Using a modem (modulator/demodulator), the analog telephone systems (Plain Old Telephone Service-POTS) digital signals at data rates up to about 30 kbps can be transmitted. However, depending upon the quality of the circuit, the maximum reliable data transfer rate may be less than half of this rate. This service is adequate for data file transfers of still images.

DDS: Telephone traffic between major exchanges is today carried in digital format through the digital data system (DDS). Switched-56 is the most common type of DDS service which provides a data rate of 56 kbps. To obtain increased data communication capacity a number of Switched-56 lines can be combined in a single channel. Six Switched-56 lines give 336 kbps, which provide realistic reproduction of detail and motion when used for applications such as video conferencing and distance learning.

The digital data system of AT&T is based on the conversion of an analog voice signal into a digital equivalent at a data rate of 64 kbps. This basic channel rate represents the bandwidths needed to encode telephone quality audio using the pulse code modulation encoding method. Channels are combined into larger units by time division multiplexing of digital voice data. If carried on conventional copper wire cables, the digital signal is denoted by *T* representing a transmission rate of 1536 kbps. T-1 and fractional T-1 service is available from service providers.

ISDN: Today, the all digital integrated services digital network (ISDN) is available for the transmission of voice and data. The common form of ISDN, BRI (Basic Rate Interface) consists of two 64 kbps data (bearer)channels and 16 kbps data control channel (2B+D) multiplexed on two wire pairs. The data channels can be combined into 128 kbps channels; for example, a codec used for desktop video conferencing. ISDN dialling and other control functions are handled in the D-channel. In some locations, ISDN, PRI (Primary Rate Interface) is available and provides 23 B-channels at 1472 kpbs and one D-channel at 64 kbps. Not all telephone ISDN services are the same and depend upon the user's equipment and the type of ISDN switch used by the service provider.

ATM: ATM (Asynchronous Transfer Mode) is a high capacity communication link between widely dispersed sites, usually connected by fibre communication channels such as OC-3 (155 Mbps) or OC-12 (622 Mbps). This network service is well suited for transmitting digital video and audio.

Table 9.2 shows the important features for various types of digital communication services.

• Table 9.2 **Digital Communication Services**

Type of Service	Media/Carrier	Data Rate
Switched-56	2 or 4 wire/T–0	56 Kbps
Dedicated line	4 wire/T–1	56 Kbps–1.54 Mbps
Frame Relay		
Data only	2 wire/ISDN-BRI	64 Kbps-128 Kbps
Data and Video	4 wire/ISDN-PRI	384 Kbps –1.54 Mbps
2B+D	2 wire/ISDN-BRI	128 Kbps
Broadband	Fibre/ATM	1.5 Mbps and higher

The primary transmission media for communication systems are copper wires, fiber optics, microwave links and/or satellite transponders. A hybrid network may rely on satellite transponders for remote facilities, fiber optics for video images and copper wires for data, signalling and control. The communication requirement is predicted on real-time/store and forward modes of operation. The real-time interactive mode requires the transmission of a large amount of information in a short time. The major emphasis is on transmission speed and bandwidth. On the other hand, store and forward tends to move blocks of data at a time and is less demanding in speed and bandwidth requirements (Lin, 1999).

9.7.7 Telemedicine Using Mobile Communication

Mobile communication and satellite communication have opened up new possibilities for mobile telemedicine in emergency situations. A typical set-up is shown in Fig. 9.18. In a moving vehicle,

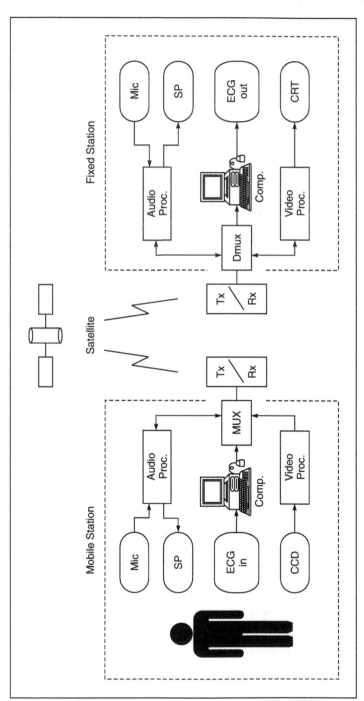

Mic   :Microphone   Video   : Video processor   DMUX : Demultiplexer
SP   : Speaker   Proc   Comp   : Computer
CCD   : Charge coupled   MUX   : Multiplexer   CRT   : Cathode ray tube
          device camera   Tx   : Transmitter
Audio   : Audio Processor   Rx   : Receiver
Proc.

➤ **Fig. 9.18** *Principle of telemedicine using mobile satellite communication*

colour images, audio signals and physiological signals such as ECG and blood pressure are obtained from the patient. These images and signals are multiplexed and transmitted to a fixed station. In the fixed station, the signals received are demultiplexed and presented to a medical specialist. Instructions from the specialist are then transmitted back to the mobile station through the communication link.

In mobile communication, the capacity of transmission link is generally limited and is typically 10 kbps–100 kbps. These capacities are far below those required for the transmission of medically significant information. For example, the transmission of a colour video signal requires a transmission capacity of 1–10 Mbps with data compression for a moving image (Shimizu, 1999). By adopting data compression on video, audio, ECG and blood pressure signals, the total capacity required is about 19 kbps, which is well within the practical capacity of mobile communication link. Table 9.3 gives transmitting information on various parameters and data reduction ratios used in mobile communication.

• Table 9.3   **Parameters and data compression ratios (After Shimizu, 1999)**

Data	Sampling	Compression Ratio	Bit Rate
Video	256 × 256 pixels/plane 8-bit RGB/pixels 1 plane/20 sec	10:1	8 kbit/sec
Audio	8 bit/sample 6000 sample/sec	4.8:1	10 kbit/sec
ECG	3 channel 8 bit/sample 200 sample/sec	8:1	600 bit/sec (3 channel)
Blood Pressure	1 sample/min 16 bit/sample	1:1	0.3 bit/sec

9.7.8   Use of Internet Resource for Telemedicine

The world wide web (WWW) is an internet resource through which information-producing sites offer hyper-linked multi-media information to the general public or in some cases restricted access to a certain group of people. Graphical browser programs such as Netscape are specially designed to access WWW resources and view their contents in text, graphics images and video. Suggestions have been made to make use of the WWW for telemedicine applications. However, it is advantageous and widely recommended to have a dedicated link as it offers higher security to the data, is more reliable due to lesser links in the channel, is better tuned for real-time applications and is an ideal solution for an integrated application to the requirements of both the transmitting as well as the receiving ends; e.g. image processing applications, video conferencing, video-microscopy, scanning, etc.

Oximeters

▶ 10.1 OXIMETRY

Oximetry refers to the determination of the percentage of oxygen saturation of the circulating arterial blood. By definition:

$$\text{Oxygen saturation} = \frac{[\text{HbO}_2]}{[\text{HbO}_2] + [\text{Hb}]}$$

where $[\text{HbO}_2]$ is the concentration of oxygenated haemoglobin
[Hb] is the concentration of deoxygenated haemoglobin

In clinical practice, percentage of oxygen saturation in the blood is of great importance. This saturation being a bio-constant, is an indications of the performance of the most important cardio-respiratory functions. It is maintained at a fairly constant value to within a few percent in an healthy organism. The main application areas of oximetry are the diagnosis of cardiac and vascular anomalies; the treatment of post-operative anoxia and the treatment of anoxia resulting from pulmonary affections. Also, a major concern during anaesthesia is the prevention of tissue hypoxia, necessitating immediate and direct information about the level of tissue oxygenation. Oximetry is now considered a standard of care in anaesthesiology and has significantly reduced anaesthesia-related cardiac deaths.

The plasma (liquid part of the blood) is a very poor carrier of oxygen. At the pressures available, only 0.3 ml of oxygen can dissolve in 100 ml of plasma, which is quite insufficient for the needs of the body. The red blood cells contain haemoglobin which can combine with a large volume of oxygen so quickly that in the lungs it may become 97% saturated forming a compound called oxyhaemoglobin. The actual amount of oxygen with which the haemoglobin combines depends upon the partial pressure of oxygen. The total quantity of oxygen bound with haemoglobin in the normal arterial blood is approximately 19.4 ml percent at a pO_2 of 95 mmHg. On passing through the tissue capillaries this amount is reduced to 14.4 ml per cent at a pO_2 of 40 mmHg. Thus, under normal conditions, during each cycle through the tissues, about 5 ml of oxygen is consumed by the tissues from each 100 ml of blood which passes through the tissue capillaries. Then, when the blood returns to the lungs, about 5 ml of oxygen diffuses from the alveoli into each 100 ml of blood and the haemoglobin is again 97% saturated.

10.1.1 In Vitro Oximetry

When blood is withdrawn from the subject under anaerobic conditions and measurement for oxygen saturation is made at a later time in the laboratory, the procedure is referred to as *in vitro* oximetry. For discrete blood samples, a spectrophotometric measurement of oxygen saturation can be made by either a transmission method or a reflection method.

Transmission Oximetry: Measurement of the degree of oxygen saturation of the blood can be made by spectrophotometric method. In spectrophotometry, the concentrations of substances held in solution are measured by determining the relative light attenuations that the light absorbing substances cause at each of several wavelengths. Applying Beer-Lambart's model (Fig. 10.1) of a light absorbing medium of

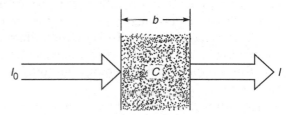

> **Fig. 10.1** *Principle of transmission oximetry*

concentration C and thickness b, the intensity of transmitted light I is related to the incident light I_O, as follows:

$$I = I_O^{-kCb}$$

where K is known as the extinction coefficient and varies as a function of the substance and the wavelength of light. The quantity KCb is called the absorbance A.

The spectral transmission characteristics of oxyhaemoglobin and reduced haemoglobin in the visible and infrared regions of the spectrum are shown in Fig. 10.2(a). It is obvious that the wavelength best suited for the measurement of oxygen saturation of blood is between 600 and 700

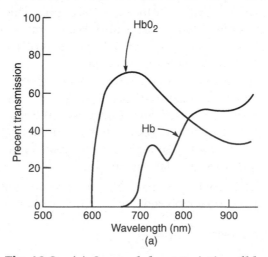

> **Fig. 10.2** *(a) Spectral characteristics of blood*

nm where the difference between the extinction coefficients of oxidized and reduced blood is the greatest. However, if the measurement is based on a given wavelength the result does not depend only on the extinction coefficient but also upon the total haemoglobin content and the thickness of the blood quantity equivalent to be found in the tissues. Generally, the total haemoglobin varies with the individual and the thickness of the equivalent changes in time with a periodicity determined by cardiac function. It is thus necessary to attempt the elimination of these factors. This is usually achieved by making measurements at two wavelengths, so that the variations in the haemoglobin concentration gets nullified.

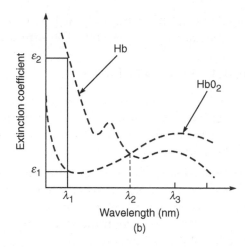

> **Fig. 10.2** *(b) Molecular extinction coefficient for haemoglobin and oxyhaemoglobin plotted against wavelength*

The extinction coefficient for haemoglobin and oxyhaemoglobin, when plotted against wavelength in the visible and near infrared regions of the spectrum, are shown in Fig. 10.2(b). It is observed that at 805 nm, the molecular extinction coefficient for fully saturated and fully reduced blood, are equal. This point is vital to the design and principle of operation of oximeters. One measurement is made at a wavelength of 650 nm (red) and the second at 805 nm (infrared). The red channel provides a signal which depends upon the amount of oxygen in the blood and the amount of blood and tissues in the optical path. The infrared channel signal is independent of oxygen saturation and carries information on the amount of blood and tissue in the optical path.

The Lambert-Beer law or the spectrophotometric technique applies only to haemolyzed blood, i.e. to blood in which the individual red cells have been destroyed and the pigments contained in the cells are homogeneously distributed in the whole solution. This is necessary to eliminate artefacts associated with multiple scattering of the measuring light from erythrocytes. Because of the high density of haemolyzed blood, very short light paths have to be used. Several designs of micro and ultramicro cells have been made for the purpose.

The calibration curve of the oximeters departs from linearity at 90 to 100% saturation. This is due to the fact that the light absorbing characteristics of oxygenated and reduced haemoglobin do not exactly follow Lambert-Beer's law in this region (Matthes and Gross, 1939). Also, the relationship of concentration to light absorption applies only in the case of monochromatic light. Oximeters which make use of filters fall far short of giving monochromatic light. Therefore, the calibration curve some what deviates from theoretical predictions.

Reflection Oximetery: Reflection oximetry is based on the scattering of light by the erythrocytes. For the light scattered from the unhaemolyzed blood sample, oxygen saturation is given by:

$$\text{Oxygen saturation} = \frac{I_r(\lambda_2)}{I_r(\lambda_1)} + b_r \qquad \qquad (i)$$

Polanyi and Hehir (1960) showed experimentally that a linear relationship exists between $I_r(\lambda_2)/I_r(\lambda_1)$ and oxygen saturation.

They computed the relationship as follows:

$$\text{Oxygen saturation} = 1.13 - 0.28 \times \frac{I_r(805)}{I_r(650)} \qquad \text{(ii)}$$

Figure 10.3 shows the schematic arrangement of a reflection oximeter. Light from a tungsten filament lamp (E) is condensed on the plane bottom of a cylindrical cuvette (F). The cuvette has a 15 mm internal diameter and contains about 2 ml of whole blood. A portion of the light scattered by the sample at an angle of about 135^0 with respect to the impinging light is condensed on two matched photoconductors (A and B). Two interference filters (C and C') limit the light reaching each cell to a narrow band centred at $\lambda_1 = 650$ nm and $\lambda_2 = 805$ nm, respectively. The ratio of the resistance of the photocells is measured by means of a conventional Wheatstone bridge. The ratio of the light intensity scattered by the sample is obtained as the inverse of this ratio.

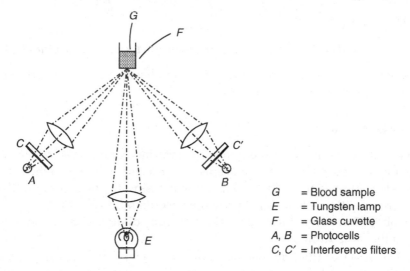

G	=	Blood sample
E	=	Tungsten lamp
F	=	Glass cuvette
A, B	=	Photocells
C, C'	=	Interference filters

➤ Fig. 10.3 *Schematic arrangement of reflection oximeter (after Polanyi and Hehir,1960)*

The optical path of a typical instrument which measures the oxygen saturation of unhaemolyzed blood on samples as small as 0.2 ml by using a micro-cuvette is shown in Fig. 10.4. Operating on the principle of reflection, the instrument employs two wavelengths (660 and 805 nm) and heparinized whole blood to measure the percentage of oxygen independent of wide haematocrit changes. The light reflected at 660 and 805 nm (isobestic point) is measured and the ratio of the two light intensities is, therefore, a direct function of the blood oxygen saturation of blood.

10.1.2 In Vivo Oximetry

In vivo oximetry measures the oxygen saturation of blood while the blood is flowing through the vascular system or it may be flowing through a cuvette directly connected with the circulatory

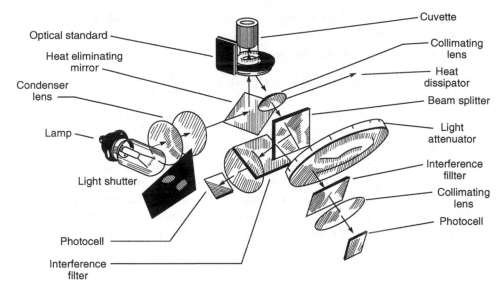

Optical standard

Heat eliminating
mirror

Condenser
lens

Lamp

Light shutter

Photocell

Interference
filter

Cuvette

Collimating
lens

Heat
dissipator

Beam splitter

Light
attenuator

Interference
fillter

Collimating
lens

Photocell

> **Fig.** 10.4   *Optical path of dual beam oximeter (Courtesy: M/s American Optical, USA)*

system by means of a catheter. The blood in this case is unhaemolyzed. Both techniques, reflection
and transmission, are utilized for in vivo oximetry.

10.2   EAR OXIMETER

Ear oximeters usually make use of the transmission principle to measure the arterial oxygen
saturation. In this case, the pinna of the ear acts as a cuvette. Blood in the ear must be made similar
to arterial blood in composition. This is done by increasing the flow through the ear without
appreciably increasing the metabolism. Maximum vasodilatation is achieved by keeping the ear
warm. It takes about 5 or 10 min for the ear to become fully dilated after the ear unit has been put
up in place and the lamp turned on.

Merrick and Hayes (1976) describe details of an ear oximeter which enables the measurement of
oxygen saturation of blood. This measurement is independent of a wide range of encountered
variables and is made without involving patients in any calibration or standardization procedure.
In brief, the technique involves measuring the optical transmittance of the ear at 8 wavelengths in
the 650 to 1050 nm range. A 2.5 m long flexible fibre ear probe connects the patient to the instrument.
The ear probe can be either held in position for discrete measurements or can be conveniently
mounted to a headband for continuous display. The resulting light transmissions are processed
digitally according to a set of empirically determined constants and the resulting oxygen saturation results are displayed in the digital form.

The instrument is based on the Beer-Lambert law. However, it is assumed that the optical
absorbers act independently and additively and that the effects of light scattering by the ear tissue
can be minimized by a proper source and detector geometry. The mathematical statement of this
law for wavelength can be written as:

$$Aj = E_{1j} C_1 D_1 + E_{2j} C_2 D_2 + ... + E_{ij} C_i D_i + ... + E_{Nj} C_N D_N$$

where A_j is the total absorbance due to N layers of absorbers of concentration C_i, thickness D_i and with extinction coefficients E_{ij}. If the measurements are made at eight wavelengths, the absorbance A_j at wavelength j can be related to the transmittance T_j in the following way:

$$A_j = -\log T_j \quad \text{where} \quad T_j = [I/I_O]_j$$

where I_O is the intensity of the light falling on the ear and I is the transmitted light intensity, all at wavelength j. The instrument is designed to measure the percentage of functional haemoglobin combined with oxygen. This can be expressed as:

$$SO_2 = \frac{C_O}{C_O + C_R} \times 100$$

where C_O is the concentration of oxyhaemoglobin and C_R is the concentration of deoxyhaemoglobin. These are two of the eight absorbers thought to be present. A working expression for saturation can be found by manipulating the set of eight simultaneous absorbance equations involving eight concentrations. This results in an expression with the following form:

$$SO_2 = \frac{A_O + A_1 \log T_1 + \dots + A_8 \log T_8}{B_O + B_1 \log T_1 + \dots + B_8 \log T_8}$$

The coefficient A_O through A_8 and B_O through B_8 are constants which were empirically determined by making a large number of measurements on a select group of volunteers. Having found the coefficient set, the instrument measures ear transmittance at eight wavelengths 20 times per second, performs the indicated calculations, and displays the results.

Figure 10.5 explains the basic operation of the instrument. The light source is a tungsten-iodine lamp that has a high output in the spectrum of interest. A lens system collimates the light beam and directs it through thin-film interference filters that provide wavelength selection. These filters

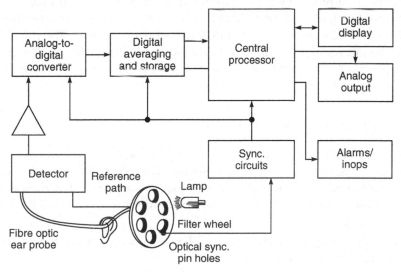

> **Fig. 10.5** *Block diagram of ear oximeter Model 47201A H.P. (Courtesy: Hewlett Packard, USA)*

are mounted in the periphery of a wheel rotating at 1300 rpm and thus cut the light beam sequentially. The filtered light beam then enters a fibre optic bundle that carries it to the ear. Another fibre optic bundle carries the light passing through the ear back to a detector in the instrument. A second light path is developed with a beam splitter in the path of the collimated light beam near the source. This path also passes through the filter wheel and then through a fibre optic bundle directly to the photodetector. So, the detector receives two light pulses for each wavelength. The processor takes the ratio of two pulses as the measured value; so readings are compensated for any changes in the spectral characteristics of the light source and optical system.

The current developed at the photodetector is only 0.5 nA or less during a light pulse. This is amplified in a high gain amplifier and then converted to a 16-bit digital form by an A-D converter synchronized with the wheel rotation. The 16-bit words are given to a digital signal averager that performs two functions. First, it averages out the noise content of the signal with a time constant of 1.6 s and secondly it serves as a buffer to hold information till it is required for computation.

Computation of percent oxygen saturation is accomplished by a 24-bit algorithmic-state machine. It uses serial processing with the program stored in ROM and the necessary coefficients of the equations stored on a field programmable ROM. The computation circuits also derive the quantity of total haemoglobin seen within the field of view of the earpiece. If this quantity is low, the instrument displays an 'Off Ear' indication. From the computational section, data is transferred in pulse-decimal form to the output circuit board where it is converted to BCD for the front panel digital display.

The patient related part consists of arterializing blood flow in the pinna by a brisk 15 s rub. Application of the probe to the ear results in a suitable display in about 30s. A built-in heater regulated to $41^{\circ}C$ maintains arterialization. Restandardization is not required when the instrument is to be used on other patients.

Measurements at eight wavelengths provide a great deal of information, which makes it possible to account for eight unknowns. This is sufficient to take into consideration the patient to patient variables and account for the various forms of haemoglobin. The procedure is simple, requiring only the storage of initial light intensities at each of the eight wavelengths. However, it is still necessary to arterialize blood flow by warming the ear, and a large ear probe incorporating fibre optics is necessary to make the system work.

▨▶ 10.3   PULSE OXIMETER

Pulse oximetry is based on the concept that arterial oxygen saturation determinations can be made using two wavelengths, provided the measurements are made on the pulsatile part of the wave-form. The two wavelengths assume that only two absorbers are present; namely oxyhaemoglobin (HbO_2) and reduced haemoglobin (Hb). These observations, proven by clinical experience, are based on the following:

   (i) Light passing through the ear or finger will be absorbed by skin pigments, tissue, cartilage, bone, arterial blood, venous blood.

   (ii) The absorbances are additive and obey the Beer-Lambert law:

$$A = -\log T = \log l_0 / I = \varepsilon D C$$

where I_0 and I are incident and transmitted light intensities, ε is the extinction coefficient, D is the depth of the absorbing layer and C is concentration.

(iii) Most of the absorbances are fixed and do not change with time. Even blood in the capillaries and veins under steady state metabolic circumstances is constant in composition and flow, at least over short periods of time.

(iv) Only the blood flow in the arteries and arterioles is pulsatile.

Therefore, only measuring the changing signal, measures only the absorbance due to arterial blood and makes possible the determination of arterial oxygen saturation (SaO_2). This is uninfluenced by all the other absorbers which are simply part of the constant background signal.

Figure 10.6 (a) shows a typical finger tip oximeter probe in use whereas Fig. 10.6(b) shows the construction of a typical pulse oximeter probe. This has two LEDs (light emitting diodes), one that transmits infrared light at a wavelength of approximately 940 nm and the other transmitting light at approximately 660 nm. The absorption of these select wavelengths of light through living tissues is significantly different for oxygenated haemoglobin (HbO_2) and reduced haemoglobin (Hb). The absorption of these selected wavelengths of light passing through living tissue is measured with a photosensor.

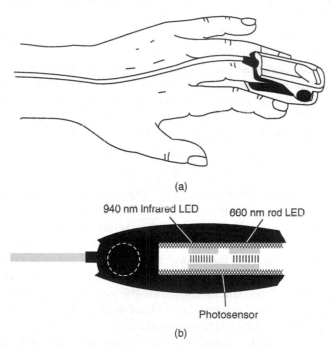

(a)

940 nm Infrared LED   660 nm rod LED

Photosensor

(b)

➤ **Fig. 10.6 (a)**   *A typical finger tip pulse oximeter probe in use*
        (b)   *Components of a pulse oximeter probe*

The red and infrared LEDs within the probe are driven in different ways, depending on the manufacturer. Most probes have a single photodetector (PIN-diode), so the light sources are generally sequenced on and off. A typical pulsing scheme of the LEDs is shown in Fig. 10.7. To

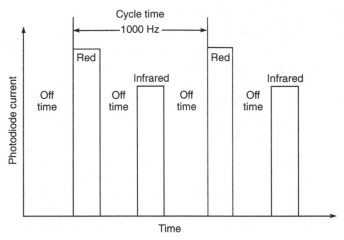

> **Fig.** 10.7 *Typical pulsing of red and infrared light emitting diodes by a pulse oximeter*

compensate for ambient light during the time when both LEDs are off, the light level is measured and then subtracted from each light channel between cycles. This minimizes the effects due to ambient conditions which may vary during monitoring. Depending on the make and model of pulse oximeters, the drive currents of LEDs, pulse widths, off and on cycles between pulses and cycle times can all vary (Ackerman and Weith, 1995).

The output of the photodiode has a raw signal represented in Fig. 10.8. There will be one signal that represents the absorption of red light (660 nm) and one that represents infrared light (940 nm). The ac signal is due to the pulsing of arterial blood while the dc signal is due to all the non-pulsing absorbers in the tissue. Oxygen saturation is estimated from the ratio (R) of pulse-added red absorbance at 660 nm to the pulse-added infrared absorbances at 940 nm.

$$R = \frac{ac\,660/dc\,660}{ac\,940/dc\,940}$$

Figure 10.9 shows the analog signal processing technique used in pulse oximeters. To simplify the diagram, the circuitry required to drive the LEDs in the sensor are omitted, and only the analog signal processing blocks between the sensor and the digital processing circuitry are shown. The signal from the sensor is a current. The first amplifier stage is a current to voltage converter. The voltage signal then goes through the following circuits: amplifiers to further amplify the signal; noise filters to remove different kinds of interference, a demultiplexer to separate the interleaved red and infrared signals; bandpass filters to separate the low frequency (dc) component from the pulsatile, higher frequency (ac) component; and an analog–digital converter to convert the continuously varying signal to a digital representation.

An advancement over the analog signal processing arrangement has been described by Reuss (2000) which eliminates analog circuitry for signal processing and replaces it with a digital signal processing the microprocessor. The output from the sensor is directly given to a high dynamic range analog-to-digital convertor followed by a microprocessor which supports the required digital signal processing. This technique offers the advantages of less circuitry, higher reliability, smaller size and lower cost.

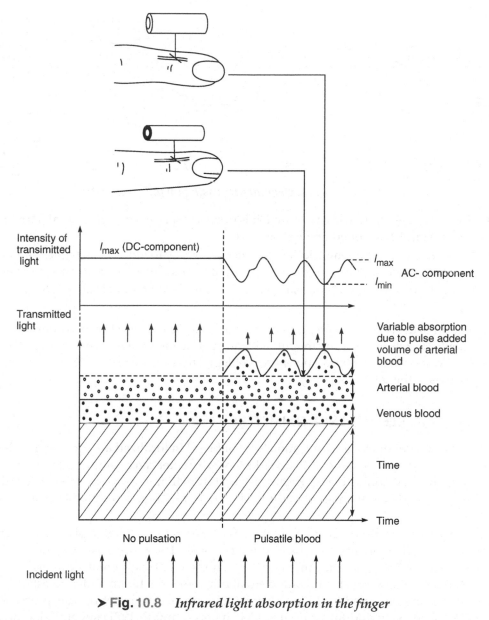

➤ Fig. 10.8 *Infrared light absorption in the finger*

An accuracy of 1% or better has been reported for the saturation range of above 80% for most transmission type pulse oximeters. Usually, the accuracy is less at lower saturation because of non-linear effects of absorption.

The pulse oximeter offers the following advantages:

- It removed the requirement of arterializing blood flow. No heating or rubbing is necessary. The measurement requires that pulsatile activity should be present, but the level is not critical.

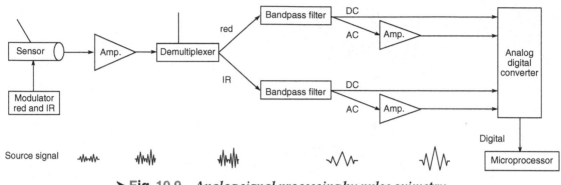

> **Fig. 10.9**   *Analog signal processing by pulse oximetry*

- Since a change in signal is measured, it is not necessary to store any initial light intensity values, simplifying operational procedures.
- The instrument can be empirically calibrated. Subject variability (skin pigmentation, thickness, tissue, sensor location, etc.) has no significant influence on the measurement.
- True arterial saturation is measured because the pulsatile signal comes from the arterial blood.

A limitation of the pulse oximeter is that ambient lights have been shown to interfere with the measurement. Therefore, covering the cuff with an opaque material is necessary to prevent such interference. Motion artifact is also a potential problem. This is because the information containing pulse activity is in the same frequency range as motion artifact.

⯈ 10.4   SKIN REFLECTANCE OXIMETERS

For the measurement of oxygen saturation level of blood in localized areas of oxygen deprived tissues on the limbs, head and torso, a skin reflectance oximeter can be employed. The instrument basically depends on monitoring backscattered light from living tissue in two wavelengths. The backscattered light data is then used for the in vivo determination of the blood's relative oxygen saturation.

Cohen and Wadsworth (1972) bring out the difficulties in the extraction of useful information from backscattered light intensity from human tissue. There are vast variations of tissue construction and optical properties among various subjects and in different locations on the same subject. Moreover, blood constitutes only a small fraction of the tissue under investigation. The fraction of volume occupied by blood is unknown, as are the exact optical properties and construction of the skin and lower tissues. Cohen and Longini (1971) suggested a theoretical solution to some of these problems. They considered human tissues to be composed of parallel semi-infinite layers of homogeneous materials. By using normalization techniques based on data experimentally collected from the living tissue under varying conditions of oxygenation and blood volume from various subjects, and processed by a computer, it was found possible to design an oximeter which was less sensitive to variations in parameters such as blood volume, skin pigmentation and construction, age, etc.

However, in comparison to transmission, the reflection pulse oximeter has poorer signal-to-noise ratio. Mendelson *et al* (1988) utilized multiple photodiodes around the light source to enhance signal level. This arrangement is shown in Fig. 10.10. A pair of red and infrared light emitting diodes are used for the light source, with peak emission wavelengths of 665 nm (red) and 935 nm (infrared). The reflected light from the skin at these two wavelengths is detected by a silicon diode. These detected signals are processed in the form of photo-plethysmographs to determine So_2.

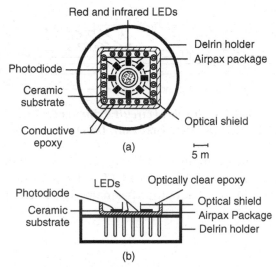

> **Fig.** 10.10 *Schematic diagram of the reflectance oximeter sensor (redrawn after Mendelson et al, 1988)*

A heater was incorporated in the sensor to warm the tissue so as to increase local blood flow. Excellent correlation in comparison with the transmission oximeter has been shown from the calf and thigh.

▶ 10.5 INTRAVASCULAR OXIMETER

For intravascular oximetry, modern instruments make use of optical fibres to guide the light signal inside the vessel and the reflected light from the red blood cells back to the light detector. For estimating SO_2, usually the reflectance at two wavelengths, one in the red and the other in the near infrared regions, are used. The relationship is given by:

$$SO_2 = A + B \, (R\lambda_1 / R\lambda_2)$$

where A and B are the constants that depend upon the fiber geometry and physiological parameters of blood (Polany and Hehir, 1960). The equation forms the basis of reflection oximetry.

Although the computation of reflectance ratios at two wavelengths has been suggested to cancel the effects of non-linearity due to scattering, effects of haemotacrit change may become important.

Therefore, currently available fiber-optic oximeters utilize more than two wavelengths to adjust for haematocrit variation (Takatani and Ling, 1994).

One of the problems in fiber optic oximeters is that damage to optical fibers result in severe measurement error. In order to overcome this problem, catheter tip type oximeters using hybrid type miniature sensors such as shown in Fig. 10.11 have been developed.

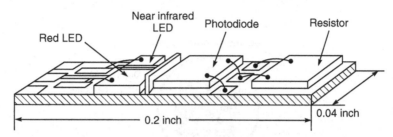

> **Fig. 10.11**   *A catheter – tip hybrid circuit oximeter for intravascular measurement (after Takatani, 1994)*

Intravasacular oximeters are normally used to measure mixed venous saturation, from which the status of the circulatory system can be deduced. Mixed venous saturation varies in reflecting the changes of oxygen saturation, cardiac output, haematocrit or haemoglobin content and oxygen consumption.

Blood Flowmeters

Blood flow is one of the most important physiological parameters and also one of the most difficult to measure accurately. This is because instruments for measuring the flow through blood vessels within the body have to meet certain stringent specifications; e.g. sensitivity and stability requirements depend upon the magnitude of flow, location and the diameter of the individual vessels. The average velocities of blood flow vary over a wide range in vessels with diameters ranging from 2 cm to a few millimetres. Besides meeting the technical requirements, the measuring system must meet the specific clinical requirement of being the least traumatic. Blood flow measurement is thus a difficult engineering and clinical problem. It is only natural that a variety of techniques have since been developed in an effort to meet the requirements of an ideal flow metering system.

There are many widely used techniques for measuring the blood flow and velocity. They are categorized into invasive (surgical) and non-invasive (through the skin). The most accurate method, obviously, is to simply sever the vessel and time the blood flow into a calibrated beaker. But this procedure is too radical for most protocols. The most commonly used blood flow meters are described in this chapter.

▪▪▪▷ 11.1 ELECTROMAGNETIC BLOOD FLOWMETER

The most commonly used instrument for the measurement of blood flow is of the electromagnetic type. With this type of instrument, blood flow can be measured in intact blood vessels without cannulation and under conditions which would otherwise be impossible. However, this method requires that the blood vessel be exposed so that the flow head or the measuring probe can be put across it.

The operating principle underlying all electromagnetic type flowmeters is based upon Faraday's law of electromagnetic induction which states that when a conductor is moved at right angles through a magnetic field in a direction at right angles both to the magnetic field and its length, an emf is induced in the conductor. In the flowmeter, an electromagnetic assembly provides the magnetic field placed at right angles to the blood vessel (Fig. 11.1) in which the flow is to be measured. The blood stream, which is a conductor, cuts the magnetic field and voltage is induced in the blood stream. This induced voltage is picked up by two electrodes incorporated in the magnetic assembly. The magnitude of the voltage picked up is directly proportional to the strength of the magnetic field, the diameter of the blood vessel and the velocity of blood flow, i.e.

$$e = CHVd$$

where e = induced voltage

H = strength of the magnetic field

V = velocity of blood flow

d = diameter of the blood vessel

C = constant of proportionality

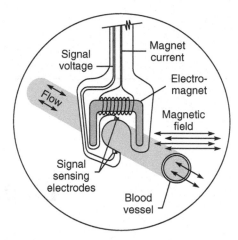

➤ Fig. 11.1 *Principle of electromagnetic flowmeter*

If the strength of the magnetic field and the diameter of the blood vessel remain unchanged, then the induced voltage will be a linear function of the blood flow velocity. Therefore, $e = C_1 V$ where C_1 is a constant and equal to CHd.

Further, the flow rate Q through a tube is given by

$$Q = VA$$

therefore,

$$V = Q/A$$

where A is the area of cross-section of the tube, hence

$$e = C_1 \times Q/A = C_2 \times Q$$

where C_2 is a general constant and is given by C_1/A. This equation shows that the induced voltage is directly proportional to the flow rate through the blood vessel.

The induced voltage picked up by the electrodes is amplified and displayed/recorded on a suitable system. The system is calibrated in terms of volume flow as a function of the induced voltage. The diameter of the blood vessel is held constant by the circumference of the hole in the probe that surrounds it.

The above relationship is true only if there exist conditions of axial symmetry and the blood velocity is considered independent of the velocity profile. Only then, the induced voltage is directly related to the blood volume flow.

Design of the Flow Transducer: In actual practice, the electromagnetic flowmeter transducer (Wyatt, 1984) is a tube of non-magnetic material to ensure that the magnetic flux does not bypass the flowing liquid and go into the walls of the tube. The tube is made of a conducting material and generally has an insulating lining to prevent short circuiting of the induced emf. The induced emf is picked up by point electrodes made from stainless steel or platinum. The flow head (Fig. 11.2) contains a slot through which the intact blood vessel can be inserted to make a snug fit. Several probes of different sizes must therefore accompany the flowmeter to match the full range of sizes of the blood vessels which have various diameters. It is naturally more difficult to construct flow

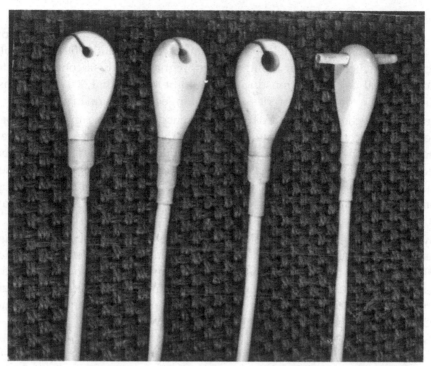

> ➤ **Fig. 11.2**　*Different types of flow heads (Courtesy: Cardiovascular Instruments Ltd., UK)*

heads suitable for use with very small blood vessels. However, flow heads having as small as 1 mm external diameter have been reported in literature.

The flow-induced voltage of an electromagnetic flowmeter is, within certain limitations, proportional to the velocity of the flow. This velocity is the average across the flow stream with an axis symmetric velocity profile. The average flow velocity appears to be 20 to 25 cm/s in arteries and 10 to 12 cm/s in veins. For designing the probe, velocity for the cardiovascular system is taken as 15 cm/s. For non-cannulated probes, a uniform magnetic field over the measuring area is so selected that it has a convenient shape and the smallest size (Cunningham et al. 1983). Iron cored electromagnets are used in probes having a diameter between 1 to 8.2 mm, and air cored electromagnets are use in diameters above 8.2 mm. Cannulated probes for extracorporeal use can have greater field strengths and magnet size as the constraint of small size is no longer present.

To obtain a reliable recording of the flow, a certain constriction of the vessel within the probe is necessary to maintain good contact. In order to limit the amount of constriction, a 20% incremental range of sizes is chosen. To protect the probe from chemical attack, it must be encapsulated in a biologically inert material having a high electrical and chemical resistance, e.g. silicone rubber. The probes can generally be sterilized by chemical means. Probe calibration is carried out in 0.9% saline during manufacture and each probe is given a calibration factor that is engraved on the connector. This factor is set on a multi-turn potentiometer to adjust the amplification to give a full scale output on the display meter.

The transducers are usually equipped with an internal ground electrode so that no external grounding is necessary. The cable from the transducer to the instrument should comprise of a teflon insulated wire completely shielded with a tinned copper braid. The entire cable is sleeved with medical grade silicone rubber tubing and impregnated with silicone rubber to minimize leakage and electrical noise. The transducers should be tested for 1,000 megaohms minimum resistance between the coil and electrodes, after prolonged immersion in saline.

Common-mode rejection is influenced by the difference between the impedances of the pick-up electrodes of a transducer and is maximum only when these impedances are identical. Electrode impedances in saline generally lie between 1 kΩ and about 10 kΩ, values of 1.5–2 kΩ being typical of platinized platinum and dull gold electrodes. The common-mode impedance of the measuring circuit should be at least 100 MΩ to ensure that variations of several hundred ohms between the impedances of a pair of transducer electrodes do not significantly affect the common mode rejection.

The early models of electromagnetic blood flowmeters employed permanent magnets which were subsequently replaced by electromagnets powered at mains frequencies. However, the tremendous interference from mains voltages at the transducer electrodes resulted in baseline drifts and poor signal-to-noise ratio. Various types of waveforms at different frequencies have since been tried to overcome these difficulties. Today, we have electromagnetic flowmeters whose magnetic coils work on sine, square or trapezium current waveforms.

The probes have an open slot on one side which makes it possible to slip it over a blood vessel without cutting the vessel. The probe must fit the vessel during diastole so that the electrodes make good contact. Probes are made in 1 mm increments in the range of 1 to 24 mm to ensure that they can be used on a variety of sizes of arteries. The probes generally do not operate satisfactorily on veins because the electrodes do not make good contact when the vein collapses. Attempts have been made to miniaturize the flow head to an extent that it can be mounted on the tip of a catheter. Such devices are needed in experimental as well as in clinical cardiology for mapping the velocity conditions of the entire circulatory system.

11.2 TYPES OF ELECTROMAGNETIC FLOWMETERS

Basically all modern flowmeters consist of a generator of alternating current, a probe assembly, a series of capacitance coupled amplifiers, a demodulator, a dc amplifier and a suitable recording device. The shape of the energizing current waveform for the electromagnet may be sinusoidal or square.

11.2.1 Sine Wave Flowmeters

In a sine wave flowmeter, the probe magnet is energized with a sine wave and consequently the induced voltage will also be sinusoidal in nature. The major problem encountered with the sinusoidal type of magnetic field is that the blood vessel and the fluid contained in it act as the secondary coil of a transformer when the probe magnet is excited. As a result, in addition to the induced flow voltage, there is an induced artefact voltage generally referred to as 'transformer voltage'.

The 'transformer voltage' is much larger than the signal or flow induced voltage and is $90°$ out of phase with it. This also causes baseline drift which necessitates high phase stability in the amplifier and demodulator circuits.

In earlier versions of sine wave flowmeters, this unwanted voltage was eliminated by injecting into the signal a voltage of equal strength, but having an opposite phase. The artefact signal is thus cancelled and only the flow induced voltage is left behind for display.

An alternative method to eliminate the transformer induced voltage in sinewave flowmeters is by using a gated amplifier. The function of this amplifier is to permit the amplification of the signals only during the portion of the cycle where flow induced voltages are maximum and the transformer induced voltages are minimum. By this method, the artefact voltage is prevented from getting amplified. This type of instrument is known as a 'gated sine wave flowmeter'.

The sine wave flowmeters require complicated electronic circuitry for the removal of transformer induced voltage from the flow induced voltage. As both the waveforms are of the same type, complete elimination of the artefact voltage becomes extremely difficult. The sine wave system, no doubt yields good signal-to-noise ratio, but imposes stringent phase stability requirements with increasing frequency.

11.2.2   Square Wave Electromagnetic Flowmeters

This differs from a sine wave flowmeter in that the energizing voltage given to the magnet is a square wave and therefore, the induced voltage is also a square wave. The square wave flowmeter has less stringent requirements of phase stability than the sine wave type as it can suppress the quadrature voltages relatively easily. Also, it is easier to control the magnitude and wave shape of the energizing current in the case of a square wave system.

The transformer induced voltage in this case is only a spike, superimposed on the beginning of the square wave flow induced voltage. Separation of these two voltages becomes easier as the amplifier can be gated only for a very short period. In a square wave flowmeter, blanking is required only during the portion when the current in the magnet is reversing and the amplifier works during the flat portion of the square wave. The square wave is amplitude modulated by the variation in blood flow and requires to be demodulated before it can be fed to a recorder.

Figure 11.3 shows the block diagram of a square wave electromagnetic blood flowmeter.

Transducer: The flow transducer consists of an electromagnet, which provides a magnetic field perpendicular to the direction of flow and lying within the field are a pair of pick-up electrodes whose axis is perpendicular to both the field and the flow axis. The electrodes may be in contact with either the flowing blood or the outer surface of the blood vessel carrying the flowing blood. The former is called 'Cannulating flowmeter' and the latter 'Cuff flowmeter'.

Preamplifier: The induced voltage picked up by the electrodes is given to a low noise differential amplifier through a capacitive coupling. The preamplifier must have a very high common-mode rejection ratio and input impedance. The preamplifier used by Goodman (1969(a)) has a CMRR of 106 dB (200,000 : 1) with a common mode input impedance of 150 MΩ. The preamplifier gain is of the order of 1000. The preamplifier also must incorporate the facility for 'probe balance' by which signals in phase with the magnet current can be selected to balance background voltages in phase

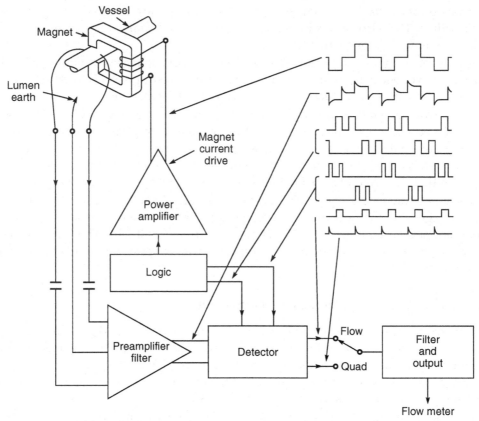

> **Fig. 11.3**   *Block diagram of a square wave electromagnetic flowmeter*

with flow voltages. A calibrating signal of 30 μV amplitude can be connected to the preamplifier with an input selector switch.

Noise voltage generated in the preamplifier is an important factor in the performance of an electromagnetic flowmeter. The noise voltage is seen as a random movement of the baseline of the recorded flow. When expressed in terms of flow, it is typically 1–2% of the full scale output of the chosen probe. For example, a 2.7 mm probe gives full scale deflection for 500 ml/min, so the noise is equivalent to 10 ml/min.

Gating Circuit: A gating amplifier helps to remove spurious voltages generated during magnet current reversal. For the flowmeter to exhibit a satisfactory base line stability, it is essential that the spurious signals produced during the magnet current reversal and those in phase with flow voltages are made negligible. The gating action is controlled by the circuit which provides an excitation current to the electromagnet.

Bandpass Amplifier: Following the gating amplifier is an active RC bandpass amplifier, which selectively passes through it the amplified square wave signal. The peak response is kept for 400 Hz. The 3 dB points are at 300 and 500 Hz. The gain of this amplifier is typically 50. The shape of the wave after this amplifier is a distorted sinusoid.

Detector: A phase sensitive detector is used to recover the signal, which is an analogue of the flow rate being measured. This type of demodulator not only offers maximum signal-to-noise ratio but also helps in the rejection of interfering voltages at frequencies well below the carrier frequency.

Low-Pass Filter and Output Stage: The demodulated signal is given to an active RC low-pass filter, which provides a uniform frequency response and a linear phase shift from 0–30 Hz. This is followed by an integrator circuit to provide an output corresponding to the mean flow. The output signal thus obtained can be put to a recorder to read the blood flow rate from the calibrated scale.

Magnet Current Drive: The excitation current supplied to the electromagnet is a one ampere peak square wave current. It is given from a source of high impedance to ensure that it remains constant for variations in magnet winding resistance of up to 5 Ω. The square wave input to the power amplifier stage which supplies current to the electromagnet is fed from a free running multi-vibrator working at 400 Hz.

Zero-Flow Reference Line: Before measurement can be made for blood flow with electromagnetic flowmeters, it is essential to accurately establish the signal corresponding to zero-flow. Although de-energizing the magnet should produce a zero reference line, unfortunately this line does not always coincide with the physiological zero-flow line. This is owing to some effects at the electrode vessel interface. An alternative method could be to occlude the blood vessel in which flow is to be measured. Several arrangements have been used to act as occluders (Jacobson and Swan, 1966). However, there is a serious objection to their use because the necessity for occlusion of the vessel, in order to obtain a zero flow reference, introduces the possibility of producing a spasm and hence a change in the blood flow.

📖> 11.3 ULTRASONIC BLOOD FLOWMETERS

There are basically two types of ultrasonic blood flow-velocity meters. The first type is the transit-time velocity meter and the second is the Doppler-shift type. For routine clinical measurements, the transcutaneous Doppler instrument has, by far, superseded the transit-time type. Therefore, most of the recent efforts have been concentrated on the development of Doppler-shift instruments, which are now available for the measurement of blood velocity, volume flow, flow direction, flow profile and to visualize the internal lumen of a blood vessel.

11.3.1 Doppler-shift Flow-velocity Meters

It is a non-invasive technique to measure blood velocity in a particular vessel from the surface of the body. It is based on the analysis of echo signals from the erythrocytes in the vascular structures. Because of the Doppler effect, the frequency of these echo signals changes relative to the frequency which the probe transmits. The Doppler frequency shift is a measure of the size and direction of the flow velocity. The principle is illustrated in Fig. 11.4. The incident ultrasound is scattered by the blood cells and the scattered wave is received by the second transducer. The frequency shift due to the moving scatterers is proportional to the velocity of the scatterers. Alteration in frequency occurs first as the ultrasound arrives at the 'scatterer' and second as it leaves the scatterer. If the blood is moving towards the transmitter, the apparent frequency f_1 is given by

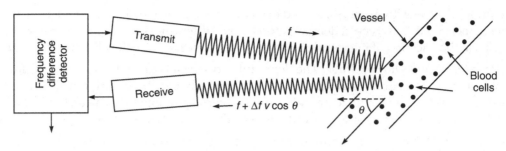

> **Fig. 11.4**　*Principle of ultrasonic Doppler-shift flow velocity meter*

$$f_1 = \frac{(C - v \cos \theta)}{C} \tag{i}$$

where f = transmitted frequency

C = velocity of sound in blood

θ = angle of inclination of the incident wave to the direction of blood flow

v = velocity of blood cells

Assuming that the incident and scattered radiations are both inclined at θ to the direction of flow as shown in Fig. 11.4,

$$f_2 = f_1 \left[\frac{C}{C + v \cos \theta} \right] \tag{ii}$$

The resultant Doppler shift

$$\Delta f = f - f_2 = f - f_1 \left[\frac{C}{C + v \cos \theta} \right] \tag{iii}$$

From equation (i), substituting f_1, we get

$$= f - \left[\frac{f(C - v \cos \theta)}{C} \right] \left[\frac{C}{C + v \cos \theta} \right]$$

$$= f \left[1 - \frac{(C - v \cos \theta)}{(C + v \cos \theta)} \right]$$

Since $C \gg v$,

$$\Delta f = \frac{2fv \cos \theta}{C}$$

$$v = \frac{\Delta f \cdot C}{2f \cos \theta}$$

This relationship forms the basis of measuring blood velocity. Depending on the application, either a signal proportional to the average instantaneous velocity or a signal proportional to the peak instantaneous velocity may be required. The peak signal is easier to obtain and has been shown to be of particular importance in localizing and quantifying the severity of peripheral vascular diseases (Johnston et al, 1978).

In order to measure absolute velocity, the angle of inclination of the ultrasonic beam to the direction of flow must be known. Several methods are available for doing so. The first uses the principle that the Doppler shift signal is zero when the ultrasonic beam is at right angles to the direction of flow. So, by finding the position of the zero Doppler signal with the probe over the vessel and then by moving it through a known angle of inclination, the angle becomes known. However, due to the separation of the transmitter and receiver, the Doppler shift frequencies are not zero. In such cases, the position at which the minimum Doppler shift frequencies are present is taken for the probe to be at right angles.

The early instruments used a continuous wave (CW) ultrasonic beam. Basically, a CW ultrasonic Doppler technique instrument works by transmitting a beam of high frequency ultrasound 3–10 MHz towards the vessel of interest. A highly loaded lead zirconate titanate transducer is usually used for this purpose. The transducer size may range from 1 or 2 mm to as large as 2 cm or more. A separate element is used to detect the ultrasound back scattered from the moving blood. The back-scattered signal is Doppler shifted by an amount determined by the velocity of the scatterers moving through the sound field. Since the velocity varies with the vessel diameter to form a velocity profile, the returned signal will produce a spectrum corresponding to these velocities.

Flax *et al* (1973) discuss the circuit blocks (Fig. 11.5) of the Doppler ultrasonic blood flowmeter. The piezo-electric crystal **A** is electrically excited to generate ultrasonic waves, which enter the blood. Ultrasound scattered from the moving blood cells excites the receiver crystal. The electrical signal received at **B** consists of a large amplitude excitation frequency component, which is directly coupled from the transmitter to the receiver, plus a very small amplitude Doppler-shifted component scattered from the blood cells. The detector produces a sum of the difference of the frequencies at **D**. The low-pass filter selects the difference frequency, resulting in audio frequencies at **E**. Each time the audio wave crosses the zero axis, a pulse appears at **G**. The filtered output level at **H** will be proportional to the blood velocity. The following two pitfalls are encountered in Doppler ultrasonic blood flowmeters. High frequency response is usually inadequate which introduces a non-linearity into the input-output calibration curve. Also, the low frequency gain is

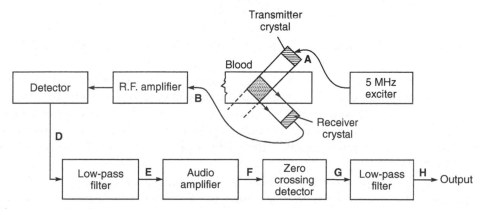

> **Fig. 11.5** *Block diagram of Doppler shift blood flowmeter (after Flax et al, 1973; by permission of IEEE Trans. Biomed. Eng.)*

normally too high, resulting in wall motion artefacts. The maximum Doppler shift has been calculated as about 15 kHz. The wall motion signal can be significantly reduced by filtering out frequencies below 100 Hz.

11.3.2 Range gated Pulsed Doppler Flowmeters

Instruments based on a CW Doppler can be better characterized as flow detectors rather than flowmeters. Even a simple measurement of the mean flow velocity requires that the angle between the sound beam and the velocity vector should be known. Baker (1970) states that recordings in literature that are calibrated in terms of Doppler frequency shift are both misleading and frequently inaccurate. The investigators did not or could not make the angle measurement required to quantify their data. Many of the difficulties associated with the CW system can be overcome if the ultrasonic source is pulsed and the Doppler shift of the returning echo is determined. If the return signal is range gated, then the distance to the moving interface (blood vessel diameter) as well as its velocity with respect to the beam can be measured.

Figure 11.6 shows the block diagram of a pulsed ultrasonic Doppler flow detector. The system consists of:

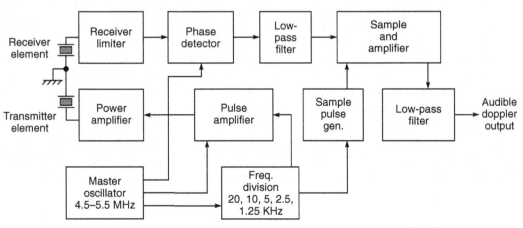

> **Fig. 11.6**   *Block diagram of a pulsed Doppler flowmeter (after Baker, 1970)*

Transmitter: It comprises a master oscillator whose frequency may vary from 2 to 10 MHz; however 5 MHz is a good compromise for detecting the blood flow in vessels 4-6 cms in depth. The choice of frequency is a compromise between the transmission attenuation losses over a given path length. The master oscillator also drives the pulse repetition frequency (PRF) ripple counter and provides a continuous reference signal to the receiver phase detector for the detection of the Doppler-shifted echoes.

The selection of PRF depends on the depth of the vessel, expected Doppler shift and attenuation characteristics of the tissue. The following values of PRFs and range depth combinations for optimum results have been suggested:

PRF (kHz)	Approximate Depth Range (cm)	Max. Δf Detectable (kHz)
25	3.0	12.5
18	4.3	9.0
12.5	6.0	6.25

The PRF pulse triggers a 1 μs one shot multi-vibrator that generates the gating signal. When the gate opens, 5 cycles of the master oscillator pass through to a power amplifier. The peak power output applied to the transmitting transducer is in the range of 10 to 30 W during the 1 μs excitation burst. With the PRF of 25 kHz, the average power applied to the transducer ranges from 0.25 to 0.75 W depending on the transducer impedance, which may be 50 to 200 Ω. The radiating area of the transducer ranges from 0.5 to 2 cm^2.

The transducers for pulsed Doppler applications should ideally have the sensitivity of the lightly backed narrow-band CW Doppler units and the wide bandwidth of the pulse echo units. The Q of transducers designed for CW applications ranges from 10 to 30 or more. These transducers ring at their resonant frequency long after the electrical signal stops. Pulse echo-transducers use tungsten epoxy backing to have a Q as low as 1.5–2.5. However, the Q is not lowered to a desirable value of about 1.5 to 5 because this would greatly decrease both the efficiency of the transmission and the sensitivity of the reception. The Q of pulsed Doppler transducers is generally kept at 5 to 15 and some ringing is thus tolerated. These transducers have a backing of aluminium powder in epoxy.

Receiver: The back-scattered Doppler-shifted signals from a blood vessel range in intensity from 50 dB to more than 120 dB down from the transmitted signal. The receivers designed for this purpose have a bandwidth of 3 MHz and gain in excess of 80 dB. This is followed by a single side band type quadrature phase detector which separates the upper and lower Doppler side bands for sensing flow direction. The detector consists of a phase-shift network, which splits the carrier into two components that are in quadrature, which means they are 90°. These reference cosine and sine waves must be several times larger than the RF amplifier output. The Doppler shift of the received echo, back scattered by the moving blood is detected by sensing the instantaneous phase difference between the echo and a reference signal from the master oscillator. If the echo at a particular range or depth contains a Doppler-shift, then the amplitude of a sample of the instantaneous phase difference will vary in amplitude exactly with the Doppler difference frequency. The envelope frequency from the phase detector is the Doppler difference frequency.

If the flow of blood is in the same direction as the ultrasonic beam, then it is considered that the blood is flowing away from the transducer. In this case, the Doppler-shift frequency is lower than that of the carrier and the phase of the Doppler wave lags behind that of the reference carrier. If the flow of blood is towards the transducer, then the Doppler frequency is higher than the carrier frequency and the phase of the Doppler wave leads the reference carrier. Thus, by examining the sign of the phase, the direction of flow can be established.

The depth at which the Doppler signal is sensed within the blood vessel and tissue depends on the delay interval between the transmitted burst and the sample gate. For a velocity of sound, in tissue of approximately 1500 m/s, the range factor is 13.3 μs delay/cm of depth. A one-shot multi-vibrator is used to develop the adjustable range delay calibrated in millimetres of depth. This is followed by a 'sample gate'. The detected Doppler shift frequency will correspond to the mean velocity averaged over the sample volume. However, the Doppler signals at the output of the sample and hold circuit are low in amplitude and contain frequency components from the transmitter pulse repetition rate and the low frequency large amplitude signals produced by the motion of interfaces such as vessel walls and the myocardium. Therefore, both high-pass and low-pass filters are used to extract the signals of interest. Once the signals are filtered with a passband of 100 Hz to 5 kHz, it is further amplified to drive an external audio amplifier or a frequency meter.

Zero-crossing Detectors: In order to measure blood velocity, a frequency meter is needed to analyze the frequency components of the Doppler signal. Nearly all Doppler velocimeters use zero-crossing detectors for this purpose. The function of the zero-crossing detector is to convert the audio frequency amplifier output to a proportional analog output signal. It does this by emitting a constant area pulse for each crossing of the zero axis from negative to positive. The output of the zero-crossing detector is a series of pulses. These pulses are passed through a low-pass filter to remove the high frequency component. The filter pass frequencies from 0 to 25 Hz in order to reproduce the frequencies of interest in the flow pulse. The zero-crossing detector is capable of measuring blood velocity to within 20% and can detect changes in velocity of about 5%. A major limitation of the zero-crossing detector used in simple flowmeters is that it cannot detect the direction of flow. A better approach is to use the quadrature-phase detector approach, which is commonly used in radar technology. This detector gives not only the velocity with which the blood flows but also its direction.

Spectrum Analyzers: Because of the limitations of the zero-crossing detector, alternative methods have been employed for the analysis of Doppler signals for velocity measurements. For example, spectrum analyzers are used to derive blood flow velocity information from Doppler signals. A spectrum analyzer processes short lengths of audio signal to produce spectral displays, which have frequency as the abscissa, time as the ordinate and spectral intensity represented by record darkening.

Spectral analysis of Doppler signals with a spectrum analzser is an off-line technique. The process of obtaining a good record involves aural recognition of the relevant Doppler wave forms, the recording of these on tape and the subsequent analysis of short sections at a time. In order to study effects extending over several cardiac cycles and also to allow the operator to optimize the recording during clinical investigation, a real-time continuous form of analysis is needed. Macpherson *et al* (1980) describe a real-time spectrum analyzer for ultrasonic Doppler signals. The analyser calculates spectral components using a type of Discrete Fourier Transform (DFT) known as the Chirp-Z-Transform.

Development of a real-time system for the three dimensional display of Doppler-shifted frequencies within an ultrasound flowmeter signal is described by Rittgers *et al* (1980). The device is based on a Fast Fourier Transform (FFT) spectrum analyzer, which provides both high frequency and time resolution over a variable range of frequencies up to 40 kHz. Signal pre-processing filters out unwanted low frequency artefacts due to arterial wall motion and automatically controls input gain to limit peak amplitudes while matched phase shift networks translate forward and reverse components for a directional output.

Ultrasound directional Doppler blood-velocity detection may be achieved using any of the three basic detection systems. These are single sideband, heterodyne and phase-quadrature detection systems. Out of these, phase-quadrature detection system is the most commonly used. Phase-quadrature detection systems operate by detecting both the amplitude and phase of the Doppler-shifted frequency.

Schlindwein *et al* (1988) used a digital signal processor chip to perform the online spectrum analysis of the Doppler ultrasound signals. It is a flexible and a programmable method, which has every aspect of the system such as window function, frequency ranges and the transform rate under software control.

11.3.3 Blood Flow Measurement through Doppler Imaging

Doppler ultrasound is not only used for the measurement of the absolute value of blood velocity and volume flow, but it also helps to have a direct visualization of the blood vessels and to study the blood-velocity/time wave form shapes over the cardiac cycle. The imaging facility helps to measure beam/vessel angle to detect the location of the required site of measurement of velocity and flow. This technique is particularly attractive in studying blood flow in carotid and femoral arteries and to visualize the bifurcation of the former.

The block diagram of an imaging instrument (Fish, 1975) using a pulsed flow detector is shown in Fig. 11.7. The probe is mechanically coupled to the position resolvers, which gives electrical outputs proportional to the movements of the probe. The position of the spot on the CRT screen corresponds to that of the sampling volume. When the sampling volume is within a blood vessel, the spot is intensified and stored when Doppler signals are received. A vessel is imaged by moving the probe over the skin surface and by adjusting the probe/sampling-volume distance until the sampling volume has been swept through the section of the vessel of interest. The three-dimensional information about the geometry of the vessel is displayed by selecting, in turn, two of the three dimensions available for display on the monitor. Thus, it is possible to construct anterior-posterior, lateral and cross-sectional scans of blood vessels.

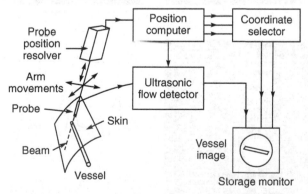

➤ **Fig.** 11.7   *Doppler imaging system-block diagram (redrawn after Fish, 1975)*

Obviously, the time needed to complete a single cross-sectional scan of a blood vessel is long (about 3 min) even after the vessel is located. An additional difficulty, if the flow detector is not direction-resolving, is that of separating the images of contiguous veins and arteries. In order to overcome these difficulties, Fish (1978) developed a multi-channel direction-resolving flow detector. Although the instrument has the same basic form as the single channel instrument, it is a 30 channel device and can determine the direction of the component of flow along the beam. The instrument can map flow towards and away from the probe and, therefore, permits separation of the images of the vein and artery when these are contiguous. The depth of the vessel can be measured by writing the position of the skin surface on the screen by intensifying a spot on the screen corresponding to the end of the probe.

The imaging facility can aid the fixing and maintenance of sample-volume/vessel geometry. Figure 11.8 shows lateral, anterior-posterior and cross-sectional views of a length of the femoral artery, together with recordings of relative velocity from three points on a diameter. The position of the 30 sampling volumes is clear from the lateral and cross-sectional views. From the images, it is quite convenient to see where the measurement is being made and to maintain the sampling volumes in position during the measurement. As it is possible to record velocity profiles over the cardiac cycle, it is possible to compute mean blood flow, peak flow, peak reverse flow by sampling measurements at intervals, say 40 ms. The system computes flow information from the velocity profiles and cross-sectional diameter of the blood vessel under investigation. The transducer

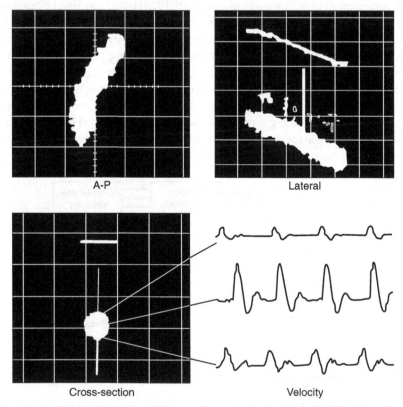

A-P   Lateral

Cross-section   Velocity

➤ **Fig. 11.8** *Lateral, anterior-posterior and cross-section views of a length of femoral artery together with recordings of relative velocity from three points on a diameter of blood vessel*

frequency is usually 5 MHz and it transmits ultrasound beam focussed at 2–4 cm, which is the approximate depth of the carotid or femoral artery beneath the skin surface. The signal processing circuit generates simultaneous forward and reverse flow-velocity channels by the use of a phase-quadrature detection system and the processor scans the instantaneous blood-velocity spectrum as it is formed. The circuit rejects low frequency signals due to arterial wall movement and spurious signals due to noise or transducer movement artefacts by adjusting the threshold for Doppler signals.

Colour Doppler sonography is based on the measurement of the local flow velocity in real-time and the display the surrounding structures in colour coded form, together with the section scan of the vessel walls. The colour, usually red or blue, indicates the direction of blood flow. The colour intensity indicates the local flow velocity. The brighter the colour intensity, the higher the velocity. The image gives an immediate overview of flow direction and flow dynamics, as well as a haemodynamic effect of alterations to the vascular structures.

One of the interesting applications of using pulsed Doppler ultrasound is to measure intra-cranial blood flow-velocities and their related pathologies. This highly sensitive technique provides a window to the brain for access to important clinical information previously unavailable with non-invasive techniques. The ultrasound probes used for blood flow-velocity range from 2 MHz to 8 MHz, and 20 MHz for micro-vascular applications. Figure 11.9 shows a transcranial Doppler system from M/s Nicolet, USA.

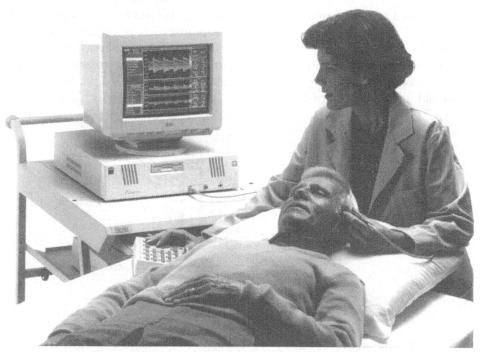

> **Fig. 11.9** *Transcranial Doppler blood flow measuring machine (Courtesy: M/s Nicolet, USA)*

▥▥▷ 11.4 NMR BLOOD FLOWMETER

Nuclear magnetic resonance principle offers yet another non-invasive method for the measurement of peripheral blood flow or blood flow in various organs. The method pertains to a quantum mechanical phenomenon related to the magnetic energy levels of the nucleus of some elements and their isotopes. For blood flow measurement work, behaviour of the two hydrogen atoms of water is studied, since blood is approximately 83% water. Due to the magnetic moment of the hydrogen atom, the nucleus behaves as a microminiature magnet which can be affected by externally applied magnetic fields. The hydrogen nuclei orient themselves to produce a net parallel alignment to a steady magnetic field. The nuclear magnets precess around the magnetic field lines until they become aligned. The angular frequency (Larmor frequency) of this precession is given by

$$W = 2\pi v = r\, B_0$$

where r is the ratio of the magnetic moment to the angular momentum (the magnetic gyro ratio) and B_0 is the density of the steady magnetic field and v is the frequency of radiation.

When the frequency of the magnetic field is near the Larmor frequency, the nuclear magnets can be rotated and their presence detected secondary to an externally applied radio frequency magnetic field. Also, a region of demagnetization can be made by disorienting the nuclear magnets with a radio frequency field very close to the Larmor frequency.

An arrangement for the measurement of blood flow using the NMR technique is shown in Fig. 11.10. The blood vessel of interest is positioned in a uniform steady magnetic field B_0. The nuclear magnets of the hydrogen atoms, which before insertion were randomly oriented, now begin to align themselves with B_0. Some begin to align parallel, whereas others commence to align anti-parallel. The alignment occurs exponentially with a time constant T_1 known as the longitudinal relaxation time. There are more nuclear magnets aligned parallel to B_0 in contrast to those aligned anti-parallel. For blood at 37°C, there are 0.553×10^{23} hydrogen nuclei per

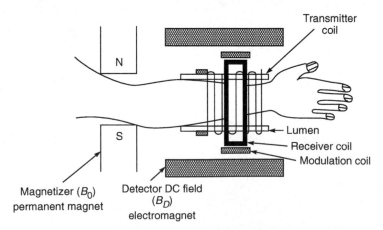

> **Fig. 11.10** *Arrangement for measurement of arterial blood flow using NMR principle. Hydrogen nuclei in the blood are magnetised by the magnetiser B_0. The nulcei are detected downstream by the receiver coil*

millilitre. For $B_o = 1000$ gauss, there are 1.82×10^{16} per millilitre net nuclei aligned. The maximum magnetization M_o due to the magnetic field B_o is given by $M_o = X_o B_o$, where X_o is the static nuclear magnetic susceptibility equal to 3.23×10^{-9} for blood at 37°C. Thus for $B_o = 1000$ gauss, $M_o = 3.23$ microgauss.

The magnitude of the magnetization can be related to either flow velocity or flow rate. Since the magnetization can be changed at some point in a magnetic field, and the change detected at some distal point, it is possible to determine the transport time between the two points. A crossed coil configuration is used to detect the level of magnetization in the limb. Two magnets are used, a strong permanent magnet B_o for premagnetization and a weaker, homogeneous electromagnet B_D for detection. A graph plotted between the magnetization at the centre of the receiver coil for typical distances as a function of velocity shows that the magnetization changes proportionally to velocity and thus a measurement of the magnetization can yield velocity information. If a receiver coil is used, the voltage induced in the coil by the magnetization is proportional to the cross-sectional area of the vessel carrying the blood. The NMR signal voltage proportional to velocity V, and multiplied by area A, will give a volumetric flow rate Q. With a homogeneous detector field B_D, the magnetization in all the blood vessels in the region is detected. In a non-homogeneous field, B_D can be tuned to detect the magnetization of selected vessels. The major shortcoming of NMR blood flowmeters is the physical size of the magnets and the sensitivity of the system needed to detect the magnetically tagged bolus of blood.

For the measurement of arterial blood flow, the blood in the upper arm is magnetized by the niagnetizer B_O (Fig. 11.10) and the magnetized blood is detected in the forearm by the receiving coil. The NMR signal waveform will be pulsatile because the arterial flow is pulsatile. For venous flow measurement, the arm is inserted in an opposite direction to that shown in the figure. Since venous flow is relatively steady, the magnetization at the receiver coil is also steady. For cerebral flow studies, a 78 cm wide, 45 cm high, 44 cm deep permanent magnet system of 540 gauss field is used. The size is large enough to accommodate a human head. NMR based flowmeters are limited in their applications to the measurement of blood flow in limbs, since the part of the body to be measured must be located inside the lumen of a cylinder.

11.5 LASER DOPPLER BLOOD FLOWMETER

A system utilizing the Doppler-shift of monochromatic laser light to measure blood flow in skin is described by Watkins and Holloway (1978). When a laser beam is directed towards the tissue under study, absorption and scattering takes place. Radiation scattered in movable structures, such as red cells, is shifted in frequency due to the Doppler effect, while radiation scattered in non-moving soft tissue is unshifted in frequency. A part of the total scattered radiation is brought to fall on the surface of a photodetector. The effective radiation penetration depth is approximately 1 mm in soft tissue and scattering and absorption take place mostly in the papilla region and the underlying corium—two dermal layers containing the capillary network of the skin.

In principle, light from a low power (5 mW) He–Ne laser is coupled into a quartz fibre and transmitted to the skin. The light is reflected from both the non-moving tissues (reference beam) and moving red blood cells (Doppler-shifted beam). The two beams are received by a plastic fibre and transmitted back to a photo-diode where optical heterodyning takes place. The heterodyned output signal which is proportional to the Doppler shift frequency is amplified and both RMS and dc values are calculated. The RMS value is weighted against the back-scattered light intensity using the measured dc value as an index of total received power. This gives the output flow velocity.

Figure 11.11 shows a block diagram of the laser Doppler system. The He–Ne laser operating at 632.8 nm wavelength is used. The laser output is coupled into the fibre using a converging lens, which results in an increased power density at the skin surface and thus enables the detection of flow in the more deeply seated veins and arteries. The receiving fibre is coupled to the photodiode through a laser line filter. The photodetector functions as a square law device and gives out current, which is proportional to the intensity of the incident light and, therefore, to the frequency of beating of the shifted and unshifted signals. The light falling on the photodetector is an optically mixed signal involving a Doppler-shifted signal back scattered from the moving red blood cells with the 'reference' signal reflected from the non-moving skin surface. The diode is connected in a configuration so that it gives wideband performance (dc–100 kHz). The amplifier is constructed using a standard operational amplifier with a low noise preamplifier. System output is obtained by taking the RMS value of the total signal, separating it from the total zero light noise, and normalizing it for total back scattered light.

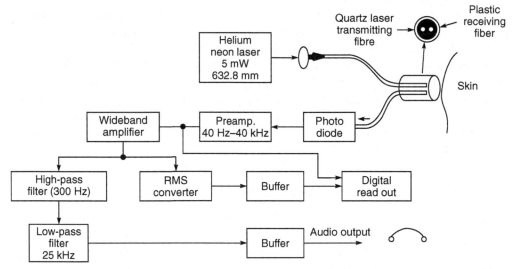

➤ **Fig. 11.11** *Block diagram of a laser Doppler system for blood flow measurement in skin (after Watkins and Holloway, 1978 by permission of IEEE Trans. Biomed. Eng.)*

An audio output of the signal before RMS conversion is also produced to 'hear' the flow pattern. It may be noted that the instrument measures an averaged red blood cell velocity and not true flow, as both the cross-sectional area of flow and the angle the incident light beam makes with each capillary are not known.

Laser Doppler flowmetry is a non-invasive technique and seems to offer several advantages like light reproducibility and sensitivity. However, its disadvantages like poor selectivity, base line instability and restriction in site of measurement are still limiting factors in its successful clinical utilization. The blood cells move through the capillaries at about 1 mm/s, which in tissue is approximately 2.10^{11} mm/s. The Doppler-shift will be the same fraction of the light frequency. To make adequate measuring possible under such conditions, the noise level of the entire system, particularly of the light source and the photo-elements, must be kept extremely low, lower than what can normally be achieved with ordinary photodetectors and low power lasers.

The laser probe is easily adapted to suit different applications. Its main part, the core, is a thin piece of stainless steel tubing (diameter 1.5 mm and length 40 mm), in which the terminating ends of the three fibres—one efferent and two afferent—are moulded together. For easier handling, the core can be inserted into a tight fitting plastic sleeve. The core can also be inserted into a thermostat head where the core-end forms the centre of a 15 mm diameter contact surface, the temperature of which can be set from 25 to 40°C.

Evaluation of the laser Doppler flowmeter has shown that a linear relationship exists between flowmeter response and the flux of red cells, with red cell velocities and volume fractions within the normal physiologic range of the microcirculatory network of the skin. It has also been found that different degrees of oxygenation influenced the Doppler signal only to a minor extent.

Cardiac Output Measurement

Cardiac output is the quantity of blood delivered by the heart to the aorta per minute. It is a major determinant of oxygen delivery to the tissues. Obviously, problems occur when the supply of blood from the heart is unable to meet the demand. A fall in cardiac output may result in low blood pressure, reduced tissue oxygenation, acidosis, poor renal function and shock. It is a reflection of the myocardial function and when taken with other measurements such as blood pressure and central venous pressure, the rational treatment of cardiac disorders becomes clearer. Stroke volume of blood pumped from the heart with each beat at rest varies among adults between 70 and 100 ml, while the cardiac output is 4 to 6 l/min.

The direct method of estimating the cardiac output consists in measuring the stroke volume by the use of an electromagnetic flow probe placed on the aorta and multiplying it by the heart rate. The method involves surgery and, therefore, is not preferred in routine applications. Another well known method for measuring cardiac output is the Fick's Method, which consists in determining the cardiac output by the analysis of the gas-keeping of the organism. Even this method is rather complicated, difficult to repeat, necessitates catheterization and, therefore, cannot be considered as a solution to the problem, though it is practised at many places even now. The most popular method group is the one applying the principle of indicator dilution.

12.1 INDICATOR DILUTION METHOD

Indicator dilution principle states that if we introduce into or remove from a stream of fluid a known amount of indicator and measure the concentration difference upstream and downstream of the injection (or withdrawal) site, we can estimate the volume flow of the fluid. The method employs several different types of indicators. Two methods are generally employed for introducing the indicator in the blood stream, viz: it may be injected at a constant rate or as a bolus. The method of continuous infusion suffers from the disadvantage that most indicators recirculate, and this prevents a maxima from being achieved. In the bolus injection method, a small but known quantity of an indicator such as a dye or radioisotope is administered into the circulation. It is injected into a large vein or preferably into the right heart itself. After passing through the right heart, lungs and the left heart, the indicator appears in the arterial circulation. The presence of an indicator in the peripheral artery is detected by a suitable (photoelectric) transducer and is displayed on a chart recorder. This way we get the cardiac output curve shown in Fig. 12.1. This is also called the dilution curve.

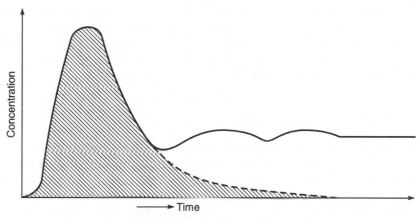

> **Fig. 12.1** *The run of the dilution curve*

The run of the dilution curve is self-explanatory. During the first circulation period, the indicator would mix up with the blood and will dilute just a bit. When passing before the transducer, it would reveal a big and rapid change of concentration. This is shown by the rising portion of the dilution curve. Had the circulation system been an open one, the maximum concentration would have been followed by an exponentially decreasing portion so as to cut the time axis as shown by the dotted line. The circulation system being a closed one, a fraction of the injected indicator would once again pass through the heart and enter the arterial circulation. A second peak would then appear. When the indicator is completely mixed up with blood, the curve becomes parallel with the time axis. The amplitude of this portion depends upon the quantity of the injected indicator and on the total quantity of the circulating blood.

For calculating the cardiac output from the dilution curve, assume that

M = quantity of the injected indicator in mg

Q = cardiac output

then $Q = \dfrac{M}{\text{average concentration of indicator per} \times \text{curve duration}} \cdot 1/s$
$\phantom{Q = \dfrac{M}{\text{aver}}}$litre of blood for duration of curve $\quad$ in seconds

$$= \dfrac{M \times 60}{\text{area under the curve}} \; 1/\text{min}$$

Suppose that 10 mg of the indicator was injected and the average concentration as calculated from the curve was 5 mg/l for a curve duration of 20 s; then $Q = 6$ l/min.

The area under the primary curve obtained by the prolongation of the down slope exponential curve to cut the time axis, encloses an area showing the time concentration relationship of the indicator on its first passage round the circulation and does not include any of the subsequent recirculations. It demands a considerable time to perform the exponential extrapolation for calculating the area. The evaluation of the dilution curve is simplified by replotting the curve on a semilogarithmic scale paper. The indicator concentration (Y-axis) is plotted on a logarithmic scale and the time (X-axis) on a linear scale. The decreasing exponential portion of the curve appears as a straight line, which is projected downwards to cut the time axis. The area under the replotted primary dilution curve is then measured either with a mechanical planimeter or by counting the square units under the curve. It can be approximated by summing the indicator concentration occurring at one second intervals from the start to the end of the curve.

▸ 12.2 DYE DILUTION METHOD

The most commonly used indicator substance is a dye. Fox and Wood (1957) suggested the use of Indocyanine green (cardiogreen) dye which is usually employed for recording the dilution curve. This dye is preferred because of its property of absorbing light in the 800 nm region of the spectrum where both reduced and oxygenated haemoglobin have the same optical absorption. While using some of a the blue dyes, it was necessary to have the patient breathe oxygen. The concentration of cardiogreen can be measured with the help of infra-red photocell transducer. Dye cuvettes of as small volume as 0.01 ml are available.

The procedure consists in injecting the dye into the right atrium by means of a venous catheter. Usually 5 mg of cardiogreen dye is injected in a 1 ml volume. The quantity used may be 2.5 mg in the case of children. A motor driven syringe constantly draws blood from the radial or femoral artery through a cuvette. The curve is traced by a recorder attached to the densitometer. After the curve is drawn, an injection of saline is given to flush out the dye from the circulating blood.

There are problems relating to the use of the indicator indocyanine green. It has been experimentally determined that above a dye concentration of approximately 20 µg/ml of blood, the optical density rises less with an increase in dye concentration than below this level (Chamberlain, 1975). Thus for optimum accuracy, the amount of dye chosen for injection should result in dye curves whose peak concentration is less than 20 µg/ml.

Figure 12.2 shows a diagrammatic representation of a densitometer which can be used for the quantitative measurement of dye concentration. The photometric part consists of a source of radiation and a photocell and an arrangement for holding the disposable polyethylene tube constituting the cuvette. An interference filter with a peak transmission of 805 nm is used to permit only infrared radiation to be transmitted. This wavelength is the isobestic wavelength for haemoglobin (Jarlov and Holmkjer, 1972) at various levels of oxygen saturation. In order to avoid the formation of bubbles, the cuvette tubing should be flushed with a solution of silicone in ether. A flow rate of 40 ml/min is preferred in order to get as short a response time as possible for the sampling catheter. The sampling syringe has a volume of 50 mi/min. The output of the photocell is connected to a low drift amplifier. It has a high input impedance and low output impedance. The amplification is directly proportional to the resistance value of the potentiometer R. A potentiometric recorder records the amplifier signal on a 200 mm wide recording paper and a paper speed of 10 mm/s.

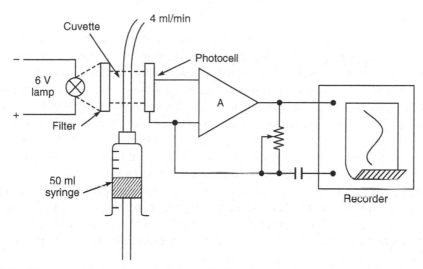

> **Fig. 12.2** *Diagrammatic representation of a densitometer for quantitative measurement of dye concentration (redrawn after Jarlov and Holmkjer, 1972; by permission of Med.& Biol. Eng.)*

In the recording of dye dilution curves, it is generally necessary that the densitometer be at some point removed from the site of interest. A catheter is used to transport the blood containing dye from the sampling site, inside the cardiovascular system, to the densitometer located outside the body. Sampling through the catheter densitometer system distorts the concentration time curve. First, the velocity of flow within the catheter is not uniform, which causes the dye to mix within the tube as it travels downstream. The mixing is a function of the flow rate and the volume of the sampling system, the viscosity of the sampled fluid and the shape of the configuration of the sampling tube. The second source of distortion is the measuring instrument itself, which may not have response characteristics fast enough to record instantaneous dye concentration as it actually occurs in the lumen. Distortion is very important when the indicator dilution method is used to measure volume since it is the measurement of the mean transit time of an indicator from the point of injection to the point of sampling, which is of interest. To reduce distortion, computer software-based corrections have been devised.

12.3 THERMAL DILUTION TECHNIQUES

A thermal indicator of known volume introduced into either the right or left atrium will produce a resultant temperature change in the pulmonary artery or in the aorta respectively, the integral of which is inversely proportional to the cardiac output.

$$\text{Cardiac output} = \frac{\text{"a constant"} \times (\text{blood temp.} - \text{injectate temp.})}{\text{area under dilution curve}}$$

Although first reported by Fegler (1954), thermal dilution as a technique did not gain clinical acceptance until Branthwaite and Bradley (1968) published their work showing a good correlation between Fick and thermal measurement of cardiac output in man. However, the technique of cannulation of the internal jugular vein and the difficulty of floating small catheters into the pulmonary artery prevented a rapid clinical acceptance of the technique.

In 1972, a report appeared in the American Heart Journal describing a multi-lumen thermistor catheter, known today as the Swan-Ganz triple lumen balloon catheter (Ganz and Swan, 1972). The balloon, located at or near the tip, is inflated during catheter insertion to carry the tip through the heart and into the pulmonary artery. One lumen terminates at the tip and is used to measure the pressure during catheter insertion. Later, it measures pulmonary artery pressure and inter-mittently, pulmonary–capillary wedge pressure. A second lumen typically terminates in the right atrium and is used to the monitor right atrial pressure (central venous pressure) and to inject the cold solutions for thermal dilution. A third lumen is used to inflate the balloon. For use with thermal dilution, the pulmonary-artery catheter carries a thermistor proximal to (before) the balloon. The thermistor is encapsulated in glass and coated with epoxy to insulate it electrically from the blood. The wires connecting the thermistor are contained in a fourth lumen (Fig. 12.3). This catheter simplified the technique of cardiac cannulation making it feasible to do measurements not only in

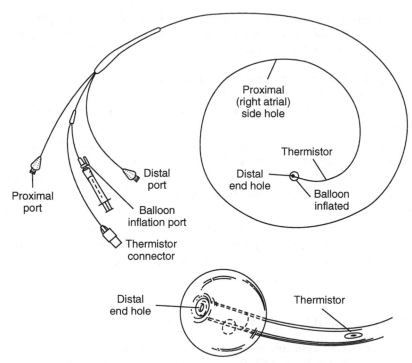

> **Fig. 12.3** *Swan-Ganz Catheter - A 4-lumen catheter. Distal Lumen - connects to transducer system to monitor (i) pulmonary artery pressure (ii) Wedge pressure with balloon inflated. Inflation Lumen - connects to balloon located approximately 1 mm from catheter-tip. Balloon inflates with 1 to 1.5 ml of air. Proximal lumen – to monitor central venous pressure or right atrium pressure. Thermistor Lumen – connects with cable to cardiac output computer.*

the catheterization laboratory but also in the coronary care unit. The acceptance of the thermal dilution technique over the past few years can only be attributed to the development of this catheter.

Figure 12.4 shows a typical cardiac output thermal dilution set up. A solution of 5% Dextrose in water at room temperature is injected as a thermal indicator into the right atrium. It mixes in the right ventricle, and is detected in the pulmonary artery by means of a thermistor mounted at the tip of a miniature catheter probe. The injectate temperature is also sensed by a thermistor and the temperature difference between the injectate and the blood circulating in the pulmonary artery is measured. The reduction in temperature in the pulmonary artery (due to the passage of the Dextrose) is integrated with respect to time and the blood flow in the pulmonary artery is then computed electronically by an analog computer which also applies correction factors. A meter provides a direct reading of cardiac output after being muted until integration is complete so as to avoid spurious indications during a determination.

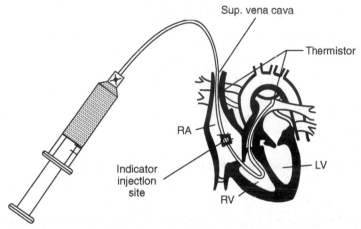

➤ Fig. 12.4 *Cardiac output thermal-dilution set-up*

The electronic computation is relatively simple, because there is no significant recirculation of the indicator in man. The calculation rests upon the integral of the inscribed curve, the resting temperature in the pulmonary artery, the temperature of the injectate, and a number of constants. Absence of the need to subtract that part of the area under the curve due to recirculation, and the ease with which an unsatisfactory curve can be detected by failure to return to the original baseline value of temperature, contribute to the internal consistency of the results.

The system calibration is based upon the use of an injection of 10 ml of 5% Dextrose solution at a temperature in the range of 18–28°C. Within this range, the injectate temperature is measured to an accuracy of ± 0.2°C, and is also displayed on a meter.

Blood temperature is measured over a range of 30 to 40°C to an accuracy of ±0.2°C. During a determination the incremental temperature is automatically derived, relative to a baseline value equal to the blood temperature, immediately before starting the determination. The incremental temperature is measured and displayed in the range 0-1°C full scale to an accuracy of ± 0.02°C.

The sensitivity of the measuring system is adjusted in such a way that a temperature variation of 0.3–0.5°C (normally encountered temperature fluctuation at the measuring point), measured by the catheter tip, gives a full scale deflection on the recorder. The total quantity injected is normally 10 ml which gives a full scale deflection with the aforesaid sensitivity of the recorder. The thermodilution curve can be recorded on an electrocardiograph machine. The normal paper speed is 10 mm/s and the normal curve amplitude is 30–50 mm.

12.3.1 Computing System

Assuming that the injectate mixes thoroughly with the blood and that negligible net heat flow occurs through the vessel wall during the passage of the blood injectate mixture between the points of injection and measurement, the heat (or coolth) injected can be equated to the heat (coolth) detected. If the volume of injectate is small as compared with the volume of the blood it mixes with, it can be shown that within a close approximation,

$$V\,D_i\,S_i\,(T_i - T_b) = Q\,Db\,Sb \int \Delta T\,dt$$

which, when rearranged gives

$$Q = \frac{V(T_i - T_b)}{\int \Delta T\,dt} \cdot \frac{D_i\,S_i}{D_b\,S_b}$$

where Q = volumetric flow
         V = volume of injectate
         T = temperature
         ΔT = incremental temperature of the blood injectate mixture
         D = density
         S = specific heat

The suffixed i and b refer to the injectate and blood respectively. The thermal dilution curve can be validated to ensure that the downslope is exponential and that the area under an acceptable curve is determined by integration. The equation for determining the cardiac output by this method thus reduces to:

$$\text{Cardiac output} = \frac{(1.08)\,(C)\,(60)\,(V)\,(T_i - T_b)}{\int \Delta T\,dt}$$

where: 1.08 is the ratio of the products of specific heats and specific gravities of 5% dextrose in water and blood

         C = 0.827 for 10 ml injectate at ice temperature (0 to 2°C)
           = 0.747 for 5 ml injectate at ice temperature
           = 0.908 for 10 ml injectate at room temperature (22 to 26°C)
           = 0.884 for 5 ml injectate at room temperature
         V = volume of injectate (ml)
         T_b = initial temperature of blood (°C)

T_i = initial temperature of injectate (°C)

$\int \Delta T \, dt$ = integral of blood temperature change (°C.s).

Figure 12.5 shows a block diagram of the thermal dilution system. The linearizing amplifier works on the principle that when a fixed resistance of suitable value is placed in parallel with a thermistor, a virtually linear relation between the temperature and resistance of the parallel combination can be obtained over a limited range of operation. The integrator responds to a ΔT signal corresponding to a maximum temperature change of typically 0.3°C with reference to a T_b baseline, which itself may suffer medium term variations of a similar amount and may additionally vary from one patient to another over a range of 30-40°C. Also, the circuit must hold the integral obtained so that the value of the cardiac output computed from it may be displayed.

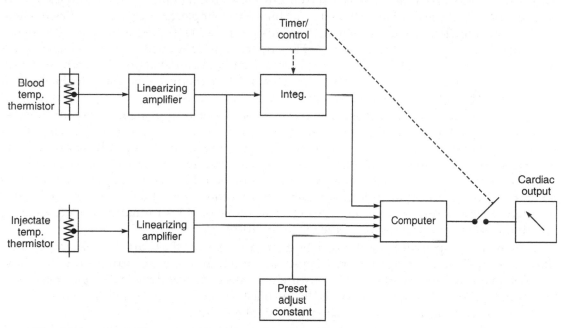

> **Fig. 12.5** *Block diagram of the processing and computing circuit of thermal dilution method (redrawn after Cowell and Bray, 1970)*

The summing and multiplication/division operations required for the evaluation of cardiac output are performed by a simple analog computing circuit. The timer control unit generates the switching signals necessary for the proper integration of the ΔT signal and for the display of the computed value of the cardiac output. The cardiac output is displayed in two ranges: 0–10 l/min and 10–20 l/min.

Philip *et al* (1984) used a resistive element in a modified Swan-Ganz catheter and energized it with a periodic electrical waveform. The resulting thermal signal is diluted by the blood flow and sensed by a thermistor in the pulmonary artery. The thermal signal is processed by a microprocessor-based instrument. This technique avoids the introduction of fluids and thereby reduces the risks of fluid overload and sepsis (Bowdle *et al*, 1993).

Neame *et al* (1977) illustrated the construction of thermistor probes intended for use in cardiac output estimations by the thermodilution method. Ideally, the probes should have infinitesimal thermal capacity and hence an infinite frequency response in order to reproduce thermal transients accurately; minimal size so as to cause no flow obstruction, produce negligible heat dissipation, a linear temperature response and an infinite working life. Bead type of thermistors are available in the smallest form and are thus suited to applications for cardiac output measurements.

Although the excellent linearity characteristics of thermocouples and their fast dynamic response, as opposed to the slower non-linear thermistor beads, make them more desirable, the inherent low sensitivity of thermocouples, however, provides a severe limitation. Compared to thermistors excited by a current below the effective self-heating range, thermocouples are two orders of magnitude less sensitive than thermistors. For example, temperature coefficient is 0.04 mV/°C for chromel-constantan as compared to 3 mV/°C for a 2000 Ω thermistor with a dissipation constant of 0.2 mW/°C excited by 0.1 mA. Flow directed precalibrated thermistor catheters are about 100 cm long and contain a thermistor located 4 cm from the distal tip. The thermistor resistance, typically, may be 7000 Ω at 36.6°C, with a temperature coefficient of resistance being 250 Ω per °C ± 5%. The injectate orifice is 26 cm proximal to the thermistor.

Basic source of inaccuracies in thermo-dilution cardiac output measurements arise with imperfect indicator mixing. Pneumatically controlled injectors are used for rapid and consistent injections. These injectors are powered by compressed air or CO_2 from a pressurized cylinder or a hospital air system. A two-way automatic valve between the injection syringe and the catheter make possible a fully automatic operation with injection and automatic refill. Injection speed is 7 ml/s via standard 7F triple luman thermistor catheter. Both injection and refill speed are usually kept adjustable. Slow return speed minimizes chances of air bubbles formation during refilling. Injection volume is also adjustable in the range 1 to 10 ml.

As blood flow is pulsatile, injection can be done in any part of the heart cycle. Therefore, time allowed for indicator mixing can vary as much as a full heart cycle. In addition, dilution curve obtained in the pulmonary artery will look different if injection was done in diastole or systole of the heart. Optimum injection moment is during diastole and, therefore, injectors are equipped with ECG synchronizer to ensure injection at the right moment.

ECG synchronizer consists of an ECG amplifier with adjustable gain. A threshold detector senses appearance of *R* wave and the interval between two adjacent *R* waves is measured and stored in a the memory. Delay of injection is set manually in a percentage of the cardiac cycle, and not in milliseconds as in the majority of angiographic injectors. The method of presetting delay of injection in a percentage of the cardiac cycle makes the synchronizer insensitive to variations in heart rate.

An electrosurgical unit emits radio frequency energy that can distrupt or scramble the thermodilution temperature curve and result in wildly inaccurate results. Such a disruption can cause major problems for the anaesthesiologist measuring cardiac output intra-operatively. The effect of high frequency interference on the measurements becomes readily apparent by an inspection of the thermodilution temperature. A visual display of the thermodilution curve is therefore necessary.

12.4 MEASUREMENT OF CONTINUOUS CARDIAC OUTPUT DERIVED FROM THE AORTIC PRESSURE WAVEFORM

It is not always possible, particularly in critically ill patients, to estimate cardiac output on a beat-to-beat basis. Thermal dilution measurements can be repeated more frequently than the other methods, which has made this method popular, but even then the measurements can only be made at certain intervals. The technique also requires the presence of well-trained operators for obtaining reliable measurements. These problems are circumvented in an instrument developed by Wessling et al. (1976a), which is based on the analysis of the pressure pulse contour illustrated in Fig. 12.6.

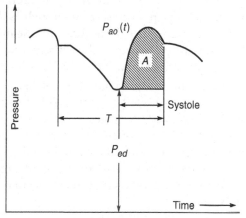

> **Fig. 12.6** *Pressure pulse contour method of measuring beat-to-beat cardiac output*

The method is based on the analysis of the aortic pressure wave, and estimates the left ventricular stroke volume from the pulse contour. The analysis of the pressure wave depends on simple hydraulic relationships between flow, pressure and time. During the ejection phase flow is ejected into the aorta, the total amount depending on the driving pressure, on the duration of the ejection period and on the impedance to flow in the aorta (Wesseling *et al*, 1974). Expressed mathematically:

$$\text{Stroke volume} = \frac{1}{Z_{ao}} \int_{T_o}^{T_E} (P_{ao} - P_{ed})\, dt$$

This equation shows that the area 'A' under the systolic portion of the aortic pressure wave is integrated. This area 'A' is divided by the patient calibrator (Z_{ao}: aortic characteristic impedance) to obtain a quantity approximating the left venticular stroke volume. Thus, changes in this area 'A' reflect corresponding changes in stroke volume. The instantaneous heart rate is derived as the inverse of the time lapse between the onset of ejection for every two consecutive pressure pulses. Cardiac output is computed for each beat as the instantaneous product of stroke volume and heart rate:

$$\text{Stroke volume (SV)} = \frac{A}{Z_{ao}}\ (\text{cm}^3)$$

$$\text{Heart rate} = \frac{60}{T}\ (\text{bpm})$$

$$\text{Cardiac output} = \frac{(SV) \times (HR)}{1000} = \frac{60 \cdot A}{1000 \cdot Z_{ao} T}\ (\text{l/min})$$

The instrument uses pattern recognition techniques to detect the onset (T_o) and the end (T_E) of ejection, and integrates the difference between the actual aortic pressure and the end diastolic pressure over that time period. The average value of Z_{ao} is found to be 0.140 for adults.

In this technique, all that is required is one pressure measurement at any site in the aorta. To measure the pressure signal, both catheter tip manometers as well as ordinary catheter manometer systems can be used. It should be ensured that pressure is measured in the aorta or near the aorta proximally in a major side branch such as the subclavian artery. Other pressure signals are much distorted and reduce the accuracy of the method. For reliable operation of the computing circuit, the resonant frequency of the catheter manometer system must be 25 Hz or higher. If no artefact distorts the pressure waveform, the resonance frequency may be as low as 15 Hz.

The cardiac output computer consists of three modules: (i) the pressure transducer amplifier, (ii) the cardiac output computer, (iii) dual numerical display and alarm limits. Quantities, which can be displayed, are mean systolic and diastolic pressure, cardiac output, stroke volume and pulse rate. Two parameters must be set initially for each patient: (i) patient age and (ii) the patient calibrator Z_{ao}. In addition, waveform display is also provided for visually checking the pressure curve shape and to ensure the correct triggering of the processing circuit. The instrument can be used to monitor trends in cardiac output or for displaying absolute values of cardiac output after calibration.

Deloskey *et al* (1978) showed a comparison of stroke volumes calculated using the Wesseling pulse-contour formula and those obtained with the electromagnetic flowmeter with several interventions introduced to vary stroke volume in one patient. The agreement between the two methods was very good (0.91) over a wide range of values for stroke volume, heart rate, mean pulmonary artery pressure and pulmonary-artery resistance.

▷ 12.5   IMPEDANCE TECHNIQUE

The technique used for the measurement of cardiac output by the impedance method is illustrated in Fig. 12.7. If p is the resistivity, the resistance of the thorax between two sensing electrodes (2 and 3) is given by

$$R_0 = \frac{\rho L}{A}$$

where L is the separation between the electrodes and A is the cross-sectional area of the thorax.

Assuming that with each ejection of stroke volume dV, the resistance decreases by dR, it can be derived that

$$dV = -\rho \left[\frac{L^2}{R_0^2} \right] dR$$

R can be replaced by Z if an ac signal is used for transthoracic impedance measurement, thus giving $dV = -\rho \, (L^2/Z_0^2) \, dz$. In this relationship, dV is the stroke volume in ml, ρ is the resistivity of the patient's blood in Ω.cm and dz is the decrease in Z_0 during a particular systolic ejection.

The stroke volume is given by the product of the initial rate of change of impedance and the time the aortic and pulmonic valves open, i.e., $dz = T(dz/dt)_{max}$ where (dz/dt) max corresponds to the peak negative value of dz/dt found during systole and T is the interval between $dz/dt = 0$ and the second heart sound.

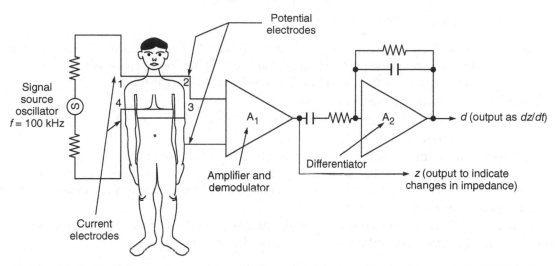

> **Fig.** 12.7   *Technique of measuring cardiac output by impedance changes (after Hill and Thompson, 1975; by permission of Med. & Biol. Eng)*

The equation used by Kubicek *et al* (1966) for stroke volume can be expressed as

$$dV = -\rho \left[\frac{L^2}{Z_0^2} \right] \cdot T \cdot \left[\frac{dz}{dt} \right]_{max}$$

The value of p is related to the patient's haematocrit and for normal values, a blood resistivity of 150 Ω·cm can be assumed.

For experimentally calculating the stroke volume, a constant current at 100 kHz is applied between electrodes 1 and 4. The resulting voltage fluctuations occurring across the thorax coincident with cardiac activity are detected at the inner pair of electrodes 2 and 3.

The basal impedance between these electrodes is found to be about 25 Ω and this diminishes by about 0.1 Ω with each systole. The voltage signal due to changes in impedance is amplified and demodulated to obtain Z. The *dz/dt* is calculated using a differentiator. A two-channel recorder is used to record *dz/dt* and the phonocardiogram. For each beat the maximum value of the *dz/dt* at systole is noted as is the ejection time from the moment the *dz/dt* tracing crosses the *dz/dt* = 0 line at the commencement of systole until the onset of the second heart sound.

The method of measuring cardiac output from transthoracic impedance plethysmograms has several advantages in clinical use, especially in monitoring each stroke volume non-invasively. For this reason, there have been many correlation studies of cardiac-output values between those measured by this method and those by other methods such as indicator dilution, Fick and pressure-gradient methods. Application of the thoracic bio-impedance technique has not been very encouraging because the correlation to thermodilution appears generally unacceptable (Bowdle *et al*, 1993). A completely non-invasive and continuous bio-impedance cardiac output meter would be a preferred device over thermodilution, but only if the problems of accuracy can be further solved.

▐▌▌▌▶ 12.6   ULTRASOUND METHOD

Ultrasound can be used to measure the velocity of blood flow in the ascending aorta by the application of the Doppler principle (Huntsman *et al*, 1983). If the area of cross-section of the aorta is known or can be measured, the blood flow rate can be calculated as follows:

Blood flow = velocity (cm/sec) × area (cm^2)

$\qquad$ = cm^3/sec = ml/sec × L/100 ml × 60 sec/min

$\qquad$ = L/min

The blood flow in the ascending aorta would be identical to the cardiac output minus the quantitatively negligible flow to the coronary arteries. However, the equation above is an over simplification because blood flow in the aorta is not constant but pulsatile and blood velocity in the aorta varies during the cardiac cycle.

Cardiac output measurement devices based on ultrasound actually measure the stroke volume during the cardiac cycle as per the relationship given below:

$$SV = CSA \times \int^{VET} V(t)\, dt$$

where $\qquad$ SV = stroke volume

$\qquad\qquad$ CSA = cross-sectional area of the aorta

$\qquad\qquad$ VET = ventricular ejection time

$\qquad\qquad$ V = blood velocity

Stroke volume is then multiplied by the heart rate to give cardiac output.

The blood velocity V calculated from the Doppler equation is as follows:

$$V = \frac{C \times F_d}{2F_o \times \cos \theta}$$

where: $\qquad$ C = speed of sound

$\qquad\qquad$ F_d = frequency shift

$\qquad\qquad$ F_o = frequency of the emitted sound

$\qquad\qquad$ θ = angle between the emitted sound and the moving object

Blood velocity in the ascending aorta can be measured by a probe that is held in the operator's hand and positioned just above the sternal notch. Blood velocity in the descending aorta can be measured (Mark *et al*, 1986) by a probe attached to an esophageal stethoscope. When compared to standard thermodilution measurement of cardiac output, the correlation coefficients range from 0.63 to 0.91 (Wong *et al*, 1990). These variations account for the relatively poor accuracy of this technique.

The area of cross-section of the aorta can be measured by an echo technique, but this is seldom done in practice. Average human dimensions for the aorta are available from nomograms that consider sex, age, height and weight. However, the individual patients may vary considerably from the average. Therefore, lack of a precise area of cross-section of the particular blood vessel poses a significant potential source of error. Due to lack of accuracy, determination of cardiac output by the ultrasound method has not become popular in clinical practice.

An esophagial ultrasound Doppler instrument for quick and continuous cardiac output and stroke volume has been introduced by M/s Deltex Medical, UK. This is shown in Fig. 12.8.

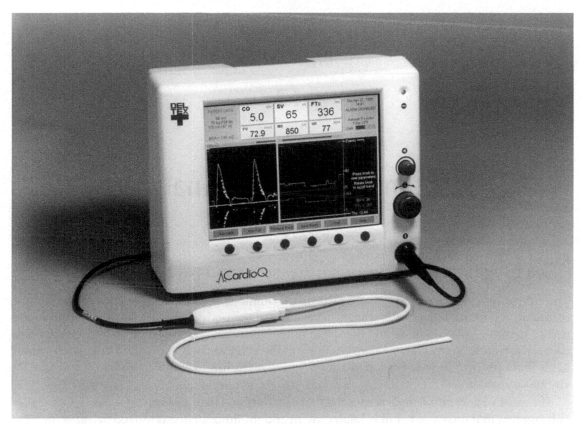

> **Fig. 12.8**   *Cardio Q – Doppler shift based cardiac output and aortic volume measurement and display system (Courtesy: Deltex Medical, USA)*

A single use, small diameter (6mm) probe is inserted into the esophagus and a continuous Doppler signal is obtained. The equipment utilizes descending aortic flow to provide a real-time assessment of the left ventricular performance. The system works on a continuous 4 MHz ultrasound frequency.

Pulmonary Function Analyzers

▶ 13.1 PULMONARY FUNCTION MEASUREMENTS

Three basic types of measurements are made in the pulmonary clinic: ventilation, distribution and diffusion. Ventilation deals with the measurement of the body as an air pump, determining its ability to move volumes of air and the speed with which it moves the air. Distribution measurements provide an indication of where gas flows in the lungs and whether or not disease has closed some sections to air flow. Diffusion measurements test the lung's ability to exchange gas with the circulatory system.

The most widely performed measurement is *ventilation*. This is performed using devices called spirometers that measure volume displacement and the amount of gas moved in a specific time. Usually this requires the patient to take a deep breath and then exhale as rapidly and completely as possible. Called the forced vital capacity, this gives an indication of how much air can be moved by the lungs and how freely this air flows.

Distribution measurements quantify degrees of lung obstructions and also determine the residual volume, which is the amount of air that cannot be removed from the lungs by the patients effort. The residual volume is measured indirectly, such as with the nitrogen washout procedure.

Diffusion measurements identify the rate at which gas is exchanged with the blood stream. This is difficult to do with oxygen since it requires a sample of pulmonary capillary blood, so it is usually done by measuring the diminishment of a small quantity of carbon monoxide mixed with the inhaled air.

Pulmonary function analyzers provide the means for automated clinical procedures and analysis techniques for carrying out a complete evaluation of the lung function or the respiratory process. The respiratory activity ensures supply of oxygen to and removal of carbon dioxide from the tissues. These gases are carried in the blood—oxygen from the lungs to the tissues and carbon dioxide from the tissues to the lungs. During quiet breathing, the ordinary intake of air or tidal volume is about 0.5 l. However, only part of this volume takes part actually in oxygenating the blood, because no exchange of gases between air and blood takes place in the mouth, trachea and bronchi. The air filling these parts is called 'Dead Space' air and in adults it typically amounts to about 0.15 l. The rest of the inspired air ventilates the alveoli and takes part in the exchange of gases. In each minute, under normal conditions, about 250 ml of oxygen is taken up and 250 ml of CO_2 is given out by the body. The average composition of atmospheric air and alveolar air is given in Table 13. 1.

• Table 13.1 *Atmospheric and Alveolar Air Composition*

Gas	Atmospheric air (%)	Alveolar air (%)
O_2	20.9	14
CO_2	0.1	5.5
N_2	79.0	80.5

It might be thought that as the ultimate function of the lungs is to exchange gas with the environment, measurement of the arterial blood gases would be sufficient to assess lung function. However, the respiratory reserve of the lungs is so considerable that in many cases, symptoms or signs do not develop until an advanced degree of functional impairment is present. Many of the available tests are very sensitive and will detect abnormalities in lung function well before lung disease has become clinically apparent.

The pulmonary function can be assessed by means of two major classes of tests. These are:

(i) Evaluation of the mechanical aspects of pulmonary function, which affects the bulk gas transport into and out of the lungs.

(ii) Evaluation of gas exchange or diffusion at the alveoli.

The ability of the pulmonary system to move air and exchange oxygen and carbon dioxide is affected by the various components of the air passages, the diaphragm, the rib cage and its associated muscles and by the characteristics of the lung tissue itself. Among the basic tests performed are those to determine the volumes and capacities of the respiratory system. These are defined as follows in Fig. 13.1.

13.1.1 Respiratory Volumes

Tidal Volume (TV): The volume of gas inspired or expired (exchanged with each breath) during normal quiet breathing, is known as tidal volume.

Minute Volume (MV): The volume of gas exchanged per minute during quiet breathing. It is equal to the tidal volume multiplied by the breathing rate.

Alveolar Ventilation (AV): The volume of fresh air entering the alveoli with each breath.

$$\text{Alveolar Ventilation} = (\text{Breathing rate}) \times (\text{Tidal volume} - \text{Dead space}).$$

Inspiratory Reserve Volume (IRV): The volume of gas, which can be inspired from a normal end-tidal volume.

$$IRV = VC - (TV + FRC)$$

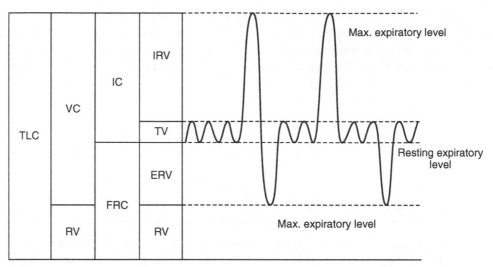

> **Fig.** 13.1   *Volume and capacities of the lungs-standardization of definitions and symbols in respiratory physiology*

Expiratory Reserve Volume (ERV): The volume of gas remaining after a normal expiration less the volume remaining after a forced expiration.

$$ERV = FRC - RV$$

Residual Volume (RV): The volume of gas remaining in the lungs after a forced expiration.

13.1.2   Respiratory Capacities

Functional Residual Capacity (FRC): The volume of gas remaining in the lungs after normal expiration.

Total Lung Capacity (TLC): The volume of gas in the lungs at the point of maximal inspiration.

$$TLC = VC + RV$$

Vital Capacity (VC): The greatest volume of gas that can be inspired by voluntary effort after maximum expiration, irrespective of time.

Inspiratory Capacity (IC): The maximum volume that can be inspired from the resting end expiratory position.

Dead Space: Dead Space is the functional volume of the lung that does not participate in gas exchange.

13.1.3   Compliance and Related Pressures

Compliance (C): Change in volume resulting from unit change in pressure. Units are l/cmH_2O.

Lung Compliance (C_L): Change in lung volume resulting from unit change in transpulmonary pressure (P_L)

Chest-Wall Compliance (C_{cw}): Change in volume across the chest wall resulting from unit change in transchest-wall pressure.

Static Compliance (C_{ST}): Compliance measured at point-of-zero airflow by interruption or breach-hold technique.

Elastance (E): Reciprocal of compliance. Units are cmH_2O/litre.

Transpulonary Pressure (P_L): Pressure gradient developed across mouth (P_{ao}) and pleural surface at lung (P_{PL}).

Transalveolar Pressure (P_{EI}): Pressure gradient developed between alveolar wall ($P_{ALV} - P_{AL}$).

Transairway Pressure (P_{RES}): Pressure gradient developed between alveoli and mouth.

Static Elastic Recoil Pressure ($R_{ST(L)}$): Pressure developed in elastic fibers of the lung by expansion.

13.1.4 Dynamic Respiratory Parameters

A number of forced breathing tests are carried out to assess the muscle power associated with breathing and the resistance of the airway. Among these are:

Forced Vital Capacity (FVC): This is the total amount of air that can be forcibly expired as quickly as possible after taking the deepest possible breath.

Forced Expiratory Volume (FEV): The percentage of the VC that can be forced out of the lungs in a given period with 'maximal exertion'. This is written as FEVT where T is usually in seconds.

Maximum Mid-Expiratory Flow (MMEF or MMF) or Maximum Mid-Flow Rate (MMFR): The maximum rate of flow of air during the middle half of the FEV spirogram. One half VC is obtained from the volume indicated by the curve between 25 and 75% VC. This is illustrated in Fig. 13.2.

Mid-Expiratory Time (MET): It is the time in seconds over which this volume is forcibly exhaled. The MMEF is calculated from $MMEF = (1/2\,VC) \times (1/MET)$

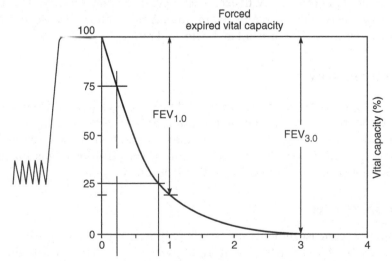

> **Fig. 13.2** *The calculation of MMEF from the FEV spirogram*

Normal values for each of these volumes and capacities have been calculated. They have been found to vary with sex, height and age. All pulmonary volumes and capacities are about 20 to 25 % less in females than in males.

A particular pattern of abnormal lung volume may occur in a particular form of lung disease and such a pattern is useful confirmatory evidence of a diagnosis made on clinical grounds. Further, serial lung function testing is of use in demonstrating progressive deterioration in function or in confirming a satisfactory response to therapy. For example, if the ratio $(FEV_1)/(FVC)$ of the volume of gas that can be exhaled forcibly in one second from maximum inspiration (FEV_1) to the forced vital capacity (FVC) is less than 70%, airway obstruction as in chronic bronchitis is likely to be present. If the FEV_1/FVC is greater than 85%, a so called 'restrictive' defect may be present. This is seen in cases of diffuse pulmonary fibrosis.

Pulmonary function tests are performed for the assessment of the lung's ability to act as a mechanical pump for air and the ability of the air to flow with minimum impedance through the conducting airways. These tests are classified into two groups: single-breath tests and multiple-breath tests.

There are three types of tests under the *single-breath* category. These are

- Tests that measure expired volume only.
- Tests that measure expired volume in a unit time.
- Tests that measure expired volume/time.

In the class of *multiple-breath* test measurements is the Maximal Voluntary Ventilation (MVV) which is defined as the maximum amount of air that can be moved in a given time period. Here, the patient breathes in and out for 15 s as hard and as fast as he or she can do. The total volume of the gas moved by the lungs is recorded. The value is mulitiplied by 4 to produce the maximum volume that the patient breathed per minute by voluntary effort.

A resting person inspires about 0.5 litre of air with each breath, with the normal breathing rate of 12 to 20 breaths per minute. With exercise, the volume may increase 8 to 10 times and the breathing rate may reach 40 to 45 breaths per minute. A respiratory disease may be suspected if these volumes, capacities or rates are not in the normal range.

ⅢⅢ▶ 13.2   SPIROMETRY

The instrument used to measure lung capacity and volume is called a spirometer. Basically, the record obtained from this device is called a spirogram. Spirometers are calibrated containers that collect gas and make measurements of lung volume or capacity that can be expired. By adding a time base, flow–dependent quantities can be measured. The addition of gas analzsers makes the spirometer a complete pulmonary function testing laboratory.

13.2.1   Basic Spirometer

Most of the respiratory measurements can be adequately carried out by the classic water-sealed spirometer (Fig. 13.3). This consists of an upright, water filled cylinder containing an inverted counter weighted bell. Breathing into the bell changes the volume of gases trapped inside, and the

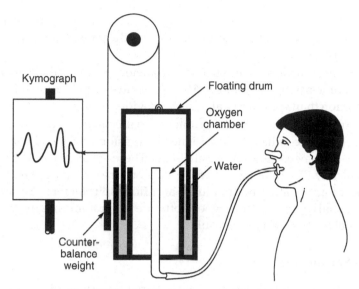

> **Fig. 13.3** *Basic water sealed spirometer*

change in volume is translated into vertical motion, which is recorded on the moving drum of a Kymograph. The excursion of the bell will be proportional to the tidal volume. For most purposes, the bell has a capacity of the order of 6–8 l. Unless a special light weight bell is provided, the normal spirometer is only capable of responding fully to slow respiratory rates and not to rapid breathing, sometimes encountered after anaesthesia. Also, the frequency response of a spirometer must be adequate for the measurement of the forced expiratory volume. The instrument should have no hysteresis, i.e. the same volume should be reached whether the spirometer is being filled or being emptied to that volume.

As the water-sealed spirometer includes moving masses in the form of the bell and counter-weights, this leads to the usual problems of inertia and possible oscillation of the bell. This can lead to an over-estimation of the expiratory volume. A suggested compensation is by the use of a spirometer bell having a large diameter and which fits closely over the central core of the spirometer, so that the area of water covered by the bell is small in relation to that of the water tank. If the spirometer is used for time-dependent parameters, then it must also have a fast response time, with a flat frequency response up to 12 Hz. This requirement applies not only to the spirometer, but also to the recorder used in conjunction with the recording device.

The spirometer is a mechanical integrator, since the input is air flow and the output is volume displacement. An electrical signal proportional to volume displacement can be obtained by using a linear potentiometer connected to the pulley portion of the spirometer. The spirometer is a heavily damped device so that small changes in inspired and expired air volumes are not recorded. The spirometers can be fitted with a linear motion potentiometer, which directly converts spirometer volume changes into an electrical signal. The signal may be used to feed a flow-volume differentiator for the evaluation and recording of data. The response usually is ± 1% to 2 Hz and ± 10% to 10 Hz.

Tests made using the spirometer are not analytical. Also, they are not completely objective because the results are dependent on the cooperation of the patient and the coaching efforts of a good respiratory technician.

There have been efforts to develop electronic spirometers which could provide greater information-delivering and time-saving capabilities. Also, there have been efforts to obtain more definitive diagnostic information than spirometry alone can provide.

Calculating results manually from the graph of the mechanical volume spirometer requires considerable time. Transducers have been designed to transform the movement of the bell, bellows or piston of volume spirometers into electrical signals. These are then used to compute the numerical results electronically. The popularity and low cost of personal computers have made them an attractive method of automating both volume and flow spirometers. An accurate spirometer connected to a personal computer with a good software programme has the potential of allowing untrained personnel to obtain accurate result.

13.2.2   Wedge Spirometer

A wedge spirometer (Fig. 13.4) consists of two square pans, parallel to each other and hinged along one edge. The first pan is permanently attached to the wedge casting stand and contains a pair of 5 cm inlet tubes. The other pan swings freely along its hinge with respect to the fixed pan. A space existing between the two pans is sealed airtight with vinyl bellows. The bellows is extremely flexible in the direction of pan motion but it offers high resistance to 'ballooning' or inward and outward expansion from the spirometer. As a result, when a pressure gradient exists between the interior of the wedge and the atmosphere, there will only be a negligible distortion of the bellows.

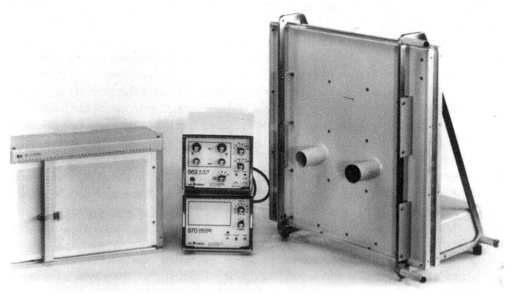

> **Fig. 13.4**   *Wedge spirometer (Courtesy: Med. Science, USA)*

As gas enters or leaves the wedge, the moving pan will change position in compensation for this change in volume. The construction of the wedge is such that the moving pan will respond to very slight changes in volume. Under normal conditions, the pressure gradient that exists between the wedge and the atmosphere amounts to only a fraction of a millimetre of water.

Volume and flow signals for the wedge are obtained independently from two linear transducers. The transducers are attached to the fixed frame and are coupled to the edge of the moving pan. One transducer produces a dc signal proportional to displacement (volume), while the other has a dc output proportional to velocity (flow).

The transducer outputs are connected to an electronics unit, which contains the power supply, an amplifier, and the built-in calibration networks.

A pointer attached to the moving pan and a scale affixed to the frame, combine to provide a mechanical read out for determining the approximate volume position of the spirometer. When open to the atmosphere and standing upright, the wedge will empty itself due to the force of gravity acting on the moving member. An adjustable tilt mechanism provides the means for changing the resting point of the moving pan to any desired volume point. An adjustable magnetic stop insures a more highly defined resting position.

Neither the tilt nor the magnetic stop has any noticeable effect on the moving pan position once it is connected to a closed system. This is primarily due to the large surface area of the pans, which serves to convert small pressures into large forces.

Thus, the relatively small forces due to gravity and the magnetic stop are overcome by a negligible rise in pressure in the patient's lungs. When gravitational return of the moving pan to the resting position is deemed undesirable, the wedge may be turned on its side so that at any point, the pan will be in a state of equilibrium. The wedge may be calibrated with a selector switch, which determines the magnitude of the calibration signal. The volume may be calibrated with a signal corresponding to either 0.5 ml or 5 l. The flow calibration signals for each particular wedge are adjusted, using special fixtures. A volume of one litre is introduced at a certain point and a flow rate of 1 l/s is introduced at another point, with the calibration signals then being adjusted to produce equal signals.

As on conventional spirometers, all standard pulmonary function tests may be performed on the wedge. X-Y recorders featuring high acceleration slew rates may be used in recording flow/volume loops.

13.2.3 Ultrasonic Spirometer

Ultrasonic spirometers depend, for their action on transmitting ultrasound between a pair of trans-ducers and measuring changes in transit time caused by the velocity of the intervening fluid medium (McShane, 1974). They employ piezo-electric transducers and are operated at their characteristic resonant frequency for their highest efficiency. Gas flowmeters generally operate in the range from about 40 to 200 kHz. At frequencies higher than 200 kHz, absorption losses in the gas are very high whereas sounds below 40 kHz are audible and can be irritating.

Ultrasonic spirometers utilize a pair of ultrasonic transducers mounted on opposite sides of a flow tube (Fig. 13.5a). The transducers are capable of both transmitting and receiving ultrasonic pulses.

In conventional ultrasonic flowmeters, pulses are transmitted through the liquid or gas in the flow tube, against and then with the direction of flow. The pulse transit time upstream, t_1, and downstream, t_2, can be expressed:

$$t_1 = \frac{D}{C - v'} \quad \text{and} \quad t_2 = \frac{D}{C + v'}$$

where D is the distance between the transducers, C is the velocity of sound propagation in the fluid and v' is the fluid velocity vector along the path of the pulses. The average gas velocity v through the flow tube is a vector of v' so that

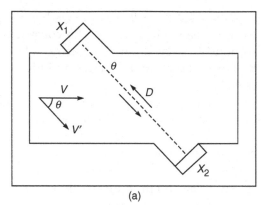

(a)

> **Fig. 13.5(a)**   *Diagram of flow tube and the position of ultrasonic transducers used in transit-time based ultrasonic spirogram*

$$v = \bar{v} \, \cos \theta$$

A frequency (f) is usually measured, which is the reciprocal of the transit times:

$$\frac{1}{t_2} - \frac{1}{t_1} = f_2 - f_1 = \frac{C + v'}{D} - \frac{C - v'}{D} = \frac{2v'}{D} = \frac{2\bar{v} \cos \theta}{D}$$

The flow velocity is:

$$\bar{v} = \frac{D}{2 \cos \theta} \left[\frac{1}{t_2} - \frac{1}{t_1} \right] = \frac{D}{2 \cos \theta} [f_2 - f_1]$$

The velocity of sound, C, does not appear in the final equation. Thus, the output accuracy is unaffected by fluid density, temperature, or viscosity. In gas flow measurements, pulmonary function tubes larger than 3 cm in diameter must be used; the single frequency systems that measure time delay directly must be able to resolve nanoseconds since the total transit delay, t, is usually measured in microseconds. This technique is not easily implemented because of the difficulty in measuring these small time differences.

In these flowmeters, disc type flat transducers are mounted in recessed wells on opposite sides of the flow tube at an angle to the flow as shown in Fig. 13.5(a). This arrangement avoids flow disruption and provides relative immunity to transducers from contaminations by flowing substances. Acoustically transparent windows are employed in liquid flowmeters to protect transducer surfaces and to improve impedance matching to the medium. For measurement of respiratory gas flows, however, the medium must be in close contact with the transducer to achieve good acoustic transmission. Consequently, the recess has to be kept open which can lead to unwanted turbulence and moisture collection.

The open tubular wells in which the crystals are mounted, by their geometry, are subject to a fluid accumulation, which interferes with or obliterates coherent ultrasonic transmission. The dead space associated with them is also undesirable. In addition to the condensation problem, the zero flow base line signal shows drift and is found to show oscillations with changes in temperature. The situation is worse when the transducers are applied to patients undergoing ventilatory assistance as highly inaccurate signals are likely to be obtained due to fluid accumulation. Blumenfeld *et al* (1975) modified the construction and designed a coaxial ultrasonic pneumotachometer. In their design (Fig. 13.5(b)), the crystals are mounted midstream in the line of flow, with their transmission axis on the centre line of the tubular housing. The principle of measurement of flow is that of measuring transit times, which is a function of the linear gas velocity and hence of flow.

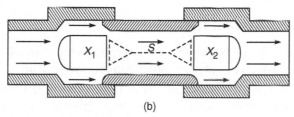

(b)

➤ **Fig. 13.5(b)** *Coaxial ultrasound pneumotachometer (after Blumenfeld et al, 1975; by Permission of Med. & Biol. Eng.)*

Although, in theory, the technique seems simple, it is fraught with several sources of error. For example, a reflected ultrasonic transmission when combined with the primary transmission, may produce artefacts of significant magnitude. Similarly, there could be alterations in the effective joint path length of flow and ultrasound transmission as a function of gas velocity and composition. The method requires further experimental work to minimize such problems. Furthermore, use of this transducer for the estimation of flow in gases of varying composition, temperature and humidity requires a correction for the corresponding variation of C (ultrasound velocity).

Plaut and Webster (1980 a) have attempted to construct an ultrasonic pneumotachometer which possesses the desirable characteristics of open tube geometry but avoids problems of flow disruption, fluid collection, reduced sensitivity and dead space which are present in the cross-the-stream and coaxial transducer configurations. They used cylindrical shell transducers with their inner surfaces flushed with the walls of the flow tube to prevent flow disruption and fluid accumulation. The axis of transmission of the ultrasound being parallel to the flow axis enhances the sensitivity of the device. Figure 13.5(c) shows the configuration of the cylindrical shell transducers and the airway. A circuit for measuring phase shift as a measure of

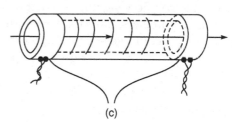

(c)

➤ **Fig. 13.5(c)** *The configuration of ultrasound pneumotachometer showing the cylindrical shell transducer and the airway (after Plaut and Webster, 1980; by Permission of IEEE Trans. Biomed. Eng.)*

transit time change due to flow is illustrated by Plaut and Webster (1980 b). They conclude that the design described by them can give accurate quantitative information when gas composition and temperature are limited to a narrow range of conditions. It can also measure qualitative parameters associated with respiration while presenting little obstruction to breathing. They however caution that poor understanding of the nature of the ultrasonic field and how it inter-acts with moving gas remains the most troublesome problem for the successful development of ultrasonic pneumotacho-meters. The other problems are the poor acoustic efficiency of ultrasonic trans-mission through gases, the wide variation in gas composition, temperature and humidity and the need for high accuracy and wide dynamic range.

ⅢⅢ▶ 13.3 PNEUMOTACHOMETERS

Pneumotachometers are devices that measure the instantaneous rate of volume flow of respired gases. Basically, there are two types of pneumotachometers, which are:

(i) *Differential manometer*—It has a small resistance, which allows flow but causes a pressure drop. This change is measured by a differential pressure transducer, which outputs a signal proportional to the flow according to the Poiseuille law, assuming that the flow is laminar. The unit is heated to maintain it at 37°C to prevent condensation of water vapour from the expired breath.

(ii) *Hot–wire anemometer*—It uses a small heated element in the pathway of the gas flow. The current needed to maintain the element at a constant temperature is measured and it increases proportionally to the gas flow that cools the element.

Pneumotachometer is commonly used to measure parameters pertaining to pulmonary function such as forced expiratory volume (FEV), maximum mid-expiratory volume, peak flow and to generate flow-volume loops. Although these devices directly measure only volume flow, they can be employed to derive absolute volume changes of the lung (spirometry) by electronically integrating the flow signal. Conventional mechanical spirometers, though more accurate than pneumotachometers, have limitations due to their mechanical inertia, hysteresis and CO_2 build-up. Pneumotachometers, on the other hand, are relatively non-obstructive to the patient and this makes them suitable for long-term monitoring of patients with respiratory difficulties. A basic requirement of pneumotachometers (PTM) is that they should present a minimum resistance to breathing. An acceptable resistance would be between 0.5 and 1.0 cm H_2O s/l. The pressure drop across the flow head at peak flow is also indicative of PTM resistance. Fleisch PTMs normally have a peak flow pressure drop of around 1.5 cm H_2O. Normal respiratory phenomenon has significant frequency components up to only 10 Hz and devices with this response should be quite suitable for most applications. More often, it is not the frequency response but the response time, which is generally specified. The response time of a typical ultrasonic spirometer is 25 ms. The dead space volume of the flow head should be as small as possible. A bias flow into the flow head is sometimes introduced to prevent rebreathing of expired air. A good zero stability is a prerequisite of PTMs to prevent false integration during volume measurements. The popular types of pneumotachometers are explained below.

13.3.1 Fleisch Pneumotachometer

Flow transducers generally used in respiratory studies are the Fleisch-type pneumotachometers. These transducers are made by rolling a sheet of thin, corrugated metal with a plain strip of metal and inserting this core within a metal cover (Fig. 13.6). Thus, these transducers are resistance elements consisting of small, parallel metal channels. This construction helps to maintain a laminar flow at much higher flow rates than would be possible for a gauze of similar area. In case of laminar flow, the pressure drop across the element is directly proportional to the flow rate of a gas passing through it.

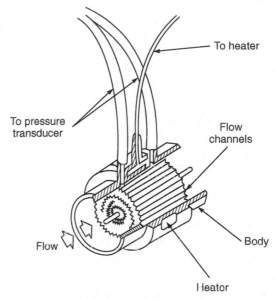

The output of the flow transducer appears as a differential pressure. To convert this pressure into an electrical signal, a second transducer is required. A capacitance type pressure transducer is used in such applications. They are more stable and less vibration-sensitive than resistive or inductive type transducers.

> **Fig. 13.6** *Construction of pneumotach transducer (Courtesy: Hewlett Packard, USA)*

At high flow rates, turbulence develops in the hose leading to the pneumotach and its response tends to become non-linear. This limits the usable range of the transducer. The relationship between pressure drop and flow is given by $DP = AV + BV^2$, where the term BV^2 introduces the non-linearity effect. This non-linearity is generally corrected electronically.

According to Poisseuille's law, the pressure developed across a pneumotach by laminar gas flow is directly proportional to the gas viscosity. The viscosity and temperature coefficient of viscosity for a variety of respiratory gases (Blais and Fanton, 1979) are given in Table 13.2.

The viscosity (η) of a mixture of gases is approximated by the equation

$$\frac{1}{\sqrt{\eta}} = \frac{X_1}{\sqrt{\eta_1}} = \frac{X_2}{\sqrt{\eta_2}} = \frac{X_3}{\sqrt{\eta_3}} + \ldots\ldots$$

where X_1 is the fraction of gas having the viscosity η_1. This necessitates the application of an automatic correction factor to the flow rate for changes in viscosity.

Hobbes (1967) studied the effect of temperature on the performance of a Fleisch head and found that the output increased by 1% for each degree C rise. He also noted that the effect of saturating air at 37°C with water vapour was to reduce the output from the head by 1.2% as compared with dry air at the same temperature. The calibration of a pneumotachograph head in terms of volume flow rate can be done by passing known gas flows through it. The flow can be produced by a compressor and measured with a rotameter type gas flowmeter.

Most respiratory parameters are reported in BTPS conditions (body temperature, ambient pressure, saturated with water vapour). This is the condition of air in the lungs and the mouth. To

• Table 13.2   *Viscosity and Temperature Coefficient of Viscosity for Respiratory Gases*

Gas	Viscosity (Micropoises) at 25°C	Temperature coefficient (Micropoises/°C)
Air	183	0.47
Nitrogen	177.3	0.43
Oxygen	205	0.52
Carbon Dioxide	148.35	0.47
Helium	200.15	0.43
Water Vapour	162	–

prevent condensation and maintain the gas under these conditions, the temperature of the pneumotach is maintained at 37°C. The heater that warms the pneumotach is electrically isolated from the metal case for patient safety, and is encapsulated so that the entire unit may be immersed in liquid for sterilization. The thermistor that senses the temperature and controls the heater through a proportional controller, is buried in the metal case.

13.3.2   Venturi-type Pneumotachometer

This type works similarly to the Fleisch pneumotachometer, but have a venturi-throat for the linear resistance element. The resulting pressure drop is proportional to the square of volume flow. They have open geometry and therefore are less prone to problems of liquid collection. Their main disadvantages are the non-linearity of calibration and the requirement for laminar flow.

13.3.3   Turbine-type Pneumotachometer

In this design, air flowing through the transducer rotates a very low mass (0.02 g) turbine blade mounted on jewel bearings. Rotation of the turbine blade interrupts the light beam of a light-emitting diode (LED). The interrupted light beam falls on a phototransistor, which produces a train of pulses, which are processed and accumulated to correspond to an accumulated volume in litres.

A special feature of this transducer is a bias air flow, applied to the turbine blades from a pump. This flow keeps the blades in constant motion even without the sample flow through it. This allows measurement of sample air flow in the range of 3 to 600 l/min in the most linear range of the volume transducer, by overcoming much of the rotational inertia of the turbine. The 'ZERO' control of the volume transducer adjusts the bias air flow to produce a train of clock pulses of exactly the same frequency as those generated by the crystal oscillator. Figure 13.7 is a diagram of the transducer.

▶ 13.4   MEASUREMENT OF VOLUME

The volume of gas flowing into and out of the lungs is a factor of considerable importance in investigations of lung function. Whilst the volume of a single breath, or the total volume expired in

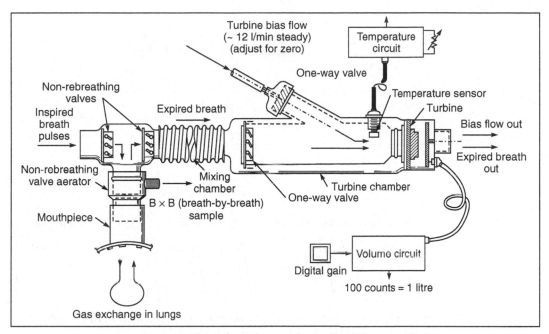

> ➤ **Fig. 13.7** *Turbine type volume transducer (Courtesy: Beckman Instruments Inc., USA)*

a given time, can be measured by continuously acting spirometers, continuous breath-by-breath measurements are often difficult. One method is to integrate the flow rate electronically and record the resulting signals. The flow rate is measured as the pressure change across a pneumotachograph head with a micromanometer whose output is a voltage proportional to the pressure difference at the manometer input, i.e.

$$V_1 = K(P_1 - P_2)$$

where K is a constant.

The output from the integrator is a voltage V_0 such that

$$V_0 = \frac{1}{RC} \int_{t_1}^{t_2} V_i \, dt$$

A simplified integrator set up is shown in Fig. 13.8 for flow and volume measurement. It consists of an 'autozero' flowmeter together with a threshold detector and an integrator. The threshold detector selects which portion of the flow signal is to be integrated and this is normally set to switch on when the flow signal moves past zero in a positive direction, and off again when the flow signal returns to zero. This means either inspiration or expiration can be measured depending on how the flowhead is connected.

When it is intended to measure tidal volume, the flow signal moves positive and continues until the flow output returns to zero, when the integrator output is reset. The display shows the size of each breath which is referred to as constant baseline. In case cumulative volume is to be measured the volume displayed after each breath is held up. The next breath integrated is added

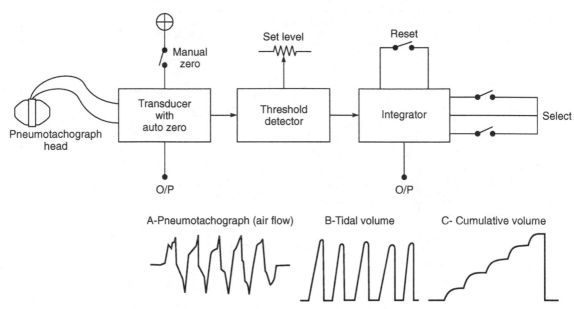

> **Fig.** 13.8   *Schematic diagram of electrospirometer for measurement of flow and volume (Courtesy: Mercury Electronics, Scotland)*

to its predecessor, thus producing a staircase pattern. The pattern can be recorded on a chart paper.

Unless extremely high quality amplifiers are used, the integrator circuit will drift and give false readings. Drift can be minimized by starting each volume from a fixed baseline on the record.

3.4.1   Flow-Volume Curve

This is a plot of instantaneous maximum expiratory flow rate versus volume. The shape of the flow-volume curve does not vary much between normal subjects of different age, size and sex, although absolute values of flow rate and volume may vary considerably. In patients with obstructive airway disease, the shape of this curve is drastically altered. For this reason, the flow-volume curve is a good early indication of abnormality. Typical MEFV curves are shown in Fig. 13.9.

There are various methods of producing the flow-volume curve. The method which has been very common in the past was to record it on the storage oscilloscope and then make a permanent record by photographing it with a polaroid camera. This procedure, obviously, is time consuming and expensive.

General purpose X-Y recorders are not fast enough to follow the rapid changes encountered in the signals while recording flow-volume curves. Therefore, special recorders have been designed to meet this requirement. For example, in the H.P. Pulmonary Function Analyzer, the recorder used has an acceleration of 76.2 m/s^2 and a slewing speed of over 0.762 m/s will result in approximately a 7.5 cm deflection. The recorder will thus be able to accurately plot a MEFV curve in which the subject reaches a peak flow of 10 l/s in less than one-tenth of a second.

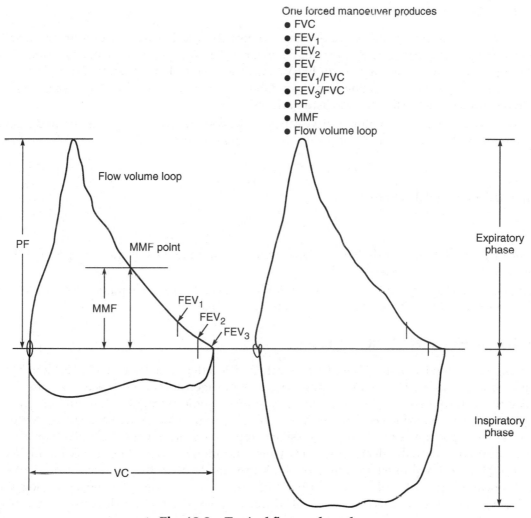

One forced manoeuver produces
- FVC
- FEV$_1$
- FEV$_2$
- FEV
- FEV$_1$/FVC
- FEV$_3$/FVC
- PF
- MMF
- Flow volume loop

Flow volume loop

PF

MMF point

MMF

FEV$_1$

FEV$_2$

FEV$_3$

VC

Expiratory phase

Inspiratory phase

> **Fig. 13.9** *Typical flow-volume loops*

A plot of the inspiratory flow-volume curve is also found to be useful in the detection of certain lung abnormalities, though it does not yield as much information about the lung mechanics as does the expiratory flow-volume curve. A useful indicator of the relative degrees of inspiratory and expiratory obstruction is the $MEF_{50\%}/MIF_{50\%}$ ratio (Jordanoglou and Pride, 1968) $MIF_{50\%}$ is maximum inspiratory flow at 50% of vital capacity.

A microcomputer is incorporated in the modern equipment to calculate the maximum spirometer value stored; FVC, the FEVI and the ratio FEVI/FVC. Besides these, some other indices were also evaluated. For example, the average flow over the middle portion of the spirogram has been the most widely accepted parameter for the early detection of increased airway resistance. A microcomputer based system facilitates automating many of such indices which are under investigation.

13.4.2   Area of the Flow-Volume

The area under the maximum expiratory flow-volume curve (AFV) is a sensitive indicator of lung function impairment. This is because within the scope of a single measurement, maximum flow at all lung volumes, the vital capacity and FEV_1 are all accessible. It has been found that AFV is the only variable capable of detecting significant inter and intra-individual differences in respect of both the immediate and the delayed asthmatic responses.

The area under the flow-volume curve can be easily computed by using a square-and-integrating circuit. In the derivation of area, the following equation is used:

$$A = \int_0^{V_T} F \, dV$$

where $dV = F \, dt$ and therefore,

$$A = \int_0^T F^2 \, dt$$

where V_T = total volume exhaled during the time. T is a time greater than the period taken to complete the respiratory manoeuvre.

A multiplier module like integrated circuit FM 1551 can be used to square the flow rate signal and then integrate it with respect to time to obtain the area.

13.4.3   Nitrogen Washout Technique

Nitrogen washout technique is employed for the indirect determination of RV, FRC and TLC. In this technique, the subject breathes 100% oxygen. A nitrogen analyzer is placed near the mouthpiece to continuously monitor the nitrogen content. Some nitrogen is eliminated on every expiration, but none is inhaled. The analyzer records nitrogen content which decreases with each successive expiration since it is progressively replaced with oxygen. The alveolar nitrogen concentration eventually decreases to 1% when an almost steady state is reached. Nitrogen washout curves are plotted with time on the X-axis and % N_2 in the expired air on the Y-axis. Theoretically, a plot of % N_2 versus expired volume will be a straight line on a semi-log paper. However, with the presence of anatomical dead space, the relationship between nitrogen concentration and expired volume is no longer exponential but is dependent upon the dead space/tidal volume ratio. To compensate for this, the dead space is automatically measured in the analyzers at the start of the test and subtracted from each breath during the test, thus yielding a washout recording unaffected by the patients breathing pattern. A typical complete multi-breath nitrogen washout test would take about 10 min. with modern instruments.

The single breath nitrogen washout test is another index of alveolar ventilation in addition to providing closing volume information. The test is performed with the subject exhaling to residual volume, making a maximal inspiration of 100% oxygen and exhaling his vital capacity slowly. Nitrogen concentration is plotted versus volume during the expiration to yield a curve as shown in Fig. 13.10. Information on different parameters as obtained in this test is shown in the diagram. For example, closing volume is an important parameter which gives an early indication of small airway disease. Closing volume increases with age since the lung loses its elastic recoil and more

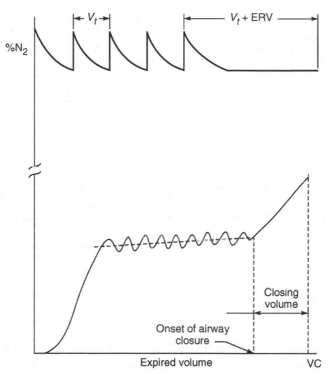

> **Fig. 13.10** *Single-breath N_2 washout curve (Courtesy: Hewlett Packard, USA)*

and more airways are closed at higher lung volumes. Closing volume is that volume exhaled from the onset of closure to the end of vital capacity. It is the first permanent upsloping departure of the curve from the straight line.

Computational techniques have been developed to eliminate the main drawbacks of traditional multiple breath nitrogen washout methods. It reduces the time required to deliver an accurate nitrogen washout test and the results are available at the same instant the subject completes this breathing manoeuvre.

13.5 PULMONARY FUNCTION ANALYZERS

A complete pulmonary function analyzer contains all the equipment necessary for testing various parameters. It comprises a nitrogen analyzer, a vacuum pump, an X-Y recorder, pneumotachs, a digital display, plumbing and valves and other electronic circuits. A simplified block diagram of the system is shown in Fig. 13.11. Modern instruments are designed to completely automate the measurements of ventilation, distribution and diffusion. The systems are designed around computers which control the procedures by opening and closing appropriate valves, measuring flow rates and the concentrations of various gases, and calculating and printing the results. An analog-to-digital converter supplies the measurement data to the computer. Inputs to the A-D

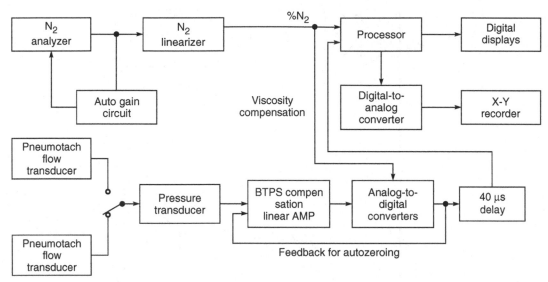

> **Fig.** 13.11   *Block diagram of pulmonary function analyzer (Courtesy: Hewlett Packard, USA)*

converter are from various measurement devices, which include a pneumotach that provides a signal proportional to the air flow for various measurements and carbon monoxide and helium analyzers for diffusion measurements. The software controlled pulmonary measurement procedure allows new programs to be added or existing ones to be modified.

Figure 13.12 is a typical example of a microprocessor based pulmonary function testing system. It makes use of a modified 'Fleisch-tube' and replaces the capillary system with a sieve which serves as a flow transducer. This enables it to achieve a large range of linearity from 0-15 l/s. This offers better precision for the recording of the flow volume loop in the range of high flows as well as of low flows. The flow proportional pressure drop over a calibrated screen resistor is led to a pressure transducer via two silicon hoses. The pneumatic value is converted into an electrical signal and fed into an amplifier whose output represents 'Flow'. The signal is integrated to volume by a voltage to frequency converter. Both signals, 'flow' and 'volume' are converted into digital form in an A-D converter and given to the processing unit.

Processing, evaluation and representation of the data is carried out by an 8-bit microprocessor. The program memory capacity amounts to a maximum of 46 k byte resident in the computer; out of this 16 k bytes are used for the software system and 30 k bytes for the measuring and organization programmes. For the intermediate storage of data, the 12 k bytes RAM is utilized. The input of characters and numbers, mainly of patient data, is done through a keyboard. The system communication is enabled by a 20-digit alphanumerical display line. A 20-digit alphanumerical printer-plotter produces a hard copy of output values. Forced expiratory curves can be presented in the form of histograms also.

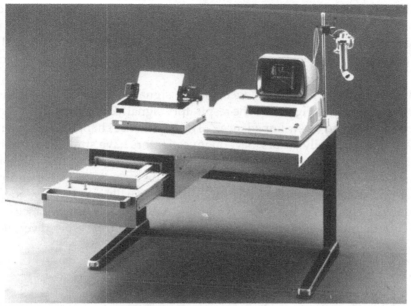

> **Fig.** 13.12   *Microprocessor based pulmonary function testing system (Courtesy: Erich Jaeger, W. Germany)*

13.5.1   Impedance Pneumograph

Impedance pneumography is an indirect technique for the measurement of respiration. It measures the respiratory volume and rate through the relationship between respiratory depth and thoracic impedance change. Impedance pneumography avoids encumbering the subject with masks, tubes, flowmeters or spirometers; does not impede respiration, and has a minimal a effect on the psychological state of the subject. Also, it can provide a continuous volumetric record of respiration. During measurement of the thoracic impedance change signal, an ac excitation is applied to the subject. Choice of the optimum frequency for recording transthoracic impedance changes accompanying respiration is dictated by two important considerations. The first is that of excitability of the various tissues between the electrodes and the second is the nature of other recordings made at the same time, notably those of bioelectric origin like the ECG. The excitation frequency employed is usually in the range of 50–100 kHz, with an amplitude of the order of one milliampere peak-to-peak and in terms of power, it is less than one milliwatt. The excitation is too high in frequency and too low in amplitude to stimulate the tissues. The signals of this frequency get attenuated in almost all biological amplifiers whose frequency response is limited to well below 10 kHz. Also, a natural rejection of bioelectric events occurs when frequencies in the tens of kilocycles range are used for the detection of impedance changes with respiration. Therefore, a tuned amplifier can be used which has practically zero response for the spectrum of bioelectric events while at the same time provides high amplification for the carrier used for impedance pneumography.

Changes in transthoracic impedance during respiration should be independent of impedance resulting from the sum of the resting thoracic impedance and the contact impedance between the electrodes and the skin. This is possible if a constant current is maintained through the subject over a large range of thoracic and electrode-skin impedance, likely to be encountered in practice.

The thoracic impedance and impedance change signal may be viewed as either polar or cartesian vectors as shown in Fig. 13.13. The respiratory signal ΔZ is shown as the change between an initial impedance Z_0 and a new impedance Z. The impedance change ΔZ is drawn much larger on the diagram than it usually occurs. Several investigators have studied the changes in transthoracic electrical impedance associated with respiration after applying low-intensity sinusoidal currents from 20 to 600 kHz. ΔZ has been shown to be essentially a change in the resistive component, virtually independent of frequency within the usually employed frequency range and correlates well with changes in the respired volume (ΔV) so that the impedance changes can be used, within certain limits of accuracy, as a quantitative measure of respired volume.

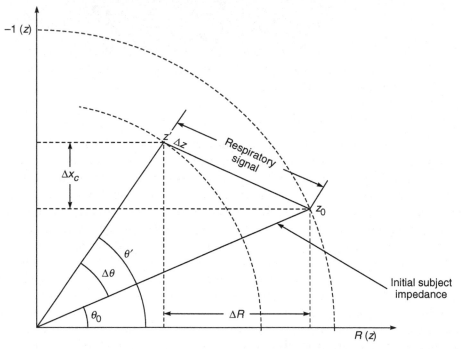

> **Fig. 13.13** *Change in impedance (ΔZ) expressed in terms of polar or cartesian vectors*

The transthoracic impedance is a function of frequency and the type and size of electrodes. Silver/silver chloride electrodes with a diameter of 9.5 mm have a typical impedance of 500–800 Ω at 50 kHz, whereas 4 mm diameter electrodes exhibit 1000–1500 Ω at the same frequency. Typically, the impedance change signal is of the order of 3 Ω/l of respiratory volume change.

Baker *et al* (1966) report that the maximum change in transthoracic impedance per unit volume of respired air *(ΔZIΔV)* was recorded from electrodes secured to the eighth rib bilaterally on the midaxillary lines. The magnitude of *dZld* V decreased rapidly with distance along the midaxillary line in both directions from the eighth rib. Also, for a given electrode location, the ratio *ΔZIΔV* was found to be essentially the same whether the lungs were inflated by spontaneous respiration or inflated by a positive pressure respirator after relaxation of spontaneous efforts following hyperventilation.

There are disadvantages in the use of the impedance pneumograph if absolute measurements of respiration are required. This is because the conversion of impedance change to lung-volume change is a function of electrode position, body size and posture.

Miyamoto *et al* (1981) studied the reliability of the impedance pneumogram as a method to assess ventilation during exercise in comparison with the standard pneumatic method. It is found that a correlation coefficient of 0.92 to 0.99 exists between tidal volume changes and impedance variation. However, it was observed that the accuracy of the impedance method is still inferior to the standard pneumatic methods. Individual as well as regional variations of the impedance sensitivity also require time consuming calibration procedures for each subject.

13.6 RESPIRATORY GAS ANALYZERS

A knowledge of the qualitative and quantitative composition of inspired and expired gas and vapour mixtures is of great importance in investigations connected with respiratory physiology, lung function assessment and anaesthesia. A number of physical methods have been utilized as the basis for a gas analyzer. Commercially available analyzers depend for their action on the measurement of quantities such as infrared or ultraviolet absorption, paramagnetism, thermal conductivity or the ratio of charge to mass of ionized molecules. There are analyzers, designed primarily for a analysis of a single component of the gas mixture whereas there are instruments such as the mass spectrometer and gas chromatograph which provide a multi-component analysis. The most common gases of interest for measurement and analysis are carbon dioxide, carbon monoxide, nitrous oxide and halothane in mixtures of respiratory or anaesthetic gas samples.

13.6.1 Infrared Gas Analysers

Infrared gas analyzers depend for their operation upon the fact that some gases and vapours absorb specific wavelengths of infrared radiation. One of the most commonly measured gas using the infrared radiation absorption method is carbon dioxide. The technique used for this purpose is the conventional double-beam infrared spectrometer system having a pair of matched gas cells in the two beams. One cell is filled with a reference gas which is a non-absorbing gas like nitrogen whereas the measuring cell contains the sample. The difference in optical absorption detected between the two cells is a measure of the absorption of the sample at a particular wavelength. Commercial gas analyzers used for measuring the amount of carbon dioxide (CO_2), carbon monoxide (CO), nitrous oxide (N_2O) or halothane in mixtures of respiratory and anaesthetic gases make use of a non-dispersive infrared (NDIR) analysis technique. A major advantage of this technique is that it is highly specific to the gas being measured and, therefore, separate pick-up heads selective to the wavelengths absorbed by the gas are necessary.

Infrared gas analyzers are particularly useful for measuring carbon dioxide in the respired air. For the measurement of gases in the respired air two types of samplers are employed—a micro catheter-cell and a breathe-through cell. The micro-catheter cell is used with a vacuum pump to draw off small volumes from the nasal cavity or trachea. Its typical volume is 0.1 ml and it is particularly useful when larger volumes could cause patient distress. The breathe-through cell accepts the entire tidal volume of breath with no vacuum assistance. It can be connected directly into the circuit of an anaesthesia machine. These instruments have a typical response time of 0.1 s and a sensitivity range of 0 to 10% CO_2.

The block diagram of an infrared gas analyzer is shown in Fig. 13.14. The solid state detector is PbSe. The chopper has a high speed of 3000 rpm and provides a response time of up to 100 ms for 90% reading. The infrared source operates at a temperature of about 815°C where it emits infrared energy optimized for the spectral bands of interest. The infrared energy source is located at the front focal plane of a parabolic reflector, so that the reflected energy from the reflector is effectively collimated. The collimated energy is chopped by the coaxial chopper, which allows the energy to pass alternately through the reference and sample tubes. Since the energy is collimated, it passes through these tubes without internal reflections so that gold foil coatings on the inside of these tubes are not necessary. The sample tube length can be selected according to the absorption strength and concentration of the sample gas. At the output end of the two tubes, a second parabolic reflector images the energy onto the detector filter assembly. The filter is a narrow bandpass interference filter with band pass characteristics matched to the absorption spectra of the gas of interest.

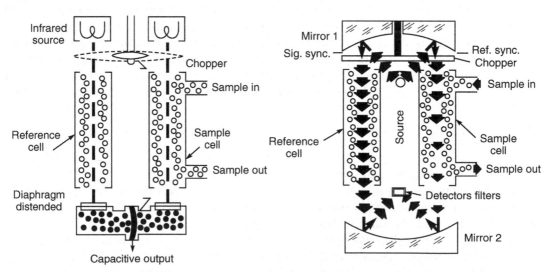

> **Fig. 13.14** *Principle of infrared gas analyzer: conventional (two sources) and improved design (with single source) (Courtesy: Infrared Industries, USA)*

13.6.2   Paramagnetic Oxygen Analyser

Oxygen has the property of being paramagnetic in nature, i.e. it does not have as strong a magnetism as permanent magnets, but at the same time it is attracted into a magnetic field. Nitric oxide and nitrogen dioxide are two other gases which are paramagnetic in nature. Most gases are, however, slightly diamagnetic, i.e. they are repelled out of a magnetic field. The magnetic susceptibility of oxygen can be regarded as a measure of the tendency of an oxygen molecule to become temporarily magnetized when placed in a magnetic field. Such magnetization is analogous to that of a piece of soft iron in a field of this type. Similarly, diamagnetic gases are comparable to non-magnetic substances. The paramagnetic property of oxygen has been utilized in constructing oxygen analyzers.

The paramagnetic oxygen analyzer was first described by Pauling *et al* (1946). Their simple dumb-bell type of instrument has formed the basis of more modern instruments. The arrangement incorporates a small glass dumb-bell suspended from a quartz thread between the poles of a permanent magnet. The pole pieces are wedge-shaped in order to produce a non-uniform field. Referring to Fig. 13.15, when a small sphere is suspended in a strong non-uniform magnetic field, it is subject to a force proportional to the difference between the magnetic susceptibility of this sphere and that of the surrounding gas. The magnitude of this force can be expressed as:

$$F = C(K - K_o)$$

where C = a function of the magnetic field strength and gradient

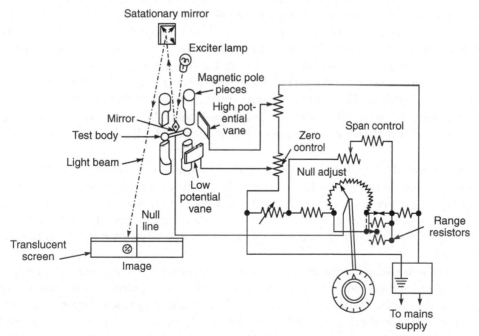

> **Fig. 13.15**   *Functional diagram of Oxygen analyzer based on magnetic susceptibility (Courtesy: Beckman Instruments Inc., USA)*

K_0 = magnetic susceptibility of the sphere

K = magnetic susceptibility of the surrounding gas

The forces exerted on the two spheres of the test body are thus a measure of the magnetic susceptibility of the sample and therefore, of its oxygen content.

The magnetic forces are measured by applying to one sphere an electrostatic force equal and opposite to the magnetic forces. The electrostatic force is exerted by an electrostatic field established by two charged vanes mounted adjacent to the sphere. One vane is held at a higher potential than the test body, the other at a lower potential. Since the glass test body must be electrically conductive, it is sputtered with an inert metal.

The test body is connected electrically to the slider of the 'Null Adjust' potentiometer R_{20}. This potentiometer is part of a voltage-dividing resistor network connected between ground and B+. Potential to the test body can be adjusted over a large range. An exciter lamp directs a light beam on to the small mirror attached to the test body. From the mirror, the beam is reflected to a stationary mirror and then on to a transluscent screen mounted on the front panel of the instrument. The geometry of the optical system is so arranged that a very small rotation of the test body causes an appreciable deflection of the image cast by the beam.

Zero control of the instrument is provided by ganged R_{13}–R_{15} setting, which changes the voltage present on each vane with respect to the ground, but does not change the difference in potential existing between them. This adjustment alters the electrostatic field. Rheostat R_{19} sets the upscale standardization point, i.e. provides span or sensitivity control. When no oxygen is present, the magnetic forces exactly balance the torque of the fibre. However, if oxygen is present in the gas sample drawn in the chamber surrounding the dumb-bell, it would displace the dumb-bell spheres and they would move away from the region of maximum magnetic flux density. The resulting rotation of the suspension turns the small mirror and deflects the beam of light over a scale of the instrument. The scale is calibrated in percentages by volume of oxygen or partial pressure of oxygen. Paramagnetic oxygen analyzers are capable of sampling static or flowing gas samples. The null position is detected by a sensitive light, mirror and scale arrangement.

Oxygen analyzers are available with continuous readout 0–25% or 0–100% oxygen. The instruments are calibrated with the reference gas specified. Standard cell volume is 10 ml and response time is about 10 s.

The recommended flow rate is 40 to 250 cc/min when the sample enters the analysis cell. Before the gas enters the analyzer, it must be pressurized with the pump and passed through a suitable cleaning and drying system. In many cases, a small plug of glass wool is sufficient for cleaning and drying functions. The entry of moisture or particulate matter into the analyzer will change instrument response characteristics. Therefore, the use of a suitable filter in the sample inlet line is recommended.

Any change in the temperature of a gas causes a corresponding change in its magnetic susceptibility. To hold this temperature of the gas in the analysis cell constant, the analyzer incorporates a thermostatically-controlled heating circuit. Once the instrument reaches temperature equilibrium, the temperature inside the analysis cell is approximately 60°C. The sample should be admitted to the instrument at a temperature between 23°C and 43°C. If the temperature is less than 10°C, then the sample may not have time to reach the temperature equilibrium before entering the analysis cell.

Calibration of the instrument consists of establishing two standardization points, i.e. a down scale and an up scale standardization point. These two points can be set by passing standard gases through the instrument at a fixed pressure, normally atmospheric pressure. First a zero standard gas is admitted and the zero control is adjusted. Then the span gas is admitted and the span control is adjusted. Alternatively, the required practical pressures of oxygen are obtained by filling the analysis cell to the appropriate pressures with non-flowing oxygen or air. If the highest point is not greater than 21% oxygen, then dry air is used to set the span point. If the point is greater than 21% oxygen, the oxygen is used to set this point.

13.6.3 Polarographic Oxygen Analyzer

Polarographic cells are generally used to measure the partial pressure or percentage of oxygen from injected samples, continuous streams or in static gas monitoring. They find maximum utility in the respiratory and metabolic laboratories. Polarographic cells are based on the redox reactions in a cell having both the electrodes of noble metals. When a potential is applied, oxygen is reduced at the cathode in the presence of KCI as the electrolyte and a current will flow. The cathode is protected by an oxygen permeable membrane, the rate at which oxygen reaches the cathode will be controlled by diffusion through the membrane. The voltage current curve will be a typical polarogram (Fig. 13.16). A residual current flows in the cell at low voltages. The current rises with the increase in voltage until it

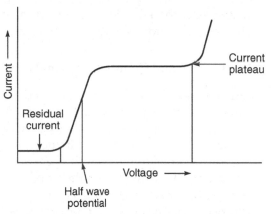

> **Fig. 13.16** *Typical polarogram showing relationship of current generated in the polarographic cell as a function of voltage*

reaches a plateau where it is limited by the diffusion rate of oxygen through the membrane. For a given membrane and at a constant temperature, this would be proportional to the partial pressure of oxygen across the membrane. When the voltage is applied in the plateau region, the current in the cell is proportional to oxygen concentration.

Polarographic cells are temperature sensitive as the diffusion coefficient changes with temperature. The temperature coefficient is usually 2–4% per degree centigrade. Therefore, temperature compensation circuits are used to overcome this problem.

Polarographic oxygen cells are used mainly for portable gas detectors, where simplicity, low cost and light weight are important. They are preferably used for measuring oxygen in liquids, especially in water pollution and medical work. The commercial oxygen analyzers use oxygen sensors, which contain a gold cathode, a silver anode, a potassium chloride electrolyte gel and a thin membrane. The membrane is precisely retained across the exposed face of the gold cathode, compressing the electrolyte gel beneath into a thin film. The membrane, permeable to oxygen, prevents airborne solid or liquid contaminants from reaching the electrolyte gel. The sensor is insensitive to other common gases. A small electrical potential (750 mV) is applied across the anode and cathode.

Although the composition of the atmosphere is remarkably constant from sea level to the highest mountain, i.e. oxygen, 21% and nitrogen 79%, there is a great difference in the partial pressure of oxygen at different altitudes. The polarographic sensor, which actually senses partial pressure, would, therefore, require some adjustment to read correctly the approximate percentage of oxygen at the altitude at which it is used. Humidity can also affect oxygen readings, but to a lesser degree. Water vapour in air creates a water vapour partial pressure that slightly lowers the oxygen partial pressure. Therefore, for precision work, it is often desirable to use a drying tube on the inlet sample line. Also, care should be taken to calibrate and sample under the same flow conditions as required for the gas to be analyzed. The range of the instrument is 0–1000 mmHg O_2 and the response time is 10 s for 90%, 35s for 99% and 70s for 99.9%. The instrument can measure oxygen against a background of nitrogen, helium, neon, argon etc. with no difficulty. The sensor is very slightly sensitive to carbon dioxide and nitrous oxide with typical error less than 0.1% oxygen for 10% carbon dioxide and 4% oxygen for 100% nitrous oxide.

13.6.4   Thermal Conductivity Analysers

Thermal conductivity of a gas is defined as the quantity of heat (in calories) transferred in unit time (seconds) to a gas between two surfaces 1 cm^2 in area and 1 cm apart, when the temperature difference between the surfaces is 1°C. The ability to conduct heat is possessed by all gases but in varying degrees. This difference in thermal conductivity can be employed to determine quantitatively the composition of complex gas mixtures. Changes in the composition of a gas stream may give rise to a significant alteration in the thermal conductivity of the stream. This can be conveniently detected from the rise or fall in temperature of a heated filament placed in the path of the gas stream. The changes in temperature can be detected by using either a platinum filament (hot wire) or thermistors. In a typical hot-wire cell thermal conductivity analyzer, four platinum filaments (Fig. 13.17) are employed as heat-sensing elements. They are arranged in a constant current bridge circuit and each of them is placed in a separate cavity in a brass or stainless steel block. The block acts as a heat sink. The material used for the construction of filaments must have a high temperature coefficient of resistance. The material generally used for the purpose are tungsten, Kovar (alloy of Co, Ni and Fe) or platinum.

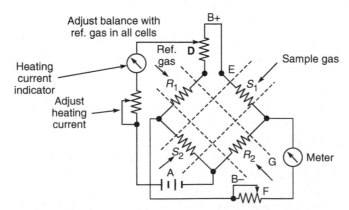

> **Fig. 13.17**   *Arrangement of a thermal conductivity based gas analyzer*

Two filaments (R_1 and R_2) connected in opposite arms of the Wheatstone bridge act as the reference arms, whereas the other two filaments (S_1 and S_2) are connected in the gas stream, and act as the measuring arms. The use of a four-cell arrangement serves to compensate for temperature and power supply variations.

Initially, the reference gas is made to flow through all the cells and the bridge is balanced precisely with the help of the potentiometer **D**. When the gas stream passes through the measuring pair of filaments, the wires are cooled and there is a corresponding change in the resistance of the filaments. The higher the thermal conductivity of the gas, the lower would be the resistance of the wire and vice-versa. Consequently, the greater the difference in thermal conductivities of the reference and sample gas, the greater the unbalance of the Wheatstone bridge. The unbalance current can be measured on an indicating meter or on a strip chart recorder.

Thermistors can also be used as heat sensing elements arranged in a similar manner as hot wire elements in a Wheatstone bridge configuration. Thermistors possess the advantage of being extremely sensitive to relatively minute changes in temperature and have a high negative temperature coefficient. When used in gas analyzers, they are encapsulated in glass. Thermistors are available which are fairly fast in response.

Thermal conductivity gas analyzers are inherently non-specific. Therefore, the simplest analysis occurs with binary gas mixtures. A thermal conductivity analyzer can be used in respiratory physiology studies to follow CO_2 concentration changes in the individual breaths of a patient. A high speed of response necessary for this purpose can be obtained by reducing the pressure of the gas surrounding the filaments to a few millimetres of mercury. The variations in the proportions of oxygen and nitrogen in the sample stream will have little effect, since they both have almost the same thermal conductivity. The effect of changes in water vapour content can be minimized by arranging to saturate the gas fed to both the sample and reference filaments.

13.6.5 N_2 Analyzer Based on Ionization Technique

With sufficient electrical excitation and at suitable pressures, certain gases emit radiation in different ways like spark, arc, glow discharge in different parts of the radiation spectrum. Measurement of the emitted radiation can help in the determination of the unknown concentration of a gas in a mixture. This technique has been utilized for the measurement of nitrogen gas, particularly in respiratory gases. The measuring technique is essentially that of a photo-spectrometer, wherein a gas sample is ionized, selectively filtered, and detected with a photocell, which provides an appropriate electrical output signal. The presence of nitrogen is detected by the emission of a characteristic purple colour when discharge takes place in a low pressure chamber containing the gas sample.

The instrument consists of a sampling head and a discharge tube. The sampling head contains the ionizing chamber, filter and the detector. The other part contains the power supply, amplifier and display system. The ionizing chamber or the discharge tube is maintained at an absolute pressure of a few torr. A rotary oil vacuum pump draws a sample and feeds it to the discharge tube. The voltage required for striking the discharge in the presence of nitrogen is of the order of 1500 V dc. The light output from the discharge tube is interrupted by means of a rotating slotted disc (Fig. 13.18) so that a chopped output is obtained. This light is then passed through optical filters having a

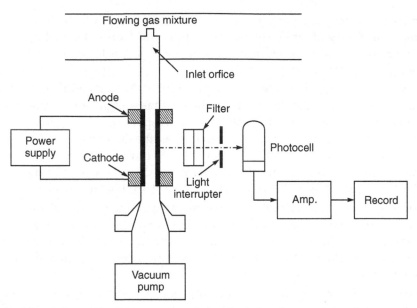

> **Fig. 13.18** *Schematic diagram showing principle of operation of nitrogen gas analyzer*

wavelength corresponding to the wavelength of the discharge. The intensity of light is measured with a photocell and an amplifier specifically tuned to the chopping frequency. The light intensity is proportional to the nitrogen concentration. The sampling rate is adjusted with the help of a needle valve, which is normally set at 3 ml/min. The vacuum system provides 600–1200 µHg. The instrument is calibrated for water saturated mixtures of nitrogen and oxygen as a reading error up to 2% can be expected with dry gases. Compensation for this error can be simply made by adjusting the sampling head needle valve, if it is desired to monitor dry gases.

Clinical Laboratory Instruments

IIID> 14.1 MEDICAL DIAGNOSIS WITH CHEMICAL TESTS

Most of the pathological processes result in chemical changes in the internal environment of the human body. These changes can generally be detected by the analysis of various samples taken from the body. The analysis not only helps in the diagnosis of various ailments but also in determining the progress of treatment and for making a prognosis. Samples taken from the body are analyzed in three different areas within the clinical laboratory set up. These are:

- *Chemistry* section deals with the analysis of blood, urine, cerebrospinal fluid (CSF) and other fluids to determine the quantity of various important substances they contain. Most of the electronic instruments in the clinical laboratory are available in this section.
- *Haematology* section deals with the determinations of the number and characteristics of the constituents of the blood, particularly the blood cells.
- *Microbiology* section in which studies are performed on various body tissues and fluids to determine the presence of pathological micro-organisms.

The most common substance for analysis from the body is blood. This is because the blood carries out the most important function of transportation and many pathological processes manifest themselves as demonstrable changes in the blood. Deviations from the normal composition of urine also reflect many pathological processes. The liquid part of the blood—the blood plasma, and the formed elements—the blood cells are analyzed during a chemical examination. The blood plasma accounts for about 60% of the blood volume and the blood cells occupy the other 40%. The plasma is obtained by centrifuging a blood sample. During centrifugation, the heavy blood cells get packed at the bottom of the centrifuge tube and the plasma can thus be separated. The plasma is a viscous, light yellow liquid, i.e. almost clear in the fasting stage.

IIID> 14.2 SPECTROPHOTOMETRY

Spectrophotometry is the most important of all the instrumental methods of analysis in clinical chemistry. This method is based on the absorption of electromagnetic radiation in the visible, ultraviolet and infrared ranges. According to the quantum theory, the energy states of an atom or molecule are defined and change from one state to another, would, therefore, require a definite amount of energy. If this energy is supplied from an external source of radiation, the exact quantity of energy required to bring about a change from one given state to another will be provided by photons of one particular frequency, which may thus be selectively absorbed. The study of the frequencies of the photons which are absorbed would thus provide information about the nature of the material. Also, the number of photons absorbed may provide information about the number of atoms or molecules of the material present in a particular state. It thus provides us with a method to have a qualitative and quantitative analysis of a substance.

Molecules possess three types of internal energy—electronic, vibrational and rotational. When a molecule absorbs radiant energy, it can increase its internal energy in a variety of ways. The various molecular energy states are quantized and the amount of energy necessary to cause any change in any one of the above energy states would generally correspond to specific regions of the electromagnetic spectrum. Electronic transitions correspond to the ultraviolet and visible regions, vibrational transitions to the near infrared and infrared regions and rotational transitions to the infrared and far-infrared regions. The method based on the absorption of radiation of a substance is known as *Absorption Spectroscopy*. The main advantages of spectrometric methods are speed, sensitivity to very small amounts of change and a relatively simple operational methodology. The time required for the actual measurement is very short and most of the analysis time, in fact, goes into preparation of the samples.

The region in the electromagnetic spectrum which is normally used in spectroscopic work is very limited. Visible light represents only a very small portion of the electromagnetic spectrum and generally covers a range from 380 to 780 mµ. The ultraviolet region extends from 185 mµ to the visible. Shorter wavelengths lie in the far ultraviolet region, which overlaps the soft X-ray part of the spectrum. Infrared region covers wavelengths above the visible range.

14.2.1 Interaction of Radiation with Matter

When a beam of radiant energy strikes the surface of a substance, the radiation interacts with the atoms and molecules of the substance. The radiation may be transmitted, absorbed, scattered or reflected, or it can excite fluorescence depending upon the properties of the substance. The interaction however does not involve a permanent transfer of energy.

The velocity at which radiation is propagated through a medium is less than its velocity in vacuum. It depends upon the kind and concentration of atoms, ions or molecules present in the medium. Figure 14.1 shows various possibilities which might result when a beam of radiation strikes a substance. These are:

(a) The radiation may be transmitted with little absorption taking place, and therefore, without much energy loss.

(b) The direction of propagation of the beam may be altered by reflection, refraction, diffraction or scattering.

(c) The radiant energy may be absorbed in part or entirely by the substance.

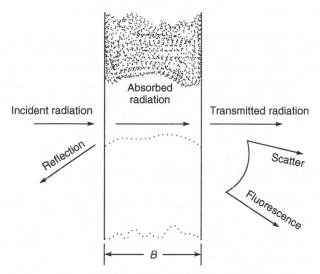

> **Fig. 14.1** *Interaction of radiation with matter*

In absorption spectrophotometry, we are usually concerned with absorption and transmission. Generally, the conditions under which the sample is examined are such that they keep reflection and scattering to a minimum.

Absorption spectrophotometry is based on the principle that the amount of absorption that occurs is dependent on the number of molecules present in the absorbing material. Therefore, the intensity of the radiation leaving the substance may be used as an indication of the concentration of the material.

Let us suppose, P_o is the incident radiant energy and P is the energy which is transmitted. The ratio of the radiant power transmitted by a sample to the radiant power incident on the sample is known as the transmittance.

$$\text{Transmittance } T = P/P_0$$
$$\% \text{Transmittance} = (P/P_0) \times 100$$

The logarithm to the base 10 of the reciprocal of the transmittance is known as absorbance.

$$\text{Absorbance} = \log_{10}(1/T)$$
$$= \log_{10}(P_0/P)$$
$$\text{Optical density} = \log_{10}(100/T)$$

According to the Beer-Lambert Law,

$$P = P_0 10^{-\varepsilon b c}$$

where c is the concentration, b the path length of light and ε the extinction coefficient. The concentration c of the substance is calculated from the absorbance, also called the extinction E, as follows:

$$\text{Absorbance } (E) = \log P_0/P = \varepsilon b c$$

In spectrophotometric measurements ε and b are nearly constant so that for a particular determination, absorbance (A) varies only with concentration (c). Therefore, if absorbance is plotted graphically against concentration, a straight line is obtained. A graph derived from the transmittance data will not be a straight line, unless transmittance (or per cent transmission) is plotted on the log axis of a semi-log paper. The constant a in the Beer Lambert law must have units corresponding to the units used for concentration and path length of the sample.

The most usually employed quantitative method consists of comparing the extent of absorption or transmittance of radiant energy at a particular wavelength by a solution of the test material and a series of standard solutions. It can be done with visual colour comparators, photometers or spectrophotometers.

▓▓▶ 14.3   SPECTROPHOTOMETER TYPE INSTRUMENTS

Figure 14.2 shows an arrangement of components of a typical spectrophotometer type instrument. The essential components are:

- A source of radiant energy, which may he a tungsten lamp, a xenon-mercury arc, hydrogen or deuterium discharge lamp, etc.
- Filtering arrangement for the selection of a narrow band of radiant energy. It could be a single wavelength absorption filter, an interference filter, a prism or a diffraction grating.
- An optical system for producing a parallel beam of filtered light for passage through an absorption cell (cuvette). The system may include lenses, mirrors, slits, diaphragm, etc.
- A detecting system for the measurement of unabsorbed radiant energy, which could be the human eye, a barrier-layer cell, phototube or photo-multiplier tube.
- A readout system or display, which may he an indicating meter or a numerical display.

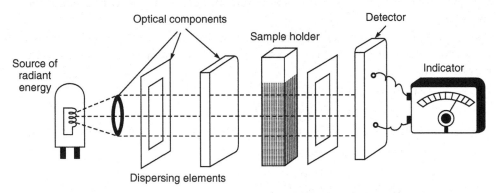

> **Fig. 14.2**   *Various components of a spectrophotometer type instrument*

14.3.1   Radiation Sources

The function of the radiation source is to provide a sufficient intensity of light which is suitable for making a measurement. The most common and convenient source of light is the tungsten lamp. This lamp consists of a tungsten filament enclosed in a glass envelope. It is cheap, intense and reliable. A major portion of the energy emitted by a tungsten lamp is in the visible region and only about 15 to 20% is in the infrared region. When using a tungsten lamp, it is desirable to use a heat absorbing filter between the lamp and the sample holder to absorb most of the infrared radiation without seriously diminishing energy at the desired wavelength. For work in the ultraviolet region, a hydrogen or deuterium discharge lamp is used.

In these lamps, the envelope material of the lamp puts a limit on the smallest wavelength which can be transmitted. For example, quartz is suitable only up to 200 mm and fused silica up to 185 mμ. The radiation from the discharge lamps is concentrated into narrow wavelength regions of emission lines. Practically, there is no emission beyond 400 mμ in these lamps. For this reason, spectrophotometers for both the visible and ultraviolet regions always have two light sources, which can be manually selected for appropriate work.

For work in the infrared region, a tungsten lamp may be used. However, due to high absorption of the glass envelope and the presence of unwanted emission in the visible range, tungsten lamps are not preferred. In such cases, nernst filaments or other sources of similar type are preferred. They are operated at lower temperatures and still radiate sufficient energy. For fluorescent work, an intense beam of ultraviolet light is required. This requirement is met by a xenon arc or a mercury vapour lamp. Cooling arrangement is very necessary when these types of lamps are used.

Mercury lamps are usually run direct from the ac power line via a series ballast choke. This method gives some inherent lamp power stabilization and automatically provides the necessary ionizing voltage. The ballast choke is physically small and a fast warm-up to the lamp operating temperature is obtained.

Modern instruments use a tungsten-halogen light source, which has a higher intensity output than the normal tungsten lamp in the change over region of 320–380 nm used in colorimetry and spectrophotometry. It also has a larger life and does not suffer from blackening of the bulb glass envelope. In the ultraviolet region of the spectrum, the deuterium lamp has superseded the hydrogen discharge lamp as a UV source. The radiation sources should be highly stable and preferably emit out a continuous spectrum.

Deuterium arc lamp provides emission of high intensity and adequate continuity in the 190–380 nm range. A quartz or silica envelope is necessary not only to provide a heat shield, but also to transmit the shorter wavelengths of the ultraviolet radiation. The limiting factor is normally the lower limit of atmospheric transmission at about 190 nm. Figure 14.3 shows the energy output as a function of wavelength in the case of a deuterium arc lamp and a tungsten-halogen lamp.

In the modern spectrophotometers, the power supply arrangements including any necessary start-up sequences for arc lamps as well as change-over between sources at the appropriate wavelength are automatic mechanical sequences. Lamps are generally supplied on pre-set focus mounts or incorporate simple adjustment mechanisms for easy replacement.

14.3.2 Optical Filters

A filter may be considered as any transparent medium which by its structure, composition or colour enables the isolation of radiation of a particular wavelength. For this purpose, ideal filters should be monochromatic, i.e. they must isolate radiation of only one wavelength. A filter must meet the following two requirements:

(a) high transmittance at the desired wavelength and (b) low transmittance at other wavelengths. However, in practice, the filters transmit a broad region of the spectrum. Referring to

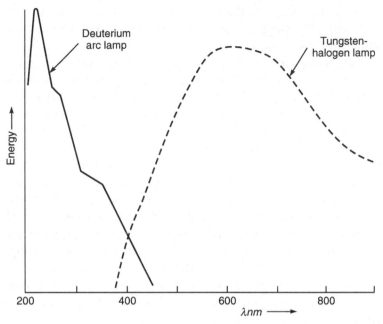

> **Fig. 14.3** *Energy output as a function of wavelength for deuterium arc lamp and Tungsten-halogen lamp*

Fig. 14.4, they are characterized by the relative light transmission at the maximum of the curve T_λ, the width of the spectral region transmitted (the half-width—the range of wavelength between the two points on the transmission curve at which the transmission value equals $1/2 T_\lambda$) and T_{res} (the residual value of the transmission in the remaining part of the spectrum). The ideal filter would have the highest value of T_λ and the lowest values for the transmission half-width and T_{res}. Filters can be broadly classified as absorption filters and interference filters.

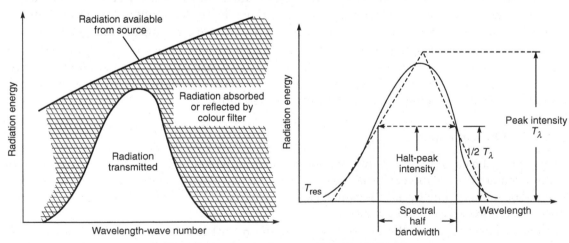

> **Fig. 14.4** *Optical properties of a light filter*

Absorption Filters: The absorption type optical filter usually consists of colour media: colour glasses, coloured films (gelatin, etc.), and solutions of the coloured substances. This type of filter has a wide spectral bandwidth, which may be 40 to 50 μ in width at one-half the maximum transmittance. Their efficiency of transmission is very poor and is of the order of 5 to 25%.

Composite filters consisting of sets of unit filters are often used. One combination set consists of a long wavelength and sharp cut-off filters and the other of short wavelength and cut-off filters combinations are available from about 360 nm to 700 nm.

Interference Filters: These filters usually consist of two semi-transparent layers of silver, deposited on glass by evaporation in vacuum and separated by a layer of dielectric (ZnS or MgF₂). In this arrangement, the semi-transparent layers are held very close. The spacer layer is made of a substance which has a low refractive index. The thickness of the diaelectric layer determines the wavelength transmitted. Figure 14.5 shows the path of light rays through an interference filter. Some part of light that is transmitted by the first film is reflected by the second film and again reflected on the inner face of the first film, as the thickness of the intermediate layer is one-half a wavelength of a desired peak wavelength. Only light which is reflected twice will be in-phase and come out of the filter, other wavelengths with phase differences would cause destructive interference. Constructive interference between different pairs in superposed light rays occurs only when the path difference is exactly one wavelength or some multiple thereof.

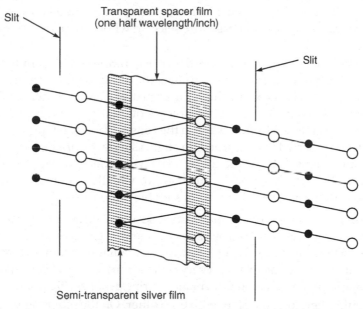

> **Fig. 14.5** *Path of light rays through an interference filter*

Interference filters allow a much narrower band of wavelengths to pass and are similar to monochromators in selectivity. They are simpler and less expensive. However, as the selectivity increases, the transmittance decreases. The transmittance of these filters varies between 15 to 60 per cent with a spectral bandwidth of 10 to 15 nm.

For efficient transmission, multilayer transmission filters are often used. They are characterized by a bandpass width of 8 nm or less and a peak transmittance of 60-95%. Interference filters can be used with high intensity light sources, since they remove unwanted radiation by transmission and reflection, rather than by absorption.

14.3.3 Monochromators

Monochromators are optical systems, which provide better isolation of spectral energy than the optical filters, and are therefore preferred where it is required to isolate narrow bands of radiant energy. Monochromators usually incorporate a small glass of quartz prism or a diffraction grating system as the dispersing media. The radiation from a light source is passed either directly or by means of a lens or mirror into the narrow slit of the monochromator and allowed to fall on the dispersing medium, where it gets isolated. The efficiency of such monochromators is much better than that of filters and spectral half-bandwidths of I nm or less are obtainable in the ultraviolet and visible regions of the spectrum.

Prism Monochromators: Isolation of different wavelengths in a prism monochromator depends upon the fact that the refractive index of materials is different for radiation of different wavelengths. If a parallel beam of radiation falls on a prism, the radiation of two different wavelengths will be bent through different angles. The greater the difference between these angles, the easier it is to isolate the two wavelengths. This becomes an important consideration for the selection of material for the prisms, because only those materials are selected whose refractive index changes sharply with wavelength.

Prism may be made of glass or quartz. The glass prisms are suitable for radiations essentially in the visible range whereas the quartz prism can cover the ultraviolet spectrum also. It is found that the dispersion given by glass is about three times that of quartz. However, quartz shows the property of double refraction. Therefore, two pieces of quartz, one right-handed and one left-handed are taken and cemented back-to-back in the construction of 60° prism (Cornu mounting), or the energy must be reflected and returned through a single 30° prism, so that it passes through the prism in both directions (Littrow mounting). The two surfaces of the prism must be carefully polished and optically flat. Prism spectrometers are usually expensive, because of exacting requirements and difficulty in getting quartz of suitable dimensions.

Diffraction Gratings: Monochromators may also make use of diffraction gratings as a dispersing medium. A diffraction grating consists of a series of parallel grooves ruled on a highly polished reflecting surface. When the grating is put into a parallel radiation beam, so that one surface of the grating is illuminated, this surface acts as a very narrow mirror. The reflected radiation from this grooved mirror overlaps the radiation from neighbouring grooves (Fig. 14.6).

The waves would, therefore, interfere with each other. On the other hand, it could be that the wavelength of radiation is such that the separation of the grooves in the direction of the radiation is a whole number of wavelengths. Then the waves would be in phase and the radiation would be reflected undisturbed. When this is not a whole number of wavelengths, there would be destructive interference and the waves would cancel out and no radiation would be reflected. By changing the angle at which the radiation strikes the grating, it is possible to alter the wavelength reflected.

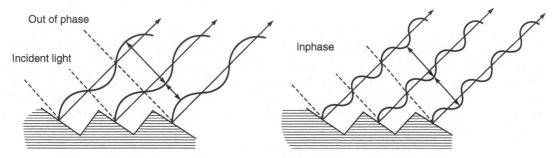

> **Fig. 14.6** *Dispersion phenomenon in diffraction gratings*

The expression relating the wavelength of the radiation and the angle (θ) at which it is reflected is given by

$$m\lambda = 2d \sin \theta$$

where d is the distance separating the grooves and is known as the grating constant and m is the order of interference.

When $m = 1$, the spectrum is known as first order and with $m = 2$, the spectrum is known as second order.

The resolving power of a grating is determined by the product mN, where N is the total number of grooves or lines on the grating. Higher dispersion in the first order is possible when there are a larger number of lines. When compared with prisms, gratings provide a much higher resolving power and can be used in all spectral regions. Gratings would reflect, at any given angle, radiation of wavelength λ and also $\lambda/2$, $\lambda/3$, etc. This unwanted radiation must be removed with filters or premonochromators, otherwise it will appear as stray light.

Most modern instruments now use a diffraction grating as a dispersing element in the monochromator, as prisms in general have a poorer stray light performance and require complex precision cams to give a linear wavelength scale. Replica gratings can even be produced more cheaply than prisms and require only a simple sine bar mechanism for the wavelength scale.

A typical reflection grating may have 1200 grooves/mm, which means the grooves are spaced at about 800 nm intervals. The grating may have a width of 20 mm or more, giving a total of at least 24,000 grooves. To obtain constructive interference across this large number of grooves with little light scattering, the spacing and form of the grooves must be accurate to within a few nanometers to give a high quality grating. Development of mechanical diamond ruling engines which give a high quality grating near to this accuracy gives rise to difficult technological problems.

Holographic Gratings: Precision spectrophotometers use holographic or interference gratings, which have superior performance in reducing stray light as compared to diffraction gratings. Holographic gratings are made by first coating a glass substrate with a layer of photo-resist, which is then exposed to interference fringes generated by the intersection of two collimated beams of laser light. When the photo-resist is developed, it gives a surface pattern of parallel grooves. When coated with aluminum, these become the diffraction gratings.

Compared with ruled gratings, the grooves of holographic gratings are more uniformly spaced and shaped smoother. These characteristics result in much lower stray light levels. Moreover, the holographic gratings can be produced in much less time than the ruled gratings. Holographic gratings used in commercial spectrophotometers are either original master gratings produced directly by an interferometer or are replica gratings. Replica gratings are reproduced from a master holographic grating by moulding its grooves onto a resin surface on a glass or silica substrate. Both types of gratings are coated with an aluminum reflecting surface and finally with a protective layer of silica or magnesium fluoride. Replica gratings give performance which is as good as the master gratings. The holographic process is capable of producing gratings that almost reach the theoretical stray-light minimum.

14.3.4   Optical Components

Several different types of optical components are used in the construction of analytical instruments based on the radiation absorption principle. They could be windows, mirrors and simple condensers. The material used in the construction of these components is a critical factor and depends largely on the range of wavelength of interest. Normally, the absorbance of any material should be less than 0.2 at the wavelength of use. Some of the materials used are:

- Ordinary silicate glasses which are satisfactory from 350 to 3000 nm.
- From 300 to 350 nm, special corex glass can be used.
- Below 300 nm, quartz or fused silica is utilized. The limit for quartz is 210 nm.
- From 180 to 210 nm, fused silica can be used, provided the monochromator is flushed with nitrogen or argon to eliminate absorption by atmospheric oxygen.

Reflections from glass surfaces are reduced by coating these with magnesium fluoride, which is one-quarter wavelength in optical thickness. With this, scattering effects are also greatly reduced.

With a view to reduce the beam size or render the beam parallel, condensers are used. These condensers operate as simple microscopes. To minimize light losses, lenses are sometimes replaced by front-surfaced mirrors to focus or collimate light beam in absorption instruments. Mirrors are aluminized on their front surfaces. With the use of mirrors, chromatic aberrations and other imperfections of the lenses are minimized.

Beam splitters are used in double-beam instruments. These are made by giving a suitable multi-layer coating on an optical flat. The two beams must retain the spectral properties of the incident beam. Half-silvered mirrors are often used for splitting the beam. However, they absorb some of the light in the thin metallic coating. Beam splitting can also be achieved by using a prismatic mirror or stack of thin horizontal glass plates, silvered on their edges and alternatively oriented to the incident beam.

14.3.5   Photosensitive Detectors

After isolation of radiation of a particular wavelength in a filter or a monochromator, it is essential to have a quantitative measure of its intensity. This is done by causing the radiation to fall on a photosensitive element, in which the light energy is converted into electrical energy. The electric current produced by this element can be measured with a sensitive galvanometer directly or after suitable amplification.

Any type of photosensitive detector may be used for the detection and measurement of radiant energy, provided it has a linear response in the spectral band of interest and has a sensitivity good enough for the particular application. There are two types of photo-electric cells; photo-voltaic cells and photo-emissive cells. These cells have been described in Chapter 3.

14.3.6 Sample Holders

Liquids may be contained in a cell or cuvette made of transparent material such as silica, glass or perspex. The faces of these cells through which the radiation passes are highly polished to keep reflection and scatter losses to a minimum. Solid samples are generally unsuitable for direct spectrophotometry. It is usual to dissolve the solid in a transparent liquid. Gases may be contained in cells which are sealed or stoppered to make them air-tight. The sample holder is generally inserted somewhere in the interval between the light source and the detector. For the majority of analyses, a 10 mm path-length rectangular cell is usually satisfactory.

In analyses where only minimal volumes of liquid samples are practical, microcells, which have volumes as small as 50 µl, can be employed. Most of the rectangular liquid cells have caps and, for the analyses of extremely-volatile liquids, some of the cells have ground-glass stoppers to prevent the escape of vapour. Studies of dilute or weakly-absorbing liquid samples, or of samples where trace components must be detected, require a cell with a long path-length. For such applications, a 50 cm path-length with about a 300 ml volume cell is employed.

Cylindrical liquid cells offer higher volume to path-length ratios than do rectangular cells, being available in path-lengths of 20, 50 and 100 mm and in volumes of 4, 8, 20 and 40 ml, respectively.

Similar to these cylindrical cells are the demountable cells that have silica windows which are easily removable. This demountable feature is especially useful for the containment of samples that are difficult to remove and clean from conventional cylindrical cells. Demountable cells are equipped with ground-glass stoppers. Figure 14.7 shows a selection of typical sample cuvettes.

ⅢⅢ▶ 14.4 COLORIMETERS

A colorimetric method in its simplest form uses only the human eye as a measuring instrument This involves the comparison by visual means of the colour of an unknown solution, with the colour produced by a single standard or a series of standards. The comparison is made by obtaining a match between the colour of the unknown and that of a particular standard by comparison with a series of standards prepared in a similar manner, as the unknown. Errors of 5 to 20% are not uncommon, because of the relative inability of the eye to compare light intensities.

In the earlier days, visual methods were commonly employed for all colorimetric measurements, but now photo-electric methods have largely replaced them and are used almost exclusively for quantitative colorimetric measurements. These methods are more precise and eliminate the necessity of preparing a series of standards every time a series of unknowns is run.

Strictly speaking, a colorimetric determination is one that involves the visual measurement of colour and a method employing photoelectric measurement is referred to as a photometric or spectro-photometric method. However, usually any method involving the measurement of colour

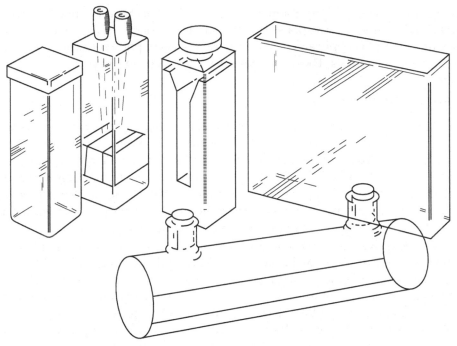

> **Fig. 14.7**   *Selection of sample cuvettes*

in the visual region of the electromagnetic spectrum (400–700 nm) is referred to as the colorimetric method.

In a colorimeter, the sample is normally a liquid. The sample compartment of a colorimeter is provided with a holder to contain the cuvette, in which the liquid is examined. Usually this holder is mounted on a slide with positions for at least two cuvettes, so that sample and reference cuvettes are measured first and a shutter is moved into or out of the light beam until the microammeter gives a full-scale deflection (100% T-scale reading). The sample is then moved into the beam and the light passing through it is measured as a percentage to the reference value.

$$\text{Sample concentration} = \text{Standard concentration} \times \frac{\text{Sample reading}}{\text{Reference reading}}$$

Colorimeters are extremely simple in construction and operation. They are used for a great deal of analytical work, where high accuracy is not required. The disadvantage is that a range of filters is required to cover different wavelength regions. Also the spectral bandwidth of these filters is large in comparison with that of the absorption band being measured. The basic component of a colorimeter are those which are shown in Fig. 14.2. They are available as single beam or double beam configurations.

14.4.1   Multi-channel Colorimeter (Photometer)

An increasing number of chemical analyses is carried out in the laboratories of industry and hospitals, and in most of these the final measurement is performed by a photometer. Obviously, it is possible to increase the capacity of the laboratory by using photometers, which have a large measuring capacity. One of the limitations for rapid analyses is the speed at which the samples can be transferred in the light path.

In a multi-channel photometer, instead of introducing one sample at a time into a single light path, a batch of samples is introduced. Measurements are carried out simultaneously, using a multiplicity of fiber optic light paths (Fig. 14.8) and detectors, and then the samples are scanned electronically instead of mechanically. The 24 sample cuvettes are arranged in a rack in a three key eight matrix. The 25th channel serves as a reference beam and eliminates possible source and detector drifts. The time required to place the cuvette rack into the measuring position corresponds to the amount of time necessary to put one sample into a sample changer.

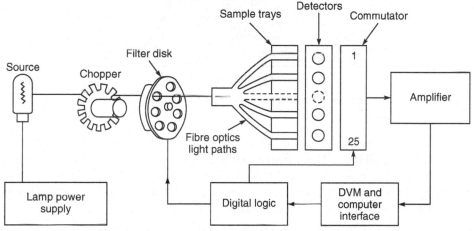

> **Fig.** 14.8   *Schematic of a multi-channel photometer*

ⅢⅢ> 14.5   SPECTROPHOTOMETERS

A spectrophotometer is an instrument which isolates monochromatic radiation in a more efficient and versatile manner than colour filters used in filter photometers. In these instruments, light from the source is made into a parallel beam and passed to a prism or diffraction grating, where light of different wavelengths is dispersed at different angles.

The amount of light reaching the detector of a spectrophotometer is generally much smaller (Fig. 14.9) than that available for a colorimeter, because of the small spectral bandwidth. Therefore, a more sensitive detector is required. A photomultiplier or vacuum photocell is generally employed. The electrical signal from the photoelectric detector can be measured by using a sensitive microammeter. However, it is difficult and expensive to manufacture a meter of the required range and accuracy. To overcome this problem, either of the following two approaches are generally adopted:

(a) The detector signal may be measured by means of an accurate potentiometric bridge. A reverse signal is controlled by a precision potentiometer, until a sensitive galvanometer shows that it exactly balances the detector signal and no current flows through the galvanometer. This principle is adopted in the Beckman Model DU single-beam spectrophotometer.

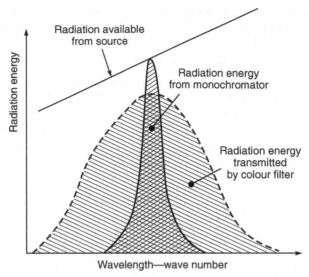

> **Fig.** 14.9 *Comparison of radiation energy from a monochromator and colour filter*

(b) The detector current is amplified electronically and displayed directly on an indicating meter or in digital form. These instruments have the advantage in speed of measurement. As in the case of colorimeters, the instrument is adjusted to give a 100% transmission reading, with the reference sample in the path of the light beam. The sample is then moved into the beam and the percentage transmission is observed.

Modern commercial instruments are usually double beam, digital reading and/or recording instruments, which can provide absorbance, concentration, per cent transmission and differential absorbance readings. It is also possible to make reaction rate studies. They can be used to include specialized techniques, such as automatic sampling and batch sampling, with the addition of certain accessories. The measurements can be generally made with light at wavelengths from 340 to 700 nm and from 190 to 700 nm with a deuterium source. In variable-slit type of instruments, the slit can be made to vary from 0.05 to 2.0 mm. The wavelength accuracy is ± 0.5 nm. The recorders are usually single channel, strip-chart potentiometric recorders. They are calibrated from 0.1 to 2.0 A or 10 to 200% *T* full-scale. The recorder used with spectrophotometers have four wavelength scanning speeds (100, 50, 20 and 5 nm/mm) and seven chart speeds (10, 5, 2, 1, 0.5, 0.2 and 0.1 inch/mm). It has a sensitivity of 100 mV absorbance units or 100 mV/100% *T*.

When scanning a narrow wavelength range, it may be adequate to use a fixed slit-width. This is usually kept at 0.8 mm. In adjustable slit-width instruments, it should be so selected that the resultant spectral slit-width is approximately 1/10th of the observed bandwidth of the sample, i.e. if the absorption band is 25 nm wide at half of its height, the spectral slit-width should be 2.5 nm. This means that the slit width set on the instrument should be 1.0 mm. This is calculated from the dispersion data, as the actual dispersion in grating instruments is approximately 2.5 nm/mm slit-width.

Spectrophotometers generally employ a 6 V tungsten lamp, which emits radiation in the wavelength region of visible light. Typically, it is 32 candle power. These lamps should preferably be operated at a potential of say 5.4 V, when its useful life is estimated at 1200 h. The life is markedly decreased by an increase in the operating voltage. With time, the evaporation of tungsten produces a deposit on the inner surface of the tungsten lamp and reduces emission of energy. Dark areas on the bulb indicate this condition. It should then be replaced.

The useful operating life of the deuterium lamp normally exceeds 500 h under normal conditions. The end of the useful life of this lamp is indicated by failure to start or by a rapidly decreasing energy output. Ionization may occur inside the anode rather than in a concentrated path in front of the window. Generally, this occurs when the lamp is turned on while it is still hot from previous operation. If this occurs, the lamp must be turned off and allowed to cool before restarting.

Wavelength calibration of a spectrophotometer can be checked by using a holmium oxide filter as a wavelength standard. Holmium oxide glass has a number of sharp absorption bands, which occur at precisely known wavelengths in the visible and ultraviolet regions of the spectrum. Holmium oxide filter wavelength peaks are given below:

Ultraviolet range with deuterium lamp : 279.3 and 287.6 nm

Visible range with tungsten lamp : 360.8, 418.5, 453.4,536.4 and 637.5 nm

The wavelength calibration can also be checked in the visible region by plotting the absorption spectrum of a didymium glass, which has been, in turn, calibrated at the National Bureau of Standards, USA.

14.5.1 Microprocessor Based Spectrophotometers

In spectrophotometry, computers have long been used, especially for on line or off line data processing. Since the advent of the microprocessor, their application has not only been limited to processing of data from analytical instruments, but has also been extended to control of instrument functions and digital signal processing, which had been performed conventionally by analog circuits. This has resulted in improved performance, operability and reliability over purely analog instruments.

A microprocessor, in a spectrophotometer, could be used for the following functions:

- *Control functions:* Wavelength scanning, automatic light source selection, control of slit-width, detector sensitivity, etc.
- *Signal processing functions:* Baseline correction, signal smoothing, calculation of % *T*, absorbance and concentration, derivative, etc.
- *Communication functions:* Keyboard entry, menu-driven operations, data presentation, warning display, communication with external systems, etc.

Figure 14.10 shows a block diagram of a microprocessor controlled spectrophotometer. The diagram shows only the post-detector electronic handling and drive systems, all controlled via a single microprocessor. Once the operator defines such parameters as wavelength, output mode and relevant computing factors, the system automatically ensures the correct and optimum combination of all the system variables. Selection of source and detector are automatically determined;

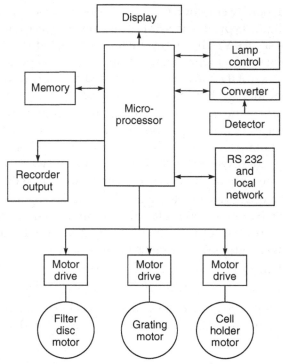

> **Fig. 14.10**   *Block diagram of a microprocessor controlled spectrophotometer*

any filters introduced at appropriate points and sample and reference cells are correctly managed in the sample area. Output in the desired form (transmittance, absorbance, concentration, etc.) is presented along with the sample identification. Secondary routines such as wavelength calibration and self-tests become available on demand.

For wavelength scanning, a stepper motor is used, which ensures accurate and fast scanning. The automatic selection of samples is also made with a motor driven system under the control of a microprocessor.

The signal from the photodetector is amplified in a preamplifier and converted into digital form in an A-D converter. The signals are differentiated into sample signal S, reference signal R and zero signal Z and stored in the memory. From these values, the microprocessor calculates the transmittance $T = (S\text{-}Z/R\text{-}Z)$ and absorbance $= -\text{Log } T$. In order to obtain R or S values within a specified range, the microprocessor provides control signals for slit-width and high voltage for the photomultiplier.

Baseline compensation due to solvent and optical unmatching of cells, which has been difficult with conventional instruments is conveniently possible in microprocessor based systems. Improvements have also been achieved in such functions as auto-zero, expanding and contracting of the photometric scale, automatic setting of wavelength as well as in ensuring repeatable and more accurate results.

The digital output from the microprocessor is converted into analog form with a D-A converter and given to an X-Y recorder as the Y-axis signal, whereas the wavelength forms the X-axis, to obtain absorption or reflected spectra. Microprocessors have also enabled the making of such measurements as higher order derivative spectra and high speed sampling and storage of fast reaction processes and for presenting processed data during and after the completion of the reaction.

PC-based spectrophotometers are now commonly available. In these instruments, wavelength drive, slit drive, filter wheel drive, source selector, mirror drive, detector change and grating drive are all carried out by stepper motors. Stepping control is effected by the computer, with the pulse frequency depending upon the individual scan speed. The computer processes the data supplied by the microprocessor and transmits it with the suitable format to the video display screen and printer/plotter.

14.6   AUTOMATED BIOCHEMICAL ANALYSIS SYSTEMS

14.6.1   The System Concepts

The development of new concepts and more advanced techniques in analytical methodology have resulted in the estimation of blood constituents as a group, whose metabolic roles are related and which collectively provide more meaningful information than the individual analyses. For instance, the group of important anions and cations of the blood plasma (electrolytes) like sodium, potassium, chloride and bicarbonate, which together with serum urea form a related set of tests—performed on patients with electrolyte disturbances. Another group consists of the analyses—protein, bilirubin, alkaline phosphatase and SGOT, which together assess liver function. The effect of this trend is in the replacement of single isolated analysis by groups of analyses, all of which are carried out routinely on each sample with highly reproducible and accurate results. With this object in view, automatic analysis equipment have been designed and put to use. Automated analysis systems are available in multichannel versions and a full description of the detailed working instructions and details of techniques for individual substances are given in the literature supplied by the manufacturers to the purchaser of their equipment. The major benefit of automation in the clinical laboratory is to get rid of the tasks that are repetitive and monotonous for a human operator, which may lead to improper attention that may cause an error in analysis. However, it may be remembered that improvement in reproducibility does not necessarily enhance the accuracy of test results, because accuracy is basically influenced by the analytical methods used. The automated systems are usually considered more reliable than normal methods, due to individual variations that may appear in handling various specimens.

The automated system is usually a continuous flow system, in which individual operations are performed on the flowing stream as it moves through the system. The end-product passes through the colorimeter, where a balance ratio system is applied to measure concentrations of various constituents of interest. The final results are recorded on a strip-chart recorder along with a calibration curve, so that the concentration of the unknowns can be calculated. The output may also be connected to a digital computer to have a digital record along with the graphic record.

14.6.2   THE SYSTEM COMPONENTS

The automated system consists of a group of modular instruments (Fig. 14.11) interconnected together by a manifold system and electrical systems. The various sub-systems are:

- Sampling unit
- Proportioning pump
- Manifold
- Dialyzer
- Heating bath or constant temperature bath
- Colorimeter/flame photometer/Fluorometer
- Recorder
- Function monitor

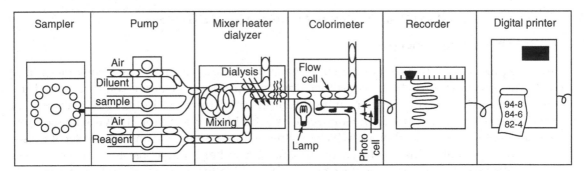

> **Fig. 14.11**   *Schematic diagram of an automated continuous flow type analysis system*

The sample to be analyzed is introduced into a stream of diluting liquid flowing in the narrow bore of a flexible plastic tube. The stages of the analytical reaction are completed by the successive combination of other flowing streams of liquids with the sample stream, by means of suitably shaped glass functions. Bubbles of air are injected into each stream, so that the liquid in the tubes is segmented into short lengths separated by air bubbles. This segmentation reduces the tendency for a stationary liquid film to form on the inner walls of the tubes and decreases the interaction between a sample and the one which follows it. The diluted samples and reagents are pumped through a number of modules, in which the reaction takes place, giving a corresponding sequence of coloured solutions, which then pass into a flow-through colorimeter. The corresponding extinctions are plotted on a graphic recorder, in the order of their arrival into the colorimeter cell. The air bubbles are removed before the liquid enters the colorimetric cell or flame emission. Details of the individual units are described below:

Sampling Unit: The sampling unit enables an operator to introduce unmeasured samples and standards into the auto-analyzer system. The unit in its earlier form consisted of a circular turntable (Fig. 14.12) carrying around its rim 40 disposable polystyrene cups of 2 ml capacity. The sample plate carrying these cups rotates at a predetermined speed. The movement of the turntable is synchronized with the movements of a sampling crook. The hinged tubular crook is fitted at a

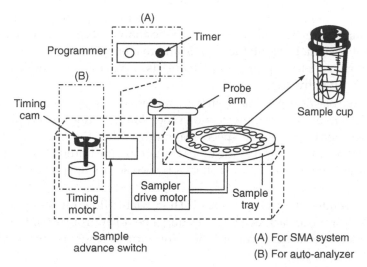

> **Fig. 14.12** *Sampler controls (Courtesy: M/s Technicon Corp., USA)*

corner of the base. The crook carries a thin flexible polythene tube, which can dip into a cup and allow the contents—water, standard or test solution—to be aspirated. At regular intervals, the crook is raised, so that the end of the sample tube is lifted clear of the cup.

Between each sampling, the crook enters a receptacle of water or other suitable wash fluid, to reduce cross-contamination of one sample with another. The ratio of sampling time to wash time is normally 2:1. The plate then rotates a distance sufficient to allow the tube, when it next moves down, to dip into the next cup. One complete rotation of the plate thus presents 40 samples. As the sample plate completes a cycle, a switch is operated, which stops the rotating action of the plate and the sampling action of the sample probe. The sampling rate can be adjusted to 20, 40 or 60 per hour. According to the above ratio, the time during which a sample is being drawn in, will be two minutes, one minute or 40s respectively. The volume of liquid taken up in most cases ranges from about 0.2 to 1.0 ml. This depends upon the rate at which the plate is run and the diameter of the pump tube.

The earlier version has been replaced by a more versatile form of the sampler, in which during the time the sample tube is out of the specimen, the crook quickly comes down into the water, and thus successive samples are separated by a column of water instead of air. This provides a better separation between them. With this sampler, the sample size may range from 0.1 to 8.5 ml. It utilizes cups of sizes 0.5, 2, 3 and 10 ml. The sample plate is kept covered to prevent evaporation, which may sometimes lead to errors up to 5%. Sampling and washing periods are controlled by a programming cam. The sample speed and sample wash cycles are selected by the markings on the cam, such as 40 and 2:1. This implies that the speed is 40 per hour at a sample wash ratio of 2:1.

Mechanical cams were used in the earlier modules of auto-analyzers to initiate and control sample aspiration and wash cycles. Modern systems use electronic timers to do the same function. These timers provide a greater flexibility in the control of the sample-to-wash ratios, which in turn allows flexibility in setting up parameters for analyses.

The Proportioning Pump: The function of the proportioning pump is to continuously and simultaneously push fluids, air and gases through the analytical chain. In fact, it is the heart of the automatic analysis system. Here, all the sample and reagent streams, in any particular analysis, are driven by a single peristaltic pump, which consists of two parallel stainless steel roller chains with finely spaced roller thwarts.

A series of flexible plastic tubes, one from the sampler, the others from reagent bottles or which simply draw in air are placed lengthwise along the platen spring-loaded platform. The roller-head assembly is driven by a constant-speed gear motor. When the rollers are pressed down and the motor switched on, they compress the tubes containing the liquid streams (sample, standard and reagents) against the platen. As the rollers advance across the platen, they drive the liquid before them.

The roller head rotates at a constant speed. The variable flow rates required in the different streams (0.15 to 4 ml/min) are achieved by selecting tubes of appropriate internal diameter, but of constant wall thickness. Since the proportions of the various reagents are fixed by the tube sizes, no measurements are needed.

Proportioning pumps are available either for single-speed or for two-speed operations. The single-speed pump has the capacitor synchronous gear head utilizing 10 rpm output shaft at 50 Hz. The two-speed pump has a non-synchronous 45 rpm motor. The slow speed in this pump is used for the ordinary working during a run and a much quicker one for filling the system with reagents before a run and for rapid washing to clear out reagents after the run. It is also utilized for rapid cleaning of the heating bath, or of the complete system, when fibrin (an insoluble protein) problems are evident and are disturbing the run. High speed is not used for analysis. Heavy-duty pumps are also available, which enable 23 pump tubes to be utilized simultaneously.

The plastic tubes are held taut between two plastic blocks having locating holes, which fit on to pegs at each end of the platen. Before beginning a run, the tubes are stretched. With use and time, the tubes loose elasticity and pumping efficiency is reduced. Therefore, each block has three sets of holes, so that the tubes can be increasingly stretched and the tension thus maintained. The tubes are replaced at the first sign of aging. In fact, they should be replaced at regular intervals to fore-stall failure. When not in use, one of the blocks is removed, so that the tubes are not kept in tension.

Actually, the sample or reaction stream is separated by air bubbles into a large number of distinct segments. The air bubbles completely fill the lumen of the tubing conducting the flow, thereby maintaining the integrity of each individual aliquot. In addition, the pressure of the air bubble against the inner wall of the tubing wipes the surface free of droplets which might contaminate the samples which follow. The proportioning includes an air bar device (Fig. 14.13), which adds air bubbles to the flowing streams in a precise and timed sequence. The air bar is actually a pinch valve connected to the pump rollers that occludes or opens the air pump tubes at timed intervals. Every time a roller leaves the pump platen—and this occurs every two seconds—the air bar rises and lets a measured quantity of air through. The release of air into the system is carefully controlled, thereby insuring exactly reproducible proportioning by the peristaltic pump.

The continuous flow analyzers make use of liquid reagents. Large volumes of reagents are stored in the systems and their quantity is adequate for the operation of the analyzer for several hours or days. Some automated systems use reagents in a dry tablet form. When required, the tablet is dispensed into a one-test reaction vessel and dissolved. The sample is then added for the

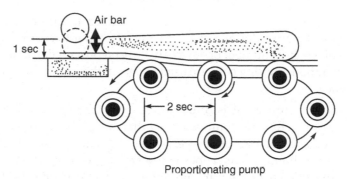

reaction to take place. This is basically a unit-dose concept, which offers several advantages like less storage space and operator time, long stability of reagents and lesser wastage.

Manifold: A manifold mainly consists of a platter, pump tubes, coils, transmission tubing, fittings and connections. A separate manifold is required for each determination and the change can be effected within a few minutes. The pump tubing and the connected coils are placed on a manifold platter, which keeps them in proper order for each test. The pump tubing are specially made, are of premeasured length and are meant to introduce all constituents of an analysis into the system. The physical and chemical properties of the tubing are extremely important in the correct functioning of the pump. It must not be so flexible as to expand beyond its normal internal dimensions on release of pressure, which may lead to variation in the flow, thereby affecting reproducibility and accuracy of the system. The tubes should be chemically inert for the constituents which are expected to flow through the tube. The constant and correct tension also provides the continual delivery of a constant volume. The inside diameter of the pump tubing determines the flow rate per minute.

Several other tubes are required to introduce reagents and to transport the specimen from one module to another. There are five types of such tubings. They are of varying sizes and are to be selected according to the requirements. These are: standard transmission tubing (Tygon), solvaflex tubing, acidflex tubing, polyethylene tubing and glass tubing.

Two types of coils are employed in the system—mixing coils and delay coils. Coils are glass spirals of critical dimensions, in which the mixing liquids are inverted several times, so that complete mixing can result.

Mixing coils are used to mix the sample and/or reagents. As the mixture rotates through a coil, the air bubble along with the rise and fall motion produces a completely homogeneous mixture. The mixing coils are placed in a horizontal position to permit proper mixing. Delay coils are employed when a specimen must be delayed for the completion of a chemical reaction before reaching the colorimeter. These coils are selected in length according to the requirements. The standard delay coil is 40 ft long, 1.6 mm I.D., and has a volume of approximately 28 ml. The time delay can be calculated by dividing the volume of coil by the flow rate of specimen plus bubbles.

Phasing: With 12 tests to be recorded on each sample and a sampling rate of 60 samples per hour, it follows that 5 s are allowed to record each steady state plateau. The reaction streams in the 12 channels and up to four blank channels must, therefore, be phased to arrive at the colorimeter in waves 5 s apart. For example, if the cholesterol stream arrives at X time, calcium must arrive at $X + 5$ s, total protein at $X + 10$ s, albumin at $X + 15$ s, etc. In order to ensure a proper sequencing for the presentation of results, a number of devices have been provided to make this adjustment an extremely simple operation. Phasing coils are used to permit the channels to enter the colorimeter in the proper sequence.

Dialyzer: In analytical chemistry, it is often necessary to remove protein cells to obtain an interference-free analysis. This is accomplished by dialysis in the auto-analyzer. The dialyzer module (Fig. 14.14) consists of a pair of perspex plates, the mating surfaces of which are mirror grooved in a continuous channel, which goes in towards the centre, on itself and returns to the outside. A semi-permeable cellophane membrane is placed between the two plates and the assembly is clamped together, similar to the kidney dialyzer. The continuous groove channel thus gets divided into two halves and the dialysis occurs across the membrane. A solution containing the substance to be analyzed passes along one-half, usually the upper one, of the channel, while the solvent that is receptive to the substance to be removed enters the other half. The substance to be separated from the sample diluent stream, will diffuse through the semi-permeable membrane by osmotic pressure into the recipient stream and the non-diffusable particles will be left behind.

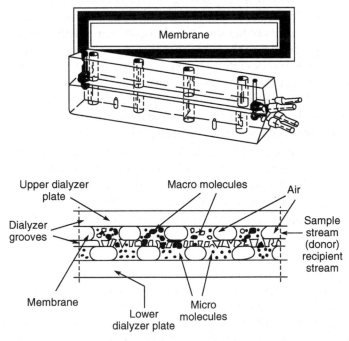

> **Fig. 14.14**   *Simplified block diagram of the dialysis process*

The cellophane membrane usually used in the dialyzer has a pore size of 4–6 μm. The rate of dialysis is stated to be dependent upon the temperature, area, and concentration gradient. For this reason, the dialyzer unit is usually immersed in a water bath maintained at a constant temperature (37± 0.1°C). The temperature is kept constant with a thermostatically controlled heater and a motorized stirrer. Both streams pass through preheating coils, before entering the dialyzer unit. The channel path is 87 inch long, which provides a large surface presentation to the dialyzing membrane. The plates of the dialyzer must be a matched set. If the plates are not a matched set, the channels may be slightly off, causing leakage, poor bubble patterns and loss of dialyzing area, which would ultimately result in the loss of sensitivity.

The quantity of solute that passes through the membrane in the dialyzer is determined by the concentration gradient across the membrane, the duration of contact of the two solutions, the area of contact, the temperature and by the thickness and porosity of the membrane. Other factors which affect the rate of transfer are the size and shape of the molecules, their electrical charge and the composition of the fluids across the membrane.

A decrease in the flow rate of the liquid streams increases sensitivity in continuous flow systems, since more concentrated samples and thinner membranes can be used. Modern dialyzers, therefore, have shallower and shorter grooves, resulting in reduced sample interaction and carry over. Membranes have been found to age with use and time due to protein deposition on their surface and therefore need periodical replacement. In the recent systems, the computer informs the operator to investigate the need for membrane replacement.

Heating Bath: On leaving the dialyzer, the stream may be combined by one or more additional reagents. It is then passed to a heating bath. This module is not used in all the tests performed by the auto-analyzer. The heating bath is a double-walled insulated vessel, in which a glass heating coil or helix is immersed in mineral oil. A thermostatically controlled immersion heater maintains a constant temperature within ±0.1°C, which can be read on a thermometer. Inside the bath, the stream passes along a helical glass coil about 40 ft long and 1.6 mm I.D., immersed in oil, which is constantly stirred. The heating bath may have a fixed temperature, as 95° or 37°C or an adjustable value. Passage through the heating coil takes about five minutes, but it would obviously vary with the rate at which the liquid is moving, which, in turn, depends on the diameter of the tubes in the manifold.

Measurement Techniques: Although automated analyzers are mostly using absorption spectro-photometry as the major measurement technique, several other alternative photometric approaches have been utilized in the recent years. These are reflectance photometry, fluorometry, nephelometry and fluorescence polarization. Use of ion-selective electrodes and other electrochemical measurement techniques are also becoming popular.

In the absorption photometers used in automated systems, the *radiant energy sources* employed include tungsten, quartz, halogen, deuterium, mercury and xenon lamps and lasers. The spectrum covered is usually from 300 to 700 nm. *Spectral isolation* is generally achieved with interference filters in most automated systems. These filters have peak transmittance of 3~80% and bandwidths of 5-15 nm. The filters are usually mounted on a rotating wheel and the required filter is brought in position under the command of a microprocessor. Some automated systems also make use of monochromators, which obviously provide greater flexibility for the development and addition of new assays. The most popular *detector* used in the automated systems is the photomultiplier although some of the recent systems also employ photodiodes.

Proper alignment of flow cells or cuvettes with the light path is as important in automated systems as in manual methods. Stray energy and internal reflections are required to be kept as low as possible, to approximately less than 0.2%. This is usually done by careful design of the wavelength isolation filters, or monochromator, or by the use of dual filters to increase rejection of stray light.

The colorimeters used in the automated systems continuously monitor the amount of light transmitted through the sample. They employ flow-through cuvettes. In the earlier designs of flow cells, the arrangement was such that as the incoming stream entered the cell, the air bubbles escaped upwards through an open vent, so that a continuous stream of liquid could fill the cell before going to waste. The flow cell size varied from 6 to 15 mm. The later designs of flow cells are all of tubular construction. This requires a much smaller volume of fluid, so that a smaller volume of sample can be used. Being completely closed, it does not require separate cleaning. Before the stream enters the flow cell, it is pumped to a debubbler, where the air bubbles are removed. The stream is then pulled through the flow cell under the action of another pump.

Some automatic continuous flow chemical analysis systems, incorporate a multi-channel colorimeter The colorimeter employs a fiber optics system, using a single high-intensity quartz iodine lamp, which is coupled to a highly stabilized power supply. This lamp passes its energy through light guides, which are connected to up to five independent colorimeter modules. Each colorimeter has two detectors, one for the sample cell and the other for the reference beam. An insulating wall is placed between the light source and the colorimeter modules, so as to maintain good temperature and electrical stability. The colorimeter contains a glass continuous-flow sample cell with an integral debubbler.

Light guides from the main light source plug into one side of the cell block and the photo-detectors are placed on the other side of the block. The filters are inserted into a slot between the light guide and flow cell. The detectors are closely matched and their output is connected to a highly stabilized bridge circuit. The output fed to the recorder is logarithmic, hence for solutions which obey Beer's Law, the recorder output is linear in concentration over the normal range of optical densities observed in a continuous-flow automatic analysis system. The measurement of sodium and potassium is carried out by a flame photometer. Fluorimetric analysis permits measurements to be made at concentrations as low as 0.01 part per billion. The fluorimeters used for automated work, like colorimeters, have flow-through systems. The continuous flow cuvette is made of Pyrex glass, which transmits light from the visible region to approximately 340 nm. For the ultraviolet region below 320 nm, quartz cuvettes are available.

Signal Processing and Data: The availability of low-cost microprocessors had a major impact on the signal processing and data handling of analytical procedures in automated systems. Real-time acquisition and processing of data, by means of specific algorithms, so that the output is immediately useful and meaningful, has become possible. Transformation of complex, non-linear standard responses into linear calibration curves have allowed automation of procedures, such as reflectance spectrometry. Specifically, microprocessors are now being used in automated methods for the following functions:

1. Complete control of the electromechanical operation of the analyzer in relation to the transfer of solutions, selection and placement of proper filters and continuous monitoring of the operation. This ensures that all functions are performed uniformly, repeatably and in the correct sequence.

2. Acquisition, assessment, processing and storing of operational data from the analyzer.
3. Providing effective communication between the analyzer and the operator through alpha-numeric display on the CRT. Some systems even monitor the equipment function and give out messages describing the site and type of problem in a malfunctioning equipment.
4. Facility to communicate to with the main-frame computer through the RS-232 interface for an integration of the instrument with laboratory information.
5. Facility to communicate over the telephone lines, using a modem, with the manufacturer's central service department, thereby enhancing ability of the on-site operator to service and repair the analyzer.

All 12 tests for each sample are reported in directly readable concentration units, on a single strip of precalibrated chart paper. Since the normal ranges for each parameter are printed as shaded areas, the physician does not have to remember the normal values. Thus each abnormality stands out clearly.

Actual measurement are made only after the analytical curve (Fig. 14.15) reaches its steady-state plateau (equilibrium condition in the system at which there are no changes in concentration with time). At this steady-state plateau, all effects of possible sample interaction have been eliminated, and the recorded signal gives a true reflection of the concentration of the constituent being measured. Herein lies the importance of segmenting each of the sample and reagent streams with air bubbles. In effect, the air bubbles act as barriers to divide each sample and reagent stream into a large number of discrete liquid segments. Equally important, the air bubbles continually scrub the walls of the tubing. This sequential wiping of the walls diminishes

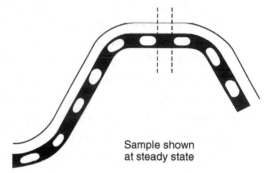

Sample shown at steady state

> **Fig. 14.15** *Typical curve showing the steady state conditions when measurements are made in a continuous flow system*

the possibility of contamination in succeeding segments of the same sample. Thus, should there be any interaction between two samples, it can easily be seen that the effects of this interaction will occur only in the first few segments of the second sample. In the middle segments, the air bubbles immediately preceding will have effectively cleansed the system and prevented further interaction. It is these middle and final segments, free from interaction, which are recorded as the steady-state plateau and appear as flat lines on the graph.

▐▐▐▷ 14.7 CLINICAL FLAME PHOTOMETERS

The flame photometer is one of the most useful instruments in clinical analyses. This is due to the suitability of the flame photometer for determining sodium, potassium and calcium, which are of immense importance in the development of the living being and are indispensable for its physiological functions. In the clinical analysis of sodium and potassium, the flame photometer gives, rapidly and accurately, numerous differential data for normal and pathological values.

The method of flame photometric determinations is simple. A solution of the sample to be analyzed is prepared. A special sprayer operated by compressed air or oxygen is used to introduce this solution in the form of a fine spray (aerosol) into the flame of a burner operating on some fuel gas, like acetylene or hydrogen. The radiation of the element produced in the flame is separated from the emission of other elements by means of light filters or a monochromator. The intensity of the isolated radiation is measured from the current it produces when it falls on a photocell. The measurement of current is done with the help of a galvanometer, whose readings are proportional to the concentration of the element. After carefully calibrating the galvanometer with solutions of known composition and concentration, it is possible to correlate the intensity of a given spectral line of the unknown sample, with the amount of the same element present in a standard solution.

A flame photometer has three essential parts (Fig 14.16). These are:

(a) *Emission System:* Consists of the following:

   (i) **Fuel gases:** and their regulation: comprising the fuel reservoir, compressors, pressure regulators and pressure gauges.

   (ii) **Atomizer:** consisting, in turn, of the sprayer and the atomization chamber, where the aerosol is produced and fed into the flame.

   (iii) **Burner:** receives the mixture of the combustion gases.

   (iv) **Flame:** the true source of emission.

(b) *Optical System:* It consists of the optical system for wavelength selection (filters or mono-chromators), lenses, diaphragms, slits etc.

(c) *Recording System:* It includes detectors like photocells, photo-tubes, photomultipliers, etc. and the electronic means of amplification, measuring and recording.

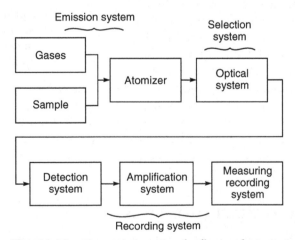

> **Fig. 14.16** *Essential parts of a flame photometer*

Dedicated instruments for the simultaneous analysis of sodium, potassium and lithium are available. In these instruments, sample handling is automatic, as the system has a turntable, which will hold up to 20 samples in cups and an automatic positive piston displacement dilutor, that dilutes the sample prior to entering the spray chamber.

Ignition and shutdown of the flame are automatic. When the calibrate button is depressed, the flame is ignited and the circuits are energized. Passing a standby button extinguishes the flame, but maintains thermal equilibrium. Twenty-four microlitres of the sample are aspirated for the simultaneous measurement of sodium and potassium, providing microsample analysis suitable for paediatric or geriatric work. After analysis, the results are displayed directly in millimoles per litre or milliequivalents per litre.

For precise and accurate determination of **Na** and **K** concentrations, use is made of the fact that lithium normally not present in significant concentrations in serum, exhibits about the same flame emission characteristics as Na and K. Lithium ions are added to the diluent used for samples, standards and controls. The lithium in the diluent is referred to as the internal standard.

The diluted sample containing the fixed known amount of lithium, in the form of a dissolved salt, is nebulized and carried by the air supply into the first of two compartments in the spray chamber (Fig 14.17). Heavier droplets fall out of the stream on to the chamber walls or separate from the stream upon striking a partition in the chamber and flow to a drain from the compartment. Propane enters the first chamber to mix with the air and sample stream, and carry it through a tubular glass bridge into a second compartment. The aerosol and propane mixture travels up from the chamber to the burner head, where the mixture is burned. Exhaust gases are vented to room air from a cover located on top of the instrument.

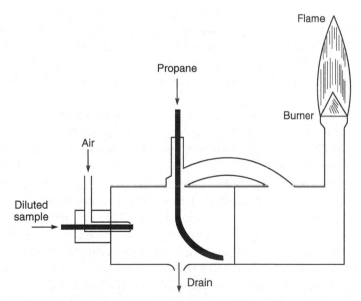

> **Fig. 14.17** *Flame spray chamber and burner of KliNa flame-photometer (Courtesy: M/s Beckman Instruments, USA)*

To provide internal standardization, the response of the sodium and potassium detector is a ratio function of the response by the lithium detector. Thus, any change in air-flow rate or fuel pressure that may affect the sample would proportionately affect the lithium detector.

The flame photometer determines and provides digital display of sodium and potassium or lithium concentrations in a sample, by responding optically and electronically to the intensity of the principal emission lines that characterize each ion as it is excited in the propane and air flame. Figure 14.18 shows the schematic diagram of a flame photometer.

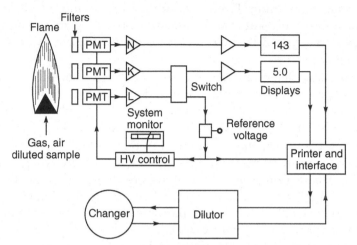

➤ Fig. 14.18 *Schematic diagram of a flame photometer*

The flame is monitored continuously by three photomultiplier detectors. Each detector views the flame through an optical filter that passes only that wavelength band which is of interest to the particular detector. The sodium detector therefore, responds only to wavelengths in a narrow band centered at 589 nm; the potassium detector responds only to wavelengths in a narrow band centered at 766 nm; the lithium detector responds only to wavelengths in a narrow band centered at 671 nm.

The system accuracy is ±0.2 mmol/litre for potassium and lithium, and ±2.0 mmol/litre for sodium. Potassium and lithium both show linearity to 20 mmol/litre, while sodium is linear to 200. In addition to the 0–20 scale, potassium may be rescaled to readout to 200 mmol/litre for convenient analysis of urine samples.

Modern flame photometers come with many useful accessories. For example, the diluter is a motor driven cam-programmed system that functions through a cycle of operations. These operations involve sample pick-up and transport to an internal mixing cup, the addition of a measured volume of diluent, mixing to ensure a properly prepared sample aliquot, coupling of the mixing cup to the spray chamber, so that sample aspiration can occur, and finally the washing and draining of the mixing cup. The diluter uses positive displacement pumps to assure exact sample dilutions in the operator selectable ratios of 50:1, 100:1 or 200:1.

The automatic changer enables automatic presentation to the diluter sample probe up to 20 successive samples. It is a turntable, which rotates stepwise to locate each sample cup under the extended and down position of the diluter sample probe. The probe is extended from the diluter, once for each sample determination. The probe tip enters the sample and a measured volume is taken for transport to the diluter mixing cup. Individual sample trays are placed on the changer turntable. Each tray can hold 20 sample cups. The sample cup could be of 0.25 ml, 0.5 ml or 2.00 ml size. The 2.0 ml size is generally recommended.

The concentration of solutions is usually expressed in parts per million (ppm) in flame photometry. This type of expression enables to make easy calculations on dilute solutions and the concentrations can be expressed in weight/weight, weight/volume and volume/volume ratios.

▚▚▶ 14.8   SELECTIVE-ION ELECTRODES BASED ELECTROLYTES ANALYZER

Over the past decade, the pH meter has been at the centre of a most important change in the field of analytical measurements due to the introduction of selective-ion electrodes. As their name implies, these electrodes are sensitive to the activity of a particular ion in solution and quite insensitive to the other ions present. As the electrode is sensitive to only one ion, a different electrode is needed for each ion to be studied. Approximately 20 types of selective-ion electrodes are presently available.

Ion-selective electrodes are classified into four major groups:

Glass Electrodes: The first glass ion-selective electrode developed is the one sensitive to hydrogen ions. Glasses containing less than 1% of Al_2O_3 are sensitive to hydrogen ions (H^+) but almost insensitive to the other ions present. Glasses, of which the composition is Na_2O–11%, Al_2O_3–18%, $Si O_2$–71% is highly selective towards sodium, even in the presence of other alkali metals. Glass electrodes have been made that are selectively sensitive to sodium, potassium, ammonium and silver.

Solid State Electrodes: These electrodes use single crystals of inorganic material doped with a rare earth material. Such electrodes are particularly useful for fluoride, chloride, bromide and iodide ion analysis.

Liquid-Liquid Membrane Electrodes: These electrodes are essentially liquid ion-exchangers, separated from the liquid sample by means of a permeable membrane. This membrane allows the liquids to come in contact with each other, but prevents their mixing. Based on this principle, cells have been developed that are selective to calcium and magnesium. These cells are used for measuring water hardeners.

Gas Sensing Electrodes: These electrodes respond to the partial pressure of the gases in the sample. The most recent of these to be developed are the gas sensing electrodes for ammonia and sulphur dioxide. Ammonia or sulphur dioxide is transferred across a gas permeable membrane, until the partial pressure in the thin film of filling solution between the glass electrode membrane and the probe membrane equals that in the sample. The resultant pH change is measured by a combination pH electrode. A potential is developed related to the partial pressure and hence the ammonia or sulphur dioxide concentration is measured.

Applications using ion-selective electrodes are many, most being time saving and simple to use. The electrodes are now used for the continuous monitoring of important electrolytes in the blood such as sodium, potassium, calcium, chloride, etc.

14.8.1   Ion Analyzers

Ion analyzers are basically pH/mV meters, which enable the operator to calculate the concentration of specific ions from the potentials developed at the ion-sensitive electrode, when dipped in sample solution. By measuring both the electrodes potential in a standard solution and in the sample solution, it is possible to calculate the unknown solution concentration by solving the following equation:

$$C_x = C_s \times 10\,\Delta E/S$$

where, C_x = concentration of the unknown solution

C_s = concentration of the standard solution

E = difference between the observed potential in the sample solution and the observed potential in the standard solution

S = electrode slope (change in electrode potential per ten-fold change in concentration)

Ion analyzers are mostly microprocessor-based instruments, which are programmed to calculate sample concentration from a set of input data, such as electrode potentials, standard concentration, slope and blank correction. The instruments measure relative millivolts, pH and concentration of specific ions. The program for direct measurement concentration is based on the Nernstian electrode response:

$$E_x = E_o + S \log (C_x + C_b)$$

where E_x = electrode potential

E_o = constant

C_b = blank concentration

The blank correction (C_b) accounts for the finite lower limit of detection of electrodes. If a solid or liquid-membrane electrode is placed in pure water, the membrane dissolves slightly, producing an equilibrium concentration of the measured ion. This concentration is a constant background for all measurements and is represented by C_b. Typical electrode response curves are generally given by the electrode manufacturers. If the sample concentration fall in the linear response region, a blank correction may not be necessary. But, if the sample concentrations are low, and fall in the non-linear region of the response curve, blank correction must be applied.

The standard calibration procedure for a specific ion meter is similar to that used to calibrate a pH meter with pH buffers. Two standard solutions are used, which are a decade apart in concentration and approximately bracket the expected concentration range of the unknown sample solution.

A block diagram of the ion analyzer is shown in Fig. 14.19. The first stage is the input buffer amplifier, which provides a very high input impedance and less than 1 pA input bias current. The electrode potentials are individually buffered by unity gain amplifiers with FET front ends. Figure 14.20 shows the input buffer stage. The two FETs are operated as source followers, each running at a constant drain current determined by its associated op-amp. The voltage at the + input of each op-amp is held constant, and therefore the drain current in the FETs will be constant. To do this, the op-amp output voltage must maintain a constant V_{GS} and must therefore follow the input voltage. The op-amps effectively serve the dual purpose of controlling the operating current of the FET and providing current gain. As with other similar circuits, the high input impedance of the buffer amplifier gets degraded by the presence of dirt, moisture or solder flux. Also, the input FET is delicate and will get destroyed by static discharge. When the inputs are not being driven by a signal, they must be grounded with shorting straps. The input amplifier is followed by a differential amplifier, before the signal is given to an A-D converter.

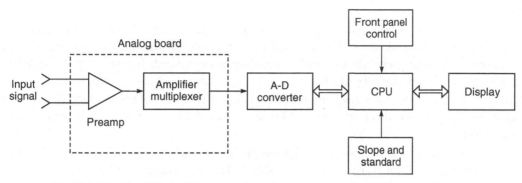

> **Fig. 14.19** *Block diagram of a microprocessor based ion-analyzer*

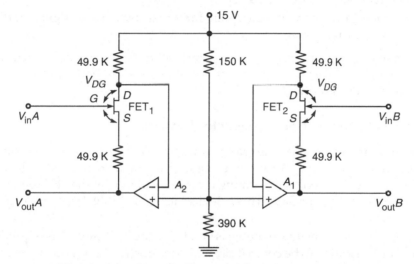

> **Fig. 14.20** *Input-buffer amplifier of an ion analyzer*

The results of the A-D converter are held in the A-D data latch by using shift-registers and the loading function is controlled by the A-D converter. The output of the latch remains in a high impedance state, until they are enabled by a signal from the control port decoder. Thus, the loading and reading of data from the A-D are independent. The microprocessor may read data from the A-D converter, regardless of the timing of the analog-to-digital conversion cycle.

The microprocessor sends and receives information through the input-output (I/O) bus. The bus is driven by only one source at a time and all other sources must be disabled, i.e. kept in a high impedance state. The bus may be driven by the CPU, A-D converter, slope switches, standard value switches and mode switches. The CPU and display receive data from the bus.

The microcomputer consisting of a microprocessor, memory interface and read-only memory (ROM) perform well-defined processing functions. Therefore, the program is stored in permanent read-only memory. Under program control, the microprocessor generates signals on the control port to select the path along which data will flow on the I/O bus. The CPU communicates with the memory and the memory interface through the microprocessor data bus. Through this bus, instructions and numerical constants flow from the memory outputs into the CPU. The memory interface performs the task of generating the address for each instruction stored in the memory. It does this by maintaining a program counter according to commands from the CPU. The timing for the microprocessor and for all signals on the bus is generated by the CPU clock.

Because of the low level of signal generated and high impedance of the ion-selective electrodes, the grounding system is designed very carefully. Usually, the ion-analyzing instruments have three grounds:

    (i) The chasis and the electrostatic shield in the power transformer are connected to the *earth ground* through the third wire of the AC line. This provides isolation from line noise

    (ii) *Digital ground* provides the return path for all the logic signals, including the microprocessor signals and the display current

    (iii) *Analog ground* provides a reference point for the electrode input signals and a return path for all analog current.

The analog and digital grounds are kept separate, so that digital signal return currents never flow through the same conductor as the analog signal returns. The earth ground is not connected to either digital or analog ground.

14.8.2 Chemically Sensitive Semiconductor Devices

Considerable effort has recently been directed towards the development of ion-sensitive electrodes based on a modification of the metal-oxide-semiconductor field transistor. In these devices, chemical sensitivity is obtained by fabricating the gate insulation of the FETs out of ion-sensitive materials, usually a polymer or SiO_2. These devices are called ISFETs (ion-selective field-effect transistors).

In these devices, the ion-sensitive material is bonded to the FET itself. This requires the material and its method of fabrication to be compatible with the substrate (high purity silicon). This very significant requirement puts a severe limitation on the use of some of the best characterized membrane materials, including ion sensitive glasses. The development of a pH-sensitive electrode by means of thick-film screening techniques has extensively been reported. This electrode retains the advantages of ion-sensitive FET transducers, but eliminates the restrictions on membrane selection and fabrication. Here, the ion-sensitive structure is physically separated from the FET. In this way the ion-sensitive membrane can be fabricated on a compatible substrate and the FET can then be attached appropriately and placed in close proximity to the ion-sensitive membrane. A hybrid electrode structure permits the incorporation of a source follower FET amplifier, directly adjacent to the pH membrane, significantly reducing response time and noise pick-up (Janata, 1989).

Figure 14.21 presents the cross section of an ISFET, which is essentially a conventional insulated gate field effect transistor, that has its metallic gate contact replaced by a chemically sensitive coating and a reference electrode. In solution, the gate region can be coated with an ion-sensitive membrane. Interaction of ions in solution with the membrane results in a change of the interfacial potential and corresponding alteration of drain current. By this technique, numerous cations and anions have been sensed (H^+, K^+, Ca^{2+}, Cl^-, I^- and CN^-). The ISFET has advantages in its small size

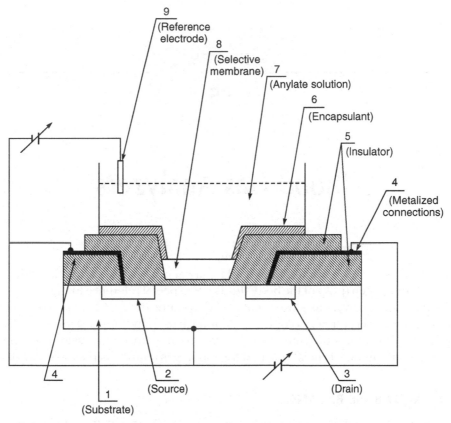

> **Fig. 14.21** *Constructional details of an ion-sensitive field-effect transistor*

(less than 1 mm^2) and low output impedance, which makes it ideal for in vivo monitoring or analysis of small sample volumes. However, problems like ion-selective coating adhesion and device encapsulaition have prevented large scale use of ISFETs.

ISFETs for up to eight different sensors have been fabricated on a single silicon chip. In addition, probes 50 µm in diameter have been fabricated for on-chip circuitry that can measure pH, glucose, oxygen saturation and pressure for biomedical applications (Webster, 1995). In addition to the ISFET sensor, integrated circuits for signal processing can also be deposited on a single chip.

Blood Gas Analyzers

Blood gas analyzers are used to measure the pH, partial pressure of carbon dioxide (pCO_2) and partial pressure of oxygen (pO_2) of the body fluids with special reference to the human blood. The measurements of these parameters are essential to determine the acid-base balance in the body. A sudden change in the pH and pCO_2 could result in cardiac arrhythmias, ventricular hypotension and even death. This shows the importance of the maintenance of physiological neutrality in blood, and consequently the crucial role that the blood gas analyzers play in clinical medicine.

▶ 15.1 ACID-BASE BALANCE

The normal pH of the extracellular fluid lies in the range of 7.35 to 7.45, indicating that the body fluid is slightly alkaline. When the pH exceeds 7.45, the body is considered to be in a state of alkalosis. A body pH below 7.35 indicates acidosis. Both acidosis or alkalosis are disease conditions widely encountered in clinical medicine. Any tendency of the pH of blood to deviate towards these conditions is dealt with by the following three physiological mechanisms: (i) buffering by chemical means, (ii) respiration, (iii) excretion, into the urine by kidneys.

The blood and tissue fluids contain chemical buffers, which react with added acids and bases and minimize the resultant change in hydrogen ions. They respond to changes in carbon dioxide concentration in seconds. The respiratory system can adjust sudden changes in carbon dioxide tension back to normal levels in just a few minutes. Carbon dioxide can be removed by increased breathing and therefore, hydrogen concentration of the blood can be effectively modified. The kidney requires many hours to readjust hydrogen ion concentration by excreting highly acidic or alkaline urine to enable body conditions to return towards normal.

Arterial blood has a pH of approximately 7.40. As venous blood acquires carbon dioxide, forms carbonic acid and hydrogen ions, the venous blood pH falls to approximately 7.36. This pH drop of 0.04 units occurs when the CO_2 enters the tissue capillaries. When CO_2 diffuses from the pulmonary capillaries into the alveoli, the blood pH rises 0.04 units to bring the normal arterial value of 7.40. It is quite difficult to measure the pH of fluids inside the tissue cells, but from estimates based on CO_2 and (HCO_3^-) ion concentration, intracellular pH probably ranges from 7.0 to 7.2.

In order to maintain pO_2, pCO_2 and pH within normal limits, throughout the wide range of body activity, the rate and depth of respiration vary automatically with changes in the metabolism. Control of alveolar ventilation takes place by means of chemical as well as nervous mechanisms. The three important chemical factors regulating alveolar ventilation are the arterial concentrations of CO_2, H^+ and O_2. Carbon dioxide tension in the blood stream and cerebrospinal fluid is the major chemical factor regulating alveolar ventilation. The carotid and aortic chemoreceptors stimulate respiration when oxygen tension is abnormally low. In fact, so many organs participate in the control of respiration that it is difficult to include all aspects in this text. The readers may like to read any standard textbook on human physiology to appreciate the mechanism of respiration control and the maintenance of physiological neutrality of the blood.

Table 15.1 Lists out the normal range for pH, pCO_2, pO_2, total CO_2, base excess and bicarbonate, all measurements made at 37°C (Gambino, 1967).

• Table 15.1   *Typical Expected Values of Blood Gas Parameters*

Parameter		Arterial or arterialized capillary blood	Venous plasma (separated at 37 °C)
pH		7.37 to 7.44	7.35 to 7.45
pCO_2	men	34 to 35 mmHg	36 to 50 mmHg
	women	31 to 42 mmHg	34 to 50 mmHg
pO_2	resting adult	80 to 90 mmHg	25 to 40 mmHg
	resting adult over 65 years	75 to 85 mmHg	
Biocarbonate	men	23 to 29 mmol/l	25 to 30 mmol/l
	women	20 to 29 mrnol/l	23 to 28 mmol/1
Total CO_2	men	24 to 30 mmol/l	26 to 31 mmol/l
(plasma)	women	21 to 30 mmol/l	24 to 29 mmol/l
Base Excess	men	−2.4 to +2.3 mmol/l	0.0 to +5.0 mmol/l
	women	−3.3 to +1.2 mmol/l	−1.0 to + 3.5 mmol/l

ⅢⅢ▶ 15.2   BLOOD pH MEASUREMENT

The acidity or alkalinity of a solution depends on its concentration of hydrogen ions. Increasing the concentration of hydrogen ions makes a solution more acidic, decreasing the concentration of hydrogen ions makes it more alkaline. The amount of hydrogen ions generally encountered in solutions of interest is extremely small and, therefore, the figure is usually represented in the more convenient system of pH notation. pH is thus a measure of hydrogen ion concentration, expressed logarithmically. Specifically, it is the negative exponent (log) of the hydrogen ion concentration.

$$pH = -\log (H^+)$$

If the number 10^{-7} represents the concentration of hydrogen ions in a certain solution, then its pH would be 7. As hydrogen ion concentration rises, pH falls because the logarithm gets smaller, and as hydrogen ion concentration falls, pH rises because the logarithm gets larger. As we deal in logarithms to base 10, a pH of 7 represents 10 times the number of hydrogen ions as does a pH of 8.

Since pure water dissociates into 10^{-7} mol/l of (H^+) and 10^{-7} mol/l of (OH^-), a pH of +7 is considered a neutral solution; a pH of +6 represents an acid, a pH of +8 an alkali. Since 10^{-6} is a larger number than 10^{-8}, the former solution has a larger hydrogenion concentration. Thus, a pH of 6 is more acidic than a pH of 8. Concentration of ions, like the concentration of atoms or molecules, is expressed in terms of mols/l (1 mol = 6.02×10^{-23} molecules, known as the Avogadro's Number. This is the number of molecules in one mole—an amount of material in grams equal to the molecular weight). Whole blood with a (H^+) of 4×10^{-8} moles/1 would have a pH of 7.4; an increase in the (H^+) to 1×10^{-7} moles/1 would correspond to a decrease in pH to 7.0.

Electrochemical pH determination utilizes the difference in potential occurring between solutions of different pH separated by a special glass membrane. If the pH of one of the solutions is kept constant, so that the potential varies in accordance with the pH of the other solution, then the system can be used to determine pH. The device used to effect this measurement is the glass electrode.

Glass Electrode: The potential (E) of the glass electrode may be written by means of the Nernst equation:

$$E = E_0 - \frac{2.3036\ RT}{F} \cdot \Delta pH$$

where E_0 = standard potential

   R = gas constant

   T = absolute temperature

   F = Faraday constant

  ΔpH = pH value deviation from 7

The above relation shows that the emf developed in the electro-chemical pH cell is a linear function of ΔpH.

Change of pH of one unit = 58.2 mV at 20°C = 62.2 mV at 40°C

The factor $-2.3036\ RTF$ is called the slope factor and is clearly dependent upon the solution temperature. With a 1°C change in temperature, the emf changes by 0.2 mV. It is also obvious that the measurement of pH is essentially a measurement of millivolt signals by special methods. Figure 15.1 shows a relationship between the pH and emf at different temperatures. The reference

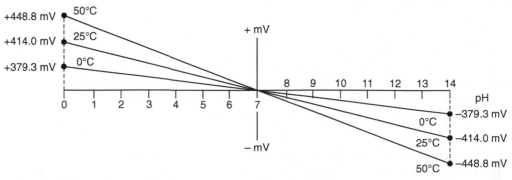

> **Fig. 15.1**   *Relationship between pH and emf at different temperatures (Courtesy: Beckman Instruments Inc., USA)*

electrode provides a constant potential against which the potential of the indicator or glass electrode is measured. An almost universally employed reference electrode is the saturated calomel electrode.

pH Measurement For making pH measurements, the solution is taken in a beaker. A pair of electrodes: one glass or indicating electrode and the other reference or calomel electrode, are immersed in the solution. The voltage developed across the electrodes is applied to an electronic amplifier, which transmits the amplified signal to the display. The pH meter is usually equipped with controls for calibration and temperature compensation.

The glass electrode exhibits a high electrical resistance, of the order of' 100–1000 MΩ. The emf measurement, therefore, necessitates the use of measuring circuits with high input impedance. Further, the high resistance of glass electrodes render them highly susceptible to capacitive pick-up from ac mains. In order to minimize such effects, it is advisable to screen the electrode cable. The screen is usually grounded to the case of the measuring instrument.

The error caused in pH measurements due to temperature effect can be compensated either manually or automatically. In manual adjustment the instrument is calibrated at 25°C. Then the control is simply set to the actual measuring temperature. By this adjustment, the output current of the amplifier gets corrected to the desired temperature. In automatic adjustment, a variable resistor which is usually a thermistor or wire wound resistance that has an approximate desired resistance temperature coefficient is inserted in the circuit. During measurement, it is placed in the test solution. The use of an automatic temperature compensator will ensure that the pH meter is operating with a correct mV/pH conversion ratio.

If it is desired to have the accuracy of a pH measurement as 0.001 pH, then the voltage must be measured with an accuracy of 0.058 mV, assuming an ideal sensitivity of 58 mV per pH unit. With a symmetrical scale on the measuring device of 6 pH units around pH 7, the maximum voltage will be ±348 mV. Therefore, the accuracy requirement of the measurement will be 0.01%. This implies that the internal resistance of the measuring device has to be 10^4 times the internal resistance of the glass membrane (Bergveld and de Rooji, 1979).

Electrodes for Blood pH Measurement: Several types of electrodes have been described in literature for the measurement of blood pH. They are all of the glass electrode type but made in different shapes so that they may accept small quantities of blood and yield accurate results. The most common type is the syringe electrode, which is preferred for the convenience of taking small samples of blood anaerobically. The small 'dead space' between the electrode bulb and the inner surface of the syringe barrel is usually filled with dilute heparin solution to prevent blood coagulation. Before making measurements, the syringe should be rolled between the hands to ensure thorough mixing.

Microcapillary glass electrodes are preferred when it is required to monitor pH continuously: for example during surgery. These types of electrodes are especially useful when a very small volume of the sample is to be analyzed.

Typically, a micro-electrode for clinical applications requires only 20–25 µl of capillary blood for the determination of pH. The electrode is enclosed in a water jacket with circulating water at a constant temperature of 38°C. The water contains 1% NACI for shielding against static inter-ference. The capillary is protected with a polyethylene tubing. The internal reference electrode is silver/silver chloride and the calomel reference electrode is connected to a small pool of saturated KCl, through a porous pin. An accuracy of 0.001 pH can be obtained with this electrode against a constant buffer. Figure 15.2 shows the constructional details of a typical blood pH electrode and the measurement set-up used in practice.

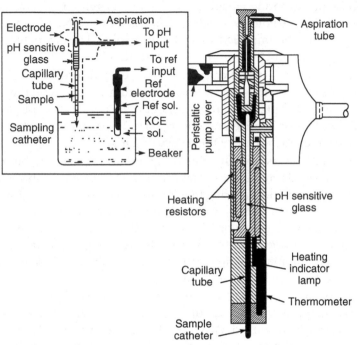

> **Fig. 15.2** *Microcapillary electrode for measurement of blood pH (Courtesy: Corning Scientific Instruments, USA)*

Quite often, combination electrodes comprising both measuring and reference electrodes offer single-probe convenience for all pH measurements. Several instruments offer the ability to measure pH in small containers with as little as 250 µl of the sample.

Ahn *et al.* (1975) bring out the drawbacks of the conventional macro-and micro-size pH electrodes when used for biomedical applications. These are due to the relatively large size of the macro-electrode and the fragility of the micro-electrode. They constructed a miniature pH glass electrode, using Corning 015 glass as the hydrogen-ion-sensitive glass. The dimensions of the electrode were –1.0 mm outside diameter and 0.25 mm wall thickness. The inner electrolyte is a solution of 0.1 N hydrochloric acid and the inner reference electrode is a silver/silver chloride electrode. The silver/silver chloride electrode is made from a silver wire (0.127 mm in diameter and 99.9% purity) by electrolytic method. A FET input operational amplifier is integrated into the pH electrode. Temperature response of the electrode was –1.51 mV/K at a pH value of 7. Evaluation of the stability of the electrode showed a 1% drift over a 7–hour operational time. The pH temperature hysteresis effects showed a 0.5 and 1.0% deviation, respectively. The response time was within 4s for a 99% response.

Effect of Blood on Electrodes Glass electrodes deteriorate if allowed to remain in contact with blood for a long time. This results in a change of the emf-pH slope. The poisoning effect appears to be due to protein deposition. Therefore, as a precautionary measure, in an apparatus where blood necessarily remains in contact around the electrode for long periods (more than 20 min.), the response must be checked frequently against buffer solutions. The poisoning effect can be reduced by putting the electrode in pepsin and 0.1 N HCl, followed by careful wiping with a tissue paper.

The pH of blood is found to change linearly with temperature in the range of 18° to 38°C. The temperature coefficient for the pH of blood is 0.0147 pH unit per degree centigrade. This necessitates the use of a highly accurate temperature controlled bath to keep the electrodes with the blood sample at 37°C ± 0.01°C. A circuit diagram for controlling temperature of the bath is shown in Fig. 15.12.

Another important point to be kept in mind while making blood pH measurements is that because of the possible individual variations in the temperature coefficient of blood pH, the method of measuring at some temperature other than 37°C followed by correction is not recommended. It is advisable to keep both the glass as well as the reference electrode at the temperature of measurement.

Buffer Solutions: Buffer solutions are primarily used for (i) creation and maintenance of a desired, stabilized pH in a solution and (ii) standardization of electrode chains for pH measurements. A buffer is, therefore, a substance which by its presence in a solution is capable of counteracting pH changes in the solution as caused by the addition or the removal of hydrogen ions. Buffer solutions are characterized by their pH value. They are available in tablets of pH value 4.7 and 9.2. Buffer solutions used in blood pH measurements are the following:

- 0.025 molar potassium dihydrogen phosphate with 0.025 molar disodium hydrogen phosphate. This solution has a pH value of 6.840 at 38°C and 6.881 at 20°C.
- 0.01 molar potassium dihydrogen phosphate with 0.04 molar disodium hydrogen phosphate. This buffer has a pH value of 7.416 at 38°C and 7.429 at 20°C.

These buffers should be stored at a temperature between 18° to 25°C. To maintain an accurate pH, the bottles containing them should be tightly closed.

▊▊▊▶ 15.3 MEASUREMENT OF BLOOD PCO_2

The blood pCO_2 is the partial pressure of carbon dioxide of blood taken anaerobically. It is expressed in mmHg and is related to the percentage CO_2 as follows:

$$pCO_2 = \text{Barometric pressure} - \text{water vapour pressure} \times \frac{\% \, CO_2}{100}$$

At 37°C, the water vapour pressure is 47 mmHg, so at 750 mm barometric pressure, 5.7% CO_2 corresponds to a pCO_2 of 40 mm.

All modern blood gas analyzers make use of a pCO_2 electrode of the type described by Stow *et al* (1957). It basically consists of a pH sensitive glass electrode having a rubber membrane stretched over it, with a thin layer of water separating the membrane from the electrode surface. The technique is based on the fact that the dissolved CO_2 changes the pH of an aqueous solution. The

CO_2 from the blood sample defuses through the membrane to form H_2CO_3, which dissociates into (H^+) and (HCO_3^-) ions. The resultant change in pH is thus a function of the CO_2 concentration in the sample. The emf generated was found to give a linear relationship between the pH and the negative logarithm of pCO_2. Although the electrode could not provide sensitivity and stability required for clinical applications, it made way for realizing a direct method for the measurement of pCO_2.

The basic construction of the electrode was modified by Severinghaus and Bradley (1958) to a degree that made it suitable for routine laboratory use. In the construction worked out by them, the water layer was replaced by a thin film of an aqueous sodium bicarbonate ($NaHCO_3$) solution. The rubber membrane was also replaced by a thin Teflon membrane, which is permeable to CO_2 but not to any other ions, which might alter the pH of the bicarbonate solution. The CO_2 from the blood diffuses into the bicarbonate solution. There will be a drop in pH due to CO_2 reacting with water forming carbonic acid. The pH falls by almost one pH unit for a ten-fold increase in the CO_2 tension of the sample. Hence, the pH change is a linear function of the logarithm of the CO_2 tension. The optimum sensitivity in terms of pH change for a given change in CO_2 tension is obtained by using a bicarbonate solution of concentration of about 0.01 mole/l. The electrode is calibrated with the known concentration of CO_2. The response time of the CO_2 electrode is of the order of 0.5 to 3 min. This electrode was twice as sensitive and drifted much less than the Stow's electrode. Figure 15.3 shows the construction of a typical pCO_2 electrode.

Further improvements in stability and response time were achieved by Hertz and Siesjo (1959). They used a dilute solution of $NaHCO_3$ (0.0001 N), which helped in reducing the response time but the drift introduced posed serious problems. The compromise between response time and drift was achieved by using a 0.001 N solution of $NaHCO_3$. Silver/silver chloride reference electrode was replaced by a calomel cell which was made an integral part of the electrode.

Severinghaus (1962) made a further improvement upon the earlier Severinghaus-Bradley electrode in the low pCO_2 range by replacing the cellophane spacer with a very thin nylon mesh. Glass fibres or powdered glass wool were also found to be good separators. He used a membrane of 3/8 mil Teflon and glass wool for the separator. Electrodes constructed in this way had a 95% response in 20 s.

Reyes and Neville (1967) constructed a pCO_2 electrode using 0.5 mm polyethylene as a membrane and used no

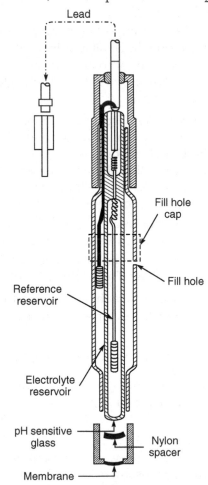

> **Fig. 15.3**   *Construction of pCO_2 electrode (Courtesy: Corning Scientific Instruments, USA)*

separator between the glass surface and this membrane. They added carbonic anhydrase to the electrolyte. The response time was found to be 6 seconds for 90% of a step change from 2% to 5% CO_2. Use of a pCO_2 electrode for the measurment of blood or plasma pCO_2 has been studied repeatedly and has been found to be accurate, precise and expedient, (Hill and Tilsley, 1973). An extension of the miniature pH electrode (Ahn *et al*, 1975) is the miniature pCO_2 electrode described by Lai *et al* (1975).

15.3.1 Performance Requirements of pH Meters Used for pCO_2 Measurement

The emf generated by a pCO_2 electrode is a direct logarithmic function of pCO_2. It is observed that a ten-fold change in pCO_2 causes the potential to change by 58 ± 2 mV. The pH versus log pCO_2 relationship is linear within ±0.002 pH unit from 1 to 100% carbon dioxide. Since 0.01 unit pH change corresponds to a 2.5% change in pCO_2 or 1 mmHg in 40 mmHg, for achieving an accuracy of 0.1 mmHg, it is desirable to read 0.001 pH unit, i.e. a resolution of 60 μv. This order of accuracy can be read only on a digital readout type pH meter or on an analog meter with expanded scale. The instrument should have a very high degree of stability and a very low drift amplifier. The input impedance of the electronic circuit must be atleast $10^{12} \Omega$.

It is essential to maintain the temperature of the electrode assembly constant within close limits. It is experimentally shown that variation in the temperature of ± 1°C produces an error of ±1.5 mmHg or about ±3% at 5 mm pCO_2. The combined effects of temperature change upon the sensitivity of the pH electrode and upon the pCO_2 of the blood sample amount to a total variation in sensitivity of 8% per degree centigrade.

Calculated Bicarbonate, Total CO_2 and Base Excess: Acid-base balance determinations are based on several calculations, which are routinely used in conjunction with blood pH and gas analysis. An accurate picture of acid-base balance can be determined from the equilibrium.

$$CO_2 + H_2O \rightarrow H_2CO_3$$
$$H_2CO_3 \rightarrow H^+ + HCO_3^-$$

which for bicarbonate has an equilibrium constant

$$K_{H_2CO_3}/HCO_3^- = \frac{[H^+][HCO_3^-]}{H_2CO_3}$$

where (H^+), (HCO_3^-) and (H_2CO_3) refer to the concentration of these substances.

Since $$H_2CO_3 = 0.03 \, pCO_2$$
and since $$pH = -\log[H^+]$$

Therefore, $$pH = pK + \log \frac{[HCO_3^-]}{0.03 \, pCO_2}$$

where pK equals 6.11 for normal plasma at 37°C. This formula is used in blood gas analysers for calculating actual bicarbonate.

Total CO_2 is calculated from the relationship:

$$[HCO_3^-] + (0.03 \times pCO_2) = \text{total } CO_2 \text{ in millimoles}/l$$

Base excess is calculated from the formula described by Siggaard-Andersen (1963).

$$\text{Base excess} = (1 - 0.0143 \times \text{Hb}) [\text{HCO}_3^-] - (9.5 + 1.63 \, \text{Hb}) \times (7.4 - \text{pH}) - 24$$

where Hb represents the patients' haemoglobin value.

Base excess is the number of milliequivalents of a strong acid or base which would be required per litre of blood to restore it to a pH of 7.400 at 37°C with pCO_2 held at 40 torr. This is usually estimated from pH and pCO_2 measurements done at 37°C in a sample of blood using Siggaard-Andersen's alignment monogram (Siggaard-Andersen, 1963).

15.4 BLOOD pO_2 MEASUREMENT

The partial pressure of oxygen in the blood or plasma indicates the extent of oxygen exchange between the lungs and the blood, and normally, the ability of the blood to adequately perfuse the body tissues with oxygen. The partial pressure of oxygen is usually measured with a polarographic electrode. There is a characteristic polarizing voltage at which any element in solution is predominantly reduced and in the case of oxygen, it is 0.6 to 0.9 V. In this voltage range, it is observed that the current flowing in the electrochemical cell is proportional to the oxygen concentration in the solution.

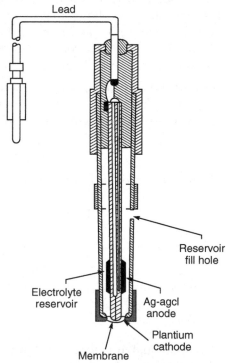

> **Fig. 15.4** *Construction of pO_2 electrode (Courtesy: Corning Scientific Instruments, USA)*

Most of the modern blood gas analyzers utilize an oxygen electrode first described by Clark (1956) for measuring oxygen partial pressure. This type of electrode consists of a platinum cathode, a silver/silver chloride anode in an electrolyte filling solution and a polypropylene membrane. The electrode is of a single unit construction and contains the reference electrode also in its assembly. Figure 15.4 shows the construction of a typical Clark-type oxygen electrode. The entire unit is separated from the solution under measurement by the polypropylene membrane.

Oxygen from the blood diffuses across the membrane into the electrolyte filling solution and is reduced at the cathode. The circuit is completed at the anode, where silver is oxidized, and the magnitude of the resulting current indicates the partial pressure of oxygen. The reactions occurring at the anode and cathode are:

Cathode reaction:

$$O_2 + 2H_2O + 4e^- \rightarrow 4\,OH^-$$

Anode reaction:

$$4Ag \rightarrow 4Ag^+ + 4e^-$$

The Clark electrode for measuring pO_2 has been extensively studied and utilized. It is found to be of particular advantage for measuring blood samples. The principal advantages are: (i) sample size required for the measurement can be extremely small, (ii) the current produced due to pO_2 at the electrode is linearly related to the partial pressure of oxygen, (iii) the electrode can be made small enough to measure oxygen concentration in highly localized areas, (iv) the response time is very low, so the measurements can be made in seconds. As compared to this, it takes a very long time if the measurements are made by chemical means.

McConn and Robinson (1963) observed that zero electrode current was not given by a solution having zero oxygen tension, but occurred at a definite oxygen tension, which they called the 'electrode constant'. So, for calibrating the electrode it was necessary to know this constant for that particular electrode. They further showed that when the straight line calibration curves (Fig. 15.5) were extended backwards, they did not pass through the origin, but intersected the oxygen tension axis at a negative value. To obtain a true zero-current (less than 10 nA), the electrolyte of the electrode is deoxygenated by bubbling nitrogen through it for about half an hour and then placing the electrode in water redistilled from alkaline pyragallol.

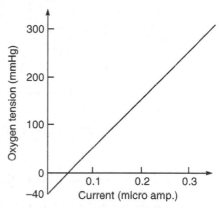

> **Fig. 15.5** *Calibration curve of pO_2 electrode (after McConn and Robinson, 1963)*

The platinum cathode of the oxygen electrode tends to become contaminated or dimensionally unstable with time and use. The result is usually an inability to calibrate and slope the electrode on any pO_2 range. The manufacturers usually recommend application of ammonium hydroxide on the tip of the electrode (10% solution), with a gentle, rotary motion using a swab. The silver chloride gets dissolved in ammonium hydroxide. It is then flushed with distilled water.

The polarographic electrodes usually exhibit ageing effect by showing a slow reduction in current over a period of time, even though the oxygen tension in the test solution is maintained at a constant level. Therefore, it needs frequent calibration. This is probably associated with the material depositing itself on to the electrode surface. The effect due to ageing can possibly be avoided by covering the electrode with a protective film of polyethylene, but it has the undesirable effect of increasing the response time.

The measurement of current developed at the pO_2 electrode due to the partial pressure of oxygen presents special problems. The difficulty arises because of the extremely small size of the electrical signal. The sensitivity (current per torr of oxygen tension) is typically of the order of 20 pA per torr for most commercial instruments. It is further subject to a constant drift and is also not independent of the sample characteristics. Measurement of oxygen electrode current is made by using high input impedance, low noise and low current amplifiers. Field effect transistors usually form the input stage of the preamplifiers.

Hahn (1969) used a field-effect transistor operational amplifier to measure small polarographic currents. The op-amp. is connected as a transresistance converter, the output of which can be read directly by a digital voltmeter. Figure 15.6 shows the circuit. The polarizing voltage is supplied by the cell B (1.3 V) and variable resistance VR_1. The standing current from the electrochemical cell is cancelled by means of VR_2, Battery B and 1 GΩ resistance. Capacitor C (100 pF) is included to limit the bandwidth of the amplifier to reduce noise and to ensure good dynamic stability.

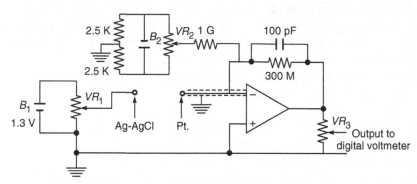

> **Fig. 15.6**   *Current amplifier for use with pO_2 electrode*

▷ 15.5   INTRA-ARTERIAL BLOOD GAS MONITORING

Arterial blood gas analysis is beneficial in the assessment and management of patients requiring mechanical ventilation and for those suffering from cardiopulmonary and other difficulties. Arterial blood gas values provide vital information about the adequacy of oxygenation, ventilation, acid-base balance and gas exchange in the lungs. *In vitro* blood gas analyzers, though commonly used, have several limitations. They require that blood be drawn and the sample analyzed, often at a distant blood gas laboratory. In addition to problems associated with blood handling, the need to send the sample to a laboratory delays results and treatment. Blood gas values can fluctuate rapidly in critically ill patients, and therefore, patient care decisions based on delayed information may be inappropriate. *In vivo* methods of blood gas monitoring have been developed to overcome these drawbacks.

15.5.1   Catheter Tip Electrode for Measurement Of pO_2 and pCO_2

Miniature electrodes are required for *in vivo* transcutaneous measurements of pO_2 and pCO_2. The electrodes must be small enough to be mounted on the catheter tip and should preferably perform measurements of more than one parameter. One such electrode capable of simultaneous measurement of both pO_2 and pCO_2 is described by Parker *et al* (1978). The device is built into the tip of a SF (1.65 mm) catheter, 40 cm in length. The electrode (Fig. 15.7) comprises a pH sensitive glass bulb at the tip of the catheter for measuring changes in pH and hence pCO_2 according to the method described by Severinghaus and Bradley (1958).

A 180 μm diameter silver cathode constitutes a pO_2 measuring electrode. The common electrode used is of silver/silver chloride. A semi-solid electrolyte is common for both the pO_2 and pCO_2 electrodes. The electrodes are dip-coated with a thin polystyrene diffusion membrane. When the device is placed in blood, water vapour diffuses through the membrane and together with the $NaHCO_3$ and NACI crystals deposited in the hydrogel film constitutes the electrolyte normally used with a pCO_2 electrode. Under these conditions, the output signal from both the pO_2 and pCO_2

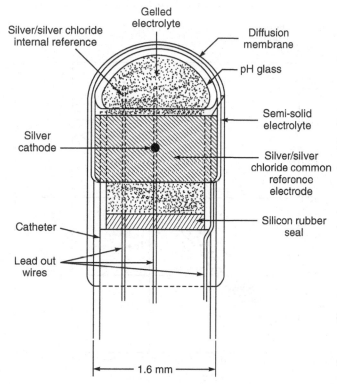

> **Fig.** 15.7 *Catheter tip electrode for measurement of pO_2 and pCO_2 (after Parker et al. 1978; reproduced by permission of Med. & Biol. Eng. and Comput.)*

electrodes is obtained. The response time for 90% response is found to be 2 min. However, the time for 100% stabilization of the outputs after a step change in pCO_2 and pO_2 in solution is pretty long.

15.5.2 pO_2 Measurement with Cutaneous Electrodes

Continuous intra-arterial monitoring of oxygen is unsatisfactory as a clinically reliable procedure. It is expensive and relatively traumatic. Oximetric methods of monitoring oxygen tension from oxygen saturation are unreliable at the pO_2 level above 50 mmHg (Scacci *et al*, 1976). Relatively simple and non-invasive methods are required to continuously monitor pO_2 to detect changes or establish trends. Eberhard *et al* (1973) concluded, after measuring oxygen tension using a Clark-type electrode applied directly to the skin, that an excellent correlation (0.98) exists between skin pO_2 and arterial pO_2 in infants and new borns (Fig. 15.8). This could provide more immediate detection of hypoxia or hyperoxia than arterial sampling.

The principle underlying the skin sensor is that since oxygen is able to diffuse through body tissue and skin, the measurement of pO_2 can be obtained indirectly by applying a Clark-type electrode sensor to the skin, heated to a constant temperature higher than the skin (44°C). Active vasodilation of the cutaneous vessels is achieved by warming the cathode and anode of the oxygen sensor to a temperature, which is higher than the normal body surface temperature. Oxygen

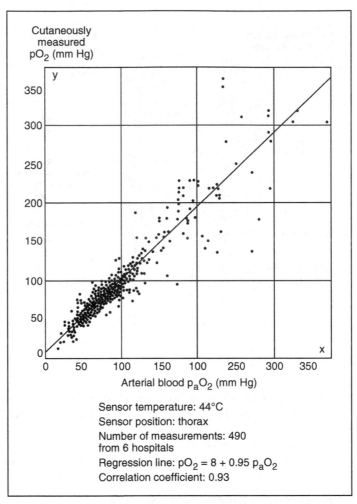

Cutaneously
measured
pO_2 (mm Hg)

Sensor temperature: 44°C
Sensor position: thorax
Number of measurements: 490
from 6 hospitals
Regression line: $pO_2 = 8 + 0.95\ p_aO_2$
Correlation coefficient: 0.93

Arterial blood p_aO_2 (mm Hg)

> **Fig. 15.8** *Relationship between central arterial pO_2 and cutaneous pO_2 of new-borns. A significant correlation (0.93) was found to exist between the two values (after Eberhard et al, 1976)*

diffuses from the arterialized capillary bed through the epidermis to the skin surface and is measured there by an electrochemical reduction at the cathode of a Clark-type sensor. The electrode is 14 mm in diameter, with a 4 mm gold cathode, silver/silver chloride anode, covered with a 6 μ thick Mylar membrane (Fig. 15.9). The electrolyte used is a solution of KCl buffered to pH 10 which has a 3 to 45 days life if kept moist. A coil of resistance wire embedded in the Sensor heats the electrode to 44°C, a temperature selected to provide good hyperemisation as well as safety for continuous application over several days. A thermistor is also imbedded in the sensor to provide a control signal for monitoring constant temperature. The response time of the electrode is 60 s for 95% response to a full-scale step change. The sensor is applied to the skin using adhesive tape. The polarogram of this sensor in the gas phase shows a stable plateau extending from –700 mV to

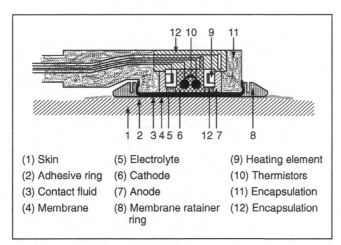

(1) Skin	(5) Electrolyte	(9) Heating element
(2) Adhesive ring	(6) Cathode	(10) Thermistors
(3) Contact fluid	(7) Anode	(11) Encapsulation
(4) Membrane	(8) Membrane rataîner ring	(12) Encapsulation

> **Fig. 15.9** *Cross-section of cutaneous oxygen sensor (Courtesy: Roche Medical Electronics Inc., USA)*

-1.1 V with an operating voltage of -900 mV. The zero-current in 100% nitrogen is less than 0.5 nA/mm^2. Eberhard *et al* (1975) explain the details of the electronic circuit used with cutaneous pO_2 sensors.

The cutaneous oxygen electrode is a satisfactory indicator of the changes in arterial oxygen tension in only those patients who have a good peripheral circulation. Skin pO_2 is probably more dependent on perfusion changes than on arterial tension. Consequently, this measurement may be more sensitive to physiologic changes than is the arterial oxygen tension.

In spite of the fact that a significant correlation between cutaneous pO_2 and arterial pO_2 exists if the sensor is heated to a temperature of 44°C, the cutaneous measurement technique cannot be regarded as a substitute for conventional blood-gas analysis via sampling of arterial blood (Eberhard and Mindt, 1976). In some situations, e.g. during marked peripheral vasoconstriction, the cutaneous pO_2 does not correlate with arterial blood pO_2, and hyper-oxemic or hypoxemic situations may not be reliably detected. Cutaneous pO_2 measurement is a supplemental technique, which helps to continuously monitor clinically significant changes in the oxygenation state of the subject, which may otherwise go unnoticed in the time interval between blood sample analysis.

15.6 A COMPLETE BLOOD GAS ANALYZER

Blood gas analyzers are designed to measure pH, pCO_2 and pO_2 from a single sample of whole blood. The size of the sample may vary from 25 μl to a few hundred microlitres. The estimations take about 1 minute. With built-in calculators, the instruments can also compute total CO_2, HCO_3 and Base Excess. A typical block diagram of a blood gas analyzer machine is shown in Figure 15.10. In this machine, separate sensors are used for pH, pCO_2 and pO_2. The instrument contains three separate high input impedance amplifiers designed to operate in the specific range of each measuring electrode. A separate module houses and thermostatically controls the three electrodes. It also provides thermostatic control for the humidification of the calibrating gases. A vacuum

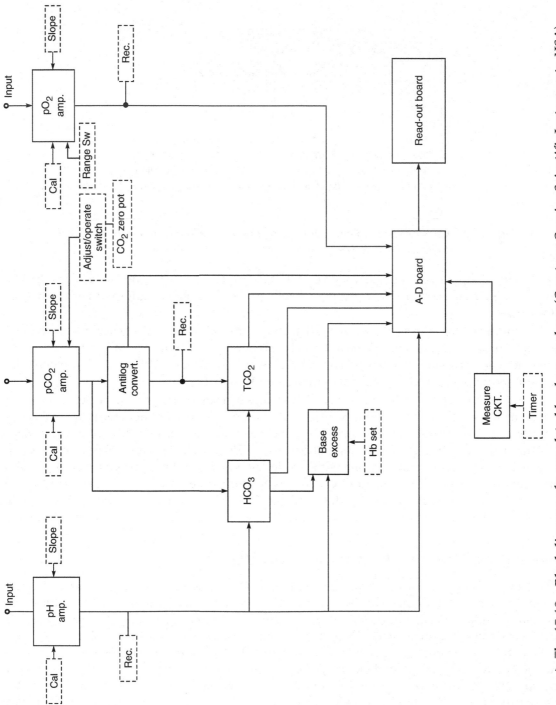

▶ **Fig. 15.10** *Block diagram of a complete blood gas analyser (Courtesy: Corning Scientific Instruments, USA)*

system provides aspiration and flushing service for all three electrodes. Calibrating gases are selected by a special push button control and passed through the sample chamber when required. Two gases of accurately known O_2 and CO_2 percentages are required for calibrating the analyzer in the pO_2 and pCO_2 modes. The gases required are: O_2 value of 12% Cal and 0% Slope and CO_2 value of 5% Cal and 10% Slope. These gases are used with precision regulators for flow and pressure control. Two standard buffers of known pH are required for calibration of the analyzer in the pH mode. The buffers that are used are 6.838 (Cal) and 7.382 (Slope). It is generally recommended that the sample chamber should control 7.382 buffer when in the standby mode.

Input signal to the (HCO_3^-) calculator (Fig. 15.11a) comes from the outputs of the pH and pCO_2 amplifiers. The outputs are suitably adjusted by multiplying each signal by a constant and are given to an adder. The next stage is an antilog-generator similar to the one used in a pCO_2 amplifier. The output of this circuit goes to an A–D converter for display. Resistance R is used to adjust zero at the output.

Total CO_2 is calculated (Fig. 15.11b) by summing the output signals of the (HCO_3^-) calculator and the output of the pCO_2 amplifier. Facilities for adjusting the slope and zero at the output are available.

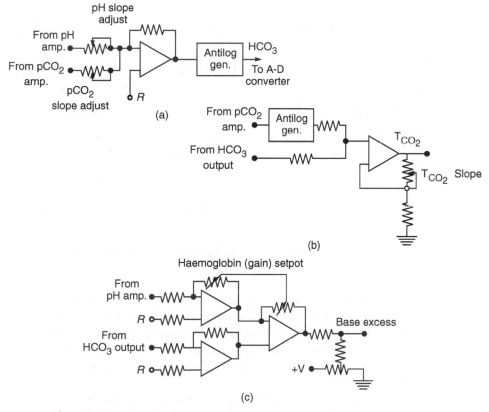

> **Fig. 15.11** *Circuit diagrams for computation of*
> *(a) Bicarbonate (HCO_3^-) (b) Total CO_2 (c) Base excess*

The base excess calculator (Fig. 15.11c) consists of three stages. In the first stage, the output of the pH amplifier is inverted in an operational amplifier whose gain is controlled with a potentio-meter (Haemoglobin value) placed on the front panel. The output of the HCO_3^- calculator is inverted in the second stage. The third stage is a summing amplifier A_3 whose output is given to an A–D converter.

The analog output of the selected parameter channel is given to the input of an A–D converter. The output of the A–D converter goes to a digital readout circuit like LEDs.

The three electrodes (pH, pO_2 and pCO_2) are housed in a thermostatically controlled chamber. It also provides thermostatic control for the humidification of the calibrating gases. The thermal block and the humidifier block heat control circuits are of the same type (Fig. 15.12). The temperature is set with a potentiometer for exactly 37°C. The heater circuit is controlled by a thermistor in the block, which acts as a sensor. As the heat increases, the resistance of the thermistor decreases. At 37°C, the thermistor is calibrated to have a resistance of 25 K.

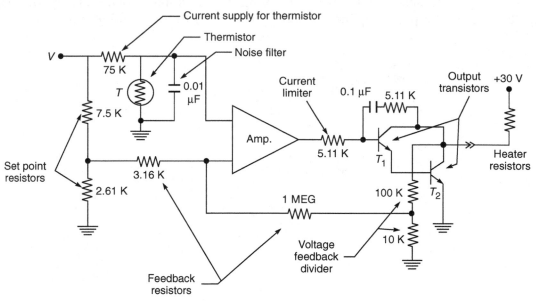

> **Fig. 15.12**   *Temperature control circuit for thermostated chamber (Courtesy: Corning Scientific Instruments, USA)*

Supposing the temperature of the block decreases, the resistance of the thermistor will increase. The increase in resistance will cause the voltage at inverting input of op-amp to become more negative. This results in the output voltage becoming more positive, increasing the base current of transistors T_1 and T_2. The increase in base current increases the collector current, which goes directly to the heater resistor on the block. As the heater resistor heats up the block, the thermistor will decrease until it returns to 25 K.

Many of the blood gas analyzers have a provision for checking the membrane of pO_2 and pCO_2 electrodes. In the check position, a potential is applied across the membrane. Any leak in the membrane of sufficient magnitude will result in a considerable lowering of the resistance may be from 100 MΩ to 500 KΩ. The change in resistance can be used to have a change of potential to switch on a transistor, which would cause a lamp to light on the front panel of the instrument. This would indicate that a new membrane is needed.

15.6.1 Fiber-optic Based Blood Gas Sensors

For *in vivo* measurements and for reliably analyzing blood gases, a small, stable, accurate and biocompatible sensor is required which could be inserted in the blood flow of an artery (Miller, 1993) through an arterial cannula and remain in place for several days. In addition, it has to be low cost so that it could be used as a disposable item. Advances in fiber-optics and the development of pH and oxygen sensitive dyes have made such a sensor possible. Blood gas analyzers based on such sensors are now commercially available.

Figure 15.13 shows the schematic diagram of a fiber-optic based blood gas analyzer (Soller, 1994). The sensors are interfaced with an electro-optic monitor. The monitor supplies the excitation light, which may be from a monochromatic source such as a diode laser or a broadband source

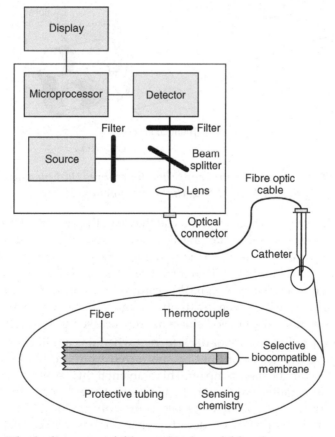

➤ **Fig. 15.13** *Block diagram of fiber optic based blood gas sensor and monitor (after Soller, 1994)*

like a xenon lamp whose light is filtered to provide a narrow bandwidth of excitation. Two wavelengths of light are provided, one wavelength is sensitive to changes in the species to be measured, while the other wavelength is unaffected by changes in the analyte concentration. The unaffected wavelength serves as a reference and is used to compensate for fluctuations in the source output and detector efficiency. The light output from the monitor is coupled into a fiberoptic cable through appropriate lenses and optical connectors. The cable is sufficiently long to permit easy patient access by allowing the monitor to be placed at a distance.

Within the sensor assembly (Fig. 15.14) are three optical fibers—one each for measuring blood O_2, CO_2 and pH. The optical-fiber is approximately 10 cm long and also has a thermocouple or thermistor wire running along side the fiber to measure temperature near the sensor tip. Temperature correction is necessary for optical blood gas sensors. The solubility of the gases namely O_2 and CO_2 in the sensing material, is a function of temperature and the optical properties of the sensing chemistry also change as the temperature varies. The fibers and the temperature sensor are encased in a protective tubing to contain any fiber fragments in case of sensor breakage.

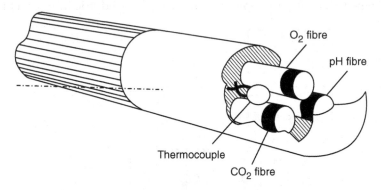

> **Fig. 15.14** *The sensor assembly contains three fibres that measure pH, pCO_2 and pO_2, bundled together with the thermocouple*

Each fiber is as thin as human hair and coated at the tip with a specific chemical dye (Fig. 15.15). When light of a known wavelength strikes the dye, the dye fluoresces, giving off light of a different wavelength. The fluorescent emission changes in intensity as a function of the concentration of the analyte (O_2, CO_2 or pH) in the blood. The emitted light travels back down the fiber to the monitor where it is converted into an electrical signal by using a solid state detector or a photomultiplier. The signal is amplified before it is given to a digitizer. Signal processing to relate the light intensity to the analyte concentration is achieved using a microprocessor and is digitally displayed.

Because a detector produces noise or dark current when it is not illuminated, accurate signal measurements require that ambient light be subtracted from the total signal. Thus, a signal measurement is made with the flash lamp off. This ambient light is subtracted from the total signal by correlated double integration circuits. Another factor that affects accurate blood gas calculations is the background fluorescence of the materials in the optical block. This value is obtained by measuring the current developed by the detector when no sensor is connected. As with ambient light, the background fluorescence is subtracted from the signal measurement.

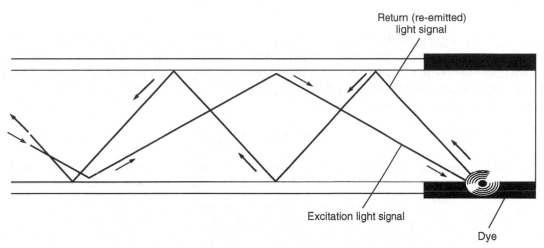

➤ **Fig. 15.15** *Within each fibre's core, excitation light reflects along the fibre toward the fluorescent dye at the fibre's tip. The dye at the tip reacts to the excitation light and analyte concentration by fluorescing. The fluorescent signal then returns in the same fibre to the monitor, which measures the intensity of the signal*

Considerable effort has gone into identifying organic molecules, which would make suitable sensors. These molecules must have a high fluorescent intensity at excitation and emission wavelengths that match the available light sources and detectors. They must be photostable, i.e. their emission properties should not change as they are continually illuminated by the excitation source. Sensors based on fluorescence quenching of organic dyes such as perylene dibutyrate have been reported for the measurement of pO_2. Oxygen sensors based on the phosphorescence quenching of metal—loporphyrins and terbium complexes have also been successfully tried.

It has been found from experimental studies that as the partial pressure of oxygen increases, the sensitivity decreases. The best sensitivity is achieved in the region of 30 to 150 mmHg, but drops off considerably with higher pO_2, making it difficult to resolve small changes in pO_2, when the oxygen partial pressure is greater than 200 mmHg. Further, at high pO_2 as the quenching increases, the light reaching the detector decreases. A compromise is thus required to be made in selecting a sensing material that provides adequate sensitivity over the required pO_2 measurement range and simultaneously offers good signal-to-noise ratio at the detector. The performance range of the sensor is normally limited to under 300 mmHg in order to achieve good sensitivity and adequate light detection.

pH sensor: Designs are based on dye molecules whose optical properties change as the pH is varied between 6.8 and 7.8. At any pH in the range of interest, both the acid and base forms of the dye molecule are present and each form has distinct optical characteristics. pH sensors have been developed which take advantage of the fact that the excitation wavelength for fluorescence emission of some dyes is different for the acid and base forms and the ratio of emission excited at these two wavelengths can be used to calculate pH. Additionally, sensors have been developed which utilize the difference in absorption maxima for both the acid and base forms of the dye.

The commonly used pH-sensitive dye is phenol red whose absorption spectra is shown in Fig. 15.16. The largest peak is observed from base form of phenol red at 560 nm and is used to measure pH because it is more sensitive to pH changes than the acid peak at 430 nm. A wavelength, which is insensitive to pH changes, is used as a reference; either a wavelength greater than 600 nm or the isobestic point at 480 nm. The relationship between pH and the base form of the dye is given by the Henderson-Hasselbalch equation:

$$pH = pKa - \log \frac{[HA]}{[A^-]}$$

where pH is the negative logarithm of hydrogen and concentration and pKa is the negative logarithm of the equilibrium constant K_a, which describes the dissociation of the acid, HA.

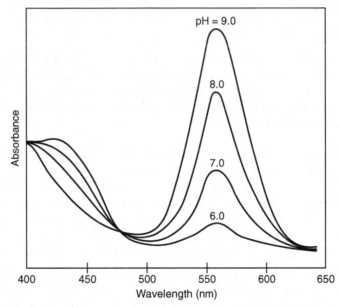

➤ **Fig. 15.16** *Absorption spectra of pH sensitive dye (Phenol red)*

One of the difficulties in designing a pH sensor is to achieve resolution of 0.01 pH units over the range of 6.8 to 7.8. An effective way to achieve this is to optimize the pKa of the dye material. This can be done through proper choice of a functional group attached to the dye molecule or by immobilizing the pH-sensitive material on a polymer with the appropriate ionic characteristics.

Most fiberoptic sensors for measuring pCO_2 use the same approach as a pCO_2 electrode. A pCO_2 sensor is fabricated by surrounding a pH sensor with a gas permeable membrane containing a bicarbonate ion (HCO^-_3) buffer. The membrane allows gaseous CO_2 and water vapour to enter the sensor, and they combine to form carbonic acid as per the following equations:

$$CO_2 + H_2O \rightarrow H_2CO_3$$
$$H_2CO_3 \rightarrow H^+ + HCO^-_3$$

The partial pressure of CO_2 can be related to the measured pH through the equilibrium constants for the above reactions and the equation is

$$pH = \log N + pK_1 - \log (K_s\, pCO_2)$$

where N = concentration of bicarbonate ion in the sensor

pK_1 = negative log of the acid dissociation constant for H_2CO_3 times the hydration constant for CO_2.

K_s = solubility coefficient for CO_2

This principle for the design of pCO_2 sensors has been implemented using both fluorescence-based and absorption-based pH sensors (Vurek *et al.*, 1983).

The methods for measuring pH, CO_2 and O_2 are similar, except that the wavelength of light used for different blood gas parameters vary. The optics is composed of three channels, each for measuring one of the parameters. Provision for calibration is made in the measuring system to compensate for individual physical variations between sensors and monitors. The calibration technique involves placing the sensor in a calibration solution, then bubbling precision mixtures of O_2, CO_2 and nitrogen (N_2) through the fluid. When equilibrium is reached, there are known partial pressure of O_2 and CO_2 in the solution. The pH is also known from the gas tensions and the chemical composition of the solution. Bubbling is repeated with a second gas mixture to provide a second calibration point. Using both calibration points, the monitor can calculate the appropriate calibration factors for that sensor.

With the development of fiber-optic based blood gas sensors, routine electrode membraning and maintenance have become history. Continuous self-monitoring provides clear and immediate information of instrument performance. The keyboard-based user interface provides advanced analytical performance and data processing capabilities. Along with measurement of the blood pH, pO_2 and pCO_2, some instruments like the AVL OPTI Critical Care Analyzer (Fig. 15.17) also include facilities for measuring other important ions such as Na+, Ka+, Ca++ and Cl⁻ in the blood. This is possible with the development of optical sensors based on fluorescence emission. All the sensors are mounted on a cassette shown in Fig. 15.18. The syringe adapter shown at the right side of the cassette allows the automatic aspiration of a sample directly from a syringe. Removing the adapter allows for direct sample aspiration from a capillary or microsampler. The sensor calibration is verified by the system automatically after insertion of the cassette. The cassette is removed after sample analysis.

The optical sensors in the cassette are designed in a way that the analytes bind with the fluorescent sensor molecule. The sensor molecule is selective for the specific analyte, i.e. the pH sensor molecule reacts only with "H^+", the O_2 sensor only with O_2 molecules, etc. The intensity of the emitted fluorescent light varies with the concentration of the ions (H^+, Na^+, K^+) or the partial pressure of the gas molecules (O_2 and CO_2) in the sample. The relationship is specific for each analyte. The corresponding calibration information for each component is encrypted in the bar code. Before the analyte can bind to the fluorescent molecule, it is made to pass through an optical isolator. The isolator prevents interference by unspecific light with the light detection system. The pO_2 sensor also makes it possible to measure and compute the total haemoglobin and oxygen saturation. The equipment works on a minimum sample size of 125 µl.

> **Fig. 15.17**   *Critical care analyser for measuring blood gases and other parameters (Courtesy: M/s AVL Medical Instruments)*

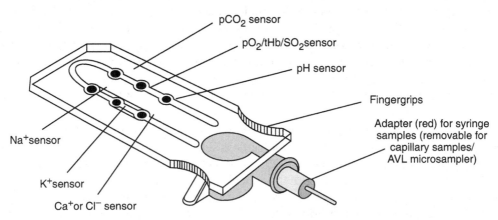

> **Fig. 15.18**   *Sensor cassette (Courtesy: M/s AVL Medical Instruments)*

With the miniaturization of the direct reading electrodes, it is possible to combine them into a single cuvette so that a complete blood gas determination could be made on a single small sample.

The introduction of the microprocessor and its use in blood gas analyzers free the medical personnel from monitoring the reaction in the electrode chamber and from the tedious chores of calculating and copying the results.

All commercial blood gas analyzers make use of the same basic electrodes and signal conditioner circuitry. The main differences between instruments manufactured by various companies are not the measurements of the parameters but the degree of automation and the technique by which the sample is presented to the electrodes.

CHAPTER

16

Blood Cell Counters

〰▶ **16.1 TYPES OF BLOOD CELLS**

Changes in the normal functioning of an organism are often accompanied by changes in the blood cell count. Therefore, the determination of the number and size of blood cells per unit volume often provides valuable information for accurate diagnosis. The blood constitutes 5–10% of the total body weight and in the average adult, it amounts to 5–6 l. Blood consists of corpuscles suspended in a fluid called plasma in the proportion of 45 parts of corpuscles (cells) to 55 parts of plasma. The percentage of cells in the blood is called the haematocrit value or packed cell volume (PCV). The majority of the corpuscles in blood are red blood cells (erythrocytes), others being white blood cells (leucocytes) and platelets (thrombocytes).

Blood cells are divided into groups according to their form and function as shown in Table 16.1.

• Table 16.1 *Blood Cell Types*

Blood cell types	Number of cells in mm^3	Mean cell volume (MCV) In μm^3	Relative proportion of different leucocyte count (differential)
1. Erythrocytes	$(4.8–5.5) \pm 1 \times 10^4$	90	
2. Leucocytes	5000–10,000		100%
(a) Neutrophils	2000–7500	450	59 ± 18%
(b) Lymphocytes	1500–4000	250	34 ± 10%
(c) Eosinophils	40–400	450	2.5%
(d) Basophils	10–100	450	0.5%
(e) Monocytes	200–800	600	4%
3. Thrombocytes	$1.5 \times 10^5 – 4 \times 10^5$	8	–

Erythrocytes (Red Blood Cells): Red blood cells have the form of a bi-concave disc with a mean diameter of about 7.5 μ and thickness of about 1.7 μ. The mean surface area of the cell is about

$134\,\mu m^2$. There are about 5.5 million of them in every cubic millimetre of blood in men and nearly 5 million in women. In the whole body, there are about 25 billion erythrocytes and they are constantly being destroyed and replaced at a rate of about 9000 million per hour. The normal red cell lasts approximately 120 days before it is destroyed.

The erythrocytes have no nucleus. They are responsible for carrying oxygen from the lungs to the tissues and carbon dioxide from the tissues to the lungs. Anaemia (reduction in the oxygen carrying capacity of blood) can develop from a change in the number, volume or Hb concentration of erythrocytes, caused by bone marrow dysfunction resulting in the poor production rate of RBCs. Since these changes are specific, the measurement of packed cell volume (PCV), the number of RBCs and the haemoglobin (Hb) are very important.

Leucocytes (White Blood Cells): Leucocytes are spherical cells having a nucleus. There are normally 5000–10,000 white cells per cubic mm of blood but their number varies during the day. They live for seven to fourteen days and there is a rapid turn over, with constant destruction and replacement.

Leucocytes form the defence mechanism of the body against infection. They are of two main types: the neutrophils and the lymphocytes. Neutrophils ingest bacteria and lymphocytes are concerned with immunological response. The number and proportion of these types of leucocytes may vary widely in response to various disease conditions. For this reason, it is important to know the total leucocyte count. The change, however, is often so small that the WBC count remains within normal limits and only the differential count would indicate any abnormality.

Neutrophils are nearly twice as big as the red cells and contain both a nucleus divided into several lobes and granules in their protoplasm. Lymphocytes are of the same size as the red cells but contain a large density staining nucleus and no granules. Monocytes are another type of leucocytes, which are twice as big as the neutrophils. They have a single large nucleus and no granules.

Thrombocytes (Platelets): Platelets are usually tiny, round, oblong or irregularly shaped cells of the blood with an average diameter of approximately $2\,\mu$. They play an important role in the blood coagulation process. There are usually 250,000–750,000 platelets in every cubic mm of blood.

16.1.1 Calculation of Size of Cells

Mean Cell Volume (MCV): It is calculated from the PCV and the number of red cells present per litre of blood. For example, if the PCV is 0.45, i.e. 1 litre of blood contains 0.45 litres of red cells and if there are 5×10^{12} red cells per litre, then

$$\text{Mean volume of one cell} = \frac{0.45}{5 \times 10^{12}} = 90\,\text{f/l} \qquad \text{f/l} = \text{Femolitres}$$

$$1\text{f/l} = 10^{-15}$$

Normal mean red cell volume is 86 ± 10 f/l. In diseased conditions, it may fall to 50 f/l or rise upto 150 f/l.

Mean Cell Haemoglobin (MCH): It is calculated from the Hb and red cell count. For example, if there are 15 g of Hb per decilitre (dl) of blood, there will be 150 gram Hb per litre of blood. Supposing the number of red cells is 5×10^{12} per litre, then

$$\text{MCH} = \frac{150}{5 \times 10^{12}} = 30\,\text{picogram (pg)}$$

Normal mean cell haemoglobin is 29.5 ± 2.5 pg. In diseased conditions it may rise to 50 pg or fall to 15 pg.

Mean Cell Haemoglobin Concentration (MCHC): It can be calculated if PCV and Hb per dl are known. For example, if PCV is 0.45 and there are 15 g Hb per dl of blood, then

$$MCHC = \frac{15}{0.45}\, g/dl$$
$$= 33.3\, g/dl$$

Mean Platelet Volume (MPV): It is the ratio of the integrated platelet volume to the platelet count and is expressed in femolitres.

Plateletcrit (PCT): It is the percentage of the total specimen volume occupied by the platelets. Information from the platelet count (PLT) and mean platelet volume is expressed by the following equation:

$$PCT\% = \frac{MPV(fl) \times (PLT) \times (10^9/l)}{10}$$

Red Cell Distribution Width (RDW): It is a numerical expression of the width of the size distribution of red cells. It is derived by analog computation. The total erythrocyte count is scanned by a continuously variable thresholding circuit. The upper threshold is moved progressively lower from a level equivalent to 360 femolitres until 20 per cent of all erythrocytes present have a size above a certain value. This is recorded as the twentieth percentile value. The lower threshold is moved downwards until it reaches a level, which 80% of all erythrocytes exceed. This is labelled as the eightieth percentile value.

The RDW index is expressed by the following equation:

$$RDW = \frac{(20th - 80th)\ Percentile\ volume}{(20th + 80th)\ Percentile\ volume} \times 100 \times K$$

The constant K is the calibration factor to produce a result of 10 for a normal population.

Platelet Distribution Width (PDW): This index is related to the size range covered by those platelets lying between the sixteenth and eighty fourth percentile. This is the conventional geometric standard deviation of the mean platelet size and is derived from the distribution curve based on the data in a 64-channel pulse height analyzer.

16.2   METHODS OF CELL COUNTING

16.2.1   Microscopic Method

The most common and routinely applied method of counting blood cells even today, particularly in small laboratories, is the microscopic method in which the diluted sample is visually examined and the cells counted. Commonly known as the counting chamber technique, it suffers from several common drawbacks. Apart from the inherent error of the system, which may be about 10%, there is an additional subjective error of ±10% entailing poor reproducibility of the results. Furthermore, the lengthy procedure involved results in the rapid tiring of the person making the examination. There is also poor time and labour utilisation.

Another problem with microscopic counting is that the data gathered by this measurement is not directly suitable for storage or for further processing and evaluation. Furthermore, the increasing number of examinations carried out in large series in busy laboratories necessitated the development of automatic instruments for counting the blood particles, with the errors of counting significantly reduced than a counting chamber. Agoston and Zillich (1971) compared the results of microscopic counting with those made by electronic counters (Fig. 16.1). It may be observed that instead of the ±20% measuring accuracy in microscopic counting, the electronic counters can provide an accuracy of ±3%.

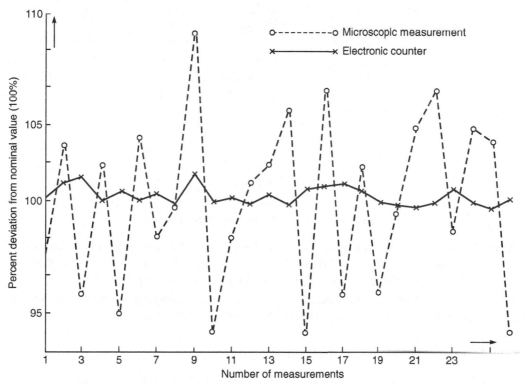

> **Fig. 16.1** *Comparison of results obtained with microscopic counting and electronic techniques (after Agoston and Zillich, 1971).*

16.2.2 Automatic Optical Method

The method is based on collecting scattered light from the blood cells and converting it into electrical pulses for counting. Figure 16.2 shows one type of arrangement for the rapid counting of red and white cells using the optical detection system. A sample of dilute blood (1:500 for white cells and 1:50,000 for red cells) is taken in a glass container. It is drawn through a counting chamber in which the blood stream is reduced in cross-section by a concentric high velocity liquid sheath. A sample optical system provides a dark field illuminated zone on the stream and the light scattered in the forward direction is collected on the cathode of a photomultiplier tube. Pulses are

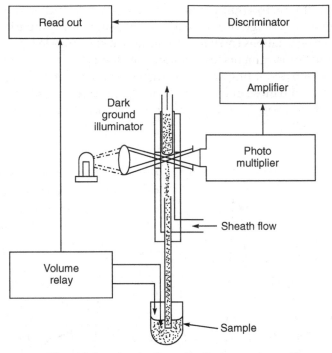

> **Fig. 16.2**   *Optical method of counting cells*

produced in the photomultiplier tube corresponding to each cell. These signals are amplified in a high input impedance amplifier and fed to an adjustable amplitude discriminator. The discriminator provides pulses of equal amplitude, which are used to drive a digital display.

Instruments based on this technique take about 30 s for completing the count. An accuracy of 2% is attainable. The instruments require about 1 ml of blood sample.

16.2.3   Electrical Conductivity Method

Blood cell counters, operating on the principle of conductivity change, which occurs each time a cell passes through an orifice, are generally known as Coulter Counters. The method was patented by Coulter in 1956 and it forms the basis of several particle counting instruments manufactured by a number of firms throughout the world. The technique is extremely useful for determining the number and size of the particles suspended in an electrically conductive liquid.

The underlying principle of the measurement is that blood is a poor conductor of electricity whereas certain diluents are good conductors. For a cell count, therefore, blood is diluted and the suspension is drawn through a small orifice. By means of a constant current source, a direct current is maintained between two electrodes located on either side of the orifice. As a blood cell is carried through the orifice, it displaces some of the conductive fluid and increases the electrical resistance between the electrodes. A voltage pulse of magnitude proportional to the particle volume is thus produced. The resulting series of pulses are electronically amplified, scaled and displayed on a suitable display.

To achieve optimum performance and to enable the relationship of change in resistance with volume of the cell to hold good, it is recommended that the ratio of the aperture length to the diameter of the aperture should be 0.75:1, i.e. for an orifice of 100 μ diameter the length should be 75 μ.

The instrument based on the Coulter principle works most satisfactorily when the average diameter of the particles ranges between 2 to 40% of the diameter of the measuring hole. Therefore, the following condition must be met for the measuring range:

$$D/50 \leq d \leq D/2$$

where d = maximum particle size

D = diameter of the measuring aperture

The lower limit of measurement in the system is governed by the noise sources involved. The noise sources include the thermal noise of the detector due to the resistance of the fluid flowing across the orifice and the noise inherent in the electronic circuits.

Particles of sizes larger than the diameter of the measuring aperture can only pass through the aperture if their longest dimension is parallel to the axis of the measuring aperture; otherwise they cause the clogging of the aperture. The upper limit of measurement is thus imposed by the increasing size of the particles. When the size of the particle approximates the diameter of the aperture, an amplitude linearity error is produced. Therefore, to count particles of different sizes, the diameter of the measuring aperture must be chosen in such a way as to meet the conditions of measurement. Simple procedures enable the extention of the range as needed via the use of successive aperture size—the total range covered is from about 0.5 microns to upwards of 500 microns. The applicability of the Coulter principle for measuring particles of various sizes is shown in Fig. 16.3.

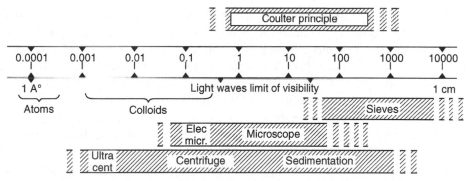

> **Fig.** 16.3 *The applicability of Coulter principle in the range from 0.5 microns to 500 microns*

16.3 COULTER COUNTERS

A wide range of particle counting instruments designed to meet a wide variety of needs in the haemotology laboratory are being commercially produced. These instruments range from the small counters used primarily for red and white cell counts in very small hospitals and clinics, to the multi-parameter microprocessor controlled instrument featuring fully automatic diluting of samples and printing of results.

Figure 16.4 shows a block diagram showing the principle of a Coulter counter. A platinum electrode is placed inside the orifice tube and a second electrode is submerged into the beaker containing the cell dilution, creating an electrical circuit between the two electrodes. Current will flow from one electrode to the other through the orifice. When the cell suspension is drawn through the orifice, cells will displace their own volume of electrolyte and cause a resistance change, which is converted to a voltage change, and is amplified and displayed.

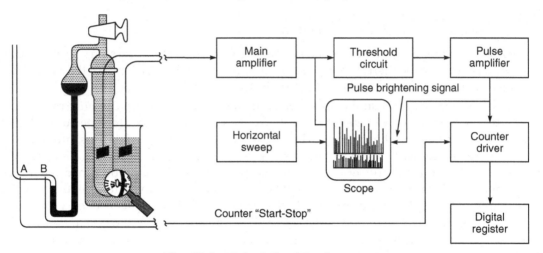

> **Fig. 16.4** *Principle of Coulter counter*

In practice, the cell suspension is drawn through the orifice by means of a mercury manometer. This manometer includes two platinum wire contacts (A and B) set through the glass walls. Contact A will start the count and contact B will stop it when precisely 0.5 ml of the dilution has passed through the orifice tube. Thus, it provides a count of the number of particles in a fixed volume of suspension. Figure 16.5 shows the sequence of building up the pulse in terms of increase in resistance at different positions of the cell with respect to the orifice.

To enable the instrument to count only those pulses, which fall within certain preset size limits, the threshold facility is required. The threshold is also necessary to enable the instrument to ignore any electronic noise, which may be present in the system. The lower threshold sets an overall voltage level, which must be exceeded by a pulse before it can be counted. The upper threshold will not allow pulses to be counted which exceed its preset level.

The Coulter counters are usually provided with an oscilloscope monitor to display the pulse information, which has passed through the amplifier, and acts as a visible check on the counting process indicating instantaneously any malfunctions such as a blocked orifice. In particular, it provides information regarding (i) relative cell size, (ii) relative cell size distribution, (iii) settings of the threshold level control, and (iv) means to check the performance of the instrument for reliability of counts. The voltage pulses produced each time a cell passes through the orifice are displayed on the oscilloscope screen as a pattern of vertical spikes.

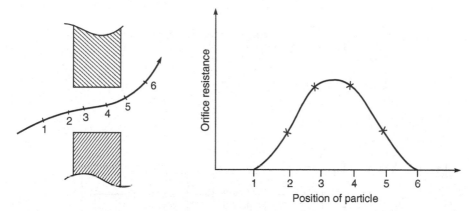

➤ **Fig. 16.5** *The sequence of building up the pulse in terms of increase in resistance at different positions the cell has with respect to the orifice*

Coulter counters also help to give an idea of the size distribution of various types of cells. It has been stated that the pulse-height is to a first approximation, proportional to the volume of the particle. Converting the pulse height into a digital number, through an A-D converter, and storing it in memory can help to obtain a plot of the number of cells as a function of their size (McGann *et al* 1982).

Taylor (1970) suggests that an aperture diameter of 100 μ would be generally useful. For such an aperture, a length of about 200 μ and flow rate of 0.04 ml/s would be optimum. The aperture can be made using ruby watch jewels bonded to a glass surface. Typically, an aperture of 100 μ diameter and 200 μ length, separating two solutions of phosphate buffered saline, has a resistance of about 25 kΩ and capacitance of 120 pF. The minimum rise time is about 5 μs. This means that the electronic circuit must have upper frequency response greater than 70 kHz. The preamplifier used in cell counters must be of very low noise preferably having noiset voltage less than 2 nA at the required bandwidth of 70 kHz.

Calibration: The calibration factor is constant for a given aperture size, electrolyte resistivity and amplifier gain setting. It is used for the conversion of threshold settings to particle volumes or their cube roots to equivalent spherical diameters. Calibration is done simply and quickly by observing the threshold for monosized particles of known diameter (adjusting the threshold level to the peaks of the single-height pulses on the oscilloscope screen). Ragweed pollen (19 micron in diameter) and polystyrene latex particles (6–14 micron in diameter) seem to meet these requirements. Of the two, polystyrene latex is preferred for calibration purposes (Thom, 1972). The particles when used seldom plug the orifice. These can be conveniently obtained in the range of 5 million particles per cubic mm.

16.3.1 Multi-parameter Coulter Counter

Figure 16.6 shows the block diagram of a multi-parameter Coulter counter. It provides the universally accepted profile of white cell count, red cell count, mean cell volume, haemotocrit, mean cell haemoglobin concentration, mean cell haemoglobin and haemoglobin. Besides this, the

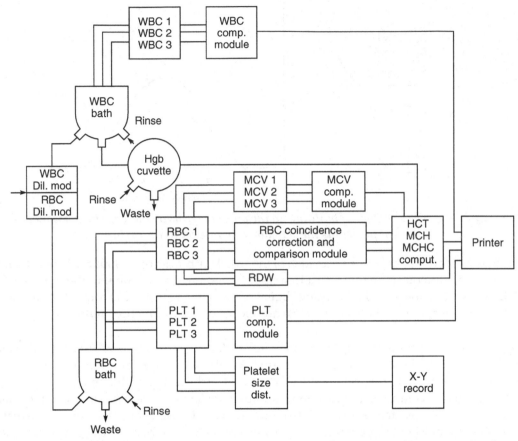

> **Fig. 16.6**   *Block diagram of multi-parameter Coulter counter (Courtesy: Coulter Electronics, USA)*

following five parameters are presented: platelet count, red cell distribution width, mean platelet volume, plateletcrit, and platelet distribution width. This instrument is microprocessor-controlled and provides the flexibility of expressing in various forms the available count and size data stored in memory.

All the directly measured parameters are measured in triplicate and the average results are displayed. All the 14 parameters are obtained from 1 ml of whole blood in 34–50 s depending on the number of platelets present. All diluting and mixing of samples is controlled by the microprocessor, which at the same time monitors the instrument and checks all the electronic circuits and pneumatics for correct functioning. A specially designed ticket is provided showing results of the measurements and computations. All this information is presented simultaneously (Fig. 16.7) so that the data can be quickly reviewed just by scanning the display. Since each distribution is an average of the information from the three measurement apertures, data from any one of the three apertures, can be examined, one after the other, or overlaid, one upon the other, for a comparative review.

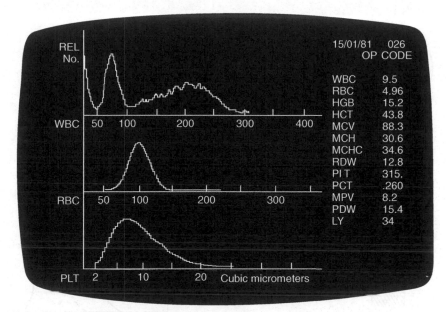

> ➤ **Fig. 16.7** *Simultaneous display of normal white cell data, red cell distribution curve on the data terminal of the model S-plus Coulter counter*

Coincidence Compensation: Coincidence error occurs when two or more particles are present in the sensing zone at the same time. This will result in the instrument detecting fewer particles than are actually present. This will also result in the instrument adding the pulses together to produce a single much longer pulse (Fig. 16.8). As the instrument detects fewer pulses than are actually present, to render the total count accurate, it is necessary to add on the pulses that have been 'lost' due to coincidence. The rate at which this happens has been mathematically determined. Model S-plus Coulter counter automatically compensates for this loss. With some instruments, a correction chart is available to allow the correct number to be determined. Under a total count of 10,000 pulses, primary coincidence is negligible and can be ignored.

Coulter counters have a serious drawback linked with the mercury manometer arrangement. The surface of the mercury gets dirty as a consequence of which the contact bordering the volume becomes uncertain, which may make the measured values uncertain. The mercury also dirties the neighbouring glass wall, which further causes uncertainty in counting. Also, the instrument can only be transported and stored after removing the mercury from it.

16.3.2 Picoscale

Based on the same principle of detecting change in conductivity in the presence of a cell in the orifice of a measuring tube, there is another cell counting instrument known by the name Picoscale. This instrument does not make use of a mercury manometer for fixing the volume thus eliminating the problems associated with its use.

This instrument is primarily meant for counting RBC, WBC and PBC and is manufactured by MEDICOR, Budapest. In this instrument (Fig. 16.9), a glass measuring tube 'C' provided with an

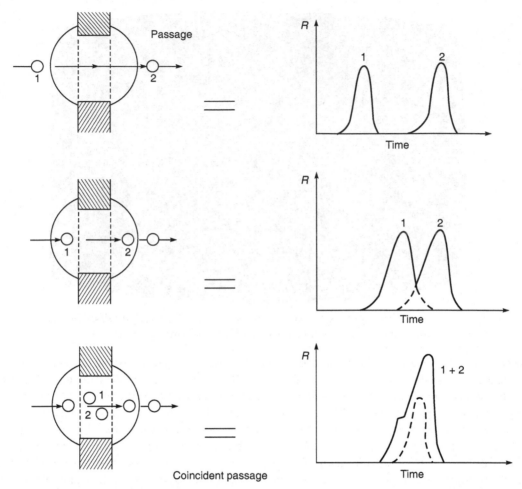

➤ Fig. 16.8 *Presence of more than one particle in the orifice resulting in loss of some valid pulses*

aperture 'A' is immersed into the suspension. The pressure difference created between the two sides of the aperture draws the suspension to flow through the aperture. This pressure difference is generated by a simple mechanical pump consisting of a syringe, a relay and other parts.

A constant current is normally passed between the electrodes E_1 and E_2. Therefore, the electric resistance of the liquid measured between these two electrodes changes rapidly when a particle having electric conductance differing from the conductance of the electrolyte passes through the aperture. This results in the generation of a voltage pulse, which is amplified in a preamplifier of high gain and low noise level. The output signal of this stage goes to a discriminator, which compares the amplitude of the pulse arriving at its input with the preset triggering level. If the input signal exceeds the triggering level, the discriminator gives out a pulse of constant shape and amplitude. These pulses go to a counting circuit for the display of the measured parameter.

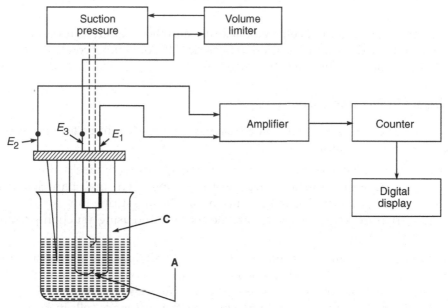

> **Fig. 16.9** *Block diagram of 'Picoscale' blood cell counter*

The measuring tube **C** is provided with a third electrode E_3 which helps to monitor the suction of a limited volume of the suspension. When the liquid level reaches E_3, the pump is changed over from the suction phase to the pressure phase. The counting process also occurs during the time the electrolyte is forced out of the measuring tube. After the liquid looses contact with the electrode E_1, the counting automatically stops and the unit becomes ready once again for the operation. The manufacturers recommend that to increase the reliability of the results, it is preferable to repeat the measurements several times, and calculate the mean value based on these measurements.

One great advantage of this instrument is that the clogging of the capillary is greatly eliminated by applying a bi-directional flow during the measurement procedure. The number of particles N in a unit volume is determined from the relation,

$$N = \frac{HLE}{V}$$

where

H = factor of dilution

L = scaling factor of the counter

V = measured volume

E = result displayed on the digital display.

For example, if the diluting factor is 63,000 (typical for red cell count in this instrument) and 520 appears on the display; L being 60, and V equal to 0.378 cm^3, then

$$N = \frac{(6.3 \times 10^4) \times (60)\,(5.20 \times 10^2)}{(3.78 \times 10^2)}$$

$$= 5.20 \times 10^6 \text{ per mm}^3$$

In other words, the solution contains 5.2 million blood cells per cubic millimetre.

The capillary diameter for red cell count is 72 µm and the dilution factor is 63,000. For white cells, the diameter is 102 µm, and the dilution factor is 630. For platelet count, the diameter of the capillary is 72 µm and a dilution of 6300 is used.

16.3.3   Errors in Electronic Counters

There are a number of errors that may occur in the electronic cell counting technique. Briefly, these errors are categorized as follows:

Aperture Clogging: Partial clogging of the measuring aperture adversely affects the results of counting. The aperture diameter becomes constricted and even small particles that were not intended for counting get included in the final count. The measuring time also becomes larger. Clogging of the aperture is cleared by cleaning the aperture and using a suction-pressure cycle by means of a pump.

Uncertainty of Discriminator Threshold: The uncertainty of triggering at threshold level and the threshold hysteresis means that particles of the same size are sometimes counted and sometimes left out. In addition to this, a threshold uncertainty error may also be introduced if the measuring capillary has a diameter larger than that rated because the passage of cells in a capillary of larger diameter will result in smaller changes in the resistance.

Coincidence Error: Coincidence error will be observed if more than one particle passes through the measuring aperture simultaneously. For the calculation of this error, a Poisson distribution is assumed whereby it is possible to determine the probability of $0, 1, 2, \ldots, n$ particles entering the measuring capillary during the time only a single count is produced. The coincidence loss depends upon the length and diameter of the aperture, the number of particles per unit volume and the dilution ratio. A correction for the coincidence loss is usually incorporated in the instruments.

Settling Error: This error arises due to the settling of the particles in the solution, with the result that the measurements show a decreasing tendency with time. If the readings are taken within 4–5 min., the settling error is less than 1%.

Statistical Error: Suppose that during a time t, n particles are detected. Then, during an additional time t, the number of particles detected will be $n \pm \sqrt{n}$. Therefore, the mean statistical error $= \sqrt{n}$ and the relative error $= \pm 100 / \sqrt{n}\%$. Assuming Gaussian distribution, the mean statistical error means that 67% tests fall into the interval $n \pm \sqrt{n}$ with 33% of the measurements greater than that. To obtain the statistical error, the instrument reading should be multiplied by the scaling factor of the counter. This will yield the value of n.

Error in Sample Volume: The measuring volume should be as accurate as possible because the magnitude of the measuring volume affects the accuracy of measurement. Greater the sample volume, the more accurate will be the result. The diameter of the measuring tube should be very accurate as it appears on the second power in the volume calculations.

Error due to Temperature Variation: Conductivity of the electrolyte increases with rising temperature. Hence, the specific resistance of the solution decreases, resulting in lowering the amplitude of the pulse generated by the particle. For accurate results, the temperature of the solution should be maintained constant.

Biological Factors: Factors like deformation of cells in the diluting solutions while passing through the aperture, the haemolyzing effect under the influence of electric fields and the presence of bacteria, etc. also introduce errors in the measurement process.

Dilution Errors: To obtain accurate measurements, the suspension containing the particles should be prepared with utmost care and with minimal error possible. The dilutions should be made accurately and the solution should be of a nature that the cells do not shrink or swell.

Error due to External Disturbances: A high intensity magnetic field in the vicinity of the instrument may interfere with the measurement. It might be necessary to screen the instrument with some magnetic material.

The sources of errors enumerated above originate from random factors and since the results too would contain random errors, different results may be observed in repeated tests. The actual value should therefore be determined with due consideration of the fluctuations, measurement losses and other circumstances.

16.4 AUTOMATIC RECOGNITION AND DIFFERENTIAL COUNTING OF CELLS

Along with the automated instruments for obtaining the erythrocyte, leucocyte and platelet counts, there has been a considerable interest in developing automated techniques for identifying and counting the different types of cells within a given class. Examples of this could be the immature red cell count, the differential leucocyte count and the recognition of normal versus malignant cells in other cell types. Various diseases affect the mechanism of blood cell formation in different ways. In particular, fractional proportion of the five major leucocytes is a sensitive measure for indicating and assisting in the diagnosis of various diseases. Furthermore, some diseases cause immature forms, which normally are present in the blood forming tissue to appear in the peripheral blood. Some diseases affect red cell morphology, and a qualitative report of red cells also forms a part of the differential count. There is, thus, a need for automation of acquisition and interpretation of data in routine clinical differential cell count. Several approaches have been employed in the pursuit of automating the techniques.

Miller (1976) describes a differential white blood cell classifier based upon a three-colour flying spot-scanner approach. It utilizes recognition parameters based on the principle of geometrical probability functions, which are generated at high speed in a dedicated computer. The system uses a conventional microscope with automatic focus and stage motion. A block diagram of the cell identification steps of the system are shown in Fig. 16.10. The system is built around a Zeiss microscope with two 15 × eyepieces and a 40 × oil immersion objective and with computer-controlled focusing. A television monitor displays the data and shows the relative position of the cells in each field. The slide is positioned precisely on the transport to allow reloading, oil is applied from a reservoir, and the objective is focused. A cathode ray tube generates a flying spot, which traverses the slide in 1 μm-wide sweeps. When a cell nucleus is located, the coordinates of the beam are immediately stored. A 24 × 32 μm window is established around the centre of the

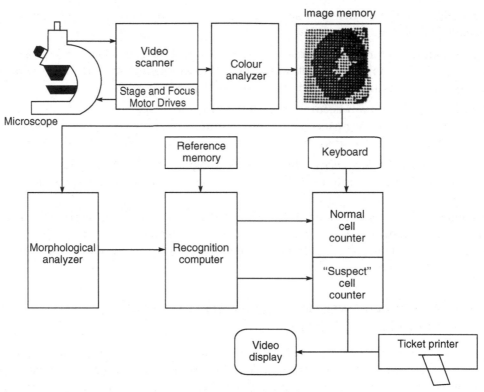

> **Fig. 16.10**   *Block diagram of cell identification system*

nucleated cell, and the spot scanner is reset to sweep the area in 0.25 μm wide passes. The light passing through the cell is split and sent through three filters to break it into its red, green, and blue components, with highly sensitive photomultiplier tubes detecting the resulting radiation. This system improves cell classification and allows wedged as well as spun slides to be used. Based on the number of grid positions and three colours, the system has over 3 million data values to use in each cell identification. These are used to first locate and then characterize the nucleii based on size and segmentation. The system is capable of determining segmented neutrophils, bands, eosinophils, basophils, lymphocytes, and monocytes, as well as abnormal cells such as atypical lymphocytes, blasts, nucleated red cells, and immature granulocytes. In addition, the system carries out the red cell morphology, evaluating size, shape, and colour, counts the reticulocytes, estimates the platelet count and plots a distribution of red cell diameters. The instrument therefore, performs all the functions that a human being sitting at a microscope would be expected to do. However, the system is much faster than a human technologist as it can process up to 100 slides per hour.

If the cell is identified as one of the six normal cell types, it is counted otherwise it is shown as a suspect cell to be identified by the operator. The same field is searched again for other nucleated cells. If no other is found, then the stage drive is incremented and the electronic search resumes in the new field presented to the system.

The automatic focus mechanism operates from high frequency video information as an error signal to position a special piezoelectric focus drive. The piezoelectric crystal adjusts the lens assembly so as to maximize the high frequency content of the video signal.

The pattern recognition program was developed from 21000 normal and abnormal cells and artefacts, from 700 different people and was stored on magnetic tape. It was observed that the degree of agreement between the instrument developed (Hematrak, registered trade mark of Geometric Data Corporation, Wayne, PA. 19087) and the technologist was 98%. The time required to perform a differential count on the instrument was small enough to yield a throughput of 40 slides per hour. An example of an automated cell recognition and counting system is that of the "diff-3" system which is illustrated below.

16.4.1 "diff-3" System

System Concept: The "diff-3" System (Perkin-Elmer, USA) is a completely automated system for the evaluation of red and white human cells and platelets. The system electronically examines conventional microscope blood smear slides and employs optical pattern recognition techniques to achieve the following:

- Counts and differentiates seven important categories of red blood cells (erythrocytes); three based on size, two on colour, one on shape and one covering red cells with nucleii (nucleated red cells).
- Enumerates white blood cells (leucocytes) and differentially classifies them into the 10 most significant medical categories and estimates their total number.
- Surveys platelets (thrombocytes) cell size and sufficiency.

The system is designed to analyze standard slides at a 35 to 40 slides per hour rate. The actual analysis task takes only 90 s. In fact, the system largely duplicates mechanically and opto-electronically, the manual procedures followed while examining blood smears with a microscope.

System Operation: The blood sample is first made into a conventional slide by a slide-spinner that ensures a monolayer of cells on the slide. About 200 μ of the sample is aspirated via the built-in pipetter-diluter and dispensed on to the standard 1 × 3 unit (25 × 75 mm) slide which has been labeled previously and put into a slide spinner unit. The slide will spin under low RPM (3200) for 0.6 s and comes to a stop. While the slide is spinning, the pipetter is flushed automatically to prevent sample carry over. The low spin speed and short spin time helps create consistently high quality slides with excellent morphology and random cell distribution. Slides are then removed by hand and taken to the slide staining unit—use of Wright's stain and an Ames Hema Tek automatic stainer helps assure the consistency of stained smears. If necessary, the system automatically compensates for normal variations in stain. After staining the slide, they are loaded into the magazine. The system keeps track of the magazine and position number to ensure correct patient identification.

Upon loading, the slides go to the blood smear analyzing part, which is an open L-shaped desk console unit. Built into the system is a microscope with a binocular eyepiece angled for easy observation by the operator; easy to read TV screens and the signal processing equipment.

Slides inserted into the smear analyzer are moved onto the stage of a binocular microscope (400× 1.0 N.A. Zeiss Optics) receiving a drop of oil for the microscope's immersion type objective lens as it moves from loading breech to the stage. Automatically, the slide is focussed for the electronic analyzer system. Light from the illuminated slide is split between the electronic pattern recognition unit and the microscope eyepiece for viewing. The microscope's motorized stage moves the slide in a programmed, back and forth raster pattern over the slide. Its movement is halted upon recognition of white cells. At each stop, the system studies isolated white and red cells and platelets.

Each white cell thus found is isolated from its surroundings via electronic image scanning and digital techniques. From then onwards, an image processor (Golay Logic Processor) takes over and carries out a total of 50 measurements on the white cells. These measurements are those commonly used in manual practice, i.e. evaluation of cell size, colour of the cytoplasm, number of nuclear lobes, etc. On the basis of these measurements, the cells of the blood are differentiated and counted. White cells, while being counted, are categorized as one of the 10 following types:

(i) segmented neutrophils, (ii) bonded neutrophils, (iii) eosinophils, (iv) basophils, (v) immature granulocytes, (vi) lymphocytes, (vii) monocytes, (viii) atypical lymphocytes, (ix) blasts, and (x) others (generally bizarre and usually less than 1% of the cells checked). In the final print-out, an estimate of the WBC count per cubic millimetre is also available.

The red cells are isolated and examined at the same time as the whites are being studied. Some 20 measurements are made on each individual red cell, so that RBCs can be classed by form and structure into five primary groups with several subdivisions (i) nucleated red blood cells: reported as NRBCs per 100 WBCs, (ii) hypochromic cells, (iii) polychromatophilic cells, (iv) size distribution—the printout indicates the relative percentage of red cells which are small, normal or large in size and (v) mis-shaper cells. These along with the spherocytes and target cells are grouped under the term 'poikilocytes'.

Platelets are examined for size and sufficiency. Giant platelets are reported as count per 100 WBCs whereas platelet count is shown per cubic millimetre. Sufficiency of platelets is estimated as low, normal or high.

Image Processor: The system uses two computers. A general purpose minicomputer operates and controls the system and makes it inherently more flexible than systems using hard-wired logic. The second computer is a special purpose pattern recognition computer, the Golay Logic Processor (Golay, 1969), which enables the system to transform cell pattern recognition information into differential results. Golay logic (Preston *et al* 1979) enables the system to 'see' a cell in much the same way as a technologist does. For example, in classifying nucleated cells, it evaluates the cell area, nuclear area, nuclear colour, nuclear clumps, cytoplasmic colour, granular colour, nuclear lobe count, and about 35 other measurements. Similarly, for red cell morphology, the system evaluates the cell area, area of central pallor, optical density of the entire cell, concavities, irregularities of shape, and about 10 other measurements.

In performing classification, it positions a nucleated cell in a $50\,\mu m \times 50\,\mu m$ field. This field is scanned as 4096 discrete data points called picture elements. The separation between data points is $0.8\,\mu m$. This low resolution scan is used to automatically adjust the fine focus and confirms that a suitable cell has been found rather than dirt or debris. Then, automatically switching to a higher resolution, a field of $25\,\mu m \times 25\,\mu m$ is scanned and another 4096 data points are examined at a resolution of $0.5\,\mu m$. Each picture element, representing a $0.4\,\mu m \times 0.4\,\mu m$ point carries a digital number representing its optical density at two different wavelengths. This picture is stored in the computer's memory (Fig. 16.11a).

➤ **Fig. 16.11(a)** *Image of a band neutrophil and surrounding red cells as stored in video picture memory*

The contents of the video memories are processed by the special purpose pattern recognition computer that examines these pictures. The process starts by generating separate, simplified pictures of separate parts of the cell such as the nucleus cytoplasm, granules, etc. The pictures are made by applying the principle that picture elements with about the same transmission are most likely located in the same parts of the cell. For example, all the very dark picture elements having low transmission values at a certain wavelength are likely to be in the nucleus, and so on. Each picture element is then turned 'off' or 'on' when it falls within an optical density range under investigation. The resulting pictures resemble line drawings, or silhouettes of parts of the cell.

The logic processor examines the 'on' and 'off' status of each picture element and its six nearest neighbours at many optical density levels. In this way, it reconstructs the image being scanned for identification. For example, a point will be judged to be an 'interior point' if the point is determined to be on and its six nearest neighbours are also determined to be on. For a background point, all seven points will be judged to be off. An edge point is recognized as a combination of on and off points. In all, 14 combinations are possible to define elements such as convex edges, concave edges, strings, branches and end points (Fig. 16.11b). The processor can examine the outline of a cell by turning on only the edge points of the cytoplasm. The picture represented in Fig. 16.11(c) would result. By turning on only interior points of the nucleus, Fig. 16.11(d) would result. Having isolated a single cell, the processor evaluates the nucleus and cytoplasm of the cell for shape, colour, texture and other elements.

In addition to a thorough examination of each cell, the processor must also ensure that only one cell is evaluated at a time. Also, the measurements of the cell area, nuclear colour, etc. are used in mathematical equations to decide the type of cell present. Comparisons are conducted on each cell and a required number of conditions are needed before classification is made into any category. If no category meets a sufficient number of conditions, then the cell will be classified as an 'other'.

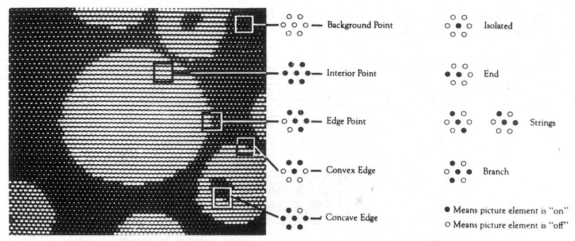

> **Fig. 16.11(b)** *Examples of points identified by the Golay logic processor*

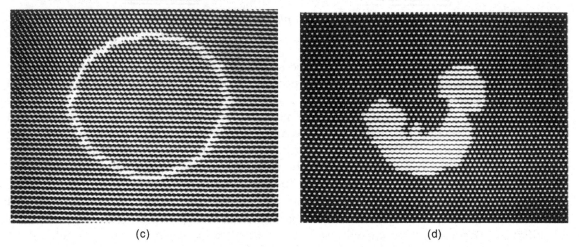

| (c) | (d) |

> **Fig. 16.11(c)** *Outline of whole cell made by turning on only edge points in picture of the whole cell*
>
> **(d)** *Picture of the nucleus generated by turning on only interior points of the necleus*

This may happen with some frequency on abnormal samples, but it happens only occasionally on normal samples.

At every observation step, the instrument shows the cells it is studying on a CRT screen, exactly as the technologist would view it in a microscope at the same time. The results of examinations are also shown upon completion of the analysis on the CRT. End results are printed in triplicate on an 8×22.5 cm ticket.

CHAPTER

17

Audiometers and Hearing Aids

ⅢⅢ▶ 17.1 MECHANISM OF HEARING

Sound waves are longitudinal waves in which the motion of each particle of the medium in which the wave is travelling, moves backward and forward along a line in the direction in which the wave is propagated. The human aural system reacts to these oscillating pressure changes and transmits them to the brain through a series of steps. Figure 17.1 shows the anatomy of the human ear. The outer ear consists of the pinna or auricle, together with the ear canal, the external auditory meatus, which is a convoluted tube, about 1 cm^3 in volume and terminates at its inner end in the

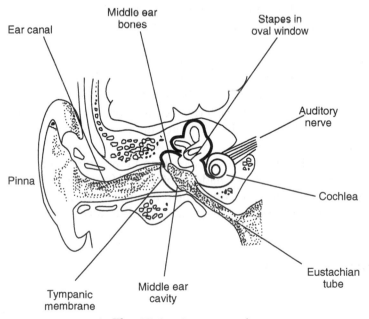

➤ **Fig. 17.1** *Anatomy of ear*

There is a rapid drop in impedance of the middle ear at high frequencies and very little of the acoustical energy fed to the ear by air conduction is transmitted to the cochlea. But bone-conducted sound by-passes the middle ear, as is shown in Fig. 17.3. This to some extent explains the different threshold shapes at high frequencies.

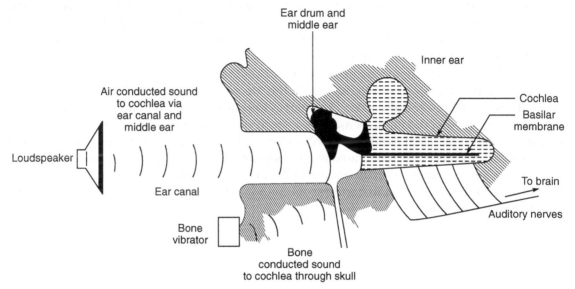

> **Fig. 17.3** *Mechanism of air and bone conduction*

17.1.2 Threshold of Hearing

The threshold pressure level of a sound is the lowest level at which an observer can discriminate between the desired sound and the noise background always present in the auditory system. In other words, it is minimal intensity on the stimulus scale, which is just barely adequate to elicit a response. The hearing threshold is not an invariable fixed intensity above which sound is always heard and below which sound is never heard. In fact, the sensitivity of the auditory mechanism is found to vary with interactions between certain physiological, psychological and physical factors. Therefore, the threshold may be regarded as an intensity range within which sound stimuli at or near the statistically determined threshold may or may not be perceived.

The base of each audiological examination is the determination of the hearing threshold. A complete audiogram, one which shows both air and bone conduction threshold bilaterally, is not only a graphic representation of the dB loss at different frequency levels for both air and bone conduction but also indicates the type and location of the hearing impairment. Thus, on the basis of the audiograms, the conduction and perceptive hearing deficiencies can be easily separated from each other.

After establishing the sensitivity of the ears to air bone sounds, it is necessary to know whether the hearing loss is due to a disorder of the hearing organ (cochlea) and connecting nerves with the brain (perceptive loss) or whether the loss arises because of reduced transmission of sound vibrations through the middle ear mechanism itself (conductive loss).

▶ 17.2 MEASUREMENT OF SOUND

Sound intensity may be defined as the amount of energy flow per unit time through a unit area perpendicular to the direction of energy flow. It is expressed as watts per square centimetre. However, the common receivers of sound are microphones, which do not measure sound intensity directly. They are sensitive to sound pressure and therefore, it is more pertinent to measure sound in terms of sound pressure which is given in dynes per square centimetre or in microbars (one microbar equals 1 dyne per cm^2). Sound pressure, for a given sinusoidal event, is related linearly to both amplitude and frequency. Sound intensity is proportional to the square of sound pressure.

In most acoustic considerations what is of importance is not the absolute magnitude of the intensity but its relative magnitude as compared to some assumed reference standard. The convenient unit for making such comparisons and to express the sound intensity and sound pressure data for all practical purposes, is the decibel (dB). The dB is 1/10 of a larger unit, the bell, named after Alexander Graham Bell. Decibel expresses the logarithm of the ratio between two sound intensities, powers or sound pressures. Since dB is merely a ratio, it is a dimensionless entity.

If I_1 and I_2 are two intensities in watts per square centimetre, then the number of decibels with which they are related can be expressed as:

$$N = 10 \log I_1/I_2$$
$$= 20 \log P_1/P_2$$

where P_1 and P_2 are sound pressures in dynes per square centimetre. Since intensity is proportional to the square of sound pressure, the constant 20 has been used in this case.

Attenuation is commonly expressed in negative dB numbers whereas amplification is given in positive dB numbers.

Use of decibels as units for comparison of intensities help to avoid all mathematical calculations except algebraic addition or subtraction of small numbers. The transmission efficiency of any medium like air, a hearing aid or an amplifier is usually expressed in dB as a gain when the output is greater and as a loss, if less. When two media are connected in series, the net efficiency is the sum of the two. For example, if a patient has a hearing loss of 70 dB and is fitted with a hearing aid with a gain of 60 dB, the net efficiency of the air-to-bone transmission system comes out to be -70 + 60 or a 10 dB loss.

Level (Volume): This is sound power measured in dB from a chosen zero level of power. The zero level is usually 10^{-16} W per cm^2. This is the sound power (intensity) in air flowing past an area of 1 sq cm at right angles to the direction of the sound, at the rate of 10^{-16} W. This is approximately the just audible power of a 1000 Hz pure tone for young healthy adults. The corresponding reference value in terms of sound pressure is 0.0002 dyne per sq cm. The smallness of both the sound pressure and sound intensity values show that the human hearing mechanism is very sensitive.

17.2.1 Transducers

In audiometry, generally employed transducers are (i) earphone, (ii) microphones, (iii) bone vibrators, and (iv) loud speakers.

Earphones: They are usually of the moving coil type. They give a reasonably flat frequency response up to 6 kHz after which their sensitivity falls rapidly. They are not specially designed for audiometric applications but for communication purposes. In their miniature form, they are used in hearing aids. When used via insert earphone and ear moulds, they provide greater acoustic power to be transferred to the small volume of the external ear. It may be noted that audiometer earphones are not interchangeable and must remain identified with a specific instrument to preserve its calibration.

In conditions of ambient noise being too high for unshielded earphones, specially designed audio cups are used. They use a fully-articulated suspension system which leaves the standard ear caps free to locate against the pinnate with normal pressure, and at the same time to enclose fully the external ears with noise-excluding shells, sealed with soft plastic cushions, to exclude background noise which will otherwise result in elevated threshold measurements.

Microphones: These are used to translate wave motion in air into electrical signal. Usual types are: (i) carbon button which changes resistance with air pressure, (ii) electrodynamic where a voltage is induced in a coil by its motion relative to a magnet, (iii) condenser where capacitance of a condenser is varied by the vibration of one of the condenser plates. High quality condenser microphones of diameters 12.5, 6.25 and 3.125 mm are currently used, depending on the frequency to be measured. For special purposes, microphones can be fitted to the ear caps and used in reciprocal arrangement to transmit sound to the ear.

Bone Vibrators: In the early days, bone vibrators were composed of a rod connected to an electromagnetically excited driving system. They were often held by hand against the mastoid process. Such units were large and cumbersome and are not in clinical use today. In the present form, bone vibrators are of the hearing-aid type in which the transduction mechanism changes the alternating current into a vibratory force through a diaphragm. The diaphragm and its basic mechanical parameters like mass, compliance and resistance are important in establishing its response characteristics. Though convenient, it is a very inefficient means of transduction and has a rather limited and peaky frequency response. The plane circular contact area of a bone vibrator is recommended to be 175 ± 25 mm^2. It is held in position by a headband.

Loudspeakers: They are used to deliver auditory stimuli, when it is not possible to have close coupling of the transducer to the ear. Obviously, the acoustic energy loss into the surroundings will be much greater than when stimulation is applied directly via an earphone. The acoustics of the test room and masking of the non-test ear are two important factors which merit special considerations while using the sound field of a loudspeaker.

ⅢⅢ▶ 17.3   BASIC AUDIOMETER

An audiometer is a specialized equipment, which is used for the identification of hearing loss in individuals, and the quantitative determination of the degree and nature of such a loss. It is essentially an oscillator driving a pair of headphones and is calibrated in terms of frequency and acoustic output. Both frequency and output are adjustable over the audio range. The instrument is also provided with a calibrated noise source and bone-conductor vibrator.

Audiometers may be divided into two main groups on the basis of the type of stimulus they provide to elicit auditory response: *pure-tone* audiometers and *speech* audiometers. A pure-tone audiometer is used primarily to obtain air-conduction and bone-conduction thresholds of hearing.

These thresholds are helpful in the diagnosis of hearing loss. Pure-tone screening tests are employed extensively in industrial and school hearing conservation programmes. Speech audiometers are normally used to determine speech reception thresholds for diagnostic purposes and to assess and evaluate the performance of hearing aids.

Screening audiometers are used to separate two groups of people. One that can hear as well as or better than a particular standard and the other that cannot hear so well. Applications of these instruments are found in industry, schools and military service.

An important application of audiometers is in industry. They help to assess the hearing function of personnel at different stages of their detection of changes in auditory acuity, identify noise susceptible persons and evaluate the effectiveness of ear protectors and noise control measures.

In conventional pure-tone audiometry, head phones are worn by the subject and a set of responses is obtained for air-conducted sounds directed to each ear in turn. A bone conductor vibrator can then be attached to the head at the centre forehead position to see whether the hearing threshold improves. If it does, then the disorder is most likely wholly or partly conductive in origin. To avoid stimulation of the ear not under test with the vibrator, it can be temporarily made deaf by introducing a suitable masking noise in the non-test ear via an earphone. A narrow-band noise centred on the pure-tone test frequency or a wide-band white noise is used for this purpose. The problem of how to recognize the need for masking and then applying the correct intensity poses a considerable difficulty.

17.3.1 General Requirements of Audiometers

Modern audiometers are solid-state instruments covering a frequency range from approximately 100 to 10,000 Hz. Some instruments produce this range in discrete octave or semi-octave steps or intervals, while others provide for continuously variable frequency over their designed range. The frequency must remain sensibly constant at a value within 1–3% of the indicated value. Where automatic recording facilities include a continuous sweep frequency, the rate of change is normally kept as one octave per minute. If an automatic recording audiometer provides fixed frequencies, then a minimum period of 30 s must be allowed at each frequency.

The test frequencies should have sufficient purity of tone or approximation to the ideal sine wave form to ensure response only to the desired fundamental frequency. The maximum harmonic distortion in pure-tone air conduction audiometry is specified as 2% for the second and third harmonic and much less at higher order harmonics. The total harmonic distortion should not be more than 3%.

The intensity range of most audiometers starts from approximately 15 dB above normal to 95 dB below normal over a frequency range from approximately 500 to 4000 Hz. The intensity range is somewhat less for frequencies below 500 Hz and above 4000 Hz. This is partly because of certain instrumental limitations imposed by the earphone or vibrator and partly due to the desire to avoid the threshold of feeling from stimulation at the lower frequency levels. The threshold of feeling is the sensation of pain or tickle in the ear, which results from sound pressures and limits the maximal sound intensity that can be tolerated by the ear. The intensity level at which the threshold of feeling is stimulated varies with frequency. For example, the threshold of feeling is stimulated at an intensity level approximately 120 dB above the normal threshold of audibility from about 500 to 4000 Hz, but at 64 Hz the threshold of feeling is stimulated by sound pressures approximately 65 dB above the normal threshold value.

The attenuation dials on the audiometers provide variable intensity or volume controls. They are calibrated in decibels usually in discrete steps, which differ by 5 dB in intensity from step to step. Auditory acuity for each frequency is thus measured in dB above or below the normal hearing-zero dB reference level for that frequency. This level is the minimal intensity at which each given frequency can be perceived by the normal ear in a noise free environment and is experimentally determined by averaging the results of measurement on a large number of normal individuals between 18 and 25 years of age.

Audiometers usually have two channels with single pure-tone generators. The first channel has pure-tone or speech output while the second channel has nominal masking. The pure-tone and speech can be switched to both channels for special tests. Channel two can have either wide or narrow-band masking. Each channel has an accurate independent attenuator output and the transducers are switched to each attenuator as required.

In the recent years, numerous audiometric products incorporating microprocessors have been introduced in the market. Such equipment offers greater convenience in calibration, test signal presentation and versatility. Automated data collection and storage are also useful features included in such equipment. It may however be noted that the audiometric measurement principles as described in the following text are not altered with the use of microprocessors and digital technology.

It is extremely important in audiometry to ensure that only the testing signal reaches the ear. Therefore, all testing must be done in a noise-free environment. Since environmental noise is difficult to control, the noise-free conditions are achieved by performing the audiometric testing in a sound isolating enclosure. Such enclosures help to attenuate all frequencies within the sensory range below the threshold of hearing of normal ears. By using double wall construction and appropriate sound absorbing material, it is common to achieve 25 dB attenuation at 125 Hz and 60 dB attenuation at frequencies between 1000 and 8000 Hz.

17.3.2 Masking in Audiometry

In the presence of monaural and asymmetrical binaural hearing losses, there is serious difficulty in obtaining accurate measures of hearing for the poorer ear. The answer to the problem is to eliminate responses from the better ear by masking in order, to shift the threshold to a higher level, permitting greater intensities to be presented to the poorer ear without any danger of cross-over. If the difference in air conduction acuity between the two ears is 50 dB or more, then it is advisable to place a masking noise over the better hearing ear while determining the threshold in the other. Masking efficiency depends upon the nature of masking sound as well as its intensity. A pure tone can be used to mask other pure tones. However, over a range of test frequencies, masking efficiency of a pure tone is low as compared to a noise composed of many frequencies, as usually provided in commercial audiometers.

Saw-tooth noise and white noise have been most commonly used for masking in clinical audiometry, but narrow band noise, i.e. a restricted frequency bandwidth of white noise is also often used. Saw-tooth noise is a noise in which the basic repetition rate (fundamental frequency) is usually that of the mains voltage and contains only those frequencies that are multiples of the fundamental. The intensity of these multiples decreases as their frequencies increase. Noises referred to as 'complex' or square waves are similar in that they are composed of a fundamental frequency and components that are multiples of it.

White noise is a noise containing all frequencies in the audible spectrum at approximately equal intensities. However, the spectrum is limited at the ear by the frequency response of the earphone, which may essentially be flat to 6000 Hz and may drop rapidly beyond. An excellent complex masking noise can be obtained by using the thermal or random electronic emission from a semiconductor diode, since it generates all frequencies simultaneously and with equal amplitude over a frequency range wider than the response of the ear.

Narrow-band noise has been used by a number of investigators in audiometric studies. It is produced by selectively filtering white noise. It has been found that narrow band noise is the most efficient masking noise in pure-tone audiometry. The masking audiograms for normal hearing subjects and the clinical results for hearing-impaired subjects show that for equal intensity levels, narrow band noise produces greater threshold shifts than do either of the other two types and thereby provides greater protection from false responses due to cross-over of the test tone.

▶ 17.4 PURE TONE AUDIOMETER

A wave in air, which involves only one frequency of vibration, is known as pure-tone. Pure-tone audiometry is used in routine tests and, therefore, it is the most widely used technique for determining hearing loss. Pure-tone audiometers usually generate test tones in octave steps from 125 to 8000 Hz, the signal intensity ranging from –10 dB to +100 dB.

Pure-tone audiometry has several advantages, which makes it specifically suitable for making threshold sensitivity measurements. A pure-tone is the simplest type of auditory stimulus. It can be specified accurately in terms of frequency and intensity. These parameters can be controlled with a high degree of precision. Speech audiometry normally allows measurements to be made within the frequency range of 300–3000 Hz. Some patients may have impaired high frequency response due to high intensity level occupational noise at 4000 or 6000 Hz. Pure-tone measurements at these frequencies prove to be a more sensitive indicator of the effect of such noise on the ear than speech tests. Changes in threshold sensitivity associated with various middle ear surgical procedures can be monitored more accurately with pure-tone than speech tests.

A pure-tone audiometer basically consists of an LC oscillator in which the inductance and tuning capacitance are of close tolerances for having a precise control on the frequency of oscillations. The oscillator is coupled to an output current amplifier stage to produce the required power levels. The attenuators used in these instruments are of the ladder type, of nominal 10 Ω impedance. The signals are presented acoustically to the ear by an earphone or small loudspeaker. The available sound pressure levels in a typical audiometer are given in Table 17.1.

▶ 17.5 SPEECH AUDIOMETER

Besides tonal audiometry, it is sometimes necessary to carry out tests with spoken voices. These tests are particularly important before prescribing hearing-aids and in determining the deterioration

• Table 17.1   *Test Tones and Signal Intensity in Audiometers*

Frequency	Pure-tone (head-phones)	Pure-tone (bone conduction)	Balance channel	Narrow band masking (head-phones)	Narrow band masking (bone conduction)
125	70	–	–	–	–
250	90	45	90	80	50
500	110	60	110	90	60
1000	110	60	110	90	60
1500	110	60	110	90	60
2000	110	60	110	90	60
3000	110	60	110	90	50
4000	110	60	110	90	50
6000	90	–	–	80	–
8000	90	–	–	80	–
Speech	110	–	110	–	–

of speech understanding of patients. Specially designed speech audiometers are used for this purpose. They incorporate a good quality tape recorder, which can play recorded speech. A double band tape recorder is preferred to interface the two channel audiometer units. Masking noise is supplied by the noise generator. The two channels supply the two head-phones or the two loud speakers which are of 25 W each.

The tape recorder has a capacity for recording a limitless variety of test material and a consistency of speech input, which cannot be obtained for live-voice audiometry in relation to test-retest repeatability. Another advantage of the tape recorded material is that the test words and sentences can be selected to cater for the widely differing needs of age, intelligence, dialect and language.

In speech audiometers, live-voice facilities are incorporated primarily for communication purposes as the inherent unreliability of live-voice speech tests may lead to serious errors. The microphone amplifier used for this purpose is a simple two stage amplifier. The frequency response characteristics of a live-voice channel should be such that with the microphone in a free sound field having a constant sound pressure level, the sound pressure level developed by the earphone of the audiometer in the artificial ear at frequencies in the range 250 to 4000 Hz does not differ from that at 1000 Hz by more than 110 dB. Also, it shall not rise at any frequency outside this band by more than 15 dB, relative to the level at 1000 Hz.

▶ 17.6   AUDIOMETER SYSTEM BEKESY

George Van Bekesy, a Hungarian scientist, designed an automatic audiometric testing method for plotting the hearing threshold based on the patient's signal. A principal feature of the method, differentiating it from conventional pure-tone audiometric techniques, is the interdependence of the patient's response and stimulus intensity: responses govern intensity and are affected by changes they introduce in it. An audiogram traced by the Bekesy method represents the absolute threshold values at all frequencies in the range tested. In addition, it shows the difference, in decibels, between levels at which the patient just hears a signal of increasing intensity and those at which he just ceases to hear the signal when its intensity is decreasing. This latter characteristic often varies significantly with the type of hearing impairment, and can aid in establishing the site of lesion within the auditory system. On the basis of the audiograms, one can easily separate the conduction and perceptive hearing deficiencies from each other.

Audiometers Bekesy are relatively simple for the patient to operate. The instrument generates a pure-tone signal, which is presented to him through an air-conduction earphone. The subject is told to press a switch when the tone is heard and to release the switch when it is not heard. This switch controls the motor-driven attenuator of the audiometer: when it is pressed, signal intensity decreases and when it is released, signal intensity increases. A pen connected to the attenuator traces a continuous record of the patient's intensity adjustments on an audiogram chart, producing a graphic representation of the subject's threshold. The test signal may be presented in a variety of ways, each suited to the investigation of a particular problem.

A block diagram of the audiometer system Bekesy is shown in Fig. 17.4. It consists basically of an electrical section and a mechanical section. The electrical section includes an oscillator and modulator circuits for the generation of the desired test signal, an automatic attenuator linked to the writing system, control circuits for the drive motors of the mechanical section and a master clock generator for the control of all timing functions via a logic control circuit. The carriage drive and the writing system with their separate drive motors constitute the mechanical section.

17.6.1 Electrical Section

Sine Wave Oscillator: This oscillator generates test signals with frequencies of 125, 250, 500, 1000, 1500, 2000, 3000, 4000, 6000 and 8000 Hz. This sequence is first presented to the left ear automatically, each tone for 30 s, and then to the right ear, the shift between the frequencies being noiseless. After both ears have been tested, a 1 kHz tone is presented to the right ear to provide a useful indication of test reliability.

Modulator: From the oscillator the test signal is fed to the modulator, where the mode of operation is selected by the 'Tone' switch, via the logic control circuit. Two models, 'Pulse' or 'Cont', are available. In the 'Pulse' mode the test signal is modulated giving a signal, which is easily recognized by the patient. In the 'Cont' mode no modulation is applied, giving a signal suitable for use, when calibrating the audiometer.

Automatic Attenuator: The signal from the modulator feeds the automatic attenuator situated on the carriage together with the writing system. The attenuator consists of a logarithmic potentiometer which has its wiper attached to the pen drive so that the attenuation of the potentiometer corresponds to the position (y-axis) of the pen on the audiogram chart. The potentiometer has infinite resolution. The attenuation range is 100 dB, thereby covering the range of hearing levels from -10 to +90 dB. When the test is initiated, the attenuator starts at its top position of -10 dB and then increases the level with a rate of 5 dB/s. Also, when the test signal switches between the ears and when retesting at 1 kHz, the attenuator decreases the signal level to −10 dB to ensure that the right ear does not receive a tone at the elevated level possibly required at 8 kHz.

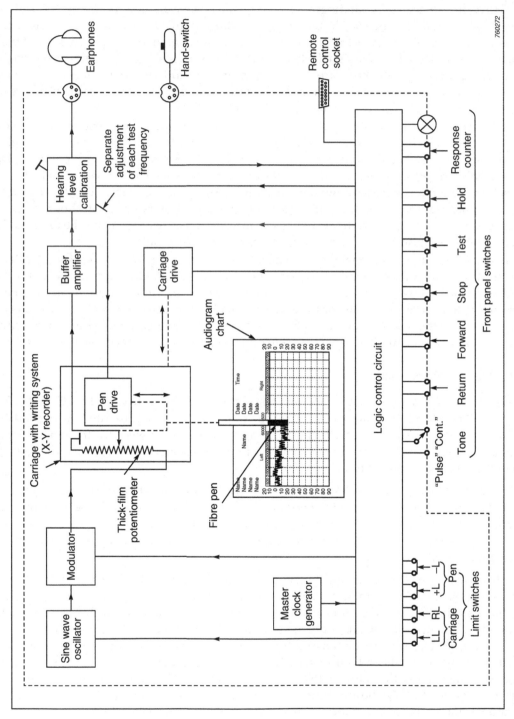

> **Fig. 17.4** *Block diagram of the audiometer system Bekesy*

Hand Switch: The pen drive is controlled via the logic control circuit by means of the hand-switch operated by the patient. Pressing the switch decreases the output from the potentiometer and thereby the level in the earphones, while releasing the switch increases the output both ways with a speed of 5 dB/s.

Buffer Amplifier and Calibration Circuit: From the attenuator the signal is fed via a buffer amplifier to the hearing level calibration circuit. The buffer amplifier isolates the attenuator from the calibration circuit in order not to affect its output. The calibration circuit consists of seven potentiometers, one for each test frequency. During calibration, the potentiometers are adjusted one at a time until the correct level, measured in a coupler, is obtained in the earphones.

Earphones: The earphones are a matched pair with distortion, typically less than 1%.

Master Clock Generator: A stable clock generator supplies the necessary signals for the control of motor speed, attenuator speed, frequency shift, modulation and other timing functions. This makes the system independent of variations in line voltage and frequency.

17.6.2 Mechanical Section

Carriage: The carriage with the writing system is driven by a stepping motor via a toothed belt. The speed and direction of rotation of the motor are automatically controlled via the logic control system. When the test is initiated and the patient indicates that he hears the signal by pressing the hand switch, the carriage moves along the X-axis (frequency axis) of the audiogram in agreement with the frequency of the test signal. When the frequency shifts, the carriage stops until the patient again, by pressing the hand switch, indicates that he hears the signal. This avoids wastage of recording space on the audiogram if a patient's hearing threshold varies from frequency to frequency or from left to right ear. When the complete test is finished the carriage and writing system return to the start position. To prevent carriage over-run, two limit switches are included in the carriage drive circuit.

Writing System: The writing system is operated by the pen drive, which is driven by a stepping motor. The pen drive moves the pen, and with it the wiper of the automatic attenuator, along the Y-axis (hearing level axis) with a constant speed corresponding to the change in attenuation of 5 dB/s. The direction of movement of the pen is determined by the position of the hand switch operated by the patient. Limit switches are also included with the pen drive.

Audiogram Chart: The audiogram is printed in standard A5 format (148 × 210 mm). The recording space is large, 0.8 dB/mm, to enable easy reading. Space is provided on the audiogram side for registration of information on the patient, audiometer, operator, etc. while the other side has space for recording the patient's medical and occupational history. Four holes in the chart give precise and automatic location of the audiogram on the chart bed.

In order to establish a more exact diagnosis applying adaptation and hearing fatigue tests, several other tests besides the pure-tone Bekesy audiometry, have been suggested and can be performed using the basic Bekesy system. For example, for carrying out the Fowler loudness balance test, a second channel is provided. The second channel has a continuously variable intensity over the range 0 to 110 dB and is calibrated in 1 dB increments.

Ⅲ⑨▷ 17.7 EVOKED RESPONSE AUDIOMETRY SYSTEM

Evoked response audiometry has been the subject of research for several years. This work has established evoked response electroencephalography resulting from an auditory stimulus above the hearing threshold. Instruments based on this principle have been found particularly suitable for determining auditory threshold in the absence of voluntary response in subjects such as infants, uncooperative adults, or animals.

The system basically comprises a conventional wide range pure-tone audiometer, which operates under the control of an automatic programmer and provides a series of auditory stimuli to the subject via either a loudspeaker or standard earphones. The EEG signal is picked up by standard electrodes placed in contact with the subjects scalp. One electrode is usually placed on the vertex, one at the post auricular area, and a third (ground) on the earlobe or forehead. The instrument stores and evaluates that part of the EEG signal, which follows each individual stimulus presentation. At the end of the programmed series of stimuli, it writes out on a paper chart a waveform that is the average response to stimuli. The presence of characteristic amplitudes and latencies in this waveform give an indication that the test intensity exceeded the subject's threshold at the test frequency. Similar trials at other intensity levels and other frequencies establish the threshold contour.

Figure 17.5 shows a block diagram of the evoked response audiometry system. It consists of the following five major subsystems.

The Tone Generator: It is a wide range pure-tone audiometer whose frequency output can be selected at 250, 500, 750, 1000, 1500, 2000, 3000, 4000, 6000 and 8000 Hz. The output power levels

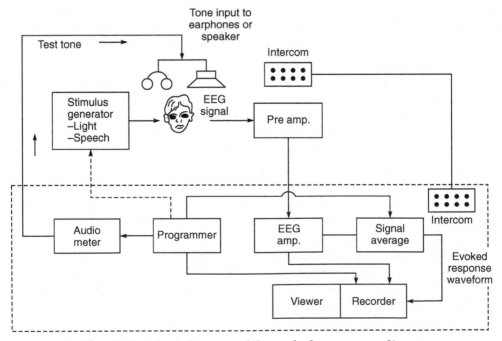

➤ **Fig. 17.5** *Block diagram of the evoked response audiometer*

are adjustable from –5 to + 110 dB in 5 dB steps. In addition to the pure tones, internally generated broad-band noise may be used as the stimulus. Provision is also made for external input from other types of stimulus. A special feature of the generator is the selectable rise/decay times for 1–100 ms. Outputs are provided for the left ear, right ear, or both. A variable intensity masking noise source is included in the generator. A power amplifier is incorporated to drive a speaker or tactile transducer. Total harmonic distortion should not exceed 2%.

EEG Amplifier: It is a conventional high gain, high impedance, low noise amplifier. The first stage of the EEG amplifier is preferably kept in a separate "preamp head" located near the subject. Its design and location minimize power line frequency pick-up. An ohm meter provided in the preamp head enables the measurement of the electrode contact impedance and thus indicates when satisfactory contact resistance of the electrodes is obtained. The EEG preamplifier gain is fixed at 200. The overall sensitivity is adjustable in steps of 10 to 1000 μV/div on the chart recorder. The amplifier also provides selectable roll-offs at the high and low ends of the spectrum of interest and a 60 Hz sharp notch filter. The low frequency roll-off points are at 0.15, 0.30, 0.60, 1.5, 3, 6, 10, 15, 30, 60 Hz at 6 dB (half amplitude), whereas high frequency points are at 1.5, 3, 6, 10, 15, 30, 60, 100 Hz at 6 dB (half amplitude) yielding a 12 dB/octave roll-off.

The Programmer: A logic device that controls the system operation in correct time sequence. It helps to have a selectable rate of stimulus presentation, stores the number of pulses that the operator chooses to constitute a run, starts the recorder at the beginning of the run, turns the audiometer tone generator on and off to provide the auditory stimuli at the proper time. It also speeds up the chart drive for the detailed signal samples, stops the recorder after providing for paper clearance, erases the signal averaging computer and clears and resets itself for the next run.

Total count in the programmer is selectable from 1–109 stimuli with a selector switch. The pulse interval is normally kept as 0.1, 0.2, 0.5, 1, 2, 5, 10 and pulse duration as 1–10,000 ms.

Signal Averaging Computer: This separates evoked responses from the normal EEG activity by ignoring those components, which are not synchronized with stimuli. Because the waveform of the evoked potential will be essentially the same every time in response to the tone presentation, and the other electrical activity will vary randomly, the evoked response "grows" in the computer memory and the noise component tends to average to zero with repeated presentations. It may be provided with either 50 or 100 averaging points depending upon the degree of resolution required. The computer includes a provision for selection of integrating time constant (5, 10, 20, 50, 100, 200, 500 s) and sweep duration or analysis time (0.1, 0.2, 0.5, 1, 2, and 5 s). A delay circuit is incorporated to select a delay between the onset of the stimulus and the start of analysis by the signal averager. The delay time is selectable as 0, 0.2, 0.5 and 1 s. The amplitude of the evoked response may be normalized to that of the on-going EEG monitoring signal using a gain control (gain variable from 20 to 50). The computer provides outputs for display either on a conventional oscilloscope or on an X-Y plotter.

Chart Recorder: It is a two channel recorder. One of the two channels is used to display the averaged response after it has been processed by the computer and the other displays unprocessed EEG. There are two event markers, one of which is activated by each gating pulse from the programmer to show the beginning and duration of each stimulus and the other is available for registering any mark at the desired instant. The chart can be driven at four different speeds (1, 5,

25, 125 mm/s), which are automatically switched by the programmer. Translucent chart paper is usually employed so that records may be compared by overlaying one on another on the illuminated opal-glass viewer.

Evoked response audiometer systems also contain a provision for 'External' mode of operation where any other type of stimulus generator can be connected into the system and controlled from the programmer. This could include narrow band noise, speech, a high frequency auditory signal for animal research or even a photic or tactile stimulator.

Modern evoked response audiometers are built around microcomputers. In these instruments, the stimulators, preamplifiers and amplifiers are all digitally controlled via the central processing unit, which automatically avoids undesirable parameters. They have no push button or dials as the parameters are varied by means of the keyboard. The parameters can be controlled to a very wide range, which would not be possible with conventional knobs and switches. Stimuli from 12 to 16 kHz are included to facilitate investigation into high frequency hearing loss and ototoxic drug effects. Texts and waveforms are displayed on a large size TV screen. A built-in chart recorder helps to make recordings under the control of the keyboard.

▶ 17.8   CALIBRATION OF AUDIOMETERS

The purpose of audiometric testing is to compare. The comparison may be between an individual's present audiogram and one taken previously, or between his audiogram and the audiograms of others. Whatever the reason for recording an audiogram, clearly the test conditions and the reference levels for the measurements must be identical for valid comparisons to be made between them. Furthermore, for valid comparisons with audiograms from other instruments, the test conditions and reference levels must be universal. Various standard institutions have published audiometric standard threshold values for several widely used combinations of earphones and artificial ears for carrying out audiometer calibrations.

Accurate calibration of audiometers is essential to ensure that the instrument produces a pure tone at the specified level and frequency, and that the signal is present only in the transducer to which it is directed. For pure tone and speech audiometers, the parameters, which are commonly checked, are frequency and intensity.

The frequency output from the audiometer is checked using either a counter-timer or an oscilloscope. A counter-timer is preferable because it gives a quick, direct reading. It is connected across one earphone, and the audiometer is set to give maximum output. The specifications for audiometers allow a tolerance of $\pm 3\%$ for frequency output from a fixed frequency pure-tone audiometer.

Sound pressure levels are best checked by an 'artificial ear' or coupler together with a sound level meter. The artificial ear consists of a condenser microphone and a 6 ml coupler. The coupler originally designed by the National Bureau of Standards, USA encloses a volume approximately the same as the volume under a earphone for a human ear. The earphone is placed squarely on the coupler with a 500 g weight. It should be ensured that it forms a good seal. The output on the sound level meter is read directly in dB and compared with the expected output per frequency. It may, however, be noted that volume displacement is only one component of acoustic impedance and, therefore, it is recognized to be a rather crude acoustic approximation of the average human ear. This is why the reference levels vary with the earphone type. The disadvantage of different reference levels is overcome with a more elaborate design of the wide-band artificial ear. A typical example of an artificial ear is that of type 4153 Artificial Ear from Bruel and Kjaer. It is designed in accordance with IEC R 318. It has a three cavity coupler and provides acoustic impedance, which closely resembles that of the human ear. Different types of couplers are available for use with artificial ears for measurements on headphones, insert type hearing aids and other earphones.

While the artificial ear is used as a standardized substitute for the human ear when making measurements using an air-conduction earphone, an artificial mastoid is used as a standardized substitute for the human mastoid. The artificial mastoid consists of a mechanical simulation of the human one, incorporating a built-in force transducer to monitor the output of the device to be calibrated. The artificial mastoid must meet standard specifications such as the American National Standard ANSI S3 13-1972 or the IEC R 373.

Audiometric tests should be conducted in rooms which are reasonably quiet. A subject under test should not be disturbed either by sounds created within the room or by those intruding from the outside. Such rooms are known as sound-treated rooms. In order to keep out extraneous noises, the outside walls of such rooms consist of heavy hard-surfaced shell. The inside is lined with absorptive material to keep reverberation low and to prevent practically any reflection.

ⅢⅢ▷ 17.9 HEARING AIDS

Hearing loss has many forms. The most common is related to the body aging process and to long-term cumulative exposure of the ear to sound energy. As one grows older, it becomes more difficult to hear. The ear becomes less sensitive to sound, less precise as a sound analyzer and less effective as a speech processor. Loss of hearing differs greatly in different individuals. Changes in the ear occur gradually over time. However, by the time the changes are manifested, it is estimated that approximately 30 to 50 percent or more of the sensory cells in the inner ear have suffered irreparable structural damage or are missing (Engebretson, 1994). Under these conditions, the only choice available for hearing-impaired individuals is to wear a hearing aid.

Hearing impairment is caused by either loss in sensitivity (loss in perceived loudness), or loss in the ability to discriminate different speech sounds or both. Loss of loudness may be due to either increased mechanical impedance between the outer ear and the inner ear or by the reduced sensitivity of the sensory organ of hearing. Loss of the discrimination ability is basically associated with damage to the sensory organ, although, other neural structures at higher levels may also be involved.

The modern hearing aid became possible with the invention of the transistor, which has enabled to develop small, power-efficient amplifier circuits that could be packed in a form that fits behind or in the ear. Even though the primary function of an hearing aid is to compensate for the loss of sensitivity of the impaired ear, in practice, it is not this simple. The ear behaves differently for soft sounds near the hearing threshold than it does for loud sounds. Therefore, a frequency response that restores normal hearing thresholds for soft sounds will not, in general be appropriate for louder sounds. Furthermore, even when speech sounds are made audible for the hearing-impaired listener, it does not follow that he/she will be able to understand speech. Hearing-impaired listeners experience more difficulty in understanding speech in background noise than normal-hearing listeners.

17.9.1   Conventional Hearing Aid

Modern hearing aids have evolved from single-transistor amplifiers to modern multi-channel designs containing hundreds and even thousands of transistors. A typical design is shown in Fig. 17.6. The basic functional parts include a microphone and associated preamplifier, an automatic gain control circuit (AGC), a set of active filters, a mixer and power amplifier, an output transducer or receiver. The total circuitry works on a battery. The use of multiple channels in this design provides different compression characteristics for different frequency ranges. Typically, the crossover frequencies of the channels and the compression characteristics can be adjusted with potentiometers. Most of the latest hearing aids are electronically programmable. The programmable parameters are downloaded from a computer-based system and stored in digital registers. The register outputs are used to switch resistor networks that control various analog circuitry. The active filters are adjusted to generally provide for low-frequency attenuation of up to 30–40 dB relative to the high-frequency response. This is because most hearing aid wearers require high frequency gain.

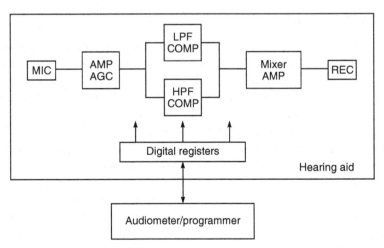

> **Fig. 17.6**   *Conventional analog type hearing aid*

The transducer in a hearing aid, which is a microphone, can be realized in an integrated form with a field-effect transistor preamplifier (Fig. 17.7). The preamplifier is housed in the metallic, microphone case to shield its input from extraneous noise. On the other hand, the receiver is an electromagnetic device, which drives a miniature diaphragm to produce acoustic output. The acoustic output is routed to the ear-mould through a flexible tubing whose frequency response can be altered to boost the high-frequency response. This is done by tapering its inside diameter from the ear mould back to the receiver port end.

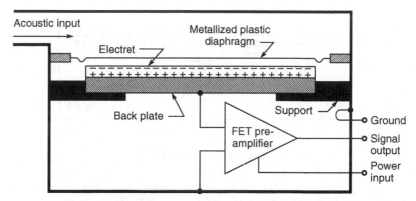

> **Fig. 17.7** *Integrated microphone and FET preamplifier (Electret microphone)*

All the electronics circuitry is packaged in a housing, which can be designed for fitting to the ear in any one of the following ways:

1. Placing all the components in a pocket-sized enclosure or box which is connected to the output transducer worn in the ear. The box can be carried in the shirt pocket or carried with a belt around the waist. With the availability of miniature-sized aids, this approach is no longer employed.

2. The components are packaged in a curved module, which is designed to fit comfortably behind the ear.

3. The most popular design is in which the total package can be put inside the outer ear.

Much depends on the performance of the filters for further reduction of the size and improvement in the working of the hearing aids. The dynamic range of an operational amplifier, which is the basic building block of an electronic filter, decreases as the three halves power of feature size. Since dynamic range of analog hearing aids is already marginally acceptable, it appears that further reduction in size to achieve increased processing complexity is not practical. The potential for greater dynamic range, with less power consumption and greater complexity in hearing aid design is feasible only with digital processing technologies.

17.9.2   Digital Hearing Aid

A typical digital hearing aid is illustrated in Fig. 17.8. The major parts are the microphone, an analog-to-digital converter (ADC), the digital signal processor (DSP), the digital-to-analog converter (DAC), the receiver and a two port memory. Essentially, sound waves picked up by the microphone and transformed into electrical signals are converted into digital form by an A-D converter. A typical microphone will have an internal noise of 20 dB SPL (sound pressure level) when referred to the input and maximum undistorted output corresponding to a signal of about 90 dB SPL. Allowing some margin for peak performance, the total dynamic range required of the ADC is 80 dB. This requirement can be achieved with a 14 bit A–D converter.

The DSP is a fixed (wired-program) digital processing device containing an array of adders, multipliers and registers which provide the fundamental operations necessary for implementing

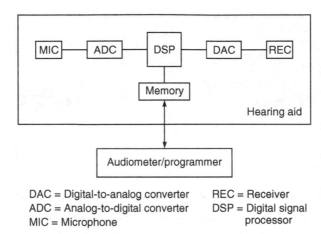

DAC = Digital-to-analog converter   REC = Receiver
ADC = Analog-to-digital converter   DSP = Digital signal
MIC = Microphone   processor

➤ **Fig. 17.8**   *Block dragram of a digital hearing aid (after Engebreston 1994)*

various digital algorithms. In a general-purpose DSP, considerable power is consumed in executing the programme instructions. Since power consumption is a major consideration in the design of hearing aids, the wired-program approach is followed. The DSP is associated with a two-port memory, which is used to store processing parameters that can be down loaded from the external programmer to the hearing aid while it is adjusted for the intended user.

The dynamic range requirements of the DAC are more severe. Some hearing impaired listeners have almost normal sensitivity at low frequencies but significantly elevated thresholds at high frequencies. Since the conversion noise generated by the DAC has a uniform spectrum and is a function of the overall output signal level, high-level high-frequency sounds can create low-frequency noise and distortion that falls above the threshold at low frequencies.

The digital hearing aids are implemented with CMOS technology, with a feature size of 1 μm or less and with an estimated power consumption of 20 μW. An estimated 10,000 CMOS inverters are required to implement 400,000 multiply-add operations for filtering, compression functions and other processing requirements.

The digital hearing aids promise to provide capabilities of superior signal processing, ease of fitting and stable long-term performance. However, they are still under development. It has often been seen that a person buys a hearing aid but does not use it because it does not help very much. The basic reason is that the impaired ear has its capacity to process speech and hearing aids are simply sound amplifiers that do not compensate for the loss of processing power. It needs to be emphasized that today's hearing aids are at an early stage of development and need to reach a highly refined stage before they can find wide spread and useful applications. The potential areas of improvement include shaping the frequency response to invert the patient's hearing loss, enhancing the signal-to-noise ratio with adaptive filtering, reducing acoustic feedback and compressing/expanding signals with minimum distortion.

17.9.3   Cochlear Implants

Sensori-neural deafness affects a large number of people throughout the world (Spelman,1999). The treatment of choice for the sensori-neural deaf is the cochlear prosthesis or cochlear implant.

Sensori-neural deafness can be caused either by cochlear damage or by damage within the auditory nerve or to the neurons of the central auditory system. The hair cells are the sensory cells that transduce mechanical motion into signals that can be recognized by auditory neurons. The auditory neurons carry information from the hair cells to the cochlear nucleus in the brainstem and, via the cochlear nucleus, to higher nuclei in the brain (Fig. 17.9).

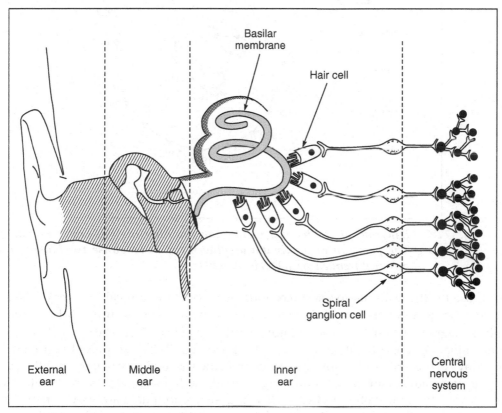

> **Fig. 17.9** *Details of cochlear part of the human ear (after Loizou, 1999)*

The normal cochlea and the associated neurons of the central auditory system provide information about both the frequency content and intensity of the auditory signal. Information is conveyed to the acoustic nerve about frequency content by the mechanically tuned properties of the basilar membrane. The inner hair cells, which connect to the vast majority of afferent neurons, are thought to be the sensory cells of the cochlea whereas the role of the outer hair cells is still under investigation. The location of hair cells along the cochlea determines their optimal response to frequency: hair cells at the apex are responsive to low frequencies, while hair cells at the base are responsive to high frequencies. The distribution of frequencies along the spiral is logarithmic (Fig. 17.10).

If the damage to the auditory system is peripheral in the inner ear, then a cochlear implant can be used. In the general design of a cochlear implant, the sound is decomposed into frequency

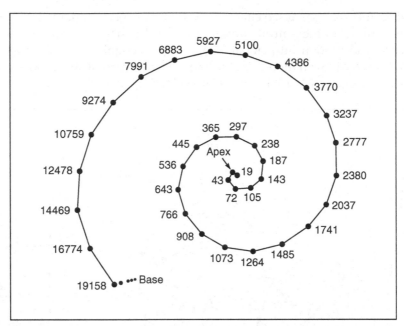

> **Fig. 17.10**   *Diagram of the basilar membrane showing the base and the apex.*
> *The position of maximum displacement in response to sinusoid of*
> *different frequency (in Hz) is indicated*

bands of use for the transmission and reception of speech, and critical features of the signals within those frequency bands are delivered to auditory neurons via an array of electrodes.

A block diagram of a generic cochlear implant is shown in Fig. 17.11. The microphone converts acoustic signals into electrical signals. The electrical signals are amplified and encoded in various ways in the block called stimulus encoder. In the vast majority of implants, the stimulus encoder is worn outside the head, producing a serially coded signal that is transmitted with a transcutaneous link, most often inductive. The link sends both data and power to an internal circuit that decodes the serial data stream and decomposes it into signals that are delivered to the current sources that drive the electrodes of the cochlear electrode array. Each electrode of the array is driven with either a pulsatile or an analog electrical signal. The signals traverse the tissues of

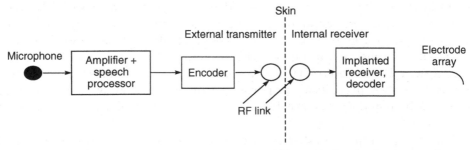

> **Fig. 17.11**   *Schematic of a cochlear prosthesis (after Spelman, 1999)*

the inner ear, usually the fluids of the scala tympani, and excite the auditory neurons. The excitation depends upon the number of intact neurons that remain, the proximity of the electrode array to the neurons, and the spatial and temporal characteristics of the current-density fields that affect the neurons.

In single channel implants, only one electrode is used. In multi-channel cochlear implants, an electrode array is inserted in the cochlear so that different auditory nerve fibres can be stimulated at different places, thereby exploiting the place mechanism for coding frequencies. Different electrodes are stimulated, depending on the frequency of the signal. Electrodes near the base of the cochlea are stimulated with high frequency signals while electrodes near the apex are stimulated with low frequency signals. The signal processor is responsible for breaking the input signal into different frequency bands or channels and delivering the filtered signals to the appropriate electrodes.

Different types of cochlear implants are available which differ in the following characteristics (Loizou,1999):

- Electrode design: Number of electrodes and their configurations
- Type of stimulation: Analog or pulsatile
- Transmission link: Transcutaneous or percutaneous
- Signal processing: Waveform representation or feature extraction

Cochlear implants have been a spectacular success story for biomedical engineers. They have been successful in restoring partial hearing to profoundly deaf people. While not all patients are able to talk on the telephone when they use their implants, a substantial number can, and all users can improve their communication skills by using lip reading. Users can hear environmental sounds such as automobile horns, knocks at the door, and sirens. Despite the maturity of today's cochlear implants, there are exciting opportunities for bioengineers to advance designs to provide better devices to the patients.

Patient Safety

Hospitals are confronted with the difficult problem of creating a safe electric environment for the care and comfort of the patients. Electric shock, burns and fire hazards result from the careless use of electricity. When electricity is relied upon to support life with devices like external pacemakers, respirators, etc. power failure is a continuous threat. Shock resulting from electric power is a common experience. Disruption of physiologic function by leakage current applied internally remains sometimes hidden and mysterious. While faulty electric cords and appliances contribute to the former, lack of concept and faulty design are responsible for the latter.

In the past few years, with the extent and proliferation of electro-medical equipment, the patient has been included as a part of an electrical system and has generated a problem of reconsidering the well established safety conventions. The patient is wired for collecting data important for diagnosis and therapy and his/her presence in the system has wholly changed the safety concepts. Currents of the order of microamperes, which were earlier ignored, have been found to be potentially hazardous.

Electric current can flow through the human body either accidentally or intentionally. Electrical currents are administered intentionally in the following cases: (i) for the measurement of respiration rate by impedance method, a small current at high frequency is made to flow between the electrodes applied on the surface of the body, (ii) high frequency currents are also passed through the body for therapeutic and surgical purposes, (iii) when recording signals like ECG and EEG, the amplifiers used in the preamplifier stage may deliver small currents themselves to the patient. These are due to bias currents. Accidental transmission of electrical current can take place because of a defect in the equipment; excessive leakage currents due to defect in design; operational error (human error) and simultaneous use of other equipment on the patient which may produce potentials on the patient circuit.

18.1 ELECTRIC SHOCK HAZARDS

It is a common experience that hazards due to electric shock are also associated with equipment other than that used in hospitals. However, the equipments used in medical practice have to operate in special environments, which differ in certain respects from the others. Some such special situations are as follows:

(i) A patient may not be usually able to react in the normal way. He is either ill, unconscious, anaesthetized or strapped on the operating table. He may not be able to withdraw himself from the electrified object, when feeling a tingling in his skin, before any danger of electro-cution occurs.

(ii) The patient or the operator may not realize that a potential hazard exists. This is because potential differences are small and high frequency and ionizing radiations are not directly indicated.

(iii) A considerable natural protection and barrier to electric current is provided by human skin. In certain applications of electromedical equipment, the natural resistance of the skin may be by-passed. Such situations arise when the tests are carried out on the subject with a catheter in his heart or on large blood vessels.

(iv) Electromedical equipment, e.g. pacemakers may be used either temporarily or permanently to support or replace functions of some organs of the human body. An interruption in the power supply or failure of the equipment may give rise to hazards, which may cause permanent injuries or may even prove fatal for the patient.

(v) Medical instruments are quite often used in conjunction with several other instruments and equipment. These combinations are often adhoc. Several times there are combinations of high power equipment and extremely sensitive low signal equipment. Each of these devices may be safe in itself, but can become dangerous when used in conjunction with others.

(vi) The environmental conditions in the hospitals, particularly in the operating theatres, cause an explosion or fire hazards due to the presence of anaesthetic agents, humidity and cleaning agents, etc.

The various factors listed above indicate that the electromedical equipment may be used in different places and under different circumstances. It is also obvious that an optimum level of safety can only be achieved when efforts are made to include safety measures in the equipment, in the installation as well as in the application.

Broadly speaking, there are two situations which account for hazards from electric shock. They are (i) gross shock and (ii) micro-current shock. In the case of gross shock, the current flows through the body of the subject, e.g. as from arm to arm. The other case is that of micro-current shock in which the current passes directly through the heart wall. This is the case when cardiac catheters may be present in the heart chambers. Here, even very small amounts of currents can produce fatal results.

18.1.1 Gross Shock

Gross shock is experienced by the subject by an accidental contact with the electric wiring at any point on the surface of the body. The majority of electric accidents involve a current pathway through the victim from one upper limb to the feet or to the opposite upper limb and they generally occur through intact skin surfaces. In all these cases, the body acts as a volume conductor at the mains frequency.

For a physiological effect to take place, the body must become part of an electric circuit. Current must enter the body at one point and leave at some other point. In this process, three phenomena can occur. These are:

  (i)   Electrical stimulation of the excitable tissues—nerves and muscles
  (ii) Resistive heating of tissue
  (iii) Electro-chemical burns and tissue damage for direct current and very high voltages.

The value of electric current, flowing in the body, which causes a given degree of stimulation varies from individual to individual. Typical threshold values of current produce certain responses where the current flows into the body from external contacts (e.g. hand to hand) and these have been investigated. A review of the principal factors involved in electrical shock hazard has been presented by Bruner (1967).

For a given voltage present on the surface of the body, the value of current passing through it would depend upon the contact impedance. The impedance of the human body to electrical currents through intact skin surfaces can vary from 1 kΩ, if the skin is damp or scrubbed to over 100 kΩ, if the skin is dry. Besides this, it depends on many other factors such as age, sex, condition of skin (dry or wet, smooth or rough, etc.), frequency of current, duration of current and the applied voltage.

18.1.2   Effects of Electric Current on the Human Body

Threshold of Perception: Bruner (1967) states that the threshold of perception of electric shock is about 1 mA. At this level, a tingling sensation is felt by the subject when there is a contact with an electrified object through the intact skin. The threshold varies considerably among individuals and with the measurement conditions. The lowest threshold could be 0.5 mA when the skin is moistened at 50 Hz. Threshold for dc current are 2 to 10 mA.

Let-go Current: As the magnitude of the alternating current is increased, the sensation of tingling gives way to the contraction of muscles. The muscular contractions increase as the current is increased and finally a value of current is reached at which the subject cannot release his grip on the current carrying conductor. The maximum current at which the subject is still capable of releasing a conductor by using muscles directly stimulated by that current is called "let-go current". The value of this current is significant because an individual can withstand, without serious after-effects, repeated exposure to his 'let-go current' for at least the time required for him to release the conductor. Currents even slightly in excess of the 'let-go current' would not permit the individual to release his grip from the conductor supplying current.

Experiments conducted by Lee (1966) on 124 males and 28 females reveal that the average value of the 'let-go current' for males was 16 mA and for females it was 10.5 mA. Based on these experiments, it is generally accepted that the safe 'let-go current' could be taken to be approximately 9 and 6 mA for men and women respectively.

Physical Injury and Pain: At current levels higher than the 'let-go current', the subject loses the ability to control his own muscle actions and he is unable to release his grip on the electrical conductor. Such currents are very painful and hard to bear. This type of accident is called the 'hold-on-type' accident, and is caused by currents in the range of 20–100 mA. These currents may also result in physical injury because of the powerful contraction of the skeletal muscles. However, the heart and respiratory functions usually continue because of the uniform spread of current through the trunk of the body. Strong involuntary contractions of the muscles and stimulation of the nerves can be painful and cause fatigue if there is long exposure.

Ventricular Fibrillation: If current comes in contact with intact skin and passes through the trunk at about 100 mA and above, there is a likelihood of pulling the heart into ventricular fibrillation. In this condition, the rhythmic action of the heart ceases, pumping action stops and the pulse disappears. Ventricular fibrillation occurs due to the derangement of function of the heart muscles rather than any actual physical damage to it. Ventricular fibrillation is a serious cardiac emergency because once it starts, it practically never stops spontaneously, even when the current that triggered it is removed. It proves fatal unless corrected within minutes, since the brain begins to die 2 to 4 min after it is robbed of its supply of oxygenated blood. Obviously, experiments involving currents likely to cause ventricular fibrillation cannot be performed on human beings. They have been performed on a variety of animals and relationships have been determined between fibrillating currents versus body weight, current magnitude and shock duration.

Sustained Myocardial Contraction: At currents in the range of 1 to 6 A, the entire heart muscle contracts. Although the heart stops beating while the current is applied, it may revert to a normal rhythm if the current is discontinued in time, just as in defibrillation. The damage is reversible if the shock duration is only of a few seconds. This condition may however be accompanied by respiratory paralysis.

Burns and Physical Injury: At very high currents of the order of 6 amperes and above, there is a danger of temporary respiratory paralysis and also of serious burns. Resistive heating causes burns, usually on the skin at the entry points, because skin resistance is high. Voltages higher than 230 V can puncture the skin. The brain and other nervous tissue loose all functional excitability when high currents pass through them. However, even in this case, if the shock duration is only of a few seconds, there is a possibility, of the heart reverting to the normal rhythmic action. The defibrillators are based for their effectiveness on this phenomenon.

The threshold of perception depends greatly on the current density in the body tissues. It may vary widely depending upon the size of the current contact. At a very small point contact, it is probable that even 0.3 mA current may be felt whereas a current in excess of perhaps 1 mA may not produce a sensation if the contacts are somewhat larger. Similarly, depending on the size of contact, the threshold of pain may also be considerably above 1 mA; probably 10 mA if the contacts are large enough. Besides the magnitude of current, the current duration and the relationship of current flow resistance are also important. Duration of less than 10 ms typically does not produce fibrillation whereas a duration of 0.1 s or longer does.

Outside the medical practice, a general limit of 500 μA is established for currents passing through the body. This is intended as a guide to acceptable leakage current levels because a current of this magnitude may give a tingling sensation, which can be disagreeable over a long period. The International Electrotechnical Commission, therefore, recommends that for medical equipment the current flowing continuously through the body should not exceed 100 μA within a frequency range of 0 to 1 kHz. In case of abnormal situations (e.g. in the case of equipment failure), the recommended maximum current is 500 μA for frequencies up to 1 kHz. Above 1 kHz, the recommended maximum increases proportionally with frequency (Fig. 18.1).

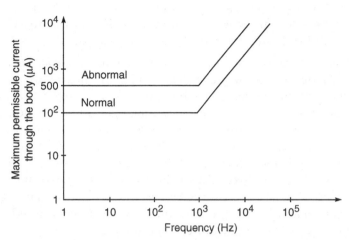

> **Fig. 18.1**   *Maximum permissible leakage currents through the body versus*
> *frequency (Courtesy: Philips Medical Systems, Netherlands)*

Gross shock hazards are usually caused by electrical wiring failures, which allow personal contact with a live wire or surface at the power line voltage. This type of hazard is dangerous not only to the patient but also to the medical and attending staff. The most vulnerable part in the system of electrical safety is the cord and plug. Physical abuse and deterioration are so common that human perception is blunted and faulty cords and plugs are continued in use. Their use can result in fatal accidents. Broken plugs, faulty sockets and defective power cords must be immediately replaced.

The commonly found fluids in medical practice such as blood, urine, intravenous solutions, etc. can conduct enough electricity to cause temporary short circuits if they are accidentally spilled into normally safe equipment. This hazard is particularly more in hospital areas that are normally subject to wet conditions, such as haemodialysis and physical therapy areas. The cabinets of many electrically operated equipment have holes and vents for cooling that provide access for spilled conductive fluids, which can cause potential electric shock hazard.

Whalen and Starmer (1967) have published comprehensive data on shock hazards arising from polarity reversal or wiring errors. The frequency with which such errors create hazards has prompted the designing of many testers for checking outlets or appliances. The cords of the instruments must be limited in length as the leakage current is a function of the length. The length should be standardized, say at 3 m. This should be shielded and a low dielectric loss insulation be used. Extension cords introduce needless risk. They should be avoided as far as possible.

18.1.3   Microcurrent Shock

The threshold of sensation of electric currents differs widely between currents applied arm to arm and currents applied internally to the body. In the latter case, a far greater percentage of the current may flow via the arterial system directly through the heart, thereby requiring much less current to produce ventricular fibrillation. Such situations are commonly encountered in hospitals; for example, the patients in the catheter laboratory or in the operating room, with a catheter in the heart, would have very little resistance to electric currents. A cardiac catheter connected to an electrical circuit for the measurement of pressure provides a conductive fluid connection directly to the heart. This makes the patient highly vulnerable to electric shock because the protection he would have had from layers of intact skin and tissue between his heart and the outside electrical environment has now been by-passed by the wire of fluid column within his heart or blood vessels.

Experiments on human beings have not been and are not likely to be conducted for the determination of hazardous threshold when internal electrodes are applied to the patient. Experiments with dogs have indicated that in some animals ventricular fibrillation can be produced with currents as low as 17 μA applied directly to the dog's heart. Based on the data collected by several researchers, 10 μA has been postulated as the safe upper limit.

The International Electrotechnical Commission's recommendations covering the safety aspects of electrical equipment for medical applications stipulate that the current that flows continuously through the heart shall not exceed 10 μA for a frequency range of 0 to 1 kHz (Fig. 18.2).

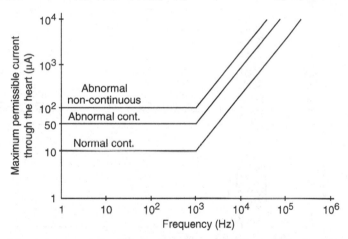

➤ **Fig.** 18.2 *Maximum permissible leakage-current through the heart versus frequency (Courtesy: Philips Medical System, Netherlands)*

In abnormal situations (e.g. in the case of equipment failure), the maximum value in this frequency range may be 50 μA for currents which can flow continuously through the patient's heart. It could be 100 μA for non-continuous current which can flow through the patient's heart in an abnormal situation. The limits of current increase proportionally for frequencies above 1 kHz.

18.1.4 Electrophysiology of Ventricular Fibrillation

Electrical current passing through the human body can prove hazardous as it can induce circulatory arrest. The primary cause of circulatory arrest induced by low voltage electric shock is ventricular fibrillation. Ventricular fibrillation is the most frequently encountered problem in patient monitoring systems. Therefore, it shall be advantageous to discuss ventricular fibrillation thresholds.

Ventricular fibrillation is produced more spontaneously when current passes directly through the heart during a specific portion of the cardiac cycle known as the 'vulnerable period'. Vulnerable period for the ventricular muscle occurs during the upstroke of the *T* wave of the electrocardiogram and even a single shock impulse lasting for less than 0.1 s could cause ventricular fibrillation if received during this period. Figure 18.3 shows the probability (in percentage) of the heart getting into fibrillation during various phases of the cardiac cycle.

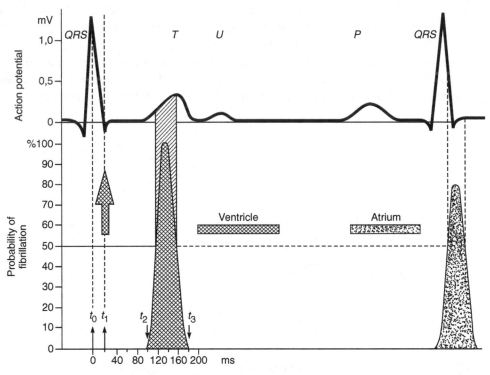

> **Fig. 18.3**   *Vulnerable zone in the ECG cycle*

Whalen *et al* (1964) showed that the risk of ventricular fibrillation in general, increases when the value of duration of the current passing through the heart increases. This risk decreases for higher frequencies. Based on the data obtained from animals, it has been found that frequencies around 50 Hz, and also direct current are relatively more dangerous as compared to currents of higher frequencies. In fact, muscle contractions and the sensation of electric shock do not occur above about 100 kHz; instead the tissues in the body get overheated.

It has been found that the minimal let-go currents occur for commercial power line frequencies of 50 to 60 Hz. For frequencies below 10 Hz, let-go currents rise because the muscles can partially relax during part of each cycle. At frequencies above several hundred cycles, the let-go current rise-sagain.

The risk of producing ventricular fibrillation is related to the current density and consequently to the electrode area. Roy (1980) compiled data from the published literature about average fibrillating current density and minimum current density and plotted these values as a function of the electrode area (Fig. 18.4). The values shown in this diagram are for currents flowing for 3 s or longer and include results from human hearts (Watson *et al*, 1973), pacing catheters as well as small bore dye-injection catheters. The figure also shows the safety margin 10 µA being the current standard.

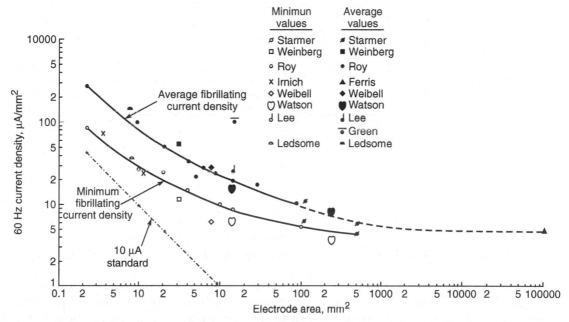

> **Fig. 18.4** *Relationship between fibrillating current densities as a function of electrode area (after Roy, 1980; reproduced by permission of Med. & Biol. Eng. & Comput.)*

The fibrillation thresholds in terms of 50 Hz current are very clear, but not so in terms of 50 Hz voltage. This is because skin contact impedance plays an important role in the determination of 50 Hz voltage. Impedance is an extremely variable factor and depends upon electrode size, electrode material, electrolyte composition and amplitude of current flowing through them. Therefore, the voltage thresholds could be considered as the minimum voltage required to drive the minimum fibrillating current through the system.

Ventricular fibrillation is probably the most common cause of death by electrocution (Dalziel and Lee, 1968). Dalziel also showed that the likelihood of an electric shock producing ventricular fibrillation increases with the duration of the shock. Also, the magnitude of the shock must be within certain limits for each duration. Figure 18.5 shows the relationship established by Dalziel (1970). It is obvious that the portion above and to the right of the electrocution threshold for adults is the most dangerous region, while the area below the 'let-go threshold' is considered safe.

Geddes *et al* (1973) reported the threshold 60 Hz (mains frequency used in USA) alternating current values required to induce ventricular fibrillation when the current is applied to electrodes

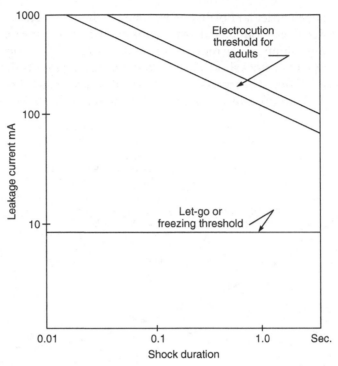

> **Fig. 18.5**   *Relationship between current and shock duration (after Dalziel, 1970)*

at different sites on the surface of animals of different weights. It was found that for a given body weight, the duration of exposure to current influences the fibrillation threshold; exposure times shorter than 1 s require more current. For a given duration of current flow, the threshold current for fibrillation is a function of body weight and electrode location. The lowest current for fibrillation was required with lead III, whereas lead I required the highest current. They concluded that for a 5 s exposure, the threshold current for fibrillation varies almost as the square root of body weight (W), the general expression being $I = KW^{\alpha}$ where α is nearly 0.5. Thus, the amount of 60 Hz thoracic current required to produce ventricular fibrillation is related to body weight, the duration of exposure to the current and the current pathway (Fig. 18.6).

The shock hazards described above automatically lead to the classification of patients in hospitals (Hill and Dolan, 1976) into three categories:

(i) A general patient who is normally not connected to any instrumentation. Such patients would not ordinarily be present in intensive care wards. They may, however, come in casual contact with electrical instrumentation.

(ii) The susceptible patients will be all those patients who are ordinarily connected to instruments like cardioscopes and other monitoring instruments. They are decidedly more susceptible to ventricular fibrillation than the general category.

(iii) A critical patient will have a direct electrically conductive path to any part of the heart. Based on this classification, the type of leakage current associated with each category can be worked out.

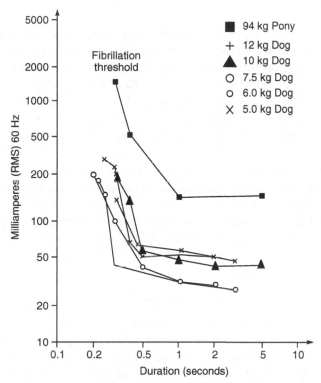

> **Fig. 18.6** *Threshold 60 Hz alternating current required to produce ventricular fibril-*
> *lation in animals of different body weights when the current is suddenly*
> *applied having a duration extending from 0.2 to 0.5 s (Geddes. et. al, 1973;*
> *reproduced by permission of IEEE Trans. Biomed. Engg.)*

▓▓▶ 18.2 LEAKAGE CURRENTS

Currents of extremely small magnitude can be fatal to a patient when a direct, localized electrical path exists to the heart. Patients in coronary care units recovering from heart diseases are specially vulnerable since their hearts are already in an irritable state. In such cases, the amount of electrical stimulation necessary to induce a life threatening arrhythmia is greatly reduced. Accidents of this nature can occur in unpredictable circumstances. For example, if a patient in a poorly grounded electrically powered bed has an external pacemaker attached to him and the doctor touching the bed adjusts a control on the pacemaker, the doctor may not feel anything but the patient may die. Accidents can even occur with safe electromedical equipment being properly used if there are defects in the wiring of power outlets.

The major source of potentially lethal currents in any instrument or equipment is the leakage current. Leakage current by definition is an inherent flow of non-functional current from the live electrical parts of an instrument to the accessible metal parts. Leakage currents usually flow through the third wire connection to the ground. They occur by the presence of a finite amount of insulation impedance, which consists of two parts: capacitance and resistance. Leakage currents due to capacitance develop because of the presence of capacities between any two conductors separated in space. Current flow shall take place if an alternating voltage is applied between them. The magnitude of the leakage current is determined by the value of the capacitance present therein. Leakage current of this type mostly originates due to capacitive coupling from the power transformer primary to other parts of the transformer or other parts of the instrument.

The resistive component of leakage current arises because no substance is a perfect insulator and some small amount of current will always flow through it. However, this type of leakage current is usually very small as compared to the capacitive leakage currents and can be safely ignored.

Instruments are generally designed so that leakage current flows to the instrument case and then to the ground via the three-wire power cord provided with the instrument. The third or the grounding wire effectively drains off leakage current to the earth. Then there will no longer be any danger for the patient or the operator. However, when the earth connection is interrupted, the leakage current can become a real danger to the patient or the operator.

18.2.1 Types of Leakage Current

Leakage currents are divided according to the current path into the following types:

Enclosure Leakage Current: The enclosure leakage current is the current which flows, in normal condition, from the enclosure or part of the enclosure through a person (or an external conductive part other than the earth connection) in contact with an accessible part of the enclosure to earth or another part of the enclosure. This current becomes significant when the person touching the equipment is connected to earth either directly or via a large capacitance.

Earth Leakage Current: The earth leakage current is the current which flows, in normal condition, to earth from the mains parts of an apparatus via the earth conductor.

Patient Leakage Current: The patient leakage current is the current which flows through the patient from or to the applied part of the patient circuits. This current does not include any functional patient current.

In almost all equipments leakage currents do flow to the ground from the ac supply operating the device. This current may not prove hazardous if there is a good grounding system to drain it away. In most of the cases, the instrument ground and the ground to which the patient is connected is the power ground. Dangerous potentials have been measured in the operating rooms between power grounds at different outlets or between power grounds and the water pipe or earth ground. Voltages as high as a few hundred millivolts have been measured between the power ground (third pin of the socket) and the water pipe nearby. These are the voltages which, if applied to a catheter positioned in the heart of a patient, could prove fatal.

We shall analyze one case in which the importance of proper grounding is brought out. Consider an electrical equipment (Fig. 18.7) connected to the power line. A leakage current of 100 µA can be assumed to be flowing in the ground wire. If the chassis of this equipment is connected to the patient who is grounded, then very little of this current will flow through him.

Let us further assume that the patient offers a resistance of $1000\,\Omega$ to the ground and the ground connection from the instrument has $1\,\Omega$ of series resistance. The current division shall be such that

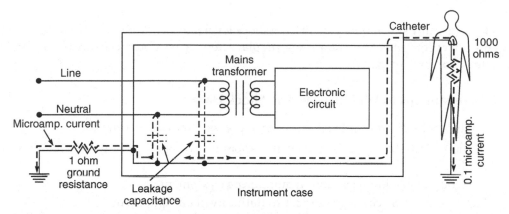

Path of leakage current in a normal case, i.e. the ground wire intact (redrawn by permission of Hewlett Packard, USA)

only 0.1 µA of current shall pass through the patient, the rest will flow to the ground. If by chance the ground connection breaks, the full leakage current will flow through the patient (Fig. 18.8). This is a very hazardous situation particularly if the current goes through internal electrodes in the vicinity of the patient's heart.

The most apparent solution to this problem of grounding failure seems to be to reduce the leakage current from the instrument to below 10 mA. Then, even if the patient were connected to the instrument case and the power cord ground breaks, there would be no hazard. Unfortunately, this is not an easy solution as it is difficult to get power transformers with low enough leakage capacitance. However, it is certainly possible to electrically isolate the patient from the input mains connections, so that even if the ground wire did break, hazardous current shall not flow through the catheterized patients.

Neither loss of ground nor increased leakage is readily evident to the user because neither will affect the functioning of the unit and by design, will not result in a shock hazard. Even if there is

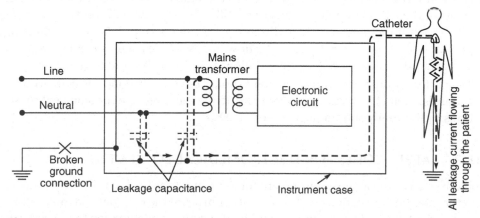

➤ **Fig. 18.8** *Path of leakage current in case of discontinuous ground (redrawn by permission of Hewlett Packard, USA)*

increased leakage and loss of ground occurring simultaneously, to a degree sufficient to pose a hazard to electrically susceptible patients, the shock level may not be perceptible to operators and would not affect equipment function. Thus, regular safety inspection checks are needed to ensure in time detection of defects.

18.2.2   Precautions to Minimize Electric Shock Hazards

The following precautions should be observed to prevent hazardous situations:

- In the vicinity of the patient, use only apparatus or appliances with three-wire power cords.
- Provide isolated input circuits on monitoring equipment.
- Have periodic checks of ground wire continuity of all equipment.
- No other apparatus should be put where the patient monitoring equipment is connected.
- Staff should be trained to recognize potentially hazardous conditions.
- Connectors for probes and leads should be standardized so that currents intended for powering transducers are not given to the leads applied to pick up physiologic electric impulses.
- The functional controls should be clearly marked and the operating instructions be permanently and prominently displayed so that they can be easily familiarized.
- Many of the portable medical equipment such as dialysis units, hypothermia units, physiotherapy apparatus, respirators and humidifiers are used with adapter plugs that do not ensure a proper grounding circuit. Particular care should be taken in such cases.
- The operating instructions should give directions on the proper use of the equipment. In fact, for electromedical equipment, the operating instructions should be regarded as an integral part of the unit.
- The mechanical construction of the equipment must be such that the patient or operator cannot be injured by the mechanical system of the equipment, if properly operated.
- A potential difference of not more than 5 mV should exist between the ground point at the outlet and the ground points at any of the other outlets and any conductive surface in the same area. If there is no voltage difference between two points or only an insignificant potential of a few millivolts exists, the flow of possible leakage current between its source and ground will be restricted to well below a level which could be dangerous.
- The patient equipment grounding point should be connected individually to all receptacle grounds, metal beds and any other conductive services. The resistance of these connections individually should not exceed 0.15 Ω.

▶ 18.3   SAFETY CODES FOR ELECTROMEDICAL EQUIPMENT

The problem of ensuring a safe environment for the patients as well as for the operators has been engaging the attention of all concerned in several countries at the national and international levels. Various countries have laid down codes of practice for equipment intended to be used in hospitals. The International Electrotechnical Commission (IEC) has brought out a fairly voluminous document, the IEC 601 (General Requirements for the Safety of Medical Electrical Equipment), to provide a universal standard for manufacturers of electromedical equipment as well as a reference manual on good safety practice. The IEC comprises over 40 countries: East and West European, Canada, New Zealand, Japan, Russia as well as USA. The intention of the common standard is that any electromedical equipment built to the standard should be completely acceptable in all IEC countries. This standard has been adopted in many countries. Adopting a common standard implies that construction of equipment shall be universally acceptable, the leakage and earth resistance paths will be assessed in identical manners and the mains leads will be coloured to the same code, etc.

Based on the IEC Document, the Bureau of Indian Standards (BIS) has issued the IS:8607 standard to cover general and safety requirements of electromedical equipment. The standard issued in eight parts, covers the following aspects:

Part I	General
Part II	Protection against electric shock
Part III	Protection against mechanical hazards
Part IV	Protection against unwanted or excessive radiation
Part V	Protection against explosion hazards
Part VI	Protection against excessive temperature, fire and other hazards
Part VII	Construction
Part VIII	Behaviour and reliability

This standard applies to all medical electrical equipment except or otherwise stated in the individual specification for the particular medical equipment for which additional or modified requirements have been specified.

Individual standards on different electromedical equipment have also been issued by the BIS. Some of the important standards issued are: Radiofrequency diathermy apparatus (IS:7583), Electrocardiograph (IS:8048), Cardiac Defibrillators (IS:9286), Diagnostic medical X-ray equipment (IS:7620), and Electromyograph (IS:8885).

18.4 ELECTRICAL SAFETY ANALYZER

A range of electrical safety analyzers are commercially available for testing both medical facility power systems and medical equipment. They vary in complexity from simple volt-ohm-meter to computerized automatic measurement systems that generate hard copies of test results. The facilities available in these testers are given below.

18.4.1 Mechanical Testing of Electrical Outlets

The power delivery point in the patient area usually consists of the outlets in the vicinity of the patient. The outlets should have three-prong wall receptecles that meet the ground retention force requirements as per the relevant standards. These force requirements are important as they ensure that plugs on medical devices do not fall out of the receptacle, possibly placing the patient in danger.

Receptacles should be tested for proper wiring, proper line voltage, low ground resistance and mechanical tension. The holding tension provided by a set of contacts in the receptacle can be measured with a spring loaded tester that measures the force required to extract the plug after it is inserted into the receptacle. The minimal mechanical retaining force for each of the three contacts in the receptacle is around 115 G or 4 oz. This gives a total mechanical holding force of ¾ lb between the plug and the receptacle. Tension testing devices are available from a number of companies that sell hospital testing equipment. It may be noted that the tension in the receptacle should not be too high as it will destroy the plug, electrical cable or the receptacle.

18.4.2   Electrical Testing of Electrical Outlets

Electrical testing of a wall receptacle should be made to determine whether power is available at the receptacle and if its polarity is correct. Proper polarity of the receptacles means that the hot, neutral and ground wires are connected to their correct positions. Miswiring of an outlet can happen during the original construction of the area or when a broken outlet is replaced. Testing of the outlet can easily be done by a receptacle tester consisting of three LEDs, which can test for 8 of 64 possible states of the outlet. The three lights have only two states (2^3) where as each of the three outlet contacts has four states (4^3)—hot, neutral, ground and open. Testing of the outlet for possible reverse wiring or loss of grounding is important as these are potentially hazardous conditions and should be corrected immediately.

Ground resistance can be measured by passing up to 1 A current through the ground wire and measuring the voltage between the ground and neutral. The resistance of neutral wiring can be tested similarly by passing the current through the neutral conductor. Ground or neutral resistance should not exceed 0.15 Ω.

18.4.3   Tests of the Grounding System in Patient Care Areas

Both voltage and impedance values have been defined with different limits for new and existing construction. The voltage between a reference ground point and exposed conductive surfaces should not exceed 20 mV for new construction. For existing construction, the limit is 500 mV for general care areas and 40 mV for critical care areas.

The impedance between the reference ground point and receptacle ground contact must be less than 0.1 Ω for new construction and less than 0.2 Ω for existing construction.

▶ 18.5   TESTING OF BIOMEDICAL EQUIPMENT

The type of tests needed for checking the safety parameters of biomedical equipments are many and varied. However, measurement of leakage current and ground resistance constitute a majority of the tests and must be carried out periodically on instruments normally applied to the patients.

18.5.1   Chassis Leakage Current Measurement

Leakage current is mostly measured from the equipment case to the ground with an ac microammeter, while the equipment is plugged in through an adapter that interrupts the ground wire, forcing the leakage current to flow through the measuring instrument. The measurement is made with the power on and off, preferably with both normal and reverse polarity on the power conductors. This helps to assess the possible hazard arising from an outlet, which is improperly wired. When several appliances are mounted together in one rack or cart, and are supplied by only one power cord, the complete rack or cart must be tested as one appliance. Leakage currents from the chassis should not exceed 500 µA for equipment not intended to come into contact with patients and should not exceed 100 µA for equipment that is likely to come into contact with the patients. These are limits on RMS currents from dc to 1 KHz.

Figure 18.9 shows a block diagram of the leakage measuring circuit. The leakage current is made to develop a potential across 1000 Ω resistance, which is shunted by a 0.15 µF capacitor. This combination simulates the patient impedance and thus helps to determine the likely leakage

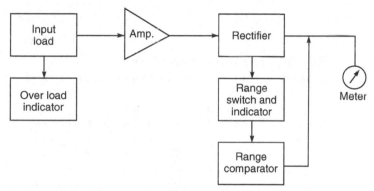

> **Fig. 18.9** *Schematic diagram of leakage current meter*

current, which would flow through the patient (Kistler and Miller, 1982). Some of the commercially available digital multimeters can be used to measure the potential developed. Such instruments must have an input impedance of at least 100 KΩ and frequency range up to 1 MHz. However, special instruments have been designed specifically for checking the leakage current on electromedical equipment. The leakage test is performed with both the power off and on (normal polarity leakage, reverse polarity leakage). These leakage currents are the maximum values that can occur with any combination of polarity or power. The test for leakage current with power off provides an opportunity for checking a short to the chassis from either side of the line. The range of leakage measurement is 0–2000 µA. Figure 18.10(a) shows the principle of a leakage current meter. Figure 18.10(b) shows a scheme for the measurement of enclosure leakage current using a leakage current meter.

The capacitor is employed to imitate the sensitivity of the heart as a function of frequency. Measurement of the enclosure leakage current should be made with a disconnected and connected ground wire. The load for the current input specified by AAMI (Association for the Advancement of Medical Instrumentation, USA) is 1 kΩ resistance in parallel with 0.15 µF capacitor having a series resistance of 10 Ω.

Another variation of current-meter circuits for measuring leakage current is shown in Fig. 18.11. This circuit has an input impedance of 1 KΩ and a frequency response that is flat to 1 kHz. It drops at the rate of 20 db/decade to 100 kHz and then remains flat up to about 1 MHz.

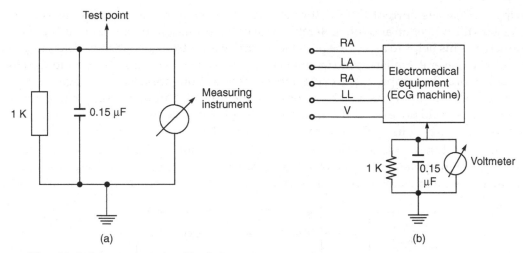

> **Fig. 18.10(a)** *Principle of leakage current meter*
>   **(b)** *Measurement of enclosure leakage current using leakage meter*

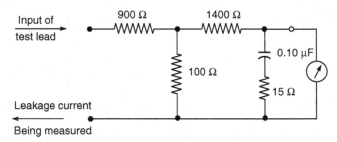

> **Fig. 18.11** *Current meter circuit suggested by NFPA (National Fire Protection Association, MA. USA)*

18.5.2 Leakage Current in Patient Leads

The patient leakage current is determined by connecting the measuring instrument between earth and one of the patient inputs. These tests are particularly required and carried out on ECG machines. The leakage current is measured under the following conditions using the standard load:

  (i) *Combined lead leakage test*–All the ECG leads are connected together so that the leakage from the combined set of leads to ground can be measured.

 (ii) *Individual lead leakage test*–Each individual lead is checked for leakage to ground independent of the remaining leads. Each lead is selected in turn and the measurements are made.

(iii) *Paired lead leakage test*–The pairs are selected by a progressive selection of the leads.

(iv) *Leakage with line voltage on leads*–This test specifically determines the amount of risk current that will flow through a patient connected to an ECG machine in the event of the patient coming in contact with the hot side of the ac supply from any source within the patient environment. This test is conducted with the ground wire to the instrument intact. For this test, line voltage is applied to the combined ECG leads. This particular test is current limited to a maximum of 1000 μA in order to prevent damage to the instrument under test, should it not have an isolated input.

Leakage current in patient leads is particularly important because these leads are the most common low impedance patient contact points. The leakage current in patient leads must be limited to 50 μA and for isolated patient leads, it should be less than 10 μA.

18.5.3   Ground-continuity Test

The integrity of the ground wire circuit is usually checked with an ohmmeter while the equipment is unplugged. This test helps to detect complete failures, but is not sensitive enough to determine the gradual development of a defect and also does not verify the capability of carrying large fault currents that could be caused by a line to enclosure short circuit. To overcome this problem, ground resistance is measured with a test current of several amperes provided either by a transformer and current limiting resistance or by a regulated dc constant current source. The high current level at which this test is made, as compared to ohmmeter testing, enables resistance measurements with enough accuracy to disclose trends, and verifies the ground circuit capability to handle fault currents large enough to actuate overcurrent devices.

The resistance between the ground pin of the plug and the equipment chassis/exposed metal objects should not exceed 0.15 Ω during the life of the equipment. During this measurement, the power cord must be flexed at its connection to the plug and at its strain relief where it enters the equipment. This test will determine if there is a break in the ground wire or if the internal device ground connection is bad or corroded. The implication of too high a ground wire resistance is that when current flows in the ground wire, the IR (Current × Resistance) drop can produce a voltage on the chassis of the equipment that may cause problems in areas with electrically sensitive patients.

Modern safety analyzers are digital reading meters incorporating facilities for measuring enclosure leakage current, patient leakage current and ground continuity tests. Besides these, they have provisions to carry out insulation tests at 500 V and isolated power system tests.

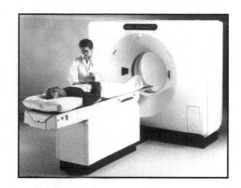

➤ PART TWO : MODERN IMAGING SYSTEMS

X-ray Machines and Digital Radiography

X-rays were discovered by the German physicist Wilhelm Konrad Röntgen in November 1895. He called the 'new kind of ray' or X-rays, X for the unknown. With these new rays, he could make a photograph of his wife's hand—showing the bones and her wedding ring. Soon afterwards, their usefulness to visualize the internal anatomy of humans was established. Today, imaging with X-rays is perhaps the most commonly used diagnostic tool with the medical profession, and the techniques from a simple chest radiography to a digital subtraction angiography or computer tomography depend on the use of X-rays.

▶ 19.1 BASIS OF DIAGNOSTIC RADIOLOGY

A radiological examination is one of the most important diagnostic aids available in the medical practice. It is based on the fact that various anatomical structures of the body have different densities for the X-rays. When X-rays from a point source penetrate a section of the body, the internal body structures absorb varying amount of the radiation. The radiation that leaves the body has a spatial intensity variation, i.e. an image of the internal structure of the body. The commonly used arrangement for diagnostic radiology is shown in Fig. 19.1. The X-ray intensity distribution is visualized by a suitable device like a photographic film. A shadow image is generated that corresponds to the X-ray density of the organs in the body section. The examination technique varies according to the clinical problem. The main properties of X-rays, which make them suitable for the purposes of medical diagnosis, are their:

- Capability to penetrate matter coupled with differential absorption observed in various materials; and
- Ability to produce luminescence and its effect on photographic emulsions.

The X-ray picture is called a radiograph, which is a shadow picture produced by X-rays emanating from a point source. The X-ray picture is usually obtained on photographic film placed in the image plane. The skeletal structures are easy to visualize and even the untrained eye can sometimes observe fractures and other bone abnormalities.

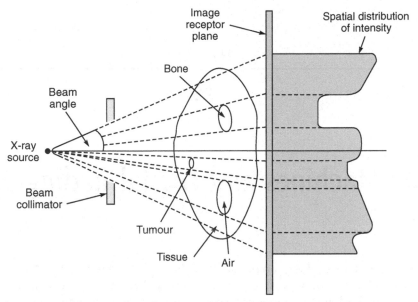

> ➤ **Fig. 19.1**   *Basic set-up for a diagnostic radiology image formation process*

Chest radiographs are mainly taken for examination of the lungs and the heart. Because of the air enclosed in the respiratory tract, the larger bronchi are seen as a negative contrast, and the pulmonary vessels are seen as a positive contrast against the air-filled lung tissue. Different types of lung infections are accompanied by characteristic changes, which often enable a diagnosis to be made from the location, size and extent of the shadow.

Heart examinations are performed by taking frontal and lateral films. The evaluation is performed partly by calculating the total heart volume and partly on the basis of any changes in shape. For visualization of the rest of the circulatory system and for the special examinations of the heart, use is made of injectible, water-soluble organic compounds of iodine. A contrast medium is injected into an artery or vein, usually through a catheter placed in the vessel. Therefore, all the larger organs of the body can be examined by visualizing the associated vessels and this technique is called angiography. The examination is designated according to the organ examined—, e.g. for coronary angiography—the coronary vessels of the heart—, angiocardiography—the heart, and cerebral angiography—, the brain.

The entire gastro-intestinal tract can be imaged by using an emulsion of barium sulphate as a contrast medium. It is swallowed or administered to diagnose common pathological conditions such as ulcers, tumours or inflammatory conditions. Negative and positive contrast media are used for visualizing the spinal canal, the examination being known as myelography. The central nervous system is usually examined by pneumography, i.e., filling the body cavities with air. It may, however, be mentioned that computerized tomography has greatly reduced the need for some of the invasive neuro-radiological methods, which involve discomfort and a certain risk for the patient.

▌▌▌▶ 19.2 NATURE OF X-RAYS

X-rays are electromagnetic radiation located at the low wavelength end of the electromagnetic spectrum. The X-rays in the medical diagnostic region have wavelength of the order of 10^{-10}m. They propagate with a speed of 3×10^{10} cm/s and are unaffected by electric and magnetic fields. According to the quantum theory, electromagnetic radiation consists of photons, which are conceived as 'packets' of energy. Their interaction with matter involves an energy exchange and the relation between the wavelength and the photon is given by

$$E = h\nu = h\frac{c}{\lambda}$$

where h = Planck's constant = 6.32×10^{-34} J s

$\qquad c$ = velocity of propagation of photons = 3×10^{10} cm/s

$\qquad \nu$ = frequency of radiation

$\qquad \lambda$ = wavelength

A vibration can be characterized either by its frequency or by its wavelength. In the case of X-rays, the wavelength is directly dependent on the voltage with which the radiation is produced. It is, therefore, common to characterize X-rays by the voltage, which is a measure of the energy of the radiation.

19.2.1 Properties of X-rays

Because of short wavelength and extremely high energy, X-rays are able to penetrate through materials which readily absorb and reflect visible light. This forms the basis for the use of X-rays for radiography and even for their potential danger. X-rays are absorbed when passing through matter. The extent of absorption depends upon the density of the matter. X-rays produce secondary radiation in all matter through which they pass. This secondary radiation is composed of scattered radiation, characteristic radiation and electrons. In diagnostic radiology, it is scattered radiation which is of practical importance.

X-rays produce ionization in gases and influence the electric properties of liquids and solids. The ionizing property is made use of in the construction of radiation-measuring instruments.

X-rays also produce fluorescence in certain materials to help them emit light. Fluoroscopic screens and intensifying screens have been constructed on the basis of this property. X-rays affect photographic film in the same way as ordinary visible light.

19.2.2 Units of X-radiation

The International Commission on Radiological Units and Measurements has adopted *Rontgen* as a measure of the quantity of x-radiation. This unit is based on the ability of radiation to produce ionization and is abbreviated 'R'. One R is the amount of x-radiation which will produce 2.08×10^9 ion pairs per cubic centimetre of air at standard temperature (0°C) and pressure (760 mmHg at sea level and latitude 45°). Other units derived from the *Rontgen* are the millirontgen (mR = 1/1000 R) and the microrontgen (μR = 10^{-6} R). The unit of x-radiation has been based on the ionization produced by the rays and not on other effects like the blackening of a photographic film due to the ease and accuracy with which ionization in the air can be measured.

The biological effects of X-rays are due to energy imparted to matter: Therefore, these effects are more closely correlated with the absorbed dose than with exposure. The unit of absorbed dose is *rad*. One *rad* is the radiation dose which will result in an energy absorption of 1.0×10^{-2} J/kg in the irradiated material. It is approximately equal to the dose absorbed by soft tissue exposed to one Rontgen of X-rays.

The Rontgen and the absorbed dose D are related as $D = f\ R$ where f is a proportionality constant and depends upon both the composition of the irradiated material and quality of the radiation beam. The value of f for air is 0.87 rad/R. For soft tissues, $f = 1$ rad/R and hence the absorbed dose is numerically equal to the exposure. However, for bone, f is larger but significantly decreases with an increase in kV. Therefore, if the contrast requirements permit, the patient's absorbed dose can be decreased by using suitably high kV.

The ionization produced by different types of radiation is not a sufficiently good criterion of biological effect. Another concept is that of the so-called *dose equivalent* (DE) H. DE is defined as the product of the absorbed dose D and a modifying quality factor *(QF)*, i.e.

$$DE = (QF)D$$

The film badge readings and radiation guides in the form of a maximum permissible dosage are expressed in *rems* or *millirems*.

In short, Rontgens express incident energy, rads give an indication of how much of this incident energy is absorbed and rems are a measure of the relative biological damage caused.

19.3   PRODUCTION OF X-RAYS

X-rays are produced whenever electrons collide at very high speed with matter and are thus suddenly stopped. The energy possessed by the electrons appears from the site of the collision as a parcel of energy in the form of highly penetrating electromagnetic waves (X-rays) of many different wavelengths, which together form a continuous spectrum. X-rays are produced in a specially constructed glass tube, which basically comprises. (i) a source for the production of electrons, (ii) a energy source to accelerate the electrons, (iii) a free electron path, (iv) a means of focusing the electron beam and (v) a device to stop the electrons.

Stationary mode tubes and rotating anode tubes are the two main types of X-ray tubes:

19.3.1   Stationary Anode Tube

Figure 19.2 shows the basic components of a stationary anode X-ray tube. The normal tube is a vacuum diode in which electrons are generated by thermionic emission from the filament of the tube. The electron stream is electrostatically focused on a target on the anode by means of a suitably shaped cathode cup. The kinetic energy of the electrons impinging on the target is converted into X-rays. Most electrons emitted by the hot filament become current carriers across the tube. It is, therefore, possible to independently set

    (i) Tube current by adjusting the filament temperature, and is it "and" or "or"?

   (ii) Tube voltage by adjusting primary voltage.

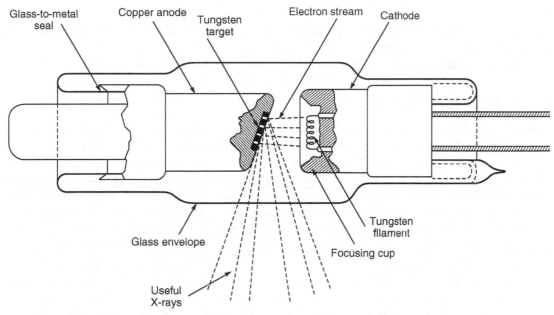

> ➤ **Fig. 19.2** *Construction of stationary anode X-ray tube*

Some X-ray tubes function as a triode with a bias voltage applied between the filament and the cathode cup. The bias voltage can be used to control the size and shape of the focal spot by focusing on the electron beam in the tube.

The cathode block, which contains the filament, is usually made from nickel or from a form of stainless steel. The filament is a closely wound helix of tungsten wire, about 0.2 mm thick, the helix diameter being about 1.0–1.5 mm. The target is normally comprised of a small tablet of tungsten about 15 mm wide, 20 mm long and 3 mm thick soldered into a block of copper. Tungsten is chosen since it combines a high atomic number (74)—making it comparatively efficient in the production of X-rays. It has a high melting point (3400°C) enabling it to withstand the heavy thermal loads. In special cases, molybdenum targets are also used, as in the case of mammography, where in improved subject contrast in the breast is desirable. The lower efficiency of X-ray production and the lower melting point make molybdenum unsuitable for general radiography.

Copper being an excellent thermal conductor, performs the vital function of carrying the heat rapidly away from the tungsten target. The heat flows through the anode to the outside of the tube, where it is normally removed by convection. Generally, an oil environment is provided for convection current cooling. In addition, the electrodes have open high voltages on them and must be shielded. The tube will emit X-rays in all directions and protection needs to be provided except where the useful beam emerges from the tube. In order to contain the cooling oil and meet the above-mentioned requirements, a metal container is provided for completely surrounding the tube. Such a container is known as a 'shield'.

Since a lot of heat will be generated by the tube, and hence this heat will cause the oil temperature to rise, the oil will expand. Being a liquid, oil is incompressible, hence a bellows, either of oil-resistant rubber or thin metal, is provided to accommodate the expansion. Due to the penetrating nature of transformer oil, particularly when it is hot, every joint on a shield has to be hermetically sealed, either soldered or sealed with a rubber gasket. Also, the shield must be made shockproof by an efficient earthing arrangement.

Stationary anode tubes are employed mostly in small capacity X-ray machines.

19.3.2 Rotating Anode Tube

With an increasing need in radiology for more penetrating X-rays, requiring higher tube voltages and current, the X-ray tube itself becomes a limiting factor in the output of the system. This is primarily due to the heat generated at the anode. The heat capacity of the anode is a function of the focal spot area. Therefore, the absorbed power can be increased if the effective area of the focal spot can be increased. This is accomplished by the rotating anode type of X-ray tubes. The tubes with rotating anode are based on the removal of the target from the electron beam before it reaches too high a temperature under the electron bombardment and the rapid replacement of it by another cooler target.

The construction of a typical rotating anode X-ray tube is shown in Fig. 19.3. The anode is a disk of tungsten or an alloy of tungsten and 10% rhenium. This alloy helps to reduce the changes in the anode track due to stress produced in the track as a result of the rapidly changing temperature. The anode rotates at a speed of 3000–3600 or 9000–10000 rpm. The tungsten disk that represents the anode has a bevelled edge that may vary from 5°–20°. Typical angles are around 15°, in keeping with the line focus principle. These design elements help to limit the power density incident on the physical focal spot while creating a small effective focal spot. With the rotating anode, the heat produced during an exposure is spread over a large area of the anode, thereby increasing the heat-loading capacity of the tube and allowing higher power levels to be used which produces more intense x-radiation.

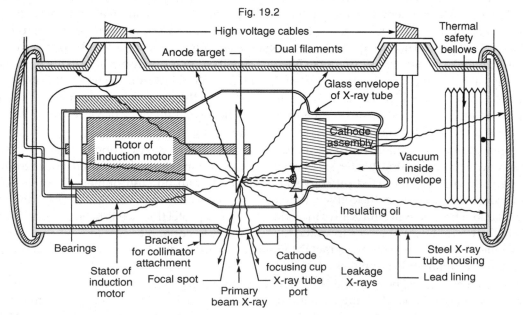

➤ **Fig. 19.3** *Constructional details of rotating anode X-ray tube*

The rotor is made from copper, either cast or from special quality rod. The molybdenum stem projecting from the rotor is either soldered or the copper of the rotor may be cast round it. The choice of molybdenum is dictated by the need for a strong metal with a melting point high enough to permit contact with a very hot tungsten disk. The anode rotation system is a high speed system. Therefore, the bearings must be properly lubricated. The high temperature environment inside the tube precludes most normal lubricants, that would have the additional disadvantage of releasing enough vapour to spoil the condition of high vacuum, which is necessary for the proper functioning of the tube. The situation has been remedied by the successful development of metal lubricants. The commonly used lubricants are lead, gold, graphite or silver. These lubricants are usually applied to the bearing surfaces in the form of a thin film (Hill, 1979).

The tube housing serves several technical purposes. It is a part of the electrical isolation between the high voltage circuits and the environment. The housing is lead-lined to keep the amount of leakage radiation below legal levels, thereby providing radiation protection for both the patient and the operator. Finally, the tube housing is an important part of the waste-heat handling system. While housings for tubes used at low mean power levels can be adequately air-cooled, it becomes necessary to provide additional cooling in case of higher power levels, which is done by circulating water through a heat exchanger contained in the tube housing or by circulating insulating oil through an external radiator. Geldner (1981) discusses electrical, thermal and load characteristics of rotating anode X-ray tubes.

Homberg and Koppel (1997) illustrate a spiral groove bearing which has several advantages over conventional anode ball bearing used in heavy duty X-ray tube assemblies. Apart from being quiet in operation, the spiral groove bearing technology permits extremely efficient cooling of the anode dish by conducting heat away into a cooling medium. This prolongs the life, even though the anode disk is operated continuously at a high speed of rotation. In contrast to ball bearings, spiral groove bearings are virtually wear and tear free. The principle of the spiral groove bearing is shown in Fig. 19.4(c). It works on the principle of the hydrodynamic wedge which is formed between the rotating and the stationary parts of the bearing. Therefore causes a 'swimming' of the rotating bearing part thus forming a gap filled with liquid metal, typically 15-20 μm, between the parts. These tubes permit anode rotation at a speed of more than 9000 rpm and are available in both two and three focus versions.

X-ray tubes are further classified on the basis of their application for diagnostic or therapeutic purposes. For diagnostic applications, it is usual to employ high milliamperes and lower exposure time whereas high kV and relatively lower mA are necessary for therapeutic uses. The description which follows relates only to the diagnostic X-ray machines.

⫸ 19.4 X-ray MACHINE

Figure 19.5 shows a block diagram of basic X-ray machine sub-systems. Basically, there are two parts of the circuit. One of them is for producing high voltage, which is applied to the tube's anode and cathode and comprises a high voltage step-up transformer followed by rectification. The current through the tube follows the HT pathway and is measured by an mA metre. A kV selector switch facilitates change in voltage between exposures. The voltage is measured with the help of a kV metre. The exposure switch controls the timer and thus the duration of the application of kV. To compensate for mains supply voltage (230 V) variations, a voltage compensator is included in the circuit.

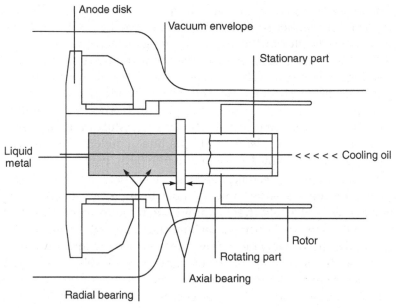

> **Fig. 19.4**   *Section through a spiral groove bearing rotating anode X-ray tube*

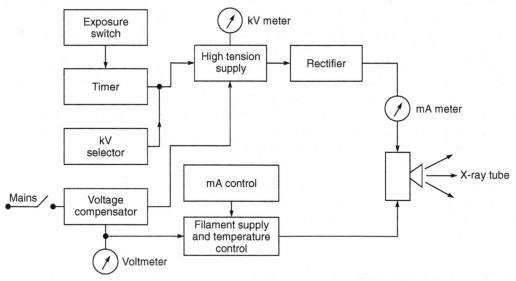

> **Fig. 19.5**   *Block diagram of an X-ray machine*

The second part of the circuit concerns the control of heating X-ray tube filament. The filament is heated with 6–12 V of AC supply at a current of 3–5 amperes. The filament temperature determines the tube current or mA, and, therefore, the filament temperature control has an attached mA selector. The filament current is controlled by using, in the primary side of the filament transformer, a variable choke or a rheostat. The rheostat provides a stepwise control of mA and is most commonly used in modern machines.

A preferred method of providing high voltage DC to the anode of the X-ray tube is by use a bridge rectifier using four valve tubes or solid state rectifiers. This results in a much more efficient system than the half wave of self-rectification methods.

19.4.1 High Voltage Generation

Voltages in the range of 30–200 kV are required for the production of X-rays for diagnostic purposes, and they are generated by a high voltage transformer. A high ratio step up transformer is used so that the voltages applied to primary winding are small in comparison to those taken from the secondary winding. Typically, the ratio would be in the range of 1:500 so that an input of 250 V would produce an output of 125 kV. Usually the high tension transformer assembly is immersed in special oil which provides a high level of insulation.

Self-rectified Circuit (One Pulse): The high voltage is produced by using a step-up transformer whose primary is connected to an auto-transformer. The secondary of the HT transformer can be directly connected to the anode of the X-ray tube, which will conduct only during the half cycles when the cathode is negative with respect to the anode or the target (Fig. 19.6). This arrangement of self-rectification is used in mobile and dental X-ray units. These machines have maximum tube currents of about 20 mA and a voltage of about 100 kV. When self-rectification is used, it is necessary to apply a parallel combination of a diode and a resistance, in series with the primary of the HT

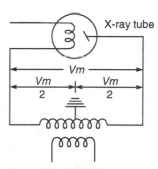

> **Fig. 19.6** *Self-rectified circuit for high voltage generation*

transformer for suppressing the higher inverse voltage that is likely to appear during the non-conducting half-cycle of the X-ray tube. This helps to reduce the cost and complexity of the X-ray machines.

Full Wave Rectification X-ray Circuit (Two Pulse): In the self-rectified units, X-rays are produced in a burst and a considerable amount of exposure time is lost during the half-cycle when the X-ray tube is not conducting. By using a full wave bridge rectifier circuit, the exposure time for the same radiation output is reduced by half in comparison to the one pulse system. This circuit produces X-rays during each half-cycle of the applied sinusoidal 50 Hz mains supply voltage as the anode would be positive with respect to the cathode over both the half-cycles (Fig. 19.7).

Full wave rectified circuits are used in medium and high capacity X-ray units which are most commonly employed for diagnostic X-ray examination.

Three Phase Power for X-ray Generation: The X-ray circuits based on a single phase supply provide a pulsating voltage to the anode. This type of voltage waveform, when used to accelerate electrons in the X-ray tube, results in the following disadvantages:

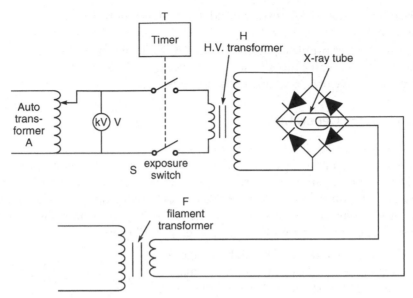

> **Fig. 19.7** *Single phase full wave rectified circuit*

- The intensity of radiation produced is lower because no radiation is generated during a large portion of the exposure time.
- When the tube voltage is appreciably lower than the peak voltage, the X-rays produced are of low energy and get mostly transformed into heat at the anode.
- A considerable part of the radiation produced is absorbed by the filter or the tube housing and patient resulting in a comparatively poor quality image.

The above-mentioned deficiencies of the single-phase systems can be overcome by using three-phase power in X-ray machines (Fig. 19.8). Three phase supply can result in steady power to the X-ray tube instead of pulsating power. The three phase equipment is more efficient than the single phase equipment of the same-rated capacity.

As with single phase X-ray systems, wherein we have one pulse or two pulses of applied voltage per cycle, a similar situation can be implemented with a three phase system. Using different types of three phase transformers and rectifier configurations, 6 pulses or 12 pulses of applied anode voltage can be obtained.

Six-rectifier Circuit (Six Pulse): Figure 19.9 shows a simple six rectifier, six-pulse circuit using three phase power supply. The primary supply is delta-connected whereas the secondary is y-connected. The secondary is connected to a six-rectifier arrangement for full wave rectification of the three phase transformer output voltage. The rectification process produces the six-pulse voltage waveform with a voltage ripple of 13% of the maximum value as compared to 100% of the single phase full wave rectified voltage system.

Twelve Rectifiers Circuit (Twelve Pulse): A further refinement of the three phase generator is to use a combination of both star and delta windings on the secondary side. They are connected as shown in Fig.19.10.

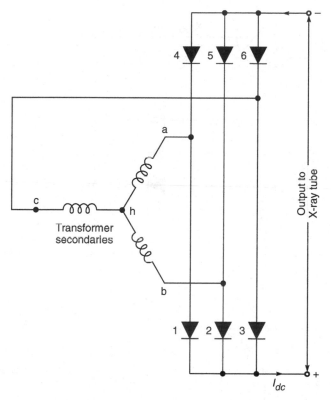

> ➤ **Fig. 19.8** *Three phase full wave bridge rectifier circuit using six diodes*

This arrangement results in reduction of the voltage ripple and an increase in the effective value of the output voltage, thereby increasing the X-ray producing efficiency of the X-ray tube.

19.4.2 High Frequency Generators

Modern X-ray machines make use of high frequency generators for producing high voltage. The single most important feature that differentiates high frequency, three-phase and single phase power is the ripple in the output. A lower ripple provides a more efficient radiation output and a shorter relative exposure for equivalent contrast radiographs. Figure 19.11 illustrates comparative wave shapes for single phase with 100% ripple, RMS three-phase with typical 6–12% ripple and RMS high frequency with typical 1–2% ripple. The high frequency used in these generators varies from 500 Hz–20 kHz.

High frequency is generated by first converting the 50 Hz power line frequency into high frequency oscillations in the converter circuit. The frequency conversion permits the use of much smaller transformers than those required with conventional equipment. A schematic diagram of a typical high frequency generator is shown in Fig. 19.12.

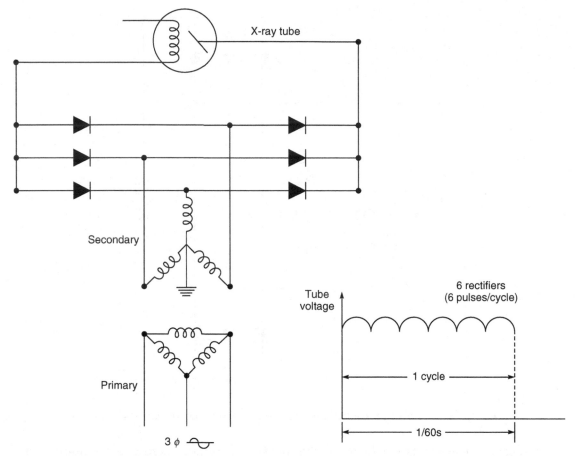

> **Fig. 19.9** *Six rectifiers, six pulses per cycle, three phase high voltage generator*

Basically, the circuit uses same switching arrangement for thyristors through a series resonant circuit, comprising L and C_1, where L is the inductance of the primary of the high tension transformer. If the circuit is switched by the thyristors at its resonant frequency, the effective current flowing through L is maximized. This primary current is transformed and rectified in the high tension circuit. The net voltage appearing across the X-ray tube is determined by the voltage present on C_2. This voltage, in turn, is supplied by the charging current and drained by the current flowing through the X-ray tube.

19.4.3   High Tension Cable

In view of the very high voltages applied to the X-ray tube, it is necessary to use special highly insulated cables for its connections to the generator. Since a typical X-ray machine may employ voltages in excess of 100 kV, the design and subsequent effect of the cable capacitance on the voltage applied to the X-ray tube must be considered.

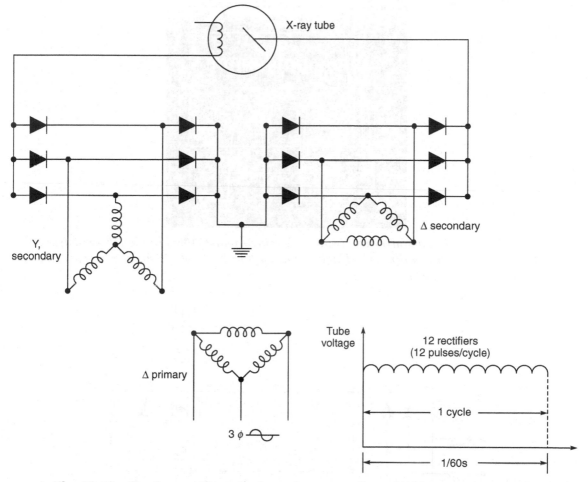

> **Fig. 19.10** *Twelve rectifiers, twelve pulses per cycle applied between X-ray tube anode and cathode*

A cross-sectional view of a high voltage cable is shown in Fig. 19.13. The centre of the cable comprises three conductors individually insulated for the low filament voltages and surrounded by semi-conducting rubber. This, in turn, is surrounded by non-conducting rubber which provides the insulation against the high voltage also carried by the centre conductors. The cable is shielded with a woven copper braiding, which is earthed, and finally covered with a protective layer, usually vinyl or some other plastic.

The metal braid is connected to the centre tap of the high voltage transformer, which is grounded. By using a centre-tapped secondary in the high voltage transformer, the voltage on each cable is reduced to half of the X-ray tube voltage relative to ground, consequently reducing the amount of dielectric required in the high voltage cables, thereby making them smaller in diameter. The grounded metal braid also serves as a safety path to ground for the high voltage, should there be a breakdown in the dielectric material for any reason.

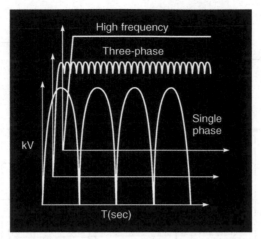

> **Fig. 19.11**   *Comparative wave shapes for single phase (100% ripple), three phase (6 to 12% ripple) and high frequency (1 to 2% ripple) high voltage generator system*

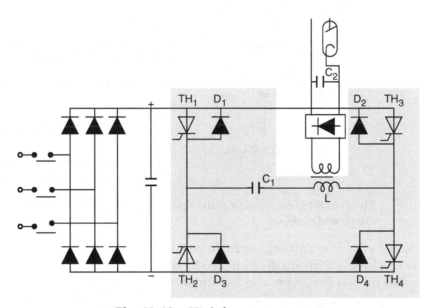

> **Fig. 19.12**   *High frequency generator*

A typical cable capacitance of high voltage cables is 130–230 pF/m. The effect of the cable capacitance is that the energy is stored during the conduction period of the rectifiers and the energy is delivered to the tube during the non-conducting period. This would change the average value of the current and voltage across the X-ray tube, which increases the power delivered to the tube.

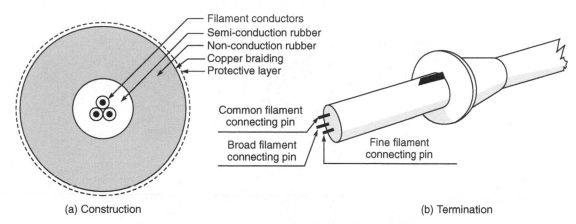

(a) Construction   (b) Termination

> **Fig. 19.13**   *Constructional details of high tension cable*

The kV meter is connected across the primary of the HT transformer. It actually measures volts, whereas it is calibrated in kV, by using an appropriate multiplication factor of the turns-ratio of the transformer. In the older types of diagnostic X-ray generators, the kV meters indicated only no-load voltage. In order to obtain the load voltage, which varies with the tube current, a suitable kV metre compensation is provided in the circuit. The kV meter compensator is ganged to the mA selector mechanically. Therefore, the mA is selected first and the kV setting is made afterwards during the operation of the machine. Moving coil meters are used for making current (mA) measurements, while for shorter exposures, an mAs meter, which measures the product of mA and time in seconds is used. Moving coil meters have now been generally replaced by digital mA and mAs meters.

19.4.4   Collimators and Grids

In order to increase the image contrast and to reduce the dose to the patient, the X-ray beam must be limited to the area of interest. Two types of devices are used for this purpose, viz. collimators and grids.

The collimator is placed between the X-ray tube and the patient (Fig. 19.14). It consists of a sheet of lead with a circular or rectangular hole of suitable size. Alternatively, it may consist of four adjustable lead strips which can be moved relative to each other. In practice, it is advisable to use the smallest possible field size. This results in a low dose to the patient and simultaneously increases the image contrast, because less scattered radiation reaches the image plane. The scattered radiation produces diffuse illumination and fogging of the image without increasing its information content, and therefore, by choosing the smallest possible field size and using a collimator, the loss of contrast due to scattered radiation is reduced.

Collimators are usually provided with an optical device, by which the X-ray field can be exactly simulated by a light field. This can be projected on the patient to ensure proper positioning of the apparatus.

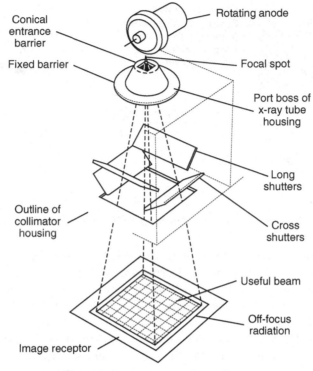

Conical entrance barrier

Fixed barrier

Rotating anode

Focal spot

Port boss of x-ray tube housing

Long shutters

Outline of collimator housing

Cross shutters

Useful beam

Off-focus radiation

Image receptor

> **Fig. 19.14** *Layout of a collimator*

Grids are inserted between the patient and the film cassette (Fig. 19.15) in order to reduce the loss of contrast due to scattered radiation. A grid consists of thin lead strips separated by spacers of a low attenuation material. The lead strips are so designed that the primary radiation from the X-ray focus, which carries the information, can pass between them while the scattered radiation from the object is largely attenuated.

Because of the shadow cast by the lead strips, the final image is striped. These grid lines do not usually interfere with the interpretation of the image. However, final details in the image may be concealed. In order to avoid this, the grid can be displaced during the exposure so that the lead strips are not reproduced in the image. Such moving grids are known as 'Bucky Grids'.

19.4.5 Exposure Timing Systems

A timer is used in X-ray machines to initiate and terminate the X-ray exposure. The timer controls the X-ray contactor which in turn, controls the voltage to the primary of the high voltage transformer. Timers vary widely in their methods of operation, starting with simple mechanical timers to microcontroller based electronic timers. The timers in older equipment were of the spring-driven hand-operated type which have now mostly been replaced by electronic timers based on the use of SCR or thyristors. However, with the variety and age of X-ray equipment presently in use, almost all of the various types of exposure timers are still in use.

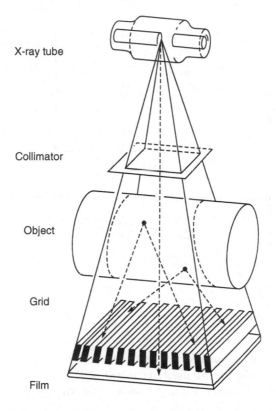

X-ray tube

Collimator

Object

Grid

Film

➤ **Fig. 19.15** *Principle of the grid. The focal radiation from the X-ray tube can pass between the lead stripes, whereas much of the scattered radiation from the object is cut off because of its direction*

With the development of high power X-ray generators, exposure times have become too short to be controlled accurately with mechanical timers. In order to meet the demand of accurate short-term timing, various types of electronic timers have been developed. The common types of electronic timers are discussed below:

RC Timing Circuits: In these timers, the length of the X-ray exposure is determined by the time constant of a simple RC circuit which is typically shown in Fig. 19.16 (a). Normally the switch 'S' is closed between the exposures. To initiate the exposure, switch 'S' is opened and the relay in the primary of the high voltage transformer circuit is closed, which begins the exposure. The capacitor starts charging and the voltage 'V'$_t$ on the capacitor is given by

$$V_t(t) = V_s(1 - e^{t/RC})$$

In this arrangement, 'C' is constant and 'R' is a variable resistor. The RC time constant can therefore be changed to get variable exposure time requirements. Fig. 19.16 (b) illustrates the capacitor voltage as a function of time for two different time settings with the help of variable resistance 'R'.

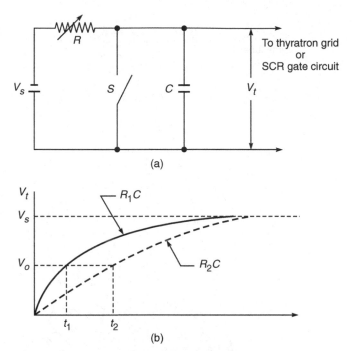

> **Fig.** **19.16** *Electronic timer*
>   (a) *RC timing circuit used to control SCR*
>   (b) *Timer RC curve. Voltage rises as a function of the resistance setting R_1 or R_2*

The voltage on the capacitor can also be used to fire a thyratron or SCR to accomplish the same function of terminating the exposure.

Digital Timers: All modern X-ray machines make use of digital timers. The latter may make use of dedicated timer ICs or conventional circuits using a reference oscillator, a counter and associated logic. The generation of a time period 'T' is based on counting out 'N' cycles of a precise frequency 'F' with the relation

$$N = T \times F$$

Referring to Fig. 19.17, the reference oscillator generates the frequency of 'F' counts per second given to an AND circuit. The value of 'N' is programmed into the counter logic. The start of the time period is controlled by a flip-flop which is set to 1 by a 'start' signal. When 'N' counts have been made corresponding to the time interval 'T', a pulse is issued that resets the flip-flop and disconnects the oscillator from the counter. The pulse output of the counter is used to trigger an SCR and to terminate the X-ray exposure.

19.4.6 Automatic Exposure Control

Radiographic practice is based on the selection of appropriate X-ray exposure factors such as patient size, shape and physical condition, examination and projection to be performed. This can

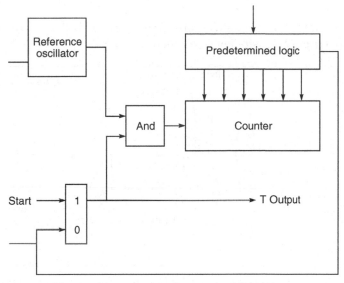

➤ **Fig. 19.17** *Circuit diagram of digital timer*

be done properly by a technologist using his or her own judgement with a manually controlled generator. However, this process has become less a matter of technologist preference and more a part of a standard protocol. Therefore, it has led to the introduction of Anatomically Programmed Radiography (APR) by combining all the primary controls of the generator and the Automatic Exposure Control. The use of machine stored parameters results in better quality of radiographs.

There are two principle methods of exposure control, one employing a photocell and the other, an ionization chamber.

In the photocell-based method, a fluorescent detector is placed on the exit side of the patient and behind the radiographic cassette (Fig. 19.18) which monitors the X-ray intensity transmitted through the film screen system. The circuit controls the X-ray exposure switch and turns the x-radiation off when a radiation flux sufficient to properly expose the film has been detected.

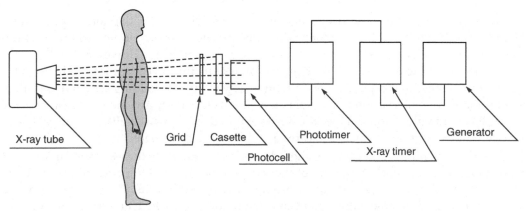

➤ **Fig. 19.18** *Automatic exposure control: photocell method*

Alternatively, an ionization chamber (Fig. 19.19) is placed between the patient and the cassette. The signal from the chamber is amplified and used to control a high speed relay which terminates the exposure, when a pre-set density level has been reached.

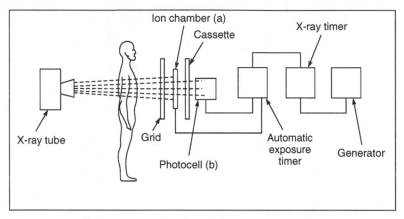

> **➤ Fig. 19.19** *Ionization method for automatic exposure control*

The basic design of X-ray generators has not changed for the last 50 years. However, there have been considerable developments in the control elements as the demand has grown for increased accuracy, better information display and greater flexibility of selection of factors. The task to be performed by the control circuits of an X-ray generator can well be performed by a microcomputer.

➤ 19.5 VISUALIZATION OF X-RAYS

X-rays normally cannot be detected or visualized directly by the human senses. Therefore, indirect methods need to be utilized to produce an image of the intensity distribution of X-rays that have passed through the body of a patient. Usually the techniques discussed below are in practice.

19.5.1 X-ray Films

X-rays which have a much shorter wavelength than visible light, react with photographic emulsions in a similar fashion as that of light. After having been processed in developing solution, a film that has been exposed to X-rays shows an image of the X-ray intensity. However, the X-ray film is relatively insensitive to X-rays. Sensitivity may be markedly improved by first producing a visible image to which the film is exposed. This is achieved by using intensifying screens consisting of a layer of fluorescent material bonded to a plastic base. The film is sandwiched between two screens and held in a light tight cassette. Thus, the film is exposed to X-rays as well as to the light from the fluorescence of the screen. Such screens are called intensifying screens.

The ideal screen fluorescent material should be colour matched to the most sensitive part of the film spectral response (blue): it should not show phosphorescence (long after glow) and it should have high absorption of X-rays without absorbing the light that it emits. The most commonly used material which satisfies these requirements is calcium tungstate which emits a broad spectrum of light of low intensity centred in blue wave length of 420 nm. In 1972, screens coated with rare-earth materials such as gadolinium (a green light emitter) or lanthanum oxybromide (a blue light emitter) were introduced commercially. The advantages of rare-earth screens lies in the speed of the system, less radiation exposure and higher resolving capacity.

19.5.2 Fluorescent Screens

In fluoroscopy, X-rays are converted into a visual image on a fluorescent screen which can be viewed directly. It facilities a dynamic radiological study of the human anatomy. The fluorescent screen consists of a plastic base coated with a thin layer of fluorescent material, zinc cadmium sulphide, which is bonded to a lead-glass plate. The image is viewed through the glass plate which provides radiation protection but allows the optical image to be viewed. Zinc cadmium sulphide emits light at 550 nm and is selected because the eye is most sensitive in the green part of the spectrum.

While keeping the radiation intensity to a safe level, the fluoroscopy image is generally faint and needs to be observed in a completely darkened room. It is therefore necessary for the radiologists become to adapt themselves to the dark for 20–30 minutes and then to view the image. Because of these inconveniences, direct fluoroscopy is not commonly used at present and has been largely superseded by X-ray image intensifiers.

19.5.3 X-ray Image Intensifier Television System

An X-ray image intensifier consists of a large evacuated glass tube with an input screen diameter ranging from 15–32 cm. The input screen converts the X-ray image into a light image. The light image thus produced is transmitted through the glass of the tube to a photo-cathode which converts the light image to an equivalent electron image. The image intensification takes place because of the very small output screen size and electron magnification in the tube.

Figure 19.20 shows the general structure of an X-ray image intensifier tube. It consists of an input screen, the surface of which is coated with a suitable material to convert X-rays into a light image. First generation X-ray image intensifiers used zinc cadmium sulphide as the input fluorescent screen. However, this material has a poor X-ray absorption efficiency. Modern X-ray image intensifiers make use of a thin layer of cesium iodide (CsI) which has the advantage of high X-ray absorption and packing density. The X-ray quanta, after getting converted to light quanta, falls on the photo-cathode in which the light quanta produce electrons. Under the influence of an electrical field, the electrons are emitted from the photo-cathode and accelerated towards the output phosphor, while being focused by the electrostatic lens system. The electrons impinging with high kinetic energy on this screen produce light quanta resulting in a much brighter and minified output image. The brightness gain is due to the acceleration of the electrons in the lens system and the fact that the output image is smaller than the primary fluorescent image. The gain is several hundred times. It not only allows the X-ray intensity to be decreased tremendously but makes it possible to observe the image in a normally illuminated room.

The output window which permits us to examine the light image presented by the output viewing screen (15–30 mm diameter) located inside the bulb near the window, is flat and allows for image transfer through large numerical aperture objective (F/0.75). Certain tubes (Thomson

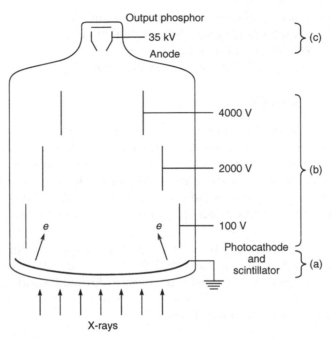

Output phosphor

35 kV

Anode

} (c)

4000 V

2000 V } (b)

100 V

Photocathode
and
scintillator } (a)

e

e

X-rays

> **Fig.** 19.20 *Constructional details of an X-ray image intensifier tube (a) input surface consists of scintillation layer and photocathode (b) three focusing electrodes with typical bias voltages (c) electrode and output phosphor*

CSF THX 475) are made with a fibreoptic output window. This permits picking up the image by mechanical coupling with a camera tube that has a fibreoptic input window (Ebrecht, 1977).

Figure 19.21 shows a system incorporating the X-ray image intensifier system which can be coupled to a closed circuit television, cine camera, photo-spot camera and video recording facilities. Although the image intensifier is a fundamental element of the chain, the final image observed on the TV monitor depends, to a large extent, on the characteristics of the TV chain, which greatly affects the signal constancy and the spatial resolution. The most commonly used TV pick-up tube is the 2.5 cm vidicon, whose target will accept a 15 mm diameter image. The sensitivity of the system depends on the conversion efficiency of the vidicon target. Available optical systems have a luminous yield of the order of 20% of green light onto the vidicon target. Also, the optical system suffers from 'vignetting' (brightness fall-off at the edges of the image). The use of a system with fibreoptic coupling solves this problem.

The combination of X-ray image intensifiers and the TV system must control the X-ray generator to produce constant density changes, despite variations in patient thickness. This is done by automatic dose rate control, also known as automatic brightness control. In the arrangement shown in Fig. 19.22 the dose control of the X-ray tube is linked to the exposure control of the camera tube and photo-camera. The intensity of the light leaving the intensifier is measured by a light sensor (LS). The exposure control circuits drive the voltage and current of the generator (XG). If the image intensifier is switched to a higher magnification, the current in the X-ray tube is

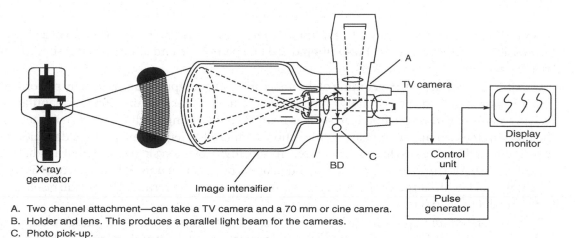

A. Two channel attachment—can take a TV camera and a 70 mm or cine camera.
B. Holder and lens. This produces a parallel light beam for the cameras.
C. Photo pick-up.
D. Electrical signal from photo pick-up. It provides the control for 70 mm and cinefluorography.

➤ **Fig. 19.21** *X-ray image intensifier system*

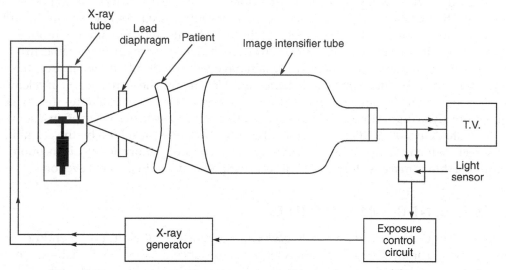

➤ **Fig. 19.22** *Automatic dose control in an X-ray image intensifier system*

increased in inverse proportion to the diameter on the X-ray screen. At the same time, the lead diaphragm output size between the patient and the X-ray tube is reduced to cut down on the area of patient exposure. Diaphragms in front of the TV camera and the photo-camera can be opened up if the light output is still not sufficient, in spite of maximum dose input.

The input screen technology of image intensifiers has improved so greatly in recent years that spatial resolution comes close to that of conventional X-ray films. But after the first step of conversion, image quality worsens because of the image intensifier electron optics, the optical lens system, and to a large extent, the TV camera. The Siemens AG 1249 line X-ray video system optimizes these steps and provides 50% better resolution as compared with the common 525/625 systems (Riemann and Marholff, 1981). With 1249 television lines and the bandwidth of 25 MHz, it is possible to achieve a vertical resolution of 2.8 line pairs/mm and a horizontal resolution of 3.0 line pairs/mm.

Cine-film or 35 mm format recording of the image intensifier output image is usually performed with a 16 or 22 cm field and is mainly employed in cardiovascular examination for quantitative measurements of certain phenomena, such as heart volume during cardiac muscle contraction. The measurement of apparent heart cross-section compared to its volume permits detection of insufficiencies that may lead to cardiac failure. The sensitivity of the system is such that a radiation dose of 20–30 μR is sufficient to produce a good image. The common optical lens system of 50 mm to 82 mm focal length can give images that are well-suited to the film format. The overall resolution is of the order of 30 lp/cm.

The use of 105-mm spot film is becoming increasingly popular, even tending to support traditional radiographic film systems for certain cardiovascular examinations. Reduction in radiation dose, high speed cine-film capacity (12 image/s), easy filing and the low cost of film are some of the advantages of using spot film.

In video fluoroscopic X-ray systems, the detector or rather the detector chain, embodies the key technology. The CCD (Charge Coupled Device) camera, introduced in some systems, is replacing the vidicon tube camera and offers significant improvements in image quality. It offers higher resolution wherein, for certain applications, 2048×2048 pixel matrices are preferable. In addition, continuous improvements in the detector chain have led to radically novel approaches which dispense with the need for an image intensifier and TV camera. The introduction of selenium—a photo-conductor with optimal properties for use in X-ray detectors—facilitates the image to be obtained directly in digital format. The next generation of X-ray systems will contain a new type of flat solid state detectors. In these detectors, the optical image is provided by the cesium iodide input screen, which is directly detected by a high-resolution amorphous silicon photo-diode matrix and a thin film transistor array. This is described in detail in the next section.

ⅢⅢ▶ 19.6   DENTAL X-RAY MACHINES

X-rays are the only media available to detect location of the teeth, their internal condition and the degree of decay at an early stage. Since the object–film distance is rather low, and the tissue and the bone thickness are limited, an X-ray machine of low power is adequate to obtain the radiograph with sufficient contrast. In practice, most dental units have a fixed tube voltage, in the region of 50 kV, and a fixed tube current of about 7 mA. The system combines the high voltage transformer and X-ray tube into a singly small case, thus greatly simplifying handling and positioning as no high voltage cables are required.

The primary winding of the transformer is fed with mains voltage via an exposure timer and the high voltage developed in the secondary windings is fed to the self rectifying X-ray tube. The complete assembly is contained in a metal case filled with special insulating oil. The X-ray tube is of special design and employs a third electrode, called a 'grid', between the anode and the cathode electrodes. The grid restricts electrons from leaving the cathode until the high voltage reaches its peak value, whereupon all electrons are released and impinge on the anode at a very high velocity. Consequently, the x-radiation generated contains fewer useless soft X-rays and more hard rays. The total radiation is, therefore, more effective and can be compared mathematically to a much higher output resulting in shorter exposure times.

▓▓▷ 19.7 PORTABLE AND MOBILE X-RAY UNITS

In many situations, portable and mobile X-ray units are necessary for X-ray patients who, for some reason, may not be able to go to X-ray departments. Such situations emerge when the patient is too ill to be moved from the hospital bed, is seriously ill at home or undergoing surgery in the theatre. Thus, the need arises to have X-ray equipments, which can be moved to the patients rather than the patient moving to the X-ray machine.

19.7.1 Portable Units

A portable unit is so designed that it can be dismantled, packed into a small case and conveniently carried to the site. The tube head is so constructed that the X-ray tube and the high voltage generator are enclosed in one earthed metal tank filled with oil. The X-ray is usually a small stationary anode type, operating in the self-rectifying mode and connected directly across the secondary winding of the transformer. The only connection required from the control desk is for the low voltage supply. The controls provided are fairly limited and include a mains voltage compensator, combined kV and current switch and time selector. As the unit is designed to be used on the domestic supply, the current must be limited to 15 A. Thus the maximum radiographic output commonly found on portable units is in the range 15–20 mA at 90–95 kV.

The tube head is mounted on a cross arm which is carried on a vertical column. The cross arm may be moved up and down this column by means of a rack and pinion drive.

19.7.2 Mobile Units

A mobile unit carries the control table and the column supporting the X-ray tube permanently mounted on the mobile base. Mobile units could be much heavier than the portable units and are capable of providing higher outputs. Mobile units provide a greater selection of mA and kV values. The high voltage generator has a full wave rectification circuit feeding a double focus rotating anode X-ray tube. Most mobile units have a radiographic output of up to 300 mA and a maximum of 125 kV. This type of output requires a main supply current of 30 A. The units are, therefore, usually fitted with a 30 A plug and special sockets need to be provided throughout the hospital for this purpose.

Because of the high current requirements of the mobile units, the mains resistance becomes a problem, specially if the mobile is to be used in different parts of the hospital. In order to ensure consistent results from one power supply socket to another, a means for mains resistance calibration is provided and must be adjusted to suit each location before making an exposure.

Where there are limitations on the electrical supply, mobile units make use of stored energy. This may be from the capacitor discharge or battery-powered invertor circuits. The former releases stored energy from the capacitor during exposure, while the latter converts energy stored in the battery.

19.7.3   Mammographic X-ray Equipment

Mammography is an X-ray imaging procedure used for examination of the female breast. It is primarily used for diagnosis of breast cancer and in the guidance of needle biopsies. The female breast is highly radiation-sensitive. Therefore, the radiation dosage during mammography should be kept as low as possible. Also, it is required to achieve better spatial resolution than other types of film/screen radiographs. In order to achieve these goals, an X-ray tube with a small focal spot size is used to minimize the possibility of geometric blur. The film/screen cassette has a single emulsion film and a single screen, and is designed to provide excellent film/screen contact.

Mammographic X-ray equipment can either be used with special film/screen cassette or as xero-radiographic units. The units intended for film/screen use have a molybdenum target X-ray tubes with a beryllium window and a 0.03 mm molybdenum filter. Radiographs are usually taken at 28–35 kV. Xero-radiographic systems use X-ray tubes with tungsten targets and about 1 mm aluminum filter. Radiographs with this technique are taken at 40–50 kV. Hence, both types of mammographic units operate at low peak voltages.

Film-based mammography has several disadvantages such as limitations in detection of micro-calcifications and other fine structures within the breast, and inefficiency of grids in removing the effects of scattered radiation. Many of these limitations can be effectively removed by using a digital mammography system in which image acquisition, display and storage are performed independently, allowing for optimization of each process. However, the availability of a suitable X-ray detector for this purpose is still a challenge that precludes the widespread use of digital mammography. Various detector technologies which are under evaluation in digital mammography are large area CCDs (Charge Coupled Devices), photo-stimulable phosphors, amorphous silicon coupled to scintillators, amorphous selenium and other solid-state devices.

▥▶ 19.8   PHYSICAL PARAMETERS FOR X-RAY DETECTORS

The physical parameters used to characterize X-ray detectors are as follows:

Detector Quantum Efficiency (DQE): The DQE describes the efficiency of a detector, i.e. the percentage of quanta for a given dose that actually contributes to the image. It is a function of dose and spatial frequency and is, by definition, effected by the various components of the system.

Dynamic Range: The dynamic range of a detector is the range from minimum to maximum radiation intensity that can be displayed in terms of either differences in signal intensity or density differences in conventional film.

Modulation Transfer Function (MTF): The MTF describes how the contrast of the image component is transmitted as a function of its size or its spatial frequency. It is expressed in line pairs per millimetre (lp/mm).

Contrast Resolution: It is the smallest detectable contrast for a given detail size that can be shown by the imaging system with different intensity (density) or the whole dynamic range. The threshold contrast is a measure for imaging of low contrast structures and is largely determined by the DQE of the detector.

ⅢⅢ▶ 19.9 DIGITAL RADIOGRAPHY

Ever since the original discovery of X-rays, film has been the preferred medium for producing medical X-ray images. This means that the same medium is employed for image acquisition, presentation and storage. Consequently, images which are produced with less than optimal quality cannot be easily manipulated to improve information retrieval.

The conventional screen film system has a moderately good detector quantum efficiency (20–30% at 60 keV) and a similarly good MTF for frequencies above 3 lp/mm. The strength of screen-film combinations lies in their high nominal spatial resolution (>3lp/mm) and the high contrast resolution at optimum exposure.

In both radiography and fluoroscopy, there are definite advantages of having a digital image stored in a computer. This allows image processing for better displayed images, the use of lower doses, avoiding repeat radiography and opening up of the possibility of digital storage with a PACS (Picture Archieving and Communication System) or remote image viewing via tele-radiology. Digitally formatted images would permit digital storage, retrieval, transfer and display of X-ray images with vast possibilities of image-related processing and manipulations, as each function can be individually optimized (Schittenhelm, 1986).

Digital X-ray imaging systems consist of the following two parts:

(i) X-ray imaging transducer or data collection; and

(ii) Data display, storage and processing.

The digitally compatible X-ray imaging transducers can be divided into the following two categories:

(i) Image intensifier TV system; and

(ii) Radiographic (film replacement) systems.

The application of X-ray image intensifier TV systems in digital X-ray imaging evolved from their use in angiography. Angiography is a diagnostic and rapidly developing therapeutic modality concerned with diseases of the circulatory system. The procedure is carried out by using a contrast material to opacify vascular structures because the radiographic contrast of blood is essentially the same as that of soft tissue. Contrast material is an iodine-containing compound which is injected through a catheter (diameter ranging from 1 to 3 mm). Radiographic images of the contrast-filled vessels can be viewed on a TV screen or are recorded by using either film or video.

The most important application of digital technology is the development of digital subtraction angiogrphy (DSA). In this technique, a pre-injection image (mask) is acquired, the injection of contrast agent is then performed, then images of the opacified vessels are acquired and subtracted from the mask. This greatly helps in contrast enhancement there by providing increased contrast sensitivity. To illustrate this, Fig. 19.23(a) represents the transmitted X-ray intensity through the cross-section of a patient. It is obvious that small contrast changes due to vessels are masked by a large anatomical background contrast change. Any attempt to amplify these small signals would

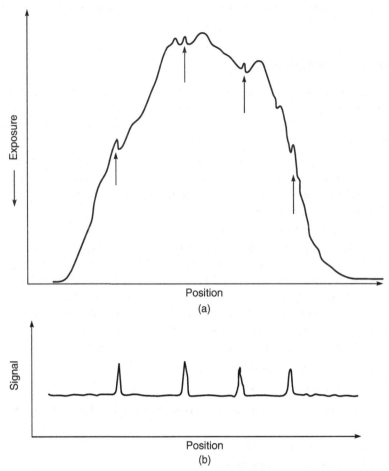

> ➤ **Fig. 19.23(a)**   *X-ray transmission cross-section of a patient with contrast*
>   *enhanced vessel images superimposed*
>   **(b)**   *Subtracted profile with uniform background to vessel image*

merely produce saturation of the display system by the large background signals. Subtraction of the constant background signal away from the contrast-enhanced signal (Fig. 19.23(b) produces a more meaningful and uniform signal. This enables the vessel signals to be amplified greatly prior to display, which improves their visibility.

Figure 19.24 shows a digital subtraction angiography system based on the use of image intensifier. The output of the video camera, which is in the analog form, is first digitized in an analog to digital converter and fed into two semiconductor memories. Theoretically, noise is added to the image due to quantization errors associated with digitization process. This additional noise can be kept to an insignificantly small degree by using a 10 or 12-bit analog-to-digital converter in order to have a sufficiently high number of digital levels.

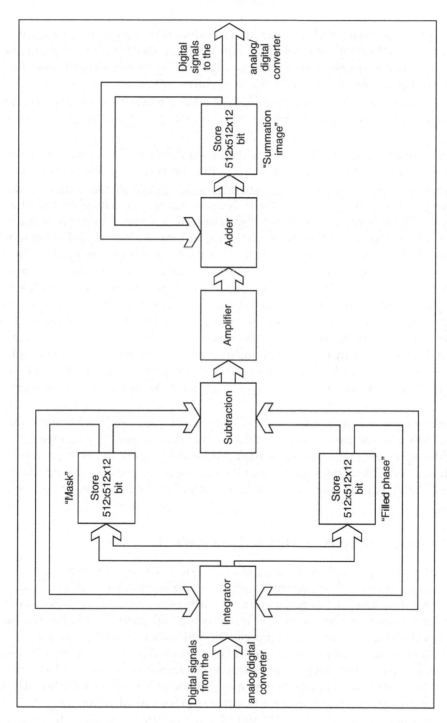

▶ **Fig. 19.24** *Block diagram of a digital video subtraction unit*

Storage of digital images is required for both online and archival purposes. Online storage is provided by real-time digital disks. Archival needs are mostly met by the storage of hard copy films generated from the digital images. Hard copy devices include multiformat cameras (laser or video) and video thermal printers. Alternatively, digital streamer tape cassettes and digital optical disks have been used to record angiographic procedures and associated images.

Film replacement digital X-ray imaging transducers make use of a number of technologies for scanning the area of interest for radiography. Moores (1987) describes these in a well-illustrated article.

Storage phosphor screens are the most widely used detectors for digital radiography. Their QE is less than that of a screen-film combination while the MTF is moderate and depends on the type of phosphor screen. The use of selenium as a detector material has long been known from xero-radiography. Here, the digital electrostatic X-ray imaging system involves the direct electrical read-out of the latent X-ray image formed by xero-radiography. This process, a thin layer of selenium on which a surface voltage has been induced is exposed to X-rays. The charges liberated by absorption of the X-ray energy migrate to the surface where they neutralize the deposited charge layer, thus forming an electrostatic image. The latent charge images thus produced are scanned by guarded electrometers. A time-dependent signal is induced, which a high input impedance pre-amplifier detects. The detected signal from the electrometers is then multiplexed, amplified and digitized. As in all digital systems, the resolution is determined by the pixel size (0.2 mm; 2166×2448 matrix) and is nominally below that of film. The dynamic range is extremely wide (1:10000) with a linear relationship between dose and signal over a wide exposure range. Xero-radiography tends to suffer from a lack of sensitivity when thicker body sections are being imaged. Therefore, its potential use may be limited to X-ray examinations of extremities of the body.

The image is displayed on the monitor immediately after acquisitions, so that the image quality can be checked while the patient is still in the examination room. Any changes in the patient position or the equipment settings can be made immediately, without having to wait until a film has been developed. The digital images can be automatically combined with the patient and the exposure data in the system and transferred online to the laser camera or the diagnostic workstation.

19.9.1 Flat Panel Detectors for Digital Radiography

The next step towards the digital integration of the classic X-ray acquisition technique is the use of electronic image detectors. The system under development is based on new X-ray detectors employing large area amorphous silicon semiconductor sensors. Amorphous (non-crystalline) silicon (a-silicon) is used in place of the classic microchip with mono-crystalline silicon, as this is necessary for achieving the large detector area. The a-Si layer is brought onto a glass carrier as a thin layer and structured into an array of sensors (photo-diodes) using conventional photo-lithographic methods. A switching element (a diode or a transistor) is allocated to each individual sensor so that the sensor can be connected to a read-out line in the column direction. The switching element are controlled via corresponding address lines in the row direction (Fig. 19.25). The signal from the individual sensors is led to pre-amplifiers, amplified and given to analog-to-digital

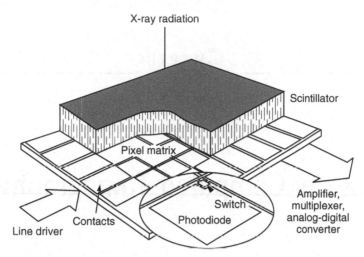

> **Fig. 19.25**   *Flat X-ray solid state detector with an amorphous silicon active readout matrix*

converters. The signals from all the sensors are read out until the whole X-ray image has been completed. As the process is electronic, it can achieve very high transfer rates. The X-ray image is displayed on the monitor of a workstation with a gray scale resolution of 12 bits.

Silicon by itself is not sufficiently sensitive for detecting X-rays in the energy range used in diagnostic radiology. For this reason, an image converter layer is applied over the layer of amorphous silicon. Generally, cesium iodide (CsI) is used as the image converter layer. This is a flourescent material, which is also used as the input screen of the X-ray image intensifier (Strotzer *et al* 1998).

The pixel size in the X-ray image is determined by the size of the sensors. In the a-Si detector, it is 143 μm × 143 μm. This allows a resolution of more than 3.5 lp/mm to be achieved, sufficient for most of the radiographic applications except for mammography. With a detector size of 43 × 43 cm, a matrix of 3000 × 3000 pixels is created on the flat a-Si detector. The flat sandwich structure of the image detector allows for a compact construction so that the flat detector can be easily integrated in the bucky table.

A flat panel detector has the potential to close the gap between digital systems already in use (CT, MR etc.) and radiography, and thus represents one more important step towards the fully digital hospital.

X-ray Computed Tomography

ⅢⅢ▷ 20.1 COMPUTED TOMOGRAPHY

There are two main limitations of using conventional X-rays to examine internal structures of the body. Firstly, the super-imposition of the three-dimensional information onto a single plane makes diagnosis confusing and often difficult. Secondly, the photographic film usually used for making radiographs has a limited dynamic range and, therefore, only objects that have large variations in X-ray absorption relative to their surroundings will cause sufficient contrast differences on the film to be distinguished by the eye. Thus, whilst details of bony structures can be clearly seen, it is difficult to discern the shape and composition of soft tissue organs accurately. In such situations, growths and abnormalities within tissue only show a very small contrast difference on the film and consequently, it is extremely difficult to detect them, even after using various injected contrast media. The problem becomes even more serious while carrying out studies of the brain due to its overall shielding of the soft tissue by the dense bone of the skull.

Various techniques have been applied in an effort to overcome these limitations, but the most powerful technique which has shown dramatic results is computed tomography, which was invented and developed by G.N. Hounsfield at the Central Research Laboratories of EMI Ltd, UK, and introduced on a commercial scale in 1972. Since then, its impact on the medical world has been as great as the discovery of X-rays itself. Despite the inherently high cost of the equipment, several thousands of these are now installed in hospitals around the world.

Tomography is a term derived from the Greek word 'tomos', meaning 'to write a slice or section' and is well-understood in radiographic circles. Conventional tomography was developed to reduce the super-imposition effect of simple radiographs. In this arrangement, the X-ray tube and photographic film are moved in synchronisation so that one plane of the patient under examination remains in focus, while all other planes are blurred. In computed tomography (CT), the picture is made by viewing the patient via X-ray imaging from numerous angles, by mathematically reconstructing the detailed structures and displaying the reconstructed image on a video monitor.

The early CT scanners were specifically designed for neuro-radiological investigations. Computed tomography enabled radiologists to distinguish, for the first time, between different types of brain tissue, and even between normal and coagulated blood. With CT images, radiologists could easily visualize the ventricles of the brain and repositories of the cerebro-spinal fluid. This capability made obsolete a rather unpleasant procedure known as 'pneumo-encephalography', in which air is pumped into the ventricles to displace the fluid and provide radiographic contrast.

The desirability of having body scanners was soon realized. The examination of the body sections, however, represents widely differing problems. Some of these problems include the movement of organs, the patient's respiratory action, the broader range of tissue densities encountered and the wide range of body sizes that have to be accommodated. In spite of these difficulties, whole body CT scanners (Fig. 20.1) with a very wide range of clinical capabilities have been made commercially available. Since respiration does not normally involve gross movements of the head, its effect on the quality of brain pictures is negligible. Consequently, a brain scanner does not need to operate at the speed of a whole body machine.

20.1.1 Basic Principle

Computed tomography differs from conventional X-ray techniques in that the pictures displayed are not photographs but are reconstructed from a large number of absorption profiles taken at

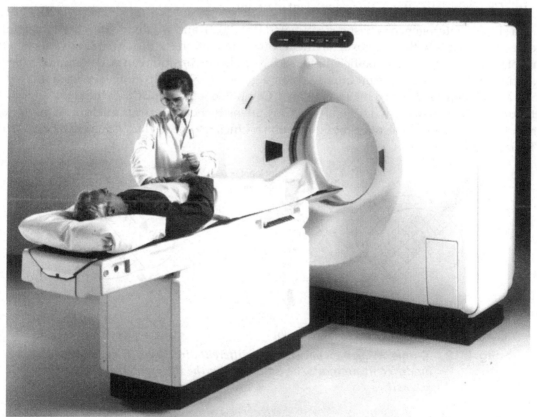

➤ **Fig. 20.1** *Whole body CT scanner (Courtesy: M/s General Electric Company, USA)*

regular angular intervals around a slice, with each profile being made up from a parallel set of absorption values through the object.

In computed tomography, X-rays from a finely collimated source are made to pass through a slice of the object or patient from a variety of directions. For directions along which the path length through-tissue is longer, fewer X-rays are transmitted as compared to directions where there is less tissue attenuating the X-ray beam. In addition to the length of the tissue traversed, structures in the patient such as bone, may attenuate X-rays more than a similar volume of less dense soft tissue. In principle, computed tomography involves the determination of attenuation characteristics for each small volume of tissue in the patient slice, which constitute the transmitted radiation intensity recorded from various irradiation directions. It is these calculated tissue attenuation characteristics that actually compose the CT image.

For a monochromatic X-ray beam, the tissue attenuation characteristics can be described by

$$I_t = Io\, e^{-\mu x}$$

I_o = Incident radiation intensity

I_t = Transmitted intensity

x = Thickness of tissue

μ = Characteristic attenuation coefficient of tissue

If a slice of heterogeneous tissue is irradiated (Fig. 20.2), and we divide the slice into volume elements or voxels with each voxel having its own attenuation coefficient, it is obvious that the sum of the voxel attenuation coefficients for each X-ray beam direction can be determined from the experimentally measured beam intensities for a given voxel width. However, each individual voxel attenuation coefficient remains unknown. Computed tomography uses the knowledge of the attenuation coefficient sums derived from X-ray intensity measurements made at all the various irradiation directions to calculate the attenuation coefficients of each individual voxel to form the CT image.

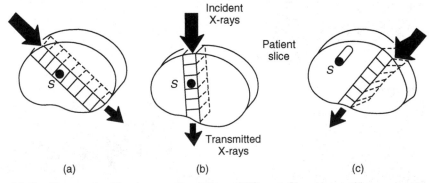

> **Fig. 20.2** *X-rays incident on patient from different directions. They are attenuated by different amounts, as indicated by the different transmitted X-ray intensities*

Figure 20.3 shows a block diagram of the system. The X-ray source and detectors are mounted opposite each other in a rigid gantry with the patient lying in between, and by moving one or both

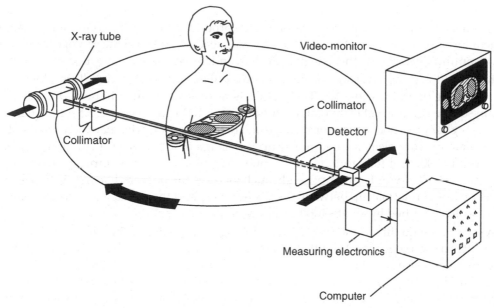

> **Fig. 20.3** *The technique of producing CT images. The X-ray tube and the detector are rigidly coupled to each other. The system executes translational and rotational movement and transradiates the patient from various angular projections. With the aid of collimators, pencil thin beam of X-ray is produced. A detector converts the X-radiation into an electrical signal. Measuring electronics then amplify the electrical signals and convert them into digital values. A computer then processes these values and computes them into a matrix-line density distribution pattern which is reproduced on a video monitor as a pattern of gray shade (Courtesy: Siemens, W. Germany).*

of these around and across the relevant sections, which is how the measurements are made. The patient lies on a motorized couch and is moved into the aperture of the gantry, with the location to be accurately determined by means of a narrow strip of light that falls on the body from the gantry and illuminates the section to be examined. From the keyboard mounted on the operating console, details such as, the patient's code, the name of the hospital, etc., are fed into the system and settings for X-ray parameters for the scan are made.

In one system which employs 18 traverses in the 20s scanning cycle, 324,000 ($18 \times 30 \times 600$) X-ray transmission readings are taken and stored by the computer. These are obtained by integrating the outputs of the 30 detectors with approximately 600 position pulses. The position pulses are derived from a glass graticule that lies between a light emitting diode and photo-diode assembly that moves with the detectors. The detectors are usually sodium-iodide crystals, which are thallium-doped to prevent an after-glow. The detectors absorb the X-ray photons and emit the energy as visible light. This is converted to electrons by a photo-multiplier tube and then amplified. Analog outputs from these tubes go through signal conditioning circuitry that amplifies, clips and shapes the signals. A relatively simple analog-to-digital converter then prepares the signals for the computer. Simultaneously, a separate reference detector continuously measures the intensity of the primary X-ray beam. The set of readings thus produced enables the computer to compensate for fluctuations of X-ray intensity. Also, the reference readings taken at the end of each traverse are used to continually calibrate the detection system and the necessary correction is carried out.

After the initial pre-processing, the final image is put onto the system disc. This allows for direct viewing on the operator's console. The picture is reconstructed in either a 320 × 320 matrix of 0.73 mm squares giving higher spatial resolution or in a 160×160 matrix of 1.5 mm squares which results in higher precision, lower noise image and better discrimination between tissues of similar density. Each picture element that makes up the image matrix has a CT number, say between −1000 and +1000, and therefore, takes up one computer word. A complete picture occupies approximately 100 K words, and upto eight such pictures can be stored on the system disc. There is a precise linear relationship between the CT numbers and the actual X-ray absorption values, and the scale is defined by air at −1000 and by water at 0.

Obviously, the quality of the reconstructed image is a matter of the differentiation between μ(X-ray attenuation coefficient) at different points and of the size of each pixel (square dots of light whose intensity varies to reflect the attenuation). The differences in μ of the various body tissues are slight and that typical tissue contains mostly elements of low atomic weight. At the photon energy employed in CT scanners, the interaction of the photons with the tissue result in the 'Compton Effect', in which the impact of an X-ray photon with an electron is accompanied by a transfer of energy and a fall in the X-ray frequency. The loss in energy is proportional to the density of electrons which results in linear relationship between tissue density and attenuations. The differential attenuation coefficient is thus well-correlated with the specific gravity. Hence, the image reconstructed as a result of computerized tomography can be considered as the mapping of densities (Table 20.1) with respect to that of water.

20.1.2 Contrast Scale

The lnear attenuation coefficient of tissue is represented by the scanner computer as integers that usually range in values from −1000 to +1000. These integers have been given the name 'Hounsfield

• Table 20.1   *Specific Gravity and Attenuation Coefficient for Various Materials*

Materials	Specific gravity	$\mu\ (cm^{-1})$	$\Delta\mu$ above water %
Water	1.00	0.205	–
Whole blood	1.034	0.214	4.3
		0.322	
Heart muscle	1.04	0.212	3.4
Fat	0.93	0.190	−7.8
Breast	0.97	0.189	−8.4
Brain white matter	–	0.215	4.8
Grey matter	–	0.213	3.9
Meningioma	1.05	0.214	4.3

units, and are abbreviated as H. They are also denoted by CT numbers. The relationship between the linear attenuation coefficient and the corresponding Hounsfield unit is:

$$H = \frac{\mu - \mu_{water}}{\mu_{water}} \times 1000$$

where μ_{water} = attenuation coefficient of water.

The CT number scale (Fig. 20.4) is defined in such a way that 0 is assigned to water and –1000 to air. The value +1000 represents highly dense materials. Since neither the human eye nor the television display system is able to differentiate all the 2000 steps in this scale, only a section of the scale is represented on the video monitor.

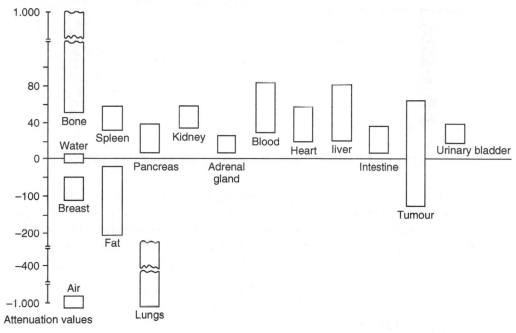

> **Fig. 20.4** *CT number scale as used in computed tomography*

Therefore, viewing systems have window level and window width controls (Fig. 20.5). These controls determine where and over what range of CT numbers will the video gray scale possibly lie. A decrease in window width enables us to see very small changes in tissue density more clearly, since the gray scale becomes spread over a smaller range and any given change in tissue density shows an increased contrast. However, it is necessary to move the window up and down the CT scale by means of the level control so that the absorption values under examination will be displayed between black and white on the monitor.

The equipment is provided with a facility for the selection of the window widths either in steps of 0, 32, 64, 128, 256, 512 and 1024, or freely. The middle position of the window *can be set between* –1024 and +1023 or between –512 and +1535.

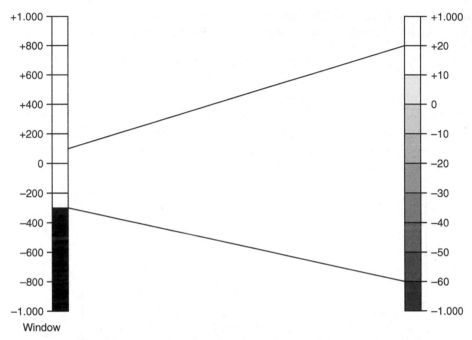

> **Fig. 20.5**   *Window level and window width control in CT*

In the relevant range of the effective radiation energy between 60 and 80 keV, the differences in the attenuation values of plexiglass and water are largely constant. The measured CT values and the known μ-values of these substances are employed in a CT number scale, and the ratio of the difference of the CT values to the corresponding difference of the values is given by the slope of the straight line:

$$\frac{I}{K} = \frac{CT_{plex} - CT_{H_2O}}{\mu_{plex} - \mu_{H_2O}}$$

where $\mu_{plex} - \mu_{H_2O} = 0.024 \ cm^{-1}$

$\mu_{H_2O} = 0.197 \ cm^{-1}$ at 66 keV

$K = 1.9 \times 10^{-4} \ cm^{-1}/CT$ value

The μ value for various tissues can thus be computed by simply multiplying the measured CT value by K.

It may be noted that this is, in fact, correct for all materials equivalent to water or plexiglass, e.g., soft tissue. But in a combination of bone and soft tissue, the actual μ values cannot be easily extracted from the CT numbers because the X-ray radiation is polychromatic, and subject to so-called beam hardening (Dichiro et al. 1978).

Most commercial CT machines have a spatial resolution of around 2 mm and almost without exception, the manufacturers claim a noise level corresponding to a 0.5% change in the X-ray absorption coefficient. Achieving this level of discrimination in a normal radiograph would require X-ray intensity differences of about 0.02% to be rendered visible on the film. (The linear absorption coefficient of soft tissue is about 0.2 cm^{-1} and the change of intensity would be $0.2 \text{ cm}^{-1} \times$ 0.2 cm X 0.5% = 0.02%). With conventional projection radiography, an intensity change of 0.02% is two orders of magnitude smaller than the maximum detectable change possible with that modality.

⚠▶ 20.2   SYSTEM COMPONENTS

All computer tomography systems consist of the following four major sub-systems:

  (i) Scanning system—This takes suitable readings for a picture to be reconstructed, and includes X-ray source and detectors.
 (ii) Processing unit—This converts these readings into intelligible picture information.
(iii) Viewing part—It presents this information in visual form and includes other manipulative aids to assist diagnosis.
(iv) Storage unit—This enables the information to be stored for subsequent analysis.

20.2.1   Scanning System

The purpose of the scanning system is to acquire enough information to reconstruct a picture for an accurate diagnosis. A sufficient number of independent readings must be taken to allow picture reconstruction with the required spatial resolution and density discrimination for diagnostic purposes. The readings are taken in the form of 'profiles'. When a plane parallel X-ray beam is passing through a required section, a profile is defined as the intensity of the emergent beam plotted along a line perpendicular to the X-ray beam. This profile represents a plot of the total absorption along each of the parallel X-ray beams. It thus follows that the higher the number of profiles obtained, the better is the resulting picture. In practice, 180 such profiles at 1° intervals are normally needed to construct a diagnostically useful picture. There are several designs of scanning gantry commercially available from various manufacturers. They use different mechanical configurations. An excellent review of the physical aspects of X-ray transmission computed tomography is given by Webb (1987).

First Generation—Parallel Beam Geometry: In the basic scanning process, a collimated X-ray beam passes through the body and its attenuation is detected by a sensor that moves on a gantry along with the X-ray tube (Fig. 20.6(a)). The tube and detector move in a straight line, sampling the data 180 times. At the end of the travel, a 1° tilt is made and a new linear scan begins. This assembly travels 180° around the patient's position. This arrangement is known as 'Traverse and Index' and was used in the earliest commercial system, the EMI MKI Brain Scanner. This procedure results in 32,400 independent measurements of attenuation, which are sufficient for the systems computer to produce an image. Obviously, this is a fairly slow procedure and requires a typical scan time of 5 minutes. It is essential for the patient to keep still during the entire scan period and for this reason, the early scanners were limited in their use to only brain studies.

Although this type of system is slow, its picture quality and hence its diagnostic utility, is exceptionally good. Scanning the brain with the early scanners provided the maximum immediate clinical benefit since traditional X-ray pictures of the brain are notoriously difficult to interpret.

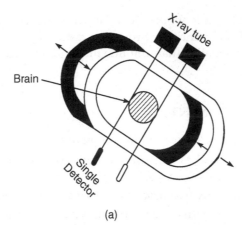

(a)

➤ **Fig. 20.6(a)**   *Scanning arrangement of the early CT machines. They made a linear traverse before taking a 1° rotation. The system employed single-source and single-detector system. It took long measuring times*

However, in order to speed up the information gathering and to achieve a reasonable patient throughput, a pair of detectors was used so that two contiguous slices could be examined simultaneously.

The slowness of the earlier brain scanners precluded the possibility of scanning areas of the body other than the head. At best, the patient could be sedated to eliminate head movement, but it is obviously not possible to eliminate respiratory movements. Therefore, there could be chances of blurring the reconstructed image caused by movement of the patient or of internal organs, which necessitates reduction of the examination period to within breath-holding times. The inherent mechanical constraints of a traverse/index system mean that each traverse must take at least 1 s. So, it was unlikely that the machines based on this principle could ever be made so fast as to scan in less than 180 s.

Second Generation—Fan Beam, Multiple Detectors: An improved version of the traverse-index arrangement consists in using a bank of detectors and a fan beam of X-rays (Fig. 20.6(b)). This system effectively takes several profiles with each traverse and thus permits greater index angles. For example, by using a 10° fan beam, it is possible to take 10 profiles, at 1° intervals, with each traverse and then index through 10° before taking the next set of profiles. Therefore, a full set of 180 profiles can be obtained with 18 traverses. This method has permitted a reduction in the scan time, and at the rate of approximately 1 s for each traverse, it has led to the systems operating in the 18–20 s range.

Third Generation—Fan Beam, Rotating Detectors: The main obstacle for a further increase in speed with the conventional computer tomographs arises from the mechanically unfavourable multiple alterations between the translational and rotational movement of the measuring system. Since the scanning of radiation absorption profiles of the object slice to be reproduced from several different projection directions is essential for the construction of a computer tomogram, the rotational movement of the radiation source cannot be dispensed with. On the other hand, the

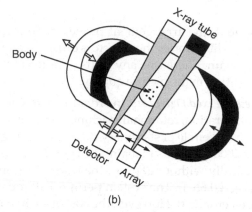

(b)

➤ **Fig. 20.6(b)** *Using a fan-shaped beam and an array of detectors, larger steps can be taken and the scanning process speeded up*

linear scanning movement can be avoided by using a sufficiently wide fan-shaped X-ray beam which encompasses the whole object cross-section, and a multiple detector system mechanically tied to the tube which permits a simultaneous measurement of the whole absorption profile in one projection direction (Fig. 20.6(c)). Also, on account of the largeness of the measuring system consisting of X-ray tube and detectors, the rotational movement must not be stepwise but continuous.

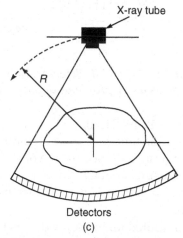

Detectors
(c)

➤ **Fig. 20.6(c)** *If the fan-beam is large, no traverse motion is needed. Only rotational movement of the scanning frame is required, thus offering considerable improvement in measuring time*

Pure rotational machines have been developed on the basis of this principle. The simplest of these has the X-ray source and detectors mounted on a common frame and rotate around the patient, usually through 360°. The system gives a wide fan beam, typically between 30° and 50°. The frame can be made to travel quite fast, so that a complete rotation takes only a few seconds.

This configuration has two major disadvantages. Firstly, it has a fixed geometry. With a fan beam set for the largest patient, the arrangement proves to be inefficient for smaller objects, particularly heads. Secondly, calibration of the detectors during scanning is not possible since the patient is always within the beam. Therefore, any drifts or faults in the detection system tend to produce a significant degradation in the picture quality.

Fourth Generation—Fan Beam, Fixed Detectors: In order to overcome the difficulties encountered in the rotating detectors configuration, rotational machines have been designed in which only the X-ray source rotates within a full circle of stationary detectors arranged around the patient. The system employs as many as 2000 detectors to maintain a good spatial resolution. The individual detectors are lined up practically without gaps, so that the radiation which has penetrated the patient is optimally used (Fig. 20.6(d)). The system permits calibration during scanning, which eliminates the problem of detector drift. However, the cost of such machines would obviously be high.

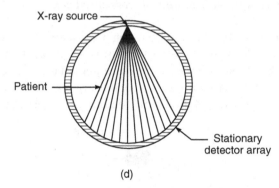

(d)

➤ **Fig.** 20.6(d)   *The x-ray tube rotates while detectors remain stationary. This arrange-*
ment overcomes many problems of pure rotational systems

Fifth Generation—Scanning Electron Beam: The 0.7 to 1 second time resolution limit of mechanical CT scanners makes phase-resolution imaging of the beating heart possible only through manipulations involving ECG triggering. The acquisition of all the cardiac phases within a single cardiac cycle can only be realized using a data acquisition system which does not contain any moving mechanical parts. One such system is the electron beam tomography (EBT) scanner (Schwierz and Kirchgeorg, 1995).

Basically, the electron beam computed tomography differs from conventional CT in terms of speed and the method of generating the X-ray. In conventional CT scanning, an X-ray tube and an X-ray detector are mounted across each other on a circular frame and rotate around the patient. In electron beam tomography, the electron beam sweeps back and forth through a magnetic field. The impact of the electron beam on a semi-circular tungsten array underneath the patient generates the X-rays and the X-ray detectors are mounted on a semi-circular array above the patient (Fig. 20.7). Because an X-ray tube and X-ray detector are heavy moving parts, weighing as much as 250 kg, it takes one second or more to take all the snapshots which are later reconstructed to form an image of one slice of the body with a conventional CT scanner. Since an electron beam can be moved back

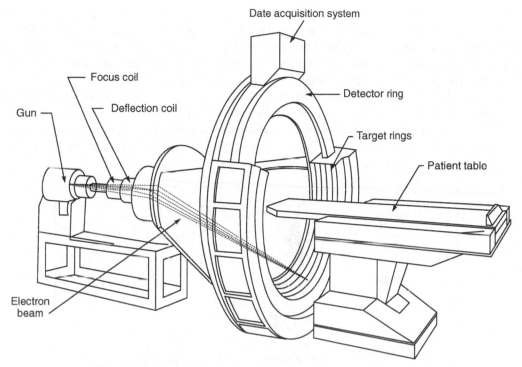

> **Fig. 20.7** *Schematic of ultrafast electron beam CT scanner*

and forth through a magnetic field very quickly, the time for scanning a slice can be of the order of 50 ms with electron beam tomography.

When combined with ECG triggering, EBCT can permit a comprehensive cardiac imaging and examination including the quantitation of flow rate over multiple heart beats. Cardiac images obtained with a conventional CT may be blurred due to motion artifact. In contrast, images of the heart obtained with electron beam tomography are precise and reproducible (Guerci and Kornhausee, 1994).

The detector array consists of two continuous ranges of 216° with 432 channels each. Luminascent crystals coupled to silicon photo-diodes are used. The scanning electron beam emitted by an electron gun is accelerated by 130–140 kV, electromagnetically focused and deflected over a target in a typical time of 50–100 ms. It was originally designed for cardiac examinations. The unit was equipped for this purpose with four anode rings and two detector rings which enabled eight contiguous slices, an area of approximately 8×8 mm, to be scanned without movement of the patient. The basic difference between an electron beam scanner and conventional units is that the patient is encircled by stationary anode rings which can thus be cooled directly. By serially scanning all four rings, a multiple-slice examination can be performed (Webb, 1987).

Spiral /Helical Scanning: This is a scanning technique in which the X-ray tube rotates continuously around the patient while the patient is continuously translated through the fan beam. The focal spot therefore, traces a helix around the patient. The projection data thus obtained allow for the reconstruction of multiple contiguous images. This operation is often referred to as helix, spiral, volume, or three-dimensional CT scanning. This technique has been developed for acquiring images with faster scan times and to obtain fast multiple scans for three-dimensional imaging to obtain and evaluate the 'volume' at different locations. Figure 20.8 illustrates the spiral scanning technique, which causes the focal spot to follow a spiral path around the patient. Multiple images are acquired while the patient is moved through the gantry in a smooth continuous motion rather than stopping for each image. The projection data for multiple images covering a volume of the patient can be acquired in a single breath hold at rates of approximately one slice per second. The reconstruction algorithms are more complex because they need to account for the spiral or helical path traversed by the X-ray source around the patient (Kalender,1993).

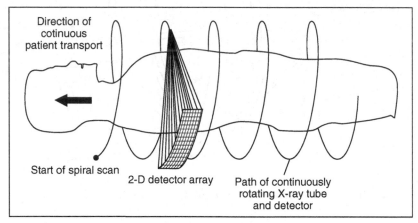

➤ **Fig. 20.8**   *The spiral CT scan principle in multislice scanning (after Theobald et al, 2000)*

Spiral CT has a special advantage in that it allows images to be reconstructed at arbitrary positions and arbitrary spacing, also resulting in overlapping. This offers a great advantage if slices at small spacings are required for the clear proof of a small lesion. The continuous acquisition of whole sections of the body, largely independent of respiration or movement, also permit the reliable localization of small lesions. Continuous data acquisition in the trunk of the body with the possibility of the reconstruction of overlapping slices could not previously be achieved (Becker *et al*, 1999).

A fundamental difference between and potential disadvantage of spiral CT as compared with conventional CT is the fact that slice sensitivity profiles are blessed by the movement of the patient in the Z direction. The degree of blurring depends upon the speed at which the patient is moved and has a corresponding influence on the spatial resolution perpendicular to the scan slice. However, this can be largely minimized by using suitable de-blurring software. In the normal case, this blurring is almost negligible if the selected table feed per 360° revolution is the same as the slice thickness (Theobald, *et al* 2000).

The SOMATOM Plus from Siemens and Toshiba 900S were the first units which offered spiral CT in 1987 and for the first time, made possible scan times of only 1 second per 360 degree scan.

Use of Slip Rings: In a conventional CT scanner, the input power is applied to the transformer, which is located separately from the gantry. The transformer steps up the voltage to the level of 80–150 kV. The high voltage is supplied by special cables, which are attached to the X-ray tube in the gantry. A sophisticated cable management system allows the tube free access for about 400 degrees of rotation in either direction . Therefore, the tube must rotate first in one and then in the opposite direction during scanning.

In third and fourth generation CT systems, it was realized that the power and signal cables would have to be eliminated as they would otherwise have to be re-wound between scans. A completely new concept to achieve this, was developed, by using self-lubrication slip ring technology, to make the electrical connections with rotating components. In the high voltage slip ring CT scanner, the input power is applied to the transformer which is located separately from the gantry. The high voltage is then connected to a ring inside the gantry. The X-ray tube has cables which are attached to metal brushes that make physical contact with the ring and transfer the high voltage to the X-ray tube. This allows for unlimited freedom of rotation in either direction for the X-ray tube. The high voltage slip rings have proven to be virtually maintenance-free and extremely reliable over several years of testing. The special arrangement of the slip rings has rendered the use of oil or gas for insulating the high voltage unnecessary and thus precluded the possible danger of a leak. In practical operation, electric power of upto 40 kW and voltage of up to 140 kV can be transmitted (Alexander and Krumme, 1988).

In an alternative arrangement, low voltage slip rings can be used to connect the input power directly to the ring inside the gantry. A small high frequency transformer is located inside the gantry at a distance of about one metre from the X-ray tube. The transformer has cables which make physical contact with the ring through metal brushes to transfer the low voltage to the transformer. The high voltage generated by the transformer is then supplied to the X-ray tube by short high voltage cables. This allows for unlimited freedom of rotation, in either direction for the X-ray tube.

X-ray Source: In CT scanners, the highest image quality, free from disturbing blurring effects, is obtained with the aid of pulsed X-ray radiation. During rotation, high voltage (120 kV) is applied at all times. A grid inside the tube prevents the electron current from striking the anode except when desired, allowing the X-rays to be emitted in bursts. As the gantry rotates, an electric signal is generated at certain positions of the rotating system, e.g., in the 4.8 second scan, 288 electrical pulses are generated at intervals of 1/60 s around the circle. Each pulse turns on the X-rays for a short period of time. The number of pulses, the pulse duration and tube current determine the dose to the patient. These factors can be selected by the operator in the same way that they are selected in conventional X-ray systems.

Since the beam is on for only a short period of time, the motion of the patient during the measurement has to be minimized to ensure that the resolution does not get degraded. For producing a fan beam, a collimator is incorporated between the X-ray tube and the patient. A filter inside the collimator housing shapes the beam intensity. Actually, in body scanners, there are two filters, one for bodies and the other for heads which are automatically selected by the computer. These filters produce an intensity variation which, when coupled with the roughly-round shape of the patient, significantly reduces the dynamic range requirements on the electronics.

The fan of X-rays extends beyond the patient diameter so that X-rays which are not attenuated can enter the detector. The intensity of these non-attenuated X-rays is measured in order to correct the data for variations in the X-ray tube output.

Two main types of X-ray tubes have been utilized for computed tomography. The first is an oil-cooled fixed anode line—focus continuous tube, which was principally, used in first and second generation CT scanners. They utilized a tungsten target with a target angle of about 20 degrees. The line focus is provided by a 2×16 mm spot. The second type of tube used in the later generations of the scanners is the rotating anode air-cooled pulsed X-ray source. These tubes have a higher power capability for exposure times in the 2–20 second range. The power requirements of these tubes are generally variable within 100–160 kV. Typical power requirements of these tubes are 120 kV at 200–500 mA, producing X-rays with an energy spectrum ranging from approximately 30–120 keV. Most systems have two possible focal spot sizes, approximately 0.5×1.5 mm and 1.0×2.5 mm. A collimeter assembly is used to control the width of the fan beam between 1.0 and 10 mm, which, in turn, controls the width of the imaged slice.

All modern systems use high frequency generators, typically operating between 5 and 50 kHz. With the production of X-rays in the X-ray tube being an inefficient process, most of the power delivered to the tube results in heating up of the anode. A heat exchanger on the rotating gantry is used to cool the tube. Spiral scanning especially places heavy demands on the heat storage capacity and cooling rate of the X-ray tube. A new X-ray tube based on liquid-metal-filled, spiral–groove bearings which allow very high continuous power, has been developed to meet this requirement. New applications such as CT angiography have become possible with these developments.

The major sources of drifting in CT scanners are variations in output of the X-ray tube and detector electronics. The reference channels included in the system correct the X-ray tube drifts. The electronics have two built-in stability circuits. The first is a switch at the input of each channel which can connect the electronic amplifiers to a battery and resistor to measure and correct for any type of electronic drift. This is done automatically by the computer. The second electronic calibration occurs on every detector channel between each X-ray pulse. Since X-rays are not present, the circuits provide zero electronic output between pulses. It may be appreciated that it is not only the number of projections and the measured data per projection which are of importance for the detail resolution that can be obtained, but also the size of the detector, the dimensions of the X-ray beam impinging upon the detectors and the path of the focal spot over the duration of the X-ray pulse, which are important factors the limit the resolving power. In principle, the factors giving rise to 'unsharpness' that have to be taken into account, are very similar to those encountered in any other X-ray imaging system.

Detectors: For a good image quality, it is important to have a stable system response and in that, detectors play a significant role. The detectors used in CT systems must have a high overall efficiency in order to minimize the patient radiation dose, have large dynamic range, be very stable with time and insensitive to temperature variations within the gantry. Figure 20.9 shows the three types of detectors commonly used in CT scanners. Fan-beam rotational scanners mostly employ xenon gas ionization detectors. The schematic diagram of the detector shows that X-rays enter the detector through a thin aluminium window. The aluminium window is a part of a chamber that holds the xenon gas, which fills the entire chamber. Only one gas volume is present so that all detector elements are under identical conditions of pressure and gas purity.

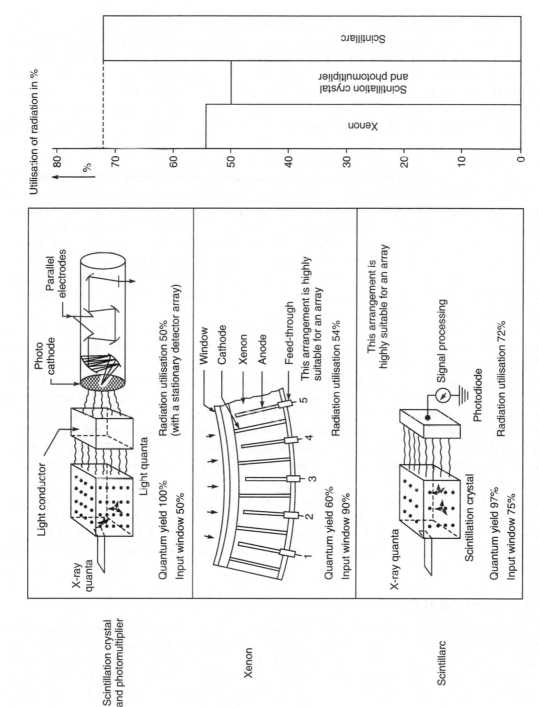

▶ **Fig. 20.9** *Three types of detectors used in computer tomography (Courtesy: M/s Siemens, W. Germany)*

The detector volume is separated into several hundred elements or cells. In a typical scanner, these cells subtend the 42 cm maximum patient diameter. There are 511 data cells and 12 reference cells for simultaneous data collection per view. The detector cells are defined by thin tungsten plates. Every other plate is connected to a common 500 V power supply. The alternate plates are collector plates and are individually connected to electronic amplifiers. X-rays which enter the gas volume between the plates interact with xenon, thus producing positive ions and negative electrons. The positive voltage accelerates the ions to the collector plate and produces an electric current in the amplifier. The resulting current through the electrode is a measure of the incident X-ray intensity.

The xenon detector is inherently a stable detector. Since the detector operates in an ionization mode rather than a proportional mode, small changes in voltage and temperature produce no measurable change in detector output. This is vastly different from photo-multiplier tubes which require almost continual calibration. The main advantages of xenon gas detectors are that they can be packed closely and that they are inexpensive. The entrance width can be as small as 1 mm. In the fixed detectors-rotating source scanners, the detectors do not have to be packed closely. Therefore, scintillation detectors are employed as opposed to ionization gas chambers. Most scintillation detectors are made of sodium iodide, bismuth germanate and cesium iodide crystals. The crystals transform the kinetic energy of the secondary electrons into flashes of light which can be detected by a photo-multiplier.

The scintillator-photo-multiplier detectors suffer from the disadvantage that the smallest commercially available photo-multiplier tube has a diameter of 12 mm. Consequently, they are employed only in translation-rotation and stationary detector arrays.

Siemens employs the SCINTILLARC detector system comprising scintillation crystals and photo-diodes in their SOMATOM machines. In this system, 520 CsI crystals, assembled with photo-diodes, are arranged on a 42° arc. In the radiation entrance plane, the detectors have very small dimensions of only 1.2 mm × 13.5 mm, thus permitting a good resolution. Owing to the fine-grid like separation of the scattered radiation collimator, high percentages (75%) of the X-ray quanta actually reach the detectors. Also, about 97% of the incident quanta can be converted into an electrical signal.

Many modern scanners use solid state detectors such as single crystal $CdWO_4$ or ceremic Gd_2O_2S, with photo-diodes which have some inherent advantages such as a higher efficiency in detecting X-ray photons. One of the next developments to be expected is the use of multi-array detectors, i.e. a number of parallel rings of solid state detectors. This will allow for faster volume scanning.

20.2.2 Processing System

Data Acquisition System: Although good detector properties are a pre-requisite for obtaining optimal image quality, the measuring electronics must have a large dynamic range to back up the detector. The dynamic range defines the ratio of the smallest, just detectable signal to the largest signal without causing saturation. The dynamic range in a typical situation is 1:4,00,000. This implies that with such systems, an optimal image will always be obtained irrespective of whether the patient is obese or thin, or whether we are concerned with bones or soft tissues.

A typical data acquisition system is shown in Fig. 20.10. It consists of precision pre-amplifiers, current to voltage convertor, analog integrators, multiplexers and analog-to-digital convertors. Data transfer rates of the order of 10 Mbytes/s are required in some scanners. This can be accomplished with a direct connection for systems having a fixed detector array. The third generation slip ring systems make use of optical transmitters on the rotating gantry to send data to fixed optical receivers.

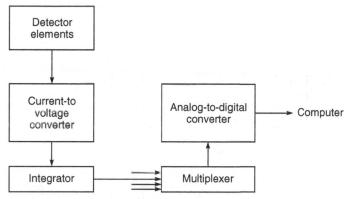

➤ **Fig. 20.10** *Data acquisition system in a CT scanner*

Processing Unit: Although for the CT image, the patient slice is divided up into numerous three dimensional voxels, the image of the slice is a two-dimensional picture in which each picture element (pixel) value corresponds to the attenuation coefficient of a voxel in the object slice. Figure 20.11 illustrates how the iterative or successive approximation method may be used to obtain an image of attenuation coefficients from the measured intensity data. Suppose the attenuation coefficients of the objects (not known before hand) in the first row is 4 and 6, and in the second row it is 1 and 8, representing the characteristics of tissue within the patients. When the object is scanned with X-rays, the sum of the values along various rays/directions are obtained. For example, for scan I, the vertical sums 5 and 14 are obtained; for scan II, the diagonal sums are 1, 12 and 6, and for scan III, the horizontal sums of 10 and 9 are obtained. This scan data will be now used to calculate the image matrix.

As the first step, the data from scan I is back-projected or distributed along the appropriate vertical column with equal weighting, by making the first estimate by placing 5/2 (2.5) in each pixel of that column. Similarly, the second column data value 14, is back-projected, giving 14/2 (7) for pixel in the second column. The matrix of the resulting image of the first iteration is next summed up diagonally and its ray sums of 2.5, 9.5 and 7 are compared with the experimental data of 1, 12 and 6 obtained from scan II. The differences of −1.5, 2.5 and −1 are back-projected with equal weighting diagonally so as to match the experimental data of the scanned object. Similarly, the resulting image matrix of the second iteration is now summed up horizontally to obtain the third iteration result. It is obvious that with more and more iterations, the image matrix matches more and more closely with the object matrix, thereby generating the image of an unknown object with the help of a computer.

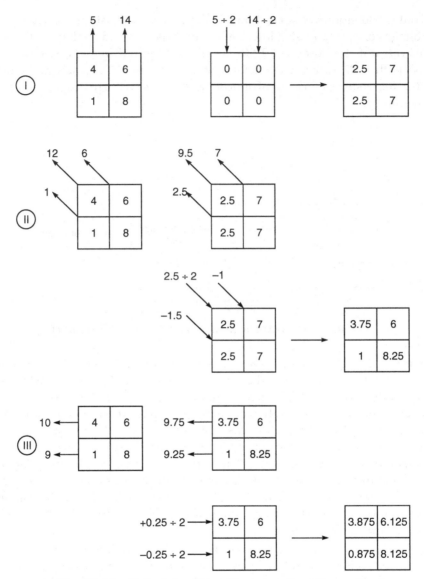

> **Fig. 20.11** *Principle of iterative reconstruction method*

The information received by the computer from the scanning gantry needs to be processed for reconstructing the pictures. The data from the gantry contains information on the following parameters:

- Positional information, such as which traverse is being performed and how far the scanning frame is along its traverse;
- Absorption information including the values of attenuation coefficient from the detectors;

- Reference information that is obtained from the reference detector that monitors the X ray output; and
- Calibration information that is obtained at the end of each traverse.

The first stage of computation is to analyse and convert all the collected data into a set of profiles, normally 180 or more. However, the main part consists of processing the profiles to convert the information which can be displayed as a picture and then be used for diagnosis. In general, the reconstruction methods can be classified into the following three major techniques:

- Back projection, which is analogous to a graphic reconstruction;
- Iterative methods, which implement some form of algebraic solution; and
- Analytical methods, where an exact formula is used. Two of these are filtered-back projection, which incorporates the convolution of the data and Fourier filtering of the image, and the two-dimensional Fourier reconstruction technique.

The method of back projection without any further processing is simple and direct. In this method each of the measured profiles is projected back over the image area at the same angle from which it was taken. At the same time, each projection contributes not only to the points that originally formed the profile, but also to all the other points in its path. This technique in fact produces 'starred' images (Fig. 20.12(a)) and blurring, which makes it totally unsuitable for providing pictures of adequate clarity for medical diagnosis.

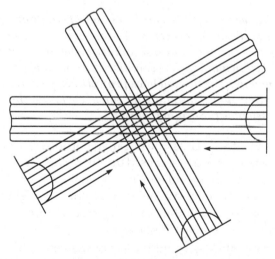

> **Fig. 20.12(a)** *By adding the back projections produced by the shadow functions, the back-projected rays are added to the reconstructed image as artefacts or unwanted points. The original circular structure is transformed into a star shaped display*

The earlier brain scanners used the iterative technique which took a succession of back projections correcting at each stage until an accurate reconstruction was achieved. The method requires several steps to modify the original profiles into a set of profiles which can be projected back to give an unblurred picture. The technique, however, tends to require long computation time.

Current commercial scanners use a mathematical technique known as convolution (Fig. 20.12(b)) or filtering. This technique employs a spatial filter to remove the blurring artifacts. This is achieved by convolving the shadow function with a filter so that each point in the projection has a negative value instead of 0, at every point other than its proper place in the projection. The resulting profiles

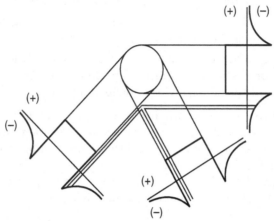

➤ **Fig. 20.12(b)**   *Filtered back projection technique of eliminating the unwanted cusp like tails of the projection. The projection data are convolved (filtered) with a suitable processing function before back projection. The filter function has negative side lobes surrounding a positive core, so that in summing the filtered back projections, positive and negative contribution cancel outside the central core, and the reconstructed image resembles the original object*

are then back-projected and added. Thus, the negative portion of each shadow function cancels out image artifacts that would otherwise be caused by other functions. Mathematically, the method of fast Fourier transform offers a powerful tool in making the required computations and special purpose high speed computers are now available to meet this requirement. The use of this method enables pictures to be reconstructed within a few seconds. Figure 20.13 shows a block diagram image reconstruction computer, used in CT scanners.

In principle, the blurring effect is counteracted in the convolution process by means of a weighing of the scan profiles. The nature and degree of the weighing is determined by the 'convolution kernel', wherein the convolution has an effect on the image structures. Thus, for example, it can be edge-enhancing, so that the bone/soft part interfaces within the skull are particularly clearly emphasized or it can have a 'smoothing effect' with the aim of producing a more uniform image structure. The 'smoothing' convolution kernel reduces image noise and such errors which, for example, can occur with motion artifacts. However, the details are more poorly resolved. The convolution kernels for the head take account of the bones forming the outer housing of the head, in such a way that the so-called 'cupping' effect is suppressed. Gilbert et al. (1981)

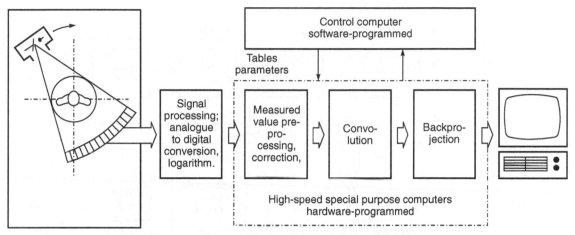

> ➤ **Fig.** 20.13 *Block diagram of the image computer. The synchronous reconstruction of the image permits the representation of the tomogram on the video monitor immediately upon completion of the scan (Courtesy: Siemen, Germany)*

review several computer software and special purpose digital hardware implementations of different forms of algorithms, either proposed or actually implemented in commercial or research CT scanners.

Computer Systems: The computer system plays a central role in CT scanning because without it, there would be no image computation and formation. The computer controls X-ray generation , gantry and table motion, data acquisition, image formation, display and storage. Usually, the CT computer system includes a microcomputer for control functions, an array processor and video memory to enable viewing of the reconstructed images. The image can be viewed on a console and a hard copy can be made on a multi-format camera. Figure 20.14 illustrates a typical computer system employed in a CT scanner. It uses twelve independent processors connected by a 40 Mbyte/s multibus configuration. A multiple array processor is used to achieve the computational speed of 200 Mflops (million floating-point operations per second). The reconstruction time from such a configuration is approximately five seconds to produce an image on a 1024 x 1024 pixel display. A multiuser and multi-tasking environment is provided by a simplified UNIX operating system.

20.2.3   Viewing System

In most of the CT systems, the final picture is available on a television type picture tube. The picture is constructed by a number of elements in a square matrix wherein each element has a value representative of the absorption value of the point in the body which it represents. This technique facilitates a much larger dynamic range than the eye can possibly have. The absorption values are displayed on a linear scale corresponding to air through tissue to dense bone, etc. Several values have been assigned to the two terminal points of the scale. For example, in some cases like the original EMI scale, air is assigned the value of –500, water the value of 0 and bones, that of +500. In the CT scanners that are presently being manufactured, the picture points are

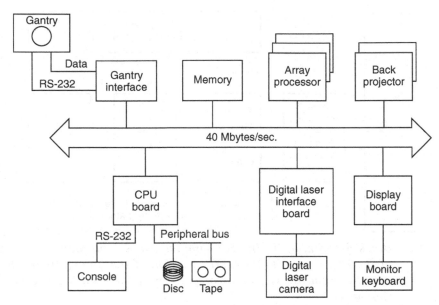

> **Fig.** 20.14 *Typical computer system organization for a CT. The system makes use of motorola family 68000, 16 bit microprocessors (Courtesy: M/s Pickers, USA)*

divided into 2000 steps. The resolution of the scale thus obtained is 1 promille difference in absorption related to the attenuation value for water.

In order to facilitate image display, it would be better if the scale could be expanded within this range. For this purpose, recourse is made to the selection of an image window. The information content of this window is spread over the representable range of colour or gray scale. As long as the original image data of the CT scan is present in the image reconstruction store of the image computer or in the image display memory of the monitor, the image window of interest can be varied in two parameters, namely window level and window width. These two parameters can be varied at will within the range of the absorption values.

Windowing is a powerful aspect of CT and shows the underlying mathematical nature of the displayed picture. This helps in defining the region of interest from where various calculations can be performed on the enclosed elements. The commonly used calculations are of the area, mean value and standard deviation which may well show an identifiable difference between healthy and diseased tissue. It is also possible to subtract one picture from another to demonstrate differences that have occurred during treatment.

20.2.4 Storing and Documentation

For subsequent processing or evaluation of a CT picture, various methods of storage are used. The picture is stored in the digital form so that the evaluation is convenient on a computer-assisted programme. For this purpose, the data carriers generally employed are magnetic disc, magnetic tape and floppy disc. Most manufacturers of CT units use the magnetic tape or floppy disc. The floppy disc provides a medium-range storage. The capacity of a bilaterally coated disc is around 20 pictures, having a matrix of 256 × 256 and depth of information of 10 bits. Floppy discs offer advantages such as ease of handling, low maintenance costs of the drive mechanism, the considerably short access time and the possibility of patient-related storage of the data carrier. For long-term storage, magnetic tapes are preferred. They are inexpensive and extremely reliable, but retrieval of the image is time-consuming and they are sensitive to environmental influences.

There are other possibilities for storing pictures which would normally only store one specific setting of window width and window level. The most common of these is a photograph which is taken from a slave monitor. It is a replica of the picture displayed on the screen at any instant. For photography, one can choose between the multiple-format camera and the 100 mm cut film camera which is available for the magazine technique. The multiple-format camera has a capacity of nine 70 mm pictures or four 100 mm pictures per sheet. A 100 mm cut film camera with the magazine technique and automatic exposure control permits the recording of up to 100 pictures without changing the magazine.

Hard copy print-outs can be made in the gray scale or in colour shades. In the gray scale picture, the shades of gray are restricted to 8 or 10. Besides this, the picture obtained is rather fuzzy and poor in contrast. The CT picture in colour can be made by using 32 colour shades using the ink-writing system, the image resolution being 256 × 256 points. A further possibility of image documentation is the print out of isodensity lines. The hard copy print-out permits the possibility of obtaining an image scaled I:1. This scale is a prerequisite to be met particularly in therapy planning. Figure 20.15 shows a set of typical CT scans.

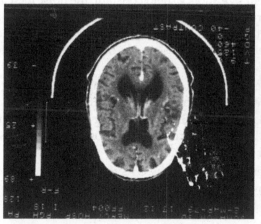 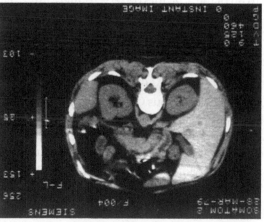

➤ **Fig. 20.15(a)** *Typical CT scan of the brain* **(b)** *CT scan of the abdominal region*

⫸ 20.3 GANTRY GEOMETRY

The CT gantry, which looks like a 'doughnut', contains X-ray tube, detection system and other associated mechanism. The patient support table allows for insertion of the patient into the doughnut hole, which is approximately 60–72 cm in diameter. Most gantries can be tilted ± 20 degree in order to obtain oblique slices. A narrow, visible light field is used to indicate where the

CT slice will be taken. An intercom system is generally available near the inside of the gantry opening to enable verbal communication with the patient.

The top of the table on which the patient is made to lie is made of carbon fibre for maximum strength and radio transparency, and is supported by a sturdy pedestal. The table provides longitudinal movement over a range of 150 cm, which enables uninterrupted scanning without repositioning the patient.

Ever since the CT technology was developed, rapid developments in computer hardware and detector technology have been witnessed. Modern CT systems acquire the projection data required for one tomographic image in approximately one second and present the reconstructed image on a 1024×1024 matrix display within a few seconds. The images represent high quality tomographic maps of the X-ray linear attenuation coefficients of the patient tissues.

The last few years have brought about relatively little change in the basic concept of CT systems. Instead development activity has been concentrated mainly on extending the range of CT applications. An example of this includes the reconstruction of three-dimensional surface images from sets of transaxial slices to assist surgical planning and follow-up assessment.

▥▶ 20.4 PATIENT DOSE IN CT SCANNERS

It is desirable to obtain a diagnostic CT scan while administering a minimal dose to the patient. Here, the radiation impinges upon the patient at a relatively high intensity in the small region of the slice width. Regions proximal to the slice are exposed to scattered radiation. In a CT system, with all of the Y factors (pixel size, algorithm, patient size and kV) held constant, the noise increases inversely as the square root of the dose. Similarly, as the algorithm and detector aperture are changed to improve the resolution, the noise of the system will go up. In other words, the dose must be increased to preserve the same noise level. In practice, it is a trade-off between resolution, noise and dose.

In CT Systems, with 360° X-ray tube movement, the skin dose around the patient is largely constant. The various CT radiographic factors are arranged in such a way that dose units of 0.25, 0.50, 1 and 2 D result. D stands for the maximum skin dose 1.3 ± 0.1 rad (J/kg) of the single slice. For the various radiographic voltages, dose units have been defined. These dose units are derived from the generator output, number of pulses and the length of each pulse.

Typical head and body doses for average patients scanned on the CT systems have been about 2.5 rads for heads and 1.0–2.0 rads for bodies. In multiple-slice operations, the overall dose profile is the sum of the individual skin dose profiles.

Nuclear Medical Imaging Systems

⫸ 21.1 RADIO-ISOTOPES IN MEDICAL DIAGNOSIS

Radio-isotopes are used in medicine both for therapeutic as well as diagnostic applications. In diagnostic practice, small amounts of radioactive chemicals, called 'tracers' (or radio-pharma-ceuticals), are injected into an arm vein or administered through ingestion or inhalation. The amount of radioactivity at different points within the patient's body, or in body fluids, is then examined by radiation detectors. Using these detectors, the amount of radioactivity can be measured within parts of organs as well as within the whole organ. The images show where biochemical processes are occurring normally and where they are occurring too slowly or too quickly.

Among the earliest procedures involving the use of radioactive tracers in medicine was measure-ment of the uptake of radioactive iodine by the thyroid gland. In the early 1940s, it was found that the rate of uptake of iodine by the thyroid greatly increased in patients with disease characterized by increased production of thyroid hormone (hyperthyroidism), a disease that led to nervousness, tremor, weight loss and, in extreme cases, even death. Other patients exhibited decreased iodine uptake (hypothyroidism) by the thyroid and had symptoms and signs of diminished thyroid function. Thus, the radioactive tracers have been in use in medical diagnostics for a long time and even today, the thyroid gland continues to be the organ most frequently examined by nuclear medicine.

In nuclear medical diagnostics, the imaging of organ functions is carried out non-invasively. In contrast to other imaging diagnostic modalities (ultrasound, X-ray, MRI), the nuclear medical examination approach is primarily function-oriented. In this case, vital processes such as blood circulation, metabolism and vitality of organs and tumours can be displayed as functional images.

The clinical use of radio-nuclide imaging depends on obtaining a suitable distribution of the radio-nuclide in the patient. The radio-nuclide is labelled to a compound which will be taken up or metabolized in some way by the human tissue to be studied. The patient receives the material, usually by intravenous injection, and after a suitable delay, which may be minutes or hours, to allow uptake in the target tissues and clearance from the blood, imaging can commence. In this way useful static images, each taking 2–10 minutes, can be produced of the bone, brain, thyroid, lung etc. Dynamic studies can be performed with the gamma camera, starting at the moment of injection and capturing frames of the data either photographically or digitally, at times ranging from minutes down to fractions of a second. Numerical analysis of such data can produce useful information on organ function, blood flow, clearance rates, etc. (Keyes, 1987).

Tecnetium-99m (Tc-99m) has proven to be the most important imaging radio-nuclide used to examine the brain, liver, lungs, bones, thyroid, kidney and heart. It combines the advantages of optimum radiation properties (emission of exclusively gamma radiation with suitable energy, short half-life of six hours) and general availability as a generator nuclide. However, TC-99m cannot be coupled with all required biologically active substances, so that with this radio-nuclide the spectrum of radio-pharmaceuticals for examinations of the organ metabolism is limited. For example, Iodine-123 labelled substances, which are used in many clinical examinations, therefore represent an important supplement to technetium studies.

21.2 PHYSICS OF RADIOACTIVITY

From the theory of atomic structure, it is known that some elements are naturally unstable and exhibit natural radioactivity. On the other hand, elements can be made radioactive by bombarding them with high-energy charged particles or neutrons, which are produced by either a cyclotron or a nuclear reactor. This process will alter the ratio of photons to neutrons in the atoms, thus creating a new unstable nucleus which could undergo radioactive decay. The extra neutron disintegrates and in the process, releases energy in the form of gamma radiation.

Radioactive emissions take place in the following three different forms:

Alpha Emissions: Alpha particles are composed of the two protons and two neutrons. They are least penetrating and can be stopped or absorbed by air. They are most harmful to the human tissue.

Beta Emissions: Beta particles are positively or negatively charged, high speed particles originating in the nucleus. They are not as harmful to tissue as alpha particles, because they are less ionizing, but are much more harmful than gamma rays.

Gamma Emissions: Like X-rays, Gamma particles constitute electromagnetic radiation that travels at the speed of light. They differ from X-rays only in their origin (Fig. 21.1). X-rays originate in the orbital electrons of an atom, whereas gamma rays originate in the nucleus. They are caused by unstable nucleus. X-rays and gamma rays are also called 'photons' or packets of energy. As they have no mass, they have the greatest penetrating capability. Gamma rays are of primary interest in nuclear imaging systems.

The energies of alpha and beta particles and gamma radiations are expressed in terms of the electron volt. One electron volt signifies the energy that an electron would acquire, if it were accelerated through a potential difference of one volt. Radioactive emissions have energies of the order of thousands or millions of electron volts. Alpha emission is characteristic of the heavier radioactive elements such as thorium, uranium, etc. The energy of alpha particles is generally high and lies in the range 2 to 10 MeV (millions electron volt). Due to the larger ionizing power of alpha particles, they can be distinguished from beta and gamma radiations on the basis of the pulse amplitude that they produce on a detector. Beta emissions have an energy range 0–3 MeV.

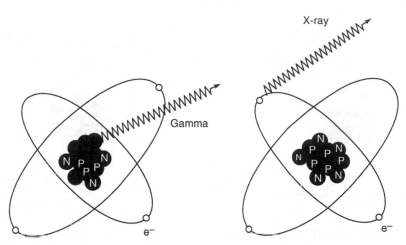

➤ **Fig. 21.1** *Difference between X-ray and gamma emissions*

There are four types of radioactive decay that produce radiation (Fig 21.2). These are:

- β^+ *(Positron Decay)*. In this, the nucleus is unstable because there are more protons than neutrons. For the atom to become stable, it must reduce the number of protons contained in its nucleus by emission of a positively charged particle. This is called positron decay (Fig. 21.2a). It occurs in man-made radioactive materials, wherein protons are added to the nucleus through bombardment using a 'cyclotron'. Cyclotrons are part of expensive equipment and are not typically available in the hospital.

- β^- *Decay (Sometimes called Negatron)*: In this the nucleus is unstable because there is an excess of neutrons. This negatively charged particle, which is an electron with a high kinetic energy, is emitted from the nucleus. This is called (-) decay. Many of the radio-nuclides used in nuclear medicine decay by the emission of a (-) particle, which in turn, triggers gamma emission (Fig. 21.2(b)).

- In *Electron Capture*: In this the nucleus captures an orbital electron from one of the surrounding energy shells. The captured electron then combines with a proton to form a neutron. During this process, energy in the form of photons or gamma rays is emitted from the nucleus, and X-rays are emitted from the electron orbits (Fig. 21.2(c)).

- *Isomeric Transition*: Sometimes, when a nucleus in an excited state begins the decay process to become more stable, it goes through more than one stage of decay. The intermediate stage is called the 'metastable' state. A nuclide in a metastable state eventually decays to a stable atom, through a process called isomeric transition (Fig. 21.2(d)). This type of decay is very important in nuclear medicine because it is the process by which Tc-99m decays. 'm' here means 'metastable'.

21.2.1 Time Decay of Radioactive Isotopes

Each radioactive isotope is characterized not only by the type and energy of radiations emitted, but also by the characteristic life-time of the isotope. This is most conveniently designated by the

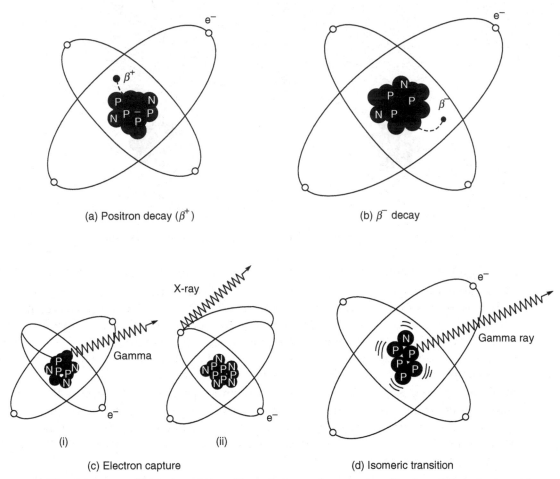

(a) Positron decay (β^+)   (b) β^- decay

X-ray

Gamma

(i)   (ii)

(c) Electron capture   Gamma ray   (d) Isomeric transition

> **Fig. 21.2**   *Radioactive decay (i) emission of gamma radiation (ii) emission of X-radiation*

half-life or half-period of the isotope. The half-period of a radioactive isotope is the time required for half of the initial stock of atoms to decay. Thus, after one-half period has elapsed, the total activity of any single radioactive isotope will have fallen to half its initial value; after two half-periods, the activity will be one-quarter its initial value and so on. After 6.6 half-periods, the activity will be 1% of the initial activity.

The half-life of a radioactive isotope is given by

$$t^{\frac{1}{2}} = 0.693 / \lambda$$

where λ is the decay constant for a particular radio-isotope. In practice, the disintegration rates are determined by counting the number of disintegrations over a certain time t^m and finding the ratio of the number of disintegrations to the time t_m.

21.2.2 Units of Radioactivity

The unit of radioactivity is *curie* and is abbreviated as Ci or ci. This was originally defined to represent the disintegration rate of one gram of radium, but it is now used as the standard unit of measurement for the activity of any substance, regardless of whether the emission is alpha or beta particles, or X or gamma radiation. When used in this way, the *curie* is defined as an activity of 3.7×10^{10} disintegrations per second. The curie represents a very high degree of activity. Therefore, smaller units such as *millicurie* (mci) or *microcurie* (μci) are generally used. Typical amounts of radioactivity used for organ scanning would be 3 mci to 25 mci.

21.2.3 Types and Properties of Particles Emitted in Radioactive Decay

Upon interaction with matter, gamma rays lose energy by three modes. (i) The photo-electric effect transfers all the energy of the gamma ray to an electron in an inner orbit of an atom of the absorber. This involves the ejection of a single electron from the target atom. This effect predominates at low gamma energies and with target atoms having a high atomic number. (ii) The Compton effect occurs when a gamma ray and an electron make an elastic collision. The gamma energy is shared with the electron and another gamma ray of lower energy is produced, which travels in a different direction. The *Compton effect* is responsible for the absorption of relatively energetic gamma rays. (iii) When a high energy gamma ray is annihilated following interaction with the nucleus of a heavy atom, *pair production* of a positron and an electron results. Pair production becomes predominant at the higher gamma ray energies and in absorbers with a high atomic number. The number of ion pairs per centimetre of travel is called *specific ionization*.

▶ 21.3 RADIATION DETECTORS

Depending upon the radiation emitted by the radio isotope of the radiopharmaceutical, a suitable detector is selected and operated under optimum conditions. Several methods are available for detection and measurement of radiation from radio-nuclides. The choice of a particular method depends upon the nature of the radiation and the energy of the particle involved.

If the radiation falls on a photographic plate, it would cause darkening when developed after exposure. The photographic method is useful for measuring the total exposure to radiation of workers, who are provided with film badges. Better methods are available for an exact measurement of the activity. These methods are the use of: (i) ionization chamber, (ii) Geiger Muller Counter, (iii) Proportional Counter, (iv) semiconductor detectors, and (vi) solid state detectors. Described below are the popular types of radiation detection methods used in modern nuclear imaging equipment.

21.3.1 Ionization Chamber

The fact that the interaction of radioactivity with matter gives rise to ionization makes it possible to detect and measure the radiation. When an atom is ionized, it forms an ion pair. If the electrons are attracted towards a positively charged electrode and the positive ions to a negatively charged electrode, a current would flow in an external circuit. The magnitude of the current would be proportional to the amount of radioactivity present between the electrodes. This is the principle of the ionizing chamber.

An ionization chamber consists of a chamber which is filled with gas and is provided with two electrodes. A material with a very high insulation resistance such as polytetrafluoroethylene is used as the insulation between the inner and outer electrodes of the ion chamber. A potential difference of a few hundred volts is applied between the two electrodes. The radioactive source is placed inside or very dose to the chamber. The charged particles moving through the gas undergo inelastic collisions to form ion pairs. The voltage placed across the electrodes is sufficiently high to collect all the ion pairs. The chamber current will then be proportional to the amount of radioactivity in the sample. Ionization chambers are operated either in the counting mode, in which they respond separately to each ionizing current, or in an integrating mode involving collection of ionization current over a relatively long period.

Figure 21.3 shows an arrangement for measuring the ionizing current. The current is usually of the order of 10^{-10} A or less. It is measured by using a very high input impedance voltmetre.

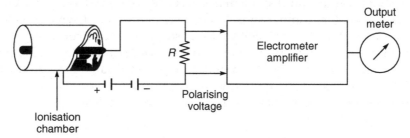

> **Fig. 21.3** *Schematic diagram for measuring ionising current using DC ionising chamber*

The magnitude of the voltage signal produced can be estimated from the fact that the charge associated with the 100,000 ion pairs produced by a single alpha particle traversing approximately 1 cm in air, would be around 3×10^{-14} coulomb. If this average charge is made to pass through a resistance of $3 \times 10^{10} \Omega$ in 1s, a difference of approximately 1 mV potential would develop across the high resistance. This voltage is a function of the rate of ionization in the chamber.

Liquid samples are usually counted by putting them in ampoules and placing the ampoules inside the chamber. Gaseous compounds containing radioactive sources may be introduced directly into the chamber. Portable ionization chambers are also used to monitor personnel radiation doses.

21.3.2 Scintillation Detector

A scintillator is a crystalline substance which produces minute flashes of light in the visible or near ultraviolet range, when it absorbs ionizing radiation. In such cases, the number of fluorescent photons is proportional to the energy of the radioactive particle. The flashes occur due to the recombination and de-excitation of ions and excited atoms produced along the path of the radiation.

The light flashes are of very short duration and are detected by using a photo-multiplier tube, which produces a pulse for each particle. A scintillator along with the photo-multiplier tube is known as a scintillation counter.

Gamma radiations cannot be detected directly in a scintillating material, because gamma rays possess no charge or mass. The gamma ray energy must be converted into kinetic energy of electrons present in the scintillating material. Thus the conversion power of the scintillating material will be proportional to the number of electrons (electron density) available for interaction with the gamma rays. Because of its high electron density, high atomic number and high scintillating yield, the scintillating material which is generally used as gamma ray detector, is a crystal of sodium iodide activated with about 0.5% of thallium iodide. For counting beta particles, scintillator crystals of anthracene are employed. Since the crystal is hygroscopic in nature, it is usually mounted in a hermetically sealed aluminium container having a glass window on the side, which is in contact with the face of the photo-multiplier.

The detector must be able to absorb a high proportion of the incident radiation and convert this energy rapidly into suitable electronic signals. Presently, thallium-activated sodium-iodide NaI(TI) scintillation crystal is used in all commercial cameras. Sodium iodide is the most versatile of all the phosphors. It has a high density that allows enough radiation to be absorbed and a high atomic number which favours photo-electric interaction. Thus a signal is generated which represents the full energy of the incident gamma ray. The activator or impurity (thallium) is necessary to provide sufficient luminescence centres. Conversion efficiency, i.e. the ratio of light output to incident photon energy, is typically 10% for 140 keV radiation. Thus a 140 keV gamma ray absorbed in the crystal produces about 4200 light photons (in the blue-green region of the spectrum where each light photon has an energy of around 3 eV). The decay time of this light flash has a half-life of approximately 0.2 ms, sufficiently fast for most clinical applications.

21.3.3 Semiconductor Detectors

There has been a great deal of development work on semiconductor radiation detectors. These detectors can be made very small and robust. Silicon and germanium crystals have been employed mainly for counting alpha and beta particles. They function in a manner similar to that of the gas ionization chamber. On absorption of radiation in the crystal, electrons and positive holes are formed, which move towards opposite electrodes under the influence of applied potential. The resulting current is proportional to the energy of the ionizing radiation.

The advantage of the semiconductor detector comes from the low energy (3 to 3.5 eV) required to produce an electron-hole pair relative to that of a gas (30 to 35 eV to produce one ion pair) or scintillation detector (200 to 300 eV to produce one photoelectron). However, its small energy gap makes it necessary to operate lithium-drifted silicon or germanium detectors at a low temperature (at the liquid nitrogen boiling point of 77°K) in order to reduce the leakage current.

21.3.4 Solid State Detectors

Solid state detectors can be made in miniature form and can be utilized as in-vivo probes for clinical and experimental applications in medicine. One such example is that of silicon and cadmium telluride detectors (1 mm diameter) which can be encapsulated and used as catheter tubes for studying blood circulation and regional pulmonary functions.

These detectors are highly sensitive to gamma rays because of the high atomic number of cadmium (48) and telluride (52). The limited thickness of solid state detectors makes them inferior to NaI (TI) detectors with respect to gamma ray detection sensitivity.

21.4 PULSE HEIGHT ANALYSER

In radioactivity measurements, the individual particles are detected as single electrical impulses in the detectors. Also, various types of detectors can be set up to operate in a region in which the particular particle produces an electrical impulse with the height proportional to the energy of the particle. The measurement of pulse height is thus a useful tool for energy determination. In order to sort out the pulses of different amplitudes and to count them, electronic circuits are employed. The instrument which accomplishes this is called a 'pulse height analyser'. These analysers are either single or multiple-channel instruments.

Figure 21.4 shows a block diagram of a single-channel pulse height analyser. The output pulses from the photo-multiplier are amplified in a high input impedance low-noise pre-amplifier. Amplified pulses are fed into a linear amplifier of sufficient gain to produce output pulses in the amplitude range of 0–100 V. These pulses are then given to two discriminator circuits. A discriminator is nothing but a Schmitt trigger circuit, which can be set to reject any signal below a certain voltage. This is required for excluding scattered radiation and amplifier noise. The upper

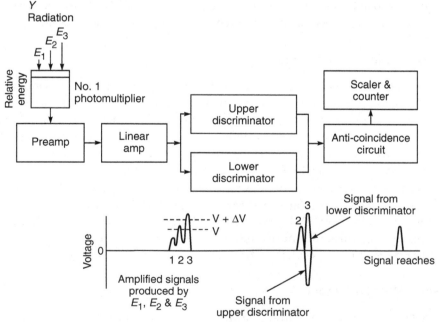

> Fig. 21.4 *Block diagram of a pulse-height analyser*

discriminator circuit rejects all but signal 3 and the lower discriminator circuit rejects signal 1 only and transmits signals 2 and 3. The two discriminator circuits give out pulses of constant amplitude. The pulses with amplitudes between the two triggering levels are counted. This difference in two levels is called the *window width*, the *channel width* or the *acceptance slit* and is analogous to monochromators in optical spectrometry.

Schmitt trigger circuits are followed by an *anti-coincidence circuit*. This circuit gives an output pulse when there is an impulse in only one of the input channels. It cancels all the pulses which trigger both the Schmitt triggers. This is accomplished by arranging the upper discriminator circuit, in such a way that its output signal is reversed in polarity and thus cancels out signal 3 in the *anti-coincidence circuit*. As a consequence, the only signal reaching the counter is the one lying in the window of the pulse height analyser. The window can be manually or automatically adjusted to cover the entire voltage range with a width of 5–10 V. Scaler and counter follow the *anti-coincidence circuit*. The scaling unit counts down the pulses from the analyser, so that they are digitally displayed. A decade system of counting is employed, which displays units, tens, hundreds time taken to record to definite number of counts, or number of counts which occur within a definite time interval.

Multi-channel pulse height analysers are often used to measure a spectrum of nuclear energies and may contain several separate channels, each of which acts as a single-channel instrument for a different voltage span or window width. The Schmitt trigger discriminators are adjusted to be triggered by pulses of successively longer amplitude. This arrangement permits the simultaneous counting and recording of an entire spectrum. A parallel array of discriminators is generally used, provided the number of channels is ten or less. If the number of channels is more than ten, the problems of stability of discrimination voltages and adequate differential non-linearity arise.

▶ 21.5 UPTAKE MONITORING EQUIPMENT

The clinical use of a radio-nuclide in medical investigations depends on obtaining a suitable distribution of the radio-nuclide in the patient. This is achieved by administering a suitable chemical substance tagged with a radio-nuclide, emitting gamma radiations. The biological system under investigation selectively assimilates the administered dose to carry out its function. Since the gamma ray gets transmitted through the body tissues, an external monitoring system can be used to detect them and provide the measurement of the chemical substance. The fraction of the chemical present in the organ at any time would indicate the functional status or what is called the uptake of the organ. The most suited gamma energy range for uptake monitoring studies is from 100 keV to 500 keV.

Figure 21.5 shows a functional block diagram of a typical gamma counting system. It basically makes use of a NaI (TI) scintillation crystal for detection of gamma rays. This is followed by the photo-multiplier which converts the scintillations into an electrical signal. The output from the photo-multiplier is given to a pre-amplifier followed by a pulse shaping circuit and a pulse height analyser. The output of the analyser drives a counter/timer which displays the information on a digital counter and a strip chart recorder.

A vital component in the system is the collimator the function of which is to exclude from the detector all gamma rays except those travelling in the preferred direction. A simple collimator

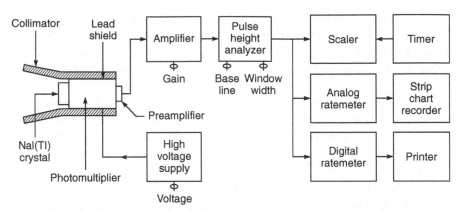

> **Fig. 21.5** *Gamma counting system for in vivo measurements*

consists of a single tapered hole in a cylindrical lead block. The sides of the lead block must be sufficiently thick to absorb the majority of the gamma rays impinging obliquely, and the detector must be surrounded by an adequate thickness of lead to effectively shield it from all gamma rays except those entering through the aperture. The shielded and collimated detector is called a probe. The probe is mounted on an adjustable support allowing it to be appropriately positioned in relation to the patient. The measurement can be carried out using the single probe or multi-probe counting system.

▐▐▶ 21.6   RADIO-ISOTOPE RECTILINEAR SCANNER

The distribution of radioactive material within an organ or part of the body is studied by using radio-isotope rectilinear scanners. The scanner is a moving detector imaging system with a block diagram shown in Fig. 21.6. Heart of the system is the detector-collimator assembly. The detector is usually a three or five inch diameter NaI crystal, situated behind a focusing collimator. This is so mounted that it can travel in a regular scanning pattern back and forth across the area of interest, so that detected and amplified signals can be plotted to give a picture or contour map of radioactivity within the organ. Usually, the detector-collimator assembly, the photo-multiplier and the pre-amplifier are housed in a single unit, which is attached to a motor-driven device. This device defines the lateral and longitudinal limits of the scan.

A single probe scanner makes use of one detector that scans the area of interest. There are dual probe scanners that have two synchronously moving, axially opposite detectors with the patient between the two detectors. The scanning can be linear or one-dimensional. In the whole body counting applications, the detector is moved continuously over the body and the counts are integrated over the entire scan.

The recording may be done either by a photographic recorder or by dot recorders. In a photographic recorder, the light flashes can be photographed on a film, from the face of a cathode ray tube. The dot recorder is most commonly used. It produces a map (Fig. 21.7) of the distribution of activity within the area of interest by recording dots or slit-like marks on paper. The dot recording mechanism consists of an electrically heated stylus to burn a small spot on a sheet of electrically

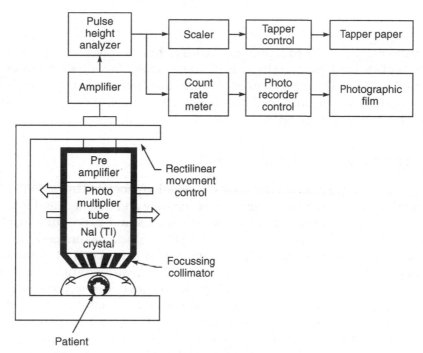

> **Fig. 21.6** *Block diagram of a typical rectilinear scanner*

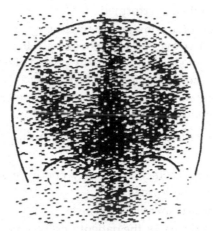

> **Fig. 21.7** *Typical scintogram using a dot recorder*

conducting paper, each time a pulse passes through the stylus. The pulses to the stylus are delivered from the pulse height analyser after scaling down the counts by an adjustable scaling factor from 1 to 256. A scaling factor of 16, for example, would mean that for every 16 counts arriving at the input of the scaling circuit from the pulse height analyser, one dot appears on the paper. This reduction in counting rate is necessary, because extremely high counting rates will drive the stylus wild. A count-rate metre is also incorporated to display or record the average count rate.

Some scanners make use of colour printing. In this technique, the maximum count rate is first established by moving the detector on the patient's body. This is then divided into six ranges, each being associated with a different colour print. As each count rate is recorded, it is allocated to one of the groups and the corresponding colour is printed. In this way, a coloured map showing the distribution of the isotope is built up.

▌▌▌▌▷ 21.7   THE GAMMA CAMERA

Gamma cameras are used to produce images of the radiation generated by radio-pharmaceuticals within a patient's body in order to examine organ anatomy and function, and to visualize bone abnormalities. The wide variety of radiopharmaceuticals and procedures used allows for evaluation of almost every organ system. In addition to producing a conventional planar image (a two-dimensional image of the three-dimensional radio-pharmaceutical distribution within a patient's body), most stationary gamma camera systems can also produce whole-body images (single head-to-toe skeletal profiles) and tomographic images (cross-sectional slices of the body acquired at various angles around the patient and displayed as a computer-reconstructed image).

The gamma camera was developed by Anger (1958). He used a large circular area of thin scintillation crystal and an array of closely packed photo-multiplier tubes to amplify and locate the gamma ray interactions in the crystal and to display the scintillations instantly on a cathode ray tube. The camera could then be used to study the rapidly changing distribution of activity, after which dynamic studies could be performed.

The gamma camera is a stationary imaging device as opposed to the rectilinear scanner in which the detector is made to move over the organ of interest. In the case of the gamma camera, the whole organ under study is viewed during the entire period of data collection. This enables fast dynamic function studies of various organs to be carried out conveniently.

Modern-day gamma cameras constitute extremely complex electronic equipment, consisting of the following functional components (Fig. 21.8).

Detector: This consists of a Collimator, crystal, photo-multiplier tubes, position localization circuitry.

Camera Electronics: This includes correction circuitry, energy analysis circuitry, counting circuit, image display and image recording device.

Briefly, the function of gamma camera is as follows:

When a photon of the radiation leaves the patient's body, it passes through the collimator and interacts with a crystal wherein its energy is converted into light. The light from the crystal is received by photo-multiplier tubes and converted into an electrical signal. The electrical signal passes through the position localization circuitry whose output consists of X and Y positional signals, and a Z or energy signal. The X, Y and Z signals are processed by special correction circuits which compensate for errors in the detection and localization of photon.

The Z or energy signal is then analysed in the pulse height analyser circuit to determine if the detected photon is within a user-specified energy range; if it is registered in the counter. The X and Y

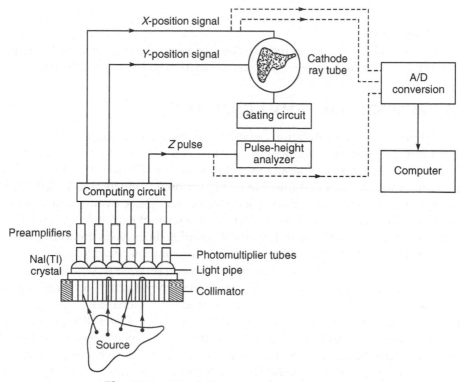

> **Fig. 21.8** *Block diagram of a gamma camera*

signals are then sent to an image recording device where they are used to position the beam of a cathode ray tube. The Z pulse then turns the beam on, causing a bright dot to appear at a location on the face of the CRT corresponding to the location in the crystal where the photon deposited its energy. This bright dot in turn exposes a film in the image recording device.

Hundreds of thousands of photons leave the patient's body and strike crystal, each causing a black spot to be formed on the film. Eventually, an image of the distribution of the radio-nuclide within the body, made up entirely of dots, is created on the film.

Since gamma photons cannot be bent by using lenses, unlike light, a collimator is used to selectively absorb unwanted radiation: only photons traveling along the desired path are allowed to pass through to the detector. The collimator is usually made up of a heavy metal absorber such as lead, with some tungsten or platinum parts. The basic types of collimators used in conventional gamma camera imaging are pinhole, parallel-hole, diverging and converging collimators.

The modern gamma camera employs a crystal of up to 500 mm diameter, typically 6.4 mm or 9.6 mm thick with an array of 61, 75 or 93 photo-multiplier tubes. The equipment has become a prime general purpose instrument for radio-nuclide imaging in routine nuclear medicine.

The collimator normally consists of a very large piece of lead with many small parallel through-holes of equal cross-section. The number of gamma rays received by any region of the crystal is directly proportional to the amount of nuclide located directly below the region. Since the gamma rays travel in all directions, only about 0.01% of the rays emitted by the labelled organ are detected and used for image formation.

A polaroid camera is mounted on the oscilloscope for photographing the build-up of about 50,000 dots on the screen. In this way, a map can be used to study distribution of activity.

21.8 MULTI-CRYSTAL GAMMA CAMERAS

A major limitation in using a scintillation camera for rapid dynamic studies is the counting losses that occur at the high count rates. The actual count rate is always greater than the registered count rate, that passes the window of the analyser. For each scintillation, the output of all photo-multipliers is summed to give a Z-signal that is proportional to the total amount of light emitted during the scintillation. The Z-pulse analyser accepts only those events which fall within the selected window. Two pulses occurring within a short span of each other are piled up and the resulting summation pulse is rejected, because its amplitude exceeds the upper window level. This results in the loss of two valid pulses, with subsequent total count loss.

The principle of operation differs slightly for multi-crystal cameras, which have a rectangular array of many small crystal detectors, each surrounded by reflective or shielding material to minimize scattering. A modified parallel-hole collimator, with bisecting lead septa within the hole of each detector, is used to increase sensitivity while maintaining overall spatial resolution. The detectors are connected to two PMTs, and as each scintillation activates only those two PMTs, and the location of the scintillation is known instantly. This type of event positioning facilitates data collection. A Lucite light-pipe array is used to couple the detectors and the PMTs, which produce a signal that is encoded into positional signals for CRT display.

With a multi-crystal data accumulation matrix, every gamma event coming from the patient and interacting in any crystal, is detected as a separate event at a unique location. If two or more events interact at the same time in different crystals, both events are discarded. This is not possible with a mono-crystal system, in which, events that occur within the scintillation decay time appear to the positioning photo-tubes as a single scintillation event and are therefore, erroneously positioned.

A computerized multi-crystal gamma camera system can accumulate high count rates (200,000 Hz), at rapid time intervals (20 s) making possible both clinical and research applications, Fig. 21.9 shows the block diagram of a typical system from M/s Baird Atomic, USA.

The detector system of this camera consists of 294 discrete crystals arranged in 14 rows and 21 columns. Each column and row is optically coupled by a lucite light guide to a photo-multiplier tube. Thus there are 35 photo-multiplier tubes. Collimation is achieved by limiting the field of view of each crystal with a single tapered hole case in lead collimators. Pulses from 35 pre-amplifiers and amplifiers serving photo-multiplier tubes pass to 35 low-discriminators, which eliminate events observed simultaneously in adjacent crystals. Pulses that are generated simultaneously in any row and column photo-tube are uniquely identified as to the location in which the interaction has taken place. *Anti-coincidence logic* rejects all pulses arising simultaneously in more than one crystal.

Since most scintillations occur in the front part of the detector, thin crystals provide better resolution by bringing the light flashes closer to the PMTs. However, the crystals allow more

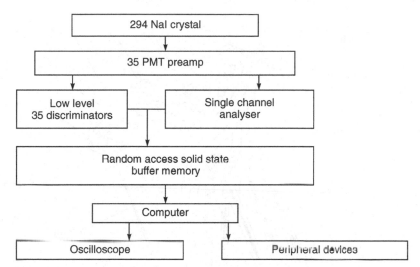

> ➤ **Fig. 21.9** *Block diagram of the multicrystal gamma camera (Courtesy: M/s Atomic Baird, USA)*

incident photons to pass through without being absorbed: therefore, the number of scintillations is reduced. The crystals of such units are 9.5 mm thick and 28 to 61 cm in diameter. Rectangular crystals are of similar thickness and range from 20 × 20 cm to 46 × 66 cm. Since sodium iodide absorbs water, a hermetically sealed aluminum housing covers the sides and front of the crystal. The back is sealed by a clear Lucite light pipe or is optically coupled directly to the face of the PMTs.

Valid events are stored in a random-access, solid-state buffer memory. Information representing data from the entire detector assembly passes into the memory of the computer. Software programs for data accumulation, correction, manipulation and display help to achieve automatic and optimum performance of the system.

The movable bed which performs a scanning motion is used to improve spatial resolution in static imaging measurements. Bed motion permits each detector crystal to scan a square area within 1.11 cm sides in 16 programmed movements of 2.78 mm each. The data observed by each of the 294 detectors at each of the 16 sites are arranged in memory as independent data points, which are utilized for image construction. Bed motion cannot be used during a dynamic study and spatial resolution may be somewhat less in these studies.

Each crystal of the detector is canned individually, has a square cross-section with 5/16 inch to a side, and is 1.5 inch long. The centre-to-centre spacing of the crystals in the array is 7/16 inch. The dimensions of the whole array are 6 × 9 inches.

The light pipe array that is required to localize each event must transmit as much light as possible in order to optimize the energy resolution of the system. The light pipe array is shown schematically in Fig. 21.10. The address of each crystal is obtained by placing two light pipes on each crystal, in such a way that one-half of the light from each scintillation event is guided down each pipe. The Y-co-ordinates are obtained by gathering the 11 rods to a 2-inch diameter

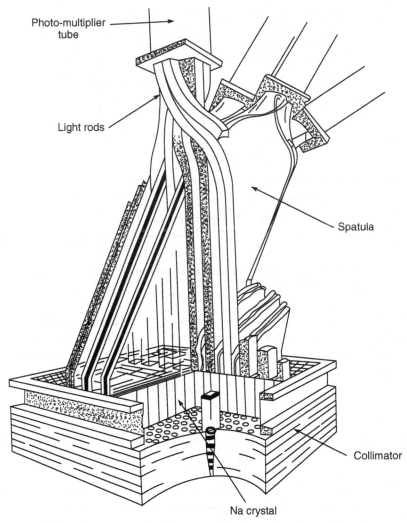

Photo-multiplier tube

Light rods

Spatula

Collimator

Na crystal

> ➤ **Fig. 21.10** *Perspective drawing of parallel hole collimator, multicrystal array, light pipe array and phototube multiplier interconnections of multicrystal camera (Courtesy: M/s Atomic Baird, USA)*

multiplier photo-tube. The Y-co-ordinates for the entire array require 14 groups of rods gathered in this manner. The X-co-ordinates are derived from 21 spatulas that are shaped and bent to fit on to the end of 2-inch diameter photo-tubes.

The main advantage of the addressing scheme is that detection and positioning are made independent. The chief disadvantage of the light-pipe scheme is that, light is attenuated in the light guides, leading to a degradation of the gamma ray energy resolution.

The principle of operation of the multi-crystal camera is illustrated in Fig. 21.11 by a simplified 2×2 crystal array. The detectors are arranged in a matrix to form orthogonal rows, A and B, and

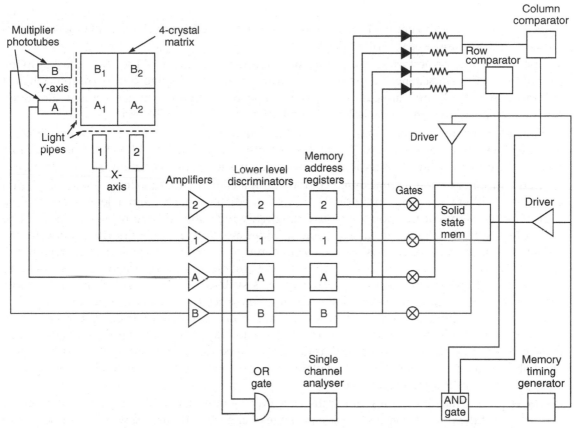

> **Fig. 21.11** *Simplified block diagram showing data accumulation principle of the multicrystal camera illustrated by a 2×2 array*

columns, 1 and 2. The light pipe array couples the detector array to row and column photo-tubes, that uniquely define the co-ordinates of the detector. If an event occurs in detector B1, the light pulse from the scintillation is directed to photo-tubes B and 1 exclusively, and their outputs are processed in amplifiers B and 1, respectively. The outputs from these two amplifiers are summed and analysed in the single-channel pulse-height analyser, and they also trigger the corresponding lower level discriminators. The discriminator outputs are led to the row and column memory address registers and these set their corresponding flip-flops, which, in turn, provide an output if only one row and one column are addressed simultaneously. An AND gate then starts the Read/Write memory cycle, if an output from the pulse-height analyser has been obtained.

The system ensures that only events that occur in one and only one detector at the correct energy range will be stored in the memory; while, numbers stored at each memory location also correspond to the number of events detected at the corresponding detector location.

The 294-crystal array can be regarded as an assembly of 294 rectilinear scanners, fixed in space with respect to each other. Whenever any one of these scanners is moved, the other 293 follow exactly the same motion. Because the detectors are spaced 11 mm apart, it is necessary for each detector to scan only an 11 mm × 11 mm area to cover the entire field of view, as all other detectors scan a corresponding area simultaneously.

The scan would be performed by a continuous motion over the 11 × 11 mm area, just as in the case of the rectilinear scanner. In practice, however, the array is displaced in 16 discrete steps of 2.78 mm in both the X and Y directions. The total number of independent image elements generated during a 16-position static study is 294 × 16 or 4704, from an area of 378 cm^2.

The modern gamma camera employs a crystal of up to 500 mm diameter, typically 6.4 mm or 9.6 mm thick, with an array of 61,75 or 93 photo-multiplier tubes, and has become the prime general purpose instrument for radio-nuclide imaging in routine nuclear medicine investigations.

Many suppliers are now marketing digital gamma cameras that perform analog-to-digital conversion, either within each individual PMT or immediately after the signal leaves the PMT. By digitizing the signal at this point, signal averaging, which affects imaging resolution, can be computer-controlled. Since digital detection provides more precise event position information, detector performance characteristics such as maximum count rate, intrinsic spatial resolution, intrinsic energy resolution, intrinsic uniformity and system sensitivity, are improved. Software control operation of digital cameras also improves system reliability and allows for remote diagnostics for servicing. New solid-state detectors constructed of cadmium zinc telluride (CZT) are under development to replace the crystal/PMT structure currently used. These solid-state CZT detectors would be used to directly convert gamma rays to electrical pulses. In addition, a curved NaI scintillation detector plate has been developed to replace 2 or 3 flat NaI detectors in single-head cameras. Because the curved plate can be positioned closer to the body, image quality and spatial resolution are improved.

The requirements for the computer system used with gamma camera are speed and stability in acquisition of the data. A comprehensive package of clinical software, in which the user's own programs can be generated and incorporated by a programming interface is required. Typical features of the hardware of modern camera computers are at least 64 Mbytes RAM, 20" Monitor with a resolution of at least 1024 × 786 for 256 colours, at least 2 Gbytes hard disk and network connection (Labmann et al, 1998).

For image output, either black and white laser printers with a resolution of at least 1200 dpi (dots per inch), colour printers (thermal transfer) or X-ray film printers are used. However, due to low film throughput compared with radiological diagnostics, the use of dry laser images is preferred in nuclear medicine in modern cameras.

The gantry and the table are operated by microprocessor-controlled motors. An open gantry design allows for the examination of standing patients or of patients in wheelchairs or hospital beds. Similarly, fully automatic or semi-automatic collimators are preferred in systems in which different radio-nuclides are frequently used.

Electrocardiographic (ECG) synchronizers are often preferred as optional equipment for gamma cameras. They are used in gated-acquisition studies to synchronize image collection with the cardiac cycle defined by ECG R wave. The beginning of the R wave triggers the ECG synchronizer to signal the start of data collection. The computer divides the interval between R waves into equal sub-divisions, usually between 16 and 32. During each cardiac cycle, data are stored in the corresponding sub-division so that a composite image of the cycle can be developed: a number of quantitative and qualitative assessments are then possible.

▥▶ 21.9 EMISSION COMPUTED TOMOGRAPHY (ECT)

Radio-nuclide tomography refers to the display of the distribution of radioactivity in a single plane or slice through the patient, just as X-ray-computed tomography attempts to display the distribution of density in a similar slice. In this technique, the three-dimensional distribution of radio-nuclide concentrations in the organ are estimated using two-dimensional projectional views acquired at many different angles about the patient. With the introduction of X-ray computed tomography and digital signal processing techniques, algorithms have become available that permit an accurate reconstruction of radio-pharmaceutical images, resulting in the development of several emission computed tomographic (ECT) systems.

For its working, ECT, depends on the measurement of an *in vivo* biochemical process, i.e. the accumulation of a radio-pharmaceutical within the body whereas transmission CT attempts to measure a physical parameter, i.e. the attenuation coefficient of X-rays. With ECT, gamma rays that are absorbed within the body before reaching the detector, or scattered within the organ and then detected, result in measurement errors that generally require compensation during reconstruction.

Emission computed tomography, provides *in vivo* three-dimensional maps of a pharmaceutical labelled with a gamma ray emitting radio-nuclide. The three-dimensional distribution of radio-nuclide concentrations are estimated from a set of two-dimensional projectional images acquired at many different angles about the patient. Several of the reconstruction algorithms are derived from the mathematical approaches used for transmission computed tomography. However, appropriate modifications have to be made to account for attenuation and photon scatter within the patient.

Emission computed tomography has developed in two complementary directions based on the type of radiation that is detected. One approach, positron emission tomography (PET), consists of the detection of annihilation coincidence radiation from positron emitter such as C-11, N-13, O-15, and F-18. When a positron (i.e., a positively charged electron) is emitted within tissue, it rapidly loses its kinetic energy in the same way that beta rays (electrons) lose their energy. The distance that the positron travels from the emission site depends on its initial energy, and typically has a range between 1 and 3 millimeters. After slowing down, the positron interacts with an electron, and both are annihilated, resulting in the emission of two 511 keV photons. To conserve momentum, the two annihilation photons are emitted in very nearly opposite directions (180°). Typically, one or more rings of discrete scintillators are used to detect the two photons (Fig. 21.12). Fast coincidence timing circuits minimize the detection of randomly occurring single events. Furthermore, collimation within the plane is not required since the emission

> **Fig. 21.12** *Principle of PET scanner (after Jaszczak, 1988)*

point essentially lies on a line determined by the two crystals that detected the two coincident photons. Collimation is usually required, however, perpendicularly to the transverse plane.

The second approach to emission computed tomography involves the detection of gamma rays emitted singly by the radio-nuclidic tracer. This approach, referred to as single photon emission computed tomography (SPECT) requires collimation within the transverse plane as well as in the perpendicular direction. SPECT uses conventional radionuclides such as Tc-99m (140 keV gamma photon) and TI-201 that are routinely used in all nuclear medicine departments.

SPECT detectors typically consist of Na(TI) scintil-lators mounted in a specially designed gantry. The system illustrated in Fig. 21.13 uses a conventional scintillation, or gamma camera that rotates about the patient to obtain a set of projectional views over 360°. These views are then used to reconstruct the regional radio-pharmaceutical concentrations within the body. Since the gamma camera obtains two-dimensional images, the entire organ of interest can be imaged with a single rotation of the camera about the patient. Although presently most SPECT systems are based on the Anger camera approach, discrete detector devices have also been developed.

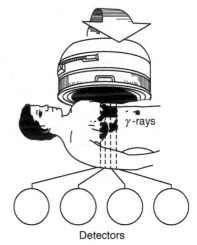

There are several techniques describing reconstruc-tion algorithms. The two broad approaches to image reconstruction consist of: 1) iterative techniques where an initial trial solution is successively

> **Fig. 21.13**   *Principle of camera based SPECT*

modified, and 2) direct analytical methods using an equation that relates the measured projections and the source distribution. There has been interest in developing iterative algorithms based on the use of prior information and the statistical nature of the measurement process [Levitan and Herman, 1987]. However, currently most commercial ECT systems use the analytical technique of back projecting the filtered projections. This technique is illustrated in Chapter 20 on computer tomography.

Although there are a few clinical applications where SPECT and PET provide similar diagnostic information, in general the applications tend to be different. These differences primarily result from the characteristics of the radio-nuclides that are used.

⑩⑩⑩▷ 21.10   SINGLE-PHOTON-EMISSION COMPUTED TOMOGRAPHY (SPECT)

Apart from some basic models and those intended only for whole-body studies, most stationary and some mobile gamma cameras can perform SPECT, a nuclear medicine technique used to create a three-dimensional representation of the distribution of an administered radio-pharmaceutical. SPECT cameras detect only radio-nuclides that produce a cascaded emission of single photons.

SPECT radio-nuclides do not require an on-site cyclotron. However, the isotopes of Tc, TI, In, and Xe are not normally found in the body. For example, it is extremely difficult to label a biologically active pharmaceutical with Tc-99m without altering its biochemical behaviour. Presently, SPECT has been used mainly in the detection of tumours and other lesions, as well as in the evaluation of myocardial function using TI-201. However, certain pharmaceuticals have been labelled with iodine and technetium and provide information on blood perfusion within the brain and the heart. Table 21.1 shows typical examples of SPECT radio-nuclides.

● Table 21.1 *Examples of SPECT Radionuclides*

Radio-nuclide	Half-life	Main-emission energy (keV)
Tc-99m	6.02	140
TI-201	73 hrs	69,71,80
I-123	13 hrs	159
In-111	2.83 days	1,71,245
Xe-133	5.25 days	81

The largest category of SPECT systems uses a single gamma camera mounted on a specialized mechanical gantry that automatically rotates the camera 360° around the patient. SPECT systems acquire data in a series of multiple projections at increments of two or more degrees. In limited-angle systems, the camera is moved a limited number of times, usually six. From the sequence of projections, an image is reconstructed by an algorithm called filtered back projection. After non-target data are mathematically filtered for each view, the reconstructed, three-dimensional image is derived from back projection, which comprises the multi-angled, two-dimensional views and projects them onto a computer monitor. The projection data are combined to produce transverse (also called axial or trans-axial) slices. Sagittal and coronal image slices can also be produced through mathematical manipulation of the data.

SPECT systems with multiple camera heads are also available (Fig. 21.14). In a dual-head system, two 180° opposed camera heads are used, and acquisition time is reduced by half with no loss in sensitivity. A triple-head SPECT system further improves sensitivity. Some suppliers also offer variable-angle dual-head systems for improved positioning during cardiac, brain and whole-body imaging. Imaging times can be decreased by using another SPECT configuration—a ring of detectors completely surrounding the patient. Although multiple camera heads reduce acquisition time, they do not significantly shorten procedure/exam time because of factors such as patient preparation and data processing.

Several approaches are being investigated to improve SPECT sensitivity and resolution. Novel acquisition geometries are being evaluated for both discrete detector and camera-based SPECT systems (Fig 21.15). The sensitivity of a SPECT system is mainly determined by the total area of the detector surface that is viewing the organ of interest. Of course, there is trade-off of sensitivity versus spatial resolution.

Kuhl [1976] recognized that the use of banks of discrete detectors (Fig. 21.15(a)) could be used to improve SPECT performance. The system (Fig. 21.15(b)) developed by Hirose et al. [1982] consists of a stationary ring of detectors. This system uses a unique fan-beam collimator that rotates in front

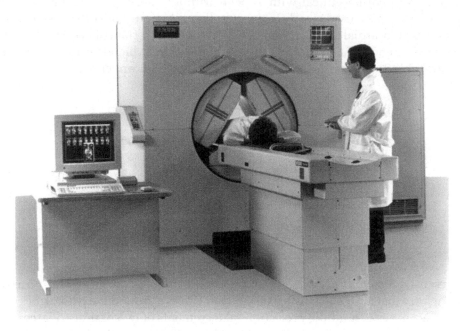

> **Fig. 21.14**   *SPECT system with multiple camera head*

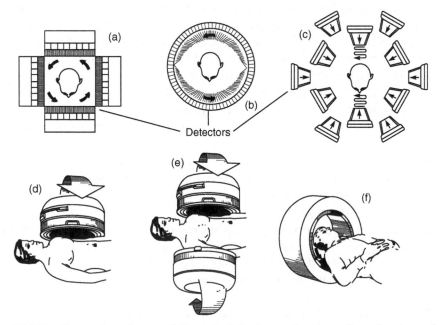

Detectors

> **Fig. 21.15**   *Examples of several discrete detector and camera-based approaches for SPECT (Jaszczak , 1988)*

of the stationary detectors. Another approach using multi-detector brain system (Fig. 21.15(c)) uses a set of 12 scintillation detectors coupled with a complex scanning motion to produce tomographic images (Moore, et al. 1984). An advantage of discrete detector SPECT systems is that they typically have a high sensitivity for a single slice of the source. However, a disadvantage has been that typically only one or at most a few non-contiguous sections could be imaged at a time. In order to overcome this deficiency, Rogers et al.[1984] described a ring system that is capable of imaging several contiguous slices simultaneously.

Camera-based approaches (Fig. 21.15(d)) for SPECT have the advantage of generating true three-dimensional images of the entire organ of interest. An obvious method to improve the sensitivity of these systems is to use more than a single camera (Fig. 21.15(e)). Lim et al. [1985] have developed a triangular configuration using three scintillation cameras. A high sensitivity system is illustrated in (Fig. 21.15(f)) which consists of an annular crystal combined with a rotation collimator. However, this device is limited to brain imaging whereas the triangular system is suitable for both brain and body imaging.

Figure 21.16 illustrates a camera-based SPECT (Jaszczak et al, 1979). A pallet, designed to minimize gamma ray attenuation, supports the patient between the two scintillation cameras. The camera separation is radially adjustable from 22 to 66 cm detector surface-to-surface. This adjustment range permits the collimators to be in close proximity to the patient for both body and brain

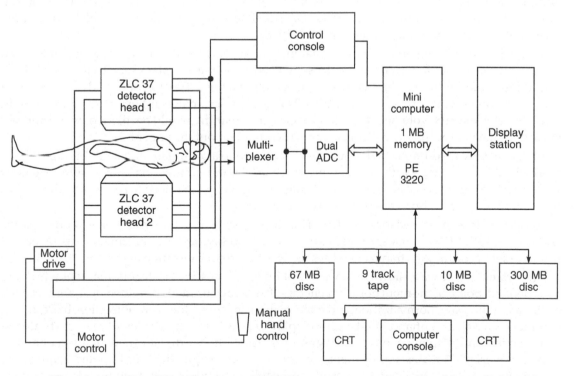

➤ **Fig. 21.16** *Simplified diagram of SPECT system consisting of dual large field-of-view scintillation cameras mounted on a rotatable gantry*

scans. The data are collected with continuous gantry motion during a 360° rotation. Acquisition times may be varied from 2 to 26 minutes. Angular samples are stored into 2° frames.

The two NaI (TI) crystals, each having a useful field-of-view of 40.6 cm, are 9.5 mm thick. Each detector crystal is optically coupled to an array of 37 photo-multiplier tubes. Detector electronics include the circuitry to compensate for positioning non-linearities and regional sensitivity variations.

During acquisition, each x-y pair of gamma ray event co-ordinates is digitized into a 128 (perpendicular to axis-of-rotation(x)) by 64 (parallel to axis-of-rotation(y)) storage array in buffer memory, together with a detector identifying bit and energy window identifying bit. Besides the primary photo-peak window, a secondary energy window is simultaneously used to record events which have undergone Compton scattering within the patient. The latter data are used to compute the body contour that is required by the attenuation compensation algorithm.

Multi-slice projection set conversion and angular framing are done in real-time by the mini-computer. The resulting projections may then be stored on disc or magnetic tape for later image reconstruction. The fast and common evaluation method for reconstruction of images in SPECT is by means of filtered back projection. However, it leads to artifacts due to its inherent properties, as a result of which pathological findings can be lost because of excessive smoothing or filtering. Iterative reconstruction helps to avoid these problems and leads to an improved evaluation of SPECT studies. However, an optimum number of iteration must be carried out and the computing time for the iteration must be kept as low as possible so that results are available in a time acceptable for clinical routine (Laβmann et al, 1998).

Parallel hole collimation is used for imaging organs such as the liver, lungs and the heart. For the brain, special fan beam collimators are used to increase system sensitivity. Using high resolution, parallel hole collimation, system sensitivity for a Tc-99 m source distributed in a 21.6 cm diameter water-filled cylinder is 54 000 (cps)/(micro-curie/m) using all axial slices. In order to determine the volume sensitivity per slice, this quantity would be divided by the number of slices that were reconstructed from a single revolution about the patient.

Image display is accomplished on a system interfaced to a computer. A 256×256 image format with 256 shades of gray with windowing and background subtraction is available. A colour monitor is also provided. The image display station is directly interfaced to a film recorder. Either transparencies or rapid-process prints may be produced.

An ECG gate is interfaced to the system. Thus it is possible to acquire multi-gated end-diastolic and end-systolic SPECT images of the heart. Coronal and sagittal sectional images are generated from the set of contiguous transverse slices using a data re-organization algorithm.

The system's software allows for a variety of image processing protocols, many of which are user-defined. Some of the more popular general software applications provided by manufacturers are image smoothing, normalization, and interpolation; image addition or subtraction; background subtraction; contrast enhancement; cyclic display of sequential images (cine) region-of-interest construction and display; curve or histogram construction and display; and creation of alpha-numeric overlays. Cardiac applications include first-pass acquisition; multi-gated acquisition; automatic edge detection: determination of end systolic and end diastolic volumes, stroke volume, cardiac output, global ejection fraction and regional ejection fraction.

▥▶ 21.11 POSITRON EMISSION TOMOGRAPHY (PET SCANNER)

Positron emission tomography is an imaging modality for obtaining *in vivo* cross-sectional images of positron-emitting isotopes that demonstrate biological function, physiology or pathology. In this technique, a chemical compound with the desired biological activity is labelled with a radioactive isotope that decays by emitting a positron, or positive electrons. The emitted positron almost immediately combines with an electron and the two are mutually annihilated with the emission of two gamma rays. The two gamma ray photons travel in almost opposite directions, penetrate the surrounding tissue and are recorded outside the subject by a circular array of detectors (Fig. 21.17). A mathematical algorithm applied by computer rapidly reconstructs the spatial distribution of the radioactivity within the subject for a selected plane and displays the resulting image on the monitor. Thus, PET provides a non-invasive regional assessment of many biochemical processes that are essential to the functioning of the organ being visualized.

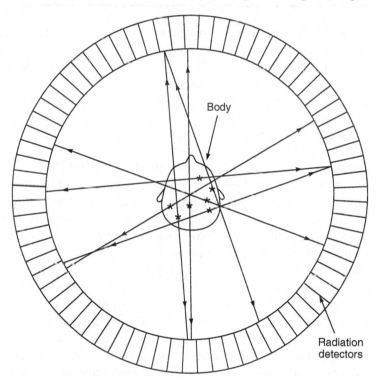

> **Fig. 21.17** *Principle of positron emission tomography (PET Scanner)*

The positron(β^+) is emitted from a proton-rich nucleus with a variable amount of kinetic energy, the maximum amount being the endpoint energy ($E\beta^+$), given for various isotopes in Table 21.2.

This energy is dissipated in the patient over a range of tissue of the order of a few millimeters. The β^+ combines with a free electron (β^-) and the masses are transmuted to two 511-keV γ rays which are emitted at $180° \pm 0.25°$ to one another to satisfy conservation of momentum as in

• Table 21.2   *Positron Emitters Commonly Used in PET*

Isotope	$E\beta^+$, (MeV)	$T^{1/2}$ (min)
^{15}O	1.74	2.07 m
^{11}C	0.96	20.39 m
^{13}N	1.19	9.96 m
^{18}F	0.65	109.77 m
^{38}K	2.68	7.64 m
^{68}Ga	1.90	68.1 m
^{82}Rb	3.35	1.27 m
^{63}Zn	2.32	38.1 m

Fig. 21.18. The variable finite range of the β^+ as well as the angular variation of about 180° are fundamental limitations to the resolution achievable with PET.

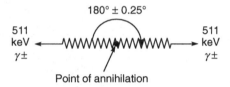

➤ **Fig. 21.18**   *The basic decay process for a positron emitter*

The compounds used and quantitated are labelled with proton-rich positron (β^+) emitters that are usually cyclotron-produced. The principal isotopes are ^{11}C, ^{13}N, ^{15}O, and ^{18}F. If the compound of interest is labelled in a known position and it maintains this positron, a PET scan permits measurement of the positron concentration ($\mu Ci/mL$) in a small-volume element within an organ or region of interest. This metabolic volume is typically 1 cm^3.

A variation of 0.25° in an angular distribution of the back-to-back 511 keV γ rays will cause a degradation of 1.75 mm at the centre of an 85-cm whole-body tomograph. Neither of the basic limitations will cause a significant loss in resolution in present-day PET designs. They are simply fundamental to the method and cannot be eliminated.

Two design types of positron-emission tomographs have been introduced, one employing opposed large-area detectors which require rotation around the patient to provide the necessary degree of angular sampling, and the other, employing multiple individual crystal detectors surrounding the patient in a circular or hexagonal array. Conventional lead absorption collimators are not required because the coincident detection of two 511 keV photons indicates the line of origin along which the photons were emitted. However, in order to reduce the random coincidence count rate, some degree of collimation is normally employed. Pulse processing needs to be much faster than with single-photon systems, to keep random coincidences to manageable proportions. With fast-response detectors and suitably fast electronics, it may be possible to use the difference in the time of arrival of the annihilation photons at opposite detectors to locate the site of positron decay and improve spatial resolution.

Figure 21.19 illustrates gantry and detector components used by Hoffman et al (1985) in a PET system. The gantry has a large opening (diameter = 65 cm) and can image both the brain and torso of an adult patient. The entire detector assembly may be tilted to obtain oblique sections. Bismuth germanate (BGO) scintillation crystals, 5.6 mm wide, 30 mm high, and 30 mm deep, are used to

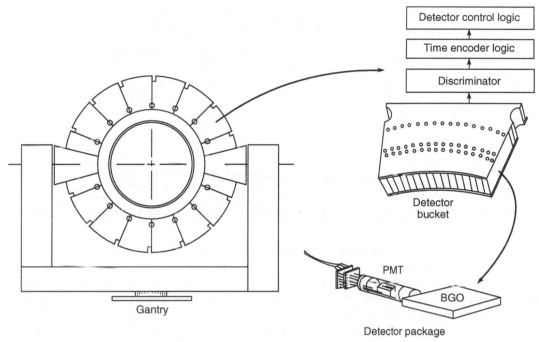

> **Fig. 21.19** *Gantry and detector modules used in PET Scanner (after Hoffman et al. 1985)*

detect the 511 keV annihilation radiation. The detectors are arranged in a circular ring geometry, with 512 detectors per ring. The system has two rings and produces three scanning planes (two direct and one cross plane). In order to facilitate replacement, the detectors are arranged in modules or buckets containing 16 detector packages. Each package contains two crystals and two PMTs. The centre-to-centre spacing of the crystals is 6.1 mm. Axially, the two rings are separated by 36 mm. Besides containing the two BGO crystals and PMTs, the bucket also contains amplifiers/discriminators and other front-end processing electronics. In order to increase linear sampling, the entire detector assembly can wobble in a small circular orbit. This wobbling procedure is used to optimize spatial resolution (Jaszczak, 1988).

The original PET scanners were constructed using a thallium-doped sodium iodide [(NaI (TI)] detector. Its high efficiency at 511 keV, ease of fabrication, and low cost made it an obvious choice in a number of initial designs utilizing discrete crystals. Its principal disadvantage in PET work was the decreasing detection efficiency caused by the trend toward smaller crystals required for high resolution while maintaining a reasonably high total system efficiency. This limitation places a practical lower limit on the resolution that is attainable with NaI (TI)-based detection systems. With the development of a new scintillator material, bismuth germanate (BGO) ($X = 79$, $p = 7.13$ g/mL), with three times the stopping power of NaI(TI), a new generation of high-resolution, high-efficiency PET scanners has become possible.

A simplified block diagram of the data acquisition system is shown in Fig. 21.20 Distributed processors are used throughout the system to maximize speed for simultaneous data collection.

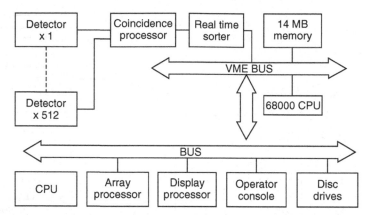

> ➤ **Fig. 21.20**   *Data acquisition system for a PET Scanner (after Hoffman et. al. 1985)*

Individual and analog detector signals are amplified and the time of interaction is then determined using a constant fraction discriminator. A time encoder converts the event into a 14 bit word containing the detector number and event time within 8 ns. This word is passed to the coincidence processor every 224 ns. The energy window is controlled automatically by the microprocessor located in each detector bucket. A threshold of 200 keV is typically used to allow for detection of gamma rays that have been scattered within a detector and escaped. The system consists of a fan beam geometry with an angular sampling of 0.7 degrees. The linear sampling is 2.9 mm.

The main processor serves to monitor and control the various processing jobs. An array processor is used to perform the primary reconstruction. Several peripheral devices, including a display processor, are attached to the system computer.

The GE 4096 Positron Emission Tomography Camera System (Fig. 21.21) is a high resolution Whole Body PET Scanner. It uses a detector ring with a diameter of 101 cm. The 4096 individual crystals of the scanner are arranged in eight rings of 512 crystals each. Onto each set of 16 crystals, two dual photo-multiplier tubes are attached providing increased positional sampling. Each crystal is made up of Bismuth Germanate ($Bi_4Ge_3O_{12}$) and is 6 mm trans-axial, 12 mm axial and 30 mm radial in size. 64 individual detector cassettes allow for easy and fast servicing.

The system provides for interleaved imaging of 30 slices in a single acquisition interval either in wobble or stationary modes. The patient port is 57 cm which allows for large patient scanning. Patient positioning is accomplished with convenient operator controls on each side of the gantry. Enhanced patient positioning is provided via +20° gantry pivot and tilt. Accurate and reproducible patient positioning is accomplished with a triple laser positioning system. The contoured, carbon fibre imaging table minimizes attenuation and additional padding allows for maximum patient comfort.

Horizontal or axial table positioning can be controlled manually from the gantry controls or by computer from the operator's console. The axial range of the table is 170 cm, the height is adjustable from 60 to 120 cm, and the maximum weight the table will support is 300 Ibs. Two, 6 mm diameter, pin-shaped[68] Ge sources (2 mCi and 10 mCi) extending over the entire axial field of view are provided for transmission measurements, adjustment of gain and coincidence timing and normalization of detector efficiencies. The pin-source is continuously sampled providing accurate

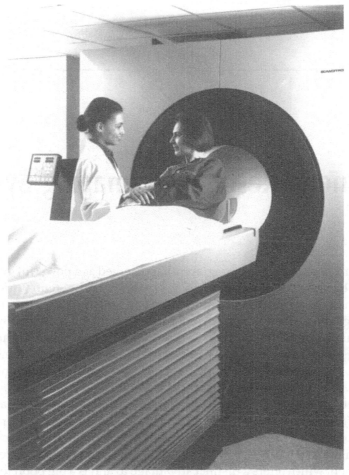

> **Fig. 21.21** *PET scanner from M/s GE Electrical, USA*

measurement of random and scattered events. The pin-sources are removable and are stored in a lead container away from the gantry when not in use.

The DAP (Data Acquisition Processor) contains a 68030 Processor and dual Intel i960 RISC Processors. The DAP controls the real-time acquisition of data for the system.

Radio-nuclide images are inherently very noisy and in comparison to most other types of images, they are of very poor quality. The radiation emitted by the object is X or gamma radiation, arising either directly from nuclear transitions or indirectly via positron emission or electron capture. In practice, surprisingly, little of radiation is used in the formation of the image, typical photon density is of the order of 10^3 cm^{-2} as compared to about 10^7 cm^{-2} in radiography and 10^{12} cm^{-2} in conventional photography. Statistical fluctuations in photon density, inherent in the radioactive decay process, are usually apparent and spatial resolution is at best currently just under 1 cm in most clinical applications.

Magnetic Resonance Imaging System

Nuclear magnetic resonance (NMR) tomography has emerged as a powerful imaging technique in the medical field because of its high resolution capability and potential for chemical specific imaging. Although similar to the X-ray computerized tomography (CT), it uses magnetic fields and radio frequency signals to obtain anatomical information about the human body as cross-sectional images in any desired direction and can easily discriminate between healthy and diseased tissue. NMR images are essentially a map of the distribution density of hydrogen nuclei and parameters reflecting their motion, in cellular water and lipids. The total avoidance of ionizing radiation, its lack of known hazards and the penetration of bone and air without attenuation make it a particularly attractive non-invasive imaging technique. CT provides details about the bone and tissue structure of an organ whereas NMR highlights the liquid-like areas on those organs and can also be used to detect flowing liquids, like blood. A conventional X-ray scanner can produce an image only at right angles to the axis of the body, whereas the NMR scanner can produce any desired cross-section, which offers a distinct advantage to and is a big boon for the radiologist.

22.1 PRINCIPLES OF NMR IMAGING SYSTEMS

Magnetic Moment: All materials contain nuclei that are either protons or neutrons or a combination of both (Show, 1971). Nuclei containing an odd number of protons or neutrons or both in combination, possess a nuclear 'spin' and a magnetic moment which has both magnitude and direction. In body tissue or any other specimen, the magnetic moments of the nuclei making up the tissue are randomly aligned (Fig. 22.1) and have zero net magnetization ($M = 0$).

When a material is placed in a magnetic field B_0, some of the randomly oriented nuclei experience an external magnetic torque which tends to

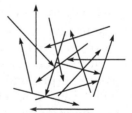

> **Fig. 22.1** *Random alignment of magnetic moments of the nuclei making up the tissue, resulting in a zero net magnetisation.*

align the individual parallel or anti-parallel magnetic moments to the direction of an applied magnetic field. There is a slight excess of nuclei aligned parallel with the magnetic field and this gives the tissue a net magnetic moment M_0. It is this differential in a magnetic moment that accounts for the nuclear magnetic resonance signal on which the imaging is based. With the magnetic moments being randomly oriented with respect to one another, the components in the X-Y plane cancel one another out while the Z components along the direction of the applied magnetic field add up to produce this magnetic moment M_0 (Fig. 22.2).

According to the electromagnetic theory, any nucleus such as a hydrogen proton which possesses a magnetic moment attempts to align itself with the magnetic field in which it is placed. This results in a precession (Fig. 22.3) or wobbling of the magnetic moment about the applied magnetic field with a resonant angular frequency, ω_0 (called the Larmor frequency) are determined by a constant γ (the magnetogyric ratio) and the strength of the applied magnetic field B_0. Each nuclide possesses a characteristic value for γ but ω_0 and B_0 are related as follows:

$$\omega_0 = \gamma B_0$$

Table 22.1 shows the NMR parameters of nuclide of interest in medicine.

Another important phenomenon of NMR is that the applied external magnetic field creates an energy absorption state from a statistical point of view. When a nucleus with a magnetic moment is placed in a magnetic field, the energy

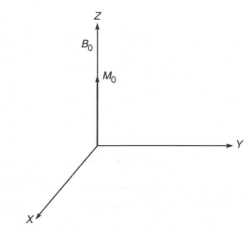

> **Fig. 22.2** *The application of external magnetic field causes the nuclear magnetic moments to align themselves, producing a net moment in the direction of the field B_0*

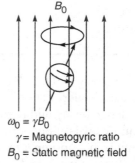

$$\omega_0 = \gamma B_0$$
γ = Magnetogyric ratio
B_0 = Static magnetic field

> **Fig. 22.3** *Precessing or wobbling of the nucleus about an applied magnetic field, with a resonant angular frequency ω_0*

of the nucleus is split into lower (moment parallel with the field) and higher (anti-parallel) energy levels. The energy difference is such that a proton with specific frequency (energy) is necessary to excite a nucleus from the lower to the higher state. The excitation energy E is given by the Planck's equation

$$E = h\omega_0$$

where h is Planck's constant divided by 2π. This energy is usually supplied by an RF magnetic field.

The spinning charged particles could be electrons, either single or unpaired, or charged nuclei such as the proton of ionized hydrogen. The ratio of excited particle to particle at rest and other

• Table 22.1   *NMR Parameters of Interest in Medicine*

Nuclei	Nuclear spin	$\gamma/2\pi$ (MHz/kg)
^{1}H	1/2	4.26
^{2}H	1	0.65
13C	1/2	1.07
15N	1	0.31
19F	1/2	4.01
31P	1/2	1.72

properties of particular nuclei determine the NMR sensitivity. Table 22.2 gives NMR frequencies and other characteristics of some of the common biological isotopes.

• Table 22.2   *Characteristics of Common Biological Isotopes*

Element	% of body weight	Isotope	NMR frequency MHz/T
Hydrogen	10	^{1}H	42.57
Carbon	18	^{13}C	10.70
Nitrogen	3.4	^{14}N	03.08
Sodium	0.18	^{23}Na	11.26
Phosphorus	1.2	^{31}P	17.24

Free Induction Decay (FID): In NMR, at room temperature, there are more protons in a low energy state than in a high energy state. The excited proton tends to return or relax to its low-energy state with spontaneous decay and re-emissions of energy at a later time 't' in the form of radio wave photons. This decay is exponential in nature and produces a "free induction decay" (FID) signal (Fig. 22.4) that is the fundamental form of the nuclear signal obtainable from an NMR system. To summarize, if in a static field, RF waves of the right frequency are passed through the sample of interest (or tissue), some of the parallel protons will absorb energy and be stimulated or excited to

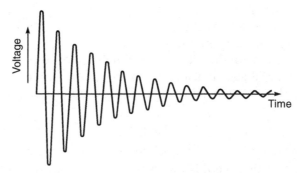

➤ **Fig. 22.4**   *Free induction decay (FID) signal obtained in NMR experiments*

a higher energy in the anti-parallel direction. Some time later, the RF frequency absorbed will be emitted as electromagnetic energy of the same frequency as the RF source. The amount of energy required to flip protons from the parallel to the anti-parallel orientation is directly related to the magnetic field strength; stronger fields require more energy or higher frequency radiation.

Excitation: If the material or tissue is now subjected to another magnetic field, say a bar magnet placed along the Y-axis, this would cause net magnetization to shift slightly from the Z-axis (B_0 magnetic field direction), through an angle x (Fig. 22.5). An alternative technique to accomplish the same result would be to apply an RF pulse at the resonant frequency of the protons in the tissue. The angle of rotation α depends on the amplitude but primarily on the length of the applied radio-frequency pulse ($\propto = KT$, where T is the pulse length in seconds and K is constant). An RF pulse of sufficient duration and power to rotate Mo through 90° is referred to as 90° RF pulse (Fig. 22.6(a)). In addition, the net magnetic moment, M now precesses with the same characteristic frequency ω_o, because the individual magnetic moments causing this shift in M are now all in phase with the applied radio waves.

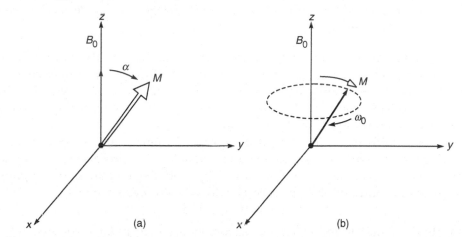

x (a) x (b)

➤ **Fig. 22.5(a)** *The magnetic moment is flipped from its equilibrium by the application of another magnetic field*

     (b) *It then precesses about the external field direction at a high angular frequency which is proportional to the field strength*

It is possible to shift the net magnetization M to any desired angle of deflection by applying the resonant frequency pulse for the appropriate amount of time. M can even be completely inverted and the corresponding pulse is called 180° RF pulse (Fig. 22.6(b)).

Emission: When the RF pulse is turned off, the net tissue magnetization begins to swing back towards the Z-axis (direction of B_o), inducing an NMR signal in the receiver coil placed perpendicular to the moving magnetic vector. In fact, individual magnetic moments begin to de-phase with one another, some apparently rotating a bit faster than the resonant frequency ω_0, and the others a bit slower. The component of M in the X-Y plane disappears as the individual magnetic

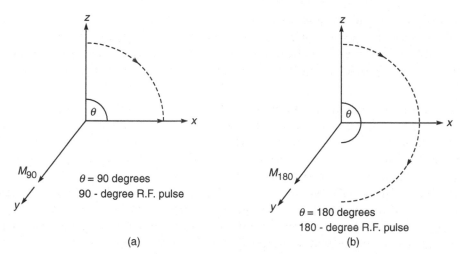

> ➤ **Fig. 22.6** *The radio frequency pulse needed to tip the vector of the net magnetic moment M through an angle of (a) 90°, (b) 180° is called 90 and 180 degree pulse respectively*

moment components on that plane cancel out each other (Fig. 22.7(a)). The amplitude (A) of the initial signal received by the coil is proportional to the magnitude of the component of M in the X-Y plane (M_{xy}) as well as various geometric factors. The amplitude of this signal decays in an exponential fashion with time t, i.e., $A = A_0 e^{-t/T2}$ where T_2 is the characteristic or average decay time for the process and I/T_2 is the decay constant.

Simultaneously, with the de-phasing decay process, there is also relaxation of the Mz component (Fig. 22.7(b)) to the pre-excitation or rest state M_0. This process is also exponential in nature with an average decay time T_1. If T_1 is sufficiently short, the resultant signal amplitude with both decay processes contributing to the observed decay of the signal would be:

$$A = A_o e^{-t/T1} \cdot e^{-t/T2}$$

It may be observed that the following two relaxation mechanisms are associated with excited nuclear spins:

(i) Relaxation time T_1 is referred to as the *spin-lattice* relaxation process as it characterizes the time for the perturbed nuclei to re-align themselves with the existing lattice structure of the host material. This is also called *longitudinal* relaxation as it is the time constant that describes the recovery of the Z component of M to its equilibrium value M_o which is along the direction of the applied magnetic field.

(ii) Relaxation time T_2 is called *spin-spin* relaxation as it indicates the time required for perturbed, in-phase spins to de-phase with respect to each other. It is also called the *transverse* relaxation process as it is relation to the decay of the component of M in the X-Y plane which is conventionally perpendicular to the Z-axis or the direction of the applied magnetic field B_0.

Transverse relaxation is faster than longitudinal relaxation so that the spin-spin relaxation time constant T_2 is always smaller than the spin-lattice relaxation time constant T_1. It is interesting

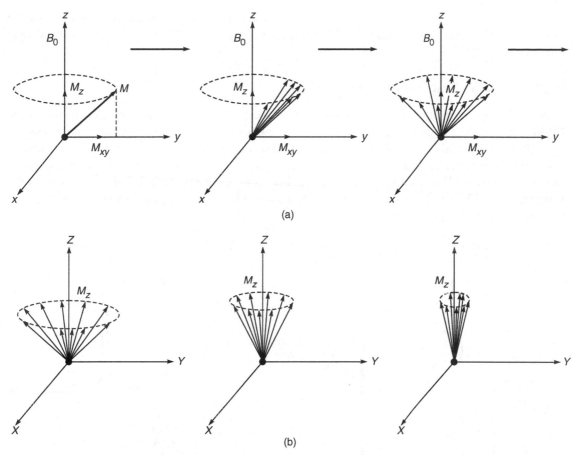

➤ **Fig. 22.7** *The decay of the magnetization*
    (a) *The signal decays with time constant T_2 to give X-Y component*
    (b) *Recovery of the magnetization to its equilibrium position parallel to the X-axis with a time constant T_1*

to note that both relaxation times (T_1 and T_2) are sensitive to the molecular structure and environments surrounding the nuclei. For example, the value of T_2 tends to increase with increased nuclear motion from micro-seconds in solids to seconds in liquids. In biological systems, typical hydrogen (^{1}H) T_2s are 0.04 to 2 s.

The value of T_1 decreases as the amount of motion that has spectral components, near the resonant frequency increases. It can vary from milli-seconds in liquids to months in solids. The range for biological tissue is typically 0.05 to 3 seconds for hydrogen (^{1}H).

Relaxation processes play a crucial role in NMR. In imaging, variations in the relaxation times among different biological tissue types provide the key contrast mechanism for anatomical discrimination. In a diseased state, the differences in relaxation times relative to the normal values can be greater than 100%, thereby providing a powerful mechanism for the detection of pathology.

Imaging capabilities of these two important parameters (T_1 and T_2) together with the proton densities of the objects thus make NMR imaging a unique, versatile and powerful technique in medical imaging.

22.1.1 Fourier Transformation of the FID

In NMR imaging systems, we are interested in sorting out what frequencies are present in the NMR spectrum and in determining the intensity of each' of the frequencies present. The FID signal is in the time domain, i.e., the magnitude is measured as a function of time and needs to be converted to frequency domain function. For this, we use Fourier transformation and determine the characteristic NMR frequencies and intensities (the NMR spectrum) that are present in the time response of a nuclear spin system subjected to an RF pulse. Figure 22.8 shows the Fourier transformation of the FID signal.

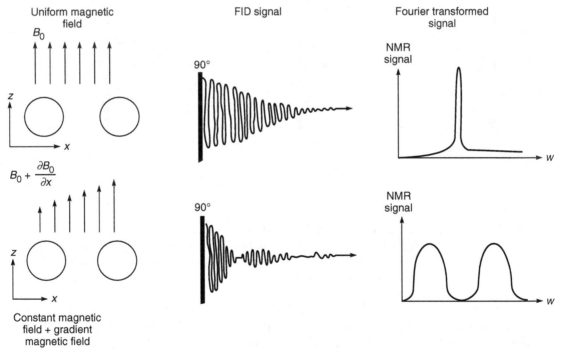

> **Fig. 22.8** *Fourier transformation of the FID signal*

The Fourier transformation $f(\omega)$ of a function of time $f(t)$ is given by

$$f(\omega) = \int_{-\infty}^{\infty} f(t)e^{-iwt}\ dt$$

where ω is the angular frequency ($\omega = 2\pi v$) and $i = \sqrt{-1}$

If $f(t)$ is the free induction decay following a pulse, $f(\omega)$ represents the ideal spectrum. In principle, instead of measuring the spectrum by conventional continuous wave methods, (which take a long time), FID can be recorded (usually in a second or so) and the same information obtained much more conveniently.

For nuclei other than protons, the FT method is especially valuable, since signals are generally very weak, a large number of repetitive scans would be needed and extremely long times are required for a conventional scan. With the FT method, the time needed for data acquisition is independent of the spectral range.

The time during which data are acquired for each FID is very important, since it may govern the resolution obtained in the Fourier transformed spectrum. The FT theory shows that an FID signal sampled for T seconds leads to lines of width $1/T$ Hz in the transformed spectrum.

However, a resolution better than that dictated by T_2 cannot be obtained by long acquisition times as an early truncation of the FID signal can result in line broadening. Another important factor which must be considered is the deterioration in signal/noise ratio as sampling time is increased, and the FID signal decays into noise. The best compromise is to acquire data for about $3T_2$ from the stand-point of resolution as well as from signal/noise ratio stand point.

Fourier transformation in NMR imaging is carried out with a small on-line computer that not only acquires data and carries out the Fourier transformation but also controls the timing of pulse widths, repetition rates and choice of various pulse sequences. The size of the computer memory for FT function is the product of the number of the data points taken per second and the time during which data are acquired for each FID. A 16K memory is adequate for most FT NMR systems.

The fact that medical NMR images can be most easily generated from the resonance of hydrogen nuclei is fortunate, because the human body is 75% water, each molecule of which has two hydrogen nuclei. Moreover, the distribution of water, along with that of various other small, hydrogen-rich molecules (for example lipids) is known to be altered by many disease states. These are of the order of 10^{22} hydrogen nuclei per gram of tissue (Bottomley, 1983) to provide a respectable population base from which the feeble NMR signal can be received. Other nuclei, such as phosphorus (^{31}P), carbon (^{13}C), fluorine (^{19}F), sodium (^{23}Na), nitrogen (^{14}N) and oxygen (^{17}O) are orders of magnitude less abundant in the body and have weaker magnetic moments.

There are two methods of measuring the NMR signal. One is the continuous wave method, in which spin responses are recorded by sweeping the magnetic field strength or exciting frequency near the resonant point by continuous radiation of electromagnetic waves for excitation. The other is the pulse method, in which the spin response is recorded after impressing radio waves that are pulse-modulated and have frequencies near the resonant frequency. The spectrum data of the frequency domain is applied on the former and the free induction decay signal of the time domain is applied in the latter. The pulse method has several advantages, such as obtaining the spectrum of wide frequency domain at one time and of the possibility of measuring parameters T_1 and T_2 directly which describe the time response of the spin system. Virtually, all NMR imaging methods use the pulse method.

22.1.2 The Bloch Equation

The Bloch equation gives a phenomenological description of the time dependence of nuclear magnetization $M(t)$ in the presence of an applied magnetic field $B(t)$. We can consider the NMR sample as a black box with $B(t)$ as the input signal or stimulus and $M(t)$ as the output signal or response. The black box is characterized by M_0, T_1 and T_2, and its behaviour is governed by the Bloch equation as follows:

$$\frac{dM}{dt} = rM_x B - \left[\frac{(M_x i + M_r j)}{T_2}\right] - \left[\frac{(M_z - M_o)}{T_1} \cdot K\right]$$

where γ is the gyromagnetic ratio and is a physical property of the nucleus of the atom. Different elements and, in fact, different isotopes of the same element exhibit significantly different gyromagnetic ratios. The gyromagnetic ratio of protons is 4.26×10^7 Hz. T^{-1} or 2.68×10^8 rad $S^{-1} T^{-1}$.

In imaging experiments, $B(t)$ consists of a static homogeneous field B_0 and orthogonal RF field $B(t)$ and a linear magnetic field gradient B_g (x, y, z). During RF pulses, $B_g(x, y, z)$ is usually negligible and can be omitted whereas between the pulses, it is $B(t)$ that is zero. The Bloch equation, therefore, can have two forms, one with a time-dependent B and the other with a time-independent B.

The co-ordinate system used in the Bloch equation is the laboratory or fixed reference frame. The k direction is taken to be parallel to B_0 where B_o is the field of the large magnet (static magnet). By convention, k is the longitudinal direction and i and j define the transverse plane.

The constant T_1 (spin-lattice time) governs the evolution of M_z towards its equilibrium value M_0 and involves the dissipation of energy from the collection of nuclei, the 'spin system' to the atomic and molecular environment of the nucleus, 'the lattice'.

The constant T_2 (spin-spin relaxation time) governs the evolution of the magnitude of the transverse magnetization, $M_x i + M_y j$, towards its equilibrium value of zero. The process can be thought of as the transverse re-orientation of the individual spins becoming de-phased, so that the sum of the transverse components of the fields of the nuclei in the collection goes to zero.

The equilibrium magnetization $M_0 k$ is the nuclear magnetization of the sample maintained at the static field for a time long compared to T_i.

Although the validity of the Bloch equation is limited, it nevertheless accurately describes the NMR imaging principle and its solution under various conditions serves as a basic 'tool kit' in understanding the magnetization behaviour of nuclei. (Hinshaw and Lent, 1983).

▸ 22.2 IMAGE RECONSTRUCTION TECHNIQUES

Over the past two decades, a variety of techniques have been proposed and demonstrated that enable the spatial discrimination and mapping of nuclear magnetic resonance (NMR) signals in heterogeneous objects. As explained earlier, for the production of images with NMR, it is necessary to differentiate between the contributions to the NMR signal originating from different regions of the sample and for this purpose, a magnetic field gradient is super-imposed on the homogeneous main magnetic field with the aid of current carrying auxiliary coils. The resonance frequency in planes perpendicular to the direction of the gradient is thus constant. If the spectrum of the NMR signals measured is analysed, then the spectral amplitude of a particular frequency corresponds to the signal contribution of all nuclear spins in an associated plane perpendicular to the applied field gradient. The entire NMR spectrum thus represents the projection of the nuclear spin density onto the direction of the field gradient.

An exciting aspect of NMR is its capability to produce medical images based on four separate characteristics of tissue. For instance, one NMR image can be constructed based on the proton density of the tissue; another can be created representing the T_1 relaxation time distribution of the same section; in a similar fashion, a third image of the section can be generated representing T_2 relaxation time distribution and yet another NMR image sensitive to flow can be reconstructed. Although, some NMR imaging equipment is capable of producing calculated T_1 and T_2 images, in actual practice, most NMR images today represent an amalgamation of the four parameters, which are combined to best depict the anatomy, or pathology of the organ or area of interest.

Other than in X-ray computed tomography, which enables only image reconstruction from projections, there are alternative scanning procedures with NMR imaging. Basically, the total imaging volume is divided into a grid matrix of rectangular cells called voxels (volume elements) along the three spatial co-ordinates. The scan is usually viewed on a cathode-ray tube represented as a two-dimensional image. The basic element usually used on most NMR scanner systems is the pixel or picture element which is an arbitrarily defined end-surface.

Several schemes have been used for the classification of the various NMR imaging methods. Figure 22.9 illustrates some of these techniques. The co-ordinate system used is such that the Z-axis is coincidental with the cylindrical axis of the sample tube.

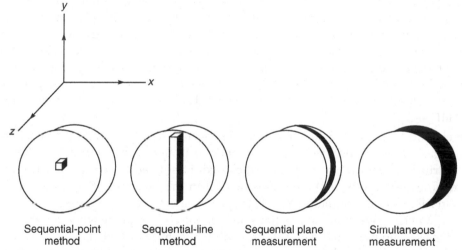

Sequential-point method Sequential-line method Sequential plane measurement Simultaneous measurement

> **Fig. 22.9** *Classification of various NMR imaging methods (i) sequential-point method (ii) sequential-line method, (iii) sequential plane measurement, (iv) simultaneous measurement. In all the methods, the Z-axis is coincident with the cylindrical axis of the sample tube*

Sequential Point Method: If the sample volume is divided into n_x, n_y and n_z volume elements (voxels) along the three respective cartesian axes, then $n_x n_y n_z$ independently measured values of the NMR signal are required to completely reconstruct the image within this volume. The method, therefore, has been termed the sequential point method.

It is observed that a continuous resonance signal could be obtained by subjecting a sample to a continuous string of intense RF pulses. The RF pulses must be near the Larmor frequency and the inter-pulse interval is short compared to the relaxation times of the sample. This pulse technique is called 'Steady State Free Precession' (SSFP). The SSFP technique produces a relatively large and continuous resonance signal and this is ideal for the sensitive-point application. Hinshaw (1976) explained the sensitive point method that requires no moving coils or gantries for moving the object. Spatial localization in these dimensions is achieved by the application of three orthogonal time-dependent linear gradient magnetic fields in the presence of a continuous string of closely spaced phase-alternated RF pulses.

The position of the sensitive point is scanned across a sample by changing the ratio of currents in each half of the appropriate gradient coil set and its size varied by altering the RF pulse repetition period or gradient strength. The NMR signal intensity is a complex function of nuclear spin density, and relaxation times T_1 and T_2.

The sensitive point method of image formation has distinct advantages. The demands made upon the uniformity of the magnetic field are much less. Also, this method is simple and direct as the inherently inexact and time-consuming computational process of image reconstruction is not required.

Sequential Line Method: In the sequential line method, volume elements along an entire selected line, for example in the Y-direction, are simultaneously observed, reducing the total number of experiments required for the complete image to $N = n_x n_z$. Thus, the method utilizes two orthogonal time-dependent magnetic field gradients, an SFP pulse sequence and signal averaging to spatially localize the NMR receiver sensitivity to a line as in the sensitive point method. The distribution of spins along the line is determined by applying a third linear magnetic field gradient in the direction of the line that is time-dependent. The NMR frequency would thus directly correspond to the spin density distribution along the line and can be obtained by the Fourier transformation of the time-averaged FID signal.

Mansfield and Morris (1982) demonstrated a line scan method which uses the irradiation cycle shown in Fig. 22.10. It has three intervals $2t_z$, t_y and t_x during which Z, Y and X gradient fields are applied. During t_z a plane perpendicular to the Z-axis is isolated by selectively saturating all the sample (magnetization of spins) lying outside of it in the presence of G_z. This is achieved by applying an RF pulse so modulated that its frequency spectrum selectively excites spins whose Larmor frequencies correspond to positions outside the chosen plane when the Z gradient G_z is applied. In the second interval t_y, a line of spins within the undisturbed plane is selectively excited by a $\pi/2$ RF pulse applied in the presence of G_y gradient. The FID resulting from excitation of the selected line is observed in the interval t_x during which the X-gradient G_x is applied. The Fourier transformation of the FID yields the NMR signal distribution along the line in the X-direction. Successive strips are scanned electronically by changing the excitation frequency of the $\pi/2$ RF pulses by computer control.

Sequential Plane Methods: Sequential plane methods permit the simultaneous observation of an entire imaging plane and for that matter $N = n_z$. The simultaneous observation and resolution of an entire plane of points requires a dynamic range capable of $n_x n_y$ pixels which places stringent demands on instrumentation and bandwidth. A number of schemes have been devised to circumvent this problem. The commonly used sequential plane methods are as follows:

Back Projection Zeugmatography: Zeugmatography, as applied to NMR imaging was described by Lauterbur (1973). This is based on the fact that the two-dimensional spatial variation or image

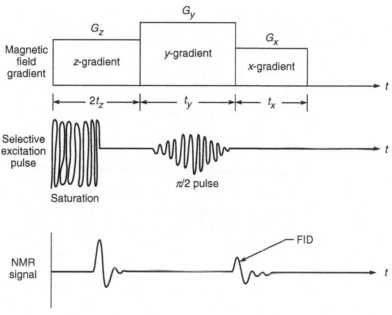

> **Fig. 22.10**   *Line-scan NMR imaging pulse sequence (redrawn from Mansfield and Morris, 1982)*

of a physical property of an object can be reconstructed from a series of one-dimensional projections of the parameter that are recorded at different orientations relative to the sample. This principle is used in X-ray computed tomography.

When applied to NMR imaging, a one-dimensional projection can be obtained by recording the NMR spectrum in the presence of a linear magnetic field gradient. Multiple projections are obtained by changing the relative orientation of the gradient and the data (proton density) processed in a computer via a standard projection reconstruction algorithm to generate an image resolved in two dimensions.

Figure 22.11 illustrates the filtered back projection technique (Bottomley, 1983). The amplitude of the NMR signal is used to assign the amount or number of nuclei present and the frequency of the signal is used to assign the spatial location. Multiple projections or angles of view are obtained by back-projecting these multiple views and the objects contained within the scanning area can be resolved. In practice, in order to image a matrix of n.n pixels, n angles of view in a π radian arc with n points (detectors) per view are required.

Another method of reconstructing an image in two dimensions is two-dimensional NMR Fourier zeugmatography suggested by Kumar et al. (1975). The technique utilizes a sequence of switched magnetic field gradients applied during the FID, combined with the two-dimensional Fourier transformation methods. Figure 22.12 shows the pulse sequence in this technique. After z-plane localization, a $\pi/2$ pulse is applied followed by the orthogonal linear gradients G_y and G_x during intervals t_y and t_x. FIDs are recorded in the interval t_x for different t_y. The two dimensional

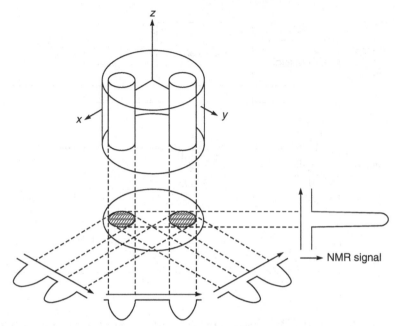

➤ **Fig. 22.11**   *Back projection reconstruction technique. A two dimensional NMR image of an object can be constructed from a series of one-dimensional projections obtained by recording NMR signals in gradients directed at different orientations (after Bottomley, 1983)*

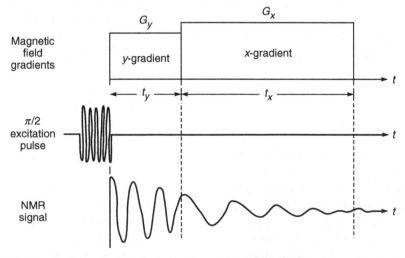

➤ **Fig. 22.12**   *Pulse sequence in two-dimensional NMR Fourier zeugmatography*

signal function contains all of the information necessary to reconstruct a two-dimensional image. Image reconstruction is affected by two-dimensional Fourier transformation of the signal function. A limitation of this method is that the signal is acquired only during part of the FID, which results in some reduction in sensitivity relative to methods such as projection reconstruction that observes the entire FID.

The spin warp imaging method is an improved technique that enables all of the NMR signals to be observed in the form of a spin echo, and the gradient pulses are varied in amplitude rather than time. Pulse sequence used for the spin-warp imaging method is shown in Fig. 22.13. A plane of spins is selected during t_2 by a selective $\pi/2$ pulse in the presence of G_z. A programmable G_y pulse is applied during t_y and n_y, spin echoes recorded in interval t_x using n_y different y-gradient pulse amplitudes from a two-dimensional Fourier transformation with respect to t_x and G_y.

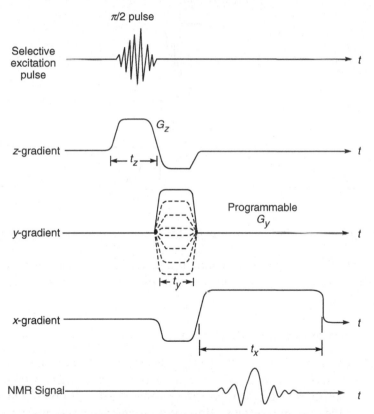

➤ Fig. 22.13 *Pulse sequence used for spin-warp imaging*

Three-dimensional Methods: In three-dimensional NMR zeugmatography, the plane localization step is omitted and instead, the gradient is re-oriented in all three dimensions. The one-dimensional projections obtained for each gradient orientation contain signal components from the entire sample. The three-dimensional image is reconstructed using a three-dimensional version of the reconstruction algorithm and displayed as a series of planes in any desired orientation.

The timing diagram for three-dimensional rotating frame zeugmatography is shown in Fig. 22.14. RF field gradients are applied in the x and y directions for variable intervals t_x and t_y. The FID is observed during t_z in a static z-gradient and a three-dimensional image reconstructed by three-dimensional Fourier transformation with respect to t_x, t_y and t_z.

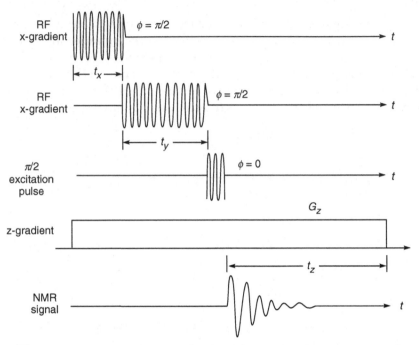

> **Fig. 22.14**   *Timing diagram for three dimensional zeugmatography*

The techniques such as planar, echo-planar and multiple-line scan selective excitation can all be extended to three-dimensional methods to multi-planer spin-echo and multi-planer multiple line scan imaging respectively.

22.2.1   Discrimination Based on Relaxation Rates

Relaxation processes can have a marked effect on the magnitude of the magnetization and hence on the intensity (contrast) of the image. This dependence of image intensity on the relaxation rates offers a unique possibility of using the relaxation rates as the discrimination factor provided that there is an appreciable variation of relaxation rates among the biological tissues, which in fact does exist.

Experimental results have shown that the proton rotation rates between the tissues of the important organs depends largely on their different water content and as such tissue types can be roughly characterized in terms of their water contents. For example, in mice, the brain has a water content of nearly 80% whereas the liver has between 66% to 70% and this has a direct effect on the relaxation time T_1. A similar trend is observed in T_2 though its absolute magnitude is much less.

The sensitivity of NMR arises from the fact that a 15% to 20% change in water concentration results in a change of about 200% in the relaxation rate.

Normally, the relaxation rates are used to modulate the individual pixel intensities rather than measuring their values. This can be done easily by making the relaxation times control the magnitude of the magnetization which induces the nuclear signal instead of allowing the magnetization to recover to its equilibrium value before each projection is taken. A number of approaches have been attempted to achieve this and the three commonly used techniques are described below:

Saturation Recovery: In a saturation recovery pulse sequence (Fig. 22.15), a series of 90-degree pulses is applied with an inter-pulse spacing t (data acquisition period) longer than the decay time T_2 and roughly the same length as T_1. When the value of t is changed, variations in T_1 in different parts of a sample will show up as differences in image intensity and a T_1 map can be generated. If t is much shorter than T_2, the signal will not decay to zero between successive pulses, creating an SSFP condition. Although images of very high quality can be generated with this technique, it is not easy to separate the individual contribution of T_1 and T_2 to the resulting image intensity.

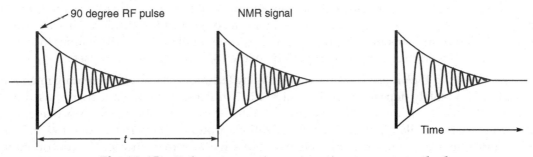

> **Fig. 22.15** *Pulse sequence in a saturation recovery method*

The saturation recovery technique allows for depressing signals from regions whose T_1 is longer than the data acquisition period and this property can be used to display fluids moving through the slice being imaged. By making the data acquisition period short relative to T_1 of the biological water, the permanently residing nuclei in the slice will be saturated and give rise to very small signals wherein the flowing fluid which is fresh from the point of NMR, will give a large signal and produce a brighter region in the image.

Inversion Recovery: 'An inversion recovery' sequence resembles the saturation recovery sequence in that T_1 variations in the sample can be exploited to achieve better contrast. In this technique, the magnetization vector is inverted by first applying a 180° pulse to the sample. T_1 relaxation preceeds during a selected pulse interval after which a 90° 'read' pulse is applied. The FID signal that follows the 'read' pulse is used to generate the image. Figure 22.16 shows the pulse sequence in the inversion recovery technique. This system involves a preparative excitation sequence involving two pulses, one of 180° and the second of 90° and during the final (90°) pulse in the sequence, a gradient is applied. The Z gradient will selectively excite a cross-sectional (X-Y) plane. Thus a single projection is taken with an amplitude that is dependent upon the proton concentration modulated by the value of T_1 due to the chemical configuration of the nuclei. Multiple experiments will be needed to reconstruct an image. Doyle and Young (1981) used a total of 180 projections for

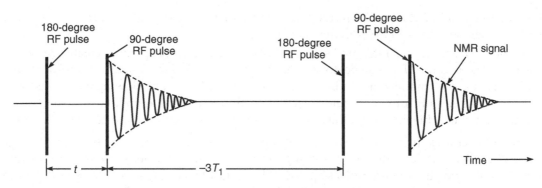

> **Fig. 22.16** *Pulse sequence in inversion recovery method*

reconstructing images. By defining a pixel intensity or colour scale in terms of the T_1 estimates, it is possible to display T_1 image.

The inversion recovery sequence generates an image of higher contrast than the saturation recovery sequence. However, it requires a longer time or reduced resolution. This is because if errors in T_1 determinations are to be avoided, a delay time equal to at least three times the value of T_1 should be allowed to elapse before repetition of the 180° and 90° pair of pulses.

Spin-echo Imaging Technique: The spin-echo imaging technique is useful for creating images that are primarily or solely dependent on T_2. Figure 22.17 illustrates the pulse sequence used in this technique. It consists in applying a 90° pulse to rotate the magnetization to the X-Y plane. This is followed by periodically flipping X-Y magnetization through 180° by means of a 180° RF pulse. Due to spin-spin relaxation, successive echoes progressively decrease in size. The contributions to the echoes from those regions with a short T_2 will decrease faster than the contributions from regions with a long T_2. Thus, by using an echo well separated from the 90° pulse, the contributions from T_2 of various values can be selectively highlighted, for example, long T_2 regions will show up brightly whereas short T_2 regions will appear black, on the NMR image.

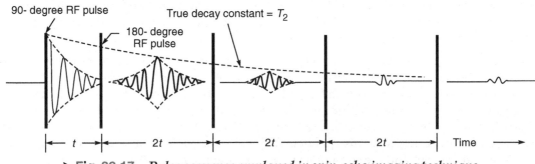

> **Fig. 22.17** *Pulse sequence employed in spin-echo imaging technique*

It may be recalled that in the saturation recovery method, the short T_1 region appears brightly whereas with spin-echo imaging the long T_2 regions are of greater intensity. Complimentary images can be constructed by modulating with T_1 on the one hand or with T_2 on the other.

22.2.2 Types of Imaging Sequences

Ever since the beginning of NMR imaging systems, two areas have dominated the field of research and intensive development. These are: (i) the extension and improvement in tissue differentiation, and (ii) the possibilities for shortening the required scan time. In the pursuit of these goals, a large number of sequences have been and are still being developed.

An NMR system would involve two processes. Firstly, spatially resolved information is encoded into a measurable signal and secondly, the spatially encoded signal is subsequently decoded to produce an image. The spatial encoding process is achieved by acquiring NMR signals under the influence of three orthogonal magnetic field gradients. There are many ways in which a gradient can influence and interact with a spin system. Figure 22.18. illustrates a generic pulse sequence and the principle of spatial encoding used in conventional NMR imaging systems (Riederer, 1988). The pulse sequence starts with the application of an RF pulse which is at the Larmor carrier frequency but modulated as shown with a sin(t)/t or "sinc" like function. The duration of the pulse is of the order of 3 ms and the amplitude is adjusted to get the desired angle of magnetization, typically 90°. The sinc modulation causes the bandwidth of the RF to be restricted to within several kHz about the Larmor frequency. Simultaneously, with the RF, a gradient (G-Slice) is applied to energize the z-gradient coil. Because of the finite bandwidth of the RF pulse, only those spins in the object within a thin slab or slice perpendicular to the z direction experience RF radiation matching their Larmor frequency. The negative lobe of G-Slice following the RF pulse is used for re-phasing of the spins within the excited slice.

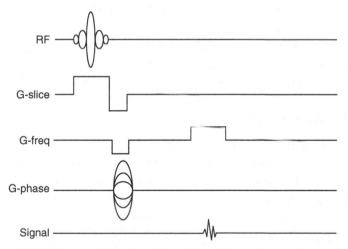

> **Fig. 22.18** *Generic pulse sequence used in magnetic resonance imaging*

After the slice is selected as explained above, the spins must be encoded spatially within the slice, i.e. along the x and y directions. The second gradient G-frequency does such encoding along the x direction. The first pulse shown is used for ensuring that all rotating spins measured in the subsequent signal are in phase. The second pulse is applied at the desired time of the signal measurement when it effectively causes the net Larmor frequency to vary linearly with position along the x-axis. The duration of the measured signal is generally 2 to 8 ms, depending upon the G-frequency gradient strength and the number of points sampled during the signal. Upon the Fourier transformation of the measured signal, the magnitude of each temporal frequency component is a measure of the net transverse magnetization at a corresponding position along the x direction.

The final gradient G-phase, performs encoding along the y direction. It is applied at an amplitude and duration such that the incremental phase accumulation provided to the transverse magnetization within the excited slice corresponds to the powers of complex exponential functions. A separate phasing coding gradient amplitude is applied during each cycle of the MR image acquisition, with each cycle corresponding to a separate exponential function. Typically, a 128 phase encoding is used in standard imaging, which in conjunction with 2000 ms repetition time, leads to scan times of about 8 minutes per slice. Several techniques have been introduced by different manufacturers to solve the dilemma of long scan times by scanning several echoes per excitation. Nitz (1996 a,b) illustrates the image sequence families which have been developed to obtain high speed imaging. The details of these sequences are beyond the scope of this chapter but a brief mention of the same is given below:

The Spin Echo Sequence Family

- Turbo Spin Echo (TSE), used by Siemens and Philips;
- Fast Spin Echo (FSE), used by General Electric;
- Rapid Acquisition and Relaxation Enhancement (RARE);
- Turbo Inversion Recovery (TIR);
- Turbo Inversion Recovery Magnitude (TIRM);
- Fluid Attenuated Inversion Recovery (FLAIR); and
- Half Fourier Acquired Single Shot Turbo Spin Echo (HASTE).

The Gradient Echo Sequence Family

- Fast Limited-Angle Shot (FLASH), used by Siemens;
- Spoiled Fast Acquired Steady State (spoiled FAST), adopted by Picker;
- Spoiled Gradient Recalled Acquisition in Steady State (spoiled GRASS or SPGR) introduced by General Electric;
- Fast Imaging with Steady State Precession (FISP);
- Fast Field Echo (FFE), used by Philips;
- Gradient Recalled Acquisition in the Steady State (GRASS), General Electric; and
- Fast Acquired Steady State Technique (FAST), Picker

Miscellaneous Type of Imaging Sequences

- Contrast Enhanced Fast Acquired Steady State Technique (CE FAST);
- Contrast Enhanced Gradient Recalled Acquisition in the Steady State (CE GRASS);

- Double Echo Steady State (DESS);
- Turbo Gradient Spin Echo (TGSE);
- Gradient and Spin Echo (GRASE); and
- Magnetization Prepared Rapid Gradient Echo (MP RAGE).

22.3 BASIC NMR COMPONENTS

The basic components of an NMR imaging system are shown in Fig. 22.19. These are:

- *A magnet*, which provides a strong uniform, steady, magnet field B_0;
- *An RF transmitter*, which delivers radio-frequency magnetic field to the sample;
- *A gradient system*, which produces time-varying magnetic fields of controlled spatial non-uniformity;
- *A detection system*, which yields the output signal; and
- *An imager system*, including the computer, which reconstructs and displays the images.

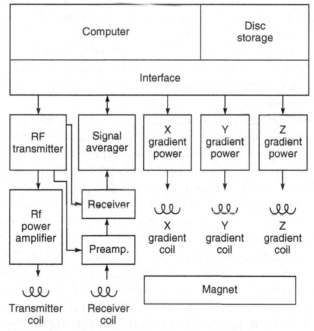

> **Fig. 22.19** *Sub-systems of a typical NMR imaging system*

The imaging sequencing in the system is provided by a computer. Functions such as gates and envelopes for the NMR pulses, blanking for the pre-amplifier and RF power amplifier and voltage waveforms for the gradient magnetic fields are all under software control.

The computer also performs the various data processing tasks including the Fourier transformation, image reconstruction, data filtering, image display and storage. Therefore, the computer must have sufficient memory and speed to handle large image arrays and data processing, in addition to interfacing facilities.

The Magnet: In magnetic resonance tomography, the base field must be extremely uniform in space and constant in time as its purpose is to align the nuclear magnets parallel to each other in the volume to be examined. Also, the signal-to-noise ratio increases approximately linearly with the magnetic field strength of the basic field, therefore, it must be as large as possible. Four factors characterize the performance of the magnets used in MR systems; viz., field strength, temporal stability, homogeneity and bore size. The effect of the magnetic field strength has been elaborated earlier. The temporal stability is important since instabilities of the field adversely affect resolution. The gross nonhomogeneities result in image distortion while the bore diameter limits the size of the dimension of the specimen that can be imaged.

Such a magnetic field can be produced by means of four different ways, viz., permanent magnets, electromagnets, resistive magnets and super-conducting magnets.

In case of the permanent magnet, the patient is placed in the gap between a pair of permanently magnetized pole faces. Permanent magnet materials normally used in MRI scanners include high carbon iron alloys such as alnico or neodymium iron (alloy of neodymium, boron and iron) and ceramics such as barium ferrite. Although permanent magnets have the advantages of producing a relatively small fringing field and do not require power supplies, they tend to be very heavy (up to 100 tons) and produce relatively low fields of the order of 0.3 T or less.

Electromagnets make use of soft magnetic materials such as pole faces which become magnetized only when electric current is passed through the coils wound around them. Electromagnets obviously require external electrical power supply.

On cost considerations, the earlier NMR imaging systems were equipped with resistive magnets. Resistive magnets make use of large current-carrying coils of aluminium strips or copper tubes. In these magnets, the electrical power requirement increases proportionately to the square of the field strength which becomes prohibitively high as the field strength increases. Moreover, the total power in the coils is converted into heat which must be dissipated by liquid cooling. For instance, at 0.2 T, the power requirement is nearly 70 kW (Oppelt, 1984) and a substantial increase of field strength above 0.2 T in resistive magnets is thus technically limited. At present, resistive magnets are seldom used except for very low field strength applications, generally limited to 0.02 to 0.06 T.

Most of the modern NMR machines utilize superconductive magnets. These magnets utilize the property of certain materials, which lose their electrical resistance fully below a specific temperature. The commonly used superconducting material is Nb Ti (Niobium Titanium) alloy for which the transition temperature lies at 9 K (–264°C). In order to prevent superconductivity from being destroyed by an external magnetic field or the current passing through the conductors, these conductors must be cooled down to temperatures significantly below this point, at least to half of the transition temperature. Therefore, superconductive magnet coils are cooled with liquid helium which boils at a temperature of 4.2 K (–269°C). The helium container with its superconductive windings is enclosed in a vacuum to keep the evaporation rate low. Internal shields cooled with liquid nitrogen prevent heating due to radiated heat passing through the vacuum vessel.

In a superconducting magnet, connection to a current supply is only necessary for energizing up to the required field strength. After this, the coils are short-circuited and require no further electrical energy. The magnetic field is temporarily stable. Due to evaporation of the liquid helium and liquid nitrogen, the monthly topping of helium and weekly topping of nitrogen is necessary. The evaporation rate in the earlier scanners was about 0.5 l/h for liquid helium and 2 l/h for liquid nitrogen. At present, many magnets now make use of cryogenic refrigerators that reduce or eliminate the need for refilling the liquid helium reservoir. Figure 22.20 shows a schematic diagram of the superconducting magnet. Because of their capability to achieving very strong and stable magnetic field strengths without any continuous power consumption, superconducting magnets have become the most widely used and preferred source of the main magnetic fields for MRI scanners.

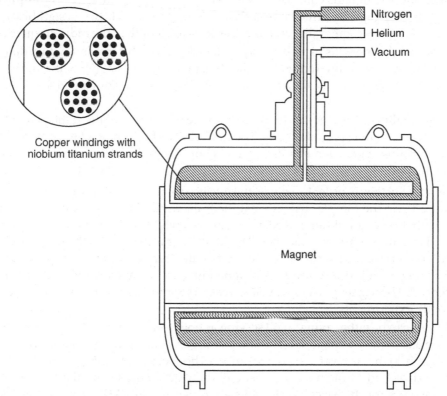

> **Fig. 22.20** *Schematic drawing of the superconducting magnet*

Superconductive magnetic resonance magnets with an open internal diameter of 1 m, as is desirable for whole body examinations, are now produced for field strengths of upto 2 T. In a typical 1.5 T magnetic field, the current required by the superconducting coils is of the order of 200 amp. The diameter of the coils is about 1.3 m and total length of the wire could be 65 kms. The magnet is operated in the persistent mode, i.e. once the current is established, the terminals may be connected together, and a constant persistent current will flow indefinitely so long as the temperature of the coils is maintained below the superconducting transition temperature.

Besides the indisputable advantages which high magnetic field strengths have on image quality, it also implies increased technical complexity, both in the installation of magnetic resonance equipment as the well as the radio-frequency technology required. The spread of the critical fringe field of a magnet is proportional to the cube root of its field strength. If with a 0.2 T magnet, the ten-fold strength of the earths field is exceeded at about 6 m distance from the centre, then at 2 T, this fringe field strength occurs at 13 m. This can seriously affect the function of nearby clinical equipment such as image intensifiers and monitors.

A field strength of 0.5 T means that a nuclear resonance frequency of 21.3 MHz is required for protons and a field strength of 2 T means that 85.2 MHz is needed. At these higher frequencies, the usual saddle-shaped antenna coils used for lower frequencies can no longer be applied, because the conductors' self-capacity is too large and travelling wave effects play a significant role. Also, the conductor length is comparable with the quarter wavelength of the radio-frequency field.

The NMR imaging systems usually incorporate magnets with a maximum flux density of 0.5 T to 1.5 T. In the system of international units (SI units), the 'Tesla' (T) is the unit of magnetic flux density. In some countries, the unit 'Gauss' (G) is also used. For conversion 1 T=10,000 G = 10 kG.

The image quality of NMR scans depends upon the uniformity of the static magnetic field and on its stability over a long period of time. The uniformity of this magnet must be at least 20 ppm within the scanning region and stability at a level of 2 ppm during short periods and under 10 ppm over long periods.

RF Transmitter System: In order to activate the nuclei so that they emit a useful signal, energy must be transmitted into the sample. This is what the transmitter does. The system consists of an RF transmitter, RF power amplifier and RF transmitting coils. The RF transmitter consists of an RF crystal oscillator at the Larmor frequency. The RF voltage is gated with the pulse envelopes from the computer interface to generate RF pulses that excite the resonance. These pulses are amplified to levels varying from 100 W to several kW depending on the imaging method and are fed to the transmitter coil. The higher power levels are necessary for the large sample volumes encountered in whole body experiments.

The RF coils can be either a single coil serving as both transmitter and receiver or two separate coils that are electrically orthogonal. The latter configuration has the advantage of reduced pulse breakthrough into the receiver during the pulse. In both cases, all coils generate RF fields orthogonal to the direction of the main magnetic field. Saddle-and solenoidal-shaped RF coils are typical geometries for the RF coils. The coils are tuned to the NMR frequency and are usually isolated from the remainder of the system by enclosure in an RF shielding cage.

For magnetic fields in the range of 0.05 to 2 T used for imaging of the human body, the resonant frequencies fall in the radio-frequency band. For example, in a field of 1 T, ^{1}H resonates at 42.57 MHz, ^{19}F at 40.05 MHz, ^{31}P at 17.24 MHz and ^{13}C at l0.71 MHz. Usually, the resonance is extremely sharp. Widths in the range of 10 Hz are typical of biological systems.

Detection System: The function of the detection system (receiver) is to detect the nuclear magnetization and generate an output signal for processing by the computer. A block diagram of a typical receiver is shown in Fig. 22.21.

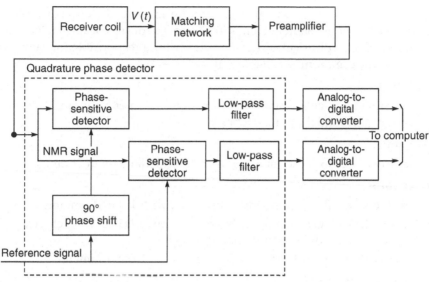

> ➤ **Fig. 22.21** *Block diagram of the NMR detection system (Adapted from Hinshaw and Lent, 1983)*

The receiver coil usually surrounds the sample and acts as an antenna to pick up the fluctuating nuclear magnetization of the sample and converts it to a fluctuating output voltage $V(t)$. Hinshaw and Lent (1983) explained that NMR signal

$$V(t) = -\frac{d}{dt} \cdot M(t,x) \cdot B_c(x) \, d_x$$

where $M(t,x)$ is the total magnetization in a volume and $B_c(x)$ the sensitivity of the receiver coil at different points in space. $B_c(x)$ describes the ratio of the magnetic field produced by the receiver coil to the current in the coil.

The receiver coil design and placement is such that $B_c(x)$ has the largest possible transverse component. The longitudinal component of $B_c(x)$ contributes little to the output voltage and can be ignored. This is because the time derivative of $M_z(t,x)$ is much less than that of the transverse component. $M_z(t,x)$ decays exponentially with time constant T_1, typically 0.1 to 1 s, while the transverse component is oscillatory with a period of, typically 0.05 to 0.2 μs.

The RF signals constitute the variable measured in magnetic resonance tomography. These are extremely weak signals having an amplitude in the nV (nano-Volt) range thus requiring specially designed RF antennas. The sensitivity of an MR scanner therefore depends on the quality of its RF receiving antenna. For a given sample magnetization, static magnetic field strengths and sample volume, the signal-to-noise-ratio (SNR) of the RF signal at the receiver depends in the following manner upon the RF-receiving antenna.

SNR ~ $K(Q/V_c)$

Where K is a numerical constant, specific to the coil geometry

$\quad$ Q is the coil magnetization factor, and

$\quad$ V_c is the coil volume.

This implies that the SNR of an MR scan can be improved by maximizing the ratio of magnetization to coil volume. Sauter et al.,1986 discuss the design of special RF coils for various applications alongwith the physical and anatomical criteria responsible for the required shapes and forms. Some of the commonly available coils are:

- **Body Coils:**
  - Constructed on cylindrical coils forms with diameter ranging from 50 to 60 cm to entirely surround the patient's body.

- **Head Coils:**
  - Designed only for head imaging, with typical diameter of 28 cm.

- **Surface coils:**
  - Orbit/ear coil: flat, planar ring-shaped coil with 10 cm diameter;
  - Neck coil: flexible, rectangular shaped surface coil (10 cm × 20 cm) capable of adaptation to the individual patient anatomy; and
  - Spine coil: cylindrical or ring-shaped coil with 15 cm diameter.

- **Organ-enclosing coils:**
  - Breast coil: cylindrical or ring-shaped coil with 15 cm diameter.
  - Helmholtz-type coil: a pair of flat ring coils each having 15 cm diameter with distance between the two coils variable from 12 to 22 cm.

Following the receiver coil is a *matching network* which couples it to the pre-amplifier in order to maximize energy transfer into the amplifier. This network introduces a phase shift ϕ to the phase of the signal.

The pre-amplifier is a low-noise amplifier which amplifies the signal and feeds it to a quadrature phase detector. The detector accepts the RF NMR signal which consists of a distribution of frequencies centred around or near the transmitted frequency w and shifts the signal down in frequency by w. By this process, the distribution of frequencies is unchanged except that it is now centred about zero.

The detector circuit accepts the inputs, the NMR signal $V(t)$ and a reference signal, and multiplies them, so that the output is the product of the two inputs. The frequency of the reference signal is the same as that of the irradiating RF pulse. The output of the phase-sensitive detector consists of the sum of two components, one a narrow range of frequencies centred at $2w_0$, and the other, a narrow range centred at zero.

The low pass filter following the phase-sensitive detector removes all components except those centred at zero from the signal. It is necessary to convert the complex (two-channel) signal to two strings of digital numbers by analog-to-digital converters. The A-D converter output is passed, in serial data form to the computer for processing.

Gradient System for Spatial Coding: Spatial distribution information can be obtained by using the fact that the resonance frequency depends on the magnetic field strength. By varying the field in a known manner through the specimen volume, it is possible to select the region of the specimen from which the information is derived on the basis of the frequency of the signal. The strength of the signal at each frequency can be interpreted as the density of the hydrogen nuclei in the plane within the object where the magnetic field corresponds to that frequency. NMR imaging methods exploit this property by way of carefully controlled, well-defined gradients to modulate the NMR signal in a known manner such that the spatial information can later be decoded and plotted as an image. Typically, the gradients are chosen with linear spatial dependence so that the NMR frequency spectrum directly corresponds to the position or even one or more spatial co-ordinate axes. The imaging methods differ mainly in the nature of the gradient time dependence (static, continuously time-depended or pulsed), and in the type of NMR pulse sequence employed.

The concept of obtaining spatial information and therefore images was given by Lauterbur (1973). He made a major advancement by super-imposing a linear magnetic field gradient on the uniform magnetic field applied to the object to be imaged. When this is done, the resonance frequencies of the precessing nuclei will depend primarily on the positions along the direction of the magnetic gradient. This produces a one-dimensional projection of the structure of the three-dimensional object. By taking a series of these projections at different gradient orientations, a two or even three-dimensional image can be produced.

There are various methods for selecting a slice, but the 'selective excitation' method is currently the most widely used method throughout the world. This method covers only the area on which radio wave pulses of the same frequency as the resonant frequency are applied, since the resonant frequency changes in the same direction if a uniform static field and a gradient magnetic field changing linearly in the vertical direction of the layer in question, are impressed simultaneously. *The characteristic of the slice will be determined by the shape of the pulse and the thickness of the slice will be determined by the width and gradient of the pulse.*

In NMR systems, for spatially resolving the signals emitted by the object, the initially homogeneous magnetic field B_0 is overlaid in all three spatial dimensions, X, Y, Z with small linear magnetic fields-gradient fields G. These gradient fields are represented in Fig. 22.22 by arrows of increasing thickness to illustrate the linear increase in magnetic field strength. These gradient fields are produced with the aid of current carrying coils and can be switched on or off as desired, both during the application of the RF energy and also in any phase of the measuring procedure.

The principle of the use of gradient field for selecting plane a is shown in Fig. 22.23. As described earlier, elsewhere, in the sensitive point method, alternating gradients are utilized. The purpose of alternating gradients is to provide a discrete but movable time variant (or sensitive) plane so that various locations may be sensed without the need to move the physical components of the sample or magnet system. With pairs of field coils and alternating complementary gradients, one can position a sensitive plane under magnetic and

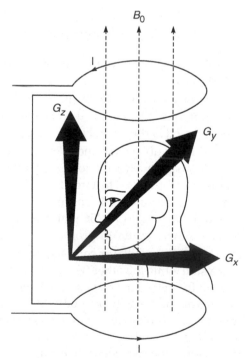

> **Fig. 22.22** *Arrangement of the field gradients*

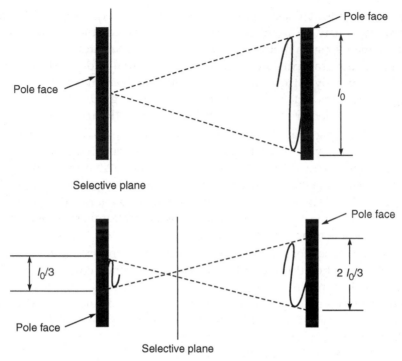

> **Fig. 22.23**   *Use of sinusoidal current amplitude to define selective plane location*

therefore, electronic control. As shown in the above figure, by controlling the relative amplitude of the signals for each coil, one can move a sensitive plane anywhere between the two gradient coils.

Several circuit arrangements for controlling magnetic field gradients are described in literature. Fitzsimmons (1982) made use of a computer to control the gradient field. A block diagram of the system used is shown in Fig. 22.24. The hardware can be broken down into four sub-systems.

The first sub-system includes the interface between the computer and the gradient control system. Its primary function is to allow the independent positioning of the three time invariant planes (X, Y and Z). The circuit is essentially a serial to parallel converter with independent reset times. This requires only six pulse lines from the host computer. The maximum number of plane positions is limited to 256 which is adequate for most imaging applications. A set of switches allows either manual or program control over each plane position.

The digital oscillator consists of a 555 timer followed by shift registers. Twelve registers shift out a sine while the other twelve shift out a cosine wave. A digital oscillator facilitates varying the output frequency over an extremely wide range through the use of a single control. Also the digital oscillator could easily be modified to allow computer-controlled stepping of the output to yield pre-set sine and cosine vectors. The 8-bit input from the interface circuit is used directly to control one attenuator while the same 8-bits are inverted to control the second attenuator. This results in two complementary sine wave outputs which can be stepped through 256 positions. The resultant current I_o is given by

$$I_o = I_{\text{coil 1}} - I_{\text{coil 2}}$$

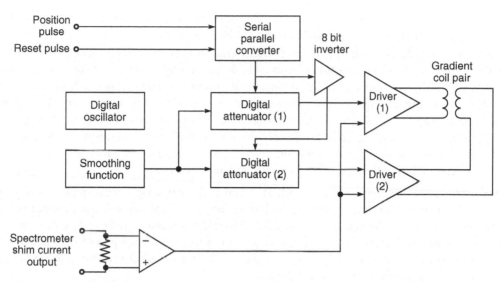

➤ Fig. 22.24 *Block diagram of gradient control system. Each X,Y and Z coil pair has its own control circuit (Redrawn after Fitzsimmons, 1982)*

The output of the attenuators is then voltage-amplified by two op amps prior to the driven circuits.

Current control requires through the shim coil that the control used to adjust the static field gradients be available for setting the DC levels upon which the alternating gradients are super-imposed. An op amp serves the differential voltage drop across a dummy load (having the same resistance as the shim coil) and produces an output which is then DC coupled to the drivers.

The high current drivers use a conventional design with a single op amp providing the input to a driver and a complimentary pair of power transistors to provide a sufficient current to the gradient coil.

In typical scanners, gradient coils have an electric resistance of about 1 Ω and an inductance of 1 mH. The gradient fields are required to be switched from 0 to 10 mT/m in about 0.5 ms. The current switches from 0 to about 100 A in this interval. The power dissipation during the switching interval is about 20 kW. This places very strong demands on the power supply and it is often necessary to use water cooling to prevent overheating of the gradient coils.

The demand on the power supply is quite high since the X and Y gradient drivers operate in the 4-6 A range. Since there are two drivers for each dimension, the total requirement is at least 20 A. The Z driver requires much less current due to the coil geometry and positions with respect to the major fields.

The software requirements pertain only to re-setting a particular gradient and then send an appropriate number of pulses to the gradient control interface. This simply requires that a register be loaded with a number corresponding to the desired plane location and that a countdown is executed outputting a pulse on each decrement of the register. In the program, the entire series of plane settings are stored in memory, so that each time the program is called by the user, it loads the pointer which selects the next number in the list, steps the gradient controller approximately, increments the pointer and returns to the main calling program.

A high degree of linearity in the magnetic field gradients is essential in order to reconstruct an accurate NMR image from projections. With well-designed coils, errors resulting from non-linear gradients will perhaps not be evident in a medical image since the image will remain clear and will not contain rigidly shaped objects or those with sharp edges for close comparison. But these gradient coils are usually designed to optimize linearity in the central region. Away from the centre, gradient linearity becomes progressively worse. Without restoration, the image will not give accurate information on the outer regions. Therefore, non-linear field gradients result in a geometrical distortion of the image reconstructed from projections.

Imager System: The imager system includes the computer for image processing, display system and control console. The timing and control of RF and gradient pulse sequences for relaxation time measurements and imaging, in addition to FT image reconstruction and display necessitate the use of a computer. The computer is the source of both the voltage waveforms of all gradient pulses and the envelopes of the RF pulses. A general purpose mini-computer of the type used for a CAT scanner is adequate for these purposes.

The computer system collects the nuclear magnetic resonant signal after A/D conversion, corrects, re-composes, displays and stores it. High speed data are sent from the system controller to the computer. An exclusive high speed computer is used to reduce the calculating time of these data. Analog-to-digital convertors with 16 bits or higher are used to produce the desired digitized signal data. During the data acquisition, information is acquired at the rate of about 800 kilobytes per second and each image can contain upto a megabyte of digital data. A specialized computer such as an array processor which is designed for the rapid performance of specific algorithms like fast Fourier transformation (FFT), is used to convert the digitized time-domain data to image data. Two-dimensional images are typically displayed as 256×256 or 512×512 pixel arrays. The images become available for viewing within about one second after data acquisition. Three dimensional imaging requires more computer processing power. The computers currently used are typically 32 bit machines equipped with upto 4 Mbyte of memory and backed by an array processor to speed up the Fourier transformation. Data storage is on high speed disks.

The reconstructed image data are transmitted to the display console by a high-capacity image memory disk. As in X-ray CT, the image is displayed on a TV monitor, either in grey scale or in colour. The display console is usually an intelligent console that can be used as an independent image processing unit in an interactive system. The screen of the high resolution monitor can be divided into four parts, if desired. It is possible to simultaneously display proton density and T_1, T_2 distribution image. A multi-format camera is used for making hard copies of the image.

The desirable features of the software are its superior operating characteristics, high speed image reconstruction and the ability to perform comprehensive image processing. The image reconstruction software is used to re-build an image, register images, display and position scanned image processing and register patients. The image display software transfers data registered in the image data file to the display console for display, sends other image data to the floppy disk or magnetic tape and also ensures data protection and editing.

The control console comprises the operation section, system control section and the display section. In the system control section, a microcomputer controls the gradient magnetic field, the high frequency pulse train (RF pulses) and the timing of A/D conversion of the signals received. The display section includes the high resolution monitor, keyboard, image memory and microcomputer for processing the image and operating keys to set the scanning conditions and to control the patient couch from the operation section, together with various panel indicators to monitor the system condition.

Contrast Enhancement: As in x-ray imaging, it is possible to artificially enhance the contrast in NMR systems. Systematic injections of paramagnetic ions and complexes that act like tiny magnets are effective in the micro-molar to milli-molar range, significantly decreasing the relaxation times of tissue water. The relaxation time differences can be accordingly converted into signal intensity differences.

The enhancement of certain tissue processes, such as blood flow, may be done by manipulating data acquisition parameters rather than by injection of contrast agents. Varying the time interval between successive data accumulations (the T_1 parameter) will selectively enhance tissues according to the respective values of T_1.

Patient Couch: The patient couch for NMR imaging applications is made of a non-magnetic material to prevent disturbing the uniformity of the magnetic field in the scanning region. The stretcher (top of the couch) is constructed for long stroke and minimum warpage. The top plate is controlled from a control panel for raising or lowering it, moving back and forth accurately. It can also be driven by remote control from the controller console and set to the scanning position or can be converted to automatic or manual feed.

▸ 22.4 BIOLOGICAL EFFECTS OF NMR IMAGING

The three aspects of NMR imaging which could cause potential health hazard are:

(i) *Heating due to the rf power:* Katinis (1982) reports that a temperature increase produced in the head of NMR imaging would be about 0.3°C. This does not seem likely to pose a problem.

(ii) *Static magnetic field:* Although no significant effects of the static field with the level used in NMR are known, Pastakia (1978) mentions about the possible side effects of electromagnetic fields. There could be a slight decrease in cognitive skills, mitotic delay in slime moulds, delayed wound healing and elevated serum triglycerides.

(iii) *Electric current induction due to rapid change in magnetic field:* It is believed that oscillating magnetic field gradients may induce electric currents strong enough to cause ventricular fibrillation. However, no damage due to NMR from exposures has been reported (Marx 1980). It is suggested that fields should not vary at a rate faster than 3 tesla/s.

▸ 22.5 ADVANTAGES OF NMR IMAGING SYSTEM

The advantages of the NMR Imaging System are:

(i) The NMR image provides substantial contrast between soft tissues that are nearly identical in existing techniques. NMR images that display T_1 and T_2 properties of tissue provide tremendous contrasts between various soft tissues, contrasts approaching 150% are possible in T_1 and T_2 images, while contrasts of only a few percent are possible between soft tissues with X-rays.

(ii) Cross-sectional images with any orientation are possible in NMR imaging systems.

(iii) The alternative contrast mechanisms of NMR provide promising possibilities of new diagnostics for pathologies that are difficult or impossible with present techniques.

(iv) NMR imaging parameters are affected by chemical bonding and, therefore, offer potential for physiological imaging.

(v) NMR uses no ionizing radiation and has minimal, if any, hazards for operators of the machines and for patients.

(vi) Unlike CT, NMR imaging requires no moving parts, gantries or sophisticated crystal detectors. The system scans by superimposing electrically controlled magnetic fields. Consequently, scans in any pre-determined orientation are possible.

(vii) With the new techniques being developed, NMR permits imaging of entire three-dimensional volumes simultaneously instead of slice by slice, employed in other imaging systems.

Ultrasonic Imaging Systems

23.1 DIAGNOSTIC ULTRASOUND

Ultrasound has become increasingly important in medicine and has taken its place along with X-ray and nuclear medicine as a diagnostic tool. Its main attraction as an imaging modality lies in its non-invasive character and ability to distinguish interfaces between soft tissues. In contrast, X-rays only respond to atomic weight differences and often require the injection of a more dense contrast medium for visualization of non-bony tissues. Similarly, nuclear medicine techniques measure the selective uptake of radioactive isotopes in specific organs to produce information concerning organ function. Radioactive isotopes and X-rays are, thus, clearly invasive. Ultrasound is not only non-invasive, externally applied and non-traumatic but also apparently safe at the acoustical intensities and duty cycles presently used in diagnostic equipment.

Diagnostic ultrasound is applied for obtaining images of almost the entire range of internal organs in the abdomen. These include the kidney, liver, spleen, pancreas, bladder, major blood vessels and of course, the foetus during pregnancy. It has also been usefully employed to present pictures of the thyroid gland, the eyes, the breasts and a variety of other superficial structures. In a number of medically meaningful cases, ultrasonic diagnostics has made possible the detection of cysts, tumours or cancer in these organs. This is possible in structures where other diagnostic methods by themselves were found to be either inapplicable, insufficient or unacceptably hazardous. Ultrasonic studies which do not involve image formation have also been extensively developed to allow the dynamics of blood flow in the cardiovascular system to be investigated with a precision not previously possible. The main limitation of ultrasound, however, is that it is almost completely reflected at boundaries with gas and is a serious restriction in investigation of and through gas-containing structures.

23.2 PHYSICS OF ULTRASONIC WAVES

Ultrasonic waves are sound waves associated with frequencies above the audible range and generally extend upward from 20 kHz. These waves exhibit the same physical properties as the audible sound waves but they are particularly preferred in situations favoured by one or more of the following reasons:

- Ultrasonic waves can be easily focussed, i.e., they are directional and beams can be obtained with very little spreading.
- They are inaudible and are suitable for applications where it is not advantageous to employ audible frequencies.
- By using high frequency ultrasonic waves which are associated with shorter wavelengths, it is possible to investigate the properties of very small structures. It is particularly true in the detection of defects where the wavelengths utilized should be of the same order as the dimensions of the defect.
- Information obtained by ultrasound, particularly in dynamic studies, cannot be acquired by any other more convenient technique.

Transmission of ultrasonic wave motion can take place in different modes. The wave motion may be longitudinal, transverse or shear. However, for medical ultrasonic diagnostic applications, the longitudinal mode of wave propagation is normally used as these waves can be propagated in all types of media, viz. solids, liquids and gases. In longitudinal waves, the particles of the medium oscillate to and fro in the direction of propagation of the wave resulting in alternate regions of compressions and rarefactions.

23.2.1 Characteristic Impedance

Characteristic impedance or the specific acoustic impedance of a medium is defined as the product of the density of the medium with the velocity of sound in the same medium.

$$z = \rho V$$

where z = specific acoustic impedance

ρ = density of the medium

V = velocity of sound in the medium.

The characteristic impedance determines the degree of reflection and refraction at the interface between two media. The percent of the incident wave energy which is reflected is given by:

$$\left[\frac{z_1 - z_2}{z_1 + z_2} \right]^2 \times 100\%$$

where z_1 = acoustic impedance of medium 1

z_2 = acoustic impedance of medium 2

provided the ultrasonic beam strikes the interface at a right angle. Beams incident on a boundary at an angle (Fig. 23.1) other than zero to the normal are reflected so that the angle of incidence and the angle of reflection are equal.

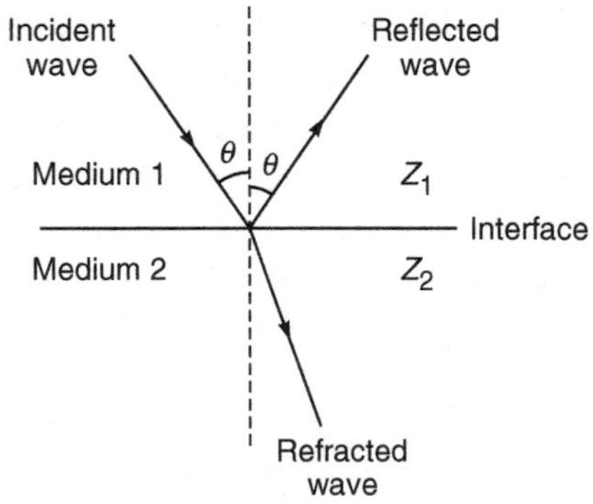

> **Fig. 23.1** *Reflection and refraction of ultrasound at an interface between two media having different acoustic impedances*

The approximate value of acoustic impedance for most of the biological materials or organs is the same. It is about 1.6×10^5 g/cm^2 s. The greater the difference in acoustic impedance, the greater the amount of reflected energy. For example, the acoustic impedance of air and tissue are 42.8 g/ cm² s and 1.6×10^5 g/cm² s respectively. This difference is so large that most of the ultrasonic energy tends to be reflected at the interface. It is for this reason that a coupling medium like olive oil or special jelly is used to minimize the energy reflection by providing an air-free path between ultrasonic transducer and skin.

23.2.2   Wavelength and Frequency

Ultrasonics follow the general wavelength and frequency relationship given by

$$V = n\lambda$$

where V = propagation velocity of sound

n = frequency or number of cycles which pass any given point in unit time

λ = wavelength, i.e., distance between any two corresponding points on consecutive cycles.

Ultrasonic frequencies employed for medical applications range from 1 to 15 MHz. This range also corresponds to radio frequencies (rf). However, there is an important basic difference between radio frequency and ultrasonic energy. Ultrasonic waves are transmitted as mechanical vibrations whereas rf energy would be in the form of electromagnetic radiations. No medium is necessary for propagation of energy and it would, therefore, pass even through vacuum; ultrasonic waves will, on the other hand, pass only through a medium.

23.2.3   Velocity of Propagation

Ultrasonic energy is transmitted through a medium as a wave motion and, therefore, no net movement of the medium is expected to occur. The velocity of propagation of the wave motion is determined by the density of the medium it is travelling through and the stiffness of the medium. At a given temperature and pressure, the density and stiffness of the biological substances are relatively constant, and, therefore, the sound velocity in them is also constant. The speed of ultrasound in m/s in various biological materials is shown in Fig. 23.2.

The knowledge of velocity of sound in a particular medium is important in calculating the depth to which the sound wave has penetrated before being reflected. If the time taken by the ultrasonic wave to move from its source through a medium, reflect from an interface and return to the source can be measured, then the depth of penetration is given by:

$$\text{Depth of penetration} = \frac{\text{Velocity of sound in the medium} \times \text{time}}{2}$$

The velocity of ultrasound in all body tissues is almost constant. Therefore, the depth of penetration can be read directly from the position of the echo pulse on the calibrated time axis of the oscilloscope trace.

23.2.4   Absorption of Ultrasonic Energy

The reduction of amplitude of ultrasonic beam while passing through a medium can be due to its absorption by the medium and its deviation from the parallel beam by reflection, refraction, scattering and diffraction etc. The relative intensity and the attenuation of an ultrasound beam is

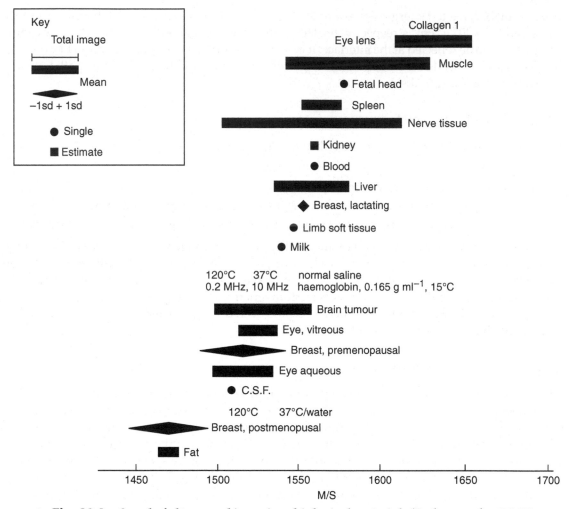

> **Fig. 23.2**   *Speed of ultrasound in various biological materials (Redrawn after P.N.T. Wells, Biomedical Ultrasound, Academic Press, London)*

expressed in decibels (dB) and the absorption coefficient α is normally quoted in dB/cm. In soft tissues, α depends strongly on the frequency and therefore, for a given amount of energy loss, the lower frequency ultrasonic signal would travel more than the higher frequency signal. Quantitatively, the average value of sound absorption in soft tissues is of the order of 1 dB/cm/ MHz. The attenuation of ultrasound in various biological materials (Wells, 1977) is shown in Fig. 23.3.

Table 23.1 shows the velocity of ultrasound, characteristic impedance and absorption coefficient in various materials.

23.2.5   Beam Width

In general, ultrasonic waves are projected in a medium as a beam. Huygen's construction may be used to determine the spatial distribution of energy in this beam, which can be conveniently split

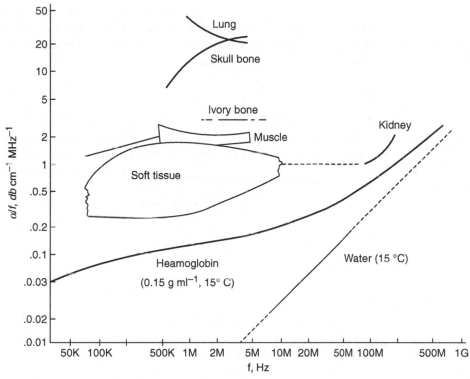

➤ **Fig. 23.3** *Attenuation of ultrasound in various biological materials (Adapted from P.N.T. Wells, Biomedical Ultrasound, Academic Press, London)*

● Table 23.1 *Velocity, Characteristic Impedance and Absorption Coefficient of Ultrasound in Various Materials*

Material	Velocity of ultrasound in $m\ s^{-1}$	Characteristic impedance, z in $kg\ m^{-2}\ s^{-1} \times 10^{-6}$	Absorption coefficient in $dB\ cm^{-1}$ at 1 MHz
Air (20°C)	343	4×10^{-4}	12.0
Water	1480	1.48	0.002
Aluminium	6220	16.5	–
Fat	1450	1.38	0.6
Brain	1541	1.58	0.85
Liver	1549	1.65	0.9
Kidney	1561	1.62	1.0
Blood	1570	1.61	0.2
Muscle	1585	1.70	2.3
Skull-bone	4080	7.80	13.0
Lens of eye	1620	1.84	2.0
Human soft-tissue	1540	1.63	0.8

into near and far fields. In the near field, within the first Fresnel zone, the beam is cylindrical with little spread. A series of maxima and minima are encountered in this region, as one travels out from the transducer which corresponds to constructive and destructive interference. The near field extends to a distance d from the transducer where

$$d = \frac{r^2}{\lambda} - \frac{\lambda}{4}$$

where r and λ are the radius of the transducer and the wavelength of the ultrasound respectively.

In the far field, the intensity of the beam reduces constantly with distance as it spreads out due to the finite size of the source. The angle of divergence within a cone of semi-angle θ about the central axis is given by

$$\theta = \frac{0.61 \lambda}{rn} \text{ in radians}$$

$$= \frac{0.61 \, V}{rn} = \frac{1.22 \, V}{nD}$$

where n = frequency

V = velocity of sound waves

D = diameter of the transducer.

For example, a cm diameter transducer, excited at 1 MHz has a near field of about 10 cm in water and a semi-angle of divergence of 3.5 degrees. The beam shape may be modified by the use of focusing elements in front of the transducer.

23.2.6   Resolution

The resolution of an ultrasound system can be defined as the system's ability to distinguish between closely related structures. In general, the resolution is divided into axial and lateral resolution.

Axial Resolution: The axial resolution is the minimal axial distance (Fig. 23.4(a)), parallel to the beam axis, at which two reflecting structures are recognized as separate structures. The axial resolution is determined by the wavelength of the transmitted pulse. This means that the smaller the wavelength, the higher the frequency and better the axial resolution.

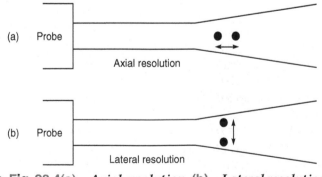

> **Fig. 23.4(a)**   *Axial resolution* **(b)**   *Lateral resolution*

Lateral Resolution: The lateral resolution is the lateral distance, in a plane perpendicular to the beam axis, at which two reflecting structures can be seen as two separate structures (Fig. 23.4(b)). The lateral resolution is determined by the shape/divergence of the ultrasound beam, produced by the probe.

23.2.7 Generation and Detection of Ultrasound

The physical mechanism normally used to generate and detect ultrasonic waves is the piezo-electric effect exhibited by certain crystalline materials which have the property to develop electrical potentials on definite crystal surfaces when subjected to mechanical strain. The converse is also true, which means that mechanical displacement is produced when electrical charges are put on their surface. The effect is demonstrated by crystals of materials like quartz, tourmaline and Rochelle salt. This phenomenon offers an excellent method for converting electrical energy into mechanical energy and vice versa.

While working with natural crystals, it is difficult to establish the appropriate axis and cut the crystal in the required form. Therefore, quartz has generally been replaced by synthetic piezo-electric materials namely barium nitrate and lead zirconate titanate. They offer several advantages because they are far cheaper to produce and are much easier to construct transducers of complex shape and large areas. They can be moulded to any shape to obtain a better focusing action for producing high intensity ultrasonic waves.

The choice of piezo-electric material for a particular transducer depends upon its applications. Materials with high mechanical Q factor are suitable as transmitters whereas those with low mechanical Q and high sensitivity are preferred as receivers and in case of non-resonance applications. Lead zirconate Titanate (PZT) crystals are much better than quartz crystals upto a frequency of about 15 MHz, because of its high electro-mechanical conversion efficiency and low intrinsic losses. The properties of PZT can be adjusted by modifying the ratio of zirconium and titanium and introducing small amounts of other substances such as lanthanum. PZT can operate at temperatures up to 100°C or higher and it is stable over long periods of time. It is mechanically strong and can be machined to various shapes and sizes. At frequencies higher than this, quartz is normally used because of its better mechanical properties. Polyvinylidene difluoride (PVDF) is another ferro-electric polymer that has been used effectively in high frequency transducers. The surface of the synthetic crystals are normally silvered for making external electrical connections.

Piezo-electric crystals are available in several shapes and the selection of a particular shape depends upon the application to which it is to be put. The application of a voltage to the transducer disc causes changes in its thickness, thereby giving rise to longitudinal waves propagated along the axis perpendicular to its face. The magnitude of the waves will be a maximum at the resonant frequency of the disc, which is determined by its thickness. Likewise, when the transducer is acting as a receiver, it will be most sensitive to ultrasonic vibrations at its own resonant frequency. Therefore, considerable attention is devoted to tailoring the dimensions of the transducer to the frequency at which it will be used.

There are three parameters that are important in optimizing transducers for various types of applications (Hunt et al. 1983). These are frequency, active element diameter and focusing. Their effects on the performance are as follows:

Frequency: With increase in frequency, the sound beam becomes more directional and the axial resolution improves. However, due to attenuation of higher frequency ultrasound waves in the tissues, the penetration decreases. For most abdominal ultrasound examinations, the frequencies used are in the range of 1-5 MHz, whereas the wavelength is in the range of 1 mm. Higher frequencies (10-15 MHz) are used for superficial organs, such as the eye, where deep penetration is not required and where advantage may be taken of the 0.1 mm wavelength to improve geometrical resolution.

The following are general rules which apply to frequency:

↑ Frequency   ↑ Axial Resolution   ↑ Lateral Resolution   ↓ Penetration

↓ Frequency   ↓ Axial Resolution   ↓ Lateral Resolution   ↑ Penetration

Frequency also influences lateral resolution by affecting beam divergence. The following rule applies, assuming all other factors remain constant.

↑ Frequency   ↓ Beam divergence   ↑ Lateral Resolution

Active Element Diameter (AED): As the transducer face diameter increases, the beam width decreases and therefore, lateral resolution improves.

↑ A.E.D.   ↓ Rate of Divergence   ↑ Lateral Resolution

The choice of which element size to use is generally based on two considerations: where on the patients' body the transducer is to be positioned, and the depth in the body to the structures of interest. While it is often desirable to use a large diameter to reduce beam divergence, this is sometimes not practical. For example, many times in echocardiography, a patient will have very narrow inter-costal spaces, necessitating the use of small diameters. In thyroid applications, the area to be scanned is irregularly shaped. Consequently, small diameters are helpful in maintaining good contact between the patient and the transducer. General abdominal and pelvic examinations usually do not have restrictions such as these, and large diameters can be used easily.

The second point to consider is the depth in the body you wish to image. For the best image detail, it is advantageous to have the ultrasonic beam as narrow as possible (minimum divergence). For superficial structures, the beam must narrow very close to the transducer face and lateral resolution at greater tissue depths is not critical; therefore, small diameter transducers are indicated. When the structures of interest lie deeper in the body (pelvic and obstetrical examinations, obese patients) large diameters are advised, since their decreased rate of beam divergence becomes important at greater tissue depths.

Focusing: Focusing a transducer is a means of minimizing the beam width and adjusting the focal zone to give optimum results for a particular examination. Acoustic lenses can be used to shape the ultrasonic beam pattern. The width of the beam can be made narrow with the result that better lateral resolution can be obtained. The focal point can be selected at different depths from the face of the transducer. The ability to select different focal points allows for the optimization of transducers for a particular type of studies. Modern transducers are internally focused and externally are of flat face.

The three important transducer parameters have been briefly touched upon. While selecting a transducer for a particular application, all three of these factors should be carefully weighed to achieve optimum performance.

23.3 MEDICAL ULTRASOUND

The use of ultrasound in the medical field can be divided into two major areas: the therapeutic and the diagnostic. The major difference between the two applications is the ultrasonic power level at which the equipment operates. In therapeutic applications, the systems operate at ultrasonic power levels of upto several watts per square centimetre while the diagnostic equipment operates at power levels of well below $100 \, \text{mW}/\text{cm}^2$. The therapeutic equipment is designed to agitate the tissue to the level where thermal heating occurs in the tissue, and experimentally has been found to be quite successful in its effects for the treatment of muscular ailments such as lumbago.

For diagnostic purposes, on the other hand, as long as a sufficient amount of signal has returned for electronic processing, no additional energy is necessary. Therefore, considerably lower ultrasonic power levels are employed for diagnostic applications. Since the absorption of ultrasound in tissue is proportional to the operating frequency which, in turn, is related to the desired resolution (ability to detect a certain size target) the choice of ultrasonic power level used is often dictated by the application. Diagnostic ultrasound are used either as continuous waves or in the pulsed wave mode. Applications making use of continuous waves depend for their action on Doppler's effect. Among the important commercially available instruments based on this effect are the foetal heart detector and blood flow measuring instruments.

There are, however, many applications where only pulsed waves can be employed. In fact, the majority of modern ultrasonic diagnostic instrumentation is based on the pulse-technique. Pulse-echo based equipment is used for the detection and location of defects or abnormalities in the structures at various depths in the body. This is possible because the time of travel of a short pulse can be measured with much greater ease as compared to continuous waves. Echoencephalograph, echocardiograph and ultrasonic scanners for imaging are all based for their working on the pulse technique.

The pulse-echo technique, basically, consists in transmitting a train of short duration ultrasonic pulses into the body and detecting the energy reflected by a surface or boundary separating two media of different specific acoustic impedances. With this technique, the presence of a discontinuity can be conveniently established and its position located if the velocity of travel of ultrasound in the medium is known. Also, it is possible to determine the magnitude of the discontinuity and to assess its physical size.

23.4 BASIC PULSE-ECHO APPARATUS

Pulse-echo technique of using ultrasound for diagnostic purposes in medical field was first attempted by making use of flaw detectors normally employed in industry for non-destructive testing of metallic structures. The basic layout of the apparatus based on this principle is shown in Fig. 23.5.

The transmitter generates a train of short duration pulses at a repetition frequency determined by the PRF generator. These are converted into corresponding pulses of ultrasonic waves by a piezo-electric crystal acting as the transmitting transducer. The echoes from the target or discontinuity are picked up by the same transducer and amplified suitably for display on a cathode ray tube. The X plates of the CRT are driven by the time base which starts at the instant when the transmitter

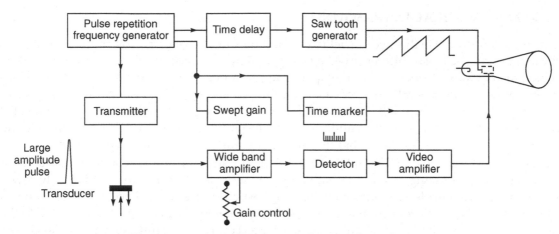

> **Fig. 23.5**   *Block diagram of a basic pulse-echo system*

radiates a pulse. In this way, the position of the echo along the trace is proportional to the time taken for a pulse to travel from the transmitter to the discontinuity and back again. Knowing the velocity of ultrasonic waves and the speed of the horizontal movement of trace on the CRT, the distance of the target from the transmitting end can be estimated.

The Probe: The transducer consists of a piezo-electric crystal which generates and detects ultrasonic pulses. The piezo-electric materials generally used are barium titanate and lead zirconate titanate. The crystal is cut in such a way that it is mechanically resonant of an increased efficiency of conversion of electrical energy to acoustic energy. It is usually one half wavelength thick for the particular frequency used.

When the transducer is excited at its resonance frequency, it will continue to vibrate mechanically for some time after the electrical signal ceases. This effect is known as 'after ringing' and destroys the precision with which the emission or detection of a signal can be timed. To reduce it, the transducer must have a good transient response and consequently a low Q is desirable. To achieve this, the transducers are normally damped. This can be done by controlling the rear surface to have a high impedance and high absorbancy of ultrasonic waves. This is to ensure that the energy radiated into it does not return to the transducer to give rise to spurious echoes. Backing material, therefore, is an important consideration in transducer construction. It is generally an epoxy resin loaded with a mixture of tungsten and aluminium powder. Backing material is made thick enough for complete absorption of the backward transmitted ultrasonic waves.

The probes are designed to achieve the highest sensitivity and penetration, optimum focal characteristics and the best possible resolution. This requires that the acoustic energy be transmitted efficiently into the patient. It is thus desirable to reduce the amount of reflected acoustic energy at the transducer-body interface. The single quarter wavelength matching layer accomplishes this by interposing a carefully chosen layer of material between the transducer's piezo-electric element and body tissue. A material with an acoustic impedance between tissue and piezo-electric ceramic is selected to reduce the level of acoustic mismatch at the transducer body interface. A uniform thickness of one-quarter wavelength for a frequency at or near the transducer's centre frequency results in higher acoustic transmission levels because of the favourable phase reversals within the layer.

The single quarter wavelength design, however, provides optimal transmission of ultrasonic energy at a particular wavelength only. This presents a problem for diagnostic pulse-echo ultrasound which is characterized by very short pulses containing a broad band of frequencies.

Also, the single quarter wavelength matching layer transducer has a face with a concave curvature. Occasionally, this can lead to air bubble entrapment or patient contact problems. Multi-layer matching (Fig. 23.6) technology overcomes these problems by interposing two layers between the piezo-electric element and body. Two materials are chosen with acoustic impedances between the values for ceramic and tissue. A stepwise transition of impedance from about 30 for ceramic to about 1.5 for tissue allows even further reduction of this acoustic impedance mismatch. The concavity can be filled with a material which is as acoustically transparent as possible, thus yielding a transducer with a hard, flat face for good patient contact, while minimally affecting the ultrasound beam.

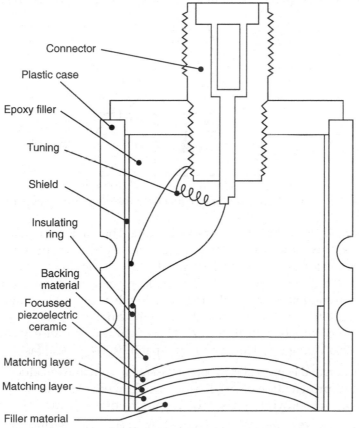

➤ **Fig. 23.6** *Multilayer matching between transducer's piezoelectric element and body tissue (Courtesy: K-B Aerotech, USA)*

Pulse Repetition Frequency Generator: This unit produces a train of pulses which control the sequence of events in the rest of the equipment. The PRF is usually kept between 500 Hz to 3 kHz.

There are several standard circuits for producing the required type of waveforms. These circuits could be the blocking oscillator or some form of the astable multi-vibrator. Mostly, the latter is preferred because the pulse duration can be more conveniently varied and the circuit does not require the use of a pulse transformer.

The width of the output pulse from the PRF generator should be very small, preferably of the order of a micro-second, to generate short duration ultrasonic pulse. It is a practice to use one astable circuit to generate a train of pulses with the required frequency and then to use them to trigger a mono-stable multi-vibrator which produces pulses of the required width. With the short pulse duration and the repetition rate of 1 kHz, only a few micro-seconds are occupied by the emission of the pulse, and the transducer is free to act as a receiver for the remainder of the time.

Transmitter: The transmitting crystal is driven by a pulse from the PRF generator and is made to trigger an SCR circuit which discharges a capacitor through the piezo-electric crystal in the probe to generate an ultrasonic signal. The circuit typically employed is shown in Fig. 23.7.

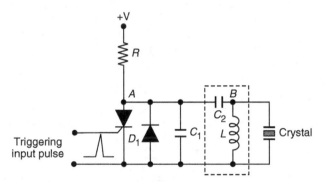

> **Fig. 23.7**   *Circuit diagram of a transmitter used in pulse-echo application*

Under normal conditions, the SCR is non-conducting. The capacitor C_1 can charge through the resistance R to the +V potential. If a short triggering positive pulse is applied to the gate of the SCR, it will fire and conduct for a short time. Consequently, the voltage at 'A' will fall rapidly resulting in a short duration, high voltage pulse at 'B'. This pulse appears across the crystal which generates short duration ultrasonic pulse. For producing a pulse with a very short duration it is necessary to use an SCR with a fast turn 'on' time and high switching current capability, which can be able to withstand the required supply voltage. SCR 2N4203 can be used because of its high peak forward blocking voltage (700 V), high switching current capability (100 A) and fast turn-on time (100 ns).

Receiver: The function of the receiver is to obtain the signal from the transducer and to extract from it the best possible representation of an echo pattern. To avoid significant worsening of the axial resolution, the receiver bandwidth is about twice the effective transducer bandwidth.

Transmitter-Receiver Matching: Ultrasonic pulse-echo systems generally use the same transducer crystal for both transmitting the ultrasonic energy and receiving the reflected echo. This permits the systems to have a compact transducer and also produces a symmetrical and well-defined beam shape. However, using a common source-receiver of ultrasound means that the sensitive input stage of the receiving amplifier must be protected from the high voltage transmission pulse. Such protection is usually provided by using a circuit shown in Fig. 23.8(a). This simple circuit has several disadvantages. During transmission, the series resistance R_s is effectively in parallel with the transducer and absorbs part of the excitation pulse. Also, if the transducer is power-matched to the receiving amplifier, then the choice of R_s is a compromise between loading during transmission and signal loss during reception. It is best to choose $R_s = R_i = R_{TD}$ (wherein R_{TD} is the transducer impedance). During reception, the presence of R_s degrades the signal-to-noise ratio due to signal attenuation; and Johnson noise and increased receiver amplifier noise due to raised source impedance.

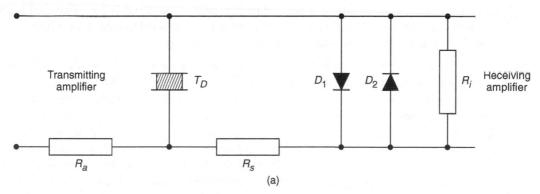

(a)

> **Fig. 23.8(a)** *Conventional circuit for input protection from large amplitude transmitted pulse in receiver*

Follett and Ackinson (1976) suggest an improved circuit to provide more effective transmitter/receiver switching. The technique has been further refined and shown in Fig. 23.8(b). The tuned transducer and input matching transformer form part of a Butterworth filter with Q of about 1.4 to reject transducer ringing. On transmission, the diodes conduct, protecting the receiver in conjunction with L_2 and effectively connecting L_2 in parallel with L_1. Since $L_2 - 4L_1$, this has negligible effect on the transducer tuning due to the low circuit Q.

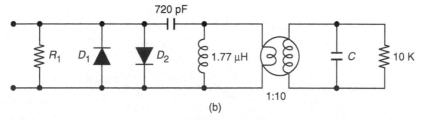

(b)

> **Fig. 23.8(b)** *Improved circuit for input matching of transmitter-receiver (after Follett and Ackinson, 1976)*

Wide Band Amplifier: The echo-signals received at the receiving transducer are in the form of modulated carrier frequency and may be as small as a few microvolts. These signals require sufficient amplification before being fed to a detector circuit for extracting modulating signals which carry the useful information. This is achieved in a wide-band amplifier, which is wide enough to faithfully reproduce the received echoes and to permit the use of different transducers operating at several different frequencies. A desirable gain of wide band amplifier is of the order of 80-100 dB. It must also have a very wide dynamic range so that the amplifier does not operate in the non-linear regions with large input signals. The amplifier must also have a low noise level to receive echoes from deep targets. In the modern instruments, the input amplifier is usually a dual gate MOSFET which is very suitable for high frequency signals and provides a high input impedance to the signals from the transducer.

Due to the wide dynamic range of echo-amplitudes that are contained in an ultrasonic image, a log amplifier is usually utilized. In this amplifier, the output voltage is proportional to the logarithm of the input voltage. By utilizing a log amplifier, one can see small relative differences in both low amplitude and high amplitude echoes in the same image.

*Swept Gain Control:*Stronger echoes are received from the more proximal zones under examination than from the deeper structures. The receiving amplifier can only accept a limited range of input signals without overloading and distortion. Abrupt changes in tissue properties that shift the acoustical impedance can cause the echo amplitudes to vary over a wide dynamic range, perhaps 40 to 60 dB. In order to avoid this, the amplifier gain is adjusted to compensate for these variations. This reduces the amplification for the first few centimetres of body tissue and progressively increases it to a maximum for the weaker echoes from the distal zone. In some instruments, segmented gain control arrangements are made to control the gain in segments on the time axis. This permits one to selectively amplify or reject echoes from different structures located at different depths. The swept gain profile can be adjusted and displayed on most of the modern instruments.

A simple technique (Fig. 23.9) of providing tissue attenuation compensation is to include a basic receiver gain control to echoes from near the skin surface. A typical control range is 0 to 60 dB. The rate of gain increase with depth is then set by a 'slope' or 'rate control' which typically provides for 0 to 10 dB/cm gain increase. The maximum gain value is reached by a 'far gain' control, also commonly with a 0 to 60 dB range (Maginness, 1979).

*Detector:*After the logarithmic amplification, the echo signals are rectified in the detector circuit. The detector employed could be of the conventional diode-capacitor type with an inductive filter

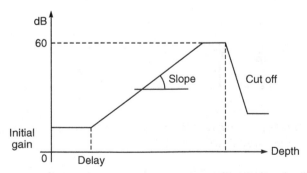

➤ **Fig.** 23.9 *A simple technique of providing tissue attenuation compensation in ultra-sound pulse echo systems*

to have additional filtering of the carrier frequency. In this rectification process, the negative half-cycles in the echo voltage waveforms are converted into positive half-cycles. This is followed by a demodulation circuit in which the fundamental frequency signal upon which the echo amplitude information has been riding, is eliminated. The output of the demodulator circuit is in the form of an envelope of the echo signal. The conventional demodulators which work well with low modulation frequencies give a much poorer performance with pulses containing only a few carrier cycles. With a peak input of 5 V, it is difficult to exceed a dynamic range of 25-30 dB. Much wider ranges of up to 40 dB can be obtained by cascade demodulation in which the individual demodulator characteristics are super-imposed to cover different input signal amplitude ranges.

The demodulator circuit may preferably employ synchronous demodulation intended for FM sound demodulation in television receivers. It consists basically of a trans-conductance multiplier with one input connected directly to the input signal and the other via a high gain limiting amplifier. The carrier frequency filtering is performed by a Gaussian filter which preserves pulse shape with minimum overshoot.

The output of the demodulator is the information desired, i.e. the amplitude of the echo signal and its time delay from the transmission pulse. The magnitude of the echo amplitude information can be controlled as it is written into the ultrasonic image. It is a great advantage to have this control due to the large variability of patient anatomy and acoustic parameters of normal tissue.

Video Amplifier: The signal requires further amplification after its demodulation in the detector circuit before it can be given to the Y-plates of the CRT. The output of the detector circuit is typically around 1 V, but for display on the CRT, the signal must be amplified to about 100 to 150 V. In addition to this, the amplifier must have a good transient response with minimum possible overshoot. The most commonly used video amplifier is the RC coupled type, having an inductance in series with the collector load. The inductance helps in extending the high frequency response of the amplifier.

The video signal is subject to some kind of compression function. This involves compression of echoes in the mid-amplitude range, denoting relatively larger sections of the output range to very weak and very strong echoes. This feature is useful in maintaining gain for weak echoes while avoiding saturation on strong echoes. With the use of digital scan converters, mid-range expand, mid-range compress, logarithmic and log-linear amplitude transfer functions can be made operator selectable.

The video signal representing the echo envelope is subjected to another processing step for 'edge enhancement'. It is obtained by partially differentiating the signal. The normal practice is to add a small fraction of the differentiated signal to the original signal. The added fraction is kept variable with an 'echo process' or 'enhancement control'.

Time Delay Unit: The time delay unit is sometimes required for special applications. Normally, the time base will begin to move the spot across the CRT face at the same moment as the SCR (Fig. 23.7) is fired. If desired, in special cases, the start of the trace can be delayed by the time delay unit so that the trace can be expanded to obtain better display and examination of a distant echo.

Time Base: The time base speed is adjusted so that echoes from the deepest structures of interest will appear on the screen before the beam has completely traversed it. Taking the speed of ultrasound in soft tissue to be about 1,500 m/s, a time of 13.3 μs must be allowed for each centimetre that the reflecting interface is below the surface. In many applications, distance markers appropriate to each time-base setting appear directly on the screen, which greatly simplifies distance measurements. Several standard circuits are available for generating the sawtooth waveform to provide a time base suitable for horizontal deflection of the spot on the CRT screen. The horizontal sweep generator is controlled by the PRF generator as the sweep starts at the moment that the transmitting pulse is applied to the transducer.

Time Marker: The time marker produces pulses that are a known time apart and, therefore, correspond to a known distance apart in human tissues. These marker pulses are given to the video amplifier and then to the Y plates for display alongwith the echoes.

Display: After amplification in the video amplifier, the signal is given to the Y plates of the CRT. CRT is not only a fast-acting device but also gives a clear presentation of the received echo signals.

There are two important controls on pulse-echo based instruments that are frequently used during a variety of examinations. These are the 'Reject' and 'Damping' controls. The reject setting controls the threshold above which the echo signal amplitude must rise to be visible on the A-scope and, as will be discussed later, to write on the B and M-mode displays. This control rejects or removes small amplitude inconsequential echoes that would otherwise produce noise signals in the displays or recordings. The damping control adjusts either the amplitude of excitation to the transmitting transducer element or the electrical load on the transducer in order to reduce the acoustical output. The effect is desirable enough to improve the echo resolution for near field interfaces, because reducing the transmitter excitation shortens the effective pulse duration.

23.5   A-SCAN

This type of scan offers only one-dimensional information. The echo signals are applied to the Y-deflecting plates of the CRT so that they are displayed as vertical blips as the beam is swept across the CRT. The height of the vertical blip corresponds to the strength of the echo and its position from left to right across the CRT face corresponds to the depth of its point of origin from the transducer.

For A-scan applications, the CRT is usually of the electrostatic deflection type. It is better to use CRT with post-deflection acceleration of the electron beam so that a very bright trace is obtained with lower deflecting voltages on the plates. The cathode ray tube should preferably be a flat face type to eliminate screen curvature error. A variable persistence scope with storage facilities would be useful for prolonged viewing.

23.5.1   Application of A-Scan

Echoencephalograph: In the normal brain, the mid-line surfaces are parallel to the flat areas of the bone near the ear. When there is a head injury, the brain gets tilted to one side or the other due to bleeding, but it still retains its normal shape. In such cases, the echoes can be easily obtained but they are placed at different distances from the probe, when the probe is placed first on one side and then on the other side of the skull (Fig. 23.10).

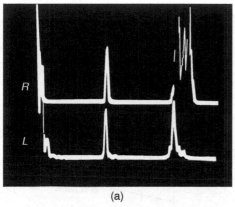 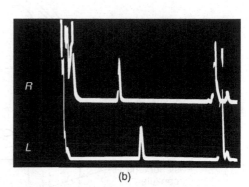

(a)   (b)

> **Fig. 23.10** *Echoes received from the brain*
>   (a) *in the normal brain, the mid-line echoes coincide for each way*
>   (b) *in the abnormal case, there is a shift between the two echoes*

Even when a tumour grows in the brain, the anato-my of the brain is gently altered and there is usually considerable tilting and displacement of the brain ventricles. Ultrasonic mid-line echo, in such cases, immediately establishes the abnormality of the brain due to its shift to one side from the centre.

The instrument for diagnosis and detection of the mid-line of the brain is called 'echoencephaloscope'. It usually incorporates a measuring range of 0–18 cm of tissue depth. The normally used frequency range is 1–3 MHz. The probe for 2 MHz with the diameter of 15–20 mm is the most common and gives a good resolution. The probe for a 1 MHz allows for deep penetration and may preferably be employed for elderly patients whose skulls are strongly calcified.

Echo-ophthalmoscope: A-mode ultrasonic technique has been found to be useful in ophthalmology for the diagnosis of retinal detachments, intra-ocular tumours, vitreous opacities, orbital tumours, and lens dislocation. It helps in the measurement of axial length in patients with progressive myopia, localization of intra-ocular foreign bodies and extraction of non-magnetic foreign bodies.

Echo-ophthalmoscopy employs a 7.5–15 MHz pencil type transducer. The transmitted pulse should be of very small width (in nanosec) and range.

In the normal eye, as is shown in Fig. 23.11, echoes can be obtained from the following structures when performing an anterior-posterior examination along the optical axis: the surface of the lid (corresponding with the zero marker that appears constantly on the trace); the surface of the cornea; the anterior lens capsule, the posterior lens capsule, the posterior wall of the globe (usually appearing as a complex of echoes originating from the retina, choroid and sclera as a unit) and retro-bulbar fat.

By using the distance markers on the trace, the depth and relative position of each structure represented in the display can be read directly from the screen or the resulting photograph. For example, in Fig. 23.11, the corneal echo appears at 2 mm and the poterior wall echo at 26 mm, giving a gross axial length measurement of 24 mm. In this eye, the depth of the anterior chamber and the thickness of the lens are each shown to be approximately 3 mm.

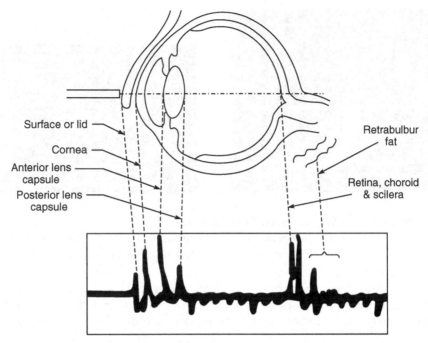

> **Fig. 23.11** *Acoustic landmarks in the normal eye along the optical axis. The electronic markers shown on the figure represent 2 mm of actual tissue depth*

➤ 23.6   ECHOCARDIOGRAPH (M-MODE)

Echocardiograph is a widely used and valuable instrument for carrying out cardiac examination and assessment of many congenital and acquired cardiac diseases. By using this instrument it is possible to detect intra-cardiac structures. The movement of these structures can also be recorded with the better resolution than with angiographic diagnostic technique. The instrument presents time-versus-motion information about heart structures on slow speeds normally used in electro-cardiogram recordings. When an ECG trace is super-imposed on the ultrasonic display, the movement of structures detected ultrasonically can be conveniently correlated with known events in the cardiac cycle. Phonocardiogram is also often recorded simultaneously.

The echocardiogram is currently the best method for the diagnosis of mitral stenosis. Echo-cardiography is also often used for the study of the aortic valve, tricuspid valve and pulmonary valve. Another very important use is in the detection of peri-cardial effusion, which is the abnormal collection of fluid between the heart and the peri-cardial sac. These examinations can be performed quickly and easily at the bedside, if necessary, without apparent risk to the patient.

In A-mode display, ultrasonic echoes produce vertical displacements of a horizontal trace on CRT. Basically, the amount of vertical displacement is proportional to the strength of the echo, and the distance along the horizontal trace represents the time of sound travel in human tissue, thereby permitting accurate measurement of tissue depth between any two echo sources. By electronically rotating the A-mode echoes 90° towards the viewer, the echoes can be presented as bright dots of light along an imaginary horizontal base line as shown in Fig. 23.12. The distance between the dots again represents time or tissue depth and the intensity of the dots represents the strength of the echoes.

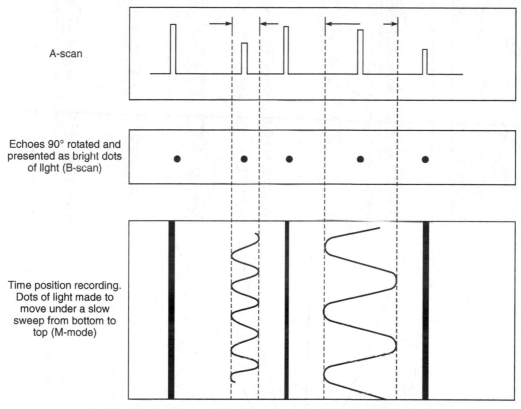

A-scan

Echoes 90° rotated and presented as bright dots of light (B-scan)

Time position recording. Dots of light made to move under a slow sweep from bottom to top (M-mode)

➤ **Fig. 23.12** *Principle of time-motion (M-mode) display*

If one of the echo sources is a moving structure, then the echo dots of light from that structure will also move back and forth. If the dots are made to move with an electronic sweep, from bottom to the top of the screen at a pre-selected rate of speed, the moving dots will trace out the motion pattern of the moving structure. This display is known as M-mode display. If a photographic film is continuously exposed to one sweep cycle of this display, a composite picture will result, providing a waveform representation of the motion pattern of the moving structure (Fig. 23.13). Alternately, thermal video printers are used for recording an echocardiograph.

Figure 23.14 shows the block diagram of an echocardiograph. Several circuit blocks are common to the general echo measuring instrument, except for the addition of a slow sweep circuit and recording arrangement.

Since the advent of two-dimensional echocardiography, ultrasonic examinations of the heart have been primarily recorded and stored on video tape. This medium is inexpensive and readily available and it is possible to record hundreds of cardiac cycles and multiple patients on a two-

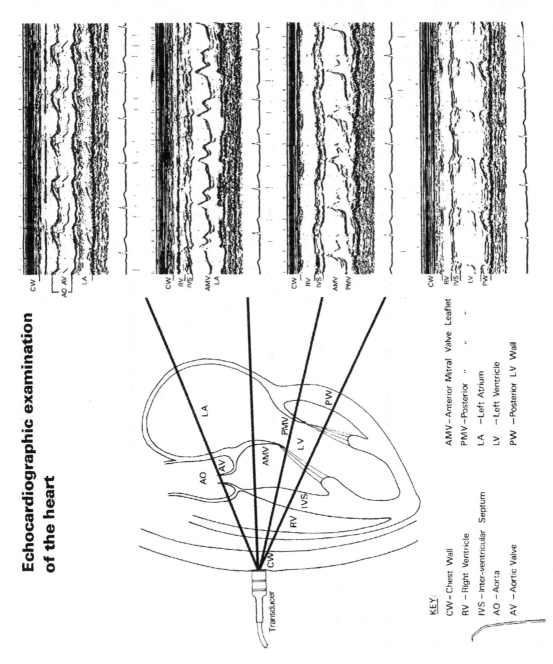

Echocardiographic examination of the heart

KEY:

CW – Chest Wall
RV – Right Ventricle
IVS – Inter-ventricular Septum
AO – Aorta
AV – Aortic Valve

AMV – Anterior Mitral Valve Leaflet
PMV – Posterior   "   "   "
LA – Left Atrium
LV – Left Ventricle
PW – Posterior LV Wall

➤ **Fig. 23.13**   *Typical M-mode display from various structures of the heart with different directions of the probe*

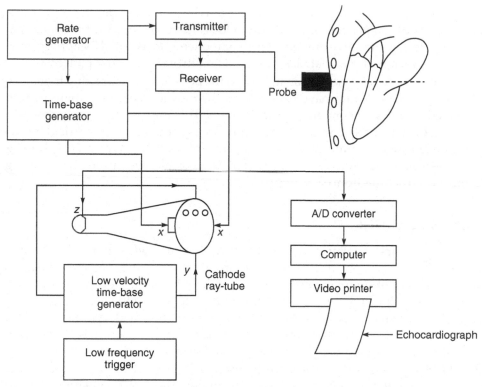

> ➤ **Fig.** 23.14 *Block diagram of an echocardiograph circuit*

hour video tape. However, many problems have arisen with video tape when used for recording echocardiogram. The major problem is that video tape technology limits the availability to quickly review the echocardiograms. Digital echocardiography which involves recording, displaying and storing an echocardiogram digitally has, therefore, largely replaced the video tape recording procedure. Also, measurements and interpretation can be done more conveniently using digital techniques of recording and manipulating imaging information.

For echocardiography, the transducer is placed between the third and the fourth ribs on the outer chest wall where there is no lung between the skin and the heart. From this probe, a low intensity ultrasonic beam is directed towards the heart area and echo signals are obtained. The probe position is manipulated to obtain echoes from areas of interest in the heart.

Pulsed Doppler echocardiography (Baker et al. 1976) depends on the sensing of blood flow velocity in contrast to M-mode echocardiography which is based on the anatomical (dimensional) properties of the heart. The technique is used as an adjunct to conventional M-mode echocardiography and frequently information obtained from the pulsed Doppler examination compliments or reinforces the M-mode procedure. In many cases, pulsed Doppler findings provide useful diagnostic information whereas M-mode findings may be nearly normal or suggestive.

The system operates on the principle of reflected ultrasound and senses flow velocity within a small 2×4 mm tear-drop shaped volume, referred to as the sample volume. The sample volume is specifically selectable within the heart and great vessels by means of a depth control setting and is subject to a variety of components of blood flow velocity; laminar, turbulent and motional components like wall motion and valve motion. These components are isolated by appropriate filters in the circuitry and each have associated audio-tonal qualities and spectral patterns.

A growing number of routine examinations and the possibility of extracting more quantitative data from the echocardiogram have necessitated the development of a computer system for semiautomated analysis of M-mode echocardiograms. The routine programme generally aims at the measurements which can be sub-divided into three groups viz.,: (i) ventricular dimensions; (ii) aorta and left atrium dimensions; and (iii) mitral valve measurements. Each group of measurements is preceded by a calibration, so that it is possible to use different recordings for the measurement of structures from each group.

23.7   B-SCANNER

Obviously the A-scope display is very difficult to interpret when many echoes are present simultaneously and often potentially useful information is wasted. A pictorial display can be conceived as a means of simultaneously presenting the echo information as well as information about the position of the probe and the direction of propagation of the sound. This is achieved in the B-scan display which results from brightness modulation with amplitude of the echoes obtained for various probe positions and orientations to produce a cross-sectional image of the object integrated by a storage display from individual scans.

Figure 23.15 represents the difference between A-scan and B-scan. Figure 23.15(a) is a hypothetical body with the probe placed upon its surface. The probe is positioned in such a way that it transmits a beam obliquely inwards. The beam encounters three interfaces in its travel. A-scan representation of this structure consists of vertical peaks (2, 3, 4) received as echoes at the receiving crystal in response to the transmitted pulse (1). The same structure in B-scan appears as light dots whose position is related to the echoing interface within the body.

In order to record cross-sectional pictures of internal structures, the ultrasonic probe is mounted on a mechanical scanner which allows movement in two directions and which links the direction and position of a B-scope time base on a CRT to those of the ultrasonic beam within the patient.

23.7.1   Types of Scans

Three types of scanning arrangements are utilized for building cross-sectional images using ultrasound. They are illustrated in Fig. 23.16. The most common scan used for abdominal studies is the linear scan (Fig. 23.16(a)). A linear scan is when the ultrasonic transducer remains parallel to the surface of the object being examined and the sound beam is perpendicular to the transducer movement. Only the location of the transducer is changed but the angle of the beam is held constant. The most common scan used in echocardiography is the sector scan (Fig. 23.16(b)). The scan is made by rocking the transducer about a fixed point such that the sound beam covers a sector. Compound scanning (Fig. 23.16(c)) is merely a combination of linear and sector scans.

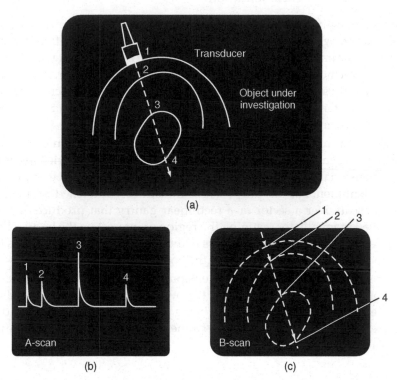

(a)

(b)   (c)

> **Fig. 23.15** *Difference between A-scan and B-scan displays*

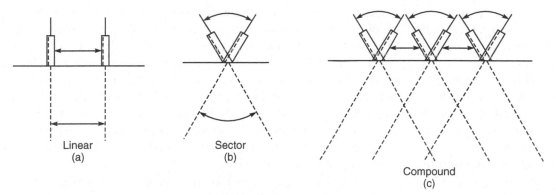

> **Fig. 23.16** *Types of scanning arrangement*
> *(a) linear scan,*
> *(b) sector scan*
> *(c) compound scan*

23.7.2 B-scanner Imaging Instrumentation

Modern ultrasonic imaging equipment is quite complex, with different manufacturers using different signal processing techniques in their equipment and offering different types of displays. However, all ultrasonic imaging systems contain the same basic building block circuits, namely: transmitter, receiver, memory and display. The transmitter and receiver circuits used in the imaging systems are similar to those which are explained under the basic pulse-echo system. However, in the case of imaging systems, the output of the video amplifier is given to the memory for scan conversion and subsequent display on a TV monitor.

B-scan is based on the fundamental information provided by echoes in an A-scan mode which is used to modulate intensity of the CRT electron beam instead of deflecting it vertically. For obtaining a 2-dimensional cross-section image (B-scan), it is necessary to know the transducer position and its orientation. For this purpose, the transducer is attached to either an articulated arm that allows a scan over a sector or a rectilinear gantry that produces a linear scan. The transducer is moved manually for scanning the region of interest on the body. The transducer is coupled by shaft encoders and position sensors to the co-ordinate generator so as to correlate the origin of the echoes from various structures of the body on the CRT scan.

B-scanning of static objects gives two-dimensional images which allow assessment of size, shape and position of the examined structures. If, however, the object is moving, the quality of the picture is degraded, the degree of degradation being proportional to the range and velocity of the movement. Therefore, fast working real time scanners have been developed to study both static and dynamic structures in the human body.

▌▌▌▶ 23.8   REAL-TIME ULTRASONIC IMAGING SYSTEMS

One of the serious limiting factors in B-scanning is the length of time taken to complete a scan. This results in blurring and distortion of the image due to organ movement, as well as being tedious for the operator. Elimination of motion artifact is important in conventional B-scanning but is critical if rapidly moving areas such as the chambers of the heart are to be made visible. Rapid scanning techniques have been developed to meet these needs. The approaches used include fast physical movement of a single transducer. Alternatively, electronic methods using arrays of transducers which can be triggered in sequence or in groups, may be utilized. In these systems, electronic manipulation allows the beam to be swept rapidly through the area of interest. Finally, instruments in which an array of transducers is combined with mechanized motion, have also been introduced. These instruments are called real-time scanners as there is negligible time delay between the input of data and the output of processed data in such systems.

A very important property of ultrasonography is the short image reconstruction time of about 20 to 100 ms, permitting the real-time scanning and observation of processes in the organs of the body. Consequently, sonography is well-suited to the fast, interactive screening of even larger organ regions and the display of dynamic processes, such as cardiology. Ultrasound section scans are, by contrast with CT and MR, largely free of kinetic blurring or motion artifacts (Haerten, 1994). On the other hand, ultrasound images are only section scans, and not whole body scans.

The real-time systems, therefore, have the following characteristics:

- Possibility of studying structure in motion—this is important for cardiac and foetal structures.
- During observation, the scan plane can be easily selected since the echo image appears instantaneously on the display.

23.8.1 Requirements of Real Time Ultrasonic Imaging Systems

The primary requirements of an ultrasonic imaging system are: high resolution, long range, adequate field of view, sufficiently high frame rate and high detectivity. However, high resolution or the ability of the system to resolve fine spatial dimensions is a key performance requirement. A resolution of 1–3 mm in all three spatial dimensions is desirable for a number of diagnostic studies like the early detection of tumours or other pathological conditions. The depth resolution (resolution along the axis) is usually governed by transducer pulse width and subsequent filtering. The round trip propagation time over a 2 mm distance is 2.67 μs. Thus, pulse widths in this range will provide the desired depth resolution. The lateral resolution (resolution along the beam diameter normal to the axis) is dominated by diffraction considerations.

The required depth range varies considerably for different anatomical studies. For example, a range of 25-30 cm is desirable for abdominal and obstetrical studies. For cardiac studies, the distance from the chest wall to the posterior heart wall is 15 cm or more. In superficial organs like the breast, thyroid, carotid and femoral arteries and in infant studies, the range of depth is 3–10 cm. Each of these depth ranges requires significantly different considerations as regards tissue absorption, specular reflection and small changes in tissue acoustic impedances. Generally high sensitivity receivers with wider dynamic range, low excitation frequencies and high transmitted powers help to extend the range. Focusing is usually necessary to achieve good resolution at the larger depths.

The field of view should be large enough to display the entire region under examination and to provide a useful perspective view of an organ of interest. When viewing small superficial structures such as the thyroid, a field of approximately 5 cm × 5 cm can encompass the desired region. For cardiac imaging, sector scans are preferable to rectilinear scans since a large structure is to be viewed through a small window. An angle of 60° is adequate to simultaneously view most of the heart. In abdominal studies, though the field of view is a matter of operator convenience, the viewed area should be at least 100 cm^2. It becomes difficult to view structures close to the surface when using sector scans because they fall in the apex of the sector. Therefore, two-dimensional arrays are used for abdominal studies.

In real-time imaging systems, the frame rate (the rate at which the image is repeated) should be rapid enough to resolve the important motions and to obtain the image without undesirable smearing. Most of these requirements are met with a frame rate of about 30 frames/s. This also satisfies the requirements for flicker-free display and is compatible with standard television formats. For some special studies, greater frame rates are often required for the data acquisition mode. These frames would then be stored and played back at about 30 frames/s to provide a slow motion presentation.

Detectivity is the ability of an imaging system to effectively capture, process and display the very wide dynamic range of signals which may, in turn, help to detect an image, lesion or other abnormal structure or process. Poor system detectivity manifests itself in lack of fidelity or picture quality which is often apparent in the visual displays of ultrasonic images.

23.8.2   Mechanical Sector Scanner

Mechanical scanning of a single transducer represents a low cost technique of extending the performance of a manually scanned system to achieve two-dimensional fields of view in real-time. In fact, the simplicity of the mechanical real-time sector scanner makes it an attractive device for medical applications. Several forms of motor-driven mechanical scanning of a single transducer have been used to achieve real-time two-dimensional imaging of the heart, the abdomen and the eye.

Schuette et al. (1976) developed a sector scanning system (Fig. 23.17) which employs a single ultrasonic transducer (2.25 MHz) through an angle that is programmable from 0 to 25 degrees. The sector scan rate is variable from 0 to 40 scans per second and scan linearity is achieved by electronic programming of the transducer position to match that of a triangular waveform.

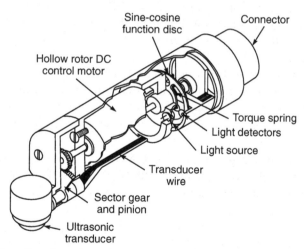

> **Fig. 23.17**   *Details of mechanical scanner with a single transducer (after Schuette et al, 1976)*

The transducer in the scanner is driven by a high performance, low inertia DC control motor. This motor incorporates a hollow-rotor armature yielding both low inertia and low inductance which are essential features for quick starting and stopping and for accurately following the rapid servo signals. The angular position of the transducer is determined photo-electrically from a sine-cosine function disc mounted on the rear of the motor shaft. Thus, the system incorporates many of the advantages of a small aperture electronically scanned array, while remaining compatible with most one-dimensional echocardiographic devices.

Holm et al (1975) constructed a mechanical scanner consisting of a rotating wheel (Fig. 23.18) on which are mounted four identical unfocused 2 MHz transducers with a diameter of 2 cm. The transducers are placed radially at 90° intervals with their fronts at the periphery of the wheel. The wheel is driven, via a Bowden cable and chain by an asynchronous motor. The rotating scanner can be used hand-held or mounted on a conventional scanning arm. With 4 revolutions/s of the transducer wheel, 16 frames/s are displayed so that 50° sector images are produced in real-time. With each transducer sweeping through 90° in 1/16 s, the 50° sector scanned during the generation of 69.4 pulses giving 69 lines in each frame or 1.38 lines/degree.

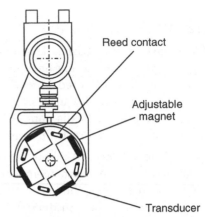

➤ Fig. 23.18 *Construction of the rotating transducer type scanner (after Holm et al, 1975)*

By super-imposing several sector scans, a large-area trapezoidal scan can be obtained. This is achieved by interposing a real-time scanner and the compound scanner. Figure 23.19 shows a sectional view of a mechanical trapezoidal scanning applicator. Three focusing indivi-dual probes rotate in a short intermediate path. Each probe covers a sector of 60° or 110°. The coupling mem-brane is shaped at the left side of the applicator in such a manner that it is also possible to take subcostal longitudinal views.

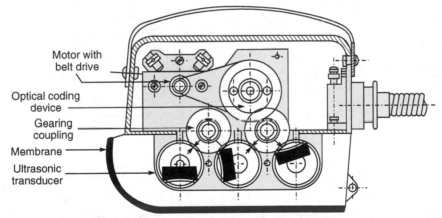

➤ Fig. 23.19 *Three transducers mounted on three separate wheels for a larger field of view*

▥▶ 23.9 MULTI-ELEMENT LINEAR ARRAY SCANNERS

Array transducers are available in several element configurations. Their arrangement determines the shape of the scan plane. For a two-dimensional image, the scanning plane is the azimuth dimension whereas the elevation dimension is perpendicular to the azimuth scanning plane. The shape of the region scanned for various array-element configurations are shown in Fig. 23.20.

Linear Sequential Arrays: The sequential linear array scans a rectangular region as shown in Fig. 23.20(a) . The scanning lines are directed perpendicular to the face of the transducer. The beam is focused but not steered. Linear array transducers are available with 512 elements in currently available ultrasound scanners. Generally, up to 128 element are selected at a time for operation. As is obvious from the diagram, the field of view with a linear array arrangement is limited to the rectangular region directly in front of the transducer.

Curvi-linear Arrays: As shown in Fig. 23.20(b) , a curvi-linear array scans a wider field of view because of the convex arrangement of the array elements. Curvi-linear or convex arrays operate in the same manner as the linear arrays and therefore, even in this case, scan lines are directed perpendicular to the transducer face.

Linear Phased Array: The arrangement of the elements is similar to the linear sequential arrays but the scanner steers the ultrasound beam through a sector shaped region in the azimuth plane. This is shown in Fig. 23.20(c). The present linear phased arrays systems have up to 128 elements and all the elements are used to transmit and receive each line of data. This arrangement has the

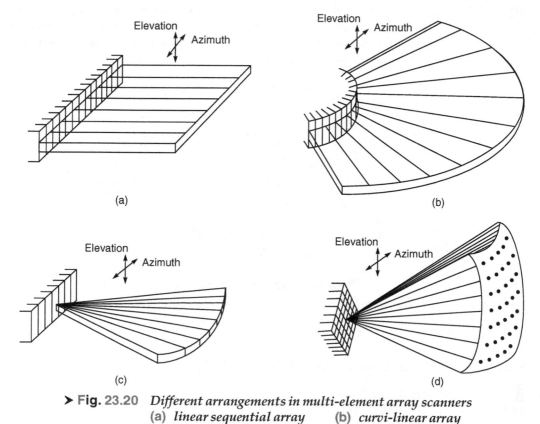

> **Fig. 23.20** *Different arrangements in multi-element array scanners*
> (a) *linear sequential array*   (b) *curvi-linear array*
> (c) *linear phased array*   (d) *phased array*

advantage that the scanned region is much wider than the footprint of the transducer, thus making them suitable for scanning through restricted acoustic windows such as in the case of cardiac imaging wherein the transducer has to look through a small window to avoid the obstructions of the ribs and lungs.

Two-Dimensional Phased Array: A two-dimensional phased array has elements in both the azimuth and elevation planes. Therefore, two-dimensional arrays can focus and steer the acoustic beam in both dimensions , thereby to produce a volumetric image as shown in Fig. 23.20 (d).

The physical shape of the various types of ultrasound probes is shown in Fig. 23.21.

Probes

A Linear Array probe 3.5 and 3.5/5.0 MHz, 12 cm
B Linear Array probe 5.0/7.5 MHz, 7.4 cm
C Linear Array probe 7.5 MHZ, 6 cm
D Curved Array probe 3.5/5.0 MHz, R75
E Curved Array probe 3.5 and 3.5/5.0 MHz, R40
F Curved Array probe 5.0 MHz, R40
G Linear Array probe 7.5 MHz, 4 cm
H Multi-angle sector probe 5.0/7.5 MHz
I Annular Array sector probe 3.5 MHz
J Annular Array sector probe 5.0 MHz
K Multi-plane endorectal probe 5.0/7.5 MHz
L Endovaginal probe 5.0/7.5 MHz
M Multi-plane endovaginal probe 5.0/7.5 MHz
N Curved Array probe 5.0/7.5 MHz, R17

> ➤ **Fig. 23.21**   *Various types of ultrasound probes used for real time scanning*

23.9.1   Linear Array Scanners

The principle of linear array scanners is described by Bom et al. (1973) and can be summarized as the use in rapid succession of a number of parallel single elements with the display of each element in the brightness mode (B-mode) at almost the same instant. The transducer used comprises 20 elements placed on a line array of 8 cm. Each element measures 4 by 10 mm and the active area of the transducer is 80×10 mm. The depth penetration is designed for 16 cm, and therefore, the cross-section covered measures about 8×16 cm.

The elements transmit a short acoustic pulse sequentially into the tissue. Returning echoes are displayed along a horizontal axis on the CRT while the vertical position of each line corresponds to the position of the respective element. A block diagram of the principal components of the apparatus is given in Fig. 23.22. In transmission, a tone burst of the resonant frequency of the crystal is fed through an electronic switch into a single crystal element. The element generates a short acoustic pulse. The echo signals arriving at this element when in the reception mode are amplified by a wideband preamplifier and fed through a switch, into the processing part of the apparatus. This cycle is subsequently repeated for all elements. A time gain compensation circuit is used to compensate for the attenuation of ultrasound by spherical spreading and absorption in tissue. The display of the image can be on any of the standard devices.

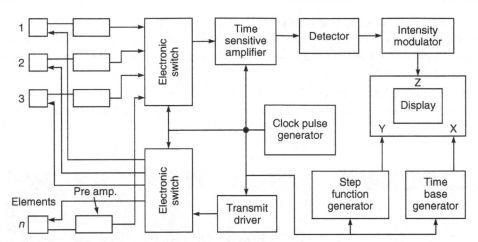

> **Fig.** 23.22   *Principle of multi-element scanning system. Block diagram shows major*
>   *components in a multi-element signal processing and display system*

In multi-element linear array transducers, the scanning action is based upon electronic beam stepping. However, the number of lines per scan is increased by using a group of elements as a composite transducer. This composite transducer may be progressively stepped, along the array in steps equal to the width of the narrow elements, by dropping the last element from the group and adding a new element to the front of the group. This produces a reasonably well-collimated beam to be moved through a series of overlapping positions along the array. Linear array applications currently available contain a large number of elements ranging from 64 to 420. They are usually activated in groups of eight, with progressive beam stepping in a sequence.

The linear array principle with beam stepping is described by Zurinski and Haerten (1978). The array applicators contain a linear array of 54 transducer elements. An initial pre-selected group of elements is actuated for the transmission of an ultrasonic pulse and receives the returning echoes. These form one line of the two-dimensional image on the monitor. By switching in an element of the selected group on one side, the image line is shifted by one half an element width. In the next step, the image line is again shifted by another half an element width by switching off an element on the other side of the group. In this way, an image is built-up which consists of 101 lines. By selecting the group width, i.e., the number of elements per group, the position of optimum lateral resolution (focus) can be located in different depth ranges. With this system, the information content is doubled as compared with conventional techniques. This is achieved by a special type of stepped group-switching and does not involve line interlacing and multiple recording. Normally, an element group is moved in the scanning direction in such a way that the first element of the group is switched off and at the same time, on the other side a new element is switched on (Fig. 23.23). The spacing between the individual scans corresponds to the element spacing (raster spacing).

The process is as follows: Firstly, a group consisting of m elements is selected (m = 2, 3 or 4). It transmits and receives in unison. A further element is then switched in without at the same time another one being switched off. This group has m + 1 elements. The central axis of the second

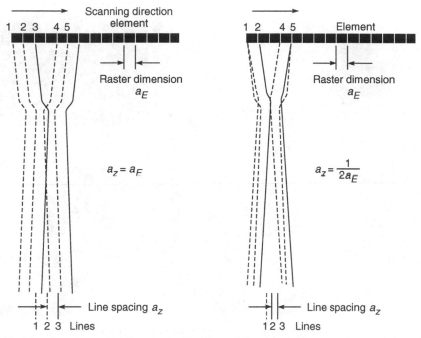

> ➤ **Fig. 23.23** *Group switching for selection of the focal distance in a multi-element scanning system (Redrawn after Zurinski and Haerten, 1978)*

group (m + 1) is shifted in the scanning direction by half the raster spacing in relation to the first group (m). After a new transmit receive cycle, the first element is switched off. The third group again has m elements and their centre axis is also shifted onwards by half a raster spacing. For the fourth group, a new element is switched in etc. The ultrasonic line sequence is certainly recorded alternately with two different group widths and sound fields but this does not impair the information. A commercial instrument based on this principle is shown in Fig. 23.24.

Delay lines are used to provide the required beam position and to shape in array transducers. To meet these requirements, the delay lines must have adequate bandwidth to preserve the pulse spectrum (2 MHz), large dynamic range (not less than 40 dB); tolerable insertion loss and adequate range (up to 10 microseconds), e.g., for 45° deflection in a 2 cm wide beam deflection array. Methods of providing electronically controlled delays include: (i) lumped constant delay lines with voltage-controlled capacitive elements; (ii) commutative delay lines where delay is determined by clock frequency; (iii) fixed delay lines, either tapped or combined in series by electronic switching; and (iv) charge coupled device (CCD) delay lines.

So simple in principle is the instrumentation associated with linear array real-time scanning that small, self-contained, hand-held battery-operated versions have been developed. This type of instrument clearly has an important potential application in the clinic, but the difficulties posed by the limitation of the CRT as a display device in portable equipment and limited resolution is presently limiting its use.

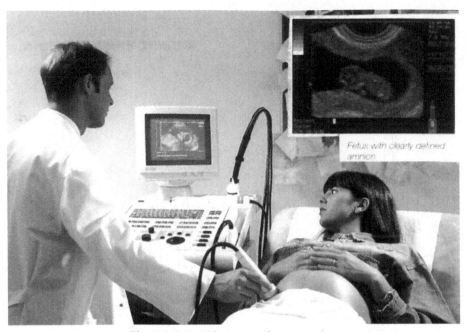

> **Fig. 23.24**   *Ultrasound scanner in use*

With rapid developments in technology, ultrasonic imaging systems are increasingly making use of digital techniques for beam forming and control, which provides better image quality than analog systems. Figure 23.25 shows a simple diagram that illustrates the difference between analog and digital ultrasound systems. In the case of the digital systems the received analog signal that comes from the transducer is digitized immediately. All the image processing and manipulation is done in the digital domain. In the case of the analog system, the digitization is done after a substantial amount of the processing has been done. The most important parameter in ultrasound image processing is the beam formation, which provides the focusing for the ultrasound beam.

In a transducer with multi-element crystals, the beam is focused and steered by exciting each of the elements at a different time so that the resulting ultrasound wave coming from each crystal will

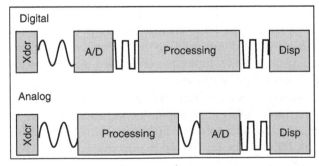

> **Fig. 23.25**   *Difference between analog and digital ultrasound systems*

arrive at the intended focal point simultaneously. Incorporating delay lines usually does this. Obviously, digital time delays would permit much greater precision in the shaping of the ultrasound beam and do not require frequent calibration associated with analog beam formers. Also, the digital beam former permits the system considerable flexibility in re-programming the size, shape, direction and intensity of the beam.

Although ultrasound resolution is affected by several factors, imaging frequency has the most direct influence. Ultrasound resolution improves in direct proportion to the frequency employed. In a typical 5-10 MHz system, the resolution cell measures roughly 0.7×0.35 mm. The result is that an anatomical structure smaller than 1 mm is likely to be missed.

Generally, the maximum imaging frequency is limited by the speed of the system's analog-to-digital converter. Conventional systems use A/D converters running at approximately 20 MHz. This limits the maximum imaging frequency to 10 MHz accordingly to the Nyquist sampling theorem. Moreover, as imaging frequencies increase, transducer design and fabrication become increasingly difficult. Further, the ultrasound penetration in the body decreases as the imaging frequency increases.

Recent advances in ultrasound transducer and electronics have enabled an extension of the ultrasound frequency for imaging beyond the conventional 5–10 MHz range. A high speed 40 MHz A/D converter has raised the Nyquist frequency, making 20 MHz imaging a possibility, to achieve a spatial resolution as fine as 70 micron. GE Medical offers a linear transducer which provides 12 MHz imaging frequency. With this system, objects smaller than 200 microns have been seen, opening an era of micron imaging. These systems have demonstrated axial, lateral and contrast resolution similar to that of Magnetic Resonance Imaging (MRI). Breast imaging is likely to see major benefits to achieve clearer visualization of the internal structure of lesions in the breast.

In such systems (Fig. 23.26), each one of the eight channels, selected in the array is amplified by means of its own amplifier and is digitized. The A/D converter used has a conversion rate of 10 ns. Afterwards, the signals of all eight channels are delayed and added in a high speed computer which carries out real-time computation. Because of the maximum transducer frequency of 7 MHz and the eight-channel operation, clock frequencies of upto 56 MHz occur. Such fast operations are carried out by ECL (emitter coupled logic) rather than by TTL (transistor-transistor logic) devices.

The linear array scanners mostly find applications in areas requiring a wide field of view such as obstetric scanning, rapid screening and orientation. On the other hand, real-time mechanical sector scanners are specifically designed for abdominal imaging for scanning organs such as the gall bladder, pancreas, and spleen, as well as hard-to-access areas of the liver and kidney and for structures such as those under the ribs or deep in the pelvis. Modern ultrasonic imaging systems sometimes incorporate both these techniques in single units in order to enhance their utility and economize on the cost by using common display for both the modes of operation.

Linear array transducers have increased versatility over mechanical transducers. Electronic scanning involves no moving parts and the focal point can be changed dynamically to any location in the scanning plane. The system is capable of generating a wide variety of scan formats. The disadvantages of linear arrays include the increased complexity and higher cost of the transducers. For high quality ultrasound images, many identical array elements are required. The

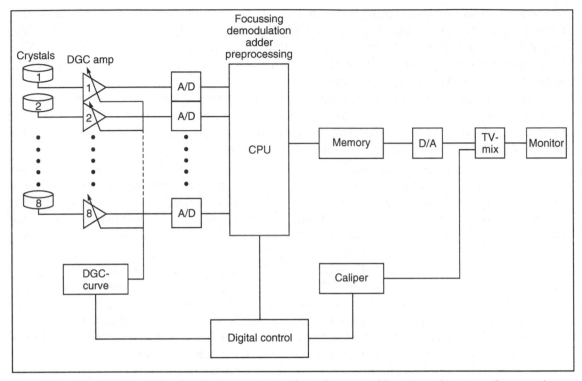

> **Fig. 23.26**   *Principle of real time computer based scanner (Courtesy: Siemens, Germany)*

array elements are typically less than a millimeter in thickness and each requires a separate connection to its own transmitter and receiver.

23.9.2   Phased Array System

Figure 23.27 shows an electronic sector scanning technique which uses a multi-element array transducer. Each element is connected to its own independent oscillator, which can be triggered at the right moment. Triggering the local oscillators subsequently with equal time intervals will cause the individual wave-fronts to construct a resulting flat wave-front, the angle of which depends on these time intervals. If time delays of the wavefronts are $d \sin \theta/c$ (where c is the ultrasonic velocity and d is the inter-element distance) with respect to the reference element, then the direction of the main beam will have an angle θ with the normal. When these time intervals are zero, it means that all elements are excited simultaneously and the result is identical to that of a conventional transducer. Electronically variable pulse-delay circuits are used for delaying the trigger pulse which permits variation of the beam direction with great rapidity.

When using a pulse repetition rate of 1000 pulses per second, a sector can be scanned 30 times per second when 32 different directions are used. In practice, 128 lines are mostly used which means that every line on the screen is repeated about 8 times per second. A sector of 90° can be easily obtained using this technique.

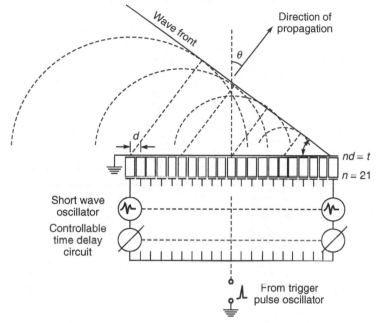

➤ **Fig. 23.27** *Beam pattern in a phased-array system*

For optimum lateral resolution, it is necessary to also make the reception of the echoes directional. Figure 23.28 shows the principle of forming a received beam with an arbitrary direction using a system of fixed and electronically variable delay lines, which compensates for the differences in arrival time on the subsequent elements of an impinging wavefront from every direction. The same control voltage controls both the direction of transmission and that of reception, in order to always be maximally sensitive in the direction in which the pulses are transmitted. The control voltage also controls the direction of the line on the CRT screen. The techniques of real-time imaging with phased array ultrasound scanners is widely used in commercial scanners.

Phased array scanners are particularly useful for cardiac ultrasound sector scanning. A transducer measuring 1.2 cm × 1.3 cm (2.25 MHz) or 1.2 cm × 0.8 cm (3.5 MHz) is commonly used. The transducer is thus small enough to be easily manipulated by hand and the beam is small enough at the transducer face to be directed through the inter-costal spaces in any desired orientation. It yields an 84° sector angle, and produces 64 shades of gray to all sector displays. The gray scale controls allow an emphasis on select echo amplitude ranges. For example, it is possible to visualize such structures better as valve orifices during studies of mitral or aortic stenosis.

23.9.3 Area Array Systems

For achieving high spatial resolution throughout a volumetric field of view, two-dimensional area arrays of transducers have been developed. In this system, an image of any place within the object can be focused on a two-dimensional transmit-receive array of piezo-electric transducers by means of an acoustic lens. The two-dimensional area array (Fig. 23.29) consists of 16 × 16 matrix of

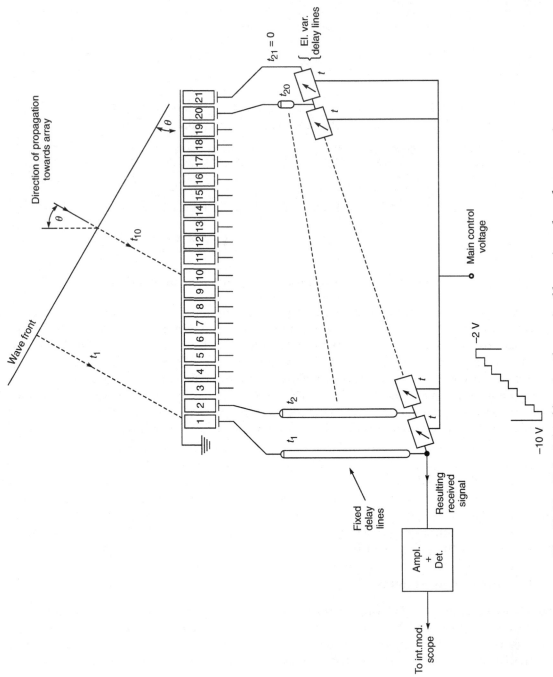

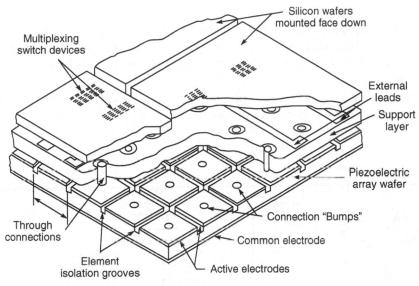

Multiplexing switch devices

Silicon wafers mounted face down

External leads

Support layer

Piezoelectric array wafer

Connection "Bumps"

Common electrode

Active electrodes

Element isolation grooves

Through connections

> **Fig. 23.29** *Construction of two-dimensional ultrasonic array using integrated circuit technique (after Meindl, 1976)*

2.25×2.25 mm transducer elements. A module of 16×16 matrices is used to form a 32×32 element array. The operating frequencies may be 2.25 or 3.5 MHz. In the time sequence, each element in the array operates in a pulse-echo mode as both the transmitter and receiver of ultrasonic energy. Accessing individual elements within this array is possible through an x-y address system consisting of row and column inter-connect lines and double-diffused metal-oxide-semiconductor transistor (DMOST) multiplex switches, co-located with each transducer element. Applying address signals to a single x-control line and a single y-control line connects the transmitter and receiver to one and only one transducer element. The piezo-electric elements and multiplexer switches are made in an integrated form. The acoustic lens provides high lateral resolution, a large depth of focus and a sector scan pattern which requires a minimal body aperture.

The area-array transducer permits operations in several different modes (Meindl, 1976). Any full row or column of elements may be scanned using either intensity modulation (B-scan) or beam deflection modulation (multiple A-scan) to produce cross-sectional displays. Electron switching from row-to-row or column-to-column at a slow rate provides the capability for online three-dimensional target visualization throughout a pyramidal volume. The range and frame rate of the system can be controlled to obtain deep body imaging in real-time. Thus accurate dynamic measurements of heart chamber volumes are quite feasible in this system. However, the area array involves a rather complex electronic circuitry considering the number of transmitters, receivers, delay lines and other circuits, etc.

23.9.4 Duplex Scanner

Duplex echo-Doppler scanning has been found to be useful for direct non-invasive measurement of peripheral arterial flow and tissue geometry. Here, the echo image is used to locate the artery and determine the position of the walls. The echo image is then used as a guide to map out the flow field with the Doppler. The incorporation of separate echo and Doppler transducers in a duplex scanner results in high resolution images due to the fact that the echo transducer beam can be aligned perpen-dicular to the artery wall, while the flow probe is at an angle appropriate for detecting Doppler shift.

Figure 23.30 shows the major components of the duplex scanner (Barbaro and Macellavi, 1978). It basically comprises the scanner head, the pulse-echo unit, the pulse Doppler unit, the duplex control unit and the display.

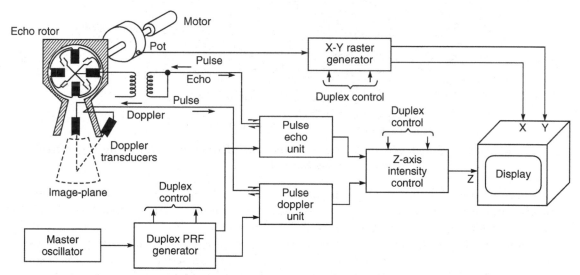

> **Fig. 23.30**   *Block diagram of duplex scanner (after Barber et al, 1974)*

Echo Rotor: Pulse-echo imaging is performed by rotating a transducer so that its beam sweeps out the image plane. The rotor contains four identical transducers but at any one time, only the one pointing in the direction of the image plane is used. The proper transducer is automatically selected by an external magnet and the reed switches shown in the rotor diagram. The proper transducer is connected through its switch to a rotary transformer to the pulse-echo unit. The rotor is driven by a DC motor and the position of the rotor is measured with a precision sine-cosine potentiometer. The peration of the system, in most respects, is therefore, independent of the speed of the motor.

The echo transducers have a 5 MHz piezo-ceramic element that is 1 cm in diameter. Each element is focused with an epoxy lens with a focal length of about 5 cm.

Doppler Transducer: There are two Doppler transducers, selectable by a slide switch mounted in the wall of the housing. With the switch in the position shown, the pulse-Doppler is connected to the transducer whose beam is contained within the image plane, and is fixed at an angle of 37.5° to the vertical. The other transducer is not in the plane of the image and is shown in projection on the diagram. Its beam intersects the image plane at a single point at an angle of 37.5° to the plane. This transducer is useful when one wishes to visualize an artery in cross-section, in which case the flow would be perpendicular to the image plane and the Doppler shift measured with the other Doppler transducer would be theoretically zero.

The pulse-echo unit contains an avalanche transistor pulser, a 5 MHz broadband amplifier with time-compensated gain, and a video detector. The pulse-Doppler unit also runs at 5 MHz. The output going to the Z-axis intensity control is normally a pulse taken from the range-gate generator.

The system runs off a 5 MHz master oscillator. The echo period is the time between each pulse from the echo PRF pulse train and is 100 µs. The Doppler period is 60 µs, a time which is shorter because the Doppler transducer is located near the skin surface. These and other logic signals control the echo and Doppler units, and the display circuitry.

The rotor potentiometer, a ramp generator and the duplex control signals are combined to generate both echo and Doppler x-y signals. The intensity of the display is quantized to four discrete shades of gray. The echo video signal is converted to a set of parallel binary signals which are re-combined with the Doppler range information, determined by the duplex control. This real-time quantization technique greatly enhances the apparent resolution of the B-mode image.

Modern duplex scanner systems are microcomputer-based which provide dual frequency pulsed Doppler with two-dimensional imaging, A-mode and gray scale M-mode. The dual frequency Doppler operates at a frequency of 3 MHz for abdominal and adult cardiology studies and at 5 MHz for peripheral vascular disease and for carotid and paediatric flow studies.

Just as echo-cardiography has greatly expanded our ability to understand spatial anatomic relationships of structures within the heart, the ability to simultaneously depict blood flow in a cross-sectional plane would greatly enhance understanding of the physiology of blood flow in the normal and diseased heart. In recent years, devices have become commercially available which literally allow one to see the blood flow through the heart alongwith echo-cardiography. The technique uses moving target indicator procedures to detect reflections from blood cells as they move through the scanned cross-section. The direction of flow is displayed in colour with one colour to display flow towards the transducers and another to display flow away from the transducer. The velocity of flow is estimated from the display by the brightness of the colour display.

23.9.5 Intravascular Imaging

Intravascular ultrasound has recently attracted considerable interest because it yields high-resolution images not only of the vessel lumen but has the unique potential of visualizing the vessel wall and pathology. Although intravascular imaging has a long history of research behind it, beginning in the early 1970s with the pioneering work by Bom et al (1972), it is only recently that it has shown potential for intra-cardiac application. However, many technical problems remain to be solved before the technique will attain full clinical utility and reliability.

The normal first order coronary is 2–5 mm in diameter, but second order vessels and atherosclerotic lumina often reach 0.1–2.0 mm. The small size of these vessels puts a limit on the size of the transducer and the acoustic power to obtain a favourable signal-to-noise ratio. The transducer is normally carried on the tip of a catheter, which should be flexible enough for safe passage into the vessels. The flexibility not only ensures that the catheter can be manoeuvred to a central and coaxial position in the vessel but also enhances both safety and image quality.

Various technical approaches have been employed to construct an ultrasound probe capable of intra-luminal examination (Nissen & Gurley, 1991). Basically, there are two fundamental approaches: one utilizing a mechanically rotated transducer or acoustic mirror, the other utilizing an electronic array.

Figure 23.31 shows an intra-luminal ultrasound catheter probe for two-dimensional imaging of the coronary arteries. The device comprises a fixed transducer element at the tip of the catheter and a rotating mirror driven by an external motor and a flexible drift shaft. By using this device, we can obtain radial cross-sectional images of the artery. After placing a guide wire in the coronary artery, the catheter is passed by guide wire control into the selected area of the artery and positioned under fluoroscopic control. The outer diameter of the catheter is typically 1.4 mm. At an ultrasound frequency of 30 MHz, high resolution images of the vascular pathology are obtained. Despite their apparent simplicity, mechanical transducers are difficult and costly to manufacture within consistent mechanical tolerances. (Bommer and Haerten, 1991).

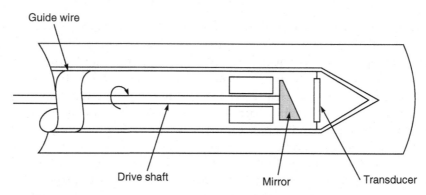

> **Fig. 23.31**　*Schematic of intraluminal ultrasound catheter device*

In another approach, very small Doppler probes have been incorporated into guide wire tips for introduction into the coronary artery. Figure 23.32 shows a schematic illustration of the Doppler guide wire. A piezo-electric element is incorporated into the tip of a 0.014" or 0.018" guide wire to produce a forward looking pulsed Doppler signal at typically 15 MHz.

The multi-element electronic array uses 1.83 mm diameter, 20 MHz intravascular ultrasound probe with no mechanical parts. The device employs an array of 64 transducer elements mounted at the tip of a 4.5 F catheter. The design permits the central lumen to accommodate a standard 0.014" guide wire to facilitate placement in the artery. The electronic signals are amplified and multiplexed by micro-miniature integrated circuits within the transducer assembly. The image reconstruction algorithm is highly complex and utilizes a principle known as a 'synthetic aperture array' and not the conventional phased array approach. With the synthetic aperture array, the multi-element transducer sends and receives ultrasound signals sequentially from each of the elements. Multiple elements contribute to the reconstruction of more than 1000 radial scan lines at 10 frames per second. For nearby pixels, only a few adjacent elements are utilized, whereas more

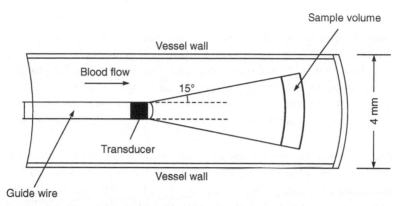

> **Fig. 23.32** *Schematic of intraluminal doppler guidewire device*

distant targets employ larger groups of elements. This variable aperture approach provides a device that theoretically remains focused from the surface of the catheter to infinity.

23.9.6 Three-dimensional Display

Ultrasonic imaging portrays qualitative information (for example, shape and physical context, and tissue type) and quantitative information (for example, organ dimensions, tissue properties such as attenuation and speed). Two-dimensional echo-cardiography allows comprehensive exploration of the structure and function of the heart by mental integration of the information from multiple cross-sectional views. Accordingly, a computer based three-dimensional reconstruction is a logical step towards improved appreciation of complex cardiac structures and reliable quantification of volumes.

Extensive research work in the field of three-dimensional echo-cardiography has been going on since the introduction of cardiac ultrasound imaging was established as a diagnostic tool. Its clinical implementation has not become widely popular because of a number of technical problems related to data acquisition and reconstruction. Specifically, these problems include spatial location of the image planes, temporal alignment of the image frames within the cardiac cycle, correction for respiratory motion of the heart, rapid data transfer, storage of large amounts of data, development of suitable algorithms for reconstruction and finally, display of three-dimensional images on a two-dimensional screen.

The most widely applied approach (Roelandt et al, 1994) for the three-dimensional image acquisition is the rotation of the scan plane around a central axis from a fixed transducer position. For data acquisition, the image plane is rotated in regular increments spanning a total angle of 180°. In this way a cone-shaped volume is scanned with its apex located at the transducer position. For each scanning position, sequential images covering an entire cardiac cycle are required and transferred either to the hard disk of the workstation or to the optical disk of the ultrasound system. Images can be stored using a video or a digital format. In order to obtain homogeneous data sets, only image sequences from cycles that fall within pre-selected limits of the R-R interval, are accepted. The final data set consists of the gray level values of all pixels with their exact spatial location and their temporal location within the cardiac cycle. For processing and reconstruction, data is later retrieved by disk by independent three-dimensional workstation. For three-dimensional display, the reconstructed volume can be dissected along any selected plane and cardiac structures of interest can be viewed from different angles. Several algorithms such as distance coding are used to create the illusion of depth. Data can be displayed as a dynamic loop sequence of an entire cardiac cycle with moving structures viewed from a selected angle. McCann et al (1988) illustrates the in *vivo* and *in vitro* techniques for three-dimensional ultrasound imaging for cardiology.

The principle of three-dimensional imaging, in a simplified form, is illustrated in Fig. 23.33 (McCann et al 1988). The transducer (5MHz) is attached to a stepper motor which rotates through 180° to obtain 100 views at 1.8° increments. One level (n) from each of the section scans is stacked into the volume image at the proper angle to generate the 'n' th section of the image. The whole sequence is recorded on videotape. A sequence of frames is selected and digitized to generate a $512 \times 512 \times 8$ bit image for each angle of view. By the process of sequential stacking of the sector scans at various positions of the transducer, the final volume image is constructed. Comprehensive user friendly software packages are now used in commercial scanners to give a three-dimensional display of scanned images.

23.9.7 Tissue Harmonic Imaging

When an ultrasound signal pulse propagates through the target media, a fundamental change occurs in the shape and consequently frequency of the transmitted signal. A sound wave propagating through the body is, in fact, a pressure wave. This pressure wave compresses and relaxes the tissue. Where tissue is compressed, the speed of sound is higher, effectively causing the top of the pressure wave to be pulled forward. As the tissue relaxes, the speed of sound becomes lower. The distortion of the waveform causes harmonics to be generated. For example, a transmitted frequency of 3.0 MHz will also return a harmonic frequency of 6.0 MHz (second harmonic) and 9.0 MHz (third harmonic). In view of the availability of harmonics in the returned echo signals combined with the advanced signal processing techniques, a new ultrasonic imaging technology called tissue harmonic imaging has recently been developed (Haerten et al. 1999).

In tissue harmonic imaging systems, the returning high frequency signal has to travel in only one direction i.e. return to the probe. The displayed image would have the benefits from the attributes of high frequency imaging and a one-way travel effect. Apart from increase in axial, lateral and contrast resolution, there is a substantial decrease in beam aberration, reverberation and side lobes.

Tissue harmonic imaging relies on the time and distance travelled and filtration of unwanted frequencies. Time is a critical factor in that the signal pulse must travel a certain distance for the change in the ultrasound frequency to occur. The harmonic effect is cumulative and builds up with increasing penetration depth. For this reason, harmonic imaging will be most efficient in the mid-to far-range before absorption limits its penetration. Similarly, the filtration of unwanted frequencies must be done to return the precise harmonic frequency necessary for optimal image quality.

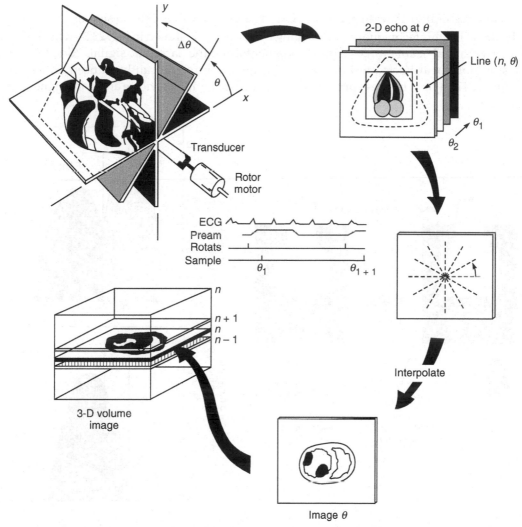

> **Fig. 23.33** *Principle of three dimensional imaging system (Redrawn after McConn et al, 1988)*

The conventional method of generating ultrasonic images based on harmonics uses a narrow band pass filter to separate the strong fundamental echo components from the harmonic components. This results in a loss of contrast resolution due to the inherent overlap of the fundamental echo with the harmonic signal, which in turn, results in both reduced spatial resolution and contrast resolution. Using phase inversion technique in which two consecutive pulses are transmitted into the body, with the phase of the second inverted relative to the phase of the first, the echoes from the linear body structure get cancelled whereas echoes from non-linear structures do not. This results in higher amplification of even harmonic components giving superior contrast and spatial resolution. A wide band pass filter is used to receive all echo signals.

The impact of this technique on the imaging performance is shown in Fig. 23.34 which compares the relative lateral beam profiles of the fundamental and the harmonic signals. Side lobes which are responsible for artifacts affecting the imaging quality are significantly reduced as compared to conventional imaging. Images appear sharper and crisper with less noise and with higher contrast resolution.

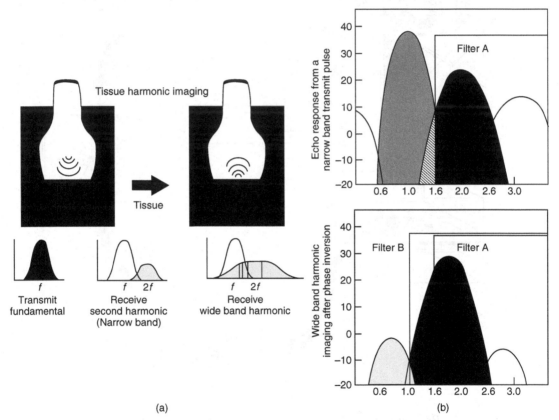

> **Fig. 23.34** *Fundamental imaging: Transmit frequency f = Receive frequency (a) Middle: 2nd Harmonic imaging-transmit frequency f, receive frequency 2f (b) Right : Wide band Harmonic imaging - transmit frequency f, receive all frequencies. (a) Echo frequency response from a narrow band transmit pulse. (i) using a high pass filter or narrow band filter (filter A) cannot effectively separate the fundamental from the harmonic signal due to the overlap of the signals. (ii) after phase inversion, the summing of the even harmonic components results in greater harmonic signal amplitude. Wider band pass filtering can now be applied more effectively (filter B). The result is improved contrast and spatial resolution.*

⟫ 23.10 DIGITAL SCAN CONVERTER

Lines of ultrasonic information are not generated in TV compatible format, consequently to use conventional TV monitors to display the image, some kind of scan conversion must be performed. The majority of scanners digitize the image information and use a digital memory as a buffer store. The memory can be updated whenever necessary and can be read to give a standard video output.

Digitization of the echo information can be performed at a number of points in the signal processing chain. The most convenient is just after de-modulation and video processing. Here, the dynamic range is about 20 dB and the required digitization rate, about 1 MHz. Both these criteria are easily met with readily available "analog-to-digital" converters. Alternatively, there is a growing interest in reducing the amount of undesirable information and digitization just after the radio frequency amplifier is becoming more popular (Halliwell, 1987). The dynamic range of 40 dB and digitization rate of 10 MHz would require expensive analog to digital converters. The advantage of digitizing at this rate is that the phase information in the carrier is not lost and frequency changes can be used to modulate the image.

Digital scan converters used in commercially available scanners provide storage arrays with picture elements 512×512. At normal viewing distances, the individual pixels are not discernible, and cosmetically the image resembles an analog image. The received echo amplitudes from each pixel are quantized (by analog-to-digital conversion) and then numbers representing these ranges of echo amplitudes are stored in the memory locations corresponding to each pixel. The echo amplitudes are usually digitized by an 8 bit A/D convertor.

Figure 23.35 shows the basic building blocks of a digital scan converter arrangement. The scan converter control receives signals of transducer position, ultrasound velocity control and TV sync. pulses and generates x and y address information which is fed to the digital memory. The echo signal is given to an analog-to-digital converter and the digitized information goes to the memory after input holding register and read/write logic. The stored image is post processed and is given to the digital-to-analog converter, which, in turn, passes it on to video section of the television monitor (Ken-Itch, Ito, 1977).

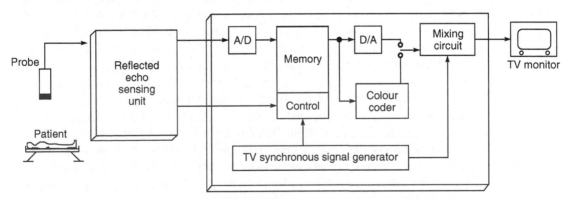

> **Fig. 23.35** *Block diagram of the digital scan conversion unit (after Ken-Itch Ito et al, 1977)*

Because of the fairly low speed of ultrasound in soft tissues and the requirement that all the echoes from the preceding pulse must be received before the next pulse can be delivered, the image line rate is limited. For penetration of 30 cm, the maximum line rate is 5 KHz. With a frame rate of 20 frames per second for non-jerky real-time imaging, the line density per frame is limited to 200. For some scanners, this trade-off between line rate, frame rate and line density produces a display with visible gaps between the lines of real ultrasonic information. Often these gaps are filled with lines calculated inside the machine by interpolation of the real information. This improvement in aesthetic appearance does not affect the diagnostic accuracy.

Digital scan converters are microprocessor-based and offer several additional benefits to the user. Alphanumeric information concerning the patient, date and equipment parameters such as transducer frequency, gain settings and location of the scan plane can be conveniently entered and displayed alongwith the ultrasound image.

23.11   BIOLOGICAL EFFECTS OF ULTRASOUND

Ultrasound methods are attractive for diagnosis because of their non-invasive nature and their apparent safety. Ultrasonic radiation in clinical use is known to be substantially less hazardous to tissues than radiations like X-rays which are in common use in some of the same applications. The apparent safety of ultrasound rests on data from experimental animal work and over 25 years of extensive human clinical experience. The experiments conducted include studies on whole animals, individual organs, isolated cultured cells and biochemical studies. Surveys of human populations which have been exposed to ultrasound, though on a limited scale, have also been taken. The types of damage sought include genetic and teratogenic effects, disorders of physiological function and inhibition of biological development (Hussey, 1975). The consensus of all these experimental results has supported the current medical opinions that no significant hazard is associated with the levels of ultrasound used in diagnosis.

Ulrich (1971) prepared a schedule of safe ultrasound dosage for experimental use on human beings (Fig. 23.36). He observed that the zone of safety for continuous wave ultrasound lay below a log/log line connecting 1 μs of 1 kW/cm^2 ultrasound, with 200 s of 100 mW/cm^2 ultrasound. An ultrasound intensity of 100 mW/cm^2 or less was also safe for at least 10,000 s. Pulsed wave ultrasound was safe for doses in which the average intensity multiplied by the total exposure time (or peak intensity multiplied by sum of pulse on-time) lay within this zone of safety. The zone of safety was valid for 0.5 to 15 MHz and for all anatomic sites except the eyes. Re-exposures were limited to 10% and 30% per year. The data in this figure provide the best estimates which can be made on the basis of the available evidence of 'safe' conditions of ultrasonic diagnosis.

Presently, it is recommended that ultrasonic diagnosis should not be permitted unless the irradiation falls within the safe conditions defined in Fig. 23.36. On the basis of Wells (1977) data, a pulse-echo system operating with pulses of 1 ms duration and 10 Wcm^{-2} on intensity, at a pulse repetition frequency of 100 s^{-1} should not be used to examine any individual patient for a total time exceeding 2.75 h. On the other hand, a Doppler detector (continuous wave) operating at an intensity of less than 0.04 Wcm^{-2} may be used without time restriction.

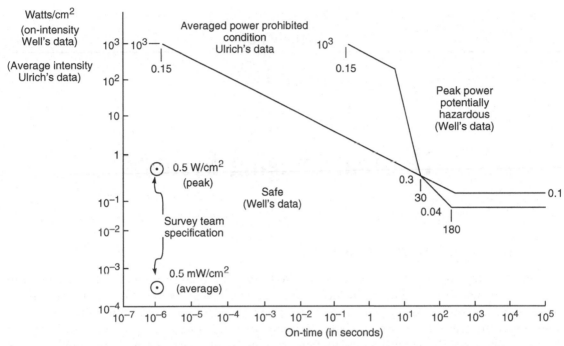

> **Fig. 23.36** *'Safezone' in ultrasonic applications at different power levels for continous and pulse-mode technique (Courtesy: National Science Foundation, USA)*

An exhaustive treatment of the subject on safety and potential hazards in the current application of ultrasound in general, and in obstetrics and gynaecology in particular has been rendered by Lele (1979). Experiments have failed to reveal any evidence that tissues have a 'memory' for expos-ure to ultrasound. Thus, unlike X-rays and other ionizing radiations, there is a little likelihood of cumulative effects of ultrasound. Also, ultrasound, because of its long wavelengths in tissues, is unlikely to cause intra-cellular focal damage such as that resulting from X-radiation.

Thermal Imaging Systems

24.1 MEDICAL THERMOGRAPHY

The medical thermograph is a sensitive infrared camera which presents a video image of the temperature distribution over the surface of the skin. This image enables temperature differences to be seen instantaneously, providing fairly good evidence of any abnormality. However, thermography still cannot be considered as a diagnostic technique comparable to radiography. Radiography provides essential information on anatomical structures and abnormalities while thermography indicates metabolic process and circulation changes, so the two techniques are complementary.

The human body absorbs infrared radiation almost without reflection, and at the same time, emits part of its own thermal energy in the form of infrared radiation. The intensity of this radiant energy corresponds to the temperature of the radiant surface. It is, therefore, possible to measure the varying intensity of radiation at a certain distance from the body and thus determine the surface temperature. Figure 24.1 shows spectral distribution of Infrared emission from human skin. In a normal healthy subject, the body temperature may vary considerably from time to time, but the skin temperature pattern generally demonstrates characteristic features, and a remarkably consistent bilateral symmetry. Thermography is the science of visualizing these patterns and determining any deviations from the normal brought about by pathological changes. Thermography often facilitates detection of pathological changes before any other method of investigation, and in some circumstances, is the only diagnostic aid available.

Thermography has a number of distinct advantages over other imaging systems. It is completely non-invasive, there is no contact between the patient and system as with ultrasonography, and there is no radiation hazard as with X-rays. In addition, thermography is a real-time system.

The examination of the female breast as a reliable aid for diagnosing breast cancer is probably the best known application of thermography. The mammary glands were the first organs that thermography was clinically applied (Lawson, 1957) to. It is assumed that since cancer tissue metabolizes more actively than other tissues and thus has a higher temperature, the heat produced

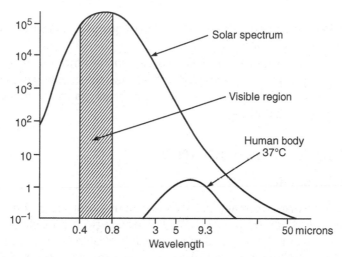

➤ **Fig. 24.1**   *Spectral distribution of infrared emission from human skin. The emission peaks at around 9 microns regardless of pigmentation*

is conveyed to the skin surface resulting in a higher temperature in the skin directly over the malignancy than in other regions. Figure 24.2 shows typical breast thermograms.

Stark and Way (1974) have been carrying out a study of women, who, in the absence of any breast pathology, have abnormal or asymmetrical thermograms. They found that the incidence of

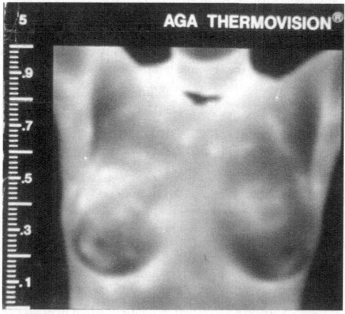

➤ **Fig. 24.2**   *Normal gray-tone thermogram of breast*

subsequent development of breast cancer in these women is upto 10 times greater than those whose thermograms have been symmetrical; therefore, thermographic examination is of considerable value in identifying a risk population.

Thermography also finds applications in the assessment and monitoring of inflammatory joint diseases, diagnosing deep vein thrombosis (Arter, 1978) and the problem of peripheral circulation. Anbar (1998) covers recent developments in clinical thermography in many fields of medicine from general surgery to ophthalmology.

24.2 PHYSICS OF THERMOGRAPHY

24.2.1 Infrared Radiation

The infrared ray is a kind of electromagnetic wave with a frequency higher than the radio frequencies and lower than visible light frequencies. The Infrared region of the electromagnetic spectrum is usually taken as 0.77 and 100 μm. For convenience, it is often split into near infrared (0.77 to 1.5 μm), middle infrared (1.5 to 6 μm) and far infrared (6 to 40 μm) and far far infrared (40 to 100 μm).

Infrared rays are radiated spontaneously by all objects having a temperature above absolute zero. The total energy 'W' emitted by the object and its temperature are related by the Stefan Boltzman formula,

$$W = \sigma \in T^4 \tag{i}$$

where W = radiant flux density and is expressed in W/cm^2

$\quad \in$ = Emissivity factor

$\quad \sigma$ = Stefan-Boltzman constant = $5.67 \times 10^{-12}\, W/(cm^2 \times K^4)$

$\quad T$ = Absolute temperature

Equation (i) shows that the amount of infrared energy emitted varies with temperature of the object. Figure 24.3 shows the spectral distribution of intensity of radiation from black bodies. The wavelength of the energy peak and the absolute temperature is given by the Wien formula,

$$\lambda_{max} = \frac{2897\,(\mu m)}{T(K)}$$

The human body has a temperature of 37°C (310 K), therefore,

$$\lambda_{max}\,(\text{human body}) \approx 10\,\mu m$$

From equation (i), Fig. 22.2, the energy density of the infrared radiation of the human body = 4.6 $\times 10^{-2}\, W/cm^2$. Assuming the surface area of the human body to be 1.5–2.0 m^2, the amount of infrared energy radiating from the whole body is approximately 700–1000 W.

24.2.2 Physical Factors

There are several physical factors which affect the amount of infrared radiation from the human body. These factors are emissivity, reflectivity and transmittance or absorption.

Emissivity: An object which absorbs all radiation incident upon it, at all wavelengths; is called a black body. A black body is only an idealized case and, therefore, all objects encountered in

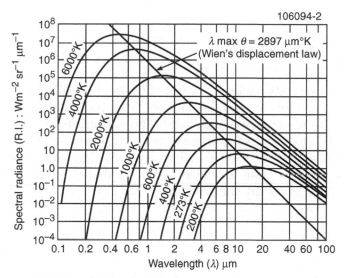

> ➤ **Fig. 24.3** *Black body spectral radiance*

practice can be termed gray bodies. We thus define the term emissivity as representing the ratio of the radiant energy emitted per unit area by an object to the radiant energy emitted per unit area of the black body at the same temperature.

$$\in = \frac{W_o}{W_b}$$

The value of $\in$ for human skin at ambient temperature is virtually unity (Watmough and Oliver, 1968) within the limit of infrared wavelengths of 3-16 μm. If a body has an emissivity that is less than one, and constant whatever the wavelength, it is called a gray body. A change in emissivity causes an error in temperature measurement, Sakurai et al. (1973) calculated that 1% of $\Delta\in$ is equivalent to approximately 0.3°C ΔT.

The spectral radiant emissivity is defined as

$$\in_\lambda = \frac{W_{o\lambda}}{W_{b\lambda}}$$

where $W_o\lambda$ is the spectral radiant power per unit area and $W_b\lambda$ is the same quantity for a black body. $\in_\lambda$ is a function of both wavelength and temperature, but the variation with temperature is usually much smaller than the variation with wavelength.

In the infrared region, the spectral radiant emissivity of most solids decreases as the wavelength increases.

Reflection: Spectral reflectivity ρ_λ is defined as the ratio of reflected power to the incident power at a given wavelength. So,

$$\rho_\lambda + \alpha_\lambda = 1 \qquad\qquad (i)$$

where α_λ is the spectral absorptance. Since

$$\in_\lambda = \alpha_\lambda$$

where $\in_\lambda$ is the spectral emissivity, then

$$\in_\lambda = 1 - \rho_\lambda. \qquad\qquad\qquad\qquad\qquad\text{(ii)}$$

Therefore, an opaque body having a low emissivity in visible light will have high reflectivity. Hardy (1939) illustrated the spectral distribution of the reflectivity ρ_λ of human skin. This is shown in Fig. 24.4. It shows that the reflectivity of human skin between the range 2 and 6 μm is 0.02–0.05. Obviously, a thermogram will be affected by strong infrared light coming from sources outside the body and getting reflected from the body. Incandescent lights emit infrared radiation and can influence a thermogram. Equation (ii) shows that if emissivity decreases, there is a corresponding increase in the reflection coefficient.

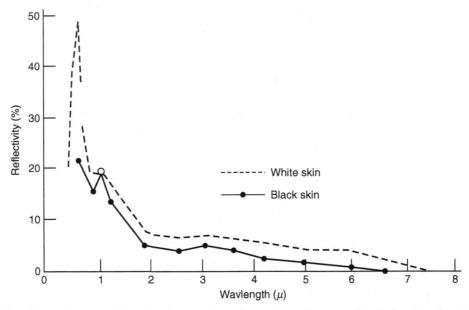

> **Fig. 24.4**   *Spectral distribution of the reflectivity of human skin (after Hardy, 1939)*

Tansmittance and Absorption of Infrared Radiation: When a semi-transparent body is placed between the surface of any radiation-emitting body and a detector, it is necessary to consider the change in emissivity related to its transmittance, reflectivity and emissivity. The presence of tape, paste or ointments have influence on thermograms.

Hardy (1939) studied the spectral absorption of infrared radiance through human skin and found that the human skin is almost opaque at the wavelengths encountered in thermography.

The background radiation of infrared rays is also found to influence the thermograms. When the infrared radiation from the background and that of the body are equal or nearly equal, it is difficult to distinguish the body from the background.

▥▷ 24.3 INFRARED DETECTORS

Infrared detectors are used to convert infrared energy into electrical signals. Basically, there are two types of detectors: thermal detectors and photo-detectors.

Thermal detectors include thermocouples and thermistor bolometers. They feature constant sensitivity over a long wavelength region. However, they are characterized by long-time constant, and thus show a slow response.

The wavelength at which the human body has maximum response is 9–10 µm. Therefore, the detector should ideally have a constant spectral sensitivity in the 3–20 µ infrared range.

However, the spectral response of the photo-detectors is highly limited. Most of the infrared cameras use InSb (indium antimonide) detector which detects infrared rays in the range 2–6 µm. Only 2.4% of the energy emitted by the human body falls within the region detected by InSb detectors. But they are highly sensitive and are capable of detecting small temperature variations as compared to a thermistor.

Another detector making use of an alloy of cadmium, mercury and telluride (CMT) and cooled with liquid nitrogen, has a peak response at 10–12 µm. The spectral response of various detectors used in thermographic equipment is shown in Fig. 24.5.

▥▷ 24.4 THERMOGRAPHIC EQUIPMENT

Thermographic cameras incorporate scanning systems which enable the infrared radiation emitted from the surface of the skin within the field of view to be focused on to an infrared detector. Most systems have a wide range of absolute temperature sensitivity ranging from 1 to 50°C. However, for most of the examinations, the gray scale is generally adjusted to represent a much narrower range of temperature, depending on the area to be examined. A full scale (black to white) temperature difference of between 5 to 10°C is usually adequate. For breast examinations, the absolute temperature range may be 25 to 35°C, for the legs 23 to 33°C and for the forehead 31 to 36°C (Carter, 1978). With the gray-scale spanning a temperature range of 5°C, it is possible to resolve temperature variations of 0.5°C on the skin. Most of the clinical changes in temperature are of the order of 1°C or more. It is not necessary to use the system on a more sensitive mode, even though some systems can theoretically resolve 0.1°C.

The equipment used in thermography basically consists of two units: a special infrared camera that scans the object, and a display unit for displaying the thermal picture on the screen. The camera is generally mounted on a tripod that is fitted on wheels.

The camera unit contains an optical system which scans the field of view at a very high speed and focuses the infrared radiation on a detector that converts the radiation signal into an electrical signal. The signal from the camera is amplified and processed before being used to modulate the intensity of the beam in the picture tube. The beam sweeps across the tube face in a pattern corresponding to the scanning pattern of the camera. The picture on the screen can be adjusted for contrast (temperature range) and brightness (temperature level) by means of controls on the display unit.

The optical system forms an important component in the scanning system. The lenses used for the infrared work are made of silicon and are anti-reflective coated which matches the spectra

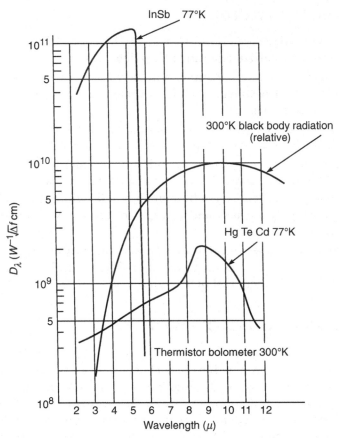

> **Fig. 24.5** *Spectral characteristics of various types of infrared detectors*

response of the scanner detector. Two standard lenses are usually available, the 12° and 20°. For normal medical use, the 12° lens is recommended. This gives a reasonable working distance to the patient and is not too close when working with extension rings. The 20° lens is preferred with working distance within 1.5 m.

Extension rings are used for close-up or macro-thermography. This gives a very high optical resolution (< 1 mm for examination of small areas). The set consists of one 12 and one 21 mm rings which can be used according to minimum distance and overall width of the object. The 'f' number of the lenses is 1.8, and the focal distance 52 and 33 mm for 12° × 12° and 20° × 20° lenses, respectively.

Figure 24.6 shows the schematic arrangement of an infrared scanner. The double-scanning movement of the plane mirror causes each spot on the patient's body to be focused in turn on the cooled detector of Indium Antimonide. The detector is mounted in a Dewar flask which is cooled with nitrogen in order to obtain optimum sensitivity and eliminate thermal noise. An optical collector system is employed in an infrared detecting system. This is due to a weak infrared input from the subject. Scanning is carried out by a reflecting mirror. The instantaneous field of view which can be scanned is decided by the optics and detector size.

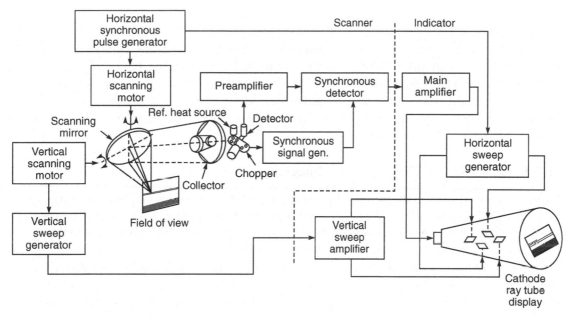

> **Fig. 24.6** *Block diagram of the scanning and displaying arrangements for infrared imaging*

The horizontal and vertical movements of the scanning mirror are controlled by individual motors. The scanning mechanism employs an arrangement for rapidly oscillating the flat mirror to give a horizontal line scan, and slowly tilting it for the frame scan. Simultaneously, horizontal synchronous pulses and vertical sweep signals for the display unit are also generated. A chopper disc interrupts the infrared beam so that an AC signal is produced from the detector. The chopping disc also interrupts light from a standard heat source, which intermittently illuminates a photocell and provides a phase-reference to a coherent detector and to a bucking circuit employed to cancel out large standing signals. The adjustment of the height and the tilt of the optical head can be controlled to have a different field of view.

The AC signal developed at the detector is amplified by the pre-amplifier. It is then rectified and fed into the bandpass filter whose central frequency is determined by the chopping frequency. The total voltage gain in the amplifier is around 120 dB. A 13 cm CRT is used for display of scanned image. Horizontal sweep, which is synchronized with the scanning motion of the scanning mirror, is displayed from top to bottom and then in the reverse order on the screen. The signals are displayed on the screen by intensity modulation which controls brightness and contrast with the strength of the signal. Each picture is built from around 150 horizontal lines in about 75 s. Temperature differences of the order of $0.1°C$ can be resolved.

In the commercial instruments based on this technique, the object is scanned in a television-like raster wherein each line consists of temperature information divided into picture elements. The thermal picture on the screen is virtually flicker-free as a result of the high frame rate of 16 frames per second. In spite of the high frame rate, temperature differences of less than $0.2°C$ can be detected. The resolving power is 140 standard lines per picture and roughly 130 elements per line. The size of the picture on the display tube is 90 × 90 mm.

The system is equipped with a 10° × 10° field of view front lens providing a minimum object size of 11.5 × 11.5 cm at a distance of 0.95 cm. By an additional extension ring, it is possible to cover an object as small as 4.6 × 4.6 cm. The front lens can be changed with a 25° × 250 field of view.

The thermal and spatial resolution of a thermographic system is determined by the optical parameters, detector performance, preamplifier's noise, the signal processing system, the picture presentation and evaluation systems.

24.4.1 Sensitivity of Thermographic Imaging System

A figure of merit for the thermographic imaging system is the noise equivalent temperature difference (NETD). This is usually called minimum temperature resolution, i.e., the temperature differential between two adjacent elements in the scene that will give a signal equal to the system noise. Thus, the smaller the NETD, the better the sensitivity.

In order to obtain the minimum value of NETD, the detector with the maximum value of integration should be used. Also, even if the scanning and optical system is the same, thermographic systems with different detectors have different NETD. The peak detectivity of InSb (photo-detective cell) is 1.3×10^{11} and that of CdHgTe, is 2×10^{9}. Therefore, InSb detector is one order of magnitude more sensitive than CdHgTe.

24.4.2 Recording Techniques

Several techniques are used for making photographic recording of the image from the screen. Adapters for 35 or 60 mm film and Polaroid cameras are available to suit different needs. For quantitative evaluation, the usual procedure is to take one photo for each isotherm level. For recording dynamic changes in thermal patterns, a 16 mm cine can be used. Cine pictures are useful when studying changes in peripheral circulation due to exercise and changes in environmental temperature. It is also possible to make a video recording on magnetic tape and play it back to the display unit for later evaluation. For colour recording, each isotherm setting is exposed through a different colour filter on a single frame of colour film. This gives concentrated and easily interpreted information on the temperature distribution. Each colour corresponds to a specific temperature.

▶ 24.5 QUANTITATIVE MEDICAL THERMOGRAPHY

For comparing the results of successive thermographic examinations, it is essential that the results are standardized and quantified. In the earlier thermographic equipment, the thermograph was recorded on a photographic film from which it was possible to quantify it by densitometry. Work with this system was limited by the long scanning time. A practical solution to this problem is the use of 'isotherms'. Differences between the various gray tones are determined accurately by means of a thermal band or isotherm. This is visualized by calibrating the electronic circuitry inside the camera, so that a particular gray level (i.e., temperature) is depicted as bright white on the screen. Thus, the isotherm is built up from white areas each indicating the selected temperature. These isotherms can be shifted to any temperature level and will register temperature differences as small as 0.2°C which are read from a scale on the monitor screen. The use of isotherms thus provides a series of points between selected temperatures providing a thermal contour map. A single isotherm could be displayed as a marker of maximum or minimum temperature, in high contrast from the surrounding temperatures. The calibration of isotherms is possible by scanning a temperature reference block.

In the modern thermographic equipment, temperature measurement is improved by providing two simultaneous isotherm functions. One isotherm could be located on an external temperature reference and the other on a point of interest. The unknown temperature is determined by difference. A convenient method of quantitative recording is the superimposition of a range of isotherms in a 7, 8 or 10 colour coded sequence between 26 and 34°C, representing the temperature range from the normal to the grossly inflamed, yellow usually represents the maximum temperature and blue the minimum. The changes can then be quantified by an integrated numerical expression of the isothermal images. For advanced medical thermography, the modern systems include facilities for both analog as well as digital analysis for automatically measuring complex temperature levels and for defining temperature areas.

24.5.1 Analog Analysis

There are many possibilities of analog analysis of the gray tone image including the following: (i) isotherm function, (ii) level analysis, (iii) sample area selector, and (iv) thermal profile analysis.

Level Analysis: This allows for greater precision in identifying the maximum, average and minimum temperature levels of selected areas within the thermal image of the patient. Level analysis is achieved by means of a thermal amplitude analyzer operating within an area selected by sample area selector control. The sample area may be tilted 45° to the left or right and is displayed as bright outlines on the screen.

Thermal Amplitude Analyser: It measures the video signal and determines the mean (average), maximum and minimum values within the desired area on the patient. This quantification of the image allows for more precise, objective and consistent analysis of even the most complex thermal patterns. Image interpretation is thus simplified. Each sample amplitude can be read out individually. Additionally reading out the difference between the two selected amplitudes can also be carried out.

Area analysis: This allows the quantification of individual isotherm level in terms of percentages of a selected region on the patient. It is a means not only to quantify the temperature uniformity, but also to quantify the temporal progression of temperature levels.

Thermal Profile Analysis: This in real-time allows for a detailed analysis of the thermal image. This type of quantized analysis allows selection of one line in the gray-tone image and the presentation of it as a single profile (A-scan). The difference in amplitude obtained along the profile gives a precise measurement of the various temperature points along the line. This makes it easy to compare two curves taken at different time intervals of the same or two different lines. In this way, asymmetries show up very clearly.

24.5.2 Digital Analysis System

In other medical fields, where complex image patterns are a regular occurrence, computers offer new opportunities for more efficient and objective analysis. Thermal images can be digitally analysed in many ways and for many reasons. Firstly, it can be used to determine numerous parameters from the image itself, highest, lowest and average temperature, or differences between one region and another or areas of various temperature contours or geometric centroids or skewnesss and so on.

Secondly and more importantly, the computer can be programmed to analyse these parameters and plot them as contours or gray tones or isothermal maps or other formats. Computers can also be used to draw conclusions from these parameters, and based on user-defined algorithms solve complex analysis problems. Alternatively, the computer may be used for automatic interpretation of the information received.

There are also more complex image computations that are frequently of interest in thermographic applications. Depending on the speed required and the complexity of the computation, the desired function may be performed by the digital image processor.

Digitizing Thermal Images: A direct CRT display of a thermogram permits only about 8 shades of gray to be distinguished between black and white, and a photographic record from a CRT face provides a permanent record of only 5 or 6 shades. Even a colour pattern has only a 10-step resolution. Thus, when a wide temperature range is measured, as sometimes encountered in clinical situations, it is difficult to distinguish small temperature differences. Besides this, absolute surface temperatures and temperature differences between two points on the body surface cannot be measured directly and precisely in a photographic thermogram. By using computer techniques or a microprocessor in a dedicated system, it is possible to have a very large range of gray levels.

Figure 24.7 shows a schematic diagram of digitizing a thermogram (Fujimasa et al, 1973). The output signal of the infrared radiation camera is recorded on a magnetic tape. Analog outputs from the camera are converted into digital thermo-profiles by the A-D converter of a computer. The

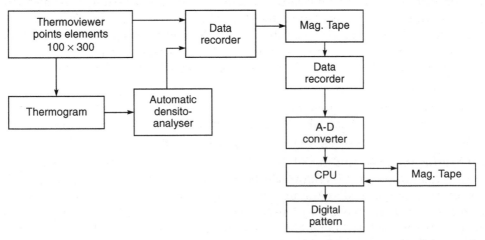

➤ **Fig.** 24.7 *Schematic diagram for digitization of a thermogram (Redrawn after Fujimasa et al 1973)*

control processing unit of the computer converts digital values into true temperature values based on the calibration data stored in the memory or magnetic tape. Alternatively, the thermographic patterns can be scanned and converted into analog electrical signals by a scanning densitometer and are converted into digital form. For a precise analysis of surface temperatures, the thermogram digitization and computers are useful in the automatic analysis of thermograms.

Computer analysis is provided as a standard option with modern thermographic equipment. This option includes not only hardware but also software. By converting the analog image into a digital form and then storing it on magnetic tape, the patients' thermal image is always available for live-like analysis, irrespective of when it was taken. There is a wide range of software available for statistical data analysis, breast thermal analysis and complex image analysis.

⁙▶ 24.6 PYROELECTRIC VIDICON CAMERA

The pyroelectric vidicon is a thermal imaging tube. This tube resembles the standard vidicon, the main difference being that the material used for the target is sensitive to infrared instead of visible light. It is a compact, sensitive and reliable device and can thus find convenient application in medical thermography.

The pyroelectric vidicon basically consists of: (i) a glass envelope; fitted with a germanium faceplate, matching the 8 to 14 μm atmospheric window, (ii) a pyroelectric target mounted on a metal backing plate that also acts as the video output electrode, (iii) an electron gun and beam shaping electrodes like those of a standard vidicon, (iv) a gas reservoir heated by a tungsten filament. The tube is surrounded by focusing and deflection coils that are identical to those used with a standard vidicon.

Incoming thermal radiation is focused by an infrared transmitting lens through the infrared transmitting face plate of the tube onto the target. The target is a triglycine-sulphate (TGS) pyro-electric target which has a high sensitivity in the 8 to 14 μm range of the spectrum. Certain types of crystals (ferro-electric) get polarized in a well-defined direction, known as the polar axis. Because the degree of polarization is temperature-dependent, heating or cooling a slice of such a crystal will create an accumulation of charge (on the faces normal to the polar axis) that is proportional to the variation in polarization caused by the temperature change. This is called the pyroelectric effect, characterized by the pyroelectric coefficient $P = db/dt$. The increase in pyroelectric coefficient with temperature is accompanied by a very rapid increase in the dielectric constant of the material that reduces the efficiency with which the pyroelectric changes can be read by the electron beam. For this reason, the maximum useful pyroelectric effect is obtained with a target temperature of about 35°C. It must never exceed 40°C, the curie point of TGS. Also, the electrical changes are produced only when the temperature of the pyroelectric material changes. So, no image output will be produced from an unchanging image. It is for this reason that the incoming radiation must be chopped or the camera panned so as to produce target temperature variations. The pyroelectric materials also exhibit piezo-electric properties. Therefore, to avoid signal degradation by micro-phonic noise, excessive vibrations should not reach the camera.

Pyricon is the name given to pyroelectric vidicon (PEV) manufactured by Thomson CSF. Figure 24.8 shows a block diagram of the electronics of a camera using the pyricon. Most of the circuit blocks are similar to the vidicon camera used in closed circuit TV systems. There are, however, two

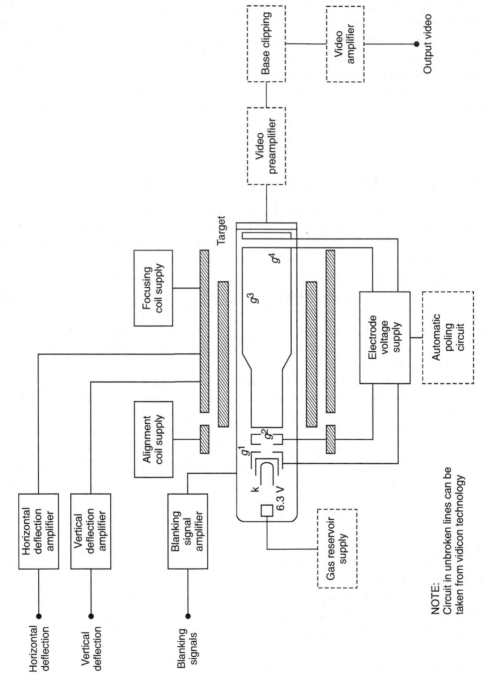

Note:
Circuit in unbroken lines can be taken from vidicon technology

▶ **Fig. 24.8**   *Block diagram of camera electronics used with Pyricon (Courtesy: Thomson CSF, France)*

special circuits; the pre-amplifier and video amplifier. The output signal current from the pyricon tube is low (0 to 50 nA). So, a high current gain, low noise pre-amplifier must be used. The amplifier should also have a bandwidth matched to the maximum spatial frequency and to the scanning standard. A typical pre-amplifier would have a bandwidth of 2.5 MHZ and a load resistance of 3.3 MΩ. The video amplifier should have an adequate gain to match the low value of the useful signal. It is generally arranged to give a gain of 10 to 60. In a typical case, a temperature change of 1°C gives a tube signal current of around 1 μA. This would give a pre-amplifier output signal of 3.3 mV. If the video amplifier has a voltage gain of 60, its output signal will be $3.3 \times 60 = 200$ mV. A base clipping circuit is provided in the pyricon camera. This is used for block level adjustment, by setting the pedestal level at the proper black level of the picture.

A pyricon camera provides 0.5°C resolution at ambient temperature. The common errors in such systems are pedestal variations noise and thermal blurring.

The PEV thermograph is silent like a TV camera and has full access to all available TV equipment and accessories.

ⅢⅢ▶ 24.7 THERMAL CAMERA BASED ON IR SENSOR WITH DIGITAL FOCAL PLANE ARRAY

Uncooled infrared technology is revolutionizing IR imaging by offering low-cost sensors with thermal sensitivity comparable to many high performance and complex cryogenically cooled infrared detectors. These sensors make use of micro-bolometer technology which involves a silicon micro-machined sensor that uses wafer-level silicon processing to fabricate a thermal sensor. White et al (1998) give details of such a sensor and that of a thermal camera based on this technology.

A bolometer is a thermal detector heated by incident radiation, resulting in temperature rise that is sensed as a change in the element resistance. The 327×245 micro-bolometer focal plane array (FPA) has micro-machined bolometer elements on a 46.25 mm pitch. Each micro-bolometer detector consists of a silicon nitride micro-bridge that lies above a CMOS silicon substrate and is supported by two silicon nitride legs. A vanadium oxide film which has an approximately 2% temperature coefficient of resistance at ambient temperatures is deposited on the bridge to form the bolometer resistor. Each of the microbolometer detectors is connected to an underlying unit cell in the silicon CMOS readout integrated circuit substrate via two holes in the passivation layer on the top of the integrated circuit. A layout of the complete integrated circuit is shown in Fig. 24.9. A unique feature of a bolometer array is pulse biasing, in which the bias necessary to measure the element resistance is applied for a very short time. This allows the basic unit cell to be simple with most of the complex circuitry located in the periphery of the chip.

The camera designed and manufactured by Lockheed Martin based on FPA is a system with functionality allocated between focal plane array and electronics to optimize performance. The analog-to-digital convertor and a number of other functions are incorporated on the FPA, while enabling the use of all digital electronics of the FPA. With the fabrication of detectors directly on silicon and application of silicon process techniques at wafer level, millions of mechanical structures can be fabricated on a single silicon wafer at a very low cost.

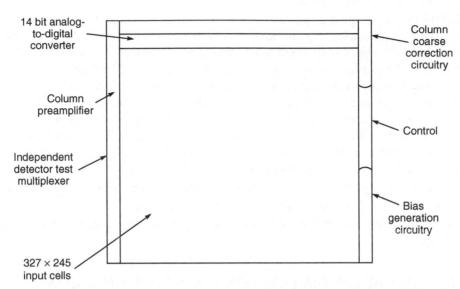

> **FIG. 24.9** *Layout of an integrated circuit used as a micro-bolometer focal plane array (redrawn after white et al, 1998). Width is approximately 17 mm*

The imaging performance of this camera shows linearity better than 0.25% over the range from 5°C to 100°C, a thermal constant of 14 msec and temperature coefficient of resistance as 1.5–2%/K. The IR absorption in the 8–12 μm range is 80%. Video output is in NTSC or PAL format.

⚙️➤ PART THREE : THERAPEUTIC EQUIPMENT

CHAPTER

25

Cardiac Pacemakers

25.1 NEED FOR CARDIAC PACEMAKER

The rhythmic beating of the heart is due to the triggering pulses that originate in an area of specialized tissue in the right atrium of the heart. This area is known as the sino-atrial node. In abnormal situations, if this natural pacemaker ceases to function or becomes unreliable or if the triggering pulse does not reach the heart muscle because of blocking by the damaged tissues, the natural and normal synchronization of the heart action gets disturbed. When monitored, this manifests itself through a decrease in the heart rate and changes in the electrocardiogram (ECG) waveform. By giving external electrical stimulation impulses (Fig. 25.1) to the heart muscle, it is possible to regulate the heart rate. These impulses are given by an electronic instrument called a 'pacemaker'.

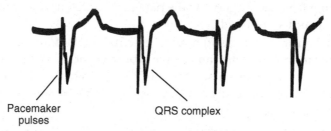

Pacemaker
pulses

QRS complex

➤ **Fig. 25.1** *Pacemaker pulses followed by QRS complex of the heart*

A pacemaker basically consists of two parts: (i) an electronic unit which generates stimulating impulses of controlled rate and amplitude, known as pulse generator, and (ii) the lead which carries the electrical pulses from the pulse generator to the heart. The lead includes the termination which connects to the pulse generator and the insulated conductors, which interface with electrodes and terminate within the heart.

A variety of pacemakers with various possibilities of operation are commercially available, each having some special advantages when used under particular circumstances. However, in almost all cases, the waveforms used for pacing are round-topped rectangular pulses of 1–3 ms duration with rates adjustable from 50-150 pulses per minute.

The electrode arrangements for use with cardiac pacemakers can be in the form of bipolar or unipolar system. In the bipolar system, two electrodes are placed on the heart for myocardial stimulation whereas in the unipolar system, one electrode is placed on the heart and the other electrode is positioned elsewhere in the body. The active electrode is connected to the negative terminal of the pacemaker so as to resemble the normal depolarization of the heart which occurs with intrinsic negative stimulation.

Cardiac pacing has reached a point wherein there are small, reliable and long-lasting units with a lot of possibilities regarding the mode of stimulation.

25.1.1   Types of Pacemakers

The classification of pacemakers into different types is based on the mode of application of the stimulating pulses to the heart. External pacemakers are used when the heart block presents as an emergency and when it is expected to be present for a short time. Internal pacemakers are used in cases requiring long-term pacing because of permanent damage that prevents normal self-triggering of the heart. In the latter case, the pacemaker itself may be implanted in the body. The patient is able to move about freely and is not tied to any external apparatus.

ⅢⅢ▶ 25.2   EXTERNAL PACEMAKERS

External pacemakers are employed to restart the normal rhythm of the heart in cases of cardiac standstill, in situations where short-term pacing is considered adequate, while the patient is in the intensive care unit or is awaiting implantation of a permanent pacemaker. Frequently, external pacemakers are used for patients recovering from cardiac surgery to correct temporary conduction disturbances resulting from the surgery. As the patient recovers, normal conduction returns and the use of pacemakers is discontinued.

The pacing impulse is applied through metal electrodes placed on the surface of the body. Electrode jelly is used for better contact and to avoid burning of the skin underneath. An external pacemaker may apply up to 80-mA pulses through 50-cm^2 electrode on the chest. This procedure is painful and therefore is used only in an emergency or a temporary situation.

The pulses may be delivered:

 (i) *Continuously*—When it is felt that the heart rate is below the pre-set value. The impulse frequency is independent of the electrical activity of the heart.

 (ii) *On demand* R-wave synchronous pacing—Normally the pacemaker is inoperative but it is activated when the heart rate falls below the normal or pre-set value. In such a situation, beat to beat examination of the time interval between two R-waves is done. When this interval exceeds the pre-set value, the pacemaker comes into operation. This technique eliminates any competition between the heart's own pacemaker and externally applied pacemaker pulses. In the R-wave synchronous mode of operation, the external pacemaker can be used to support an implanted unit shortly before re-implantation or shortly after initial implantation to secure pacing.

Pacing with external pacemakers through the chest requires a maximum of 150 V pulses across an impedance of the order 1 kΩ. However, external pacing has the disadvantage that the electrodes eventually tend to burn the skin and the electrical pulses become painful. Also, each impulse causes an uncomfortable contraction of the thoracic muscles around the area of the electrodes.

The stimulating pulses can be preferably applied to a heart through a pacing catheter passing through a vein and connected to the heart. This is called internal pacing. The pacing current required is much less than when it is applied through the chest. The voltage output of internal pacemakers is about 0–15 V and the available output current ranges from 1–20 mA. For long-term pacing by this method, the electrodes can be inserted percutaneously into the myocardium of the left ventricle.

The most common temporary pacemaker used for short-term pacing is a demand type rather than a fixed-rate type. This is because normal A-V conduction may occur intermittently, following complete A-V block and therefore be present during a pacemaker implant, especially when it is used prophylactically. A bipolar or unipolar catheter electrode is inserted through the jugular or 'cephalic vein, into the right ventricle, preferably the apical region, using an X-ray image intensifier television system.

For temporary pacing, external pulse generators (Fig. 25.2) are usually battery-powered. The generator can provide demand or asynchronous pacing. With their calibrated controls, they can be used to determine both stimulation thresholds and approximate R-wave potentials sensed by the electrodes of a pacing lead system. They provide a pacing rate adjustable from 30 to 180 pulses per minute and constant current output from 0.1 to 20 mA regardless of lead, to a maximum of 18 V as measured at output terminals across 500 Ω. The pulse width is around 2 ms. The maximum sensitivity is 1.5 mV for a 40 ms, sine-squared pulse. They are provided with defibrillator shock damage protection to 400 Ws. At maximum sensitivity, they are designed to reject mains frequency interference (50 or 60 Hz) up to a level of about 50 mV. The refractory period is generally in the range of 200-300 ms, initiated by the emission of a pacing pulse or the sensing of an R-wave. They are designed to operate on a 9-V transistor radio battery (e.g. Mallory TR 146X or Eveready E 146X or equivalent) which may give about 500 h of working life.

Pulse generators for external placement are marketed by a large number of manufacturers. They can be worn in the pocket or strapped to a limb or to the torso. They are designed to function in the asynchronous (fixed rate) and synchronous (demand) modes. Some units also operate in a synchronous triggered mode. This feature is useful for temporary pacing in the presence of an inhibiting non-cardiac signal such as one emanating from an implanted pulse generator which is producing pulses but not capturing the myocardium.

The electronic circuit of a pacemaker consists of two parts, namely the impulse-generating circuit and the output circuit. The impulse-forming circuit determines the frequency and duration of the impulses. This is usually a multi-vibrator circuit with adjustable rate and fixed pulse width. The output circuit determines the shape and amplitude of the impulse. The block diagram of the circuit used in external pacemakers is the same as that for implantable pacemakers and is shown in Fig. 25.7.

There are three types of pacemakers based on the type of output waveform (Fig. 25.3). These are:

Voltage Pacemakers: Voltage pacemakers are those in which the current in the circuit is determined by the available voltage during the entire duration of the impulse. The voltage output from the

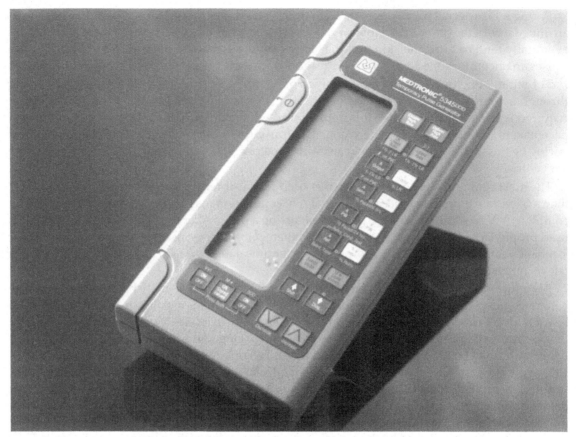

> **Fig. 25.2**   *Medtronic model 5345 external pacemaker*

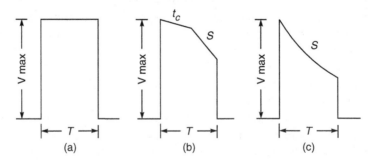

> **Fig. 25.3**   *Impulses from three different types of pacemakers as seen on oscilloscope*
> **(a)**   *constant current type pacemaker*
> **(b)**   *current limited voltage pacemaker*
> **(c)**   *voltage pacemaker*

pacemaker remains constant and changes of resistance in the circuit will influence only the current.

Current Pacemaker: In current pacemakers, throughout the impulse, the current in the circuit is determined by the internal resistance of the pacemaker.

Current Limited Voltage Pacemakers: This is primarily a voltage circuit, but the maximum current in the circuit is limited, preventing too large a current impulse to circulate when there is a low resistance in the electrode circuit. With these pacemakers, during the first part of the impulse, the current in the circuit is determined by the internal resistance of the pacemaker (constant current type) but during the second part of the impulse, the current in the circuit is determined by the voltage available (constant voltage version).

The generators include a condenser in the circuit which stores enough charge to provide pacing for 15-30 s without a battery, to allow battery changeover without interruption of delivery of pulses. However, since all units function in excess of the period for which temporary pacing is likely to be required, it is advisable to change the battery before use with the next patient. Indicators for sensing and pacing are generally provided through appropriate visual indications. The provision of a battery check facility is desirable.

Most pacemaker circuits are protected against defibrillating voltage. Furthermore, they revert to asynchronous mode rather than inhibiting completely in the presence of continuous electrical interference. The range of pulse frequency and energy output is almost similar in all commercial pacemakers and is adequate to meet a variety of the usual clinical situations. Most pulse generators have terminals that accept a wide variety of electrodes, due to the provision of adjustable pin vices.

Battery life of external pulse generators will vary depending on rate and current output settings. At typical rate and output settings of 70 ppm and 10 mA respectively, the battery life should be around 500 h. When the unit is stored for extended periods, the battery should be removed from the battery cartridge.

25.3 IMPLANTABLE PACEMAKERS

The implantable pacemaker, alongwith its electrodes, is designed to be entirely implanted beneath the skin. Its output leads are connected directly to the heart muscle (Fig. 25.4). The pacemaker is a miniaturized pulse generator and is powered by small batteries. The circuit is so designed that the batteries supply sufficient power for a long period. Since the pacemaker is located just beneath the skin, the replacement of the pacemaker unit involving relatively minor surgery has become a routine procedure.

For any implantable circuit, the basic requirements are:

- The components used in the circuit should be highly reliable;
- The power source should be in a position to supply sufficient power to the circuit over prolonged periods of time;
- The circuit should be covered with a biological inert material so that the implant is not rejected by the body; and
- The unit should be covered in such a way that body fluids do not find a way inside the circuit and thus short-circuit the batteries or result in other malfunctioning of the circuit.

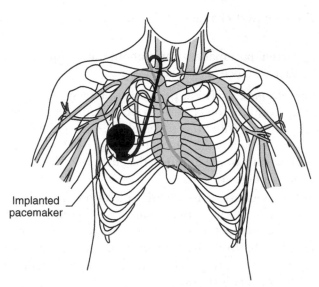

> ➤ **Fig.** 25.4   *The implantable pacemaker placed in the pectoral region. The lead enters the right external jugular vein immediately superior to the medial-end of the clavicle*

From 1968 to 1972, the average life of most pacemakers was 22 months and they were too large (Furman, 1969). With important advances in many areas, particularly in power sources, miniaturization of electronic circuits and hermetic encapsulation, the average life of a pacemaker in 1974 was around 31 months. From 1975 onwards, the life of pacemakers was increased to more than five years. Some of the present-day pacemaker manufacturers even provide a life-time of pacing performance warranty, though under certain conditions.

25.3.1   Types of Implantable Pacemakers

Depending upon the clinical requirements, different types of implantable pacemakers (Fig. 25.5) are utilized. Besides the fixed rate units, the most widely used (97%) implanted pacemakers are the R-wave controlled units (R-wave inhibited and in some cases R-wave triggered). There has been extensive development of other types of pacemakers such as atrial demand, bi-focal demand, and the programmable that enable external programming of such parameters as impulse frequency, impulse duration, output current/voltage and refractory period.

Fixed Rate Pacemaker: This type of pacemaker is intended for patients having permanent heart blocks. The rate is pre-set, say at 70 bpm. The rate can be varied externally in implanted units by magnetically actuating a built-in relay. Since the fixed rate pacemaker functions regardless of the patients' natural heart rhythm, it poses a potential danger because of competition between the patients' rhythm and that of the pacemaker.

Demand Pacemaker: These pacemakers (Fig. 25.6) have gradually almost replaced the fixed rate pacemakers because they avoid competition between the heart's natural rhythm and the

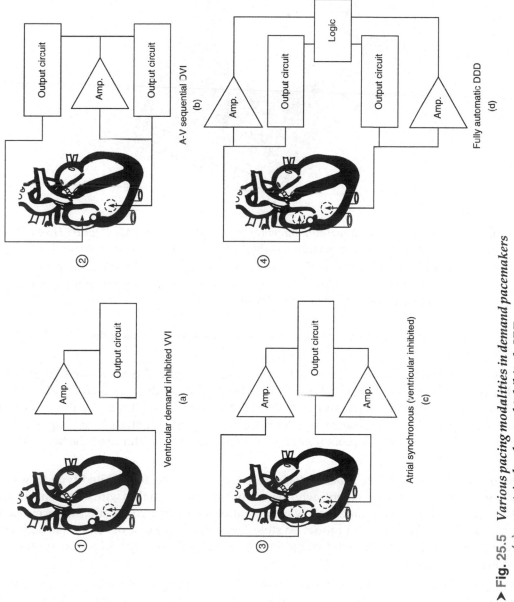

➤ **Fig.** 25.5   *Various pacing modalities in demand pacemakers*
   (a)   *ventricular demand inhibited : VVI*
   (b)   *A-V sequential DVI*
   (c)   *atrial synchronous (ventricular inhibited), VD T/I*
   (d)   *fully automatic DDD*

> ➤ **Fig. 25.6**   *Commercially available pacemakers (Courtesy: M/s Medtronic, USA)*
> *medtronic micro minix implantable cardiac pacemaker-one of the smallest*
> *in the world. Only 1 9/16 inches high and 1 1/4 inches wide, the pulse*
> *generator weighs 0.6 of an ounce. Designed for full programmability via*
> *telemetry and for pacing either the atrium or the ventricle of the heart*

pacemaker rhythm. The demand unit functions only when the R-R intervals of the natural rhythm exceed a pre-set limit.

R wave Triggered Pacemaker: The ventricular synchronized demand type (*R* wave triggered) pacemaker is meant for patients who are generally in heart block with occasional sinus rhythm. The pacemaker detects ventricular activity (*R* wave of ECG) and stimulates the ventricles after a very short delay time of some milliseconds. If there is sinus rhythm, the stimulating impulse will occur in the ventricular de-polarization. If there is asystole, the unit will stimulate the heart after a pre-set time.

Ventricular Inhibited or R Wave Blocked Pacemaker: The ventricular inhibited type (*R* wave blocked) pacemaker is meant for patients who generally have sinus rhythm with occasional heart block. The circuitry detects spontaneous *R* wave potentials at the electrodes and the pacemaker provides a stimulus to the heart after pre-set asystole. However, in the case of ventricular activity, the *R*-wave does not trigger the output circuit of the pacemaker but blocks the output circuit and no stimulation impulse is given to the heart.

Atrial Triggered Pacemaker: This is a *R* wave triggered or atrial triggered pacemaker. The pacemaker detects the atrial de-polarization and starts the pulse forming circuits after a delay so that the impulse to the ventricles is delivered after a suitable *PR* interval. The major advantage of this pacemaker is its ability to provide maximum augmentation of cardiac output at changing atrial rates to meet various physiological requirements.

Dual Chamber Pacemakers: These devices are capable of treating the majority of those patients who suffer from diseases of the sino-atrial node by providing atrial stimulation whenever needed. In these devices, both the atria and the ventricles are sensed and stimulated as needed while maintaining proper synchronization of the upper and lower chambers. Rate–adaptive features available with dual chamber pacemakers include the automatic adjustment of stimulus intensity and gains for the various sensing channels.

25.3.2 Classification Codes for Pacemakers

With rapid developments taking place in the implantable pacemaker technology, it was felt necessary to develop a standard nomenclature to facilitate identification of the type and functions of the pacemaker. Initially, a three-letter code devised by the Inter-Society Commission for Heart Disease Resources (ICHD) was introduced. The code format is given below:

First letter	*Second letter*	*Third letter*
A,V, or D	*A, V, O, or D*	*O, I , or T*
Indicates chamber paced	Indicates chamber sensed	Indicates pulse generator, mode of response
A = Atrium		I = Inhibited
V = Ventricle		T = Triggered
D = Dual (both chambers)		O = Not applicable

The first of the three letters identified the stimulated chamber, the second letter identified the chamber sensed by the pacemaker and the third letter identified the response to the sensed signal. For example; the earliest fixed rate was when pacemakers stimulated the ventricle, but had no sensing function. Thus, they were termed VOO. Similarly, the first demand pacemakers stimulated the ventricle, but if they sensed an R-wave, they inhibited the pacemaker function for one cycle. Thus, they were termed VVI.

The development of dual-chamber and multi-programmable pacemakers necessitated a more detailed nomenclature, resulting in revision of the three-letter code to five-letter codes, which was jointly adopted by the North American Society of Pacing and Electro-physiology (NASPE) and the British Pacing and Electro-physiology Group (BPEG). Table 25.1 shows the classification of the 5-letter pacemaker identification terminology.

In the five-letter code, the first letter indicates the chamber or chambers that are paced. The second letter shows those chambers in which sensing takes place. The third letter reveals how the pacemaker will respond to a sensed event. For example; the pacemaker will 'inhibit' the pacing output when intrinsic activity is sensed or will 'trigger' a pacing output based on a specific, previously sensed event. The fourth letter describes the degree of programmability of the pacemaker but is typically used to indicate that the device can provide rate response. The fifth letter is reserved specifically for anti-tachycardia functions.

• Table 25.1   *The NASPE/BPEG Generic (NBG) Pacemaker Code*

I	II	III	IV	V
Chamber(s) paced	Chamber(s) Sensed	Response to sensing	Programmability, rate modulation	Antitachyarrhythmia function(s)
0 = None	0 = None	0 = None	0 = None	0 = None
A = Atrium	A = Atrium	T = Triggered	P = Simple Programmable	P = Pacing
V = Ventricle	V = Ventricle	I = Inhibited	M = Multiprogrammable	(antitachyarrhythmia)
D = Dual (A+V)	D = Dual (A+V)	D = Dual (T+I)	C = Communicating	S = Shock
S = single	S = single		R = Rate modulation	D = Dual (P+S)
	(A or V)	(A or V)		

Note: Positions I through III are used exclusively for antibradyarrhythmia function (after Bernstein et al., 1987).

25.3.3   Ventricular Synchronous Demand Pacemaker

Figure 25.7 shows a functional block diagram of a ventricular synchronous demand pacemaker. The pulse generator has two functions, viz., pacing and sensing. Sensing is accomplished by picking up the ECG signal. In the case of dual-chamber pacing, the P wave is also sensed. Once the ECG signal enters the sensing circuit, it is passed through a QRS bandpass filter. This filter is designed to pass signal components in the frequency range of 5-100 Hz, with a centre frequency of 30 Hz.

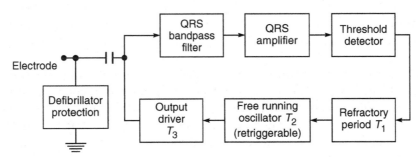

➤ **Fig. 25.7**   *Block diagram of a ventricular synchronous demand pacemaker*

This is followed by an amplifier and threshold detector which is designed to operate with a detection sensitivity of 1–2 mV. Sensitivity of this order ensures reliable detection of cardiac signals sensed on the electrodes which typically have amplitudes in the 1–30 mV range depending on the electrode surface area and the sensing circuit loading impedance (Brownlee and Tyres, 1978). A refractory period (T_1) is necessarily incorporated to limit the pulse delivery rate, particularly in the presence of electromagnetic interference. It is meant to prevent multiple re-triggering of the astable multivibrator following a sensed or paced contraction. The free-running multivibrator provides a fixed rate mode with an interval of T_2 via the output driver circuit. The output pulses of a length T_3 synchronous with input signals that fall outside the sensing refractory period T_1 are thus delivered at the stimulating electrodes.

Most of the commercial pacemakers employ a single defibrillation protection diode, while others employ a symmetrical two-diode type protection. The diodes in this case are placed back-to-back. The symmetrical protectors minimize detection of high-level, high-frequency, pulsed electro-magnetic interference artifacts, and thus raise the noise detection threshold.

Two types of stimuli are produced by the output circuit of the pacemaker. These are the constant voltage and constant current type. Constant voltage amplitude pulses are typically in the range of 5 V with a duration of 500 to 600 ms. The lithium-iodine cell voltage is 2.8 V, and therefore if pulses of about 5 V are required at the output of the pacemaker, a voltage doubler circuit is employed. Constant current amplitude pulses are typically in the range of 8 to 10 mA with pulse duration ranging from 1.0 to 1.2 ms. Pulse rates range from 70 to 90 beats per minute in synchronous pace-makers whereas those that are not the fixed rate type typically provide rates ranging from 60 to 150 beats per minute. Pacemakers usually provide for a refractory period in the range of 400 to 500 ms.

For most pulse generators, the pacing circuit impedance is 5 k to 20 kΩ, whereas a demand pulse generator sensing amplifier is generally a high input impedance device with a typical value of 5 MΩ. The amplifier must detect intrinsic cardiac activity in the millivolt region and yet withstand pacing circuit output pulses that are several volts in amplitude. Also, the input impedance of the sensing circuit and output impedance of the pacing circuit appear in parallel, and the net impedance seen by the load is not more than the output impedance of the pacing circuit.

Sensing and pacing are achieved with one of two configurations: bipolar and unipolar. In bipolar, the anode and cathode are close together, with the anode at the tip of the lead and cathode, a ring electrode about 2 cm proximal to the tip. Both the electrodes are located on or in the heart tissue. In unipolar system, only one pole (usually negative) is located on or in the heart tissue. The positive electrode is located remotely, generally on the body of the pacemaker generator itself.

The atrial-synchronous pacemaker is a more complex device which is designed to replace the blocked conduction system of the heart. As explained earlier, the heart's intrinsic pacemaker located in the SA node, initiates the cardiac cycle by stimulating the atria to contract and then providing a stimulus to the AV node which, after some delay, stimulates the ventricles. If the SA node is able to stimulate the atria, the P wave of the ECG corresponding to atrial contraction can be detected by an electrode implanted in the atrium and used to trigger the pacemaker in the same way that it triggers the AV node.

25.3.4 Programmable Pacemaker

A programmable pacemaker consists of two parts: the external unit which generates programmed stimuli which is transferred to an internal unit by one of the several communication techniques. Figure 25.8 shows a functional block diagram of the programming interface. The commonly used methods of transmitting information are: (i) magnetic—an electromagnet placed on the surface of the body establishes a magnetic field which penetrates the skin and operates the pacemaker's reed switch, (ii) radio-frequency waves—the information can be transmitted over high frequency electromagnetic waves which are received inside the body by an antenna. The antenna is usually in the shape of a coil housed within the pacemaker, (iii) acoustic-ultrasonic pressure waves from a suitable transducer placed over the skin, can penetrate the human body. They are received by a suitable receiver in the pacemaker which carries out the desired function. Out of all these methods,

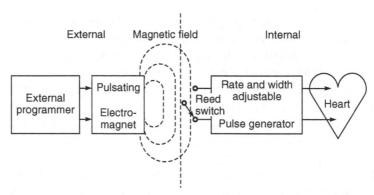

> **Fig. 25.8**   *Functional block diagram of programming interface*

the magnetic field method is the most widely used because of its simplicity and minimal power requirements.

An essential requirement of programmable pacemakers is that they should be immune to accidental programming from naturally occurring energy sources. To meet this requirement, the information is usually coded and the pacemaker contains a decoding mechanism to recognize proper information. This programming security code method makes it practically impossible to reprogram an implantable pacemaker through extraneous random magnetic fields.

Figure 25.9 is a detailed block diagram of the multi-programmable pacemaker. It may be considered as being comprised of the three systems. System I controls the main timing functions of the pulse generator and carries the rate limiter, the pulse output circuit and the stimulating function of the electrode. Operating as directed by the programmable control circuit, this system generates output pulses at the programmed rate, width and amplitudes unless over-ridden by System 2. System 2 carries the sensing and signal discriminating function of the circuit. Comprising the sensing function of the electrode, an RF filter, a signal amplifier and comparator, this system identifies signals of cardiac origin and, where appropriate, sends an inhibit signal to System 1. System 3 carries the programmable control circuit, the data validate circuit, the reed switch and the master timing crystal. This system effects program recognition, storage, and execution as well as control of the battery and various test sequences.

Under normal operating conditions, the timing control circuit periodically triggers the output circuit causing the emission, at the programmed rate, of stimulation pulses of programmed pulse width and amplitude. The period between each trigger signal is scrutinized by the rate limiter and in the unlikely event of component failure causing a rate increase, the limiter holds the rate of stimulation to less than 180 bpm.

Upon receipt of inhibit signals from System 2, the timing control circuit compares their time of arrival against the programmed refractory period. Signals arriving within the refractory period are ignored. Signals arriving outside the programmed refractory period when zero hysteresis is programmed, reset the timing control circuit thus inhibiting the output circuit. When hysteresis is programmed the arrival of a signal outside the refractory period causes resetting of the timing control circuit with the period of escape before the next stimulation of the output circuit becoming the programmed basic interval plus the programmed hysteresis period. Should an inhibit signal

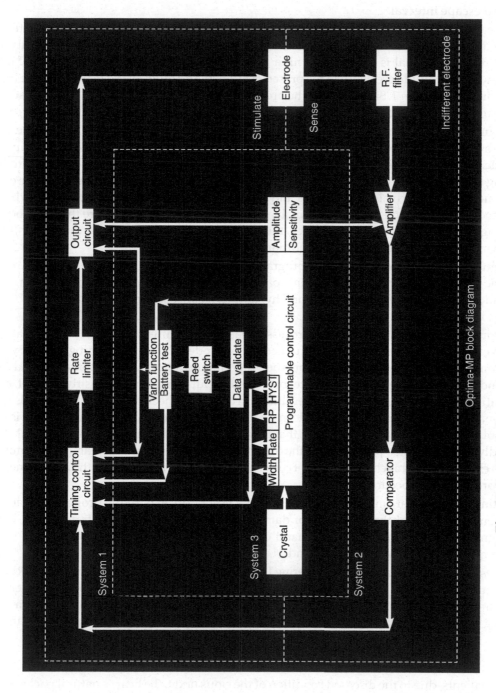

> **Fig. 25.9** *Block diagram of a multi programmable pulse generator*

be received from System 2 during the period, the timing control circuit resets to again offer the increased escape interval.

A signal detected by the electrode is filtered of high frequency components by the RF filter. The signal is then selectively amplified by an amount determined by the programmed sensitivity level and the resultant signal is compared with a preset level at the comparator. Signals of either polarity with magnitude greater than the preset level, enable an input signal to be fed to the timing control circuit of System 1. Signals below the preset level are ignored. In this way greater assurance is given to inhibitions occurring only with the detection of cardiac signals.

System 3, acting on programming signals transmitted to the reed switch, directs the pulse generator to function as programmed. The reed switch by opening and closing in sympathy with magnetic pulses emitted by the programmer, feeds signals to the data validate circuit. After first verifying that the reed switch was closed for a minimum period of 300 ms, the data validate circuit checks the speed of arrival of the signals and executes a code validation check. Only when all these checks are satisfied is the programmable control circuit directed to store the new code. By activating appropriate electronic switches, the programmable control circuit directs Systems 1 and 2 to implement the new program conditions.

The pulse generator is non-invasively programmable by means of a programmer which emits a magnetic code, that enables the following parameters to be altered: rate, pulse width, pulse amplitude, sensitivity, refractory period and hysteresis. All measurements are taken at 37°C with 500 Ω resistive load.

The programmer contains a microprocessor-based transmitter/receiver that operates by inductively coupling pulse-position modulated, binary coded data from the programmer via the programming wand to the pulse generator. The programming information is contained in a 20-bit command code specifying the desired rate, pulse width, pulse amplitude, sensitivity level, mode of operation, a pulse generator model identification code and check codes. If an attempt is made intentionally or unintentionally to include in a programming command a parameter that is not a feature of the pulse generator being used, that parameter will simply remain at its nominal value. All validly reprogrammed parameters in the command will be implemented. Part of the command code is a check code. If this check code is not correct, the command is rejected by the pulse generator, and no programming occurs.

The timing of the transmission is precise. Crystal oscillators in both the programmer and the pulse generator control the frequency of data exchanges. Each data bit is transmitted within approximately 1.0 ms. The entire command code and its complement are transmitted within approximately 40 ms or 1/25th of a second.

At present most manufacturers, are offering PC-based programming units rather than dedicated proprietary instruments. These systems are more flexible and more easily updated when new devices are released. Also, time-efficient programming plays an important role in the productivity of pacing clinics which may provide follow-up for a very large number of patients per year.

25.3.5 Rate-responsive Pacemakers

In some patients, due to the diseased condition of the sinus node, the heart's natural pacemaker is not able to increase its rate in response to metabolic demands. Although the synchronous pacemakers can meet some of the physiological demand for variation in heart rate, these devices cannot replicate the functions of the heart or meet the demands of the body during stressful activities such as exercise. A new type of pacemaker which makes use of a control system has been developed. A sensor is used to convert a physiological variable in the patient to an electrical signal that serves as an input to the controller circuit, which can determine whether any artificial pacing is required or not. Figure 25.10 shows a block diagram of a rate responsive pacemaker.

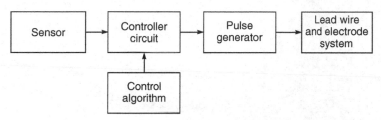

> **Fig. 25.10**   *Block diagram of a rate-responsive pacemaker*

Several sensor-based rate-responsive systems were developed during the late 1970s and 1980s, including: blood pH, respiratory rate, vibration or motion, blood temperature, and QT interval. Today, the majority of pacemakers are rate-responsive, incorporating one or more sensors. The most common sensor is the activity sensor which uses piezo-electric materials to detect vibrations caused by the body movement. Some of the other sensors which are under various stages of development are blood pH, blood temperature, stroke-volume, oxygen saturation and minute ventilation. The sensor can be placed within the pacemaker itself or located at some other place in the body. It may be noted that each of the physiological variables requires a different control algorithm for the control circuit.

Minute volume (tidal volume× number of respirations) is a fine physiological indicator, because it responds to the rapid change in metabolic demand via increases in tidal volume and respiratory rate. It provides better rate-responsive characteristics and is used in many rate-responsive pacemakers. Figure 25.11 shows 'Legend' range of pacemakers from M/s Medtronic, USA.

25.3.6   Packaging of Implantable Pacemakers

Pacemakers are now based on hybrid circuit technology because of its ability to compress complex electronic circuitry to a very small size. Hybridization telescopes the size requirements of discretely packaged transistors, integrated circuits, diodes, resistors and capacitors onto a single, compact package. The hybrid circuit and the lithium-iodide power source are hermetically sealed separately and then both of them are sealed within the stainless steel case, thus providing complete and enduring isolation from the corrosive body environment. The pulse generator is hermetically sealed by tungsten inert gas welding. The case is manufactured either of titanium or of 316 L stainless steel and tested to a standard helium leak rate of 2×10^{-8} cc/s at one atmosphere. The steel enclosure also serves as the indifferent electrode for pacing and sensing, and functions as a shield against electrical interference. A specially designed, hermetic, poly-crystalline ceramic feedthrough seal utilizing a 99.99% pure platinum conductor makes the electrical connection to the pulse generator through the hermetically sealed enclosure.

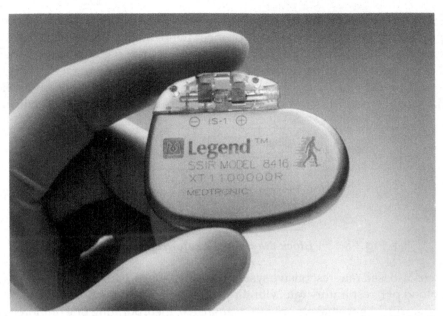

➤ **Fig. 25.11**   *"Legend" range of rate responsive pacemaker (Courtesy: M/s Medtronics)*

The materials exposed to the subcutaneous environment are medical-grade epoxy, medical-grade silicon rubber urethane (Stokes et al. 1979), acetal co-polymer and type 316L stainless steel, consisting approximately of 66.9% iron, 17% chromium, 12% nickel, 2.5% molybdenum and traces of carbon, manganese, and silicon. All these materials have a long history in medical implants, and have been found to be tissue-compatible.

25.3.7   Power Sources for Implantable Pacemakers

The life of a pacemaker is determined by the current consumption of the electronic circuit and the available energy in the unit. The first clinical application of an implantable cardiac pacemaker (Elmquist and Senning, 1960) used nickel-cadmium rechargeable cells. These pacemakers were only marginally successful and were, therefore, abandoned. It appears that the service life of rechargeable cells with recharging was not significantly longer than that of primary cells.

Mercury Batteries: The first American implantable pacemaker developed in 1960 by William Chardack and Wilson Greatbatch, used mercury oxide zinc batteries with 1200 mAh. This battery produces 1.35 V. Depending upon the pacemaker, three to five batteries were used to power a pacemaker.

For a very long time, the zinc-mercury battery has been the preferred power source for implantable cardiac pacemakers. Nevertheless, this battery has been documented as being responsible for most pacemaker failures (Ruben, 1969). The failure mechanisms included dendritic mercury growth, zinc oxide migration, leaky separators and corroded welds. Most of the failure modes are traceable directly or indirectly to the mobile, corrosive liquid electrolyte which is sodium hydroxide.

Biological Power Sources: Several investigators (Racine and Massie, 1971; and Schaldach, 1971) have experimented with biological power sources. These have mostly been galvanic cells using body fluid as the electrolyte. However, in most cases, the cell eventually seems to become permanently electrically isolated from its environment and become inoperative, thus making it hazardous to carry the concept to clinical application.

Nuclear Batteries: The most commonly used nuclear cells (Greatbatch and Bustard, 1973) utilized plutonium 238 with a half-life of 87 years. The energy liberated by the total decay of I g of Pu 238 with a power density of 0.56 W/g will be 780 kWh. If it is assumed that this prime energy could be converted into electric power at an efficiency of 1% and then, if we need only to provide a pulse power, about 20 mg of plutonium would be required.

Even though the radiation exposure to the patient from the nuclear batteries is judged to be insignificant, opponents of radio-isotopic fuel sources seemed to be concerned that the mere number of nuclear pacemakers in use will make catastrophic dissemination of particulate plutonium more likely.

Lithium Cells: The long-life lithium-iodine battery powered pacemaker represents a significant advance in pacemaker technology. The lithium battery is solid-state and consists of an anode of metallic lithium (Li) and a cathode of molecular iodine (I_2) bonded in complex form to an organic carrier. The solid electrolyte consists of crystalline lithium-iodide (LiI). The following reaction takes place:

$$2Li + I_2 = 2\,LiI + e$$

The relative ease with which the single outer shell electron is lost from the atoms of the alkali metals family makes them the most active of all metals. Lithium has the highest electrochemical equivalent of any alkali metal. It is, therefore, the most energetic anode material and is ideal for use in high energy density batteries. An anode current collecting screen is pressed between two layers of lithium, forming the anode assembly (Fig. 25.12). The battery develops a voltage of 2.8 volts, which is stepped up to 5 V in the circuitry. No gas is evolved from the simple cell reaction; therefore, the lithium cell can be hermetically sealed in a welded stainless steel enclosure.

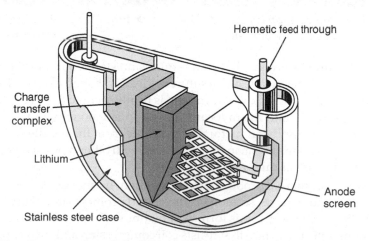

> **Fig. 25.12** *Constructional details of lithium iodine battery (Courtesy: Wilson Greatbatch, USA)*

The lithium iodide battery shows a continuous, but gradual drop in voltage over a period of years, due to the slow increase in the internal resistance. Once the output voltage has fallen to 3.3 V, producing a 6 bpm decline in pulse rate, replacement of the pulse generator is indicated. This happens when internal resistance becomes 35 to 40 kΩ. With the battery's energy capacity at 4.14 Ah, a service of more than 10 years is expected.

25.3.8   Type of Leads and Electrodes

The electrodes for delivering stimulating pulses can be connected either on the outside or inside wall of the heart. Electrodes connected to the outer wall of the heart muscle are called myocardial electrodes (Fig. 25.13);. electrodes which are connected to the inner side of the heart chamber are known as endocardiac electrodes (Fig. 25.14). Endocardiac electrodes are inserted through a suitable vein, preferably the jugular vein, and pushed directly into the heart. This method offers an advantage in that open heart surgery is not necessary for the replacement of the myocardial electrode.

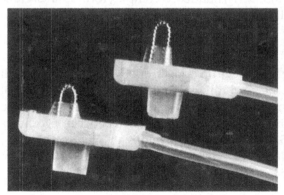

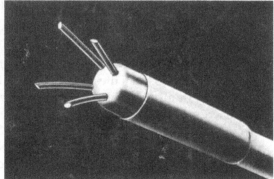

➤ **Fig. 25.13** *A typical loop type myocardial electrode*

➤ **Fig. 25.14** *A special unipolar intracardial type electrode*

Sutureless Leads: Sutureless myocardial lead requires no stab wound or sutures for electrode placement and support. The electrode of the sutureless lead is shaped like a cork screw and rotates into the myocardium with two clockwise turns and is firmly secured to it. The lead is mounted on a special handle for rotation into the myocardium. Supplied along with the lead is a tunneller which is used to release the lead conductor and electrode from the handle. Tissue damage from electrode insertion appears to be less than that associated with other types of myocardial electrodes, resulting in minimal fibrosis around the electrode tip.

The electrode has a 3.5 mm electrode depth penetration. Its surface area is 6.6 mm^2. The large electrode surface area causes fewer sensing problems and promotes reliability. The lead length is available as 54 or 35 cm. The electrode is formed of platinum (70%) and iridium (30%) which give it a strong corrosion-resistant surface that resists coil compression and distortion. The electrode's 5 mm coil diameter spreads the potential stress over a large surface area and promotes a secure attachment of the electrode with tissue. Bio-compatible silicon-rubber provides sheath to both the electrode lead and the head.

Figure 25.15 shows tined lead for transvenous applications. It incorporates four small, soft, silicon rubber tines (straight strands) at the distal tip to protect against lead dislodgement by anchoring the lead in the interstitial spaces formed by trabeculae. The ring-tip electrode design optimizes efficient current distribution. Clinical studies show that ring-tip acute pacing thresholds are approximately 40% lower than those found in other low threshold leads.

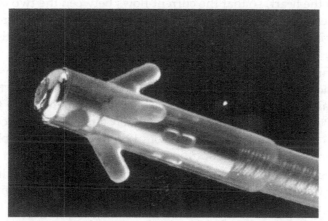

> **Fig. 25.15**   *Tined lead for transvenous applications (Courtesy: Cardiac Pacemakers Inc, USA)*

An electrode provides two functions in a demand pacemaker. It senses the heart's electrical activity and stimulates the heart. Any displacement in the electrode position manifests itself in a decrease in the R-wave amplitude and a reduced slew-rate (rate of change of voltage with time). A typical value of slew rate is 3 V/s. In practice, it is found that the slew rate is a much more sensitive indicator of electrode placement than R-wave amplitude. Pacemaker sensing amplifiers employ bandpass filters to discriminate between R-wave and T-wave. A consequence of this is that R-waves of smaller slew rate are also attenuated and therefore, it is likely that an R-wave with an amplitude exceeding the R-wave sensitivity of the sensing amplifier may not be sensed. This is of critical importance in cases of low amplitude R-waves (under 5 mV) where even moderate attenuation could lead to sensing problems.

A sensing electrode has certain impedances associated with it. The degree to which sensing impedance affects pacemaker sensing is dependent on the ratio of the input impedance of the sensing amplifier to the sensing impedance. Typically, most pacemakers have an input impedance of 20 kΩ or greater. Any leakage path between the pacemaker terminals will be in parallel with the input impedance and will lower it. For example, a leakage path of 2 kΩ between terminals will result in 1800 Ω. A small surface area electrode with a sensing impedance of 2.5 kW will create an R-wave attenuation of 58%. Therefore, small surface area electrodes having larger sensing impedance will have minimal effect in case the input impedance of the amplifier is high and the leakage is negligible (Amundson, 1977).

Two types of electrode systems are commonly used, viz., bipolar and unipolar. In the unipolar system, one electrode is inside or on the heart and is the stimulating electrode, and the second electrode (indifferent electrode) is usually a large metal plate attached to the pulse generator. The indifferent electrode is much larger in size than the pacing electrode. The current in this case flows between the pacing electrode in the heart and the indifferent electrode via the body tissue. The batteries are so arranged that the pacing electrode is negative (cathode) and the indifferent electrode is positive (anode).

In the bipolar electrode system, both electrodes are approximately of the same size and both are placed inside or on the heart, so that the current flows between the two electrodes. The pulse generator is so attached that the distal electrode, at the tip of the catheter, is negative and the proximal electrode ring is positive. Both the unipolar and bipolar configurations have advantages and disadvantages.

A typical unipolar lead used for permanent stimulation of the heart consists of two parts: the electrode and the lead. The proper negative heart electrode is represented by the platinum-iridium (Pt-Ir) catheter tip and has a length of 2 mm and diameter of 3 mm. The tip is connected to a connecting lead, a Pt-Ir specially wound wire completely embedded in silicon rubber. The outer diameter of the lead is 3 mm and its length is about 125 cm. The silicon-rubber core in the spiral gives the lead an excellent mechanical sturdiness.

The indifferent electrodes are usually made up of stainless steel whereas it is desirable to make the active electrode from platinum-iridium alloy to avoid damage which may be caused due to electrolysis by body fluids when used over long periods.

Once the pulse generator and the lead are in place, the electrode-tip which comes in contact with the endocardium or myocardium creates a dynamic process involving physiological changes at the electrodes. A fibrous tissue encapsulates the electrode which increases the stimulation threshold. This process continues for approximately one month during which period, the electrode stabilizes within the fibrous capsule. The increased threshold is due to reduced current density in the most immediate active heart cells (Lindemans, et al, 1975). As a result, significantly more efficient electrodes have been developed. The optimal electrode surface is around 11 mm^2 (Myers and Parsonnet, 1969).

Porous Tip Electrode: A porous tip lead (Fig. 25.16) provides a high stability lead for the endocardial method of pacemaker implant. The immediate tissue in growth secures the porous lead tip to the endocardium. The porous lead has an 85–90% porous platinum-iridium tip that stabilizes the position quickly with little endocardial irritation. It is observed that shortly after implant, the tissue in growth begins to knit the lead tip to the endocardium. One month after the implant, the entire electrode is filled with tissue (Amundson, 1979).

The porous lead has a tip of 7.5 mm^2 and thus requires less current to pace, reducing battery consumption as compared to conventional leads. The tip also provides a larger sensing area because of its porosity—as much as 50 mm^2. Increased contact with electrolytes reduces polarization loss—14% for this tip as compared to 30% loss for solid tip leads with the same pacing surface area. Since the porous electrode reduces polarization energy losses, a larger R-wave is sensed resulting in greater sensing reliability.

Steroid-eluting Electrode: Electrodes have evolved from large-surface-area (30–40 mm^2) platinum-iridium electrodes to relatively small-surface-area (4–12 mm^2) electrodes, composed of various new materials such as iridium oxide-coated titanium, titanium nitride, platinum black, pyrolitic

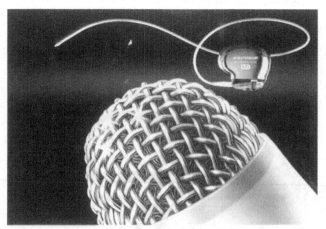

> **Fig. 25.16** *Porous tip lead (Courtesy: Cardiac Pacemakers Inc. USA)*

carbon and others. In order to stimulate the heart, an important consideration is the density of the current at the electrode tissue interface. This is influenced by several factors, such as the surface area of the electrode the amount of fibrotic encapsulation, electrical material, pulse width and pulse amplitude. Extensive studies have been carried out to help control some of these factors, including the use of an anti-inflammatory drug (Glucocorticosteroids such as dexamethasone sodium phosphate) in the tip of the electrode, known as steroid-eluting electrode (Stokes et al., 1983). In this arrangement, about 1 mg of corticosteroid is contained in a silicone core which is surrounded by the electrode material (Fig. 25.17). The leaking of the steroid into the myocardium occurs slowly over several years and reduces the inflammation that results from the lead placement. It also retards the growth of the fibrous sack that forms around the electrodes which separates it from the viable myocardium. As a result, the rapid rise in acute thresholds that is observed with non-steroid leads over the first few weeks, post-implant, is nearly eliminated. The result is a considerable reduction in energy requirements for consistent stimulation of the heart so that a smaller battery can be used to achieve the same longevity.

25.3.9 Problems with Leads and Electrodes

The leads are implanted in the highly hostile environment of the human body. They are constructed to be small in diameter, but still have to be very flexible, which evidently reduces their reliability. Leads may fail due to a variety of reasons, which may include body rejection, fibrotic tissue formation, displacement, erosion, medical complications, migration through body tissue, breach of the lead insulation or failure by breakage. Due to various reasons, manufacturers usually disclaim any warranties with respect to leads.

Pacemaker lead problems can be categorized as follows:

$$
\text{Technical problems} \longrightarrow
\begin{cases}
\text{broken conductors} \\
\text{broken insulation} \\
\text{poor interface with pulse generator}
\end{cases}
$$

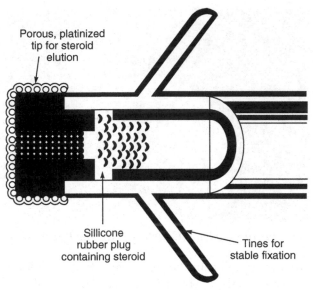

Porous, platinized
tip for steroid
elution

Sillicone
rubber plug
containing steroid

Tines for
stable fixation

> **Fig. 25.17**   *Steroid elution electrode*

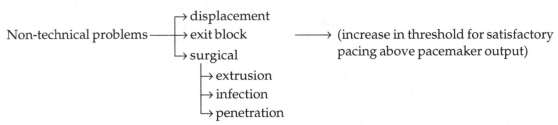

Non-technical problems ──┬→ displacement
                          ├→ exit block ────→ (increase in threshold for satisfactory
                          └→ surgical   pacing above pacemaker output)
                                   ├→ extrusion
                                   ├→ infection
                                   └→ penetration

Technical reliability, therefore, is only one aspect of the lead problem. Because of the faults in pacemaker leads, the concept of redundancy systems or of the fail-safe lead has been introduced. In the fail-safe lead, there is a secondary conducting pathway along the lead so that if the primary conductor fails, sufficient current still flows via the secondary conductor to maintain pacing. The secondary conducting medium might be a liquid, a gel or a paste. Again, this medium must be radio-opaque so that insulation failure can be detected by X-ray examination. An alternative could be a conductive coating on the inside surface of the lumen.

Special fixation mechanisms such as hooks and more recently, coils or helical designs, have considerably decreased the cases of lead dislocations. Efficiency has also improved with the decrease in the electrode surface area. Early implantable cardiac leads had large polished electrodes, as large as 85 mm^2. These had relatively high thresholds, so that stimulations at outputs such as 5 V, 1.8 ms were common. Because of the electrode's large size, pacing impedance was also very low, about 250 Ω. These leads typically drained about 120 uA current. The recently introduced electrodes with a special tip design, such as the ring shape electrode, have a surface area of 5–8 mm^2. The 8 mm^2 polished electrode requires most of the patients to be paced at 5 V, but now 0.5 ms is recognized to be an adequate pulse duration. With modern 2.7 V LiI batteries, the 8 mm^2 polished electrode drains only about 11 uA current at 72 beats per minute.

Traditionally, the feedthrough from the circuitry to the connector terminal has been the weakest link in pacemaker design. Therefore, it should have an excellent mechanical integrity to help assure electrical continuity and hermeticity. The connector boot material is made of transparent silastic thus enhancing visibility and allowing easier verification of proper lead connection. A highly reliable enclosure is ensured by laser welding the feedthru' connector to the outer titanium shield. Fabrication of the feedthru' employs the technique of' metalizing the ceramic insulator with either titanium or platinum to help eliminate the possibility of corrosion which could lead to hermetic failure.

There has been a significant reduction in the size of connectors. The diameter of the connectors used earlier was between 5 and 6 mm, and pin lengths were as long as 25 mm. The International Standard Number 1 (IS.1) now established by the pacing industry specifies a connector diameter of 3.2 mm and pin length of 5.0 mm.

Bipolar leads, which have two electrodes positioned in the heart, are designed with a coaxial connector requiring only a single receptacle, resulting in improvement in the size of bipolar pacemaker connectors. Some dual-chamber devices originally required four holes 5–6 mm in diameter; but now require only two holes with in-line connectors that are 3.2 mm in diameter.

Cardiac pacing electrodes have improved significantly over the last 40 years. The chronic stimulation thresholds have decreased from about 3 V, 1.8 ms in the 1960s to about 0.5 V, 0.5 ms in the 1990s. Pacing impedances of 1000 Ω are the state-of–the art today as compared to 250 Ω for some of the early electrodes. Consequently, we have smaller, longer lived pulse generators without sacrificing patient safety. This has been achieved in part by the use of bio-compatible materials, optimizing electrode size and surface structure, and the use of steroid elution (Stokes, 1996).

25.3.10 Reliability Aspects of Cardiac Pacemakers

There are unique and challenging problems of designing and implanting cardiac pacemakers. The most significant of these is the potential irritability to tissues for chronic applications. Some materials may function satisfactorily for a few days or weeks before chemical or physical interactions induce inflammation or rejection. Partiality reacted or incompletely cured plastics, epoxies or rubbers invariably evoke biological responses. Intra-muscular and subcutaneous tests generally serve as indicators of tissue compatibility.

Implantable cardiac pacemakers must operate at a somewhat elevated temperature in a perfectly stirred, extremely corrosive, oxygen-rich environment of 100% humidity. It is difficult to imagine a worse environment for devices that must operate reliably, 24 hour a day, for years on end. Although high reliability is a common attribute in all medical equipment, it is more so in implantable pacemakers. Since it is not possible to achieve reliable functioning with preventive or service maintenance, such reliability must be designed in. In fact, the value of the pacemaker has little relation to its price and depends far more on its failure rate. This is both because of the danger to the patient's life from a pacemaker failure and the monetary cost of replacing it, which may well be several times the price of the pacemaker itself.

A design objective of a failure rate of 0.15% per month at a confidence level of 90% has been suggested for pacemakers. This means that out of 100 implanted pacemakers, one device might be expected to fail in seven months, or three in 21 months. The 90% confidence level means that in 10 such clinical series, nine would demonstrate a failure rate equal to or less than 0.15% per month. In the past, premature battery failure accounted for up to 90% failures. Marked improvements in pacemaker reliability have been achieved with the use of lithium-iodide battery powered systems. In fact, the electrode lead has replaced the battery as the weakest point in the pacemaker system.

▮▮▮▶ 25.4 RECENT DEVELOPMENTS IN IMPLANTABLE PACEMAKERS

Developments in cardiac pacing technology have greatly reduced the size of pacemakers, while improving their longevity and reliability, and increasing their sophistication in terms of programming and other automatic features. Reductions in the size of pacing systems have been largely due to improvements in power sources, increased circuit integration, hybrid packaging, and the development of smaller leads and lead connectors. The use of sophisticated microprocessors and large amounts of memory have transformed some modern pacemakers into implantable computers allowing them to store significant amounts of intra-cardiac data. A trend towards the use of sensor technology has enabled pacemakers to provide rate response, taking the place of a damaged sinus node.

Today's pacemakers are all hermetically sealed as a result of the widespread use of lithium iodide batteries. The circuitry therefore, no longer has to be sealed from the power source, which is important in terms of electronic packaging. Hermetically sealed pacemakers, usually within stainless steel or titanium housing, allow the internal environment to be kept to a very low humidity level, virtually eliminating this source of failure.

Advances in microelectronics have resulted in developments of sophisticated pacing circuitry, beginning with the use of two transistors in the first implantable pacemaker in 1958. Integrated circuits were first used in a pacemaker in 1971 and were quickly adopted as the standard for pacemaker design. Hybridization has contributed to the miniaturization of modern pacemakers. Hybrids are an assembly of semiconductor chips, miniature resistor/capacitor chips and film resistors/capacitors which are interconnected by an insulating ceramic substrate. Now chips are placed above one another, facilitating very dense packaging involving double-sided hybrids.

Interconnections between ICs that carry high-frequency signals require more current due to losses as a result of capacitance in wire connections. There is a requirement to provide more circuit integration, in order to decrease current drain and the overall size of the package to house circuits. Whereas ICs in 1980 had device geometry of the order of several microns, ICs today used in pacemakers have device geometry at the sub-micron level (Sanders and Lee, 1996).

Further, complementary metal oxide (CMOS) integrated circuits with low power consumption and high reliability have replaced bipolar/I^2L technology. Digital and analog functions are designed onto the same chip, and techniques such as switched-capacitor filters are used to eliminate external hybrid circuit components and resistor trim operations. Other advances include the use of flex circuits and spot-welded connections from the circuit to the battery post or to the feed through and lead connectors.

Microprocessors offer a high degree of flexibility and allow a wide range of products to be developed with new software and offer the possibility of faster revision of pacemaker function. Very small, energy-efficient microprocessors are now commercially available to pacemaker manufacturers. In today's pacemakers, a single IC contains the microprocessor, memory, output circuitry and telemetry, plus other features. Reducing the pacemaker circuit to a single chip has decreased current drain and size, allowing a significant reduction in volume over earlier units.

As pacing functions have become increasingly complex, it has become standard practice to build circuitry in a more general, computer-based architecture, with functions specified in software. This provides a greater degree of freedom from making any changes in hardware and allows non-invasive modification of software after implant. A software-based pacemaker consists of a telemetry system, decoder, timing circuit, analog sensing, and output circuitry, and analog rate-limiting circuitry, with a microprocessor acting as controller (Fig. 25.18). Two kinds of memory are involved: read-only memory (ROM) and random access memory (RAM). Software instructions are stored in RAM, with programmable settings such as pacing rate, pulse amplitude, pulse width, sensing gain, etc. kept in RAM. The ROM circuitry is designed to check for the error-free flow of information, conduct internal self-testing routines during each pacing operation and switch to a back-up pacing system if errors are detected; thus reducing the possibilities of software errors causing anomalous pacing behaviour. Data such as serial number, patient identification and diagnostic information are stored in RAM, alongwith more complex features.

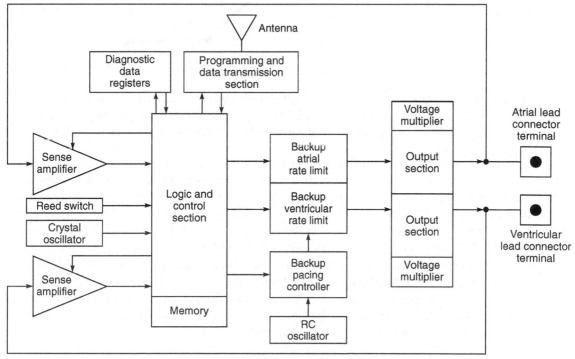

> **Fig. 25.18**   *Block diagram of a typical modern pacemaker (after Sanders and Lee, 1996)*

Modern pacemakers are extremely sophisticated and highly programmable. Weighing about 25 grams, they can pace and sense both chambers of the heart and adjust their rate by tracking intrinsic atrial activity or responding to input from a sense. Figure 25.19 shows the decreasing size of modern pacing devices. Information on a wide range of diagnostic data such as status of the battery, lead system, electronics, pacemaker/patient interaction and output of the sensor becomes available through telemetry.

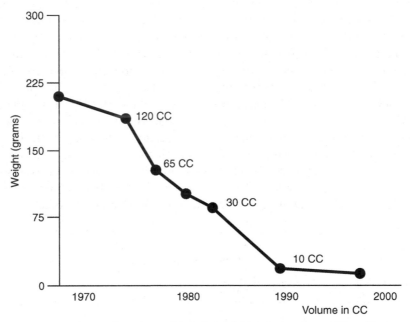

> **Fig. 25.19**   *Trends in the size of implantable pacemakers. The diagram shows dramatic decrease in weight and volume which has begun to plateau at around 25 g and 10 cc of volume (Adapted from Sanders and Lee, 1996)*

There are numerous failure mechanisms that can lead to catastrophic consequences for the patient. In such cases, the pacemaker monitors these mechanisms and automatically switches to the built-in back-up system. For example, some pacemakers are equipped with a bipolar verification function that tests the integrity of the bipolar lead every pacing cycle. In the event that a high anodic resistance is encountered, the pacemaker automatically reverts to unipolar pacing, using the can as anode thereby preventing a serious medical problem.

Another type of automatic back-up feature protects against failure due to the logic circuitry. This circuit redundancy, which involves a voltage-dependent timing circuit (or RC oscillator) provides basic pacing support in the event of microprocessor failure, software errors, crystal failures, or other disruptions. If the back-up circuit does not detect a cardiac event for 2.8 seconds, it assumes control and provides basic pacing support.

▥▷ 25.5 PACING SYSTEM ANALYSER

Pacing system analysers are useful in the operating room or catheterization laboratory during pacemaker surgical procedures. The analyser can help to determine optimum voltage and pulse width thresholds with the resultant current flow thus helping to determine the stimulation thresholds. Likewise, the R-wave amplitude of the heart sensed by the implanted electrode is displayed thus helping to determine the sensitivity threshold. The analysing system has two independent functions:

- A synchronous pulse generator with variable rate, pulse duration and output voltage— These parameters help the physician to determine the stimulation threshold.
- A digital measuring system helps to test the operation of the pulse generator as well as the lead. The system provides a digital read-out of the pulse amplitude, pulse rate and pulse width available at the output of the pulse generator. This series of test capabilities is useful when verifying the proper operation of a pulse generator to be implanted and when trouble-shooting for a suspected malfunction in an implanted system.

For pre-implantation sensitivity testing of ventricular synchronized and ventricular inhibited pacemakers, it is necessary to simulate the R-wave available at the endocardial or myocardial electrode. The sine-squared waveform has been chosen for maintaining pacemaker standards.

Cardiac Defibrillators

▥▶ 26.1 NEED FOR A DEFIBRILLATOR

Ventricular fibrillation is a serious cardiac emergency resulting from asynchronous contraction of the heart muscles. This uncoordinated movement of the ventricle walls of the heart may result from coronary occlusion, from electric shock or from abnormalities of body chemistry. Because of this irregular contraction of the muscle fibres, the ventricles simply quiver rather than pumping the blood effectively. This results in a steep fall of cardiac output and can prove fatal if adequate steps are not taken promptly.

In fibrillation, the main problem is that the heart muscle fibres are continuously stimulated by adjacent cells so that there is no synchronised succession of events that follow the heart action. Consequently, control over the normal sequence of cell action cannot be captured by ordinary stimuli.

Ventricular fibrillation can be converted into a more efficient rhythm (Fig. 26.1) by applying a high energy shock to the heart. This sudden surge across the heart causes all muscle fibres to contract simultaneously. Possibly, the fibres may then respond to normal physiological pacemaking pulses. The instrument for administering the shock is called a defibrillator.

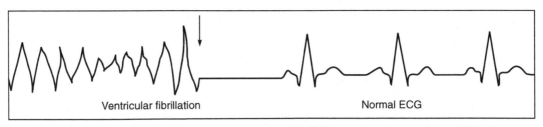

Ventricular fibrillation   Normal ECG

➤ **Fig. 26.1** *Restoration of normal rhythm in fibrillating heart as achieved by direct current shock (arrow) across the chest wall. The horizontal line after the shock shows that the cardiograph was blocked or disconnected for its protection during the period of shock*

The shock can be delivered to the heart by means of electrodes placed on the chest of the patient (external defibrillation) or the electrodes may be held directly against the heart when the chest is open (internal defibrillation). Higher voltages are required for external defibrillation than for internal defibrillation.

26.2 DC DEFIBRILLATOR

In almost all present-day transthoracic defibrillators, an energy storage capacitor is charged (Lown et al., 1962) at a relatively slow rate (in the order of seconds) from the AC line by means of a step-up transformer and rectifier arrangement or from a battery and a DC to DC converter arrangement. During transthoracic defibrillation, the energy stored in the capacitor is then delivered at a relatively rapid rate (in the order of milliseconds) to the chest of the subject. For effective defibrillation, it is advantageous to adopt some shaping of the discharge current pulse. The simplest arrangement involves the discharge of capacitor energy through the patient's own resistance (R). This yields an exponential discharge typical of an RC circuit. If the discharge is truncated, so that the ratio of the duration of the shock to the time constant of decay of the exponential waveform is small, the pulse of current delivered to the chest has a nearly rectangular shape. For a somewhat larger ratio, the pulse of current appears nearly trapezoidal. Rectangular and trapezoidal waveforms have also been found to be effective in the trans-thoracic defibrillation and such waveforms have been employed in defibrillators designed for clinical use (Schuder et al. 1980).

The basic circuit diagram of a DC defibrillator is shown in Fig. 26.2. A variable auto-transformer T_1 forms the primary of a high voltage transformer T_2. The output voltage of the transformer is rectified by a diode rectifier and is connected to a vacuum type high voltage change-over switch. In position A, the switch is connected to one end of an oil-filled 16 micro-farad capacitor. In this position, the capacitor charges to a voltage set by the positioning of the auto-transformer. When the shock is to be delivered to the patient, a foot switch or a push button mounted on the handle of the electrode is operated. The high voltage switch changes over to position 'B' and the capacitor is discharged across the heart through the electrodes.

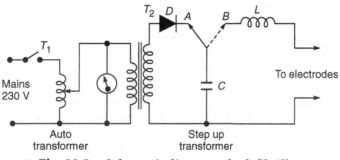

> **Fig. 26.2** *Schematic diagram of a defibrillator*

In a defibrillator, an enormous voltage (approx. 4000 V) is initially applied to the patient. It has been shown by various investigations that although short-duration pulses (as low as 20 µs), can affect defibrillation, the high current required impairs the contractility of the ventricles. This is overcome by inserting a current limiting inductor in series with the patient circuit. The disadvantage of using an inductor is that any practical inductor will have its own resistance and dissipates part of the energy during the discharge process. In practice, a 100 mH inductor will have a resistance of about 20 Ω. The energy delivered to the patient will, therefore, be only 71% of the stored energy.

The inductor also slows down the discharge from the capacitor by the induced counter voltage. This gives the output pulse a physiologically favourable shape. The shape of the waveform that appears across electrodes will depend upon the value of the capacitor and inductor used in the circuit. The discharge resistance which the patient represents for the defibrillating pulse may be regarded as purely ohmic resistance of 50 to 100 Ω approximately for a typical electrode size of 80 cm^2. The shape of the current/time diagram of the defibrillating pulse remains largely unchanged in the above resistance range, except for a change in amplitude which depends on the resistance. A typical discharge pulse of the defibrillator is shown in Fig. 26.3 (curve 1). The most common waveform utilized in the RLC circuit employs an under-damped response with a damping factor less than unity. This particular waveform is called a 'Lown' waveform. This waveform is more or less of an oscillatory character, with both positive and negative portions. The pulse width in this waveform is defined as the time that elapses between the start of the impulse and the moment that the current intensity passes the zero line for the first time and changes direction. The pulse duration is usually kept as 5 ms or 2.5 ms.

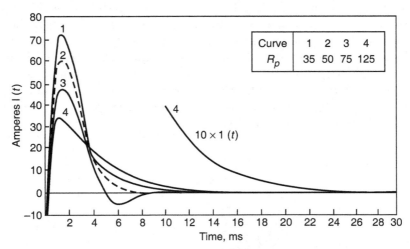

> **Fig. 26.3** *Current waveform I(t) versus patient impedance R_p for a typical damped sine wave defibrillator (C = 32 uF, L= 35 mH, RC = 67)*

Damped sinusoidal waveforms (DSW) are used in over 90% of the defibrillators currently available in the market and make use of capacitors 30–50 μF. They have critical damping resistances of 40–70 Ω, hence produce overdamped waveforms for almost all patients.

The diagram shows the current waveforms for patient resistance of 35, 50,75 and 125 Ω, for a typical defibrillator R_C = 67 Ω (critical damping resistance) charged to deliver 360 J to a

50 Ω patient. The actual delivered energy ranges from 335 J at 35 Ω to 405 J at 125 Ω (Charbonnier, 1996).

There are two general classes of waveforms: mono-phasic and biphasic. Monophasic waveforms use escalating high levels of energy delivered in one direction through the patient's heart whereas a biphasic waveform delivers energy in both directions. The biphasic waveform is preferred as it has proven in clinical studies to defibrillate more effectively than other types of waveforms.

It has been found experimentally that the success of defibrillation correlates better with the amount of energy stored in the capacitor than with the value of the voltage used. It is for this reason that the output of a DC defibrillator is always calibrated in terms of watt-seconds or joules as a measure of the electrical energy stored in the capacitor. The instrument usually provides output from 0–400 Ws and this range provides sufficient energy for both external and internal defibrillation.

Energy in watt seconds is equal to one half the capacitance in farads multiplied by the voltage in volts squared, i.e., $E = 1/2\,CV^2$. If a 16 microfarad capacitor is used, then for the full output of 400 Ws to be available, the capacitor has to be charged to 7000 V.

For internal defibrillation, energies up to 100 Ws are usually required whereas higher energy levels are necessary for external defibrillation. The DC defibrillator cannot be used for rapidly repeated shocks because it requires about 10 s to recharge the capacitor.

The adjustment of the desired energy can be done either continuously or in steps. A continuous adjustment has the advantage that the energy can be adjusted for each desired value. In this arrangement, an adjusting control with or without calibration is provided. In an alternative arrangement, the user has to keep a push button pressed-in until the desired energy level is reached. The wattsecond-meter functions, in this case, a control and an adjustment instrument at the same time. The energy level in this case can be selected through a step switch or push-button selector switch. In most of the instruments, a meter is provided which directly indicates the quantity of energy whereas in others, a lamp lights up at the moment the adjusted energy level is reached.

The amount of energy which the defibrillator actually delivers to the patient is of more relevance. This can be a determining factor as to whether or not a defibrillation is successful. The delivered energy can be estimated by assuming the value of a load resistance which is placed between the electrodes and thus simulates the resistance of the patient. Usually, values of 50–200 Ω are taken as patient's resistance at external defibrillation and 25–50 Ω at internal defibrillation. Most defibrillators will deliver between 60% and 80% of their stored energy to a 50 Ω load.

When the defibrillator is charged and not fired the instrument is a potential source for danger. In some instruments, the capacitor is automatically discharged internally through a resistance when it is not fired, say within 5 min.

26.2.1 Defibrillator Electrodes

The electrodes for external defibrillation are usually metal discs about 3-5 cm in diameter and are attached to highly insulated handles. Most of the conventional electrode systems are circular, a little concave with sharp rims and an insulated back-side. For internal defibrillation when the chest is open, large spoon-shaped electrodes are used. Usually, large currents are required in external defibrillation to produce uniform and simultaneous contraction of the heart muscle fibres. This current not only causes a violent contraction of the thoracic muscles but may also result in occasional burning of the skin under the electrodes.

The external electrodes contain safety switches inside the housings and the capacitor is discharged only when the electrodes are making a good and firm contact with the chest of the patient. This precludes the possibility of an accidental shock to the operator and the risk of burns to the patient. A number of instruments are provided with electrodes which have a spring contact. When these electrodes are adequately pressed on the thorax, the contacts close and the defibrillator is fired. In this way, burns due to poor electrical contact between electrodes and the skin are prevented. However, if the operator is not aware of the presence of the spring, there is a risk that the defibrillator is discharged internally after operating the firing control while no energy is delivered to the patient.

Flat paddle surfaces do not always conform to the body thus reducing contact areas. Proper paddle sizes for different size patients are not quickly and easily available. In order to meet these requirements, pre-gelled and self-adhesive electrodes have been introduced. Such electrodes are commercially available.

Chest wall impedance plays an important part in efficient defibrillation. The factors determining this impedance include the size of the electrodes, paddles, the energy of the discharge, the number and the time interval between previous counter-shocks and the interface material that is used between the paddle electrode and chest wall. Electrode gels are usually employed to reduce contact impedance of the interface and for that the impedance of the gel itself must be very low.

26.2.2 DC Defibrillator with Synchronizer

For the termination of ventricular fibrillation, a DC defibrillator of the type described above is used to restore synchronised working of the heart with the pacemaker of the body. But for termination of ventricular tachycardia, atrial fibrillation and other arrhythmias, it is essential to use a defibrillator with synchroniser circuit. Unlike ventricular fibrillation, there is less direct risk for the patient with auricular fibrillation. In this case, the pump action of the ventricles still exists. However, at defibrillation of a heart in auricular fibrillation, the shock may bring the ventricles into fibrillation. There is, however, a period in the heart cycle in which the danger is least. Defibrillation must take place during that period. This is called Cardio-version (Jones and Tover 1996). In this technique, the ECG of the patient is fed to the defibrillator and the shock is given automatically at the right moment.

Thus, the function of the synchroniser circuit is to permit placement of discharge at the right point on the patient's electrocardiogram. The application of the shock pulse during the vulnerable T wave is avoided, otherwise there is a likelihood of producing ventricular fibrillation. With the synchroniser unit, the shock is delivered approximately 20 to 30 ms after the peak of the R wave of the patient's ECG.

The synchroniser unit contains within it, an ECG amplifier which receives the QRS complex of the ECG and uses this to trigger a time delay circuit. After an interval of the desired delay time (approximately 30 ms), the defibrillating capacitor is discharged across the chest through the electrodes.

The electrocardiogram of the patient is simultaneously monitored on a cardioscope. The moment the discharge takes place, the synchroniser unit produces a marker pulse on these monitoring instruments to show the instant where the counter-shock has occured in the ECG cycle. The defibrillator with synchroniser unit is normally preferred in coronary care units for use in cardiac emergencies.

For effective and efficient results, the output of the defibrillating circuit is kept isolated or floating. In a floating circuit, the total energy is always contained between the two electrodes. There is no loss of energy to extraneous grounds and a high efficiency is therefore maintained. Moreover, there is no direct path to ground and the danger of shock by accidental contact with the patient during the period of defibrillation does not exist. This provides safety to the attending medical personnel.

The portable defibrillators (Fig. 26.4) operate on rechargeable batteries and therefore make use of DC-DC converters for stepping up the voltage required for charging the storage capacitor. The maximum energy delivered by portable defibrillator is generally $300\,\Omega$ delivered into $50\,\Omega$ load which is equivalent to about 400 Ws of stored energy.

> **Fig. 26.4** *Portable external defibrillator, ECG monitor and recorder (Courtesy: M/s Physio Control, USA)*

Since the patient impedance can vary considerably, it has a strong effect on defibrillation effectiveness. Hence knowledge of peak current, patient impedance and actual delivered energy will greatly enhance the ability of the operator to assess and improve defibrillation effectiveness. Patient impedance and delivered energy can be directly determined from the peak discharge current if the stored energy and the defibrillator circuit parameters (C, L, R) are known. These measurements are automatically made in the modern equipment. A simplified diagram of the discharge control and recording circuitry of this type of defibrillator is shown in Fig. 26.5. (Bennett and Jones, 1982).

Before the discharge, the operator selects the desired energy to be delivered into $50\,\Omega$ load. The microprocessor determines the corresponding value of stored energy taking into consideration the defibrillator internal resistance and the patient impedance. The corresponding storage capacitor voltage V ($E_{stored} = 0.5\,CV^2$) is sensed and regulated by the microprocessor. The discharge current

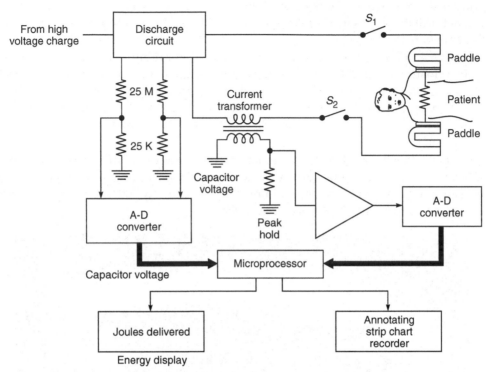

> **Fig.** 26.5 *Block diagram of the discharge control and recording circuitry of a micro-processor based defibrillator monitor (Adapted from Bennett and Jones, 1982)*

passes through a current-sensing transformer placed in the circuit. The use of a sensing transformer provides ground isolation for the patient circuit, yielding a simple method for measuring the discharge current. The transformer provides a voltage that is peak-detected and recorded by the microprocessor. The microprocessor takes the measured peak discharge current and uses this value along with the stored energy to determine patient impedance and delivered energy. The microprocessor then drives a digital display and annotates peak current, patient impedance and delivered energy on a strip chart recorder. A poor paddle contact warning indicator is also activated when the patient impedance exceeds 100 Ω. The ECG monitor has an automatic gain setting, baseline (offset) restore, 60 Hz filter and an effective heartbeat detector.

26.2.3 Automatic or Advisory External Defibrillators

An important development in the field of defibrillators has been the development and successful use of smart automatic or advisory external defibrillators (AEDs) which are capable of accurately analysing the ECG and of making reliable shock decisions. They are designed to detect ventricular fibrillation with sensitivity and specificity comparable to that of well-trained paramedics, then deliver (automatic) or recommend (advisory) an appropriate high energy defibrillating shock.

AEDs require self-adhesive electrodes instead of hand-held paddles for two reasons. Firstly, the ECG signal acquired from self-adhesive electrodes usually contains less noise and has higher quality, hence, it allows a faster and more accurate analysis of the ECG and, therefore, facilitates better shock decisions. Secondly, "hands-off" defibrillation is a safer procedure for the operator, especially if the operator has little or no training. Figure 26.6 shows a typical AED from M/s Agilent Technologies.

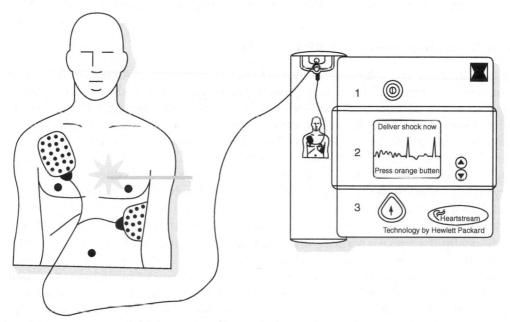

> **Fig. 26.6** *Automatic external defibrillator (Courtesty: Agilient Technologies USA)*

It was initially thought that different self-adhesive electrode designs were needed for defibrillation, pacing, and monitoring. However, studies have shown the feasibility of hybrid designs which produce multi-function electrodes with acceptable performance for all three applications. Multi-function self-adhesive electodes are now commonly used with defibrillator-monitor-pacer instruments.

A critical factor in the safety and performance of an automatic external defibrillator is the ability of the device to accurately assess the patient's heart and make an appropriate therapy decision. The defibrillator performs this evaluation by sensing electrical signals from the patient's heart via electrodes and using a computerised algorithm to interpret the electrical signals. The algorithm used in the modern defibrillators looks at four key indicators to determine whether a rhythm is shockable or non-shockable. These four indicators are heart rate, conduction (flow of electrical waves through the heart indicated by the width of 'R' wave), stability (repeatability of QRS complexes) and amplitude (magnitude of the heart's electrical activity).

An automatic external defibrillator optimised for infrequent use by both first responders and untrained bystanders has been introduced by the Agilent Techologies Heartstream. It is small, light (less than 4 lb) and virtually maintenance-free. While it is on standby for long periods, the device automatically self-tests its electronic circuitry every day and periodically performs an internal discharge and recalibration. The device is powered by a long-life disposable lithuim battey with enough capacity for 75 discharges and one year of self-test. It uses a low energy biphasic waveform with dynamic compensation which accommodates a wide range of patient impedances. The device uses comprehensive voice prompts to alert the user when it detects artfact (electrical signal present in the ECG that is unrelated to the heart signal).

▥▶ 26.3 IMPLANTABLE DEFIBRILLATORS

The use of automatic implantable defibrillators (AID) is recommended for patients who are at high risk for ventricular fibrillation. The AID was commercially introduced by Cardiac Pacemakers (CPI), USA in 1985, following three years of clinical testing (Thomas, 1988). Once the clinical benefit of the implantable defibrillator was proven and clinically accepted, rapid developments in technology were facilitated by the use of integrated circuits to reduce the device size, while enhancing functionality of the device. A modern implantable defibrillator is an implanted computer which stores recordings of the patient's heart signals and collects extensive therapy history and diagnostic data files to aid the physician in individualizing device behaviour for each patient. Less than 70 cc's in volume and with over 30 million transistors, these implantable devices draw less than 20 μA during years of constant monitoring of the patient's cardiac status.

Additionally, the device is hermetically sealed, bio-compatible, and able to survive 500 G's over a temperature range of –30°C to 60°C (Warren et al, 1996). These devices allow the physician to non-invasively programme the therapy rate threshold. Devices available today combine a defibrillator to deliver high energy to very fast and erratic heart rates, with a pacemaker to provide therapies to both increased and decreased heart rates. Other additions to these devices are more sophisticated algorithms for rhythm classification (Warren et al, 1996), and storage of patient's heart signals.

An implantable defibrillator is continuously monitors a patient's heart rhythm. If the device detects fibrillation, the capacitors with in the device are charged up to 750 V. The capacitors are then discharged into the heart which mostly represents a resistive load of 50 Ω and to bring the heart into normal rhythm. This may require delivery of more than one high energy pulse. However, most devices limit the number of high energy shocks to 4 or 5 during any single arrhythmic episode. The shock duration for efficient defibrillation is approximately 4-8 ms which results in the delivery of approximately 30 –35 J at 750 volts.

There are reports that appropriate biphasic waveforms (Figure 26.7) are more efficient and probably safer than monophasic waveforms and produce successful defibrillation at lower energies. Of course, a small capacitor must be charged to a higher voltage to store the required energy. The discharge current pulse duration (full width at half maximum current) is approximately $2.5 \, (LC)^{1/2}$, it is in the range of 2.5–3.5 ms for most defibrillators and does not increase very much with patient impedance. Based on studies by several research workers (Winkle et al, 1989), a biphasic truncated exponential waveform has become the standard waveform, among all implantable defibrillator manufacturers.

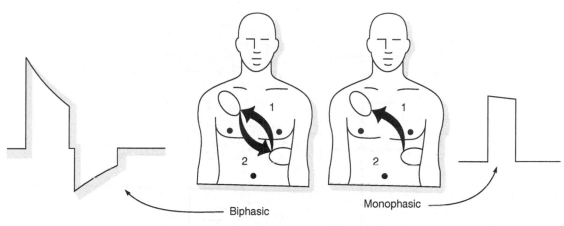

Biphasic

Monophasic

> **Fig. 26.7** *Monophasic and biphasic waveform*

Implantable defibrillator systems have three main system components: the defibrillator itself (AID), the lead system, and the programmer recorder/monitor (PRM). The AID houses the power source, sensing, defibrillation, pacing, and telemetric communication system. The leads system provides physical and electrical connection between the defibrillator and the heart tissue. The PRM communicates with the implanted AID and allows the physician to view status information and modify the function of the device as needed.

Programmer Recorder/Monitor (PRM): The PRM is an external device that provides a bidirectional communications link to an implanted AID. This telemetry link is established from a coil which is contained within the wand of the PRM, to a coil which is contained within the implanted device. This telemetry channel may be used to retrieve real-time and stored intracardiac ECG, therapy history, battery status, and other information pertaining to device function. A number of combinations of programmable therapy and detection options are available and it is not unusual to alter these prescriptions dozens of times over the life of the implant.

Programmers are capable of maintaining continuous communication with the pulse generator at a data rate of 2–4 K b/s. This supports transmission of one to four channels of telemetered electrocardiograms (intracardiac signals) and other device or patient data. Sampling rates of individual signals are from 256 to 512 8-bit samples/s, and data integrity is maintained through error checking algorithms.

Leads: Until recently, the defibrillating high energy pulse was delivered to the heart via a 6 cm×9 cm titanium mesh patch with electrodes placed directly on the external surface of the heart. Sensing was provided through leads screwed in the heart. This approach required an invasive surgical approach to provide access to the heart. The modern implantable defibrillators make use of a single transvenous lead with the multiple electrodes inserted into the right ventricle for ventricular pacing and defibrillation.

Pulse Generator: Major sub-systems of the implanted pulse generator are shown in Fig. 26.8. It has a microprocessor which controls overall system functions. An 8-bit device is sufficient for most systems. ROM provides non-volatile memory for system start-up tasks and some program space,

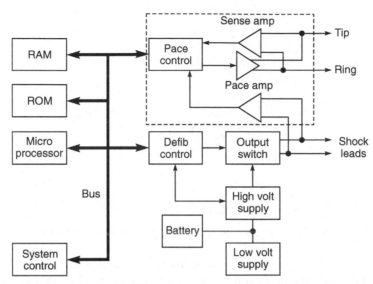

> **Fig. 26.8** *Implantable defibrillator system architecture (after Warren et al.,1996)*

whereas RAM is required for storage of operating parameters, and storage of electro-cardiogram data. The system control part includes support circuitry for the microprocessor like a telemetry interface, typically implemented with a UART-like (universal asynchronous receiver/transmitter) interface and general purpose timers.

The power supply to the circuit comes from lithium Silver Vanadium oxide (Li SVO) batteries. Digital circuits operate from 3 V or lower supplies whereas analog circuits typically require precision nanoampere current source inputs. Separate voltage supplies are generated for pacing (approximately 5 V) and control of the charging circuit (10-15 V).

High power circuits convert the 3–6 V battery voltage to the 750 V necessary for a defibrillation pulse, store the energy in high voltage capacitors for timed delivery, and finally switch the high voltage to cardiac tissue or discharge the high voltage internally if the cardiac arrhythmia self-terminates. The major components of these circuits are the battery, the DC to DC converter, the output storage capacitors, and the high power output switches.

Commercially available implantable defibrillators all utilize lithium SVO cells, with the most common configuration being two connected in series to form an approximately 6 V battery. Unlike 2.8 V lithium iodide (LI) pacemaker cells which develop high internal impedance as they discharge (up to 20,000 Ω over their useful life), SVO cells are characterized by low internal impedance (less than 1 Ω) over their useful life. The output voltage of SVO is higher than LI ranging from 3.2 V for a fresh cell to approximately 2.5 V when nearly depleted.

DC to DC converter used to convert the 6 volt battery voltage to 750 V is of classical configuration. They are operated at as high a frequency (in the range of 30–60 KHz) as practical to facilitate the use of the smallest possible core.

The storage capacitors are typically aluminum electrolytics because of the high volumetric efficiency and working voltage required. Most designs utilize at least two such capacitors in series to achieve the 750 V working voltage required for defibrillation. Since the load resistance is in the range of 20–50 Ω, peak currents of the order of 40A are common in output circuits.

Proper sensing of electrical activity in the heart requires precise sensing and discrimination of each of the components that comprise the intracardiac signal to differentiate tachycardia and ventricular fibrillation. Also, the sense amplifier must also be immune to both physiological and external sources of interference. From an electrical standpoint, the sense amplifier must be able to operate properly over a range of rates from 30 to in excess of 360 bpm. Additionally, the amplifier must be able to quickly and accurately respond to the widely varying intracardiac signals presented during arrhythmia. Jenkins and Caswell (1996) gave a detailed description of detection algorithms in implantable cardioverter defibrillators.

Heart signals of interest lie in the range of 100 μV to 20 μV, and as a result must be amplified prior to being processed. If the signal being sensed drops suddenly in amplitude, which is common during fibrillation, the automatic gain control circuit increases the gain so that the fibrillation signal continues to use the systems dynamic range. A dynamic threshold circuit may then be used to sense R-waves or fibrillation. Sensing circuitry is active through out the entire life of an implantable defibrillator, so minimizing power consumption of these circuits has a beneficial impact on device longevity. Therefore, CMOS or BICMOS IC technologies are the best suited to low power designs.

▧▶ 26.4 PACER—CARDIOVERTER—DEFIBRILLATOR

The vast majority of cardiac arrest patients suffer from tachyarrhythmias which generally develop into ventricular fibrillation. However, a smaller percentage of cardiac arrest victims suffer from extreme brady arrhythmias which require pacing. Hence, it is logical to have a multi-function defibrillator, capable of external pacing as a standard feature. Therefore, most manual defibrillators currently in the market offer both demand and asynchronous (fixed rate) external pacing facility.

Makino et al. (1988), describe an implantable defibrillator with high-output pacing function after defibrillation. A block diagram of the experimental device is shown in Fig. 26.9. It is composed of five battery-powered units: sensing circuit, high voltage converter, switching circuit, defibrillation

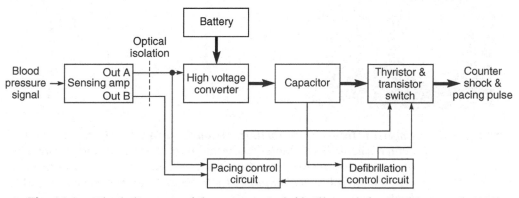

> **Fig. 26.9** *Block diagram of the automatic defibrillator (after Makino et. al, 1988)*

control circuit, and pacing control circuit. The heartbeat signal, which is detected by a catheter-type heartbeat sensor, is amplified for heartbeat monitoring. The absence of a heartbeat for 3.5 s causes the fibrillation detecting circuit to deliver the turn-on signal which then switches on the high voltage converter. At a predetermined voltage level (800 V), the thyristor switch allows the capacitor to discharge its current through the right ventricular electrode. After defibrillation, high-output demand pacing is activated by using the residual energy in the output-capacitor. The pacing rate and pulse width are controlled by the pacing control circuit, and the heartbeat signal is used for demand function. The electrical behaviour of defibrillation and pacing electrodes is discussed by Kina and Schimpf (1996).

Normally, in the cardiac pacemaker, ECG is used for demand pacing function. However, in a situation when the ECG is induced through catheter electrode immediately after defibrillation, the detection circuit will saturate for several seconds. To overcome this problem, a simple blood pressure sensor (electret condenser-microphone) was used to detect a heartbeat without the interference of the stimulation current. The outside of the sensor was covered with adhesive and polyurethane for electrical isolation and waterproofing. The system employs a specially designed multi-element electrode in which the defibrillation electrode area is 180 m^2 and the pacing electrode area is 22 mm^2. Figure 26.10 shows the Medtronic PCD system (Pacer-Cardioverter-Defibrillator).

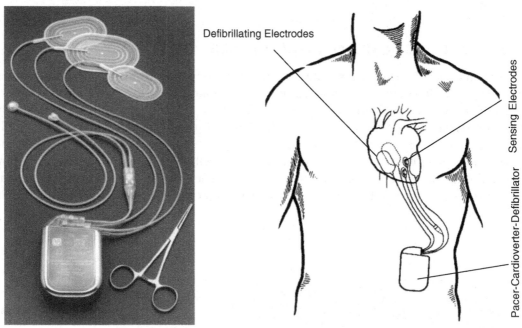

> **Fig. 26.10**   *Pacer-cardioverter-defibrillator (Courtesy: M/s Medtronic, USA). The device is usually implanted in the abdominal area. The pulse generator receives information from two sensing leads and delivers precisely programmed electrical impulses to the heart through two or three patch leads*

▶ 26.5 DEFIBRILLATOR ANALYSERS

A defibrillator analyzer is basically meant to measure the energy content in the discharge pulse. It works on the principle that the energy contained in a pulse of arbitrary shape and time duration is given by

$$E = \int_0^T e(t), i(t)\, dt \tag{i}$$

where E = energy in watt-seconds
$\quad$ $e(t)$ = voltage as a function of time,
$\quad$ $i(t)$ = current as a function of time,
$\quad$ T = time duration of the pulse.

When the voltage exists across a fixed resistance, the energy dissipated in the resistance is given by

$$E = \frac{1}{R} \int_0^T [e(t)]^2\, dt \tag{ii}$$

where R is the resistance of the load.

Equation (ii) shows that a defibrillator analyzer circuit should consist of the blocks shown in Fig. 26.11. The defibrillating pulse is applied across a standard $50\,\Omega$ load and the voltage developed across it is given to a squaring circuit. The squaring circuit consists of a four-quadrant multiplier followed by an operational amplifier. The output of this device is a current which is proportional to the product of the two inputs. In the squaring mode, the two inputs are connected together so that the output is a square of the input voltage. The operational amplifier acts as a current to voltage converter producing an output voltage which is proportional to the output current, from the multiplier.

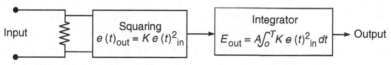

> **Fig. 26.11** *Basic block diagram of defibrillator energy meter*

Besides these basic blocks, the defibrillator analyzer contains other circuits for controlling the measurement operations. For example, the pulse integrator circuit integrates the squared function pulse during the time that a pulse is present. During the remainder of the display time, it acts as an analog storage element. The voltage stored in the pulse integrator is read on the energy meter calibrated in watt-seconds.

The instruments based on the above scheme are meant to measure precisely the pulse-energy and are designed primarily for verifying or calibrating the output energy of all defibrillators, the output pulse waveforms may be Lown or Trapezoid. They measure delivered energy into $50\,\Omega$ load when the user places the paddles of the charged defibrillator against the input contact plates.

Besides determining the value of the delivered energy from a Lown or trapezoidal defibrillator, it is often necessary to have a display of the energy waveform. This can be done by storing the waveform in digital memory and getting it out with a time expansion through a digital-to-analog converter for convenient recording on a standard ECG machine.

Instruments for Surgery

◗ 27.1 PRINCIPLE OF SURGICAL DIATHERMY

High frequency currents, apart from their usefulness for therapeutic applications, can also be used in operating rooms for surgical purposes involving cutting and coagulation. The frequency of currents used in surgical diathermy units is in the range of 1–3 MHz in contrast with much higher frequencies employed in short-wave therapeutic diathermy machines. This frequency is quite high in comparison with that of the 50 Hz mains supply. This is necessary to avoid the intense muscle activity and the electrocution hazards which occur if lower frequencies are employed. The power levels required for electro-surgery are below the threshold of neural stimulation provided that the diathermy frequency is in the radio-frequency range. When the frequency is at least 300 kHz, both the faradic and the electrolytic effects are largely eliminated during the flow of current through the human tissue. This then allows the exclusive utilization of the thermal effect in high frequency surgery providing both the applications for cutting and coagulation.

For their action surgical diathermy machines depend on the heating effect of electric current. When high frequency current flows through the sharp edge of a wire loop or band loop or the point of a needle into the tissue (Fig. 27.1), there is a high concentration of current at this point. The tissue is heated to such an extent that the cells which are immediately under the electrode, are torn apart by the boiling of the cell fluid. The indifferent electrode establishes a large area contact with the patient and the RF current is therefore, dispersed so that very little heat is developed at this electrode. This type of tissue separation forms the basis of electro-surgical *cutting*.

Electro-surgical *coagulation* of tissue is caused by the high frequency current flowing through the tissue and heating it locally so that it coagulates from inside. The coagulation process is accompanied by a grayish-white discoloration of the tissue at the edge of the electrode.

The term *'fulguration'* refers to a superficial tissue destruction without affecting deep-seated tissues. This is undertaken by passing sparks from a needle or a ball electrode of small diameter to the tissue. When the electrode is held near the tissue without touching it, an electric arc is produced, whose heat dries out the tissue. Fulguration permits fistulas and residual cysts to be cauterized and minor haemorrhages to be stopped.

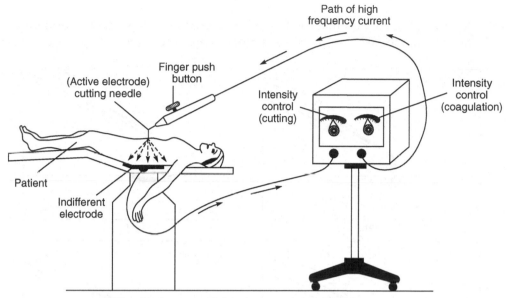

> **Fig. 27.1**   *Principle of surgical diathermy machine*

In *desiccation*, needle-point electrodes are stuck into the tissue and then kept steady. Depending upon the intensity and duration of the current, a high local increase in heat will be obtained. The tissue changes due to drying and limited coagulation. Figure 27.2 shows various types of electrosurgery techniques that are commonly employed in practice.

The concurrent use of continuous radio-frequency current for cutting and a burst wave radio-frequency for coagulation is called *Haemostasis* mode.

Different types of waveforms have been used to produce different effects for surgical procedures. Figure 27.3 shows typical waveforms found in commercial equipment for cutting and coagulation. The cutting current usually results in bleeding at the site of incision, whereas the surgeon would require bloodless cutting. The machines achieve this by combining the two waveforms shown in Fig. 27.3 (e). The frequency of this blended waveform is generally the same as that used for cutting current.

The use of high frequency current offers a number of important advantages. The separation of tissues by electric current always takes place immediately in front of the cutting edge and is not caused by it. Electric cutting therefore, does not require any application of force. Instead it facilitates elegant and effortless surgery. The electrode virtually melts through the tissue instantaneously and seals capillary and other vessels, thus preventing contamination by bacteria. A simplified method of coagulation saves valuable time since bleeding can be arrested immediately by touching the spot briefly with the coagulating electrode. A high frequency apparatus is regarded as standard equipment in the operating theatre and it is kept ready for all surgical interventions even if high frequency surgery is not intended so that use can be made of its superior method of electro-coagulation.

Biological tissue can only be cut when the voltage between the cutting electrode and the tissue to be cut is high enough to produce electric arcs between the cutting electrode and the tissue. The temperatures produced at the points at which the electric arcs contact the tissue like microscopic flashes of lightning are so high that the tissue is immediately evaporated or burned away. A voltage of approximately 200 V_P is required in order to produce the electric arc between a metal

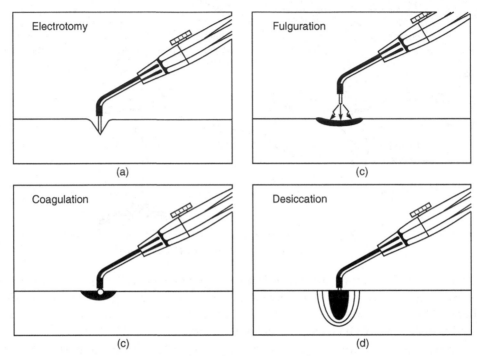

> **Fig. 27.2**   *Various types of electro-surgery techniques commonly employed in practice*

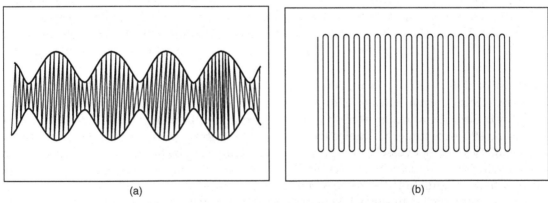

> **Fig. 27.3**   *Types of waveforms generated by surgical diathermy machines:*
> **(a)**   *Cut waveform generated by electron tube circuit, showing RF modulated by 100 Hz,*
> **(b)**   *Cut waveform generated in a solid state diathermy machine*

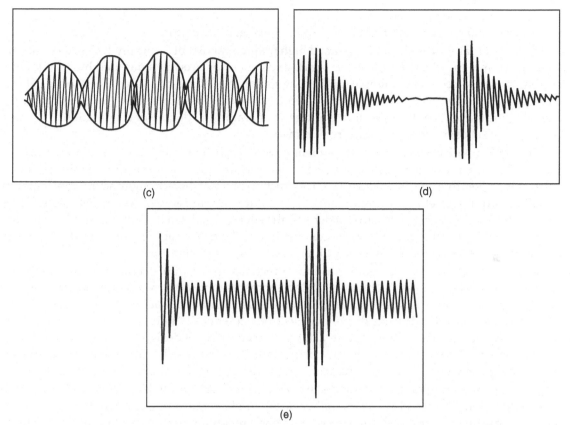

> **Fig. 27.3** *Types of waveforms generated by surgical diathermy machines:*
> (c) *Coagulate waveform generated by spark gap generator*
> (d) *Coagulate waveform generated in a solid state diathermy machine*
> (e) *Blend waveform generated in a solid state diathermy machine*

electrode and biological tissue. If the voltage is less than 200 V_P, the electric arcs cannot be triggered and the tissue cannot be cut. The voltage suitable for cutting biological tissue ranges between 200 V_P and 500 V_P. If the voltage rises above 500 V_P, the electric arcs become so intense that the tissue is increasingly carbonized and the cutting electrode may be damaged. A visible arc forms when the electric field strength exceeds 1 kV/mm in the gap and disappears when the field strength drops below a certain threshold level.

Biological tissues are coagulated by thermal means if the requisite temperature is maintained at around 70°C.

27.2 SURGICAL DIATHERMY MACHINE

Basically, a surgical diathermy machine consists of a high frequency power oscillator. The earlier types of diathermy machines consisted of spark-gap oscillators whereas the current practice is to use thermionic valves or solid-state oscillators. A majority of the earlier units have access to both these power sources, viz. an RF generator and a spark-gap generator.

The RF generator provides an undamped high frequency current (typically 1.75 MHz) which is suitable for making clean cuttings. The spark-gap generator produces damped high frequency current which is specifically suitable for the coagulation of all kinds of tissues. The mixing of both these currents signifies one of the most important possibilities for use in electro-surgery. By blending the currents of the tube and spark-gap generator, the degree of coagulation of wound edges may be chosen according to the requirements.

Whilst the detailed waveform and frequency spectrum used varies from one manufacturer to the other, the basic concept requires a high temperature arc, possibly exceeding 1000°C (Dobbie, 1969) at the operative site. In practice, the cross-section of the arc is extremely small, considerably less than 1 mm diameter, leading to a high current density in the arc. As the heating effect is proportional to the square of the current density, the effect is localized to form the arc. Other factors affecting the rise in temperature are the composition of the tissues and the magnitude of cooling provided by the local blood flow or any other heat transport system.

Despite the fact that surgical diathermy machines have been routinely employed for over half a century, the most significant technological developments have occurred within the past decade only. Solid-state generators have replaced a substantial number of vacuum tube and spark-gap units. Disposable, self-adhering dispersive electrodes (generally known as 'ground pads') are now widely used in place of the large area buttplate. A number of design features have enhanced safety. These features include a variety of circuit integrity monitors like dispersive electrode cable continuity, patient circuit continuity and alternate-path current monitors, among others.

The frequency of operation of solid-state diathermy machines is 250 kHz to 1 MHz. They deliver 400 W in 500 Ω load at 2000 V in the cutting mode and around 150 W in the coagulation mode. In coagulation, the burst duration is 10–15 s and repetition frequency of the burst is 15 kHz.

Figure 27.4 shows a block diagram of the solid-state surgical diathermy machine. The heart of the system is the logic and control part which produces the basic signal and provides various timing signals for the cutting, coagulation and haemostasis modes of operation. An astable multi-vibrator generates 500 kHz square pulses. The output from this oscillator is divided into a number of frequencies by using binary counters. These are the frequencies which are used as

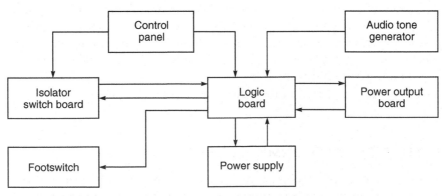

➤ **Fig. 27.4**   *Block diagram of solid state electro-surgical unit*

system timing signals. A 250 kHz signal provides a split phase signal to drive output stages on the power output circuit. A 15 kHz gating signal produces the repetition rate for the three cycles of the 250 kHz signal which make up the coagulating output. The pulse width of this output is set at about 12 µs.

The 250 kHz signal used for cutting is given to power output stage where it controls the push-pull parallel power transistor output stage. The output of this high power push-pull amplifier is applied to a transformer which provides voltage step-up and isolation for the output signal of the machine. In order to meet the high power requirements, as much as 20 transistors are used in a parallel Darlington circuit. However, the power output amplifier circuitry varies considerably among the different commercial equipment. The modern machines employ both bi-polar junction transistors and power metal oxide-semiconductor field-effect transistors (MOSFET) in a cascade configuration or the use of a bridge connection of MOSFETs.

In order to facilitate identification of each mode of operation, the machines incorporate an audio tone generator. The tone signals are derived from the counter at 1 kHz (coagulation), 500 Hz (cutting) and 250 Hz (haemostasis). The isolator switch provides isolated switching control between the active hand switch and the rest of the unit. A high frequency transformer coupled power oscillator is used in which isolated output winding produces a DC voltage. The load put on the DC output by the hand switch is reflected back to the oscillator, accomplishing isolated switching. There is a provision to interrupt the power output if so desired.

Besides these basic functional circuits, logic circuits are used to receive external control signals and to operate the isolating relays, give visual indications and determine the alarm conditions. The logic circuits receive information from the foot-switch, finger switch and alarm sensing points. A thermostat is sometimes mounted on the power amplifier heat sink. In case of over temperature, it becomes open-circuited, signalling an alarm and interrupting the output.

The output circuit in the diathermy machine is generally isolated and carefully insulated from low frequency primary and secondary voltages. Blocking capacitors serve to effectively prevent any low frequency from appearing in the output circuit, and the isolated output reduces the possibility of burns due to an alternate path to ground. Complaints of electrical shock during surgery can almost always be attributed to muscle contractions of the patient. This is caused by the rectification of the high frequency energy at the junction of the active electrode and the tissue in the presence of an arc, which is the actual means of performing electro-surgery. This phenomenon is observed most when operating in a site of sensitive nerve tissue. There is, however, no danger to the patient or to the operator due to this action. On the other hand, anyone in close proximity to the radio-frequency carrying cables or electrodes will have some energy induced into his body. If by chance, he touches the metal cabinet of the surgical unit or any other conductive surface, current will flow through his body, resulting in a spark at the point of contact. It is advisable to avoid contacts with conducting surfaces by those who happen to be near the machine or cables.

The gases used in anaesthesia tend to settle near the floor. Therefore, the construction of the foot switch should be such that no explosion should occur in the atmosphere surrounding this switch caused by the operation of the electrical contacts within the switch.

Solid-state machines mostly incorporate an independent bi-polar RF generator for microsurgery procedures offering a fine output power control. The output waveform is a damped sinusoid at a repetition frequency of 30 kHz. The power output across 300 Ω is about 20 W, with peak to peak voltage of 1500 V open circuit.

27.2.1 Automated Electro-surgical Systems

Haag and Cuschieri (1993) illustrated (Fig. 27.5) that with a conventional electro-surgical unit, there is a considerable fluctuation of the output voltage throughout the 3-s period of the cut. The cause of this undesirable fluctuation is linked to the following factors:

- *Size and Shape of the Cutting Electrode:* The conditions are different for the generator if, for instance, cutting is performed with electrode of large surface area or with a fine needle.
- *Type and Speed of Cut:* The cutting quality is determined by the speed with which the electrode is moved (quick or slow) and by the type of cut (superficial or deep)
- *Different Tissue Properties*: The tissue itself has a strong influence on the quality of the cut. For example, in tissues with a high resistance such as fat, the output voltage is increased whereas in tissues with a low electric resistance, such as nerves and blood vessels, the output voltage may drop significantly.

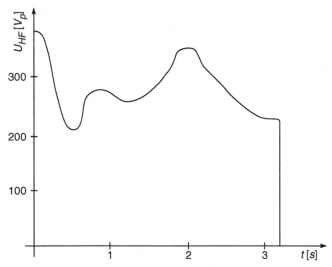

> **Fig. 27.5**   *Cut made with a conventional electrosurgical unit. The incision is uneven with variable coagulation depth along its length due to considerable fluctuation of the voltage secondary to the source resistance of the machine*

The variations in the output voltage due to the above factors considerably affects the quality of the cut. At times, the maximum output voltage can become so high (above 600°C) that severe carbonization occurs. Conversely, the minimum value of the output voltage can become so low (below 200°C) that cutting action is not achieved. In order to overcome this problem, microprocessor-controlled automated systems have been developed so that the output voltage or the spark intensity remain constant. Such a system has been introduced by M/s ERBE, Germany in their electro-surgery machine shown in Fig. 27.6.

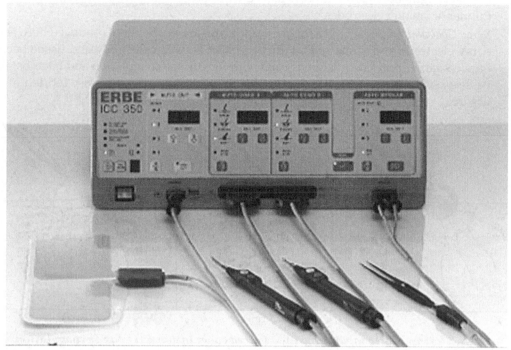

> **Fig. 27.6** *Solid state electro-surgical diathermy machine (Courtesy: M/s ERBE, Germany)*

In this machine, the variables—current, tissue resistance, voltage and spark intensity—are registered by means of an inbuilt sensor system and then processed as defined output signals. The automatic control operates on two different criteria:

– *Voltage control:* whereby the selected voltage is controlled and held constant.
– *Spark control:* by which the selected spark intensity is held constant.

The control of spark intensity is relatively complex because of its non-linear nature. It has been revealed that a number of parameters are directly proportional to the spark intensity and these are used to pre-select and maintain constant these non-linear variables. The design of the control system ensures that the cutting quality is independent of size and shape of the electrode, the type and speed of the cut and the varying tissue properties.

Apart from ensuring a good quality of the cut, the microprocessor-controlled machine also provides the following coagulation modes:

– *Soft coagulation (Fig. 27.7(a)):* In this, no electric arcs are produced between the coagulation electrode and the tissue during the entire coagulation process to prevent the tissue from becoming carbonized. Soft coagulation is recommended in which coagulation electrodes are brought in direct contact with the tissue to be coagulated.

- *Forced Coagulation (Fig. 27.7(b)):* This is characterized by the fact that electric arcs are intentionally generated between the coagulation electrode and the tissue in order to obtain deeper coagulation than could be achieved with soft coagulation, particularly when using thinner or smaller electrodes.
- *Spray Coagulation (Fig. 27.7(c)):* In this, electric arcs are deliberately produced between the spray electrode and tissue so that direct contact between electrode and tissue becomes unnecessary. Spray coagulation is used both for surface coagulation and haemostasis of vessels not directly accessible to coagulation electrodes, such as those hidden in bone fissures.

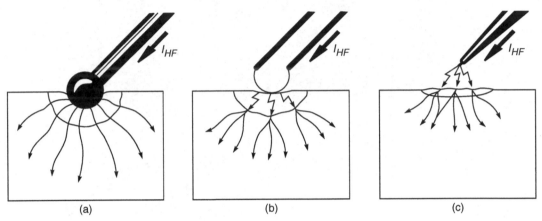

(a) (b) (c)

➤ **Fig. 27.7(a)** *Soft coagulation* **(b)** *Forced coagulation* **(c)** *Spray coagulation*

The microprocessor-controlled machines provide numerous safety features such as an error detection system, an error signalling system and an error storage system. If a certain error arises, it is immediately displayed and the cause of the error can be readily determined. A variety of safety features which help in reducing the risks in high frequency surgery for both the patient and the users are provided in the machines. These features include low frequency leakage current monitor, high frequency leakage current monitor, output error monitoring, time limit monitoring, operating errors and neutral electrode safety system.

Modern surgical diathermy machines are programmable and user-friendly. For instance, frequently used standard settings can already be programmed by the manufacturer before delivery and individual customized settings can easily and swiftly be programmed later. Some machines have a power peak system that delivers a very short power peak at the beginning of electro-surgical cutting to start the cutting arc. Thereafter, average power can be limited to relatively small amounts, which signifies an improvement in protection against unintentional thermal tissue damage. Continuous monitoring of current and voltage levels and making automatic adjustment under the control of a microprocessor provides for a smooth cutting action throughout the procedure.

27.2.2 Electro-Surgery Techniques

The electric current can flow only if the electric circuit is complete (closed). In terms of current flow, there are basically two types of electro-surgical techniques: the mono-polar and the bi-polar technique.

Mono-polar Technique: In the monopolar technique the current flows from the active electrode through the patient to the neutral electrode (patient plate) from which it returns to the generator. The cutting or coagulating effect depends on the contact area between the mono-polar active electrode and the tissue, which is very small compared with the contact area between the patient plate and patient's skin.

Bi-polar Technique: In this technique, two electrodes are used. The current in this case flows through the tissue between the tips of the two electrodes and returns to the generator without passage through the patient. The bipolar surgery is not only safer than monopolar but is also more precise since the current only flows locally at the specific site where it is actually required for heat generation. In addition, the risk of inadvertent burning of the patient at the patient plate is very low. Therefore, the bi-polar technique is becoming a method of choice wherever possible.

27.2.3 Electrodes Used with Surgical Diathermy

The bi-polar technique is used in most of the applications involving the use of surgical diathermy. The high potential terminal of the diathermy is connected to the cutting electrode which is mounted in an insulated handle. The cutting electrodes are available in a variety of shapes (Fig. 27.8), the choice depending upon the nature of application. Lancet electrodes are normally used for cutting applications whereas specific types of needle electrodes are preferred for epilation and desiccation. Loop electrodes are employed for exsecting (or opening up) channels and extirpating growths, etc. The active electrodes for coagulation purposes are of ball type (Fig. 27.9) or plate type. In electro-surgery, the surgeon must be able to switch the high frequency current on and off himself. This can be done with a finger-tip switch in the electrode handle or a foot switch.

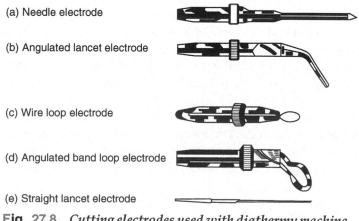

(a) Needle electrode

(b) Angulated lancet electrode

(c) Wire loop electrode

(d) Angulated band loop electrode

(e) Straight lancet electrode

➤ **Fig. 27.8** *Cutting electrodes used with diathermy machine*

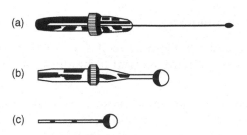

> ➤ **Fig. 27.9** *Coagulation electrodes of different shapes and sizes*

Consequently, the electrodes are used either with electrode handles with finger-tip switch, or with handles without finger-tip switch with the latter being used in conjunction with foot switches.

The low potential terminal of the radio frequency output leads is connected to the indifferent or dispersive electrode. This electrode consists of a lead plate (15 × 20 cm) wrapped in a cloth bag, soaked in saline solution and strapped onto the patient's thigh. An alternative arrangement is to use a flexible non-crumpling stainless steel sheet plate without any covering. Good contact is established with the film of perspiration rising between the plate and the patient's body. Quite often, a liberal amount of conductive paste like ECG paste is applied to the plate. This gives excellent electrical contact and removes the need to keep a wet gauze pad. However, problems may arise if the paste is not cleaned from the plate after use as it may form a hard insulating layer.

An alternative approach is to use capacitively coupled plates in which no direct contact is made between the metal of the indifferent electrode and the patient's skin. The electrode comprises a large sheet of thin metal sandwiched between two sheets of neoprene, which formed a capacitor with the patient's body. This capacitor allows an easy path for the passage of the high frequency diathermy currents. It, however, adds to the problem of introducing burn hazard by providing alternative current paths when other equipment with grounded patient connection is used.

Many manufacturers offer disposable indifferent electrodes and many advantages are claimed for these compared to conventional metal plates. One type of disposable dispersive electrode is like a disposable ECG electrode with a gel-soaked sponge backed by metal foil and surrounded by foam and pressure-sensitive adhesive.

There is no universal agreement over the safe effective area of the indifferent electrode necessary for diathermy current to exit without causing a rise in skin temperature. The American National Fire Protection Association manual (NFPA 1975) carries the suggestion that the plate area should be 1 cm^2 per 1.5 W applied power, leading to an area of 267 cm^2 for a 400 W unit. In Britain, the recommendation calls for 180 cm^2 plate area. Indifferent electrodes currently available range from 50 to 200 cm^2.

The most common reason for faulty performance of an electro-surgical unit is improper placement of the indifferent electrode. This electrode must be placed in firm contact with a fleshy portion of the patient and as near as possible to the operating site. Poor contact or excessive distance from the operating site causes a considerable loss of energy available for the actual surgical procedure.

Equipments having rated output powers exceeding 50 W are generally equipped with a circuit that will interrupt the supply voltage to the equipment's output and sound an audible alarm whenever an interruption in the indifferent electrode circuit occurs. This is achieved by measuring the electrical impedance between different electrodes and the patient. If the high frequency rises above the limit permitted for the electrical impedance, an alarm will be generated and the output disconnected.

⚙️➤ 27.3   SAFETY ASPECTS IN ELECTRO-SURGICAL UNITS

The risks associated with electro-surgery fall into four main categories viz. burns, electrical interference with the heart muscles (ventricular fibrillation), the danger of explosions caused by sparks and electrical interference with pacemakers and other medical electronic equipment.

Burns: The predominant hazard associated with electro-surgical units is burns caused by excess current density at a rate other than that at which it is meant to be present. The burn usually occurs at the dispersive electrode because of failure to achieve adequate contact. The injury can also occur because an unintended current pathway may be created. In the latter case, the lesion usually occurs at a point where the patient is inadvertently touching a grounded object and contact is made over a small area of skin.

The risk of burns also exists in the presence of moisture, i.e., the accumulation of prepping agents, blood or other fluids around the indifferent electrode can give rise to small, highly conductive areas. Burns resulting from small conductive areas between the limbs can be prevented by means of dry cloth placed between them.

During surgery, the output power of the electro-surgical unit should not be increased if the desired surgical effect is not obtained. Abnormal power settings indicate that something is wrong and that the fault must be identified. In particular, the indifferent electrode and all cables and connectors should be thoroughly checked. It is advisable to carry out surgical work with the power setting as low as possible, to reduce the risk of burns. Besides this, the active electrode, when not in use, should be placed well clear of the patient. This is to avoid its activation in case the foot switch is inadvertently pressed.

High Frequency Current Hazards: Another serious hazard associated with the use of surgical diathermy machines is the possible electrocution of the patient from faulty mains operated equipment, when one side of an electrical circuit is connected to earth. In order to provide protection against mains current electrocution, a capacitor (RF earthed) is generally included between the indifferent lead and earth.

The output configuration plays an important role in the RF current circuit. There are three technical approaches. In the earthed output system, the indifferent electrode is connected conductively to protective earth (Fig. 27.10a). The earth referenced system uses a capacitor to connect the indifferent electrode to earth (Fig. 27.10b). In the isolated system, the return electrode is floating, i.e., there is no intentional connection to earth (Fig. 27.10c). The value of the capacitor is such that while providing a very low impedance to the high frequency diathermy current, it offers a higher impedance to the mains frequency. This approach also offers only a partial solution to a complex problem.

Modern solid-state machines usually have RF isolated patient circuits. This implies that ideally RF current may take only one path, i.e. from active electrode through the patient to the indifferent

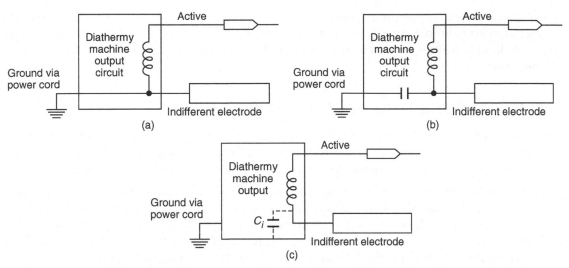

> **Fig. 27.10(a)**   *Output circuit configuration in which the indifferent circuit is directly*
>   *connected to the mains earth*
>
>   **(b)**   *Use of capacitor in the output circuit. This permits RF currents to flow*
>   *to earth through the diathermy machine. However, it effectively*
>   *blocks the passage of low frequency currents (50 Hz)*
>
>   **(c)**   *RF isolated output circuit configuration. Obviously, there is no direct*
>   *connection of the indifferent circuit to ground. The RF leakage current*
>   *is due to stray capacitance within the machine*

electrode. Since there is no earth connection, there is no propensity for the RF current to take any earth pathways which may unintentionally develop. However, due to RF leakage pathways inherent in the equipment and leads, no machine can be considered as completely isolated. The degree of RF leakage current is thus a measure of the degree of isolation of a particular machine. The lower the leakage current, the better the isolation. With the current technology, RF leakage figures of around 100 mA are generally achieved.

Of the three types of electro-surgical output systems—earthed, earth-referenced and isolated—only the last two are recommended by IEC (1978). For surgical applications in which the danger of ventricular fibrillation cannot be excluded, electro-surgical units of the isolated output type (type CF) should be used as they offer the best protection against fibrillation. Earth-referenced systems, type BF are recommended for most general applications.

The voltages of the power transformer in a surgical diathermy machine are high enough to cause serious injury. Therefore, when checking voltages, it is advisable to take adequate care. Also, caution should be taken to avoid damage to the test equipment due to high voltages and high currents.

Explosion Hazards: In operating theatres, danger zones can develop through the use of cleansing agents such as ether and alcohol, and by using explosive anaesthetic gas or mixtures with oxygen. The sparks associated with the use of surgical diathermy can cause a dangerous explosion if proper precautions are not taken. The use of non-explosive anaesthetics such as nitrous oxide, fluothane or halothane is recommended to prevent sparks generated during electro-surgery. If flammable gases are used, it is important that the electro-surgical unit be located outside the zone in which the anaesthetic is used. In addition, the foot-switches of the electro-surgical unit should always be explosion-proof.

Some diathermy machines are fitted with automatic anti-explosion devices which make the sparks occurring at the active electrodes innocuous. When the foot-switch is actuated or the finger-tip switch in the electrode handle is operated, this device causes a stream of nitrogen to emanate from the electrode handle to form a protective cloud around the cutting and coagulating electrode before the high frequency generator is switched on. Hence the explosive gas mixtures in the immediate vicinity of the electrode cannot ignite. An automatic control is incorporated in the unit which ensures that the high frequency current is not switched on until the active electrode is surrounded by protective gas. This is achieved by using an electrically heated thermistor in the handle which gets sufficiently cooled by the flow of protective gas. This ensures that an adequate stream of gas is emanating from the handle.

▏▶ 27.4 SURGICAL DIATHERMY ANALYSERS

The verification of output power of an electro-surgery unit is important to assure the continuing efficacy of the equipment and to promote consistency in the settings of output level controls. This can be done with the help of a load consisting of non-inductive resistors and current metre suitable for the frequencies and waveform measured.

Electro-surgical devices are currently capable of delivering up to 500 W of power at fundamental output frequencies ranging from 0.3 to 3 MHz. In some units, significant harmonic components exist beyond 5 MHz (Wagner and Phillips, 1980). Even at these relatively low frequencies, precise measurements of output power requires that adequate consideration be given to the radio-frequency nature of the output waveforms.

Figure 27.11 shows the principle of operation of the electro-surgical analyser. It is known that the root-mean-square value of time varying repetitive current is given by:

$$I = \left[\frac{I}{T} \int_0^T [i(t)]^2 \, d \right]^{\frac{1}{2}}$$

where I = RMS value

T = period of the current

$i(t)$ = current as a function of time

If the repetitive time-varying current is a sinusoidal waveform, the relationship between the half-cycle average and the RMS value can be shown to be 1.1, and the RMS value can be determined by measuring the half-cycle average current and multiplying by 1.1.

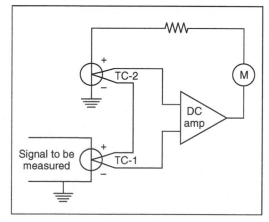

▶ Fig. 27.11   *Principle of operation of surgical diathermy analyser*

A thermal converter consisting of a heater element, a thermocouple, and an evacuated glass envelope is used to measure the output power. The heater element varies in temperature as a function of the RMS current through it. The thermocouple is in electrically insulated from, but thermally coupled to the middle of the heater element. It produces a voltage that is a function of the heater temperature. The evacuated glass envelope eliminates the loss of heat by convection and increases the sensitivity of the thermal converter.

The output of the thermal converter is proportional to the square of the applied voltage. This output when applied directly to a meter would give rise to a non-linear scale. Also, the readings are likely to vary with ambient temperature. By using two matched thermal converters, one heated by the signal to be measured and the other heated by dc signal, these problems can be conveniently overcome. The two thermocouple outputs are connected in series opposed. This combined signal is fed to the input of a high-gain DC amplifier. The output of the amplifier is fed back and drives the heater of the second thermal converter.

The operation of the system consists in applying the signal to be measured to the heater of the first thermal converter. This raises the temperature of the heater, and a DC voltage is generated by the thermocouple. This voltage is amplified by the amplifier and causes a current to flow in the heater of the second thermal converter. This, in turn, raises the temperature of the heater and thermocouple. The voltage generated by the second thermocouple subtracts from the voltage of the first thermocouple and reduces the voltage applied to the amplifier. This process continues until the two heaters are at approximately the same temperature and the thermocouples are generating approximately the same voltage, thus balancing the system.

Since the two thermal converters are a matched pair, and are at the same temperature, both heaters must have equal RMS currents flowing through them. The DC current flowing through the second thermal converter can be measured easily. Hence, the RMS value of the signal applied to the first thermal converter has also been measured.

Laser Applications in Biomedical Field

IIIID> 28.1 THE LASER

The term 'laser' has been coined by taking the first letters of the expression "light amplification by stimulated emission of radiation". It is an extension of maser (microwave amplification by stimulated emission of radiation) to the optical region of the electromagnetic spectrum. Although an amplifier, as suggested by the abbreviation, the laser is invariably used as a generator of light. But its light is quite unlike the output of conventional source of light. The laser beam has spatial and temporal coherence, and is monochromatic (pure wavelength). The beam is highly directional and exhibits high density energy which can be finely focused.

Lasers are presently used for a variety of applications in the medical field. In some cases, the techniques have already been tested clinically and are well-established; while in others, they are still in the research stage. The medical use of lasers is suitable where there is a favourable interaction between the laser radiation and the human tissue. The success of this interaction is dependent on radiation wavelength, the ability of the tissue to absorb this wavelength, delivered power on treatment area, total energy incident on tissue and the area treated. Lasers have been especially successful in the following spheres of medical treatment viz. treatment of detached retina; coagulation in diabetic retinopathy (coagulation of lesions in the retina); neuro-surgery (treatment of tissue in the skull and spine); gastro-entrology (treatment by coagulation of the lower gastrointestinal tract); dermatology (removal of skin imperfections by laser irradiation); and ear, nose and throat surgery. It may, however, be appreciated that in all these fields, lasers have either replaced the methods hitherto practised, or are used as effective back-ups to these methods.

The high radiance, monochromaticity and spatial and temporal coherence make the laser a unique light probe for non-invasive applications. The information contained in laser light reflected or scattered by structures can be detected and analysed for diagnostic purposes. The most widespread medical application for laser technology in medicine has occurred in ophthalmology. This is due to the easy accessibility of the human eye, its transparency and the absorption properties of its internal tissues. Lasers are in routine clinical use for many therapeutic and diagnostic purposes and their application has become the standard of care in the treatment of many eye diseases. Millions of patients have had their vision preserved or restored through laser treatment. Thompson et al. (1992) provided an overview of clinical and research applications for lasers in ophthalmology, while Marcus (1992) reviews the use of laser based photo-dynamic therapy of human cancer.

Table 28.1 shows typical laser characteristics and their medical applications (Cayton, 1983).

• Table 28.1 *Characteristics of Lasers Applied in Medicine*

Laser	Wavelength μm	Solid or Gas	Typical power (Watt)		Type of beam	Applied Field
			Continuous Wave	Peak		
Argon	0.49 0.52 (visible)	Gas	5	100	Continuous tunable pulsed	Neurosurgery, ophthalmology, general surgery, gynaecology, dermatology, biological research
Helium-Neon	0.63 1.15 3.39 (visible)	Gas	0.1	2	Continuous	Diagnostic applications like study of light, permeability of blood containing tissues, laser holography, etc.
Krypton	0.35 (ultraviolet)	Gas	5	100	Continuous	Ophthamology and for general diagnostic use
Ruby	0.69 (visible)	Solid	5	50	Pulsed	Ophthamology, dermatology
CO_2	10.6 (infrared)	Gas	200	75,000	Continuous	Neurosurgery, general surgery dermatology, gynaecology
Nd-YAG	1.06 (infrared)	Solid	50	1,000	Q-switched continuous	Neurosurgery, dermatology, gynaecology
Excimers			(Pulse energy: milli-Joules)			
ArF	193 (ultraviolet)	Gas	30	400 (mJ)	Pulsed	
KrF	249 (ultraviolet)	Gas	45	550 (mJ)	Pulsed	Surgery
XeCl	308 (ultraviolet)	Gas	30	200 (mJ)	Pulsed	
XeF	350 (ultraviolet)	Gas	20	275 (mJ)	Pulsed	

28.1.1 Principle of Operation of Laser

Until the advent of the laser, all sources of optical radiation were essentially hot bodies with the radiation energy spread over a broad band of frequencies. The total energy radiated by these sources may be quite high but the power per unit wavelength is rather low. Even the laboratory sources of monochromatic light such as spectral lamps are unable to give more than a few milliwatts per angstrom. Besides this, the atoms and ions in a hot body radiate independently of each other and the phase of the emitted radiation changes randomly in space and time. In other words, the emitted radiation is not coherent. Light from hot bodies is not confined to any specific direction and it cannot be collimated without a considerable loss in intensity.

According to the quantum theory of energy, each atom has known characteristic energy value. If another form of energy such as heat, light or electricity is used to stimulate an atom, its electrons are displaced and raised to a high level or excited state. Once in an excited state, the atom holds the energy for only 10^{-8}s. After this time, the energy is spontaneously released in the form of diffused light and the electrons return to their original resting state. The spontaneous release of energy is observed in conventional lighting which is diffused in all directions and is comprised of many wavelengths which collectively appear as white light to the human eye.

The laser action depends upon the phenomenon of stimulated emission. Consider a single atom A (Fig. 28.1a) in an excited state which can come back to its normal or ground state by emitting a "photon" or a light quantum whose frequency is related to the excitation energy E by the well-known equation

$$E-h\upsilon$$

where h is Planck's constant, υ is frequency of emission.

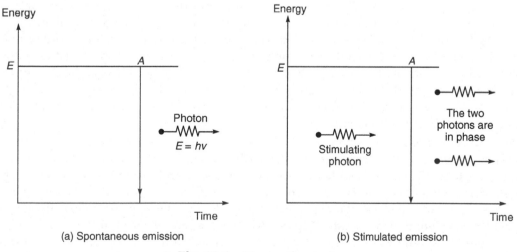

(a) Spontaneous emission   (b) Stimulated emission

> **Fig. 28.1** *Types of emission*

This corresponds to the phenomenon known as spontaneous emission. If during the period the atom is still excited, it can be stimulated to emit if it is struck by an outside photon having precisely the energy of the one that would otherwise be emitted spontaneously. So the stimulating photon or wave is augmented by the one released by the atom (Fig. 28.1b). The significant fact is that the photon upon release falls exactly in 'step' or 'in phase' with the photon that stimulated its release. So it is possible to realize a laser in terms of synchronization of a large number of excited atoms so that when they work together, they produce a powerful coherent wave.

Since most atoms are in the ground state, their absorption is generally far more likely than emission. However, if population inversion could be obtained (i.e. with more atoms in the excited state), an incident photon of the correct frequency could trigger stimulated emission causing an avalanche of coherent photons. The incident wave could continue to grow so long as the scattering processes were few and the population inversion could be maintained.

To achieve this, it is necessary to have an active medium (Fig. 28.2) in which atoms are kept in an excited state and stimulated by an outside photon to emit light in a particular direction. The process by means of which a medium is activated is called 'pumping'. This entails injecting electromagnetic energy into the medium at a wavelength different from the stimulating wavelength.

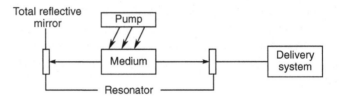

> **Fig. 28.2**   *Main elements of a laser*

The active medium is usually enclosed in a resonator box with highly reflecting walls. The photons released by the stimulated emission undergo multiple reflections and result in a coherent wave of growing strength. The laser output is obtained if the resonator box is transparent to the emitted laser beam. In order to collect the number of high energy photons accumulating within the system, a double-mirrored resonating chamber is used to reflect the light beam so that the rays of light are super-imposed as a single, high-density energy beam. The high energy stored within the resonating chamber can then be directed through the partially reflective mirror by releasing the shutter in a precisely controlled manner.

The directional properties of the laser beam can be attributed to the physics of the stimulated emission process which restricts the emission of the stimulated photon in the same direction as that of the exciting photon. The coherence of the laser beam is related to both the temporal phase correlations of the electromagnetic field which comprises the laser beam and the different positions in space over which these correlations remain constant. The laser is restricted to emit only at one particular wavelength of a very small spread which lies at the centre of the band of frequencies encountered in spontaneous emission. These frequencies, in turn, cause emission at the same frequency so that an extremely narrow beam divergence is achieved. The laser beam is intense because the rate of emission of energy is much higher in the laser than in a hot body.

A laser's properties are determined by three primary considerations: the gain medium, the pumping mechanism and the resonator design.

There are three classes of gain media: gas, liquid and solid. Gas lasers exhibit narrow wavelength regions where there is an appreciable optical gain. This is due to their sharp spectroscopic transitions. Liquid lasers have broad regions for optical gain corresponding roughly to their fluorescence. Solid-state lasers can have either narrow or broad gain regions depending upon the nature of the fluorescence.

The pumping mechanism can be classified as optical and electronic. In optical pumping, a coherent laser can be used for exciting the laser medium to its excited state. Arc lamps and tungsten lamps are generally employed in continuous lasers while flash lamps are used in pulsed lasers. In electron pumping, a discharge is created in the gain medium which excites the population inversion. Both violent discharges like sparks and more gentle discharges like glow discharges are employed. The pumping mechanism largely determines whether the laser is pulsed or continuous.

The resonator provides the means to control the laser by adjusting the losses experienced by the cavity. High losses are ensured for the undesirable modes and low losses for the modes that are desired. If wavelength selective devices like gratings, prisms or etalons are present in the cavity, it is possible to provide low losses for a selected band of frequencies. It is even possible that only one mode has a low loss. Such single mode lasers have very narrow bandwidth and are capable of very high resolution.

The resonant cavity plays an important part in laser operation. Photons which do not propagate nearly along the cavity axis tend to be lost, quickly passing out of the sides of the medium, which accounts for the high degree of collimation of the laser beam. Although the medium acts to amplify the wave, the optical feedback provided by the cavity converts the system into an oscillator.

In addition to the axial modes of oscillation, which correspond to the standing modes set up along the cavity, transverse waves can also be sustained. These are known as TEM_{1j} modes (transverse, electric and magnetic) whereas the subscripts i and j are the integer number of transverse nodal lines in the X and Y directions across the emerging beam.

Energy decay within a cavity is expressed in terms of the Q-factor (quality factor). If the cavity is disrupted by, say, displacing the mirror, laser action generally ceases. This can, however, be done deliberately to delay oscillation within the cavity. This is known as Q-switching.

28.1.2 Types of Lasers

Several types of lasers are in use including solid-state, gaseous and semiconductors. They are classified according to the two fundamental modes of operation, namely, the pulsed operation such as is achieved with the ruby and neodymium glass, and the continuous wave operation (CW) such as is achieved with helium-neon, argon, krypton, carbon dioxide lasers. Each of these two classes has specific areas of application in the medical field. The quantification of energy in case of CW and pulsed lasers is shown in Fig. 28.3.

ⅢⅢ▷ 28.2 PULSED RUBY LASER

Historically, the first operational laser was developed by T. H. Maiman in 1961. It employed a ruby crystal as the active medium. Ruby is aluminium trioxide in which about 0.05% of the aluminium atoms have been replaced by chromium atoms. In a typical pulsed laser system, a high voltage

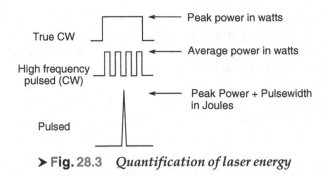

> **Fig. 28.3**   *Quantification of laser energy*

transformer (4000-10000 V) is connected to a bank of capacitors and pulse-forming inductances. This network is further connected to a xenon flash lamp (Fig. 28.4) which may be straight or helical. Pumping of the ruby crystal is done by this flash lamp which has proper arrangements for condensing and focusing light on the crystal. A high voltage trigger pulse is used to cause flash lamp ionization. The light absorbed by the chromium atoms raises them to a high level of excitation wherein, when they come down to the metastable state, they are stimulated to emit by a stray photon emitted by spontaneous decay. With the help of the resonators, a strong coherent wave of light is built up. The wavelength of this emission is 6943Å (0.69 μm) which lies in the red region of the spectrum.

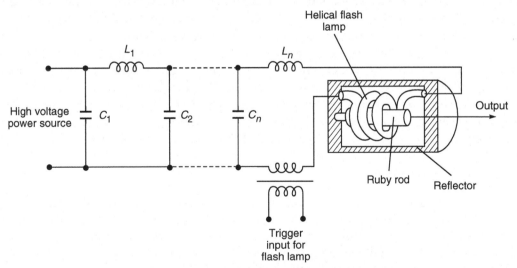

> **Fig. 28.4**   *Schematic diagram of a pulsed ruby laser*

The ruby laser is usually operated in short bursts or pulses because in continuous wave operation, it gets heated up and results in upsetting the distribution of atoms in different quantum states and physical damage to the crystal. The pulse duration is extremely small, of the order of a millionth of a second, and is capable of producing a peak power of several million watts and energy densities which easily exceed 15,000 J/cm^2.

Ruby laser finds applications in medical treatment where high energy pulses are required. It is an ideal energy source for effecting retinal coagulation since it emits light at a wavelength which is in the region of maximum absorption of light by the retina and choroid. Besides, the ruby laser produces a high power pulse making it suitable for producing lesions in time spans of the order of a millisecond.

A successful application of the laser in the medical field is in the treatment of retinal detachment. This procedure can be done in a minute or two as an outpatient procedure. No anaesthesia is required as the patient does not experience pain. The laser light travels through the cornea, pupil, lens, vitreous humour and strikes against the choroid, heating it at the spot. This heat, also transmitted back to the retinal layer results in the retina being "welded" to the choroid.

A typical photo-coagulator using a ruby laser consists of a hand-held laser head fitted with a conventional ophthalmoscope, and a power supply unit. The laser emits light at a wavelength of 0.69 μm in a burst of pulses lasting about a millisecond. As the amount of energy required to achieve coagulation varies from patient to patient, the release of energy from the laser is regulated and can be varied from 0 to 100 millijoules.

28.3 ND-YAG LASER

The Nd-YAG (Neodymium doped-Yttrium aluminium garnet) laser is very similar to the ruby laser in construction. These lasers have become very popular in recent years because of their very high output energies, repetition rates and wavelength outputs. The active element is a Nd-YAG crystal optically pumped by two krypton arc lamps. The wavelength of emission is 1.06 mm.

Neodymium has a four-level transition. The laser transition is more easily inverted with respect to the intermediate transition level than the three-level transition of ruby. It is thus much easier to achieve a population inversion in Nd-YAG because it is measured relative to another excited state instead of the ground state as in ruby lasers. The laser arrangement consists of a Nd-YAG rod placed within an elliptical cavity. The pumping of the laser rod is generally done by a pulsed or continuous discharge tube whereas pumping at low levels of CW power output is achieved by flash lamps. The oscillator is Q-switched to obtain very high, reproducable peak power that can be efficiently doubled, tripled and quadrupled in frequency. The pulse widths and repetition rates are typically 10-20 ns and 10 Hz respectively. These characteristics are almost ideal for a number of applications.

A typical commercially available Nd-YAG laser consists of a power supply and laser head fitted with light guide and focusing handpiece. The system operates on 380 V,20 A, 3 phase supply. The power supply unit contains the thyristor control for the lamp current and a control circuit for automatic switching and monitoring of the laser. Due to high power requirements, the lamp fittings and the YAG crystal are cooled by a single circuit cooling system safeguarded by flow monitors and magnetic valves. Approximately 10 L per minute water at a pressure of 3.5 bar is required for cooling. With these inputs, the maximum power output available out of the light guide is approximately 70 W.

In order to meet varied requirements of modern medical technology, a light guide is necessary which transmits the laser energy without major losses and at the same time, is sufficiently flexible to permit handling of delicate work in various fields like endoscopy, urology, neuro-surgery, ENT, gynaecology, dermatology, dental surgery and general surgery. With the flexible light guides, it is possible to replace the cumbersome articulated arm systems for delivery of the laser beam.

The effect of the Nd-YAG laser beam depends on the radiation dose (laser power times duration) per unit of area. Laser power, duration and the area of irradiation can be varied according to requirements. The infrared Nd-YAG laser beam penetrates relatively deeply into the tissue (approximately 2 to 3 mm) and is absorbed there. The essential biological effects produced depend upon the dose of radiation and include dehydration (cellular contraction), protein coagulation, thermolysis (thermal decay of molecules) and tissue vaporization (incision).

Wyman et al.(1992) illustrate the use of Nd-YAG laser for interstitial photo-coagulation for destroying solid tumour tissue in the brain, head and neck, liver, breast and pancreas. Approximately, 2 Watt of laser energy at 1064 nm is directed into the tumour through one or more optical fibres implanted interstitially. The method has the advantage of precisely controlled localized destruction of deep-seated tumours with minimal invasiveness. Marcus (1992) presented a review of the use of Nd-YAG laser for photo-dynamic therapy of human cancer, whereas Thompson et al.(1992) detailed out the therapeutic and diagnostic application of lasers, in general, and Nd-YAG laser in particular, in ophthalmology.

28.4   HELIUM-NEON LASER

This laser employs a gaseous active medium in which the atoms of one gas (neon) pump themselves up through collisions with the excited atoms of another (helium). A discharge tube (Fig. 28.5) containing 10 parts of helium and one part of neon is maintained at a very low pressure of approximately 1 mm of Hg. For continuous laser beam, pumping is achieved by an electrical discharge in the gas by radio frequency excitation.

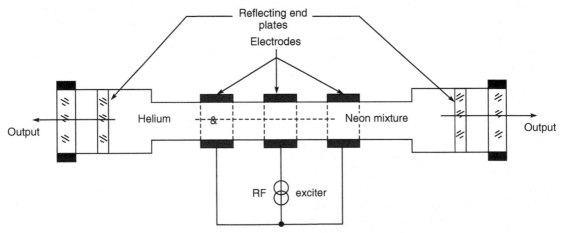

> **Fig. 28.5**   *Schematic diagram of a helium-neon laser*

The two electrons of helium can have either parallel or anti-parallel spins. In the former case, they possess an energy level which is considerably higher than the ground level. A certain number of this type of electrons in the metastable state do exist in the unexcited helium gas. Collisions between metastable helium atoms and unexcited neon atoms result in energy transfer to the neon so that the number of neon atoms in the excited level will increase. When these atoms return to their ground state, they do so by several transitions which lead to output in the optical region.

In contrast to ruby laser, a Helium-Neon laser can operate at several wavelengths. Besides this, the line width of the radiation emitted is much smaller than the ruby laser. It can operate continuously but the power output is limited only to a few milliwatts.

He-Ne Laser can be used for the measurement of visual acuity and is very helpful to the ophthalmologist in deciding about the necessity of performing cataract surgery on the patient. The laser power required for this application is $1\ \mu W/cm^2$ at the retina. This application is based on interferometry where in the spatial coherence of lasers is used to form interference fringes on the retina. These are dark lines with variable spacing and orientation whose formation is largely insensitive to the optical clarity of the intervening media. The spacing of the fringes is varied by adjusting the angle of two interference beams and the patient indicates whether or not they are visible, facilitating prediction of the patient's acuity after cataract surgery.

The scanning laser ophthalmoscope has been developed for viewing the retina and its supporting structures including blood vessels, nerve bundles and underlying layers. Here, a laser beam about 1 mm in diameter at the eye's pupil is focused to a 10 μm spot on the retina. The beam is scanned over the retina and at any instant only a 10 μm spot on the retina is illuminated and the instant may last only 100 ns. Further applications of the scanning laser ophthalmoscope include its use as a therapeutic device (photo-coagulator) and a retinal eye tracker.

▥▶ 28.5   ARGON LASER

Figure 28.6 shows a block diagram of the argon ion laser. The heart of the system is the plasma tube wherein a very high current discharge passes down the barrel of the tube through argon gas. The discharge ionizes the argon gas and also populates the excited ion states which are involved in the lasing. A strong electrical current is forced to flow through argon gas within the tube, each end of which has a mirror. The current excites the argon atoms to a higher energy level and some of them begin to emit light spontaneously (Labuda et al. 1965). As this light is reflected back and forth

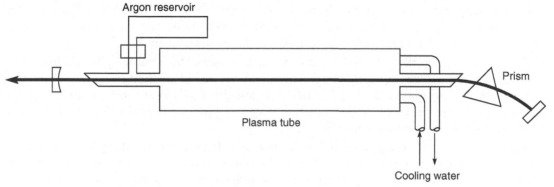

➤ Fig. 28.6   *Schematic representation of argon gas laser*

between the mirrors, it stimulates additional argon atoms to emit light by forcing them to lower energy levels. Eventually, a steady state of light amplification is reached. The mirror at one end of the laser tube is a partial mirror that permits a portion of the light to exit from that end as a laser beam. Since very high current densities are required to both ionize and excite the inert gas like argon, it necessitates large power consumption and substantial cooling arrangement. An 18 W argon ion laser will typically require 38 kW of electrical power and comparable cooling requirements. Therefore, the outside of the plasma tube is water-jacketed to provide for a cooling system.

At low laser powers (below 1 W), quartz tubes have been used but at higher powers, the laser tube must be capable of withstanding very high temperatures, and therefore, tubes made up of segmented beryllia (BeO) or graphite rings are employed. Brewster windows with external dielectric coated optics are used to obtain a high reflectivity at the short output wavelength.

The argon ion laser normally employs DC discharge at currents of around 10-100 Å to excite laser action, though induction-coupled RF discharges have also been used in some cases. An axial magnetic field is used to stabilize the DC discharge on the tube axis. The high current densities result in heating of the inside bore of the plasma tube so that appreciable amounts of incoherent black body radiation are also emitted by the laser. This background must be removed by filters or monochromators if measurements involve low light levels.

Argon ion lasers are characterized by several spectral lines evenly spaced in frequency by C/2L where C is the speed of light and L is the cavity length. These individual lines correspond to the different longitudinal modes of the cavity. Output over a range of wavelengths of 458 and 515 nm at power outputs up to several watts at about 0.05% overall efficiency can be obtained at lengths of about 2.2 m. A small etalon put into the cavity can be adjusted to select a single longitudinal mode, thereby converting the laser into a single mode laser with a very narrow bandwidth.

A major disadvantage is that the cathode erosion is severe at high currents. It limits the tube life which must be replaced occasionally and such replacement cost about one-fourth of the initial cost of the laser itself.

28.5.1 Argon-Ion Laser Photo-coagulator

The argon wavelength is highly absorbed by the red colour. Since biological tissue has both blood vessels and haemoglobin, which are red in colour and capable of absorbing the argon beam, its optical energy can be converted into thermal energy upon contact with such media. This would result in photo-coagulation of the blood protein and micro-haemostasis. Thus contiguous, similar soft-tissue may be spot-welded and made to adhere by the laser beam.

The argon wavelength is in the blue-green colour range of the visible light spectrum and is naturally transmitted through clear fluids without conversion of the light energy into heat. It also interfaces well with glass light fibres, thus allowing the argon beam to be coupled with the fibreoptic light system. This property greatly increases its manoeuvrability as a hand-held light scalpel or instrument used with operating microscope.

Figure 28.7 shows how argon laser is extremely effective as a source for photo-coagulation. Absorption in the ocular media (curve 1) has a 'window' from the visible spectrum to the near infrared. Energy in this part of the spectrum is transmitted to the retina with little loss in the media. The window corresponds to the peak absorption of the pigment epithelium and choroid combined

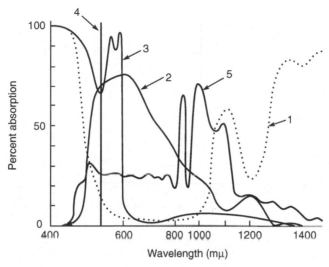

> **Fig. 28.7** *Principle of use of argon laser as a source for photo-coagulation*

(curve 2). Fortunately haemoglobin also has its peak absorption in the same part of the spectrum (curve 3). The specific blue/green argon light (line 4) is ideally matched to the absorption characteristics of the eye. Very little absorption occurs in the media while maximum absorption occurs in the blood pigment epithelium and choroid. By contrast, the continuum of the xenon gas emission (curve 5) covers the whole window extending into the near infrared where absorption in the media occurs. It may be noted that emission peaks do not correspond to absorption peaks in the fundus. These characteristics explain why argon laser coagulation can be achieved with up to 10 times less power than xenon. Argon can thus provide comparable coagulation with less overall heating in the globe, an important consideration when performing any retinal surgery.

Argon ion laser photo-coagulator is more suitable for photo-coagulation of the retina since the output of the ruby laser is not so effectively absorbed by blood vessels (L' Esperarance, 1968). The argon ion laser offers the advantage of a smaller affected zone typically 0.15-1.5 mm, with low laser energy requirements producing less damage to healthy tissue than with xenon arc light sources.

Control of Gastric Haemorrhage by Photo-coagulation Using Argon Ion Laser: Considerable clinical experience has accumulated in the use of argon lasers in ophthalmological procedures involving photo-coagulation. By scaling the power requirements of ophthalmological practice, a power level of about 10 W has been found suitable for the control of gastric haemorrhage. The argon laser beam is transmitted via a quartz fibre wave guide inserted into a gastro-intestinal fibreoptic endoscope. With advances in quartz fibre technology, fibreoptic delivery of an adequate dosage of visible or near IR radiation is practical. A gastric photo-coagulator which can be inserted into the biopsy channel of a conventional flexible endoscope for non-invasive control of gastric bleeding is shown in Fig. 28.8. The system employs a water-cooled argon laser capable of delivering more than 13 W continuous radiation when operating with all standard atomic lines. The system makes use of a special purpose beam-splitter to provide an 'aiming' beam for the endoscopist at a

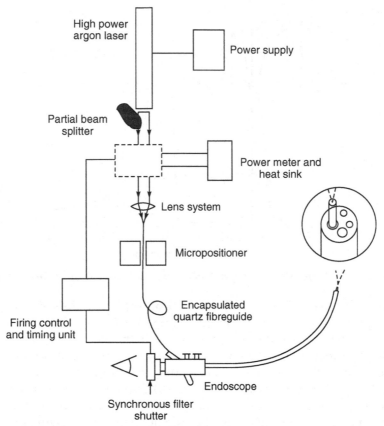

> **Fig. 28.8**   *Block diagram of fibre-optic gastric photo-coagulator*

pre-set and well-controlled operating power level. At the same time, the laser power can be adjusted and monitored prior to firing. The high power beam is transmitted to the flexible quartz wave guide by triggering an electronically controlled solenoid. The solenoid shutter can be adjusted in 0.1 second steps from 0 to 760 s. The quartz fibre is encapsulated in a laminated sheath of non-toxic polyethylene with an overall diameter of less than 2 mm. This provides physical protection and enables easy transport through the biopsy channel of a conventional endoscope. With approximately 14 W of laser radiation available, about 10 W can be deposited distally. Pilot studies with experimental animals have shown that the system can be easily adopted and operated within the conventional endoscopic procedure. Rapid haemostasis was observed in experimental animal studies using this technique.

Although the endoscopic photo-coagulator system has been found to be functionally reliable on animal experiments, this has some limitations (Silverstein et al., 1976). The major problem is that it is difficult to control severe bleeding (more than 2 cc/mm) as the technique can photo-coagulate only low to moderate bleeding ulcers quite well. By absorbing the blue-green argon beams, the overlying red blood shields the underlying bleeding vessels, thus preventing effective haemostasis at the vascular level. Another problem is that the fibre can be damaged if blood is splashed onto the tip during high power operation. Kimura et al. (1978) designed a hybrid optical fibre catheter to overcome these limitations. They found that if a jet stream of gas (CO_2) were blown on the bleeding site while the laser radiation was being emitted, the overlying blood would be blown away and the laser radiation would reach the vascular level. Thus, it will be more effective in achieving haemostasis than laser radiation alone. Also, the gas flow helps to prevent any contamination of the output tip if it were directed around the optical fibre during the photo-coagulation process. The construction of the catheter is detailed by these workers.

Silverstein et al. (1976) caution that though the argon laser photo-coagulating system works adequately from the technical viewpoint, many questions remain to be answered. Studies are necessary to optimize the combination of power, divergence angle and tip proximity. It is necessary to evolve a mechanism to slow down active arterial bleeding to give the laser a chance to coagulate the vessel. It also remains to be determined as to what type of bleeding from what size vessels is suitable for laser therapy.

ⅢⅢ▷ 28.6 CO_2 LASER

The CO_2 laser provides a means of bloodless surgery. It causes a thin layer of heat coagulated tissue immediately around the treatment site while the cells beyond this site remain untouched, undisturbed and begin the healing process promptly. Cauterization of vessels smaller than 0.5 mm results in dry-field, almost bloodless surgery. Post-operative oedema—experienced in cryosurgery and diathermy—is minimal in laser surgery. As a result, healing is faster with a minimum of tissue swelling and scarring and with less post-operative pain and discomfort.

The laser destroys tissue by vaporizing cells. Tissue, which contains 80-90% water, when exposed to laser energy, is disrupted by steam formation within the cells. Larger cellular components are blown free by the rapid vaporization and may be heated to incandescence in the plume. The area of laser destruction shows no evidence of tissue combustion. As vaporization takes place at atmospheric pressure, the intra-cellular temperatures never exceed 100°C. This relatively low temperature, coupled with poor conductivity of tissues for heat, means that there is a minimal damage to areas contiguous to the destroyed tissue. Of clinical importance is the coagulation effect of the laser beam which leads to collapse of the tissue and thus to occlusion and sealing-off of vessels opened during surgery.

CO_2 laser has been extensively used for surgical applications. The major characteristic that makes it suitable for this application is that it is a high-power continuously operating laser and that its wavelength of operation is in the infrared region at 10.5 μm, a wavelength that is almost completely absorbed by most biological tissues. This high degree of absorption in combination with the high power, the continuous mode of operation and the fact that the laser beam can be focused to very small spot size, permit the application of radiation to tissue in dosages sufficient to cause rapid burning and vapourization of a well-localized volume of tissue. The purely thermal effect of this radiation overcomes the major difficulty encountered in surgery with other high power pulsed or Q switched lasers such as the ruby and neodymium lasers. With these lasers, it is found that the disruptive effects of the radiation on tissues are not localized.

CO_2 laser is a mixture of carbon dioxide, nitrogen and helium (Polanyi et al., 1970) with the CO_2 as the active, energy emitting gas. The CO_2 laser has a high conversion efficiency of the order of

15% as compared with the ruby which has an efficiency of about 1–2%. It operates with a simple cooling system of tap water and is relatively inexpensive. The beam can be reflected or focused by appropriate mirrors and lenses, thus lending itself to passage through a surgical manipulator. The normal operation of the laser is multimode. However, by inserting an annular diaphragm into the resonator, a single mode output can be obtained and the focal spot reduced to about 0.5 mm. The operating microscope can be used in conjunction with the CO_2 laser as an aid to good visualization of tissues undergoing treatment.

Figure 28.9 shows CO_2 laser (System 450, Coherent Ltd.) whose block diagram is shown in Fig. 28.10. The laser head contains the CO_2 laser which produces the invisible infrared (10.6 micron) beam used for surgery. A continuously adjustable power up to 25 W is available. A small helium-neon laser provides a visible red (0.8 mW, 6328 Å) aiming beam. This beam provides accurate visualization of the operating area for precise laser removal of pathological tissue.

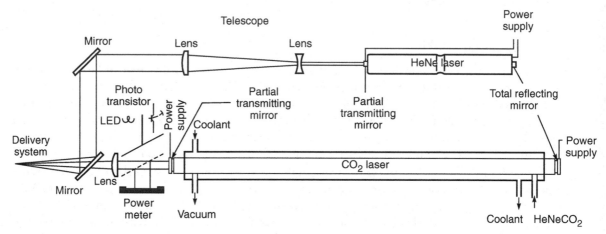

> **Fig. 28.9**   *Block diagram of a CO_2 laser with He-Ne laser for aiming the beam*

The most common applications of CO_2 laser are in microsurgery, where the laser beam is focused to be parfocal with the operating microscope. The coherent monochromatic beam can be exactly focused on an area approximately 1 mm diameter and is controlled through the optical system connected microscope or colpscope. A very precise micro-manipulator directs the laser light to the treatment area. In general surgery, the laser can be directed through a surgical handpiece which allows large or easily accessible areas to be treated efficiently. For areas that are difficult or impossible to reach by conventional methods, the laser is used in conjunction with an endoscope. In many diseases of the nose, oral cavity, nasopharynx and the tracheo-bronchial tree, lasers have become the preferred surgical instrument. The laser is now increasingly used as a primary treatment in selected cancers of the larnyx, pharynx and oral cavity. Andrews and Moss (1974) illustrated the techniques used in laser surgery of the larynx. They stated that by its precision, absence of bleeding, control of depth and area of destruction, reduction of tissue reaction and preservation of normal tissue has extended the quality of operative procedures. Strong et al. (1973) made use of continuous wave CO_2 laser in the management of patients with lesions in the

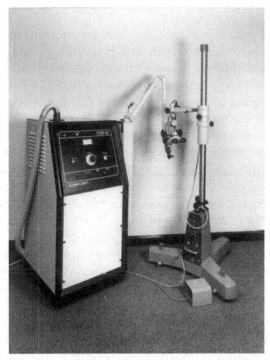

➤ **Fig. 28.10** *CO_2 laser system (Courtesy: Coherent Ltd., UK)*

upper aero-digestive tract. Strong et al. (1975) reviews the progress of the use of CO_2 laser in otolaryngology. The use of CO_2 laser in gynaecology has been particularly attractive. Previous to laser surgery, the gynaecologist was forced to sacrifice large areas of normal tissue in order to remove a little pathologic tissue. With the laser surgery, small areas of tissue can be removed even in the most inaccessible anatomic regions. Jordan (1977) suggests that the treatment with CO_2 laser will allow most patients with pre neoplastic disease of the cervix and vagina to be treated on an outpatient basis. Baggish and Dorsey (1981) used CO_2 laser for treating carcinoma in situ of the vulva. It was established that laser therapy is precise and results in rapid healing without scar formation. Stafi et al. (1977) have reported a cure rate in excess of 90% in laser treatment of cervical and vaginal neoplasia.

Goodale et al. (1970) used a carbon dioxide laser via a stiff, open endoscope to reduce bleeding from experimental gastric erosions in dogs. A CO_2 laser has also been used for brain surgery with good results (Polanyi et al. 1970). Mockwitz et al., (1975) used CO_2 laser for cutting arteries and bones. The use of lasers offers the possibility of cutting bones without undesirable massive destruction or dislocation, such as often cannot be avoided in classical surgical techniques.

Laser surgery is subject to certain limitations. For example: the operative field must be dry; the presence of blood, cerebro-spinal fluid, saliva or the like would make it impossible to proceed with tissue destruction until they have been removed. Also, the target lesion must be clearly visible at all times. The lesion can be viewed in a stainless steel mirror, if necessary and treated with laser but the mirror must be kept free of smoke and steam.

28.7   EXCIMER LASERS

The excimer lasers operate primarily in the ultraviolet spectral region. They use mixtures of rare gases such as argon, krypton, or xenon with halide molecules such as chlorine and fluorine. The molecules like ArF, KrF, and XeCI are strongly bound only in excited states. The ground state of these molecules, which are called excimers, is characterized by small dissociation energies. In these lasers, electrical pumping creates items and ions that combine to form excimer molecules. Since the ground state of these molecules is essentially empty because of rapid dissociation, a population inversion between the excimer state and the ground state is easily obtained. Transitions between these states occur in the ultraviolet region, and therefore, excimer lasers emit wavelengths shorter than 350 nm. The most common excimer lasers are argon-fluoride (193 nm), krypton-fluoride (248 nm), xenon-chloride (308 nm), and the xenon-fluoride (351 nm).

Most excimers are pulsed, but the frequency can be high enough in some cases to be considered as a continuous wave. The radiative life time of the excimer level is very short, of the order of 10 ns. Consequently, these lasers typically produce short pulses of energy with durations of 10 to 50 ns and repetition rates of 2000 pulses per second. The power levels in the excimer lasers range from a few millijoules to thousands of joules in pulsed mode. Average powers up to 200 watts can be obtained from commercial excimer lasers. They are relatively efficient with 1-5% efficiency and produce useful energy in the ultra-violet region.

A major advantage of an excimer laser is its intense ultra-violet beam which can vapourize tissue with almost no heat transfer into surrounding tissue. The two most important applications of excimer lasers in medicine are improving vision by controlled ablation of the cornea with ArF (193 nm) excimer laser and removal of artherosclerotic plaque from arteries with XeCl (308 nm) excimer laser. Kochevar (1992) studied the biological effects of excimer laser radiation at these two wavelengths.

One of the areas of great clinical interest is laser angioplasty to open atherosclerotic arterial narrowings in peripheral and coronary arteries. Continuous wave lasers such as argon and Nd-YAG were initially used to create an opening, but because of thermal injury they have had a medically unacceptable high incidence of restenosis. Specially designed excimers have been shown to cause minimal thermal injury (Haller and Wholey, 1992). However, the current laser systems have not yet been shown to offer any clear advantages for the treatment of any group of patients with occlusive arterial disease.

28.8   SEMICONDUCTOR LASERS

Semiconductor lasers are highly efficient laser light emitting devices which are extremely small. They are constructed from several semiconductor material systems, most notably the gallium arsenide/aluminum gallium arsenide system and the indium phosphide/indium gallium arsenide phosphide system. The first semiconductor laser device were made from chips of gallium arsenide. The gallium arsenide was grown such that p-n junction or diode was formed. Laser diodes which contain a medium (semiconductor material) are pumped electrically and have a resonator. Rather than using mirrors, manufacturers employ the differences in the index of refraction between semiconductor layers to form the resonator.

The early devices suffered from a lack of coherence due to the wide spectral bandwidth and low power output. These drawbacks have been overcome and the techniques used today for producing semiconductor lasers are identical to the technology which is used for manufacturing electronic devices. They can be mass produced with a comparable reliability as standard electronic components. With this link in processing technology, electronic devices can be integrated with semiconductor lasers on the same waveform. The most commonly available semiconductor laser is made from AIGaAs material. The emission wavelength can be varied from 700 to 900 nm by varying the Ga/Al composition. The laser diodes can be packaged individually or for greater power, can be packaged in arrays. Most diode lasers are continuous wave, but there are also pulsed versions, depending on the application.

Laser diodes have been used in therapeutic applications and are being used as pumps for solid-state lasers. They can be used with advantage as photo-coagulators (Balles et al. 1990). They offer greatly decreased size, cost and reduced maintenance. AIGaAs diode lasers emitting at 805 nm with output power 1–3 W have been used for the treatment of retinal vascular disease. There have been a few prototype blue laser diodes reported, but it will be some time before they are available for medical applications. The technical disadvantages of the diode laser include limited power output and a more divergent beam cone angle than the argon and krypton lasers. Diode lasers and diode-pumped solid-state lasers with multiwatt power output at wavelengths that are desirable for various clinical applications, are likely to be available to replace the more expensive and bulky argon, krypton and dye lasers for ophthalmic applications.

▮▮▮▶ 28.9   LASER SAFETY

A serious problem in the use of lasers for surgical applications is that of ignition and special precautions must be taken to avoid anaesthetic gases which are flammable or explosive. Oxygen must be handled by exercising reasonable care in using the laser. Laser dangers can include fires to unprotected drapes and endotracheal tubes, wherein oxygen-enriched environments increase the chance of fire.

Safe laser practices include wearing protective eyewear that is made for the wavelength being used. This eyewear must be labelled for the wavelength and optical density. There is a possibility that the laser beam could impinge on a flat metal surface and by chance, be reflected as a focused beam. Therefore, the operating room personnel should wear glasses to protect their eyes. For the same reason, the patient's eyes should be protected by protective eyewear or wet towels over the eyes.

The facility must also have a committee in charge of laser safety, to ensure that all safe laser practices are followed, and to monitor and enforce the control of laser hazards. Laser safety standards have been worked out such as ANSI-Z136.1, American National Standard for safe use of lasers, and ANSI-Z136.3, for the safe use of lasers in health care facilities. These standard used together should be followed by any facility that has a laser of any type.

Physiotherapy and Electrotherapy Equipment

⟫ 29.1 HIGH FREQUENCY HEAT THERAPY

Physical stimulus commonly employed in the practice of physiotherapy is in the form of heat, either by simple heat radiation or by the application of high frequency energy obtained from special generators. The use of high frequency energy in thermotherapy has the advantage of considerable penetration as compared with 'simple' heat application. Thus, with high frequency energy, deeper lying tissues, e.g. muscles, bones, internal organs, etc. can be provided heat.

High frequency heat therapy is based on the fact that the dipole molecules of the body are normally placed randomly. Under the influence of an electric field, they rotate according to the polarity of their charge in the direction of the field lines (Fig. 29.1). The positively charged end of the dipole then orients itself to the minus pole and the negatively charged end to the plus pole. Since the polarity of the electric field alternates, a micro-heating effect results from the continuous re-alignment of the molecules.

High frequency energy for heating is obtained by various ways. It may be from the short-wave therapy unit making use of either the condenser field or the inductor field method. Microwaves and ultrasonic waves are also used for heating purposes in special cases.

⟫ 29.2 SHORT-WAVE DIATHERMY

The term 'diathermy' means 'through heating' or producing deep heating directly in the tissues of the body. Externally applied sources of heat like hot towels, infrared lamps and electric heating pads often produce discomfort and skin burns long before adequate heat has penetrated to the deeper tissues. But with the diathermy technique, the subject's body becomes a part of the electrical circuit and the heat is produced within the body and not transferred through the skin (Yang and Wang, 1979).

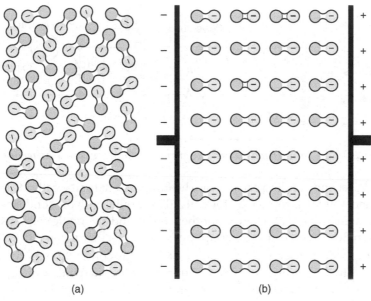

(a)   (b)

> **Fig. 29.1** *Principle of high frequency heat therapy units*
> (a) *The dipole molecules of the body are at first ordered completely randomly*
> (b) *Under the influence of an electric field, they rotate according to the polarity of their charge in the direction of the field lines*

Another advantage of diathermy is that the treatment can be controlled precisely. Careful placement of the electrodes permits localization of the heat to the region that has to be treated. The amount of heat can be closely adjusted by means of circuit parameters. The heating of the tissues is carried out by high frequency alternating current which generally has a frequency of 27.12 MHz and a wavelength of 11 m. Currents of such high frequencies do not stimulate motor or sensory nerves, nor do they produce any muscle contraction. Thus, when such a current is passed through the body, no discomfort is caused to the subject. The current being alternating, it is possible to pass through the tissues currents of a much greater intensity to produce direct heating in the tissues similar to any other electrical conductor.

The method consists in applying the output of a radio frequency (RF) oscillator to a pair of electrodes which are positioned on the body over the region to be treated. The RF energy heats the tissues and promotes healing of injured tissues and inflammations.

Circuit Description: The short wave diathermy machine consists of two main circuits: an oscillating circuit, which produces a high frequency current and a patient circuit, which is connected to the oscillating circuit and through which the electrical energy is transferred to the patient.

Earlier models of diathermy machines employed single-ended or push-pull power oscillators operating from unfiltered or partially filtered power supplies. They usually made use of a valve circuit, a typical example of which is shown in Fig. 29.2. Transformer T_1, the primary of which can be energized from the mains supply, is a step-up transformer for providing EHT for the anode of

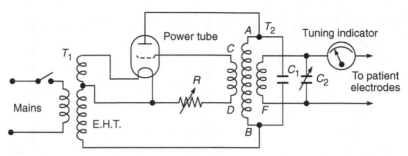

> **Fig. 29.2**   *Simplified circuit diagram of a short-wave diathermy unit*

the triode valve. A second winding can provide heating current for the cathode of the triode valve. The tank (resonance) circuit is formed by the coil AB in parallel with the condenser C_1. The positive feedback is generated by coil CD. There is another coil EF and a variable condenser C_2 which form the patient's resonator circuit due to its coupling with the oscillator coil AB.

The anode supply of such a circuit is around 4000 V. The conduction in the triode takes place during the positive half-cycle and the high frequency is generated only during this period. More usually, the supply voltage is rectified before supplying to the anode of the oscillator valve. In such a case, the oscillations produced are continuous and more power thus becomes available. In order to ensure that the oscillator circuit and the patient's resonator circuit are tuned with each other, an ammeter is placed in series with the circuit. The variable condenser C_2 is adjusted to achieve a maximum reading on the meter, the needle swinging back on either side of the tuned position. The maximum power delivered by these machines is 500 W.

A thermal delay is normally incorporated in the anode supply which prevents the passage of current through this circuit until the filament of the valve attains adequate temperature. The patient circuit is then switched on followed by a steady increase of current through the patient. A mains filter is incorporated in the primary circuit to suppress interference produced by the diathermy unit itself.

There are several ways of regulating the intensity of current supplied to the patient from a short-wave diathermy machine. This can be done by either (i) controlling the anode voltage, or (ii) controlling the filament heating current, or (iii) adjusting the grid bias by change of grid leak resistance R_1, or (iv) adjusting the position of the resonator coil with respect to the oscillator coil. However, the best way of finely regulating the current is by adjusting the grid bias, by putting a variable resistance as the grid leak resistance.

Automatic Tuning in Short-wave Diathermy Machines: Any short-wave therapy unit would give out the desired energy to the patient only if and as long as, the unit is correctly tuned to the electrical values of the object (part of the body). Therefore, tuning must be carefully carried out at the beginning of the treatment and continuously monitored during the treatment. There is a possibility of the tuning getting affected due to unavoidable but involuntary movements of the patients and the resultant fall of dosage.

In order to overcome the problem of making tuning adjustments during the course of treatment, an additional circuit is fitted in the machine. The RF current in the patient circuit changes a capacitor to a voltage, whose polarity and magnitude is a measure of the detuning of the patient circuit. This voltage accordingly moves a servo-motor, adjusting the tuning capacitor so that resonance is restored. Figure 29.3 shows a diathermy machine with automatic tuning in operation.

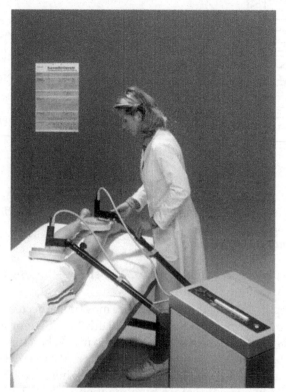

> **Fig. 29.3** *Short-wave therapy unit with automatic tuning in use (Courtesy: Siemens, Germany)*

Application Technique of Short-wave Therapy: The pattern of tissue heating is greatly affected by the method of short-wave diathermy delivery. The two most common forms of application include the capacitor plate method and the inductive method (Fig. 29.4).

In the capacitor plate method, the output of the short-wave diathermy machine is connected to metal electrodes which are positioned on the body over the region to be treated. These electrodes are called 'PADS' in the terminology of the diathermy. These pads or electrodes do not directly come into contact with the skin. Usually layers of towels are interposed between the metal and the surface of the body. The pads are placed so that the portion of the body to be treated is sandwiched between them. This arrangement is called the 'Condenser Method' (Fig. 29.4(a)) wherein the metal pads act as two plates while the body tissues between the pads as 'dielectric' of the capacitor. When the radio frequency output is applied to the pads, the dielectric losses of the capacitor manifest themselves as heat in the intervening tissues. The dielectric losses may be due to vibration of ions and rotation of dipoles in the tissue fluids (electrolytes) and molecular distortion in tissues

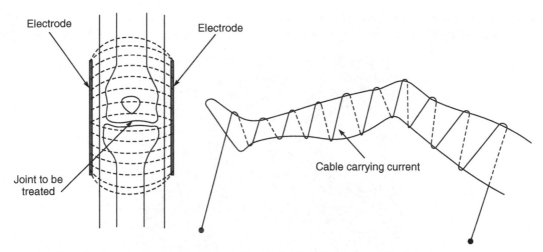

> ➤ **Fig. 29.4** *Method of applying electrodes in short-wave diathermy treatment*
> (a) *condenser method*
> (b) *inductive method*

which are virtually insulators like fats. Figure 29.4a shows the use of air-spaced condenser electrodes for short-wave diathermy treatment.

Alternatively, the output of the diathermy machine may be connected to a flexible cable instead of pads. This cable is coiled around the arm (Fig. 29.4(b)) or knee or any other portion of the patient's body where plate electrodes are inconvenient to use. When RF current is passed through such a cable, an electrostatic field is set up between its ends and a magnetic field around its centre. Deep heating in the tissue results from electrostatic action whereas the heating of the superficial tissues is obtained by eddy currents set up by a magnetic effect. This technique is known as 'inductothermy'.

Although most short-wave diathermy machines have an output power control, yet there is no indication of the amount of converted and absorbed heat within the body tissues. Therefore, the intensity of treatment is dependent on the subjective sensation of warmth felt by the patient.

29.2.1 Diapulse Therapy

A severe limitation of diathermic machines is that they direct continuous high frequency radio waves, and if a high enough output of energy is sustained for even a brief time, they can cause burns. As a consequence, the wattage has to be lowered to tolerable limits. Also, the heat resulting from diathermy has many contra-indications and limitations because the heat limits the amount of energy that can be used. The increase in energy output, while avoiding the dangers of heat, has been achieved in a machine called 'Diapulse'.

Diapulse apparatus also works at 27.12 MHz, the frequency of short-wave diathermy machine. However, the energy is delivered in the form of pulses of 65 µs with an interval between pulsations at a maximum setting of 1600 µs. The rate of pulsations is adjustable in steps from 80 to 600 pulses per second. Even at this setting, power is provided no more than 4% of the total time that the machine is in operation. The peak instantaneous wattage can be varied from 293 to 975 W. The effect of the rest periods is to reduce the output to a maximum average of only 38 W. The result is an intermittent, relatively athermic, electrotherapy. Therefore, despite the high energy pulses, all the heat is dissipated during the rest period, and as a consequence, there is no danger of burns or hyperthermic complications. The depth of penetration depends upon the peak energy delivered, which is adjustable on the machine. Tuning adjustment is provided to obtain maximum efficiency at each wattage setting.

The physiological effects of diapulse are not fully understood. Cells in vitro become aligned within a diapulse field as iron filings follow the lines of force around a magnet, and the orientating effects of the electromagnetic waves seem to be important. Perhaps the most exciting possibility with pulsed-electromagnetic waves is that they may enhance the rate at which peripheral nerves, particularly the smaller diameter fibres, re-generate.

29.3 MICROWAVE DIATHERMY

Microwave diathermy consists in irradiating the tissues of the patient's body with very short wireless waves having frequency in the microwave region. Microwaves are a form of electromagnetic radiation with a frequency range of 300-30,000 MHz and wavelengths varying from 10 mm to 1 m. In the electromagnetic spectrum, microwaves lie between short waves and infrared waves. The most commonly used microwave frequency for therapeutic heating is 2450 MHz corresponding to a wavelength of 12.25 cm. The heating effect is produced by the absorption of the microwaves in the region of the body under treatment.

Microwave diathermy provides one of the most valuable sources of therapeutic heat available to the physician. However, in many conditions, though the therapeutic effects of microwave diathermy are similar to short-wave diathermy, yet in others, better results are obtained by using microwave.

The technique of application of microwave diathermy is very simple. Unlike the short-wave diathermy where pads are used to bring in the patient as a part of the circuit, the microwaves are transmitted from an emitter, and are directed towards the portion of the body to be treated. Thus, no tuning is necessary for individual treatments. These waves pass through the intervening air space and are absorbed by the surface of the body producing the heating effect.

Production of Microwaves: Microwaves are produced by high-frequency currents and have the same frequency as the currents which produce them. Such currents cannot be produced with oscillators using ordinary vacuum tube valves or solid-state devices. A special type of device called 'magnetron' is used for the production of high frequency currents of high power.

The magnetron consists of a cylindrical cathode surrounded by an anode structure that contains cavities opening into the cathode-anode space by means of slots. The output energy is derived from the resonator system by means of a coupling loop which is forced into one of the cavities. The energy picked up on the coupling loop is carried out of the magnetron on the central conductor of a co-axial output tube through a glass seal to a director. The director consists of a radiating element of antenna and a reflector which directs the energy for application to the patient. The electrical current is transformed into electromagnetic radiation on passing through the antenna.

The reflector then focuses this electromagnetic energy and beams it to the tissues where it is subsequently absorbed, reflected or refracted, according to the electrical properties of the tissue. Tissues of lower water content (i.e., subcutaneous) are penetrated to a greater depth but little is absorbed, whereas tissues of high water content (i.e., muscle) absorb more of the electromagnetic energy but allow little penetration.

The output power of a magnetron depends upon anode voltage, magnetic field and the magnitude and phase of the load impedance to which the magnetron output power is delivered. Therefore, the cable used to carry the energy from the magnetron to the director is always of a definite length for a particular frequency. A part of the energy fed to the magnetron is also converted into heat in the anode on account of the collision of the electrons with the anode so that the output energy is considerably less than the input energy. The efficiency of a magnetron is usually 40 to 60%. The heat produced at the anode must be removed which is usually done by using water or air as a means of cooling.

Schematic Diagram of a Microwave Diathermy Unit: The essential parts of a microwave diathermy unit are shown in Fig. 29.5. The mains supply voltage is applied to an interference suppression filter. This filter helps to bypass the high frequency pick-up generated by the magnetron. A fan motor is directly connected to the mains supply. The fan is used to cool the magnetron.

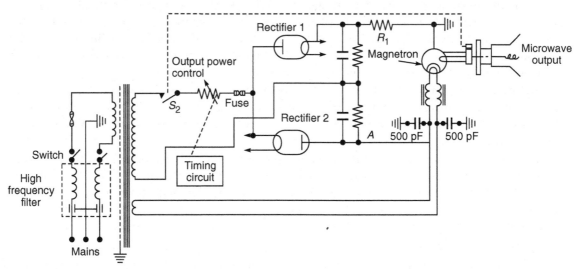

> **Fig. 29.5**   *Simplified circuit diagram of a microwave diathermy machine*

The Delay Circuit: It is necessary for the magnetron to warm up for 3 to 4 minutes before power may be derived from it. A delay circuit is incorporated in the apparatus which connects the anode supply to the magnetron only after this time elapses. The arrangement is such that a lamp lights up after 4 minutes indicating that the apparatus is ready for use.

The Magnetron Circuit: The magnetron filament heating voltage is obtained directly from a separate secondary winding of the transformer. The filament cathode circuit contains interference-suppression filters. The anode supply to the magnetron can be either DC or AC. A DC voltage is obtained by a full wave rectifier followed by a voltage doubler circuit. A high wattage variable resistance is connected in series which controls the current applied to the anode of the magnetron.

When using AC, the voltage is applied to the anode of the magnetron through a series connected thyratron so that the AC voltages of both tubes are equal in phase. By shifting the phase of the control grid voltage with respect to the phase of the anode voltage, the amount of current through the magnetron can be determined and thus the output power can be varied. The phase shift can be achieved by using a capacitor resistor network.

Safety Circuits: There are chances of the magnetron being damaged due to an excessive flow of current. It is thus protected by inserting a fuse (500 mA) in the anode supply circuit of the magnetron. The protection of both the patient and the radiator is ensured by the automatic selection of the control range depending on the type of the radiator used.

The considerable interference produced by the apparatus necessitates the use of large self-inductance coils in the primary supply. Since the cores are likely to become saturated in view of the small dimensions, the coils are split up and fitted in such a way that no magnetization occurs.

Excessive dosage can cause skin burns and in all cases, the sensation experienced by the patient is the primary guide for application. The skin should be dry as these waves are rapidly absorbed by water. The duration of irradiation generally ranges from 10 to 25 minutes.

29.4 ULTRASONIC THERAPY UNIT

Ultrasonics are used for therapeutic purposes in the same manner as a short-wave diathermy machine is used. The heating effect in this case is produced because of the ultrasonic energy absorption property of the tissues. The property of specific heat distribution in tissue and the additional effect of a mechanical component have given rise to a number of special therapeutic applications of ultrasonics. The effect of ultrasonics on the tissues is thus a high speed vibration of micro-massage. Massage as a modality in physical medicine has been used in the treatment of soft tissue lesions for centuries. Ultrasonic energy enables this massage to be carried out, firstly to a greater depth than is possible manually, and secondly at times (in acute injuries) when pressure cannot be exerted by hand because of intolerable pain caused to the patient. The thermal effects of ultrasound are dependent on the amount of energy absorbed, the length of time of the ultrasound application and the frequency of the ultrasound generator. The electrical power required in most of the applications is usually less than 3 W/cm^2 of the transducer area that is in contact with the part of the body to be treated.

Ultrasonic generators are constructed on the piezo-electric effect. A high-frequency alternating current (e.g., 0.75-3.0 MHz) is applied to a crystal whose acoustic vibration causes the mechanical vibration of a transducer head, which itself is located directly in front of the crystal. These mechanical vibrations then pass through a metal cap and into the body tissue through a coupling medium. The therapeutic ultrasonic intensity varies from 0.5 to 3.0 W/cm^2. Applicators range from 70 to 130 mm in diameter. The larger the diameter of the applicator, the smaller would be the angle of divergence of the beam and the less the degree of penetration.

Circuit Description: The equipment required for ultrasonic therapy is electronically very simple. Figure 29.6 shows the block diagram. The heart of the system is a timed oscillator which produces

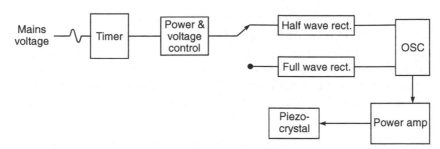

> ➤ **Fig.** 29.6   *Block diagram of an ultrasonic therapy unit*

the electrical oscillations of the required frequency. The oscillator output is given to a power amplifier which drives the piezo-electric crystal to generate ultrasound waves. Power amplification is achieved by replacing the transistor in typical LC tuned Colpitt oscillator by four power transistors placed in a bridge configuration.

The delivery of ultrasound power to the patient is to be done for a given time. This is controlled by incorporating a timer to switch on the circuit. The timer can be a mechanical spring-loaded type or an electronic one, allowing time settings from 0 to 30 minutes.

The output of the oscillator can be controlled by either of the following two methods:

- Using a transformer with a primary winding having multi-tapped windings and switching the same as per requirement;
- Controlling the firing angle of a triac placed in the primary circuit of the transformer, and thereby varying the output of the transformer.

The power output in case of triac controlled machines can be continuously varied from 0 to 3 watts/cm^2.

The machine can be operated in either continuous or pulsed mode. A full-wave rectifier comes in the circuit for continuous operation. The mains supply is given to the oscillator without any filtering. The supply voltage is therefore at 100 Hz which causes the output 1 MHz to be amplitude modulated by this 100 Hz. In pulsed mode, the oscillator supply is provided by the half-wave rectifier and the oscillator gets the supply only for a half cycle. Thus the output 1 MHz is produced only for one half of the cycle and is pulsed.

The transducer may be barium titanate or lead zirconate titanate crystal, having 5-6 cm^2 effective radiating area. The length of the cable connecting the transducer with the oscillator is of critical dimension and should not be altered. In front of the crystal lies a metal face plate which is made to vibrate by the oscillations of the crystal. Ultrasonic waves are emitted from this plate. The crystal has a metal electrode pressed against its back surface by a coiled spring. Voltage is applied to the crystal via this electrode. The front diaphragm is grounded and provides a return path for the excitation voltage.

Dosage Control: The dosage can be controlled by varying any of the following variables.

- Frequency of ultrasound;
- Intensity of ultrasound; or
- Duration of the exposure.

The frequency is involved because the absorption of ultrasonics by the tissues is a frequency-dependent phenomenon. The question as to which ultrasonic frequency to use has been the subject of much investigation and thought but it has been established that a frequency of approximately 1 MHz is the most useful. The amount of energy absorption in the human tissue has been measured experimentally and in soft tissue, a reduction of 50% occurs with a 1 MHz ultrasonic transmission at a depth of 5 cm. The higher the frequency the quicker the energy loss and thus with a transmission of 3 MHz, this reduction of 50% occurs at a depth of only 1.5 cm. Below a frequency of 1 MHz, the beam of ultrasonic energy tends to diffuse and no efficient treatment can be expected. A frequency in the range of 800 kHz to I MHz is, therefore, most widely adopted.

Unlike the operation of a short-wave therapy unit, no tuning is necessary while the treatment is in progress. The operating frequency is also not very critical and may vary to the extent ± 10%.

The output power of an ultrasonic therapy unit can be varied continuously between 0 and 3 W/cm^2. The calibrated positions are marked in steps. The steps indicate the average value of intensity monitored in terms of electric power converted into acoustic energy.

Standard values concerning dosing in ultrasonic therapy have been established on the basis of experimental studies. In order to achieve maximum therapeutic efficiency, it is necessary to ascertain the correct ultrasonic intensity and duration of application for a given indication. To make matters simpler, some instruments are equipped with a 'dose tabulator' from which the data concerning dosing can be taken at a glance. In this table, a dose mark is given for every indication (disease) and all that is required is to set a pointer appropriately to ensure that the apparatus is providing the correct output intensity.

Apart from continuous and mains frequency modulated modes, ultrasonic output can also be modulated by any other frequency. The idea behind a pulsed operation is that the predominant effect of ultrasound is not the heating effect but the direct mechanical effect (micro-massage). The thermal effect is reduced by repeatedly interrupting the supply of energy through brief pauses.

Application Technique: There are several ways for applying ultrasonics to the body. The probe can be put in direct contact with the body through a couplant provided the part to be treated is sufficiently smooth and uninjured. In case a long area is to be treated, the probe is moved up and down, and for small areas it is given a circular motion to obtain a uniform distribution of ultrasonic energy. If there is a wound or an uneven part (joints etc.), the treatment may be carried out in a water bath. This is to avoid mechanical contact with the tissues which may damage an already injured surface. It should be ensured that air bubbles are not present either on the probe or the skin. For this treatment any vessel with warm water would be suitable. The part of the body to be treated is rubbed with alcohol or soaped. The probe is moved over the area to be treated but held at a distance of about 1–2 cm from the area under treatment. This method is not generally preferred because of the difficulty of controlling the exact amount of dosage.

29.5 ELECTRODIAGNOSTIC/THERAPEUTIC APPARATUS

29.5.1 Electrodiagnosis

If a normal muscle or motor nerve is stimulated with a current of adequate intensity, it results in its contraction. When there is disease or injury of a motor nerve or muscle, alterations are liable to occur in their response to electrical stimulation. The changed electrical response may be of considerable help in the diagnosis of certain diseases affecting them. Quantitatively, these changes manifest themselves in that a higher or lower current intensity than normal is required to bring about a muscle contraction. It is, therefore, possible to determine the degeneration and regeneration processes in nerves and the muscle system by the use of the stimulation current technique.

Intensity-Time Curve (i-t Curves): In order to examine the conditions of excitability and to obtain a good picture of the degeneration and regeneration process of neuro-muscular units, modern stimulation current diagnosis plots the so called i-t curves based on the intensity of the stimulus and its duration. These curves are determined by means of rectangular and triangular pulses in such a manner that the threshold values are measured at progressively decreasing stimulation durations. The i-t curves have characteristic shapes and deviations from the standard form lead to an indication of the state of the tissues.

In order to plot such curves, the tissue (muscle, nerve) to be examined is first stimulated with long impulses (usually of 1 s pulse duration and then with shorter and shorter impulses, (down to say, 0.05 ms). For each impulse duration, the current intensity is adjusted until the stimulation threshold has been exceeded and the effect of the stimulation detected. Obviously, the current intensity has to be increased. The impulse duration is usually varied in stages such as 1000–300–100–10–3–1–0.05 ms and the control of current is effected by a continuously adjustable resistance.

Figure 29.7 shows typical shapes of i-t curves for an intact neuro-muscular unit of normal excitability when rectangular impulses are used. It also shows the shape of the curve for a totally denervated muscle having an advanced state of degeneration and required excitability.

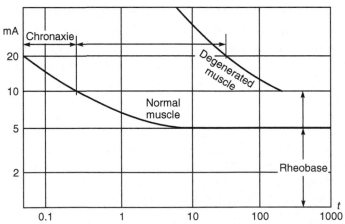

> **Fig. 29.7** *Typical intensity time curves of a normal muscle and degenerated muscle. The curve shows that decreasing excitability with progressive degeneration requires extended stimulation times and increased current strength for achieving successful stimulation*

With degenerated muscle, the curve obtained is shifted to the right and upwards. The intermediate stages of degeneration and regeneration are characterized by curves lying in between these two limits.

The chronaxie and rheobase can be easily read from the i-t curves. The rheobase is the minimum intensity of current that will produce a response if the stimulus is of infinite duration, in practice an impulse of 100 ms being adequate for estimating this. The chronaxie is the minimum duration of impulse that will produce a response with a current of double the rheobase. For example if the rheobase is 6 mA, the chronaxie is the duration of the shortest impulse that will produce a muscle contraction with a current of 12 mA.

i-t curves with exponentially progressive current impulses can be drawn in the same way as rectangular impulses. The two curves differ considerably. A typical characteristic of these curves is that the stimulation threshold first decreases when the pulse duration is reduced and the rheobase is also missing. This is due to the phenomenon of the accommodability of the neuro-muscular units.

Accommodation: Accommodation is the property of a neuro-muscular unit of being able to respond less strongly to a slowly increasing current impulse. In other words, the units exhibit a lower excitability and a higher stimulation threshold. The importance of accommodation from the diagnostic point of view lies in the effect that it gives an indication of the presence or alteration of a state of degeneration.

In the representation of i-t curves, the determination of the accommodation consists of the comparison between the 100 ms points of the rectangular i-t curves and of the triangular i-t curves and is, in a way, analogous to the determination of the chronaxie, which is essentially a comparison of the two points of the rectangular i-t curves.

The types of waveforms required for electrodiagnosis are:

- Galvanic current for qualitative and quantitative determination of the galvanic excitability (rheobase and chronaxie);
- Rectangular pulses for checking nervous conduction as a control of functioning, also of prognostic importance;
- Exponentially progressive current for checking the accommodability or its loss as a symptom of the degree of degeneration and for prognosis of the re-innervation of totally denervated muscles; and
- Faradic current for qualitative and quantitative determination of the faradic excitability.

29.5.2 Electrotherapy

Electrotherapy, employing low-volt, low-frequency impulse currents, has become an accepted practice in the physiotherapy departments. The biological reactions produced by low-volt currents have resulted in the adoption of this therapy in the management of many diseases affecting muscles and nerves. The technique is used for the treatment of paralysis with totally or partially degenerated muscles, for the treatment of pain, muscular spasm and peripheral circulatory disturbances, and for several other applications.

Although some of the principles upon which low-volt therapy depends have been known since the end of the last century, it is only in recent years that it has started being widely used with the availability of safe and simplified apparatus required for the purpose.

Different types of waveforms are used for carrying out specific treatments. The most commonly used pulse waveforms are discussed below.

Galvanic Current: When a steady flow of direct current is passed through a tissue, its effect is primarily chemical. It causes the movement of ions and their collection at the skin areas lying immediately beneath the electrodes. The effect is manifested most clearly in a bright red coloration which is an expression of hyperaemia (increased blood flow). Galvanic current is also called direct current, galvanism, continuous current or constant current.

Galvanic current may be used for the preliminary treatment of atonic paralysis and for the treatment of disturbance in the blood flow. It is also used for iontophoresis, which means the introduction of drugs into the body through the skin by electrolytic means. In general, the intensity of the current passed through any part of the body does not exceed 0.3 to 0.5 ma/sq cm of electrode surface. The duration of the treatment is generally 10–20 minutes.

Faradic Current: Faradic current is a sequence of pulses with a defined shape and current intensity. The pulse duration is about 1 minutes with a triangular waveform and an interval duration of about 20 minutes. Faradic current acts upon muscle tissue and upon the motor nerves to produce muscle contractions. There is no ion transfer and consequently, no chemical effect. This may be used for the treatment of muscle weakness after lengthy immobilization and of disuse atrophy.

Surging Current: If the peak current intensity applied to the patient increases and decreases rhythmically, and the rate of increase and decrease of the peak amplitude is slow, the resulting shape of the current waveform is called a surging current.

The main field of application of the Faradic surge current is in the treatment of functional paralysis. The surge rate is usually from 6-60 surges per minute in most of the instruments. The ratio of interval to the duration of the surging is also adjustable so that graded exercise may be administered. This type of current is usually required for the treatment of spasm and pain.

Exponentially Progressive Current: This current is useful for the treatment of severe paralysis. The main advantage of this method lies in the possibility of providing selective stimulation (Fig. 29.8) for the treatment of the paralysed muscles. This means that the surrounding healthy tissues even in the immediate neighbourhood of the diseased muscles are not stimulated. The slope of the exponential pulse is kept variable.

Biphasic Stimulation: The cell recovery from the effect of a stimulus current can be hastened by the passage of a lower intensity current of opposing polarity over a longer period so that the net quantity of electricity is zero. Such type of combination of positive and negative pulses is called biphasic stimulation. In a typical case, the stimulating pulse may be followed by a pulse of opposite polarity of one-tenth the amplitude and 10 times the width. Biphasic stimulation also helps to neutralize the polarization of the recording electrodes in case silver-silver chloride electrodes are not used. This means that there are no electrolytic effects, nor are any macroscopic changes affecting either the skin or the electrodes observed. Also, there is reduced muscle fatigue, since each current

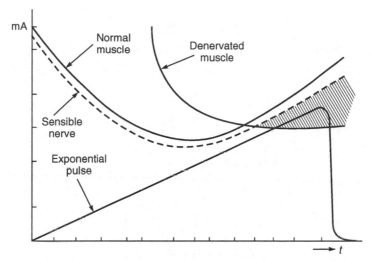

> **Fig. 29.8** *Principle of selective stimulation of the denervated musculature. Selective stimulation of the denervated muscle without irritation of the sensible receptors is possible in the shaded area of the graph.*

pulse is immediately followed by an opposite current phase of the same magnitude. The stimulation current intensity required during treatment is less as compared with monophasic currents.

Monophasic current forms, however, retain their importance in electro-diagnostic evaluation since the necessary pulse shapes are defined mono-phasically.

29.5.3   Type of Apparatus

Several types of commercial units are available which give specific output waveforms for specific applications. However, the trend is in favour of having a versatile apparatus which gives output current waveforms to cover the whole range of electro-diagnostic and therapeutic possibilities. The output waveforms generally required are shown in Fig. 29.9.

Another important consideration is that the apparatus must be either of constant voltage or constant current type. An instrument of indefinite intermediate principle can lead to unreliable results. Moreover, the apparatus must give reproducible and well-defined impulses that must correspond to the value set on the dials. For clinical practice, maximum tolerance permitted in the pulse parameters is 15%. The instrument generally has a floating output and incorporates an isolation transformer in the output.

The typical specifications of an electro-diagnostic therapy unit are as follows:

- Galvanic current up to 80 mA, ripple less than 0.5% as constant current or surging current with adjustable surge frequency from 6 to 30 surges per minute
- Exponentially progressive current pulse sequences with continuously variable pulse duration from 0.01 to 1000 ms and independently adjustable interval duration of 1 to 10,000 ms. The pulse form can be set continuously between triangular and rectangular forms;

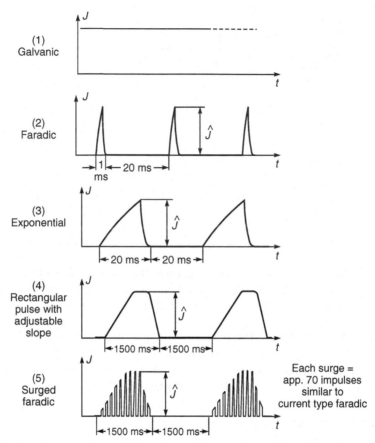

> **Fig. 29.9** *Current waveforms normally employed in electrodiagnosis and electrotherapy*

- Faradic surging current with 25 surges per minute, upto 80 mA. Precision and constancy of the values set better than ±10%; peak current measurement facility. Constant current circuit, both poles earth-free.

29.5.4 Functional Block Diagram Description

Figure 29.10 shows the block diagram of a versatile electro-diagnostic therapeutic stimulator. It makes use of a variable rate multi-vibrator (M1) to set the basic stimulus frequency. The output from the free running multi-vibrator triggers a monostable multi-vibrator (M2) circuit which sets the pulse width. The output pulse from the monostable provides an interrupted galvanic output whose rate as well as duration can be independently controlled. Another astable multi-vibrator produces short duration pulses called faradic currents. Faradic currents are usually modulated at the frequency set by the multi-vibrator M1, in a mixer circuit (M4). Since the modulation of Faradic pulses takes place with a slow rate of increase and decrease, the output of M4 is surged Faradic currents. By integrating the output of M2, the interrupted galvanic pulses can be modified to have

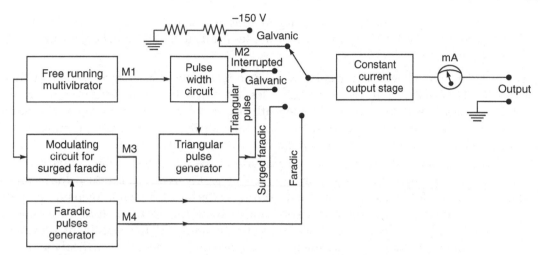

> **Fig. 29.10** *Schematic diagram of a diagnostic/therapeutic stimulating unit*

an exponential rise and fall. The shape of these pulses is similar to a triangular waveform. Galvanic current is also made available by suitably tapping the DC supply.

Finally, any one of the waveforms can be selected through a selector switch and fed either to an emitter-follower stage in order to provide a low output impedance constant voltage output or to a high output impedance constant current stage. Usually the output impedance of a constant voltage stimulator is of the order of 100 Ω and that of a constant current type is greater than 100 kΩ.

The output of a diagnostic/therapeutic stimulator is kept floating, i.e. it is isolated from earth. The usual method is to have an isolation transformer at the output of the stimulator. This transformer has floating terminals and is fitted with an electrostatic shield to reduce capacitive coupling with the earth. Another method of isolation of the output from earth is by the use of a radio-frequency output stage.

The two methods have been widely used for providing isolation of the stimulator output, but they have some drawbacks. The simple transformer cannot transmit square waves without distorting the waveform and the method of radio-frequency is rather complex. Isolation can also be provided through the opto-isolation technique.

The question as to which type of stimulation impulses, whether constant voltage or constant current type, should be preferred in carrying out electro-diagnostic studies is still a matter of choice. However, most of the present-day instruments are of the constant current type because for a given electrode geometry, constant current stimulation will provide better reproducibility for a wide variation in preparation impedance. The advantages of constant current therapy are detailed below:

- The current flow is largely constant irrespective of the patient's resistance. The selected current intensity remains constant, even if the resistance in the tissue between the electrodes should vary, as a result of, say, changes in the blood circulation during treatment, or after previous therapy.
- The current waveform is applied, and distortion-free, since micro-voltages between the electrode and the skin have no influence.

- Current therapy avoids accompanying symptoms such as irritating stimulatory sensations between electrodes by applying electrodes firmly to the skin and keeping them in one position.

In case of constant voltage, the current flow is dependent on the resistance of the patient, that is, if the electrical resistance of the tissue increases, less current will flow, and vice versa. Irritating stimulatory sensations do not occur, even beneath electrodes that are not firmly applied or are moved over the skin, for example, during the search for a trigger point. This operating mode is recommended for the combined use of stimulation current and ultrasound.

Modern electro-diagnosis/therapy units are microprocessor-controlled which make possible a number of automatic sequences in selecting the type and quality of waveform. Also, the facility for automatic self-test is followed by the automatic setting of the basic program. Acoustic and visual signals provide information on the various operating situations. Operating errors are indicated on visual displays. A built-in service advice helps identify faults. All this intelligent information reduces the demands made on the therapist, who can then devote more time to the patient.

29.5.5   Interferential Current Therapy

Interferential electrical stimulation is a unique way of effectively delivering therapeutic currents to tissue. Conventional TENS (Transcutaneous Electrical Nerve Stimulator) and neuro-muscular stimulators use discrete electrical pulses delivered at low frequencies of 2–200 Hz. However, interferential stimulators use a fixed carrier frequency of 4,000 Hz and also a second adjustable frequency of 4,001–4,400 Hz. When the fixed and adjustable frequencies combine (heterodyne), they produce the desired beat frequency or interference frequency (Fig. 29.11). Interferential stimulation is concentrated at the point of intersection between the electrodes. This concentration occurs deep in the tissues as well as at the surface of the skin. Conventional TENS and neuro-muscular stimulators deliver most of the stimulation directly under electrodes. Thus, with interferential stimulators, the current perfuses to greater depths and over a larger volume of tissue than other forms of electrical therapy. When current is applied to the skin, the capacitive skin resistance decreases as pulse frequency increases. For example, at a frequency of 4,000 Hz (interferential range), the capacitive skin resistance is 80 times lower than with a frequency of 50 Hz (in the TENS range). Thus, interferential current crosses the skin with greater ease and with less stimulation of cutaneous nociceptors allowing greater patient comfort during electrical stimulation. In addition, because medium-frequency (Interferential) current is tolerated better by the skin, the dosage can be increased, thus improving the ability of the interferential current to permeate tissues and allowing easier access to deep structures. This explains why interferential current may be most suitable for treating patients with deep pain.

Interferential currents are produced by using two-channel stimulators and four electrodes. Each channel produces a sinusoidal, symmetrical alternating current at a high frequency (2000–5000 Hz). The electrodes are used in a quad-polar arrangement and the AC frequencies are set at slightly different frequencies but at similar amplitudes. The currents from the two waveforms interface with each other in the tissue, giving constructive (when the two waveforms add to each other) or destructive (when the circuits tend to cancel each other) interference.

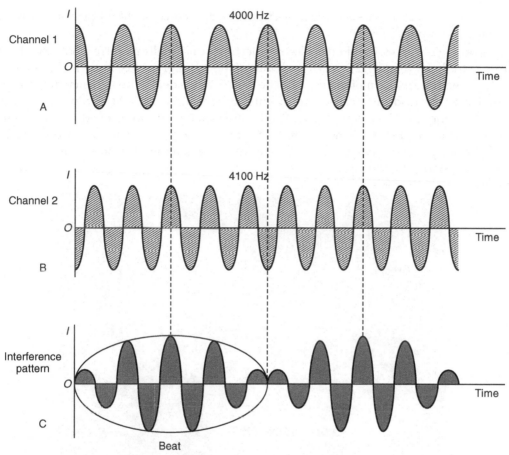

➤ Fig. 29.11 *Principle of generation of interference currents*

The therapeutic current is the beat frequency. Neither the carrier frequency nor the variable frequency alone has any therapeutic value. The major benefit of interferential currents is that the therapeutic current is generated within the tissue, and therefore the treatment is possible at deeper structures. In addition, it is possible to increase the amplitude of the stimulus that can be tolerated and increase the comfort of the pulse as it passes through the skin.

29.5.6 Types of Electrodes for Electro-diagnostic/Therapeutic Applications

Two methods of electrode systems are in common use. The mono-polar technique makes use of small active stimulation electrode. The indifferent or dispersive electrode is of larger area and is placed near to the active electrode. This technique is used for testing of the galvanic and Faradic excitability and for determining the chronaxie.

For diagnostic purposes, a ball or plate electrode which is provided with a small thick muslin strip is mounted on a special handle. The handle carries a finger-tip switch to facilitate convenient control of output. Similarly, a small metal electrode can be secured on the motor point, particularly for therapeutic applications.

For recording i-t curves, the bi-polar electrode technique is usually preferred. Both the electrodes are fixed to the body (Fig. 29.12), so that the hands of the operator are free to operate the apparatus. The active electrode in this method need not be a small as we deal with higher current intensities and small area electrode may cause unpleasant heat sensations. Suitably sized metal sheets are used as electrodes in this system. The electrodes are fastened to a moistened pad of about 1 cm thickness and 1 cm wider than the electrode sheet on all sides. The material used for pads is of good absorbancy and ordinary water can be used to moisten the electrodes. The electrodes are held in position by rubber straps.

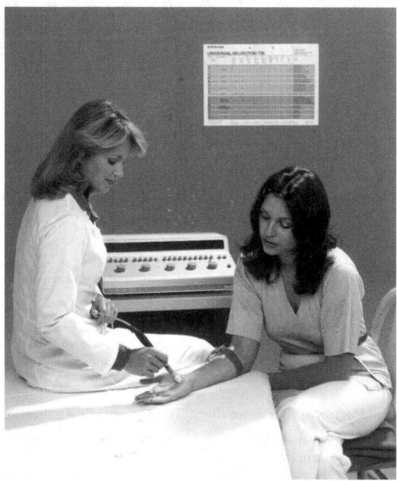

➤ **Fig. 29.12**   *Precision stimulation response measurement for plotting i-t curves using electrodiagnosis and electrotherapy apparatus*

The stimulation current therapy can also be administered with suction electrodes. The current is conducted directly to the patient through the suction electrodes; the negative pressure necessary for adhesion is generated in the electrodes themselves by a compressed air flow. This method prevents liquids, dirt and bacteria from being drawn out from the electrodes into the pump/hose system of the unit. The adhesion strength can be adjusted to suit the application. Suction electrodes have been used by M/S Siemens in their therapy unit (model NEOSERV 824).

Standard texts on muscle stimulation emphasize that successful muscle stimulation can only be achieved if the activating currents are properly applied. Physiotherapists are trained to understand all about motor points and how to apply stimulation through these points.

29.6 PAIN RELIEF THROUGH ELECTRICAL STIMULATION

29.6.1 Transcutaneous Electrical Nerve Stimulator (TENS)

Pain is man's oldest enemy and for centuries, medicine has searched for an innocuous, non-destructive, non-invasive, well-tolerated and effective way of relieving pain that is both efficient and practical. In the past few years, several workers have reported their success in using electrical impulses to block the pathways of the transmission of pain. The impulses are produced in a battery-powered pulse-generator to which a pair of electrode-tipped wires can be attached. Applied to the skin overlying any painful area of the body, these electrodes provide continuous, mild electrical stimulation. These signals seem to jam the pain signals travelling along the nerve pathways before they can reach the brain. The result is analgesia, often for hours after stimulation ends. The pain control is explained by:

- The Gate Control Theory which suggests that by electrically stimulating sensory nerve receptors, a gate mechanism is closed in a segment of the spinal cord, preventing pain-carrying messages from reaching the brain and blocking the perception of pain; and
- The Endorsphin Release Theory which suggests that electrical impulses stimulate the production of endorphin and enkaphalins in the body. These natural, morphine-like substances block pain messages from reaching the brain, in a similar fashion to conventional drug therapy, but without the danger of dependence or other side-effects.

The electrical impulses required for electrotherapy to treat the pain are provided by an instrument called TENS (Transcutaneous Electrical Nerve Stimulator). Investigations on a great variety of electrical impulse parameters have indicated that two waveforms, the square wave and the spike wave are optimally and equally effective in relieving pain. Most stimulators feature adjustable settings to control the amplitude (intensity) of stimulation by controlling voltage, current and the width (duration) of each pulse. Electrodes are placed at specific sites on the body for treatment of pain. The current travels through the electrodes and into the skin stimulating specific nerve pathways to produce a tingling or massaging sensation that reduces the perception of pain.

Typically, the stimulator is based around a 500 ms spike pulse, having an adjustable amplitude of 0 to 75 mA and an adjustable frequency of 12 to 100 pulses per second. Instruments having similar specifications except that they produce square waveform, have a pulse frequency range of 20–200 Hz, pulse width from 0.1 to 1.0 ms and pulse amplitude of 0–120 V with maximum output current as 25 mA. The instrument powered by three standard flashlight batteries of 1.5 V each gives about 100 hours of continuous operation. Transcutaneous or skin surface application of electrical stimulus is accomplished by application of the conducting pads to various trigger-zone areas, acupuncture sites or even peripheral nerves. Skin irritation at the site of electrode application is diminished by the use of carbonized rubber electrodes applied with a tincture of Benzoin interface.

The skin electrode system must be designed so as to minimize impedance variations with motion, to conform to the body surface to provide a uniform impedance across the surface of the electrode and to have an adequate surface area. The adequate surface area can be determined keeping in view the peak square-wave current at the threshold of thermal damage as a function of the electrode surface area. The thermal damage threshold varies widely with skin impedance, which is a function of skin preparation.

Transcutaneous electrical nerve stimulation (TENS) electrodes are commonly moulded from an elastomer such as silicon rubber, loaded with carbon particles to provide conductance. Conformability is achieved by making the electrode thin. Useful carbon-loaded silicon rubbers have a minimum resistivity near 10Ω cm. A thin electrode may exhibit an impedance which is not negligible as compared to the impedance of the interface and tissue under it. Thus, the design of an electrode with the required conformability and current distributing properties becomes a compromise in electrode geometry and material properties. The frequency-dependence of the electrode performance also has to be considered since the impedance between the electrode and subcutaneous contains capacitance.

29.6.2   Spinal Cord Stimulator

Spinal cord stimulation is a term relating to the use of electrical stimulation of the human spinal cord for the relief of pain. This is accomplished through the surgical placement of electrodes close to the spinal cord, either with leads extending through the skin, or chronically, with the leads connected to an implanted source of electrical current. The applied electrical impulses develop an electrical field in and around the spinal cord, which then causes depolarization or activation of a portion of the neural system resulting in physiological changes.

The stimulus source provides stimulation pulses at frequencies ranging from 10 to 1500 Hz, with pulse widths from 100 to 600 µs and controllable amplitude from 1 to 15 mA delivered into a load from 300 to 1500 Ω. These parameters can be controlled when one is using an implant that derives power and control through RF coupling from an externally power unit. Figure 29.13 shows the Medtronic Spinal Cord stimulation system, which has an implantable pulse generator and a hand-held programmer.

Since a spinal cord stimulator is not a life-support system, there is no hazard associated with a stimulator failing to provide an output. However, patients using ventricular inhibited or triggered pacemakers should not be exposed to nerve stimulation. Also, a patient entering a pulsed radio frequency field of the frequency to which the receiver is tuned would be in danger of having his stimulator activated by the field. Spinal cord stimulation has been shown to be of great benefit to some patients with multiple sclerosis and other neurological diseases; it is expected that the technique would be applied more and more in the near future.

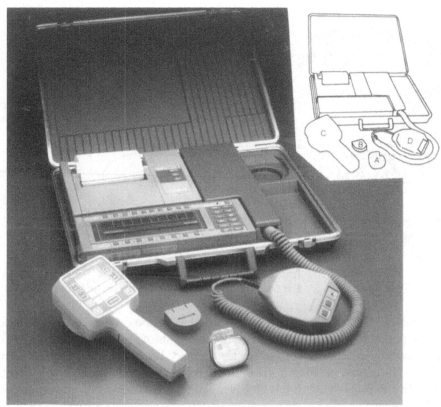

> **Fig. 29.13** *Spinal cord stimulation system (Courtesy: M/s Medtronic, USA)*
>   **(A)** *Implantable pacemaker which includes electronic circuitry and power source*
>   **(B)** *Magnet used by the patient to control the pulse generator's impulses within parameters set by the clinician*
>   **(C)** *Hand held programmer which communicates with the pulse generator*
>   **(D)** *Desk top programming console to programme the pulse generator and provides printouts of information from it*

The treatment of idiopathic scoliosis (lateral curvature of the spine) by electrical stimulation is described by Leonard (1980). The apparatus used for this purpose consists of an implanted radio receiver (Fig. 29.14(a)) and an external transmitter with an appropriate antenna (Fig. 29.14(b)). The transmitter is designed to generate pulses for muscle contraction lasting 1.6 seconds with a rest period of 9 seconds between contractions. The actual stimulation is not a single pulse but rather a burst of pulses consisting of individual pulses 220 μs wide, repeated 33 times every second. For transmission through the skin, the pulse bursts are modulated with a carrier frequency of 460 kHz.

The receiver is a passive device designed to receive only signals from the transmitter. It demodulates the signals and conducts them through the leads into the appropriate muscles to

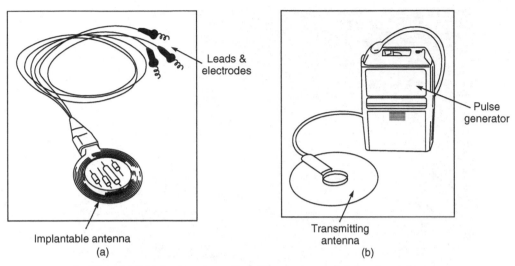

Leads & electrodes

Pulse generator

Implantable antenna
(a)

Transmitting antenna
(b)

> **Fig. 29.14(a)** *Implanted radio receiver with leads*
>   **(b)** *External transmitting unit with an antenna*

produce stimulation. The receiver circuit is embedded in an epoxy disc coated with silicon rubber for tissue compatibility. The receiver is attached to three leads of platinum-iridium wire terminating in platinum corkscrew electrodes. The electrodes are placed over appropriate para-spinal muscles during surgery. The receiver is placed in a subcutaneous pocket on the convex side of the curve. The transmitting antenna is a flat disc which is taped on the skin over the subcutaneous receiver by disposable adhesive.

This technique is considered as an exciting development in the field of scoliosis management, particularly of young children. For a given geometry, the effect of electrical stimulation on nerve tissue is determined by the charge per stimulus. Thus, in order to maintain a certain stimulus level despite a possibly varying electrode impedance the amount of charge per pulse delivered to the electrode must be kept constant. This can be achieved either by discharging a capacitor which has been charged to a given voltage or by driving the electrode by a current source thus supplying a constant current for a certain amount of time.

29.6.3   Magnetic Stimulation

A problem with electric stimulation is that it is painful (Hallett and Leonardo, 1990). The pain is not very different from that induced by the stimulation of peripheral nerves, but it is sufficient to limit its clinical acceptability. It has been shown by Barker et al (1985) that it is possible to stimulate both the nerve and brain magnetically. A magnetic pulse is generated by passing a brief, high-current pulse through a coil of wire. The technique has an advantage in that the stimulation is almost painless. Although a large number of studies have been carried out to study the effectiveness and safety of magnetic stimulation, the technique is still experimental and regulated in countries like the USA.

29.7 DIAPHRAGM PACING BY RADIO-FREQUENCY FOR THE TREATMENT OF CHRONIC VENTILATORY INSUFFICIENCY

Sarnoff et al. (1948) introduced the term electro-phrenic respiration to describe the contraction of the diaphragm following electrical stimulation of the phrenic nerve. He experimented extensively with electrical stimulation of the phrenic nerve (Fig. 29.15) and demonstrated that subaximal electrical stimulation of only one phrenic nerve could affect normal oxygen and carbon dioxide exchange. The technique is now well-developed and radio frequency phrenic nerve pacemakers are commercially available.

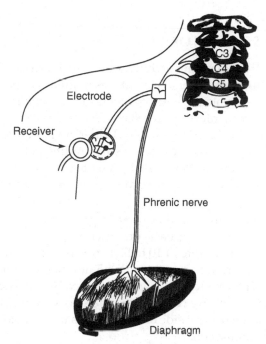

> **Fig. 29.15** *The diagram shows placement of receiver and electrode for phrenic nerve stimulation*

The apparatus is a radio-frequency-coupled stimulator composed of a small external transmitter and a passive receiver capsule implanted subcutaneously in the patient. The energy required for stimulation is transmitted via the external battery-operated transmitter through the closed skin to the internally implanted receiver capsule. Multi-strand stainless steel wires covered with silicon rubber deliver the stimulus from the implant to the nerve-cuff placed around the phrenic nerve. A small antenna composed of several loops of wire taped to the patient's skin over the receiver capsule provides the electromagnetic coupling that is necessary for the operation. In order to eliminate the problem of antenna motion with respect to the implant's causing changes in the amplitude of the current stimulus applied to the nerve, a pulse width telemetry system is used. The current stimulus supplied to the nerve is 150 μs in duration. The amplitude of the pulse applied to the nerve is a function of the duration of the applied radio-frequency carrier, not of the amplitude of the carrier.

Noshiro and Suzuki (1978) stress the importance of synchronization of respiratory rhythm with electrical stimulation of the phrenic nerve. Synchronization is of clinical importance because ventilation cannot be fully performed if the remaining spontaneous breathing in assisted respiration is asynchronous with a respirator. It was observed by these workers that synchronization occurs only within the limited range of a stimulation period. Talonen et al. (1990) illustrate neuro-physiological and technical considerations for the design of an implantable phrenic nerve stimulator.

29.8 BLADDER STIMULATORS

The micturition reflex controls emptying of the bladder and guards against urinary tract infection and possible kidney malfunction. When it is faulty, as it often is in paraplegia, chronic infection can occur which may be life-threatening. Such patients mostly require a catheter either constantly or intermittently to maintain an infection-free state. Electronic devices have been used for direct electrical stimulation of the detrusor muscle of the bladder but it produced undesirable side-effects because of the necessarily high current sent to the surrounding pelvic structure. A technique has been developed to activate the micturition reflex by remote electronic stimulation of a permanently implanted spinal electrode, with which the paraplegic is able to empty the bladder completely without the use of a catheter.

The stimulating device is similar to that used in phrenic nerve stimulation. However, the stimulus provided is in the form of a biphasic pulse with a pre-set pulse width of 0.2 ms, a pulse intensity of 0.5 to 25 V and a pulse rate of 10 to 50 Hz. The sacral cord electrode consists of two insulated platinum wires, 2.5 mm in length, with 1.5 mm bared conical tips, mounted 2.5 mm apart on an epoxy strip. The two lead out wires are flexible, silastic-coated and are made of stainless steel and connect to a receiver with a circumference of 3 cm. The receiver is placed in the subcutaneous tissue on the left or right side of the patient's waist. Voiding usually begins 10 to 15 seconds after the onset of stimulation.

29.9 CEREBELLAR STIMULATORS

Cerebellar stimulation has been found to give a favourable clinical effect in modifying or inhibiting intractable epilepsy. Stimulation to the cerebellum is provoked by transcutaneous inductive coupling, through an antenna fixed subcutaneously on the chest. It is delivered through four pairs of platinum discs fixed on a plate of silicon-coated mesh. The electrodes are applied beneath the tentorium to the cortex of an anterior lobe of cerebellum or directly to the posterior lobe. In some cases, electrode bearing plates are placed on both the anterior and posterior cerebellar cortex. The parameters of stimulation vary from person to person, and are adjusted depending upon the effects noted in individual cases. Rectangular pulses of 1 ms width, with a rate of 7-200 Hz and an intensity of 0.5–14 V are generally used for the stimulation.

Haemodialysis Machines

▶ 30.1 **30.1 FUNCTION OF THE KIDNEYS**

The main function of the kidneys is to form urine out of blood plasma, which basically consists of two processes: (i) the removal of waste products from blood plasma, and (ii) the regulation of the composition of blood plasma. These activities not only lead to the excretion of non-volatile metabolic waste products but are also largely responsible for the remarkable constancy of the volume, osmotic pressure, pH and electrolyte composition of the extra-cellular body fluids.

The human body has two kidneys which lie in the back of the abdominal cavity just below the diaphragm, one on each side of the vertebral column (Fig. 30.1(a)). Each kidney consists of about a million individual units, all similar in structure and function. These tiny units are called *nephrons* whose structure is shown in Fig. 30.1(b). A nephron is composed of two parts—a cluster of capillary loops called the glomerulus and a tubule. The tubule runs a tortuous course and ultimately drains via a collecting duct into the funnel-shaped expansion of the upper end of the ureter, i.e. the tube which conveys urine from the kidney to the bladder. The mechanism by which the kidneys perform their functions depends upon the relationship between the glomerulus and the tubule.

The kidneys work only on plasma. The erythrocytes supply oxygen to the kidneys but serve no other function in urine formation. Each substance in plasma is handled in a characteristic manner by the nephron, involving particular combinations of filtration, re-absorption and secretion.

The renal arteries carry blood at a very high pressure from the aorta into the glomerular capillary tuft. The blood pressure within the glomerular capillaries is 70–90 mm of mercury. The blood flow through the capillary tuft is controlled by the state of contraction of the muscle of the arteriole leading to the tuft. The fluid pressure within the tuft forces some of the fluid part of the blood, by filtration, through the thin walls of the capillaries into the glomerulus and on into the tubule of the nephron. The glomerular filtrate consists of blood plasma without proteins. The total amount of glomerular filtrate is about 180 litres per day, whereas the amount of urine formed from it is only 1–1.5 *l*. This means that very large amounts of water, and other substances, are re-absorbed by the kidney tubules. The re-absorption is partly an automatic process, because the absorption of water is accurately controlled by the anti-diuretic hormone of the pituitary gland, in relation to the body's

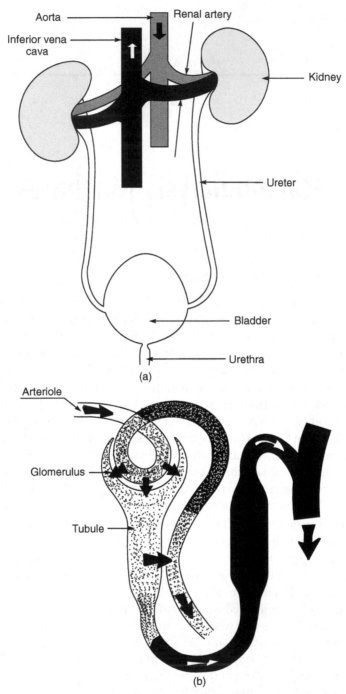

> **Fig. 30.1(a)**   *Kidneys and their connection to other organs of the body*
>   **(b)**   *Structure of nephron*

need for water. The absorption of electrolytes such as sodium and potassium is partly controlled by the supra-renal gland and the concentration of others, like chloride and bicarbonate, is related to the acid-base balance. Some of the re-absorption from the glomerular filtrate is also a passive, automatic process of diffusion depending upon pressure gradients. This applies to water itself, and its electrolytes like sodium, potassium, chloride, calcium and bicarbonate.

There are other substances such as urea, phosphates and sulphates which are the waste products of metabolism. They are unwanted by the body. The tubules are selectively porous to substances of importance to the body and impermeable to the unwanted. Therefore, the unwanted substances cannot diffuse back into the plasma and thus a large proportion is excreted in the urine. As the filtrate passes down the tubules, the concentration of waste products rises steadily and the specific gravity of normal urine varies from 1.015 to 1.030 as compared with 1.010 for glomerular filtrate. Other substances in the glomerular filtrate such as glucose and amino acids show little tendency to diffuse through the tubular walls and are returned to the plasma by a process of active re-absorption.

The total blood flow through the kidneys is about 1200 ml/min. The total extra-cellular fluid amounts to about 15 litres. The blood plasma and the extra-cellular fluid are in equilibrium with each other and, therefore, an amount of blood equivalent to all the extra-cellular fluid can pass through the kidneys once every 15 minutes. The water and electrolyte content of the blood plasma and, therefore, indirectly of the extra-cellular fluid are closely controlled by the kidneys. Kidneys also play an important role in maintaining the acid-base balance.

30.1.1 Changes in the Body Fluids in Renal Disease

The symptoms and signs of profound renal malfunction are known as uremia, meaning urine in the blood. Since most urinary contents are water-soluble, they reach high concentrations in blood, and result in deranged body parts and their physiology.

Severe cases of acute renal failure are characterized by the virtual cessation of urine formation. It is obvious that the kidney cannot exert any appreciable excretory or regulatory function. In such patients, even when protein foods are omitted completely from the diet, the production of urea and other nitrogenous waste products continues because of the metabolic breakdown of the body's own tissue proteins. These products, when retained, result in a progressive rise in their level in the plasma. Furthermore, since acids are formed by the oxidation of the sulphur and phosphorus contained in protein, the plasma bicarbonate will fall. Partly because of the acidosis and partly because of cellular and tissue breakdown, potassium is liberated into the extra-cellular compartment and the plasma potassium level rises. Dangerous changes in the volume and distribution of body water may also occur in acute renal failure.

Chronic renal failure results in changes in the body fluids which occur due to a progressive decrease in the number of functioning nephrons. With the decrease in functional nephrons, the clearance of urea, creatinine and other metabolic waste products will decrease proportionally. In consequence, the plasma concentrations of these substances will rise. A reduction in the glomerular filtration rate (GFR) will also lead to an increase in the plasma concentrations of substances (e.g., phosphate and sulphate) which are removed from the blood by filtration and partially re-absorbed by tubules. However, since tubular absorption of these substances also declines with a falling

GFR, both their clearances and their plasma concentrations usually remain unchanged until the GFR has fallen to about 25% of the normal rate. At this stage, there is also a fall in the plasma bicarbonate.

With the decline in the number of functioning nephrons, the kidney also becomes less effective as a regulatory organ. Accordingly, a patient with chronic renal failure is liable to develop either deficiencies or excess of body fluid water and sodium. However, the volume and sodium concentration of the extra-cellular fluids often remain virtually unchanged until the renal function is very seriously impaired.

Uremia is the clinical state resulting from renal failure. The signs and symptoms of uremia are extremely diverse and frequently appear to point to disease of other organs. Uremia affects every organ of the body as certain substances that accumulate in uremia are clearly toxic. Middle molecules with molecular weights of 1000-3000 are a major contender for the villainous role of the uremic toxin.

⬛⬛▶ 30.2   ARTIFICIAL KIDNEY

Intermittent treatment with a mechanical device like the artificial kidney (dialyzer) will reduce the accumulation of waste products and water and thus the blood concentrations of the toxic substances are returned to normal levels. By effectively removing these materials from the blood, the dialyzer temporarily replaces the function of the natural kidneys and is able to keep the patient close to normal condition. Essentially, an artificial kidney is a dialyzing unit which operates outside the patient's own body. It receives the patient's blood from the cannulated artery via a plastic tubing. The dialysate is an electrolyte solution of suitable composition and the dialysis takes place across a membrane of cellophane. The return of the dialyzed blood is by another plastic tube to an appropriate vein.

The dialyzing membrane has perforations which are extremely small (Fig. 30.2) and are invisible to the naked eye. Waste products in the blood are able to pass through these minute perforations into the dialysate fluid from where they are immediately washed away. The perforations in the dialysis membrane have an average diameter of 50 Å with an estimated range of 30 Å to 90 Å. The waste products pass through the membrane because of the existence of a concentration gradient across the membrane. The dialysate fluid is free of waste product molecules and, therefore, those in the blood would tend to distribute themselves evenly throughout the blood and the dialysate. This movement of waste product molecules from the blood to the dialysate results in cleaning of the blood.

The volume of body fluid cannot be controlled by dialysis. Instead, ultra-filtration across the membrane is employed. For this, a positive pressure is applied to the blood compartment or a negative pressure established in the dialysate compartment. Either way, fluid—both water and electrolytes—will move from the blood compartment to the dialysate, which is subsequently discarded. The degree of ultra-filtration depends both on the pressure difference across the membrane and the ultra-filtration characteristics of the membrane.

The artificial kidney is thus simply a membrane separation device that serves as a mass exchanger during clinical use. It is unable to perform any of the synthetic or metabolic functions of the normal kidney and, therefore, cannot correct abnormalities that result from the loss of these

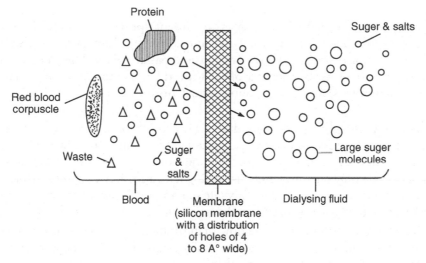

> **Fig.** 30.2 *Principle of dialysis in the artificial kidney*

functions. The only use of the artificial kidney in replacing renal function, therefore, is the transfer of noxious substances from the blood to the dialysate, so that they might be eliminated from the body. The physical processes by which solutes move across the dialysis membrane are discussed by Henderson (1976).

The application of repeated dialysis as a definitive treatment of the permanent loss of kidney function requires a simple method for obtaining repeated access to the patient's circulation over a period of years. This is achieved either by the Scribner Teflon-Silastic arterio-venous shunt or by using a subcutaneous arterio-venous fistula, surgically created by anastomosis of the radial artery to an adjacent branch of the cephalic vein in the forearm. The fistula can then be used for dialysis by insertion of wide-bore needles through the skin into the veins. The fistulas have proved to be essentially non-clotting, rarely become infected and help to keep the patient free from an external device between dialyses.

The maximum number of dialyses are performed for the treatment of chronic renal failure. This is maintenance dialysis, replacing the excretory functions of the kidney for the duration of the patient's life, or until he receives a transplant. Acute renal failure is the next major indication. Dialysis for acute renal failure is necessarily undertaken only for a few days or weeks until the patient's own kidneys recover.

> ## 30.3 DIALYZERS

The dialyzer is the part in the artificial kidney system in which the treatment actually takes place and where the blood is freed from the waste products. It is the meeting point of two circuits, one in which the blood circulates and the other in which dialysis fluid flows.

Dialyzers, in routine clinical use, may be classified according to three basic design considerations: coil, parallel plate and hollow fibre type. Each type of dialyzer has certain optimum operating requirements. The rate of clearance of substances such as urea, creatinine, etc. from the blood during passage through an artificial kidney is dependent upon the rate of the blood flow. As the flow rate falls, there is a disproportionate fall in clearance. At high flow rates, there is little advantage in further augmentation of the blood flow. The rate and pattern of the dialysate flow also influence overall performance in respect of clearance of waste products. Almost all commercial dialyzers use cellulosic type membranes, the most common being Cuprophan (cupro-ammonium regenerated cellulose).

The removal of waste products during dialysis is proportional to the concentration gradient across the membrane. In order to effect the maximum gradient, the concentration of waste products in the dialysate should be maintained at zero. This is achieved in most currently employed machines by using the dialysate only once and then discarding it. In addition, counter-current flow through the artificial kidney is used so that the dialysate enters the kidney at the blood exit-end where blood concentration of waste products is at the lowest level.

It is desirable for the resistance to blood flow in the dialyzer to be as low as possible, eliminating the need to employ a blood pump. In addition, the design of the blood compartment should be such that all the blood can be easily and completely returned to the patient at the end of dialysis. The design must effect an optimum, thin film of blood going through the dialyzer without streaming under perfused areas of membrane surface. Similarly, there must be optimum mixing in the dialysate compartment, effected via the membrane support structure.

30.3.1 Parallel Flow Dialyzers

The parallel flow dialyzer has a low internal resistance which allows adequate blood flow through the dialyzer with the patient's arterial blood pressure, eliminating the need for a blood pump. The dialyzing surface area of a parallel flow dialyzer is about 1 sq m. At a blood flow rate of 200 ml/min and a dialysate flow of 500 ml/min, the urea and creatinine clearance is about 80 and 64 ml/min. The rigid supports used in parallel flow dialyzers permit negative pressure to be created on the dialysate side of the membrane for ultra-filtration. The water is ultra-filtered at a rate of 9.2 ml/min with a negative pressure of 130 mmHg. This rate is 1.8 ml/min without negative pressure. The dialysate flows continuously at 500 ml/min in a direction counter current to the blood, permitting exchange to take place throughout the dialyzer.

The KIIL dialyzer has earlier been the most commonly used form of parallel flow dialyzer. It consists of (Fig. 30.3) three polypropylene boards with dialyzing membranes laid between them. The boards are held firmly with a frame on the top and bottom and are fastened by a series of bolts on the side. A rubber gasket runs along the periphery of the boards inner surface to prevent blood and dialysate leakage. The dialysate enters through a stainless steel port and is distributed to grooves running across the end of the board both above and below the membrane of each layer. After flowing down longitudinal grooves in the boards, it is collected and flowed out at the opposite end of the board. The outside measurements of a KIIL dialyzer are $125 \times 40 \times 16.5$ cm.

The KIIL dialyzer is not disposable. It needs to be cleaned and re-built after each dialysis operation. With this type of dialyzer, a single-pass body temperature dialysate passes through the dialyzer once before going to the drain to obtain higher operational efficiency and to minimize bacterial infection.

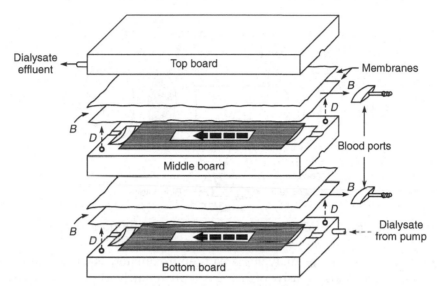

> **Fig. 30.3** *Constructional details of KIIL dialyzer showing plates separated out*

Several modifications have been introduced in the basic KIIL system. The parallel grooves have been replaced by pyramidal grooves which allow multiple point support for the membranes. This arrangement provides greater clearance of urea and creatinine under the same flow conditions because of increased surface area.

30.3.2 Coil Hemodialyzer

A coil hemodialyzer comprises a tubular membrane placed between flexible support wrapped around a rigid cylindrical core. The coil is immersed in a dialyzing bath. The tubular membrane can be of cellophane or cuprophane. The average wall thickness of cellophane membrane is 20–30 μm and that of cuprophane in the range of 18–75 μm. The coil membrane supports are woven screens or unwoven lattice. Usually, the 'twin-coil' is made with three layers of woven, polyvinyl chloride-coated fibre glass screen separated by four narrow strips of the same material, which are sewn into place with cotton thread. Coil dialyzers are available with several design variations, which include the type of membrane, the membrane support, the number of blood channels (1, 2 or 4), the width of the blood channels (38–100 mm) and surface area (0.7–1.9 m^2). Coil dialyzers can be pre-fabricated because of their simple design. They are characterized by high dialysate flow rates and high resistance to blood.

30.3.3 Hollow Fibre Haemodialyzer

The hollow fibre haemodialyzer (Fig. 30.4) is the most commonly used haemodialyzer. It consists of about 10,000 hollow de-acetylated cellulose diacetate capillaries. The capillaries are jacketed in a plastic cylinder 18 cm in length and 7 cm in diameter. The capillaries are sealed on each end into a tube sheet with an elastomer. The capillaries range from 200–300 mm internal diameter and a

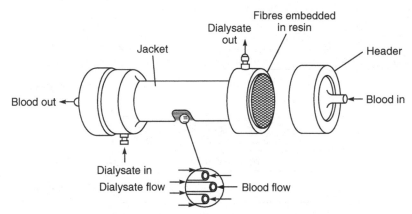

> **Fig.** 30.4   *Constructional details of hollow-fibre dialyser*

wall thickness of 25–30 μm. The dialyzing area is approximately 9000 cm^2/unit. The primary volume with blood manifolds exclusive of tubing is approximately 130 ml. The blood is introduced and removed from the hemodialyzer through manifold headers. The dialysate is drawn through the jacket under a negative pressure around the outside of the capillaries counter-current to the blood flow. The dialyzers are disposable.

Disposable dialyzers offer the advantages of a reduction in infection risk and reduced operator set-up time. The dialyzer sterilization procedure also gets eliminated. However, the use of disposable dialyzers is an expensive procedure. This has necessitated the development of 'methods of cleaning dialyzer cartridges so that they may be re-used' (Deane et al., 1978; Wing et.al.,1978). However, there are several difficulties associated with the practice of dialyzer re-use. It is a time-consuming and unpleasant procedure and requires additional technical skills.

30.3.4   Performance Analysis of Dialyzers

The dialyzer performance can be compared in terms of their clearance of urea and creating priming volume, residual blood volume, ultra-filtration rate, convenience of handling and cost, etc.

Clearance: The overall performance of the dialyzer is expressed as the clearance, analogous to that of a natural kidney. It represents that part of the total blood flow rate through the dialyzer which is completely cleared of solute. Uremic patients carry a number of toxic solutes in their blood, which are generated daily. Despite the uncertainty as to which solutes and how much should be removed, the performance of a dialyzer is generally assessed for a spectrum of molecular weight solutes. The molecular weight of urea is 60, of creatinine, 113, Vitamin B_{12}, 1355 and of insulin, 5200.

The clearance of urea and creatinine is measured at clinically useful blood flow rates and standard dialysis fluid addition or flow rate. It is calculated as

$$\text{Clearance} = \frac{\text{blood flow rate}}{A + \text{Bx blood flow rate}}$$

(where A and B are constants) with 95% confidence limits of the mean by least square approximation. The blood flow rate is measured by bubble transit time over a two-meter track using the mean of three measurements. Urea and creatinine concentrations are measured in the plasma from 1 ml sample of heparinised blood. Usually, the blood flow is maintained between 75–300 ml/min and dialysis fluid flow rate at 500 ml/min.

The performance of hemodialyzers is usually compared by employing the dialysance curve which is a graph of dialysance versus the blood flow rate.

$$\text{Dialysance } D = \frac{Q_b \, (C_{bi} - C_{bo})}{C_{bi} - C_{di}} \qquad (1)$$

where Q_b = blood flow rate

C_{bi} = blood solute concentration at the dialyzer inlet

C_{bo} = blood solute concentration at the dialyzer outlet

C_{di} = dialysate solute concentration at the inlet.

Figure 30.5 and equation (1) show that dialysance rises rapidly at low blood flow rates and tends to stabilize at high flow rates. Therefore, for comparing the performance of the dialyzer, it is important to specify the blood flow at which the measurements are made. The equation also shows that more waste is likely to be removed early in the treatment when the blood dialysate gradient is high.

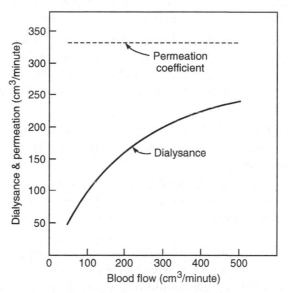

> **Fig. 30.5** *Dialysance vs blood flow*

The dialysance is also used to calculate the clearance, by the following relationship:

$$\text{Clearance} = \frac{\text{Dialysance}}{1 + \dfrac{\text{Dialysance}}{\text{dialysis fluid addition rate}}}$$

It may be noted that the clearance may vary with time despite quasi-steady state conditions. If the dialysate is re-circulated, its solute concentration increases, which effectively reduces the concentration driving force. For a given blood flow rate, the clearance is greater for the smaller molecular sized constituents. This is due to less membrane resistances and higher liquid diffusion coefficients for smaller-molecular weight solutes. On the other hand, the contribution of the membrane resistance to the overall value becomes greater with the increase in solute molecular weight. Bobb et al. (1971) suggested 'square meter hour' concept for increased removal of middle molecules by either increasing dialysis time or increasing membrane area. They illustrated that inadequate removal of solutes of 300–2000 daltons (middle molecules) might be associated with the neurological dysfunction in chronic uraemic patients. Keeping in view this study, attempts have been made to design dialyzers with large surface area (2.5 m^2) and development of more permeable membranes. Another method of comparing the performance capacity of dialyzers is given by:

$$\text{Performance capacity} = K_s A = \frac{A}{R_s} \tag{2}$$

where　A = surface area

K_s = permeability coefficient

R_s = mass transfer resistance, i.e., reciprocal of permeability.

Mass transfer resistance is composed of resistance due to a blood film layer, resistance of the membrane itself and the dialysate film resistance layer. Equation (2) is independent of blood flow rate and can be considered as a measure of dialyzer performance.

Ultra-filtration Rate: The fluid removal during dialysis (ultra-filtration) takes place due to hydrostatic and osmotic transmembrane pressure gradients. The rate of fluid removal due to hydrostatic pressure effects depends upon the specifications of the dialyzer in terms of mass-transfer coefficient and surface area. It, however, has a linear function of the transmembrane pressure (TMP) gradient. It is given by

$$\text{Mean transmembrane pressure} = \tfrac{1}{2}\,[P_{BI} + P_{BO}] - 1/2\,[P_{DI} + P_{DO}]$$

where PB_i and PB_O are the blood inlet and outlet pressures and PD_I and PD_O are the dialysate inlet and outlet pressures respectively.

The pressure losses generated by blood and dialysate flows in their respective flow paths should be small. This ensures that the local transmembrane pressure (ΔPm) will not vary excessively from the mean pressure. High values of ΔPm can result in deformation of the membrane and its possible rupture.

In actual clinical practice, only one blood pressure and one dialysate pressure are normally measured. Therefore, in order to obtain reasonable control over ultra-filtration, the pressure loss in the blood and dialysate compartments should be known as a function of the respective flow rates. The pressure drop (difference between blood inlet and outlet pressures, ΔPb) across a dialyzer is directly proportional to the length of the passage and the viscosity of the fluid and inversely proportional to the number of blood passage and some function of their cross-sectional area. Blood passage width may change with changing blood compartment pressures. Therefore, the relationship of pressure drop to blood flow is not linear at increased flows which are accompanied by increased pressures that can cause a widening of the blood passage and a decrease in $\Delta P_b/Q_b$. The viscosity of the blood is not constant as the blood is an anomalous fluid. Its viscosity tends to increase with haematocrit and decrease in small passage. The area of cross-section is important as with small areas, an inordinately high pressure is required to yield a given flow or flow is reduced to very low levels for a given pressure source.

Residual Blood Volume: Residual blood volume can be measured after an 800 ml saline wash in. The fluid remaining in the dialyzer and lines is circulated through a 0.1 litre bottle of 0.04% ammonia solution for 10 mm. The residual blood volume is calculated from the formula.

$$\text{Residual blood volume} = \frac{U(1000 + \text{volume of dialyzer and lines in ml})}{200S}$$

where U = the haemoglobin concentration of the re-circulated fluid,

  S = the haemoglobin concentration of a sample of arterial blood taken at the end of dialysis and diluted 1: 200 with 0.04% ammonia.

Residual volumes of 1.8 to 6.3 ml are quoted in the literature depending upon the dialyzer type and washback volume (Hartitzsch et al. 1973).

Priming Volume: The volume of blood within the dialyzer is known as priming volume. It is desirable that this should be minimal. Priming volume of present day dialyzers range from 75 to 200 ml, depending on the membrane area geometry and operating conditions. Requirement of low priming volume permits the use of patients own blood to prime the circuit without serious hypo-volemic effects. This is particularly significant in the case of long-term dialysis therapy.

Extra-corporeal blood volumes become important in those dialyzers which require priming. Priming is usually accompanied at relatively low pressures. Recent innovations have considerably reduced the extra-corporeal volume and a saline prime is frequently used.

Pyrogenicity: Pyrogen reactions are rare with all disposable dialyzers. However, they are known to exist with KIIL dialyzers but at rates well lower than 1%.

Leakage Rate: Blood-to-dialysis-fluid leak with the KIIL dialyzer is found to be 3% (Rastogi et.al. 1969), but it varies with the dialyzer, the batch of membrane and the skill of the operator. The leak rate from all cuprophan coils is, however, high (Easterling et al. 1969).

▶ 30.4 MEMBRANES FOR HAEMODIALYSIS

The efficiency of dialysis is determined by the permeability characteristics of the semi-permeable membrane. The ideal membrane should possess high permeability to water, organic metabolites and ions, and the capability of retaining plasma proteins. The membrane should be of sufficient wet strength to resist tearing or bursting and non-toxic to blood and all body cells. Since a fresh membrane is required for each dialysis process, it should be inexpensive to produce.

Virtually all artificial kidneys presently in use, employ cellulosic membranes. Such membranes operate as sieve-type membranes allowing the passage of solutes through micro-holes. Therefore, selective sieving of the blood is based upon the size, shape and density of the solute.

Cupraphan (trademark of Enka Glanzstoff, Germany) is the commonly used membrane for haemodialysis. It is a membrane consisting of natural cellulose and is considered puncture-proof, and of high tenacity and elasticity. During haemodialysis, different substances of varying molecular weight are to be removed. The specific membrane permeability values and their dependency on substances with increasing molecular weight are shown in Fig. 30.6. The good permeability on the

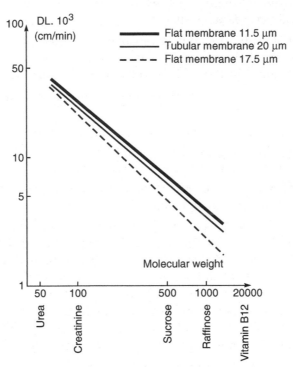

> ➤ **Fig. 30.6** *Relationship between permeability (DL) and molecular weight for different thicknesses of the membrane*

substances with middle molecular weight is obvious and substances with molecular weights of over 10,000 are retained.

Cupraphan is a moisture-sensitive cellulose hydrate membrane whose reaction during processing and whose functional value depends upon the water content. If it varies from the fixed standard values, consequently, it exhibits dimensional instability pertaining to a change in dimension and swelling as well as in the handling property during assembly of the dialyzer. Wetting with water results in a three-dimensional change in the length of the cupraphan membranes. The increase in thickness for all types amounts to a factor of 1.9 during the transition from the normally conditioned to the wet state. Glycerine is added to the membrane as a humectant and plasticizer for smooth processing. The water content of the membrane balances with the humidity level of the surroundings. It should be noted that during all phases of processing, room conditions should be around 35% relative humidity at 23°C which correspond with the equilibrium humidity of the membranes.

For applications in hollow fibre dialyzers, cupraphan fibres are used. With their small internal diameter, it is possible to design dialyzers with a low priming volume, making it possible to combine a large surface area with a low priming volume. The number of fibres in a dialyzer can be as high as 16,000 giving a density of fibres as 1000 per cm^2. Cupraphan hollow fibres are particularly suitable for the dialyses of solutes in the middle molecular weight range (500–2000).

The high solute permeability in this range is not associated with an extremely high water permeability.

30.5 HAEMODIALYSIS MACHINE

A haemodialysis machine is used for the production of warm dialysate which is then circulated through an external dialyzer assembly. It also controls the cycling of the blood from the patient through the artificial kidney (dialyzer) and back to the patient. It continuously monitors and controls all important parameters, automatically halting treatment in the event of parameters going out of pre-set limits. The haemodialysis machine performs five basic functions. It (i) mixes the dialysate, (ii) monitors the dialysate, (iii) pumps the blood and controls administration of anti-coagulants, (iv) monitors the blood for the presence of air and drip chamber pressure, and (v) monitors the ultra-filtration rate.

The machines are designed to be totally adjustable to meet individual therapy requirements. The machine pumps and controls the flow of blood from the patient through the dialyzer at a pre-determined rate and pressure to ensure effective clearances and fluid removal in a specified time period. Some machines also provide an ultra-filtration rate meter that measures the ultra-filtration rate in kilograms per hour. This allows the operator to efficiently and accurately calculate, predict and control fluid removal during dialyses. Figure 30.7 shows a block diagram of a haemodialysis machine.

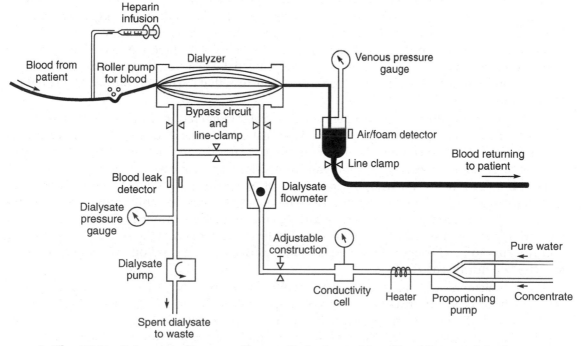

➤ **Fig. 30.7** *Schematic diagram of haemodialysis machine. Blood flow path is shown in solid black line and the dialysate circuit as open lines*

Proportioning Pumps: Mixing large amounts of dialysate from dry chemicals is time consuming and laborious. When re-circulated, glucose-containing dialysate results in the rapid growth of bacteria unless changed at frequent intervals. Single-pass proportioning systems using liquid concentrate avoid bacterial overgrowth. Proportioning systems of earlier designs used motor-driven positive displacement pumps. Water and concentrate pumps were driven by one shaft from the same motor, simultaneously delivering a fixed ratio (35:1) of water and concentrate into a mixing chamber.

Incoming water under controlled pressure can drive proportioning pumps, thus eliminating the need for a motor and permitting a smaller, quieter system. Incoming water activates a piston with an attached shaft, causing expulsion of a fixed proportion (35:1) of water (piston) and concentrate (shaft) from different chambers. A steady flow of dialysate is achieved by adding an additional concentrate chamber to the other side of the water chamber so that the back and forth movement of the piston and shaft expels water and dialysate continuously.

The proportioning system could be of the fixed ratio or variable ratio type. In a fixed ratio type, the ratio of concentrate and water is determined by the ratio of two pump piston diameters and these are mechanically fixed in their relationship. Proportioning is, therefore, fixed also, and is generally 34:1. In the variable ratio type system, a variation of ± 5% on the standard 34:1 dilution ratio is possible. .

Dialysate Temperature Control and Measurement: The dialysis is normally done at the body temperature. The temperature of the dialysate is, therefore, monitored and controlled before it is supplied to the dialyzer. In case the dialysate gets over-heated, the system should stop the flow to the dialyzer and pass it to the bypass. Dialysis at temperatures lower than the body temperature is less efficient and requires re-warming of the blood before its return to the patient's body. Temperatures in excess of 40°C tend to damage components of the blood. A temperature control system is used to raise the temperature of the dialysate to the required temperature which can be varied from 36 to 42°C. A secondary safety cut-out ensures that the heaters are switched off if the temperature exceeds 43 °C.

Two types of circuits can be used for effecting control of temperature: (i) A bi-metallic thermostat which would connect or disconnect supply to the heater coil depending upon the temperature of the dialysate, and (ii) A completely electronic single-term proportional controller which makes use of a thermistor for sensing the temperature and a triac for control of power to the heater. A typical circuit for such a system is shown in Fig. 30.8. Normally, the uni-junction transistor is 'off' until the capacitor C charges to a point of breakdown voltage. When this occurs, the transistor conducts and the capacitor is discharged through the pulse transformer T. The triac thus gets a triggering pulse and switches on the heaters. The triac switches off at the end of each half-cycle and remains so until triggered once again. Since a triac conducts in both directions, it can be switched on during each half-cycle.

The thermistor has a negative temperature coefficient. With an increase in temperature from the set value, its resistance decreases, thereby reducing the rate of charge of C. Therefore, the frequency of charge and discharge (oscillations) reduces and less power is delivered to the heaters which results in a reduction in temperature. With this method, it is possible to control the temperature with an accuracy of 0.2°C. The temperature can also be controlled by varying resistance R_2 and therefore, any temperature can be set with the help of this control.

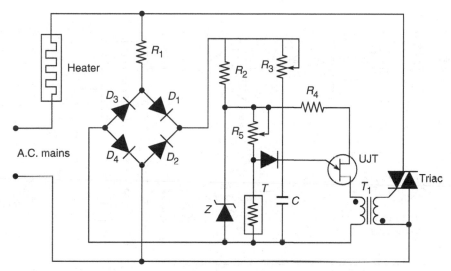

> **Fig. 30.8** *Simplified circuit diagram for controlling dialysate temperature*

A thermistor connected in one arm of a Wheatstone bridge may be used as a sensor of temperature of the dialysate in the header tank. The output of the bridge can be amplified in a differential amplifier and displayed on' a panel meter. The amplified signal would also operate alarm circuits in case the temperature of dialysate crosses the preset limits.

Some machines have facilities for automatic sterilization. Sterilization is carried out by passing water at 85° to 90°C, through the total hydraulic system. In these machines, sequence interlocks ensure that a dialysis cannot be started without sterilization and that minimum requirements, for adequate sterilization are achieved before the dialysis phase can be selected.

In the modern microprocessor-based haemodialysis machines, the temperature monitor and control circuitry generate a signal that the CPU utilizes to generate display of the fluid temperature and to control the heaters. A dual element heater assembly, which has a 150 W and a 300 W element, is used to bring the fluid up to and to maintain the operating temperature. When the fluid temperature rises to within 2.5°C of the preset temperature (between 35°C to 39°C), the 300 W heater is turned off and the 150 W heater is used to maintain the set temperature (Fig. 30.9).

As a secondary fault monitor, the system incorporates a mercury type temperature switch which is normally open when the dialysate temperature is below 40°C (± 0.5°C) and closed when the dialysate temperature exceeds 40°C ((± 0.5°C). In case the fluid temperature exceeds 40°C, the microprocessor removes the heater elements. In addition to the thermistor probe and temperature switch, the enabling of the heaters is dependent upon the flow rate. The microprocessor reads the flow pulses and determines if there is adequate flow within the system. If the flow is inadequate, the heater elements are disconnected.

The flow is measured using a flow-thru transducer which produces a precise number of pulses per unit of flow (26,000/liter or 108 pulses/second at 250 ml/min). This is achieved by monitoring the rotation of a disk which contains light reflective white spots. Light pulses from the rotating disk are transmitted by internal fibre optics. The sensor assembly includes a light source and a

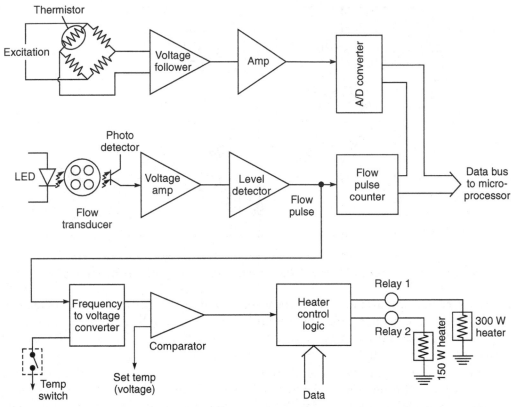

> **Fig. 30.9** *Temperature monitor and control circuit using multiple heaters*

photo-transistor to provide the optical coupling with the sensor. The pulses generated by the flow transducer are amplified, filtered and counted to determine the flow rate. These pulses are also used to control the flow rate in the hydraulic circuit. This circuit supplies a computer-generated variable drive signal to the dialysate pump and flow rate feedback signal to the CPU.

Conductivity Measurement: The conductivity of the dialysate being produced is continuously monitored by a conducting cell, to verify the accuracy of proportioning. The result is usually displayed as a percentage deviation from the standard. In practice, a fluctuation about the mean reading will occur and conductivity will normally be maintained with 1 %. If an alarm occurs due to the conductivity not remaining within limits, an alarm is given. The effluent pump motor will be switched off automatically which effectively prevents further circulation of dialysate through the dialyzer, and dialysate production will be by-passed to the drain.

The composition of the dialysate is checked by comparing the electrical conductivity of the dialysate with a standard sample of the dialysate. Proper temperature compensation is essential as the conductivity of the dialysate changes by about 2% for every 1°C change in temperature. Figure 30.10 shows a block diagram of the conductivity measuring system. It comprises a 1.5 kHz oscillator which drives a bridge circuit, one arm of which contains a conductivity cell. The conductivity cell is of the flow type and is mounted directly downstream of the header tank. In order to provide a fast response to changes of solution temperature which would otherwise considerably affect the conductivity measurements, a temperature compensation thermistor is placed in another arm of the bridge. The output from the bridge, *after* amplification, is capacitively coupled to a phase-sensitive detector where its phase is compared with the phase of the 1.5 kHz oscillator output. The magnitude and phase of the output from the phase-sensitive detector determine the direction and amount of deviation from the pre-set value.

An alternative arrangement for conductivity measurement is to make use of a conductivity flow cell consisting of two probes as shown in Fig. 30.11. Each probe consists of four ring–electrodes. Two of the electrodes provide a fixed

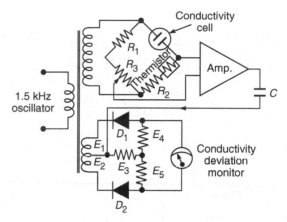

> **Fig. 30.10** *Simplified circuit diagram for monitoring conductivity of dialysate*

current through the solution, while the other two electrodes sense the differential voltage generated due to the solution conductance. The current source consists of two parts : a constant amplitude oscillator (10 kHz) and a voltage-to-current converter. The constant amplitude oscillator (A) on probe 1, and oscillator on probe 2, have a positive feedback to ensure oscillation and a negative feedback to set the frequency and the output amplitude. The voltage-to-current converters provide an output of approximately 1.0 mA. The output of the current source is transformer-coupled to the two current electrodes to minimize leakage current through the solution.

The two sensing electrodes are AC-coupled to a differential amplifier, U-1 on probe-1 and U-2 on probe-2. The output is then AC-coupled into an AC to DC converter whose output is given to the A/D converter, after making offset and gain adjustments. The output of the A/D converter is given to the system microprocessor.

Table 30.1 gives typical dialysate composition.

In practice, the mEq/l of sodium, calcium, chloride, potassium, magnesium and acetate are added to obtain the total ionic content of the dialysate in mEq/l. For example: for total ionic content of 270 mEq/l, the conductivity is 12.9 milli-ohms and for 304 mEq/l, it is 13.8 milli-ohms.

Dialysis must never commence unless it is known that the conductivity circuit calibration and concentrate in use are both correct for the intended dialysis. Therefore, it is recommended that once per month a sample of the dialysate from the machine's dialysate outlet connector be analyzed in a laboratory to check conductivity monitor calibration.

Dialysate Pressure Control and Measurement: Negative pressure upon the dialysate is created by the effluent pump. The effluent pump is a fixed flow, motor-driven gear pump. A small plastic housing encloses stainless steel gears driven by an electric motor. Pressures between zero and maximum are available by adjustment of a needle valve mounted on the machine panel. A relief valve (pre-set to suit the type of dialyzer) limits the maximum negative pressure available, thus minimizing the risk of a burst in the dialyzer membrane which may be caused by high transient

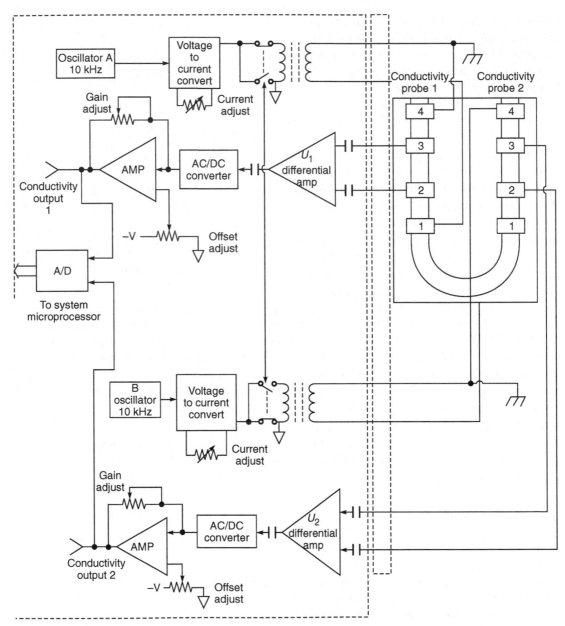

> **Fig. 30.11** *Conductivity monitoring using two probes*

pressures. Pressure adjustments should not produce any significant change in the flow rate. The pressure is measured by a strain gauge transducer connected immediately downstream of the dialysate return side. Pressures within the range 0 to –400 mmHg are generally made available and any value can be adjusted in this range.

• Table 30.1   *Composition of Dialysate*

Constituent	Concentration (mEq/l) (g/l)
Sodium (Na$^+$)	130.00
Potassium (K$^+$)	1.34
Calcium (Ca^{++})	3.30
Magnesium (Mg^{++})	1.00
Chloride (Cl$^-$)	101.00
Acetate	35.00
Lactate	1.30
Glucose	1.80

The pressure indicated on the gauge will be the dialysate pressure on one side of the dialyzer membrane. On the other side of the membrane will be the venous pressure. The effective pressure across the membrane, which is so important in consideration of filtration and weight control, will be the algebraic sum of the dialysate pressure and venous pressure. If the pressure goes beyond the alarm limits, the effluent pump is switched off automatically and dialysate production by-passed to drain by way of the header tank overflow and the waste funnel.

Venous Pressure Measurement: Venous pressure is normally measured at the bubble trap. A length of tubing connects the trap to a small plastic housing to which a strain gauge transducer is attached. The sensor diaphragm is fragile and should not be roughly handled. The elevation above the floor of the point of connection to the blood line produces a small change in the pressure reading. For maximum accuracy, the sensor connection should be maintained in the same altitude during dialysis, preferably with the luer connector downwards to prevent blood reaching the diaphragm in the event of a leak. If the venous pressure passes beyond one of the alarm limits, power to the blood pump will be isolated and the blood pump, if in use, will cease to operate.

Blood Leak Detector: In a dialysis machine, a thin membrane separates the patient's blood from the dialysate. Normally, the pressure on the blood side of the membrane is maintained at a much higher level than the pressure on the dialysate side. This is necessary to minimize the time required for the dialysis procedure. Besides this, in order to reduce the total time required for dialysis, the membrane area is made as large as possible. Therefore, these two conditions of a high pressure differential across a large fragile membrane may result in a leak in the membrane. In fact, even a relatively small tear in the membrane can result in major blood loss in a very short time, with the consequent immediate threat to the patient.

Blood-to-dialysate leaks usually occur at the beginning of dialysis and can be detected by examination of the effluent from the dialyzer. However, it is not rare to have instances wherein blood leaks have been found to start several hours after setting up dialysis. The detection of blood leaking through imperfection in the membrane into the dialysate is best achieved by monitoring the effluent from the dialyzer for changes in transmission of light resulting from the presence of haemoglobin. If there is any blood leak across the dialyzer membrane, it can be detected by using a photo-electric transducer.

The dialysis membrane leak detector (Rhodine and steadman, 1976) basically examines the light absorption of the dialysate at 560 nm, the absorption wavelength for haemoglobin. This spectral tuning of the system makes it sensitive and stable, and reduces false alarm situations. An LED is available which has a peak spectral emission at 560 nm with a spectral line half width of 27 nm.

Figure 30.12 shows a block diagram of the blood leak detector. In order to minimize drift over a period of several hours required for the dialysis, a chopped light system with AC amplifiers is employed. Chopping is achieved by driving the LED with a square wave of current. The light is detected with a cadmium sulphide (CdS) photo-conductive cell. This has peak response at 565 nm. After amplification of the AC response signal, an absolute value circuit provides a signal whose peak value is proportional to the received 560 nm light. The peak value is compared to a reference voltage which is pre-set. When the peak value falls below the selected threshold, visual and audible alarms are activated.

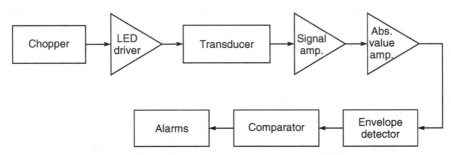

➤ **Fig. 30.12** *Block diagram of blood leak detector using LED as a light source (after Rohdine and Steadman, 1976)*

Blood leak detectors are liable to give false alarms when used over a long period of several weeks. This is the result of a gradual build-up of contaminants on the lenses of the LED and CdS cell. This needs gradual change in the setting of the threshold. Careful cleaning of the transducer can, however, restore the original threshold. This does not materially affect system performance since in almost all machines, the threshold is set at the beginning of each dialysis procedure.

Blood leak level, for normal operation, is set at 25 mg of haemoglobin per litre of dialysate. The maximum setting detects blood leaks at the rate of 65 mg/I of dialysate. If a blood leak is detected, the effluent pump is switched off automatically and dialysate production by-passed to drain by way of header tank overflow. The blood pump is de-energized and, if in use, ceases to operate.

Ultra-filtrate Monitor: The ultra-filtrate monitor circuit is used to monitor the amount of fluid removed from the patient and in conjunction with the negative pressure, to control the rate at which it is removed. This circuit generates a signal that the CPU utilizes to generate display of the total UF(ultra-filtrate). The CPU also uses this signal to calculate the TMP required to maintain the UF rate required by the operator.

$$\text{UF Rate (L/hr)} = \frac{\text{Total fluid removal required (litres)}}{\text{Treatment time (hours)}}$$

This is done by measuring the amount of fluid removed during a measured time period by the amount of fluid removed. This calculation helps determine the coefficient (K) of the dialyzer.

Dialyzer K = Total UF (6 min period)/TMP (Avg.)

The CPU then divides the required UF Rate by the dialyzer K factor to determine how much TMP is needed to achieve this UF rate.

Required TMP = Required UF Rate (L/hr)/dialyzer K

The CPU will then subtract the measured blood pressure from the TMP calculated to determine how much negative pressure is required to achieve the calculated TMP.

Blood Pressure = Venous Blood Pressure + Arterial Blood Pressure/2

Negative Pressure Required = TMP – Blood Pressure

If the calculated negative pressure is less than 30 mmHg or greater than 350 mmHg, the CPU will generate an alarm signal.

Figure 30.13 is a block diagram of the ultra-filtrate monitor. The load cell and associated electronics are used to monitor the weight changes of the fluid in the reservoir during the haemodialysis treatment. The load cell utilizes a strain gauge that produces a differential resistance proportional to the applied force. An excitation of 10V DC is supplied to the strain gauge bridge from a reference source. The differential output from the strain gauge bridge is typically 13.3 mV for a 10 kg load. The differential input is first connected to the instrumentation amplifier which gives a gain of 100

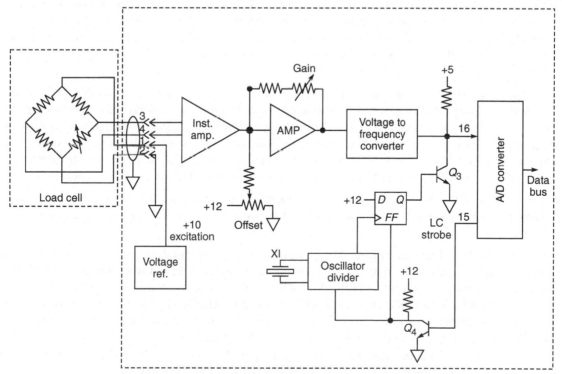

> **Fig. 30.13** *Block diagram of ultrafiltration monitor*

and produces a single-ended output. The following amplifier stage includes provisions for offset and gain adjustments. The weight is represented at this stage by a DC voltage. It is changed to a proportional frequency by a voltage-to-frequency converter. The pulses corresponding to the weight are then counted and given to the microprocessor.

Flow Meter: The flow is measured with a flow meter which comprises a stainless steel float, inside a glass tube held by plastic connectors. All fluid returning to the machine from the dialyzer passes through the flow meter. The dialysate flow rate is fixed at a nominal 500 ml/min. Since it is generally positioned downstream of the dialyzer, it is possible to observe a large blood-to-dialysate leak by the discoloration of the fluid in the flow meter tube.

Effluent Pump: Effluent pumps are available in several design configurations. The more common types are: the diaphragm type, gear type and magnetically coupled. The diaphragm type pumps are not preferred because they give problems due to diaphragm fatigue when operated over long periods. Therefore, in the modern machines, either the gear type or magnetically coupled pumps are preferred.

Blood Pump: The blood pump used in dialysis machines is usually of the peristaltic type. It is designed to give blood flow at a rate of 50 to 350 ml/min.

Bubble Trap: Air embolism is a serious hazard in dialysis. Air may be sucked in due to inadequate flow in the line in the pumped dialysis system. Alternatively, air may be transferred from the dialysate. It is for this reason that dialysis equipment includes provision for adequate de-aeration of the dialysate. The venous return flow circuit usually incorporates a bubble trap to diminish the chances of air embolism. The level of blood in the venous return bubble trap may be monitored by a photo-electric cell. Some manufacturers use the ultrasound method for detecting the presence of air in the blood line.

Heparin Pump: These pumps are usually of the plastic syringe type, having a capacity of 30 cc. The delivery of heparin from the pump is calibrated in cm/h. The pump is driven by a stepper motor and a drive screw mechanism. This drives the plunger of the syringe into its barrel which produces the pumping action. The stepper motor speed is determined by the computer based on the heparin flow rate required. The machines are generally provided with the facility to accommodate commonly used syringe sizes. The speed of the stepper motor is monitored using an optical encoder.

Blood Pressure Monitor: The blood pressure monitor circuit is used to monitor the arterial and venous blood pressures. Two separate pressure transducers of the strain gauge type are used for this purpose.

Computer System: The heart of the computer system is a microprocessor. The microprocessor operates with a clock frequency of 1 MHz or higher which is derived from a crystal controlled oscillator. The program is stored in EPROM with a total capacity of around 24 K bytes. The system includes a RAM circuit (2 K × 8 bits memory) powered by a built-in battery. A watch-dog circuit monitors the performance of the microprocessor by checking the presence of a strobe signal. The CPU strobe signal is activated only when the program is running.

Monitoring and control equipment forms an essential part of the haemodialysis system as it helps in maintaining the most optimum conditions during dialysis. In other words, it ensures a safe clinical procedure against any potential hazards. Incorrect composition and temperature of the dialysate, blood loss due to disconnection or membrane leakage and the formation of air embolism in the blood are some of the principal hazards which, when developed, require immediate remedial steps for the safety of the patient.

30.5.1 Multi-patient Dialysate Delivery System

The dialysate fluid can either be prepared prior to dialysis and stored in large tanks, or continuously prepared in a proportioning system. The latter can be done either in individual machines or prepared centrally and supplied to bedside units. The former types are generally preferred as single patient machines can be used even for home dialysis. In multi-patient dialysis systems, the dialysate is prepared by mixing a small quantity of the concentrated dialysate and water automatically and monitored for electrolyte concentration. It is also heated to reach the appropriate temperature. The bedside units contain dialysate pressure and blood leak monitors. The prepared dialysate flows through the plumbing system fed from the central dialysate delivery system.

▐▐▐▶ 30.6 PORTABLE KIDNEY MACHINES

Kidney machines are manufactured by several manufacturers. Over the years there have been efforts to develop machines for unsupervised home dialysis. At the same time, portability of the machine has been another consideration in the design of machines. The bulk of the machines is due to the dialysate preparation system and the electronic monitoring and control unit. The dialyzers have been reduced in size to a considerable extent. Kolff et al. (1976) report the development of a wearable artificial kidney (WAK) which uses a 20 l dialysate tank and 250 g of activated charcoal for a dialysate regenerating system. This has been possible with the development of battery-operated small-size pumps. In the WAK system, a single pump is used to pump both blood and dialysate. This is accomplished by using silastic tubing ventricles with unidirectional valves, giving a pulsatile flow. The pump operates on re-chargeable 12 V nickel cadmium battery. The dialyzer used with the machine is hollow fibre dialyzer (1.4 m^2 dialysis surface). The venous line bubble-catcher is attached to the outside of the dialyzer. In Fig. 30.14 the blood circuit is on the left and the dialyzing fluid circuit to the right. The dialysate is drawn from the dialyzer into the pump canister, past an accumulator which is a reservoir to expand and contract with the pulsation of the pump, then to an ultra-filtration valve which creates a negative pressure on the onflow line to the dialyzer and finally to the inflow of the dialyzer. The accumulator and ultra-filtrate reservoir do not function when the tank is in the circuit but are essential for use without the tank. Blood flow through the machine is simple, coming from the patient into the blood pump ventricle, to the inflow of the dialyzer, out of the dialyzer to the bubble-catcher and returning to the patient.

Current dialysis equipment is bulky and dialysis patients find it inconvenient. The development of really portable WAK can be expected some time in the future. A portable haemodialysis system (Portalysis 101) developed at the Lodge Moor Hospital, Sheffield, UK is shown in Fig. 30.15. The haemodialysis is carried out for 6 hours requiring two changes of dialyzing fluid and employs hollow fibre dialyzer. The total machine weighs 16 kg and measures $570 \times 412 \times 228$ mm. The system is built in a suitcase for convenience of transportation.

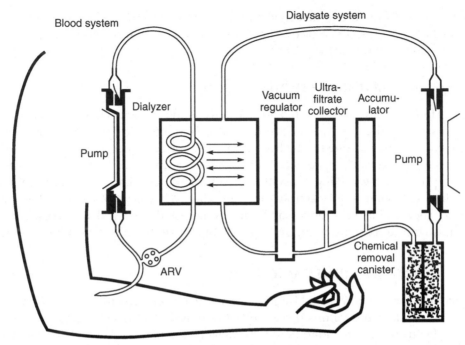

> **Fig. 30.14**   *Wearable artificial kidney flow diagram (after Kolff et al, 1976)*

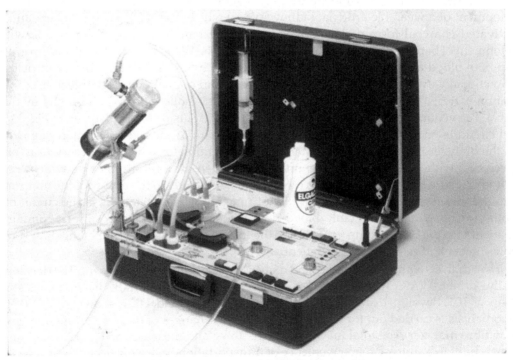

> **Fig. 30.15**   *Portable dialysis machine*

The principle of operation of this machine is very simple. The dialyzing fluid is prepared in a collapsible 20 litre container by connecting a fresh water supply to the inlet of the preparation circuit. The water is purified by passing it through a disposable denomination cartridge and checked for quality before being heated in a temperature-controlled chamber. The 41°C water passes out of the preparation circuit and into the mixture container via a sachet of dialyzing fluid concentrate. The machine senses by weight when the container has the required quantity of mixture and shuts down automatically.

Lithotriptors

31.1 THE STONE DISEASE PROBLEM

Urinary stone disease is one of the oldest diseases known to man. It is caused when the urine, for various reasons, becomes super-saturated with particular salts which may then crystallize out of solution forming a stone-like substance. Of the four main types of stones, the most common (about 70%) are salts of calcium, comprising either calcium oxalate or calcium phosphate or combinations of both salts. A further 20% are known as matrix stones and consist of magnesium ammonium phosphate, usually associated with infection and calcium phosphate and about 9% consist of uric acid and 1% of cystine (Coleman et al, 1987).

The solubility of stones depends very much on the type of stone. Calcium oxalate stones are impossible to dissolve in situ. Similarly, magnesium ammonium phosphate stones are difficult to dissolve and frequently recur unless all fragments are completely removed. Attempts to dissolve the stone by direct irrigation via a percutaneous catheter involve a significant risk of infection and may lead to renal failure.

Stone diseases of the urinary and biliary tract are rather common. They can cause very intense pain and may ultimately lead to renal failure through infection of the urinary tract. Many forms of treatment have been tried with surgery and interventional techniques, greatly increasing the success of treatment and reducing the risks in the process. Surgery is resorted to when stones are unlikely to pass out and when infection and pain cannot be controlled. The form of surgical intervention choosen would depend upon the type, size and site of the stone. The search for less invasive and more efficient methods has concentrated on the use of mechanical and acoustic energy for the destruction of concrement and has resulted in the establishment of various forms of lithotripsy (Greek for "stone grinding").

In early experiments for the contactless destruction of gallstones and stones of the urinary tract, continuous wave ultrasound was used. With intensities of up to $18\,\text{W}/\text{cm}^2$ for up to 30 minutes at frequencies from 100 kHz to 1.5 MHz, effective destruction of stones was achieved. However, it resulted in serious damage when applied to living tissues. Therefore, these experiments were discontinued due to the risks involved in the procedure (Reichenberger, 1988). A non-invasive approach to disintegrate renal and ureteric calculi that finally succeeded was the use of focused acoustic shock-waves originating outside the patient. The first publication describing this idea is a report by Hausler and Kiefer (1970) and a patent application from Dornier Medical Systems GmbH, Germany. It took almost ten years of in-vitro and animal investigations, mainly in co-operation with the Dornier System GmbH of Germany and the Klinikum Großhadern at Munich, before the first patient treatment could be performed. The methods used and results obtained are summarized by Chaussy et al, (1980).

Within only a few years, the method of extra-corporeal shock-wave lithotripsy gained worldwide acceptance and is now a preferred method for the treatment of kidney and ureter stones. Extra-corporeal shock-wave lithotripsy (ESWL) is a non-invasive treatment intended to disintegrate calculi of the urinary tract by means of focused sound waves, thus making it possible to eliminate them by natural routes. From the introduction of the original Dornier lithotripter until today, millions of patients have been treated the world over with this technique.

Ⅲ▶ 31.2   FIRST LITHOTRIPTOR MACHINE

The first machine from Dornier medical systems, pioneers in the field of lithotripsy consists of a large tub of warm water with the underwater electrode (spark plug) in the ellipsoidal reflector at the base of the tub. The water provides an acoustic coupling to the patient so that acoustic waves generated in the water penetrate the tissue and are not reflected from the skin. Since a single acoustic wave generated by the lithotriptor is of high amplitude compared with those used in normal medical diagnostic ultrasound, it propagates in a characteristic way. The shock-wave is produced when an electrical discharge occurs between two electrodes in water (Fig. 31.1). The energy deposited in the water from the electrical discharge produces a bubble of very hot, rapidly expanding plasma which subsequently collapses after emitting a shock-wave. This wave, which expands out from the electrode gap, is focused using a hollow, hemi-ellipsoidal brass reflector. Since an ellipse has two foci, a spherical wave initiated at one of the foci will be focused after reflection at the second. The point of shock-wave generation is at point P_1 of an ellipsoid, half of which constitutes the reflector at the base of the water bath tub. Through the focusing of the shock-waves, the energy is concentrated at the second focus P_2 of the ellipsoid, where the volume of the shock-wave energy corresponds to that of a fingertip, approximately 15 mm^3. In the Dornier machine, this point is about 15 cm above the upper edge of the ellipsoid.

The patient is moved, partially submerged in the bath tub, on a hydraulically operated gantry until the stone is accurately positioned at the second focus of the ellipsoid by using bi-planar fluoroscopy. The positioning system of the Dornier lithotripter uses two independent X-ray systems with separate axes. The orthogonal X-ray beams, from X-ray tubes positioned under the bath tub, are viewed by two image intensifiers, resting on the patient's lower abdomen, and the resulting images are displayed on two monitors. The crossing point of the X-ray beam axes is the second focus of the shock-wave reflector, the stone is correctly centred when it appears at the same position on both monitors which is indicated by cross-wires of each X-ray system.

The stone is possibly broken up by stress and shear forces generated in it by a series of shock-waves, though other mechanisms may also be involved including a phenomenon known as 'cavitation'. Kidney stones can generally support a compression stress up to about 8 MPa

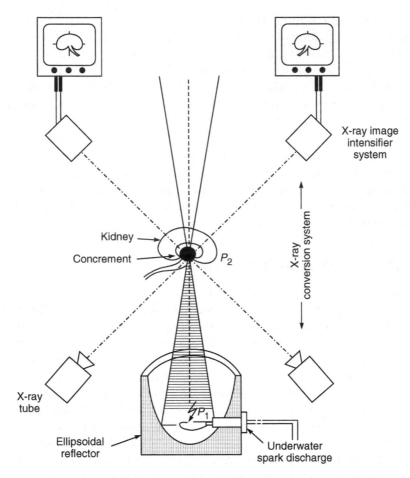

> **Fig. 31.1** *Construction for extracorporally induced destruction of kidney stone with 2 integrated X-ray positioning system (after Chaussy et al., 1980)*

(megapascal) and a tensile stress up to about 0.6 MPa. The peak positive pressure (compression) in water due to a single shock-wave has been measured at around 40 MPa at the focus. This is followed rapidly, in about 5 seconds, by a peak negative pressure (de-compression) of approximately 10 MPa. Although it is not possible to measure the pressures in tissue, yet it is expected that these peak pressures will be at least 50% smaller due to the attenuation of the wave produced by the tissue. It is therefore likely that part of the stone will be fragmented by compression as the wave enters the stone. Pressure waves of less than about 8 MPa will be able to travel in the stone and may subsequently be internally reflected on meeting an interface between stone and tissue. A reflection of this type leads to tension within the stone and its consequent fragmentation.

The size of fragments resulting from internal reflections is closely related to the decay constant of the pressure waveform; the fragments being smaller for a faster fall-off in pressure after the peak pressure. An internally reflected wave of peak positive pressure, 8 MPa, with an exponential pressure decay with a time constant of $2\,\mu s$, will give fragments of the order of 1 mm in thickness.

It is possible to estimate the velocity of the fragments due to the same pressure waveform and this is found to be of the order of 1 m/s. Clearly, there may be some advantage in keeping the time constant of the pressure waveform small (<2 µs) to reduce both the energy and size of stone fragments. It is possible that considerably smaller peak pressures may be useful in breaking up the stone simply by tension produced in the stone by internal reflection of a suitable pressure wave, though this would probably require the use of multiple shock-waves.

▶ 31.3   MODERN LITHOTRIPTOR SYSTEMS

First generation electro-hydraulic lithotriptors had several major disadvantages. The early machines were relatively expensive to install and operate, and required dedicated facilities and treatment. Besides, using the systems was painful and required general anaesthesia, which resulted in-prolonged in-patient stays and higher overall costs.

Modern machines incorporate clinical advantages over their predecessors. The integration of a variety of reflector sizes and control over voltage and power output allows for greater ease of use as well as customization of treatment parameters for increased treatment efficiency and decreased discomfort for the patient.

Shock-wave lithotripsy machines currently in the market vary in terms of several operational factors such as the energy source, the focusing system and stone localization system. In general, the main components of a lithotriptor system are:

- Focused shock-wave source;
- Means for acoustic coupling of the shock-wave to the body;
- Imaging modalities for stone localization and therapy control;
- A patient table with either the table or the shock-wave source movable in three dimensions;
- System for the measurement of physiological variables and their monitoring; and
- Trigger generation and control system.

Figure 31.2 shows the schematic diagram of a lithotriptor system. Figure 31.3 is a view of a clinical site showing a lithotriptor system in use.

31.3.1   Focused Acoustic Shock-wave Source

Focused acoustic shock-waves are necessary for extra-corporeal lithotripsy for which focal pressures of 10 to 100 Megapascal (100-1000 bars) acting in a cigar-shaped focal volume with a across-sectional diameter of approximately 2 to 8 mm and a length of about 25 to 50 mm, are required depending on the design of the particular device. These shock-waves are generated by an emitter outside the body and transmitted as pulsed longitudinal waves through a fluid coupling medium and the body tissue to the target, the concrement to be destroyed. Shock-waves are unharmonic and non-linear acoustic phenomena, characterized by an extremely steep change in pressure amplitude, the shock front. It is generally accepted that an ideal shock-wave for extra-corporeal lithotripsy shows a shock front only in the compressional part of the pulse up to a peak pressure, followed by decay (Fig. 31.4). Due to general physical principles, this compression is accompanied by a rarefaction. The characteristic parameters of the shock-wave are the peak value

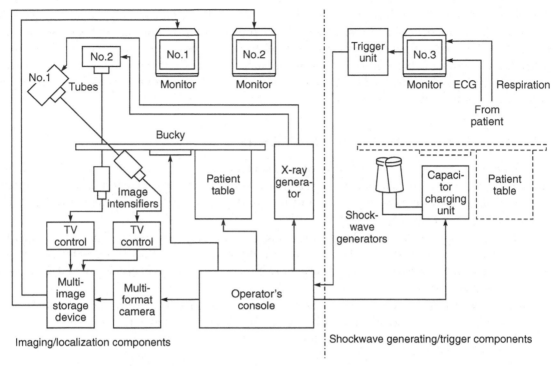

> **Fig. 31.2**   *Schematic diagram of a lithotriptor system with biplane X-ray imaging*

> **Fig. 31.3**   *View of a lithotriptor system (Courtesy: M/s Medispec Ltd., Israel)*

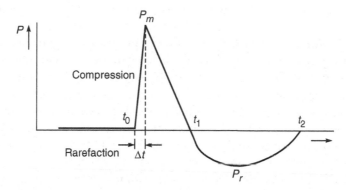

> **Fig. 31.4** *Schematic of an acoustic shock-wave pulse*

of the pressure and duration of compression and rarefaction, and the rise time of pressure Δt. In order to achieve the best results, the shock-wave must be developed with a sufficiently high positive pressure amplitude and a low negative-pressure amplitude.

Shock-waves in lithotripsy need to be strongly focused in order to keep the area of interaction with tissue or concrements restricted to the pre-determined region of interest. Any tissue in front of, behind or adjacent the target area should be left unaffected. The technical answer to this medical requirement is to use large aperture systems which spread the shock-wave energy over a wide skin entrance area. Simultaneously, the system concentrates the acoustic energy precisely to a small focal volume with a cross-sectional diameter of a few millimeters. Modern systems use aperture angles of 80 to 90 degrees to provide these favourable field parameters. Depending on the type of indication, focal distances up to 16.5 cm are available for the treatment of deeply lying structures like the kidney and ureteral stones.

To-date there is no standard to define the 'strength' of shock-waves. However, for comparison and dosage purposes, the 'energy flux density' measured by milli joule per mm^2 (mJ/mm^2) turns out to match fairly well with clinical efficiency. Of clinical importance is the ability of shock-wave devices to precisely adjust the delivered shock-wave energy. Electro-hydraulic or spark gap technology cannot provide shock-wave pulses of precisely defined energy due to the statistical nature of spark gap formation. Peak focal pressures of spark gap systems may show a pulse to pulse variability exceeding 50% whereas a state-of-the-art electromagnetic system provides a repetition accuracy better than 3%.

The three basic types of shock-wave sources for lithotripsy are:

1. Plasma explosion method;
2. Electromagnetic system;
3. Piezo-ceramic system.

These excitation sources are coupled with the following focusing methods:

(a) Ellipsoidal reflector;
(b) Focusing with an acoustic lens; and
(c) Self-focusing source.

It may be appreciated that wave excitation and focusing methods must be matched to one another.

Plasma Explosion Method: Shock-wave generation by plasma explosion over a high-voltage spark gap in water, was the pioneering method with which the first Dornier extra-corporeal shock-wave lithotriptors were operated. Even today, some of the systems still operate with this method.

In this method, a capacitor is discharged across two opposing electrodes placed at the first focus of a partial ellipsoid of rotation in a bath tub. A conducting plasma channel is formed between the electrodes and expands with supersonic velocity. The resulting compression wave in the water is a shock-wave with a steeply rising front. The initial velocity of the radial propagation of the shock-wave is significantly higher than the normal speed of sound in water, but rapidly slows down towards it. Due to non-linear absorption, the peak pressure of the shock-wave is reduced more rapidly with distance than predicted for a small amplitude wave. The resulting shock-wave can be focused by aid of a semi-rotational ellipsoidal reflector as shown in Fig. 31.5 (a).

Maximum pressure amplitudes of 2000–3000 bar are used in lithotripsy, even though calculi have been crushed at one-tenth of this level. When treating calculi, it is not a question of applying the maximum pressure from the beginning, but rather of applying pressure which is sufficient to achieve success within a reasonable time period.

The soft-tissue path lying within the focal length between shock tube lens and its focus is naturally always of a different length. This is compensated for by a water-filled bellows applied at a constant pressure on the patient's skin. Wet gel disks are placed between the surface of the bellows and the skin, or a coupling gel is used. Shock-wave release is generally triggered by pre-selected respiratory gating. A calculus which is carried along with respiratory movements can be hit with nearly 100% accuracy and tissue not containing calculi does not get into the shock-wave focus unnecessarily.

In the first generation machines based on this principle, the patient was immersed in water (bath tub). In the second generation machine, the reflector contains water contained by a rubber membrane. The sparks generated in an aqueous medium cause a local generation of vapour, in the form of a primary shock-wave which spreads in a spherical pattern. These primary shock-waves are reflected by the metal wall of the reflector, which focuses them on point F_2 as shown in Fig. 31.5 (b).

The Nova System made by Direx Medical System Ltd., USA is provided with two types of reflectors: 'Standard' and; 'small' with openings of 240 mm and 181 mm respectively. These produce focal spots of different sizes. The focal spot of small reflectors measures $5 \times 5 \times 17$ mm, whereas that of standard reflectors measures $3 \times 3 \times 14$ mm. Shocks produced by means of standard reflectors are less painful, because of the larger area of contact with the patient; however, they are less effective.

In the plasma explosion system, the electrode tip wears away with each discharge, thereby increasing the distance between the tips with use which may become too great to create a spark. Electrodes, therefore, must be replaced in such a situation.

An improvement over the electro-hydraulic method is the new electro-conductive shock-wave generator used in SONOLITH 4000$^+$ from TECHNOMED Medical Systems (France). In this arrangement, (Fig. 31.6) the shock-wave is generated in a highly conductive electrolytic solution

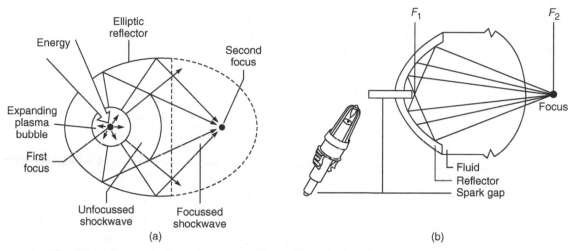

> **Fig. 31.5(a)** *Exploding plasma emitter with elliptic reflector*
> **(b)** *Shock wave generation. Plasma explosion across a spark gap under water, focusing with ellipsoidal reflectors*

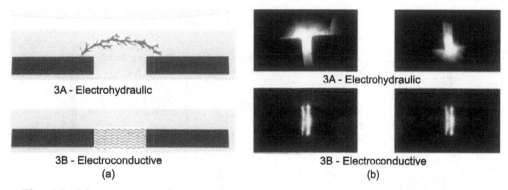

> **Fig. 31.6(a)** *Energy path at F_1*
> **(b)** *Discharge formation between the anode and cathode (i) electrohydraulic (ii) electroconductive*

and the discharge is forced to follow a precise and a reproducible pathway. Shock-waves are repetitively and accurately created at the same geometrical location (F_1) and at a stable shot-to-shot intensity. At the same time, the electro-acoustic efficiency is greatly improved due to faster release of energy. The shock-waves are focused via a shallow ellipsoidal reflector whose geometrical parameters are specially designed for electro-conductivity technology, reducing pair level. At F_2, focal volume dimensions and pressure dispersion are minimized whereas the energy concentration is maximized on the stone and disintegration efficiency is optimized. The technique gives perfect linearity between energy settings and disintegration efficiency (Flam et al, 1994).

Using an integrated hydrophone (Fig. 31.7), the actual delivered pressure of each shock is recorded/displayed, allowing the physician to manage the therapy directly through pressure rather than KV settings.

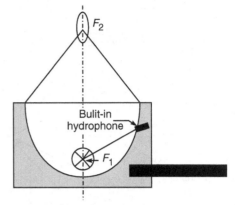

Electromagnetic System: Electromagnetic (also known as electro-dynamics) shock-wave genera-tion system is shown schematically in Fig. 31.8. The construction of the shock tube and a cross-section of single-layer helically wound coil is also shown. The coil and a flat electrically insulated metal membrane placed on it form the actual oscillation generator. When an electrical current pulse is sent through the coil, it produces a rapidly increasing magnetic field intensity which induces eddy currents in the homogeneous metal membrane. The eddy currents also produce a magnetic field which, according to the law of

> **Fig. 31.7** *Use of hydrophone for monitoring acoustic wave delivered pressure*

electromagnetic induction, is opposed to that of the coil. The membrane is repelled and transmits the released mechanical impetus to the water column. A wave travels from the membrane, becoming a shock-wave after passing through the focusing lens (Pfeiler et al, 1989).

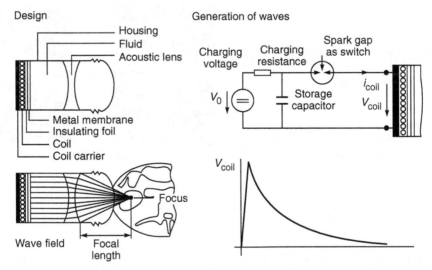

> **Fig. 31.8** *Shock-wave generation—electromagnetic system with acoustic lens. The electrical pulse from the coil forms a magnetic field, inducing eddy currents in the metal membrane. The membrane is consequently repelled*

The collecting lens is a concave lens, in contrast to lenses used in optics, since water has a higher sound refraction index than the material of the lens. The current pulse in the coil is generated by the discharge of a high-voltage capacitor charged to 10 to 20 KV which is connected with the coil through a controlled spark gap (Fig. 31.9).

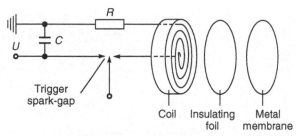

> **Fig. 31.9** *Schematic diagram of an electromagnetic acoustic source (after Pfeiler, 1989).*

An improved configuration of an electromagnetic shock-wave device utilizes a cylindrical coil arrangement with a parabolic reflector which provides significant improvements over flat coil arrangements with lens focusing. The cylindrical wavefront is focused virtually without energy loss by a rotational parabolic reflector. It simultaneously provides the appropriate space on the central axis for implementation of either in-line X-ray localization or inline ultrasound transducers. By the technique of so called 'schlierenoptics', a time sequence of shock-wave fronts propagating towards the focal point can be visualized. Cavitation bubbles are displayed which are claimed to play a significant role in stone fragmentation as well as possibly in orthopaedic applications.

Piezo-ceramic System: The piezo-electric principle operates by simultaneously driving several hundred piezo elements mounted on a spherical dish, thus providing self-focusing spherical waves. The arrangement is shown in Fig. 31.10.

In contrast to the electro-dynamics shock tube, the useful wave field is focused with a lens immediately after its formation. The wave migrating to the left is dissipated in an acoustic sink to avoid disturbing reflections. The high voltage

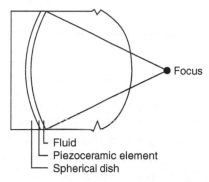

> **Fig. 31.10** *Shock-wave generation—piezo-electric system with focusing oscillator geometry. The spherically curved oscillator ceramic can be composed from a mosaic of small, flat oscillating elements*

pulses can be generated in the same way as those in the electromagnetic system. However, the piezo-electric system operates with a lower high voltage as compared to the electro-dynamic system.

In systems presently in clinical use for lithotripsy, up to 2000 piezo-ceramic elements (Fig. 31.11) are arranged on the inner surface of a supporting spherical segment. When a voltage step is applied, e.g., by a capacitor discharge, the length of elements changes due to the piezo-electric effect, their surface is displaced and an acoustic pulse is emitted. The elements, made for example from PZT-ceramic, are designed for high power output for use in lithotripsy. They need a high electric breakdown strength and a specially designed backing to increase their efficiency and to reduce

> **Fig. 31.11**   *Inside view of lithotripsy treatment head using piezo-electric ceramic elements (Courtesy: EDAP International S.A., France)*

unwanted secondary pulses. The impedance of the electrical network and the time-function of the voltage step have to be matched to the electro-acoustical characteristics of the ceramic elements so that mainly one pulse with high intensity and a low amplitude rarefaction is emitted.

Some systems offer an assortment of easily exchangeable reflectors allowing the physician to adopt treatment to the patient's specific needs. This feature provides the capability of matching more precisely both the size and consistency of stones, as well as the patients' level of pain tolerance. Once stones are identified on the image workstation, the physician may decide which reflector is best suited for the particular treatment.

Coupling: A medium is used for coupling shock-waves to the human body to minimize the presence of air to provide undisturbed propagation of the acoustic pulse and to obtain good matching of acoustic impedance to the acoustic impedance of the human skin to minimize reflection of the acoustic pulse at the skin.

An open bath tub provides optimal acoustic coupling, but restricts positioning and handling of the patient. However, it necessitates the management of large volumes of clean water and adequate safety measures to reduce electrical hazards to the patient.

Closed systems with an elastic coupling membrane also need water circulation and preparation, but demand a significantly lower volume and isolate the patient from the source. The membrane material is such that it matches the acoustic impedance of water. Coupling to the skin is provided by an additional layer of an oily gel similar to the gels used in ultrasound imaging.

31.3.2   Imaging

X-ray fluoroscopy as well as ultrasound B-scan are used as sole imaging procedures in lithotripsy. For pre-treatment diagnosis and for immediate and long-term control after the shock-wave treatment, X-ray imaging is undisputedly the optimal modality. For optimal fluoroscopy, two X-ray generators and image intensifiers are used at different imaging planes. The radiological image appears on the high resolution computer screen mounted on the control panel. Using digital image control, the computer can enhance the contrast of the calculus, and can also save an entire series of images on the magnetic media during the treatment. Advantages of ultrasound B-scan are the freedom from ionizing radiation and the possibility of online imaging. The drawbacks are the necessity to employ personnel trained well in ultrasound imaging to permanently watch the procedure, lower spatial resolution than with X-rays and the difficulty in presentation of concre-ments in the ureter.

In systems with integrated ultrasound imaging, an ultrasound sector probe (3.5 MHz) is centrally positioned in the shock-wave source and directed towards the focus. As it displays only one image plane, it has to be rotated to scan the whole volume. Improved imaging and easier handling result when a second ultrasound probe is available.

A lithotriptor intended for universal lithotripsy must be supplied with both X-ray and ultrasound imaging facility. Such a system with an electromagnetic shock-wave head with integrated sonographic array enables the progress of the therapy to be followed visually and continuously along the shock-wave path, without stress to the patient. In the shock-wave head, the shock-wave generating oscillatory system has a recess in its centre. The sonographic probe is located in this recess and can rotate about its own axis and be adjusted along its axis.

The calculus is searched for in the suspected area by rotating the sonographic probe. The calculus becomes visible when it lies in the plane of the ultrasound image. The entire shock-wave head is then adjusted so that the calculus lies within the target point indicated on the monitor, corresponding to the shock-wave focus.

Besides helping to locate the calculus, the ultrasound image also permits observation during shock-wave application and consequently helps control the successful hits. In order to compensate for small changes in the location of calculi, patients once positioned, can be treated after movement of the shock-wave source and sonographic probe within the shock-wave head, without changing the coupling of the shock-wave head. The combination of an under-table shock-wave head with X-ray orientation and an over-table shock-wave head with integrated ultrasound imaging offers a maximum of lithotripsy applications.

31.3.3   Patient Table

In most systems for the treatment of kidney and urinary stones, shock-wave sources are arranged below the structures supporting the patient. For aiming the focus at the concrement, either the table with the patient is moved or, after an approximate positioning, the shock-wave source is adjusted. With two symmetrically arranged shock-wave sources, the patient lies in a supine position independent of whether the left or right kidney is treated. The patient table is motorized that enables movement in all three directions. The table can be controlled directly from the control panel. It provides the ability to position the patient in three co-ordinate axes and provides access to the patient's lumber area. The table has an opening permitting the patient's lumber area to contact an oil-coated membrane that acoustically couples the patient with the water system.

31.3.4 Monitoring and Trigger Generation

With the original Dornier lithotriptor, wherein shock-wave treatment was performed either under general anaesthesia or with regional anaesthesia, monitoring and anaesthetic control was necessary. With growing experience and the development of improved lithotriptors, anesthesia is only indicated by demand of the patient or when additional interventional measures are necessary. Monitoring of the heart is nevertheless advisable because even when treatment is performed with a lithotriptor, circulatory reactions, such as extra-systoles are found to occur in a low percentage of the patients. In these cases, triggering of the shock-waves by the heart cycle is used as a preventive measure. Any ECG monitor that provides a 1 volt TTL R-wave sync. or defibrillation sync output signal can be used for the required synchronization of the shock-waves with a patient's heartbeat.

With respiration, the kidney, and thus the concrement, is displaced periodically. Efficiency of the treatment of kidney stones can be improved and the patient exposed to fewer shock-waves when the triggering of the shock-waves is gated by the respiratory cycle, whereby expiration is normally preferred. The treatment typically takes about 30 to 90 minutes, not including patient preparation and associated measures.

The treatment consists of about 1500 shocks average (range, 300–2000), after which the stone is reduced to 1 to 2-mm fragments that pass spontaneously down the ureter, usually within 30 days. The shocks are generally given in series of 100-150, followed by fluoroscopic verification that the stone or large particles are in the "second focus". After ESWL treatment, patients are instructed to drink enough liquids to make two quarts of urine a day and to store their urine for one week to collect stone particles for analysis. Haematuria, which occurs after about 200 initial shocks in all patients, usually ceases within a few hours to 1-2 days. Most patients may then resume full activity two days later.

▪▪▪▶ 31.4 EXTRA-CORPOREAL SHOCK-WAVE THERAPY

Valchanov and Michailov (1991) applied extra-corporeal shock-wave therapy for applications in orthopaedics for the first time in humans. Later several groups have successfully used shock-waves of different energy levels for orthopaedic pain treatment (Schleberger and Senge, 1992).

Unlike lithotripsy devices for fragmentation of body concrements like kidney and ureteral stones, an extra-corporeal shock-wave device for orthopaedic and traumatologic purposes needs to fulfil different needs.

Flexible Shock-wave Applicator: Since the area of application is not limited to the abdominal region but may be spread all over the total body from shoulder to foot sole, the shock-wave applicator needs to be very flexible for comfortable positioning and precisely adjustable for easily following the actual point of highest pain sensation.

In-line Ultrasound Localization: Although fine targeting is done by interaction of patient and operator, for coarse localization, ultrasound imaging of the region of interest is required. In-line ultrasound transducer integrated in the shock-wave head provides best targeting accuracy and easy access to virtually all possible anatomical regions.

Targeting Manipulator: As mentioned above, the actual point of highest pain sensation is followed by frequently re-positioning of the focal zone. Small displacements of the therapy focus are recognized by the patient and guided by verbal interaction with the operator. Precise manipulation of the focal position requires an adequate mechanical configuration.

Energy Dosage and Dynamic Range: Depending on different indications and individuals, the appropriate energy level and the just tolerable pain sensation during shock-wave treatment needs to be individually adjusted and kept constant unless willingly changed. Energy levels from 0.03 to 0.5 mJ/mm^2 have proven to be safe and effective for most indications.

A popular device (Fig. 31.12) which matches well with all the above-stated requirements is available from STOREZ Medical, Co., Switzerland. It is a compact and transportable device and is

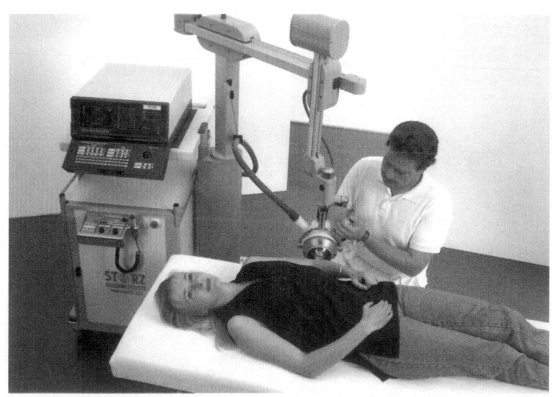

➤ **Fig. 31.12** *Shock-wave head with integrated ultrasound imaging. The sonographic probe is arranged centrally in the shock wave oscillation system. (Courtesy: M/s Storez Medical Co. Switzerland)*

characterized by a very flexible shock-wave head mounted to an articulated arm which allows almost unrestricted mobility to cover all zones of interest. The cylinder configuration of the shock-wave source is ideally suited for coaxial integration of a high resolution 7.5 MHz ultrasound sector transducer. Due to in-line integration of the transducer into the coupling cushion, automatically a stand-off effect is provided for enhancing superficial image quality in orthopaedic ultrasound imaging. For general purposes, the ultrasound may optionally be equipped with several linear or curved array transducers which are commonly favoured in orthopaedics.

The treatment is carried out in an outpatient way. Either without any medication or with the aid of subcutaneous analgesic drugs, 1500-2000 shock-wave pulses are applied to the point of highest pain sensation. At a single point, approximately 50-100 pulses are required to reduce pain sensation. Shock-waves are released with rates of 1,1.5 and 2 pulses per second, including targeting and re-positioning a treatment session lasts 20-40 minutes. Well in accordance with the reflex control theory, usually 1-5 treatment sessions provide a good treatment result. The applied energy levels are 0.03 to 0.25 mJ/mm^2.

Anaesthesia Machine

▶ 32.1 NEED FOR ANAESTHESIA

Surgical methods of treatment consists mainly of operations which are normally carried out under some form of anaesthesia. Anaesthesia serves the following two functions:

1. It ensures that the patient does not feel pain and minimizes patient discomfort; and
2. It provides the surgeon with favourable conditions for the work.

When anaesthesia is given so that the patient loses consciousness, it is called 'general anaesthesia'. In general anaesthesia, the anaesthetic agent is administered to the body so that it reaches the brain via the blood stream. The usual method is 'inhalation anaesthesia' in which gaseous anaesthetic agents are introduced via the lungs. Examples of such agents are diethyl ether, chloroform, halothane, cyclopropane and nitrous oxide (N_2O, laughing gas). During anaesthesia, not only is the anaesthetic administered in the required amount but also oxygen. Any excess carbon dioxide is also eliminated. In the superficial stages of anaesthesia, the patient can breathe for himself—spontaneous ventilation. At a greater depth of anaesthesia, it may be necessary to support the patient with artificial ventilation known as controlled ventilation.

32.1.1 Delivery of Anaesthesia

The anaesthetic delivery system consists of an anaesthesia machine, a patient breathing circuit, a ventilator and airway equipment.

The *machine* comprises a gas supply—delivery unit and an anaesthetic vapourizer.

The *breathing circuit* consists of a closed loop of breathing tubing, containing two uni-directional breathing valves and an Adjustable Pressure Limiting (APL) valve, a CO_2 absorber, a means for venting excess gases (scavenging), a humidifier, and a collapsible reservoir bag.

A *mechanical ventilator* is used for positive pressure ventilation.

The *airway management equipment* includes the mask and endotracheal tube, which interface the patient with the breathing circuit.

ⅢⅢ> 32.2 ANAESTHESIA MACHINE

An anaesthesia machine is a device which is used to deliver a precisely known but variable gas mixture including anaesthetic and life-sustaining gases to the patient's respiratory system. Generally, a variable concentration gas mixture of oxygen, nitrous oxide and anaesthetic vapour like ether or halothane is obtained from the machine and is made to flow through the breathing circuit to the patient. It is composed of two subsystems (Fig. 32.1): (i) The gas supply-delivery unit, which consists of tubing and flowmetres interconnected in parallel; and (ii) The anaesthetic vapourizer, which is used to produce an anaesthetic vapour from a volatile liquid.

32.2.1 Gas Supply System

Gases are provided to the anaesthesia machine from either a pressurized hospital central supply or small storage cylinders attached to the machine.

Centralized Supply: Centralised supply systems consist of bulk or cylinder storage for main and reserve supply, control equipment including valves and pressure regulators, a distribution pipeline, and numerous supply outlets. The system is so designed and operated that the necessary supply of gases (oxygen and nitrous oxide) is always available. The gas supplied by the hospital is regulated and maintained at 275-345 kPa (40-50 psi) at the wall outlet.

Gases are supplied to the anaesthesia machine inlet from the central system via a flexible hose connected to the operating room wall outlet. In order to prevent interchanging the gas supply wall outlet with the incorrect anaesthesia machine inlet, for example, nitrous oxide for oxygen, non-interchangeable connectors are used at each end of the hose. The two types of non-interchangeable connections most commonly used are the Diameter Index Safety System (DISS) and non-inter-changeable quick couplers. Each type of connection incorporates a male and female end that is specially designed for each type of gas. In addition to the connector design, colour-coded hoses for each specific gas are utilized.

Gas Cylinders: A second gas supply source is the cylinders located in yokes attached to the anaesthesia machine. This supply can be utilized as either the main source when a central gas supply does not exist, or a reserve when central gas supply is available.

Yoke: Each anaesthesia machine has at least one yoke for an oxygen cylinder but most are provided with two. In addition to oxygen, most machine designs include a nitrous oxide yoke. In order to prevent incorrect placement of a tank into the wrong yoke, two pins located in the yoke must fit into corresponding holes drilled into the tank neck. The placement of these pins and corresponding holes is unique for each gas. This identification system, which is referred to as the 'Pin Index Safety System', has been standardized to prevent the accidental fitting of a wrong cylinder to the yoke.

Pressure Regulator: Machine pressure regulators reduce cylinder gas pressures to 275 kPa (40 psi) before the gas flows through the machine. The regulator has one high-pressure inlet, one high-pressure outlet and two-low pressure outlets. The high pressure inlet is connected with the cylinder through a non-return valve. The non-return valve prevents the flow into an empty cylinder or back into the central piping system and also enables its removal and replacement when the reserve cylinder is turned on without interrupting the supply of gas.

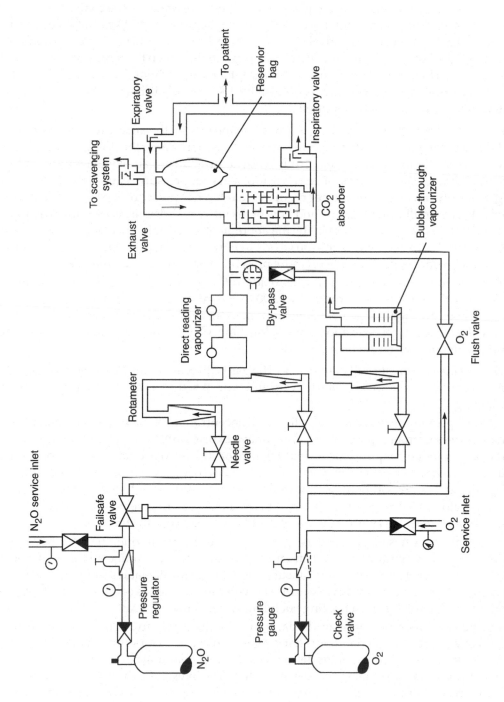

▶ **Fig. 32.1** *Schematic diagram of an anaesthesia machine and patient breathing circuit*

Pressure Gauge: Pressure gauges are attached to the cylinders to indicate the contents of the gases in the cylinders. For oxygen, the operating range of the gauge is 0 to 150 kg/cm^2. Whenever the new oxygen cylinder is hooked up and taken in line, the indicator should be above this mark. With the gradual usage of the gas, the reading would drop gradually, when the indicator shows that the pressure has fallen below the minimum level of acceptance, the cylinder should be refilled. If for any reason, the pressure gauge shows a reading above 150 kg/cm^2 during use, the cylinder should be disconnected immediately and replaced.

Fail Safe System: From the supply, the gas flows into the inlet of the anaesthesia machine and is directed through the pressure safety system (fail-safe system) towards the flow delivery unit. The pressure safety system will not allow nitrous oxide to flow unless an oxygen supply pressure exists in the machine. The fail-safe system consists of a master pressure regulator valve located in the oxygen supply line. From the master regulator, a reference pressure is provided to the slave regulator valve controlling the pressure and flow of the nitrous oxide line. When sufficient oxygen pressure of 275 kPa (40 psi) is present in the master regulator, the reference pressure enables the slave regulator valve to open and for nitrous oxide to flow. Unfortunately, pure nitrous oxide can be delivered with only oxygen supply pressure present; oxygen flow is not required.

Regulations now require oxygen-nitrous oxide ratio safeguards, which need a minimum continuous low flow of oxygen varying from 200 to 300 mL/min, as indicated by the low-flow rotameter. In newly designed machines, ingenious mechanical devices prevent the delivery of gas mixtures with an oxygen concentration below a low limit. Oxygen-nitrous oxide ratios vary from 25:75 to 30:70, depending on the manufacturer.

Gas Delivery Units: From the fail-safe system, the gas is directed to the flow delivery unit. Two methods have been used to accomplish delivery and control of the gas mixture: gas proportioning and gas mixing.

In a gas proportioning system, the delivered concentration of each gas constituent is the function of a pre-determined, precisely controlled ratio of proportionality which is independent of the total gas flow. For example, for a desired mixture of 70% nitrous oxide and 30% oxygen, the metered ratio of mass delivery will always be 7:3, regardless of the total flow rate. Concentration is only a function of the proportional relationship between constituents. It does not rely on setting individual gas flows. An oxygen-nitrous oxide bleeder used in a manner similar to the oxygen—air blenders commonly used with mechanical ventilators performs this function.

Most current anaesthesia machines use gas mixing. In this technique, the flow rate of each constituent is independently controlled and measured by a delivery unit consisting of a needle valve and a rotameter. The needle valve functions as a flow controller and a means of turning the gas on and off. The rotameter is a variable orifice flowmetre and consists of a transparent tube with a tapered internal diameter and a floating bobbin flow indicator.

During the administration of anaesthesia, it may be necessary to fill the patient breathing circuit with oxygen at a rate higher than what the gas delivery unit can supply. For example, such a situation exists any time the patient is disconnected to the breathing circuit. This higher flow of oxygen is supplied via the oxygen flush valve and line. The oxygen flush system provides a high flow ranging from 35 to 75 L/min at a high pressure (20-45 psi, 270-590 kPa) directly into the patient breathing circuit.

Each gas has a specific delivery unit. These units are connected in parallel and exhaust into a common manifold prior to leaving the machine. The final concentration and total flow determined by mixing the component flows are dependent functions and subject to the accuracy of the control and measurement equipment.

32.2.2 Vapour Delivery

The various liquids that possess anaesthetic properties are too potent (strong) to be used as pure vapours. They are thus diluted in a carrier gas such as air and/or oxygen, or nitrous oxide and oxygen. The device that allows vapourization of the liquid anaesthetic agent and its subsequent admixture with a carrier gas for administration to a patient is called a 'vapourizer'. Vapourizers thus produce an accurate gaseous concentration from a volatile liquid anaesthetic. The anaesthetic vapour can then be safely added to the previously metered oxygen and nitrous oxide as the mixture leaves the mixing manifold.

Vapourizers are available in one of the two basic designs: the flowmetre controlled or the concentration-calibrated. In either device, the anaesthetic vapours are picked up from the vapourizer by a carrier gas consisting either of pure oxygen or an oxygen-nitrous oxide mixture that bubbles through or passes over the liquid. The liquid surface area to gas interface is designed to ensure the most efficient vapourization process.

As a result of vapourization, a drop in liquid temperature is produced. As the liquid temperature decreases, a thermal gradient is established between the liquid and the surroundings. This results in a decrease in the quantity of the vapour produced. In order to maintain the performance of the vapourizer, the temperature drop is minimized or prevented by the incorporation of a thermal source. This is achieved by using a water bath or surrounding the vapourizing liquid with a heating element. These devices may also control the temperature of the carrier gas entering the vapourizer.

The materials selected for vapourizer construction require both a high specific heat and high thermal conductivity. Materials with high specific heats will change temperature more slowly and maintain an appropriate thermal inertia. The higher the thermal conductivity, the higher the conduction of heat from the surroundings. Because of its availability and lower cost, copper has been one of the most common materials used. Although not ideal, copper has a moderate specific heat and a high thermal conductivity. Early vapourizers were accordingly called "copper kettles".

In order to provide a stable and predictable concentration of anaesthetic vapour, the vapourizers include a suitable method of obtaining calibrated dilution of vapour to avoid administration of too powerful volatile anaesthetic agents to the patient. This can be done by several means and the vapourizers are accordingly classified into various categories discussed below.

Variable Bypass Vapourizer: Here the carrier gas flow from the flowmeter is split into two streams in a known ratio: one stream which is called 'chamber flow', flows over the liquid agent while the other stream goes through the bypass path and does not enter the vapourizing chamber. The final concentration can be controlled by varying the splitting ratio between the vapourizer gas and the bypass gas using an adjustable valve (Fig. 32.2).

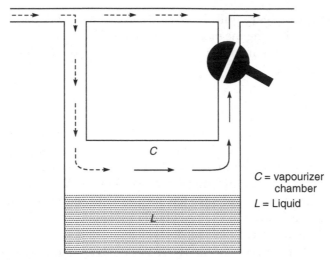

C = vapourizer
    chamber
L = Liquid

> **Fig. 32.2**   *A schematic diagram of a variable bypass vapourizer. A flow splitting valve that can be rotated to alter the relative diameters of the vapourizer and bypass channels*

The splitting ratio of the two flows depends on the ratio of resistances to their flow, which is controlled by the concentration control dial and the automatic temperature compensation valve. Usually, less than 20% of the gas becomes enriched-saturated with vapour and more than 80% is bypassed, to rejoin at the vapourizer outlet. The output of current variable bypass vapourizers is relatively constant over the range of fresh gas flows from approximately 250 mL/min to 15 L/min. The output of vapourizers is linear at the ambient temperature (2°-35°C) due to automatic temperature compensating devices that increase carrier gas flow as the liquid volatile agent temperature decreases. Also, they are composed of metals with high specific heat and thermal conductivity. Check valves are provided to prevent back pressure effect on the vapourizer from the breathing circuit due to positive pressure ventilation.

Measured-flow Vapourizers: In these devices, the anaesthetic agent is heated to a temperature above the boiling point (so that it behaves as a gas) and is then metred into the fresh gas flow (Fig. 32.3).

Various anaesthetic agents have widely different potencies and physical properties and hence require vapourizers constructed specifically for each agent. They are thus 'agent-specific'. They are only calibrated for a single gas, usually with keyed filters that decrease the likelihood of filling the vapourizer with the wrong agent.

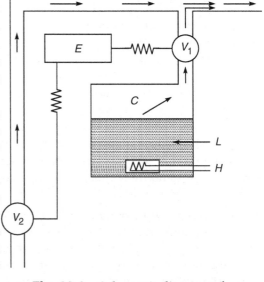

> **Fig. 32.3**   *Schematic diagram of a measured flow vapourizer*

Vapourizers are provided with various safety related inter-locks which ensure that:
- Only one vapourizer is turned on;
- Gas enters only the one which is on;
- Trace vapour output is minimized when the vapourizer is off;
- Vapourizers are locked into the gas circuit, thus ensuring that they are seated correctly; and
- Other important safety features are followed including keyed filters and secured mounting to minimize tipping (tilting) which may obstruct the working of the valves.

32.2.3   Delivery System

Patient Breathing System: The function of a patient breathing system is to deliver anaesthetic and respiratory gases to and from the patient. It describes both the mode of operation and the apparatus by which inhalation agents are delivered to the patient. The breathing system may be

1. Rebreathing Type: This refers to re-breathing of some or all of the previously exhaled gases, including carbon dioxide and water vapour.
2. Non-rebreathing Type: In this a fresh gas supply is delivered to the patient and re-breathing of previously exhaled gases is prevented. Usually, non-rebreathing type systems are applied in practice. This is achieved by using:
- Non-rebreathing uni-directional valve;
- Carbon dioxide absorption system; involving;
  - Uni-directional (circle) system; and
  - Bi-directional (to-and-fro) system.

Figure 32.4 shows the principle of a non-rebreathing system which uses uni-directional non-breathing valve. Fresh gas entering the inspiratory part is either sucked in by the patient's

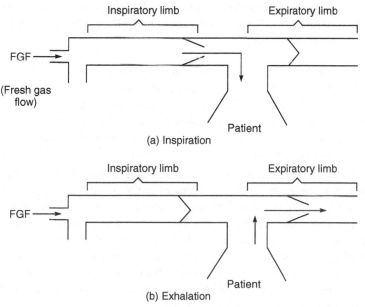

> **Fig. 32.4**   *Non-breathing valve*

inspiratory effort or blown in during controlled ventilation. The non-rebreathing valve is so designed that when it is open to admit inspiratory gas, it does not permit the flow from the expiratory part to get through it. When the patient exhales, the reverse happens, as the inspiratory valve is occluded and the expiratory valve is opened to allow expiratory gases to escape. The inspiratory system usually includes a rubber bag of two-litre capacity which acts as a reservoir for fresh gas. The reservoir bag is refilled with fresh gas during the expiratory phase. It can also be compressed normally to provide assisted or controlled ventilation. The fresh gas supply is linked to a length of corrugated breathing hose (minimum length -10 cm with an internal volume of 550 ml). This represents slightly more than the average tidal volume in an anaesthetized adult breathing spontaneously. This is, in turn, connected to a variable tension, spring-loaded flap value for venting off exhaled gases. This valve is located as close to the patient as possible, and is called an APL (adjustable pressure limiting) valve. The APL valve works as a pop-off valve to ensure that the patient is not subjected to the surges in the gas supply. When the gas encounters resistance from the patient, the excess gas pops out. The arrangement is shown in Fig. 32.5. In this case, carbon dioxide elimination is achieved by the flushing action of the fresh gas introduced with the breathing system, rather than by separation. Obviously, this system retains the potential for re-breathing of carbon dioxide when the fresh gas flow rates are reduced.

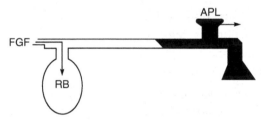

> **Fig. 32.5**   *Mapleson breathing system used with spontaneous breathing FGF (Fresh Gas Flow), APL (Adjustable Pressure Limiting) valve, RB (Reservoir Bag)*

Circle System: The circle is the most popular breathing circuit and is a closed loop of large-bore, low-pressure tubing divided into an inspiratory and an expiratory limb. Contained within this loop are two uni-directional valves, a CO_2 absorber, circuit gas venting (scavenging), an adjustable pressure-limiting valve, reservoir bag, and airway management equipment including masks and endotracheal tubes. The patient is connected to the absorber by two corrugated hoses, one inspiratory and the other expiratory.

Fresh gas is introduced proximal to a uni-directional inspiratory breathing valve. During inspiration, gas moves through the absorber from either the reservoir bag or ventilator bellows and inspiratory valve into the inspiratory limb of the circuit. The pressure difference between the inspiratory and expiratory limbs keeps the uni-directional expiratory valve closed. During exhalation, the pressure differential reverses. The inspiratory valve closes and the expiratory valve opens, allowing the exhaled gas to flow into the reservoir bag or ventilator bellows and absorber. The APL or the pop-off valve enables the anaesthetist to control the circuit volume and pressure by the regulation of gas venting from the circuit. Circuit exhaust is either carried into the room or collected by a gas scavenging system.

Uni-directional breathing valves are available in several designs. The disk valve is the most common in modern systems. This valve consists of only one movable part, a flat disk. The disk is made of either plastic or metal and is held against the valve seat by either gravity or a mechanical spring. The valves are placed in transparent devices so that their action may be observed.

The APL is designed to regulate circuit pressures by manually adjusting the spring tension against a disk. When circuit pressure overcomes the valve resistance, the disk is lifted from its seat and gas is allowed to exhaust from the circuit. The circuit volume and pressures throughout the delivery of anaesthesia are continuously observed so that the APL valve can be appropriately adjusted.

The absorber contains a carbon dioxide absorbent (soda lime) in a closed container. Soda lime is used in the form of granules so that they have a large volume and large surface area. Thus, the expired air remains in contact with the soda lime for a relatively long period of time, increasing the efficiency of absorption. Granules of an optimum size are selected as, too large a size leads to poor contact and poor absorption, while too small a size clogs the soda lime bed and causes resistance to the gas flow. The exhaled gas is made to flow through the absorber where the CO_2 is removed. The remaining gas is mixed with fresh gas flowing from the machine and re-breathed via the inspiratory limb. When the soda lime gets consumed, its colour changes from pink to yellowish.

The reservoir or breathing bag is highly compliant, with an easily expanded volume. The bag allows the accumulation of gas during exhalation so that a reservoir is available for the next inspiration. It provides a means for visually monitoring the spontaneous breathing pattern of the patient and buffers increases in breathing circuit pressure. The bag also provides a means that can be used to manually ventilate the patient. Either passive or active scavenging systems are utilized in removing the circuit exhaust. In passive scavenging, anaesthetic waste gases are vented directly into the existing room ventilation systems. The tubing connects either the ventilator or the APL valve exhaust port of the breathing circuit to the hospital ventilation system. In active methods, the anesthetic circuit connects directly to a high—flow vacuum system via an appropriate interface.

32.2.4 Humidification

Dry gases supplied by the anaesthesia machine may cause clinically significant desiccation of mucus. This may contribute to retention of secretion and the mucus flow may cease. Lung compliance will consequently fall. Therefore, air or anaesthetic gases need to be humidified.

Absolute Humidity: This is the maximum mass of water vapour which can be carried by a given volume of air (mg/L). This quantity is pre-dominantly determined by temperature. Warm air can carry much more moisture.

Relative Humidity (RH): This is the percentage of the amount of humidity present in a sample, as compared to the absolute humidity possible at the sample temperature.

It is ideal to provide gases at body temperature and 100% RH to the patient's airway. The humidification measures that are commonly employed include heated airway humidifiers, nebulizers and heat and moisture exchangers.

(i) In the *heated humidifiers*, the air passes over the surface of the heated water and vapourization takes place. The temperature of the water is thermostatically controlled.

Preferably, two thermostats in series are used, so that if one thermostat fails, the other would still cut off the electric supply before a dangerous temperature is reached. The temperature sensor is usually placed near the patient-end of the delivery tube so as to ensure the maximum efficiency.

(ii) *Nebulizers* are used to supply moisture in the form of droplets. A jet of air or gases may be used to entrain water drawn from a reservoir (Fig. 32.6). As the water enters the jet, it is broken up into a large number of droplets, i.e. it is nebulized. Nebulizers based on this principle are also used in some ventilators. In ultrasonic nebulizers, water is broken into droplets by continuous bombardment of ultrasound energy which vigorously vibrates the water.

(iii) *Heat and moisture exchangers* are based on the principle of conserving the patient's own heat and moisture without external energy or water supply. The fibre-packed cartridges contain a moisture-absorbent material that absorbs the patient's exhaled water and heat. During inspiration, the dry, inspired circuit gas flows through the warm, moisturized absorbent where it is warmed and humidified. Fibre cartridges are not as efficient at warming as humidifiers. However, they do significantly retard patient heat loss.

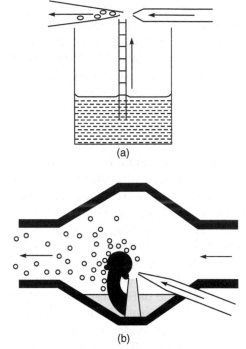

> **Fig. 32.6**   *Principle of the nebulizer*
> (a)   *Employing the bernoulli effect, a jet of air may be used to draw a liquid up a small tube from a reservoir and to entrain it as droplets*
> (b)   *The droplets may be made to impinge on an 'anvil', so causing them to be broken up into still smaller droplets*

For protection of patients from infection, clean or sterile disposable breathing circuits and bacterial filters have been advocated and widely used to reduce post-operative respiratory infections.

Although simple in design, breathing circuits can be sources of many problems. The most common and serious problem is the potential for disconnection at any of several locations. Numerous investigators have shown that 10-15% of preventable mishaps result directly from airway leaks and disconnects. The causes for leaks and disconnects include poor fit due to incorrect size, incorrect shape taper connections, inappropriate fabrication of materials, thermal expansion, broken fittings and the absence of a locking device.

32.2.5 Ventilators

An integral component of the anaesthetic delivery system is the ventilator. The ventilator provides a positive force for transporting respiratory and anaesthetic gases into an apneic patient. The ventilators provide positive pressure ventilation at a controlled minute volume (Tidal volume, Rate). They operate either electronically or mechanically with pneumatic or electric power source.

Most of the currently used ventilators consist of a bellows contained within another housing. The bellows communicate directly with the breathing circuit and causes a pre-selected volume of gas to flow into the patient. The flow of gas into the circuit results from collapsing the ventilator bellows by pressurizing the surrounding gas volume contained within the bellows housing.

The ventilator is either located within the mainframe of the anaesthesia machine or is attached as an accessory unit. The outlet of the ventilator connects directly to the patient breathing circuit of the anaesthetic delivery system at the location and in place of the breathing reservoir bag. The ventilator thus functions as a controller for both ventilation and circuit gas supply by replacing the functions of the reservoir bag and APL valve.

32.2.6 Patient Circuit

The patient circuit consists of black corrugated anti-static rubber tube, a chrome plated tube fitting (T joint), a two litre re-breathing bag, a Heidbrink valve and a face mask with an elbow fitting. The face mask is designed to fit the patient's face perfectly without any leaks and yet to exert the minimum of pressure which might depress the jaws and cause respiratory obstruction.

▶ 32.3 ELECTRONICS IN THE ANAESTHETIC MACHINE

Any delivery system is expected to meet accurately and safely, the patient's varying requirements for respiratory and anaesthetic gases. The system must be able to monitor the function of the delivery system itself and the effect of the anaesthesia on the patient. Also, during the entire procedure the machine performance should not only be monitored and controlled, but its status should be continually assessed and recorded. In order to meet these requirements, the impact of electronics on the design and functioning of the anaesthetic machine has been phenomenal.

The totally pneumatic anaesthetic machine still has many merits, which include its being easy to understand and easy to maintain, as also its cheaper cost and reliability. However, certain problems are encountered which directly affect the performance and safety of a pneumatic anaesthetic machine. This is overcome through the utilization of newer technologies and automation of instrumentation and anaesthetic delivery. Microprocessor-based anaesthetic equipment facilitates improvements in:
- Gas supply and proportioning systems;
- Breathing circuits;
- Gas scavenging and humidification devices; and
- Ventilators.

The use of microprocessor technology allows us to fully integrate control and safety functions and protects the patient from: gas supply failure, electrical supply failure, hypoxic mixtures, disconnections, vapourizer function, excessive airway pressure exhaled minute volume outside pre-set limits, oxygen or volatile agents outside the pre-set limit, end-tidal CO_2 outside present limits and technical failure. All abnormal conditions cause an alarm to appear on the monitor panel, which also displays the nature of the fault.

Computer application with appropriate data processing inputs and outputs, automation and integration of machine functions and record-keeping are increasingly becoming possible. Ergonomically designed machines with easy to read and interpret display systems are also being commonly used.

Anaesthesia has a profound effect upon all physiological systems. Most of these effects are deleterious, and therefore, it is important to know how the human body is affected by anesthesias. With a view to increasing patient safety and achieving a good degree of risk management, all systems affected by anaesthetic drugs must be monitored. This is done by using monitoring equipment with visible and audible alarms as illustrated in Chapter 6.

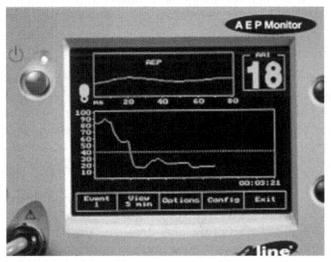

> **Fig. 32.7**   *AEP (Auditory Evoked Potential) monitor to measure individual's level of consciousness during general anaesthesia. (Courtesy: Alars Medical Systems, U.K)*

Since individual responses to a particular dose of anaesthetic vary considerably, it is advisable to measure the effect of the anaesthetic on a patient's level of consciousness. One method to do so is to measure Auditory Evoked Potentials (AEP), which is a neuro-physiological indicator of the changes in the level of consciousness during anaesthesia. This is an electrical signal contained within the EEG, which is obtained by delivering an auditory stimulus to the patient's acoustic nerve. The fast extraction of the complex AEP signal, the brain's response to the auditory stimulus of the acoustic nerve is obtained by mapping the signal and establishing an index which is developed as a graphic curve and a single number on the monitor screen (Fig. 32.7). This index which is calculated from a proprietary mathematical modelling method, quantifies the level of anaesthesia. For example, typically, if this index is higher than 60, the patient is awake, and decreases in line with decreasing level of consciousness (loss of conscious typically occurring below 30). Re-usable head-phones apply stimulation to the acoustic nerve to obtain AEP, which is then measured by a set of three disposable electrodes; two electrodes are applied in the forehead and one behind the ear to estimate the level of consciousness in a fast and non-invasive manner.

Ventilators

⯈ 33.1 MECHANICS OF RESPIRATION

Respiration is the process of supplying oxygen to and removing carbon dioxide from the tissues. These gasses are carried in the blood, oxygen from the lungs to the tissues and carbon dioxide from the tissues to the lungs. The gas exchanges in the lungs are called external respiration and those in the tissues are called internal respiration. There is a very delicate balance between the absorption and excretion of oxygen and carbon dioxide in the lungs and tissues, and this balance is maintained by the respiratory or breathing activity.

The organs of respiration are shown in Fig. 33.1. They are typically divided into the following parts:

Conducting Section: This includes the nasal cavities, pharynx, larynx, trachea, bronchi and bronchioles. These organs are thick-walled and do not participate in the gas exchange to capillaries.

Respiratory Section: This includes respiratory bronchioles, alveolar ducts and alveolar sacs. These contain thin walls and permit gas exchange to blood capillaries.

Both sections function through the muscles of respiration consisting of diaphragm and intercostal/chest muscles, ribs, and sternum. The chest is formed by twelve pairs of ribs joined together by muscles and connective tissue. It forms a closed cavity sealed off from the outside air except for a flexible non-collapsible tube, the trachea, which leads up to the larynx, nose and mouth. Just below the level of the collar bones, the trachea divides into right and left divisions or the bronchi.

Bronchi branches into about 20 non-symmetrical branches and then leads to bronchioles, each having a small diameter of about 1 mm. The bronchioles become progressively smaller until finally they lead into the alveoli where the exchange of gasses between the blood and lungs takes place. At the alveoli, only two thin layers of cells separate the air from the blood and gasses can diffuse freely between them. As it traverses the alveolar capillaries, the haemoglobin of the blood takes up oxygen and carbon dioxide passes out of the blood into the alveolar spaces. Each alveolus is extremely tiny (0.2 mm diameter), but because of the very large number of the alveoli present

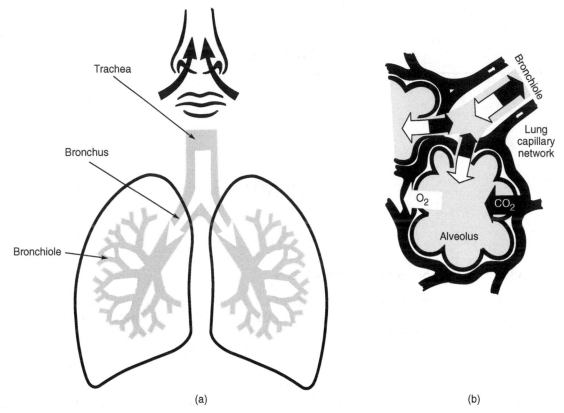

(a)   (b)

➤ **Fig. 33.1(a)**   *Organs of respiration* **(b)**   *Diagram of alveolar gas exchange*

(300 million), they comprise about 70 m^3 of the surface area. This gives rise to a total lung capacity of 3.6 to 9.4 litres in the adult male and 2.5 to 6.9 in the normal female.

The lungs consist of two cone-shaped spongy organs that contain the alveoli (air sacs) that trap air for gas exchange with the blood. The lungs are covered by a smooth glistening membrane called the pleura, which turns back at the root of each lung and covers the inner surface of the chest wall. Normally, there are two layers of pleura: a moist membrane, the visceral pleura which covers the lung surface and the parietal pleura which lines the thoracic cavity. The fluid-lined space between two membranes accounts for easy slippage between the lung and chest walls during breathing. The lungs are normally stretched or expanded against the resistance of elastic fibres' inspiration and expiration. The two sets of muscles involved are the diaphragm (a thin sheet of muscle which separates the thorax from the abdominal cavity), that moves up and down and the inter-costal muscles (surrounding the thoracic cavity), that move the rib cage in and out. Inspiration results from contraction of the diaphragm and the inter-costal muscles, whereas expiration results from their relaxation. There is no active participation of the lungs in the movements. The rate and depth of breathing are controlled from the brain in the medulla region. The control impulses reach respiratory muscles via the spinal cord, control the contraction of the diaphragm and raise the ribs to increase thorax cavity. Also, the changes in the metabolism due to different types of chemical reactions in the organs of the body regulate the process of respiration. The sensors known as chemo-receptors also regulate the breathing process. The chemo-receptors are directly influenced by the concentration of O_2, N_2, CO_2, carbonic acid, temperature and the flow rate of blood.

If breathing stops for more than five minutes, death or permanent damage will almost certainly occur. This may happen in many conditions such as asphyxia, carbon monoxide poisoning, drowning and electric shock, and artificial respiration is then essential.

33.2 ARTIFICIAL VENTILATION

For reduced breathing or respiratory failure (insufficiency), mechanical devices or respirators are used in hospitals. These devices provide artificial ventilation, supply enough oxygen and eliminate the right amount of carbon dioxide, maintain the desired arterial partial pressure of carbon dioxide ($PaCO_2$) and desired arterial oxygen tension (PaO_2).

Mechanical aids for manual artificial ventilation consist of a mask, breathing valve and self-filling bag (Fig. 33.2). The mask, which is of soft rubber or plastic, is held firmly over the patient's mouth and nose so that it fits tightly. The breathing valve serves to guide the air so that fresh air or air enriched with oxygen is supplied to the patient and expired air is conducted away. The bag is squeezed with one hand and functions as a pump. It is self-expanding and fills automatically with fresh air or oxygen when the patient breathes out.

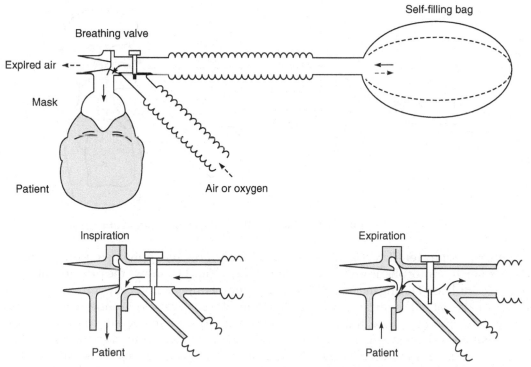

> **Fig. 33.2** *Mask, breathing valve and self-filling bag for artificial ventilation*

▥▹ **33.3 VENTILATORS**

When artificial ventilation needs to be maintained for a long time, a ventilator is used. Ventilators are also used during anaesthesia and are designed to match human breathing waveform/pattern. These are sophisticated equipment with a large number of controls which assist in maintaining proper and regulated breathing activity. For short-term or emergency use, resuscitators are employed. These depend upon mechanical cycle operation and are generally light-weight and portable.

The main function of a respirator is to ventilate the lungs in a manner as close to natural respiration as possible. Since natural inspiration is a result of negative pressure in the pleural cavity generated by the movement of the diaphragm, ventilators were initially designed to create the same effect. These ventilators are called *negative-pressure ventilators.* In this design, the flow of air to the lungs is facilitated by generating a negative-pressure around the patient's thoracic cage. The negative-pressure moves the thoracic walls outward, expanding the intra-thoracic volume and dropping the pressure inside the lungs, resulting in a pressure gradient between the atmosphere and the lungs which causes the flow of atmospheric air into the lungs. The inspiratory and expiratory phases of the respiration are controlled by cycling the pressure inside the body chamber. However, because of several engineering problems impeding the implementation of the concept and the difficulty of accessing the patient for care and monitoring, negative pressure ventilators have not become really popular.

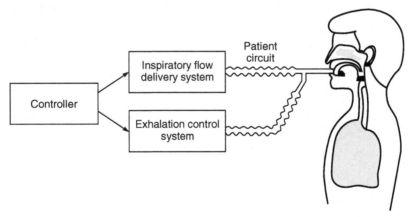

> **Fig. 33.3** *Functional diagram of a positive pressure ventilator*

Positive-pressure ventilators generate the inspiratory flow by applying a positive pressure—greater than the atmospheric pressure—to the airways. Fig. 33.3 shows the principle of a positive pressure ventilator. During the inspiration, the inspiratory flow delivery system creates a positive pressure in the patient circuit and the exhalation control system closes the outlet to the atmosphere. During the expiratory phase, the inspiratory flow delivery system stops the positive pressure at the exhalation system and opens the valves to allow the exhaled air to the atmosphere. Positive pressure ventilators have been found to be quite successful in treating patients with a wide range of pulmonary disorders.

Positive pressure ventilators operate either in mandatory or spontaneous mode. In spontaneous breath delivery, the ventilator responds to the patient's effort to breath independently. Therefore, the patient can control the volume and the rate of respiration. Spontaneous breath delivery is used for those patients who are on their way to full recovery but are not completely ready to breathe from the atmosphere without mechanical assistance. In contrast, when delivering mandatory breaths, the ventilator controls all parameters of the breath such as tidal volume, inspiratory flow waveform, respiration rate and oxygen content of the breath. Mandatory breaths are normally delivered to the patients who are incapable of breathing on their own.

In general, most ventilators in clinical use employ positive pressure during inspiration to inflate the lungs with mixture of gasses (air, oxygen). Expiration is usually passive, though under certain conditions, pressure may have to be applied during the expiratory phase in order to improve arterial oxygen pressure.

33.4 TYPES OF VENTILATORS

Anaesthesia Ventilators: These are generally small and simple equipments used to give regular assisted breathing during an operation.

Intensive Care Ventilators: Intensive care ventilators are more complicated, give accurate control over a wider range of parameters and often incorporate 'patient triggering facility,' i.e. the ventilator delivers air to the patient when the patient tries to inhale.

33.5 VENTILATOR TERMS

Lung Compliance: The compliance of the patient's lungs is the ratio of volume delivered to the pressure rise during the inspiratory phase in the lungs. This includes the compliance of the airways. Compliance is usually expressed as litres/cm H_2O.

Lung compliance is the ability of the alveoli and lung tissue to expand on inspiration. The lungs are passive, but they should stretch easily to ensure the sufficient intake of the air.

A ventilator and other parts of the breathing circuit also have compliance and some of the delivered volume is used to compress gas or expand gas in these parts.

The compliance of a patient's lungs is the ratio of pressure drop across the airway to the resulting flow rate through it. It is also expressed as cm H_2O/litres (pressure drop/flow rate).

Airway Resistance: Airway resistance relates to the ease with which air flows through the tubular respiratory structures. Higher resistances occur in smaller tubes such as the bronchioles and alveoli that have not emptied properly.

Mean Airway Pressure (MAP): An integral taken over one complete cycle expresses the mean airway pressure (Fig. 33.4).

Inspiratory Pause Time: When the pressure in the patient circuit and alveoli is equal, there is a period of no flow. This period is called inspiratory pause time (Fig. 33.5).

Inspiratory Flow: Inspiratory flow is represented as a positive flow above the zero line (Fig. 33.6).

Expiratory Flow: Expiratory flow is a negative flow below the zero line (Fig. 33.6).

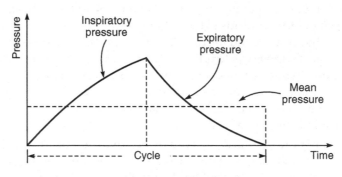

> **Fig. 33.4**   *Mean airway pressure*

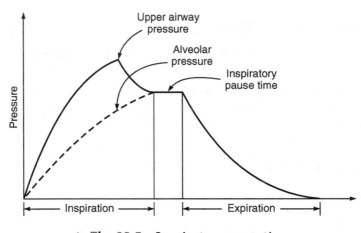

> **Fig. 33.5**   *Inspiratory pause time*

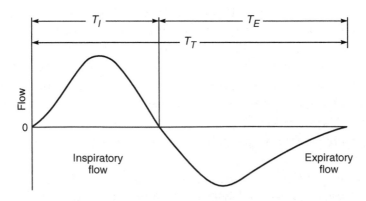

> **Fig. 33.6**   *Inspiratory/expiratory flow, airway pressure and volume-time diagrams, where the flow is approximately sinusoidal*

Tidal Volume: Tidal volume is the depth of breathing or the volume of gas inspired or expired during each respiratory cycle. It can be calculated by multiplying the flow rate (l/sec) setting by the set inspiratory time (seconds). Calibrated tidal volume settings range from 0.010 litre to 4.8 litres. If the flow is set at 0.6 l/s and inspiratory time is set at 1 sec, the tidal volume is = 0.6 litres.

Minute Volume: This refers to volume of gas exchanged per minute during quiet breathing. Minute volume is obtained by multiplying the tidal volume by the breathing rate.

Respiration Rate: This is the number of breaths per second. It represents total respiratory rate of the patient. In the assist-control mode and SIMV (Synchronized Intermittent Mandatory Ventilation) mode, the ventilator measures the previous four breaths and shows the average total rate, which is prescribed rate plus the additional breaths taken by the patient.

Conventional Mechanical Ventilation (CMV): This provides the force which determines the tidal volume (V_T) at a respiratory frequency (f) to achieve the desired minute ventilation (VE)

$$VE - V_T \times f$$

Intermittent Mandatory Ventilation (IMV): This allows the insertion of a variable time delay between each breath.

Inspiratory Expiratory Phase Time Ratio (I : E Ratio): This signifies the ratio of inspiratory interval to expiratory interval of a mandatory breath. This ratio is normally limited to 1:1, i.e. the inspiratory time should not exceed 50% of the total ventilator cycle time as set by the breath/minute control. Inverse I:E ratio is prevented.

Synchronized Intermittent Mandatory Ventilation (SIMV): It represents a combination of machine ventilation and spontaneous breathing. SIMV enables the patient to breathe spontaneously in regular prescribed cycles, with the mechanical mandatory ventilation strokes providing a minimum ventilation during the remaining cycles.

Synchronized Intermittent Mandatory Ventilation delivers a prescribed tidal volume and respiratory rate. The patient may then breathe spontaneously in between the delivered breaths (Fig. 33.7).

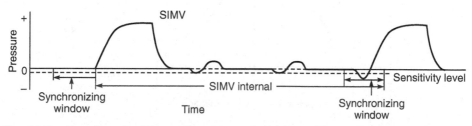

> **Fig.** 33.7 *Synchronized intermittent mandatory ventilation. It is activated simultaneously only when the patient's breath is detected during the last quarter of the set total breath cycle*

Sigh Volume: One sigh breath is 150% of the set tidal volume.

Patient Circuit: This includes a set of tools collecting the patient airway to the outlet of a ventilator.

Oxygen Percentage (F_1O_2): In all ventilatory modes, oxygen is delivered during the inspiratory phase and the percentage (F_1O_2) is adjustable from 21 to 91%.

Peak Airway Pressure: It is the highest level of pressure reached over several breaths.

Spontaneous Ventilation: This is a ventilation mode in which the patient initiates and breathes from the ventilator at will.

Bias Flow: In bias flow, mixed gas from the mixer is directed through the patient circuit in-between mechanical breaths. Bias flow stabilizes baseline pressure for spontaneously breathing patients and decreases the response time of the demand valve.

Sensitivity: It is used to detect spontaneous effort by the patient, in order to trigger mandatory ventilation with the set Respiration rate.

Mandatory Minutes Volume Ventilation (MMV): This operating mode applies mandatory ventilation only if spontaneous breathing is not yet sufficient and has fallen below a pre-selected minimum ventilation. Unlike SIMV, the mandatory strokes are not applied regularly but only in cases of insufficient ventilation.

Controlled Mandatory Ventilation: This term refers to mandatory ventilation of patients who are not able to initiate or respire on their own.

Assisted Spontaneous Breathing (ASB): It refers to the pressure support of insufficient spontaneous breathing.

Positive End Expiratory Pressure (PEEP): PEEP is a therapist-selected pressure level for the patient airway at the end of expiration in either mandatory or spontaneous breathing. PEEP is used to increase the end-expiratory lung volume (EELV) or prolong expiration with a potentially similar effect on the EELV (Fig. 33.8).

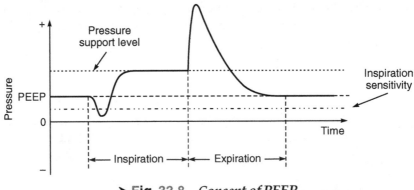

> **Fig. 33.8**   *Concept of PEEP*

Continuous Positive Airway Pressure (CPAP): CPAP is a spontaneous ventilation mode in which the ventilator maintains a constant positive pressure, near or below PEEP Level, in the patient's airway while the patient breathes at will.

Assist/Control Ventilation: During this process, a positive pressure breath is delivered with each patient's spontaneous inspiratory effort to reach the trigger level setting. In volume controlled assist control, tidal volume is determined by flow and inspiratory time settings. If the patient does not trigger the ventilation, it automatically delivers breaths according to the set rate.

Pressure Relief Valve: It determines the maximum pressure that can be reached in the patient circuit during spontaneous mechanical and manual ventilation. It is adjustable from 0-100 cm/ H_2O and functions in all modes.

⁣▶ 33.6 CLASSIFICATION OF VENTILATORS

Ventilators can be classified in terms of various methods. Discussed below are the general criteria for systematic listing and description of these classifications.

33.6.1 Based on the Method of Initiating the Inspiratory Phase

Controller: A ventilator which operates independent of the patient's inspiratory effort. The inspiration is initiated by a mechanism which is controlled with respect to time, pressure or another similar factor. Controlled ventilation is required for patients who are unable to breath on their own.

Assistor: A ventilator which augments the inspiration of the patient by operating in response to the patient's inspiratory effort. A pressure sensor detects the slight negative pressure that occurs each time the patient attempts to inhale and triggers the process of inflating the lungs. Thus the ventilator helps the patient to inspire when needed. A sensitivity adjustment provided on the equipment helps to select the amount of effort required on the patient's part to trigger the inspiration process. The *assist* mode is required for those patients who are able to breathe but are unable to inhale a sufficient amount of air or for whom breathing requires a great deal of effort.

Assistor/Controller: A ventilator which combines both the controller and assistor functions. In these devices, if the patient fails to breathe within a pre-determined time, a timer automatically triggers the inspiration process to inflate the lungs. Therefore, the breathing is controlled by the patient as long as it is possible, but in case the patient should fail to do so, the machine is able to take over the function. Such devices are most frequently used in critical care units.

33.6.2 Based on Power Transmission

Direct Power Transmission: A ventilator which delivers the gas directly from the source of compressed gas to the patient (Fig. 33.9(a)).

Indirect Power Transmission: A ventilator which has separate patient and power systems (Fig. 33.9(b)). The pressure in the power system determines the flow rate.

33.6.3 Based on Pressure Pattern

Positive-Atmosphere: A ventilator which produces a positive pressure in the patient's lungs during inspiration, with an end expiratory pressure that is equal to the atmospheric pressure. In this mode, the mean airway pressure is always higher than the atmospheric pressure and the patient normally breathes spontaneously with this mode of operation (Fig. 33.10(a)).

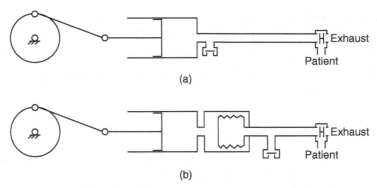

> **Fig.** 33.9(a) *Direct power transmission*
>   (b) *Indirect power transmission*

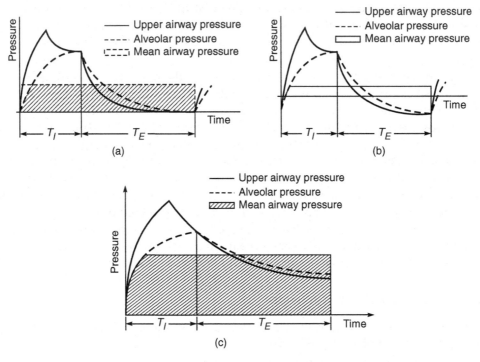

> **Fig.** 33.10 *Pressure-time diagram for*
>   (a) *Positive-atmosphere pressure pattern*
>   (b) *Positive-negative pressure pattern*
>   (c) *Positive-positive pressure pattern*

Positive-Negative: A ventilator which produces a positive pressure in the patient's lungs during inspiration and below atmospheric pressure in the airway during part of expiratory phase (Fig. 33.10b). A positive-negative pressure pattern results in a low mean airway pressure.

Positive-Positive: A ventilator which produces a positive pressure in the patient's lungs during inspiration, with an end expiratory pressure that is greater than the atmospheric pressure (Fig. 33.10c). In order to obtain an end expiratory pressure that is greater than the atmospheric pressure, it is necessary to start the inspiratory phase before the airway pressure reaches the atmospheric pressure.

33.6.4   Based on the Type of Safety Limit

Volume Limited: A ventilator in which pre-determined volume cannot be exceeded during inspiration. Volume limit normally refers to tidal volume.

Pressure Limited: A ventilator designed in such a way that predetermined pressure cannot be exceeded during inspiration.

Time Limited: A ventilator in which predetermined phase time cannot be exceeded. It limits the expiratory phase time if the patient does not initiate the inspiratory phase and is common to ventilators used for assisted ventilation.

33.6.5   Based on Cycling Control

Cycling control of a ventilator is the device which determines the change from the inspiratory phase to the expiratory phase and vice versa. The cycling of a ventilator may be based upon different factors such as pressure, volume, time and the inspiratory effort made by the patient. The common types of cycling controls are described below.

33.6.6.1   Cycling from Inspiration to Expiration

Volume Cycled: A ventilator which starts the expiratory phase after a preset tidal volume has been delivered into the patient circuit. This device normally has a pressure over-ride valve so that if, while the machine is in the process of administering the set volume, the pressure exceeds a pre-determined maximal value, the ventilator will cycle whether or not the appropriate volume has been administered.

Pressure Cycled: A ventilator which begins the expiratory phase after a preset pressure has been attained.

Time Cycled: A ventilator which initiates the expiratory phase after a preset time period for the inspiratory phase has passed.

33.6.6.2   Cycling from Expiration to Inspiration

Pressure Cycled: A ventilator which begins the inspiratory phase after a pre-set end expiratory pressure has been attained.

Time Cycled: A ventilator which initiates the inspiratory phase after a preset time period for the expiratory phase has passed.

Patient Inspiratory Effort Cycled: A ventilator which starts the inspiratory phase in response to the inspiratory effort.

33.6.7　Based on the Source of Power

Pneumatic: A ventilator powered by compressed gas.

Electric: A ventilator powered by an electrical device such as an electric motor, or similar gadget.

ⅢⅢ▷ 33.7　PRESSURE-VOLUME-FLOW DIAGRAMS

In order to understand the performance of a ventilator, it is necessary to be familiar with the pressure-time, flow-time and volume-time diagrams. The ventilated system consists of the patient circuit, the airway and the alveoli, each having its own compliance. After the start of the inspiratory phase, a certain gas volume is delivered into the system, resulting primarily in an increase of pressure in the patient circuit and subsequently, in a flow through the airway. During the inspiratory phase, the airway pressure and alveolar pressure increase gradually with the airway pressure always being higher than the alveolar pressure.

The equal pressure of the patient circuit and alveoli determines the end of the inspiratory flow and beginning of the expiratory flow, due to the fact that the pressure in the patient system is allowed to decrease. The expiratory flow is determined by the difference between alveolar pressure and pressure in the patient circuit. It may thus be noted that:

- – an airway pressure higher than the alveolar pressure characterizes an inspiratory flow; and
- – an airway pressure lower than the alveolar pressure characterizes an expiratory flow.

It may be observed that it is necessary to provide for a time delay (pause time) between the cycling of the ventilator and the change from inspiratory flow to expiratory flow in the airway. During this pause time, the flow becomes zero when the alveolar pressure equals the airway pressure and constant volume is maintained in the lungs. Ventilators producing a pause time during inspiration or expiration have certain advantages over ventilators without such a pause and are therefore preferred over the latter. Figure 33.11 shows pressure, flow and volume pattern in a ventilated system with and without pause.

ⅢⅢ▷ 33.8　MODERN VENTILATORS

The current and future trends in critical care ventilatory management demand precise flow, pressure and oxygen control for application to both adult and paediatric patients. In addition, patient monitoring and rapid, understandable alarms are extremely important for timely care of the patient. This has become possible by making use of computer technology in the ventilators to achieve a wide range of functions and controls.

Modern ventilator machines consist of two separate but inter-connected systems: the pneumatic flow system and an electronic control system. Figure 33.12 shows a block diagram of a typical ventilator.

The pneumatic flow system enables the flow of gas through the ventilator. Oxygen and medical grade air enter the ventilator at 3.5 bar (50 psi) pressure through built-in 0.1 micron filters. The normal operating range is 2 to 6 bar or 28 to 86 psi. These gasses enter the air/oxygen mixer where

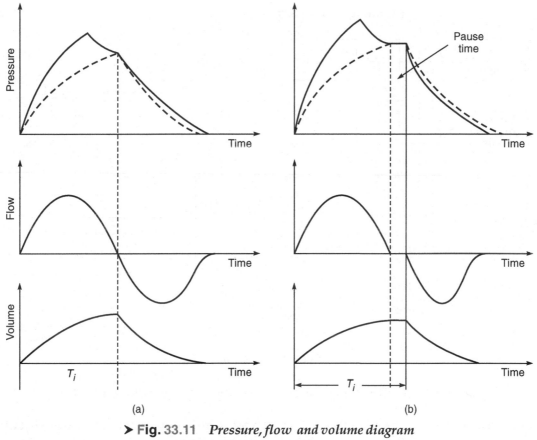

> **Fig. 33.11** *Pressure, flow and volume diagram*
> **(a)** *without pause time*
> **(b)** *with pause time*

they combine at the required percentage and reduced in pressure to 350 cm H_2O. The gasses then enter a large reservoir tank which holds about 8 litres of mixed gasses, when compressed to 350 cm H_2O. An electronically controlled flow valve proportions the gas flow from the reservoir tank to the patient breathing circuit. In some ventilators, an air compressor is used in place of a compressed air tank. The primary objective of the device is to ensure proper level of oxygen in the inspiratory air and deliver a tidal volume according to the clinical requirements.

As the gasses leave the ventilator, they pass by an oxygen analyser, a safety ambient air inlet valve and a back-up mechanical over pressure valve. The ambient valve provides the patient the ability to breathe room air when the ventilator fails or the pressure in the patient circuit drops below –10 cm of H_2O. In the patient breathing circuit is a bi-directional flow sensor to measure the gas flows. The exhaled gasses exit through an electronically controlled exhalation valve located at the ventilator. With the introduction of microprocessors for control of metering devices, electro-mechanical valves have gained popularity. The microprocessor controls each valve to deliver the desired inspiratory air and oxygen flows for mandatory and spontaneous ventilation. A high

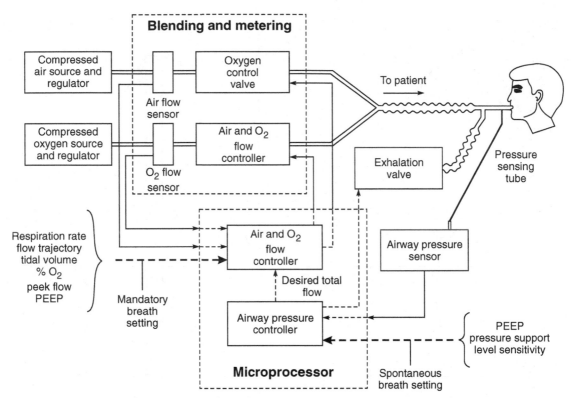

> ➤ **Fig. 33.12**   *Block diagram of a microprocessor controlled ventilator*

pressure valve is used to provide safety in case the pressure in the patient circuit exceeds 110 cm of H_2O.

The electronic control system may use one or more microprocessors and software to perform monitoring and control functions in a ventilator. These parameters include setting of the respiration rate, flow waveform, tidal volume, oxygen concentration of the delivered breath, peak flow and PEEP. The PEEP selected in the mandatory mode is only used for control of exhalation flow. The micorprocessor utilizes the above parameters to compute the desired inspiratory flow trajectory. The system consists of monitors for pressure flow and oxygen fraction. The sensors are connected to electronic processing circuits which makes them available for digital readouts. The signals are also compared with pre-set alarm levels so that if they fall outside a pre-determined normal range, alarms are sounded.

The pressure sensors are normally of semiconductor strain gauge type placed in a bridge configuration. For measurement of fraction of oxygen in the inspired air, a fuel cell type oxygen sensor is used. This sensor generates a current proportional to pO_2. As this sensor is temperature-sensitive, compensation for its operating temperature is included in the circuit. Usually, a thermistor is used to carry out this function. The flow sensor usually consists of a variable orifice and by measuring the pressure drop across the variable orifice, the patient flows can be calculated.

Ventilators are life saving equipment and therefore need regular maintenance and calibration which should be carried out as per the instructions of the manufacturers.

⬛⬛⬛▶ 33.9   HIGH FREQUENCY VENTILATORS

A new technique for ventilating patients at frequencies much higher than the respiration rate has recently been introduced. This method has been shown to improve CO_2 wash out and provide adequate oxygenation without the requirement for high inspiratory pressures (Chan and Greenough, 1993). The key principle in this technique is to provide tidal volumes equal to or smaller than the dead space, at very high rates.

In conventional positive pressure ventilation, CO_2 elimination is directly controlled by the amount of applied minute ventilation. However, it is known that mean airway pressure is the parameter that best correlates with improvement in oxygenation. Gas transport during conventional ventilation is attributed to two basic mechanisms: (i) convection or flow of gas through the conducting airways, and (ii) molecular diffusion of gasses into the alveoli and pulmonary capillaries. The tidal volume (V_T) applied to the patient at the Y-piece can be divided into the volume used to ventilate the dead space (V_D) and the alveolar volume (V_{Talv}). Only the alveolar volume takes part in the gas exchange process. Therefore,

$$V_{Talv} = V_T - V_D$$

The portion of the tidal volume used to ventilate the dead space does not take part in capillary gas exchange and is therefore wasted. To overcome the problem of wasted ventilation in conventional ventilation, the inspiratory pressure is increased in order to increase the total tidal volume. Unfortunately, however, this also increases the mechanical stress on the lung and has been associated with various traumas. High frequency ventilation has been shown to provide adequate alveolar ventilation and oxygenation without the requirement for high inspiratory pressures (Hamilton et al., 1983).

High frequency (HF) ventilators are now commercially available, the most popular being the Babylog 8000 from M/s Drager, Germany. The ventilator generates high frequency rate from 5 to 20 Hz (300 to 1200 pulse/minute). Although several methods are available to generate the high frequency pressure waves, the Babylog 8000 makes use of an oscillating diaphragm mechanism. This mechanism is computer-controlled and can precisely determine the shape of the pressure swings and I:E ratio.

An alternative method of achieving HF ventilation is based on the jet principle in which a small diameter tube is passed down a tracheal cannula and is either terminated at its distal end or extended into the trachea itself. Short pulses of higher pressure oxygen are introduced into the airway through the cannula at frequencies well above the normal respiration rate. This technique has the disadvantage of forcing volume into the patient and then leaving the patient to exhale passively, which may lead to some trapped volume inside the lung increasing the mean lung pressure. This problem is overcome by ensuring that the pressure during the exhalation phase is negative with respect to the set PEEP.

Boynton et al. (1984) found that combined high frequency ventilation and conventional ventilation facilitated gas exchange in certain critically ill neonates. Blanco et al. (1987) reported that the use of high frequency ventilation in combination with conventional ventilation produced a significant improvement in gas exchange at a lower airway pressure.

The commercial HF ventilators operate at frequencies of between 5 to 20 Hz. Inspiration-to-expiration ratios can usually be varied from 1 : 1 to 1 : 4. The wave shape can be sinusoidal or rectangular. Like conventional ventilators, HF ventilators are usually microprocessor-controlled and provide an integrated system featuring high frequency, triggered and conventional ventilation.

ⅢⅢ▷ 33.10 HUMIDIFIERS, NEBULIZERS AND ASPIRATORS

Apart from ventilation, humidification of the breathing gas plays a leading role in the intensive care of patients. The main task of a humidifier is to replace humidity in the upper air passages which has been lost by intubation. The humidity should be as close to 100% as possible, or speaking in terms of water, the absolute content per litre breathing gas should be more than 30 mg, regardless of environmental conditions. Therefore, in order to prevent damage to the patient's lungs, the air or oxygen applied during respiratory therapy must be humidified. Thus, all ventilators include arrangements to humidify the air, either by heat vapourization (stream) or by bubbling an air stream through a jar of water.

When water or some type of medication suspended in the inspired air as an aerosol is to be administered to the patient, a device called a nebulizer is used. In this device, the water or medication is picked up by a high velocity jet of air/oxygen and made to impact against one or more baffles to break the substance into controlled-sized droplets which are then applied to the patient via a respirator. More effective and efficient nebulizers are based on the use of high intensity ultrasound energy which vibrates the substance (water or medication) to produce a high volume of minute particles. Ultrasonic nebulizers do not depend upon breathing gas for operation and thus therapeutic agents can be conveniently administered during ventilation procedure.

Aspirators are often included as part of a ventilator to remove mucus and other fluids from the airways. Alternatively, a separate suction device may be utilized to achieve the same purpose.

CHAPTER

34

Radiotherapy Equipment

Cancer is one of the leading causes of death in the world. It is presently managed using one of or a combination of three methods of treatment: surgery, chemotherapy and radiation therapy. More than half of all cancer patients receive radiation therapy either as a primary or adjunctive treatment. The number of radiotherapy patients is likely to increase substantially because the percentage of elderly in the population is increasing, which means that the incidence of cancer will also increase. Radiotherapy with a good success rate and lower total cost of treatment, is considered as a preferred method in cancer management among patients.

34.1 USE OF HIGH VOLTAGE X-RAY MACHINES

Shortly after their discovery in 1895, X-rays were used to treat cancer. These early X-ray devices were of limited value in treating many types of cancer as the penetration of the radiation was inadequate to treat deep-seated tumours without doing significant damage to the healthy normal tissues overlying the tumours. Before 1951, radiation therapy was carried out almost exclusively by X-ray machines operating at tube voltages in the range of 400 KV. Such machines produce X-ray beams having a broad spectrum of X-ray energies with an average of one-third or less of the maximum. Thus, a 400 KV machine would correspond to a single energy of about 133 KeV.

Around the same time, higher energy radiation machines designed for use in high-energy physics research, such as the resonant transformer and Van de Graaff generator were put into medical use. These devices used direct acceleration methods to achieve energies of 1 or 2 million electron volts (MeV). Although they provided more penetrating X-rays, they were cumborsome devices to use and were difficult to operate and maintain. As a result, they did not receive wide acceptance.

34.2 DEVELOPMENT OF BETATRON

More popular amongst these devices was the betatron, which was used for cancer treatment in the 1950s. The betatron produced high-energy X-ray beams as well as several electron beams of various energies. Some betatron units provided X-ray and electron energies as high as 45 MeV. The great variety of treatment beams from the betatron brought a new dimension to radiotherapy. Because of their greater penetration through thick body sections, high-energy X-ray beams from the betatron were found to be well-suited for treating tumours of the trunk and pelvis. Betatron electron beams proved to be particularly useful because they deposit most of their radiation dose within a few centimetres of the surface and allowed treatment of superficial lesions while sparing the normal tissue and critical structures beneath the tumour.

However, despite their advantages, betatrons had serious drawbacks that limited their applications. Higher energy betatrons (25 MeV or greater) were large, heavy, unwieldy and cumbersome devices. They were difficult to manoeuvre around the patient due to their heavy weight and large size. Lack of a full 360 degree rotation of the treatment unit around a stationary patient made it difficult to achieve accurate patient set-ups during the course of treatment.

More compact betatrons of about 18 MeV though were relatively easy to manoeuvre, but they could produce only very low X-ray output. The maximum field size was also limited to about 20×20 cm. For larger fields, the treatments had to be given at an extended distance, further diminishing the radiation output. This low output, together with the difficulty in patient positioning generally limited the throughput in terms of number of patients that could be treated in a day. Finally, betatrons were complex devices, often requiring a technically knowledgeable person to be present on-site to maintain the unit. With these drawbacks, the betatrons did not become popular and their use remained confined only to the larger treatment facilities.

34.3 COBALT-60 MACHINE

In the early 1950s, the cobalt unit was introduced into medical use. Cobalt is a hard metallic substance and has an atomic number of 27, an atomic weight of 58.933 and a mass density of 8900 kg m^{-3}, and melts at about 1500°C. It was a simple, compact and reliable low-energy radiation treatment device using a pellet of radioactive cobalt isotope as a source of radiation. The cobalt unit made practical the widespread use of radiation therapy, not only in the larger treatment centres and university hospitals, but at hospitals and clinics throughout the world. The increased use of radiation as a therapeutic modality stimulated interest providing more compact sources of even higher energy radiation for treatment.

The radioactive isotope ^{60}Co does not exist in nature. It is man-made and is produced when the stable isotope of cobalt, ^{59}Co, is bombarded by neutrons in a nuclear reactor.

It has a relatively long half-life (5.26 years), and when it decays, it yields two γ rays with energies of 1.17 and 1.33 MeV respectively. These are produced in equal number and can be approximated by their average, 1.25 MeV, to form radiation that has high penetration in matter. These properties makes ^{60}Co unique as a source of radiation for cancer treatment. Cobalt units are mechanically and electrically simple devices and following their introduction, they rapidly became the standard machine for treatment of nearly all types of cancers.

Very soon after the cobalt unit became available commercially, and the production of cobalt sources and cobalt units expanded to such an extent that, for 30 years, more radiotherapy was carried out with cobalt-60 than with all other types of radiation combined. Cobalt machines have the tremendous advantage of producing a completely predictable, steady, reliable beam of relatively high-energy radiation, which is also relatively easy to repair when required.

Production of ^{60}Co Source: Almost any material placed within the neutron radiation field of a nuclear reactor will become radioactive. In the case of ^{60}Co, the activity produced is determined by the neutron flux density in the reactor, the neutron capture cross-section, the amount of Cobalt-59 inserted into the reactor, and the length of time it is left there.

A simple radiation treatment typically involves an absorbed dose at the tumor of 2.0 Gy (200 rad). Because of attenuation in the tissues and various other factors, this would imply an exposure of, say, 200 R. The irradiation part of the treatment should not last for much more than 1 ½ minutes., and this would require a source activity of very nearly 4000 Ci. In order to attain this level of radioactivity, the cobalt must be left in the reactor for a long time as while ^{60}Co is being formed it is also decaying. The resulting activity would be the sum of that which is being produced and the amount that decays.

Detailed mathematical calculations would show that to produce a source of strength 4000 Ci, the cobalt would have to remain in the reactor for over two years. However, as time has passed, reactor fluxes have increased. This has allowed the irradiation time to be shortened and the source size to be made smaller.

In actual practice, sources are not irradiated as solid cylinders. Rather, they are made up into a capsule on demand from stocks or pellets that were pre-irradiated to a selection of specific activities. Pellets are contained in a pair of stainless-steel containers. The pellets are loaded into the cylinder alongwith the spacers which are inserted to hold the pellets in position, and finally this cylinder, when capped, is inserted into another cylinder and cold-welded shut. All of these operations are carried out remotely in a hot cell. Finally, the source is shipped in a well-protected and shielded container to be loaded into a cobalt unit.

34.3.1 Contractional Details of the Cobalt Machine

A cobalt machine consists of the following major sub-systems: (Fig. 34.1):

1. Cobalt Source Head
2. Head Mounts;
3. Collimotor;
4. Treatment Table; and
5. Control Console and Safety Interlocks.

Cobalt Source Head: The heart of the system is the cobalt source. It is placed near the centre of a large, lead-filled steel container (Fig. 34.2). A number of methods have been devised for moving the source from the "off" to the "on" position. The two commonly used methods are shown in Fig. 34.3. In Fig. 34.3(a), the source is mounted

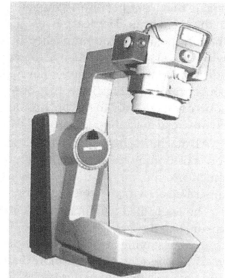

➤ **Fig. 34.1** *Various sub-systems in a cobalt machine*

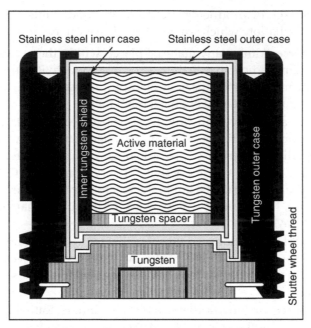

Stainless steel inner case   Stainless steel outer case

Inner tungsten shield

Active material

Tungsten outer case

Tungsten spacer

Shutter wheel thread

Tungsten

> **Fig. 34.2**   *Design layout of cobalt source*

in a heavy metal like tungsten wheel that is rotated through 180 degree to carry it from the 'off' position to the 'on' position. The transfer between 'on' and 'off' positions takes place in less than two seconds for pinpoint exposure accuracy. The motor drive acts against a heavy duty torque spring that instantly returns the source to the 'off' position for normal exposure termination and in the event of an electrical power failure. An electrical means for instantly returning the source to the 'off' position is accomplished by depressing an emergency push bar on the metre control which reverses the direction of the motor so that the source is driven back to its 'off' position. In addition, a large hand wheel is permanently monitored on the source head facial plate for manual source return. In Fig. 34.3b, the source is mounted in a sliding plug or drawer which carries the source from the 'off' to the 'on' position. This method has become the most commonly used one in commercial machines.

All machines must be arranged so that they are fail 'safe', that is, the source must be held in the 'on' position by the continuous application of a force so that if the power fails, it must return quickly to the 'off' position. This is usually provided by a strong spring and both the methods described above carry this provision.

The lead-filled container is 25 cm thick in all directions from the source. Its design should ensure that the leakage radiation coming through its thickness would not cause an over-exposure to anyone staying at its surface for prolonged periods of time. This would imply, for example, a yearly dose equivalent of no more than 50 mSv (or ~500 mrem) at a distance of 1 m from the source. The sievert is the unit of dose equivalent. One sievert will result in the same biological effect as 1 gray (Gy) of coventional X-rays. If we assume a maximum source strength of 10,000 Ci, and again use the exposure rate constant of 1.31 $Rm^2 h^{-1} Ci^{-1}$, and assume that 1 R corresponds to a dose

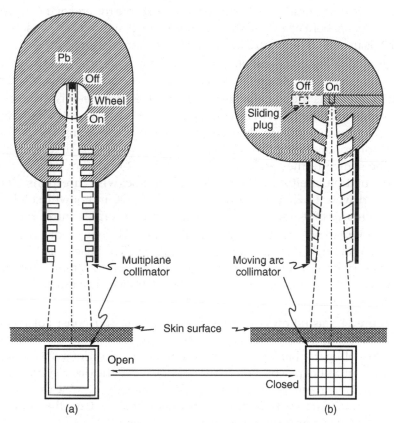

> **Fig. 34.3** *Mechanism for movement of the cobalt source in "on" and "off"*
> *positions*
>
>   **(a)** *The source is mounted in a wheel which rotates to carry the source*
>   *from a shielded "off" position to the "on" position*
>
>   **(b)** *The source is mounted in a sliding drawer or piston which moves the*
>   *source from "off" to "on" position. The diagrams also show a multi-*
>   *plane collimator with moveable bars to define the beam size*

equivalent of 0.01 Sv, this would imply a thickness of about 23 half-value layers. The half-value layer in lead for cobalt radiation is about 1.1 cm, and this calculation would imply a thickness of about 25 cm. In fact, 20 to 25 cm is about the thickness of the heads of most cobalt units. The source is 9000 Rhm ^{60}Co that provides 234 R/min at the axis of rotation. The transmitted radiation level does not exceed 10 mR/h at 100 cm from source at any point, nor should it exceed 2.0 mR/h overall average with beam off.

Mounts: There are two basic ways of mounting and 'porting' radiation treatment units. In the earlier machine design, the head of the unit is held in a yoke which can be moved up and down, or back and forth, and can be rotated about the axis. The unit is equipped with a treatment applicator which is mounted at the end of the collimator. The motions of the mount allow the unit to 'point' over a wide range of directions and enable the operator to place the end of the treatment applicator against the skin of the patient at a prescribed location. The distance from the source to the skin of the patient (SSD) is usually fixed at 80 cm, and the focus of the set-up is the surface of the patient.

In this case, the head mount is suspended from a set of rails attached to the ceiling. In other cases, the unit is mounted on a column with vertical motion and a pivot for head rotation.

The alternative mount is the so-called isocentric or fixed source-axis-distance (SAD) mount. The head, encased in a streamlined plastic cover, is mounted on a gantry that can rotate about a horizontal axis. The gantry supports the yoke-mounted biplane source head at one end and the radiation barrier or optional counter-weight at the opposite end. It rigidly maintains an 80 cm distance from the source to the axis of rotation located 120 cm above floor. The gantry has a continuous 360° rotation in either clockwise or center-clockwise directions. Both source head motions are motor driven at 0.25 rpm. The drive consists of a DC motor with solid state control and dynamic braking circuits that assure uniform rotation and accurate control of the gantry motion.

The gantry is mounted on a vertical support column which consists of a rugged heavy-gauge steel frame with 2.5 cm thick base and cap anchoring plates. The base plate is secured to the floor with two conventional anchoring bolts. The carriage and hanger assembly provide vibration-free movement alongwith the vertical guide rails. Control of the vertical movement is provided by a convenient pendant switch.

Collimators: Figure 34.3 also shows two types of collimators. Both consist of a set of bars that can produce a radiation beam with a rectangular cross-section. The diagrams at the bottom show an end-on view of the collimator bars in the open and the closed positions. The collimator rotates at an angle of 360° about its beam axis, and rotatory knobs lock the desired position. An indicator scale is calibrated in 1° increments. The field size is continuously variable to permit the shaping of square or rectangular fields from 3×3 cm to 35 cm $\times$ 35 cm at 80 cm from the source. The collimator vanes are motor-driven and adjust the field size.

The collimator is constructed of four sets of flat, inter-leaved lead vanes with angulated inner tungsten trees for continuously variable field sizes. As the vanes move, the beam defining tungsten trees automatically angulate to follow the divergence of the beam. The built-in system projects light field on the patient's skin to indicate precisely the radiation field at the 50% geometric penumbra line. A built-in optical device projects an easy-to-read scale on the patient's skin to facilitate a direct read-out. It indicates 60 to 100 cm in 1 cm increments. When the desired number is superimposed on cross hairs, the patient's skin is at the desired distance from the source.

Treatment Table: The patient lies on a couch which can be raised or lowered or moved sideways so that the tumour is positioned on this axis such that for any angle of the gantry, the beam will pass through the tumour. The axis of rotation is a fixed distance from the source, and the size of the beam is specified by its size at the axis. The focus of attention is now at the tumour rather than the surface. The mount is called isocentric because the axis of rotation of the gantry intersects with the central axis of the beam (axis of rotation of the collimator), so that they both rotate about the same centre. In addition, the couch can usually be rotated about a vertical axis, also passing through the isocentre. The travelling counter levered table top permits precise patient positioning prior to treatment with fixed or moving radiation beam. Longitudinal and transverse table top travels are manually controlled. Vertical elevation is motor-driven. In addition, the entire table pivots 180° about the vertical beam axis.

The carriage assembly consists of aluminium casting mounted directly below the table top. It includes bearing assemblies and positioning controls. Ball bearings ride in hardened steel tracks for longitudinal travel. Linear ball bearings move on hardened steel shafts for transverse travel. The patient couch is provided with multiple safeguards to protect the patient and the therapist.

Control Console and Safety Interlocks: The control console is placed outside the treatment room and facilititates control of various functions. It permits the selection of treatment techniques such as rotation, oscillation, skip-scanning or multi-portal indexing. The direction selector establishes clockwise or counter-clockwise rotation of the gantry. An exposure timer counts down as treatment time diminishes and is displayed by a four-digit LED readout. Similarly, the count-up exposure time is displayed with the progress of the treatment. An emergency button interrupts source-drive current for torque-spring return of some to 'beam off' condition with momentary actuation. Similarly, an entrance door interlock permits interconnection to the safety switch at the entrance door. Exposure automatically terminates if the entrance door is opened while the source is "ON". A 'safety key switch' controls the source transfer mechanism to prevent unauthorized use.

The cobalt unit is mechanically simple and its output is totally predictable and reliable. Sources with sufficient strength to enable short treatment times can easily be produced. The beam characteristics are well-known and relatively easy to measure. It is also easy to make special filters and beam modifiers for individual treatment needs.

Because of the source decay, sources must be renewed at intervals of around five years, but this procedure is quite straightforward and its expense is more than offset by the low maintenance cost of the machine.

▶ 34.4 MEDICAL LINEAR ACCELERATOR MACHINE

The conventional radiotherapy linear accelerators that produce X-ray and/or electron beams were introduced into clinical practice during the 1960s, and achieved widespread acceptance in the clinical community during the 1970s. The linear accelerator is a precise, reliable treatment instrument, with a wide range of capabilities. Today, most radiotherapy treatments are conducted using conventional radiotherapy linear accelerators.

Strictly speaking, the term 'linear accelerator' only applies to that part of the system wherein electrons are accelerated to the required level of energy. However, in this treatment, the term is used to describe the whole system used for radiotherapy treatments.

The accelerator is designed to deliver a mega-voltage X-ray beam with the characteristics necessary for modern radiotherapy techniques. In physical appearance, the linear accelerator machine is comprised of three major components: (a) gantry and stand, (b) treatment couch, and (c) control console.

The treatment beam is generated by a linear accelerator wave guide located in the gantry and produces a flattened photon beam of energy ranging from 4 to 20 MeV. The accelerator guide is a standing-wave design with side coupling cavities permitting a resonant standing wave condition which minimizes power loss in the accelerator and increases the efficiency of power absorption by the injected electrons. The accelerator guide is shielded to lower the leakage radiation to less than 0.1% of the primary beam intensity. The accelerator structure is colinear with the photon beam that it produces. The accelerator utilizes a tungsten target and an electron gun with an operating voltage not exceeding 35 kV.

The entire accelerator assembly is mounted in a rotating gantry (Fig. 34.4) with 100 cm target-axis distance. The gantry also contains the X-ray collimating system. Digital displays of the gantry rotation angle, collimator opening and collimator rotation angle are mounted on the gantry, at the axis of the rotation. The stand supports the gantry by means of a precision bearing mount for accurate isocentric gantry rotation. The stand contains electronic circuitry for motor control and the high voltage power supply for the microwave source and electron gun.

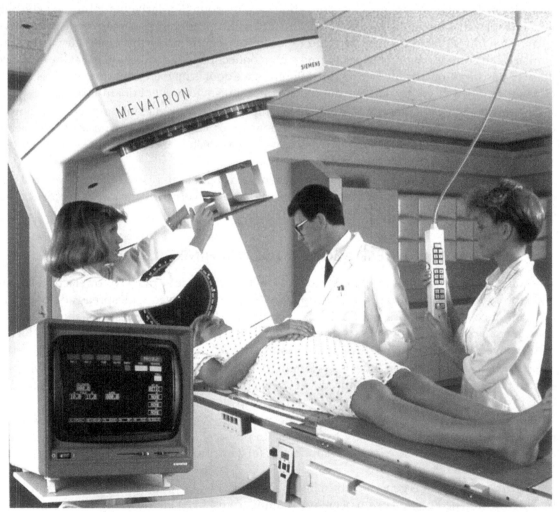

➤ **Fig. 34.4** *Linear accelerator machine in use (Courtesy: M/s Siemens, Germany)*

The collimation system is comprised of a primary collimator and two pairs of movable secondary collimators. The secondary collimators move approximately normal to the edge of the radiation field. The collimator system also contains a range finder and field defining light for determination of the target-skin distance and field size at the skin surface.

The photon beam flattening filter and two independent sealed transmission ion chambers, for measuring integrated dose and dose rate, are located above the movable collimators. The integrity of the dosimetry system is checked electronically before every treatment. Each ion chamber is designed to independently stop treatment when the desired integrated dose has been delivered. The ionization chamber also contains special electrode configurations to monitor beam symmetry in two orthogonal beam axes. The ionization chamber, in combination with other electronic circuitry, monitors beam energy.

The control console is designed to control the operation of the machine. Treatment is controlled at the console through switch settings for the dose rate, the integrated dose to be delivered, treatment time duration, and stop angle and dose per degree for arc therapy. Lighted indicators show the status of equipment interlocks, and displays are provided for the dose rate, integrated doses for both dose channels, time, gantry angle, arc therapy and rotation direction (clockwise or counterclockwise) and operating mode, either fixed or arc therapy.

The treatment couch has four motions which are motorized with variable speed motors, and are controlled from a hand-held pendant connected to the treatment couch. The longitudinal and lateral motions of the treatment couch top can also be controlled via a manual over-ride. The couch top can be rotated 180° in order to position either end section.

The Accelerator: The heart of the radiotherapy linear accelerator machine is the accelerator. All accelerators have four major components: the modulator, electron gun, RF power source, and accelerator guide which are connected as shown in Fig. 34.5.

The electron accelerator is a wave guide structure which is energized at microwave frequency, most commonly at 3000 MHz. The microwave radiation is supplied in short pulses, a few microseconds long. These pulses are generated by supplying high voltage pulses of about 5 kV from the modulator to the microwave generator, which is most commonly a magnetron valve. In higher energy accelerators, a klyston valve is used as the microwave power source. The electron gun is also pulsed so that high velocity electrons are injected into the accelerating wave guide at the same time as it is energized. The electron gun and accelerating wave guide system have to be evacuated to a pressure such that the mean free path of electrons between atomic collisions is long compared with the electron path through the system.

Modulator: The primary function of the modulator circuit is to supply high voltage pulses to the microwave generator. It steps up the input mains three-phase power supply (380V-440V) to about 50 kV prior to its rectification to DC. Specialized circuitry within the modulator produces high-voltage pulses at the rate of a few hundred pulses per second to synchronously power the electron gun and RF power source. The modulator contains a thyratron which is a high-power switching device needed to direct the high-voltage pulses generated by the modulator to the electron gun and the RF power source. The pulse repetition frequency is determined by the pulse generator which controls the thyratron grid. The peak voltage, the peak power and the mean power required from this circuit are determined by the working conditions of the microwave generator. The dose rate

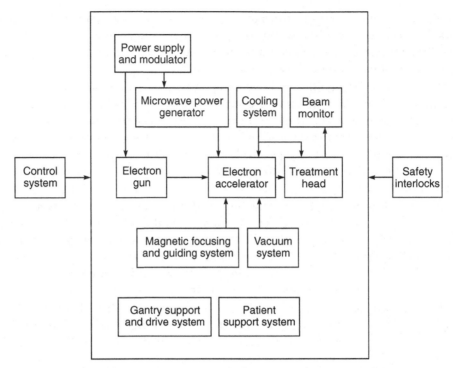

> **Fig. 34.5** *Sub-systems of a linear accelerator machine*

from the accelerating guide is regulated by controlling the pulse repetition frequency. The pulse length used is typically 3-6 μs.

The modulator may be located either in the gantry or the gantry supporting stand, or in a separate cabinet that can be located at some distance from the accelerator. Electrical connections to the electron gun and the RF power source are made through high-voltage cables.

Electron Gun: The electron gun is pulsed by the modulator and injects pulses of electrons of a few micro-seconds duration into the accelerator guide at energies of about 15-40 keV. The electrons are subsequently accelerated in the accelerator guide to the required energy level. The electron gun can either be a diode device with direct or indirect heating of the cathode, or a triode device in which the grid can be used to obtain control of the injected electron current in the electron mode. All three designs are used in commercial accelerators. All electron guns employ either a heater or filament, which eventually burns out and requires replacement.

RF Power Source: The RF power source is either a magnetron or a klystron. Klystrons are generally used in high-energy accelerators and magnetrons in low- or medium-energy accelerators. These devices employ a number of RF cavities either in a circle (magnetron) or in a straight line (klystron). An electron beam from a cathode is used to excite RF power in these cavities. The amplified RF power is fed into a wave guide—a special hollow metallic tube used to transport microwaves, that is connected to the accelerator. The RF power source, also pulsed by the modulator, provides high-frequency electromagnetic waves (3000 MHz) that accelerate the electrons injected from the electron gun down the accelerator guide. The electrons injected into the accelerator guide are captured and bunched by the accelerating electric field at exactly the optimum phase of the RF wave cycle so that they get accelerated.

Accelerator Wave Guide: A charged particle travelling along the axis of a series of conducting tubes which are connected to an alternating voltage gets accelerated and acquires energy as it passes through each gap between the tubes. A system using a radio frequency supply and medium atomic number particles formed the basis of early linear accelerators. The technology was not found to be practical because the high velocity attained by the particles would require very long flight tubes when radio frequency was used. However, at microwave frequencies, it became possible to accelerate electrons to energies of several million electron volts.

In practice, the accelerator structure or wave guide is made up of a number of specially shaped, copper microwave resonant cavities that have been brazed together to form a single structure. The length of the accelerator wave guide will vary from about 30 cm to 2.5 m. depending on the final elect-ron energy to be achieved and the type of structure utilized. A number of different accelerating structures have been employed in medical accelerators. While the shape of the individual resonant cavities and method of injecting the RF power into the accelerator guide will vary depending on the manufacturer, all accelerator structures are of two basic types: travelling wave (TW) or standing wave (SW).

However, the standing wave system accelerates the electrons in a field of constant amplitude, while the field in a travelling wave system is attenuated as it moves along the guide, the former will give a higher electron energy in the same length of guide for a given microwave power level. This means that where the length of the accelerating guide is a critical design factor, a standing wave system has a definite advantage. But this advantage is applicable only to accelerators operating at a lower MeV range. For energies above 4 MeV, the guide length becomes too long and therefore, the difference in accelerating wave guide length does not remain a decisive factor between the standing wave and travelling wave systems.

The accelerating electrons tend to diverge, partly by mutual repulsion but mainly, because the electric field in the wave guide structure has a radial component. They can be focused back onto their straight path by the use of a co-axial magnetic field, which is supplied by coils which themselves are co-axial with the accelerating wave guide. There are also additional coils which steer the electron beam in such a way that it emerges from the accelerator structure at the required position and direction.

Mechanical tolerances in the construction of accelerating guides are of the order of 0.01 mm. Both because of resistive losses in the guide walls and because some electrons may strike the structure, their accelerator guide will heat up in operation. Consequently, thermal expansion may result in significant changes in dimensions. A water cooling system for the wave guide is provided in the form of a water jacket through which temperature-controlled water is circulated at a pre-determined rate. The wave guide structure is typically 15 cm in outer diameter, with a length of one to three metres and a weight of several hundred kilograms.

Vacuum conditions need to be created in the accelerator wave guide so that the electrons being accelerated should not be deflected by collisions with gas atoms. For this purpose, an ion pump is used which has a working range of 10^{-3} to 10^{-8} torr.

Cooling System: The temperature of certain parts in the machine is critical for efficient operation. In particular, the temperature of the accelerating guide structure and the microwave valve has to be controlled because dimensional changes associated with thermal expansion will significantly change their characteristics. The X-ray target also needs to be cooled. A cooling system based on circulating water and operated using a thermostat is generally employed.

Treatment Head: The treatment head contains the X-ray target and filtering system, the beam monitor detectors and the beam defining system (Fig. 34.6).

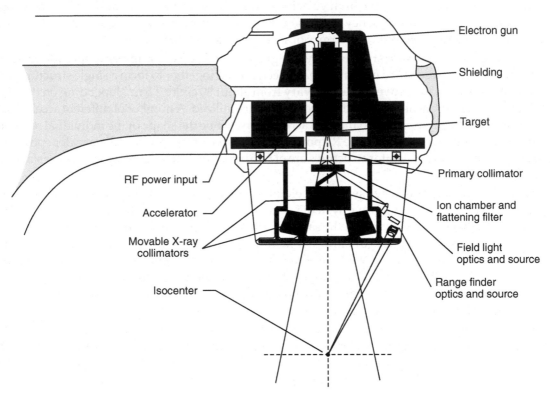

> **Fig. 34.6**   *Schematic of accelerator and collimator sub-systems*

Once the electrons have been accelerated to the correct energy level, they either impinge on a metal target and produce X-rays through atomic collisions in the metal in the treatment head or can be used directly for treatment. After X-rays are produced in the target, they emerge in a forward directed lobe whose intensity distribution must be flattened for clinical use. This is accomplished by interposing a circulary symmetric-shaped metal absorber, called a flattening filter in the path of the X-rays. The filter is a cone-shaped device which differentially absorbs the radiation towards the beam centre, i.e. it substantially reduces the dose rate at the beam centre so as to give a uniform dose over the whole area of interest. If the electron beam is to be used for treatment, it will emerge from the vacuum system through a thin window into the treatment head, where it is monitored and, if necessary, scattered to give the required field coverage.

Collimator: Most modern treatment units have two sets of adjustable collimator jaws that allow a rectangular field of irradiation to be adjusted to approximate the target area to be irradiated. The collimator jaw system is part of the collimator assembly. The collimator jaws should typically absorb 99% or more of the radiation and the opening should be variable to 40×40 cm at the isocentre to facilitate large field treatments. The collimator jaws should be able to close completely for certain quality control procedures, but need only be adjustable to an opening as small as 2×2 cm for most clinical use. The entire collimator assembly can be rotated about an axis that passes through the centre of the treatment field and the isocentre, the point in space where the gantry axis of the accelerator intersects the collimator axis of rotation. The ability to rotate the collimator allows the treatment field to be rotated, if required.

Usually, each jaw of a collimator pair of jaws moves synchronously and symmetrically about the collimator axis. In some modern accelerators, at least one jaw of a collimator pair of jaws can also move independently of the other jaw to provide asymmetric fields. The collimator assembly also contains a light source that projects a light field onto the patient to define the entry position of the radiation field during the set-up of the patient. An optical range finder projects the distance from the target to the patient surface.

All linear accelerators manufactured today are isocentric treatment units. The mechanical stability of the gantry structure and radiation head components during rotation generally assures that the beam axis passes within ± 2 mm of this isocentre point. This means that the tumour may be centred at the isocentre and the treatment unit rotated without requiring that the patient be repositioned. Thus, multiple fixed fields can be delivered with great accuracy and efficiency.

Wedge Filters: The accelerator is equipped with wedge filters and blocking trays, and for units capable of electron beam therapy, electron applicators. Wedge filters are metallic wedge-shaped devices that, when placed below the flattening filter, selectively absorb X-ray radiation, resulting in a dose distribution tilted from normal incidence by an amount nominally equal to the angle of the wedge filter inter-posed in the beam. To provide full clinical use, the wedge angle should be variable from 15° to 60° and cover a field size of 20 cm in the wedged direction and at least 25 cm in the non-wedged direction. While smaller wedged field sizes are often used, wedged fields of at least this width and length are commonly needed to treat breast carcinoma.

Dual Energy/Dual Beam Machines: Some treatment units employ the dual X-ray energy linear accelerator. These units provide two X-ray energies, a low-energy X-ray beam at 6 MeV and a higher energy X-ray beam of at least 10 MeV alongwith a range of electron beams, from 4 to 22 MeV. These units are capable of treating most patients. However, low-energy linear accelerators are still in demand because the majority of patients needing radiation therapy require low-energy X-rays. Also, they are less expensive initially and cost less to maintain than dual energy units, because of their inherently simpler design.

Higher energy units typically require higher power RF sources and longer and more expensive accelerator structures. Consequently, they require more complex circuitry and dosimetry systems. More expensive treatment rooms are also required because of the extra concrete shielding in the walls and ceiling needed to absorb the higher energy X-rays. Also, for X-ray energies above 10 MeV, the treatment room must be protected for neutrons, which are produced by these very energetic X-rays. Neutron protection is generally accomplished by building a special type of entry maze and putting about 10 cm of polyethylene neutron shielding on the treatment door. Some manufacturers equip their treatment units with beam stoppers that absorb some of the radiation and reduce the cost of the treatment room.

Dose Rate: While beam energy is generally the prime consideration in selecting a treatment unit, the dose rate is also important. Dose rates should be at least 200 cGy/min for all energies. Higher dose rates are often desirable for treatments at extended distances or to treat children or unco-operative or infirm patients quickly. The higher the dose rate, the shorter the treatment time, and the smaller the probability of patient motion during treatment.

Arc Therapy: For some treatments, it is desirable to produce radiation while the gantry is rotating. This type of treatment is called arc therapy and can be either an X-ray or an electron treatment. For flexibility in arc therapy, rotation in both directions is desirable. For electron arc therapy doses of up to 8 cGy/$^\circ$ are needed so that the irradiation may be accomplished in a single rotation. For X-ray arc therapy, generally 3 cGy/$^\circ$ is adequate. It is usually important for the dose during the rotation to be uniform. This is accomplished by rotating the gantry at a constant speed and adjusting the dose rate dynamically during rotation to maintain a uniform dose at every gantry angle. Naturally, the quality of the beam, its flatness, stability, precision and reproducibility are important in rotational beam delivery.

Methods of Mounting the Accelerator: There are two methods which have been used for moving the radiation beam with respect to the patient. They are shown in Fig. 34.7.

These are:

(a) Accelerator in line, with treatment beam—applicable only in low-energy machines, as the length of the accelerator structure for higher energy machines may not permit 360° rotation, particularly below the treatment couch.

(b) Electron accelerator parallel to axis of rotation of gantry—this requires the electron beam from the wave guide to be bent through 90° or 270° on to the X-ray target. Accelerating guides several metres long can be used. The gantry can be conveniently rotated through 360°.

Gantry: The accelerating wave guide, the focusing and steering coils, the treatment head and radiation shielding are mounted on the horizontal arm of the gantry which rotates about the horizontal axis. The gantry is supported on a vertical stand which is firmly fixed to a frame embedded in the floor. The support is provided by a slewing ring, a large diameter thrust bearing which can support the moments about both the horizontal and vertical axes. The vertical stand and the main support are box structures.

The rotation to the gantry is provided by a drive shaft which is connected to the servo drive motor via a large diameter pinion or toothed wheel, which is driven from the main gear box. Alternatively, a DC motor may be used whose speed can be controlled with a thyristor-based speed controller. The gantry speed is continuously variable from 0.1 to 1 rpm. The gantry rotates through a total useful range of 360°, in both clockwise and counter-clockwise directions.

Control Console: The accelerator is controlled by a compact console located outside the treatment room. While different manufacturers will incorporate different features in their control systems,

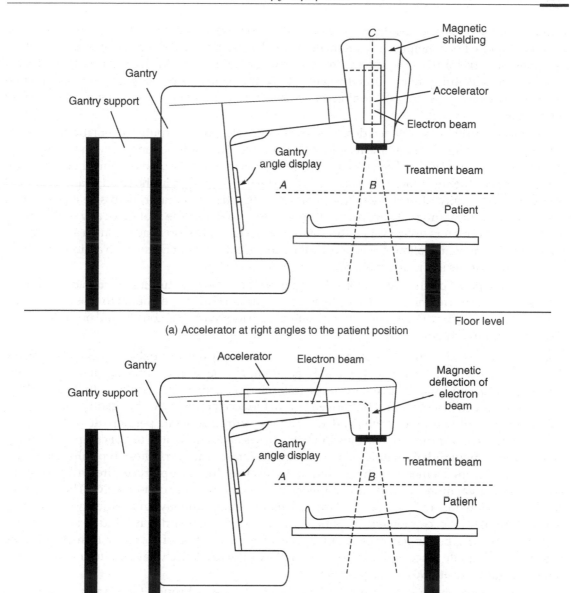

(a) Accelerator at right angles to the patient position

(b) Accelerator at parallel to the patient position

> **Fig. 34.7** *Layout design of accelerator with respect to the patient*

all consoles enable the operation of the unit, allow selection of the dose to be given for the treatment, and contain interlock circuitry designed to protect both the patient and the treatment unit.

Besides turning the radiation on and off, some control consoles permit remote mechanical movement of the linear accelerator. In addition to adjusting field size from the control console, it is often desirable to be able to adjust other mechanical parameters without having to re-enter the treatment room. For example, measurements taken during the initial machine calibration, and the on-going quality control procedures, are more convenient if the field size can be adjusted from the console. The ability to position the gantry from the control console is also useful in certain arc therapy techniques, for positioning the gantry from outside and in making quality control measurements.

The machine interlock system is designed to promote safety during routine use of the equipment. Interlocks are designed to shut the machine off in the event of malfunctions in the cooling system, vacuum system, modulator, high voltage power supply, line voltage or in certain other important components of the machine. The machine is interlocked such that in the fixed beam treatment mode, motion of any of the accelerator parameters (gantry, collimator, field size or treatment couch) will stop treatment. In arc therapy, an interlock to detect the motion of any of the above accelerator parameters except for gantry rotation is designed to stop irradiation. In addition, the circuitry of the machine allows inclusion, by the user, of external interlocks such as door switches, emergency shutoffs in the treatment room, and warning lights.

A separate panel on the control console contains monitoring circuits and certain other adjustments. The monitoring circuits allow physicists or other technical personnel to monitor various equipment parameters. Emergency off pushbuttons are located on the console and on both sides of the stand and treatment couch.

Patient Couch: The motion of the treatment couch is accomplished through controls located on the side of the couch or through a device called a pendant. The pendant is either suspended from the ceiling or attached by a flexible cord to the treatment couch. Some pendants that control the motions of the couch also control the movements of the accelerator (field size, gantry rotation, and collimator assembly rotation), while other systems require a separate pendant device to control these functions. In the modern machines, the patient set-up is automated under computer control. Since the treatment radiation times with modern accelerators are very short, typically less than 1 or 2 minutes, improvements in the operational aspects of the set-up procedure will permit the accelerator to treat more patients and thus make the treatments more cost-effective. The treatment couch should have an adequate range of travel, laterally, longitudinally, and vertically. Sometimes, to accommodate an exceptionally tall patient, a head extension board may be added to the couch to increase its length. It is generally felt that 25 cm of lateral motion of the couch from its mid-line position is adequate since it allows 100 cm source-to-skin distance treatment for most patients when the gantry is at a 90 degree position.

The treatment couch also has a rotational axis that passes through the isocenter. Since it is much easier to rotate the treatment couch than to try to re-adjust the angle at which the patient is lying on the couch, a few degrees of rotation is usually adequate to assure that the patient is accurately and appropriately aligned. Finally, the couch must go low enough to conveniently allow patients to get on and off and so that non-ambulatory patients can be easily transferred from a hospital trolley or wheel chair to the treatment couch. The couch should go high enough to permit at least a 40 × 40 cm field posteriorly.

Patients are properly positioned and aligned on the treatment couch by using the light field that simulates the radiation field, and by lasers that are located on the walls and ceiling of the treatment room. The lasers project a 2-mm spot, thin lines, or small crosses at the isocenter. These lines or spots intersect with patient landmarks that have been placed on the patient during the simulation process. Sagittal lasers are used to project a line on the patient along the length of the couch. Alignment of the patient by this line will assure that the patient is perpendicular to the plane of the rotation of the unit, a very critical alignment for arc therapy and multiple field isocentric treatments.

It may be observed that the range of technologies used in the linear accelerator system is very wide. The accelerator system requires relatively high power electronics, of the order of megawatts, while on the other hand, the dose monitoring system has to measure currents of the order of 10^{-12} A. The use of microprocessor technology in controlling the operation of the machine and safety interlocks have facilitated reliable systems. In addition, the gantry, the patient support system and the beam defining system all call for high quality and precision mechanical engineering.

Automated Drug Delivery Systems

▶ 35.1 INFUSION PUMPS

In many medical applications, intra-venous (IV) fluids and drugs need to be infused over a period of time, which could be several minutes, hours and days. The most common method of doing this is by manual injection of bolus doses using syringes by manually setting the drip rate of gravity-feed intra-venous infusion sets. The application of infusion delivery devices continues to grow, extending to patient–controlled analgesia, home therapy, chemotherapy, implantable drug pumps (such as insulin delivery pumps) etc. For meeting the exacting requirements of these applications in terms of flow rate of the fluids in a safe and effective manner, the pumps are becoming smaller and smarter. The use of microprocessor technology has allowed the systems to provide performance and functionality that were unattainable only several years ago and most new systems are designed for easy addition of new features through simple software improvements.

The volumetric infusion pump is generally used to deliver larger volumes of fluid (> 60 cc) from a bag or bottle. This pump technology evolved from the desire to improve on the flow rate control provided by the simple, mechanical roller clamp. The first improvement was the electronic drop counter that could quickly determine the drip rate of IV delivery. Soon there after came a device which could control the drop rate automatically. It wasn't long before manufacturers found ways to better control volumetric accuracy with a wide array of special tubing sets based on rotary peristaltic, linear peristaltic, modified syringe, and optical measurement mechanisms. The capability has advanced to the point that most pumps can provide a flow rate which does not vary more than ± 5% over a range of rates from 0.1 to 999 mL/hr (Evans, 1995).

Once the accuracy issue was addressed, other aspects of infusion therapy were tackled. Patient safety concerns such as excessive delivery pressure, air-in-line, and inadvertent free-flow were taken into account in the system design. Control of secondary medications, variable delivery patterns (ramp up/down stepping etc.), dose-rate calculations and even in-line mixing of medications are other refinements resulting from the changing requirements of new drug therapies.

35.1.1 Hospital Systems

The most common application of infusion devices is to maintain appropriate fluid levels in the patient. Fluid therapy is used in the management of patients during and after surgery, for treatment of burns, and in treating dehydration in paediatric patients. The therapy involves the controlled infusion of plasma expanders, usually at a rate determined by the patient's fluid balance.

Infusion systems are also commonly used to intravenously supply nutrients to support life and to maintain growth and development in paediatric patients. Because no feedback variable is available to provide control signals, these systems are operated open loop.

Continuous drug infusion is also used for such widely different applications as delivery anaesthetics during surgery, chemotherapy for cancer, oxytoxic agents for inducing labour and anti-arrhythmic drugs for patients in the coronary care unit.

35.1.2 Ambulatory Applications

The largest group of potential users of infusion devices are ambulatory type 1 diabetics. Conventional therapy, consisting of one or two daily injections of insulin, allows substantial fluctuations in blood glucose levels. Since evidence is accumulating that the complications of diabetes may be reduced by improved control of blood glucose, much effort is aimed at establishing therapeutic regimens that reduce blood glucose fluctuations.

Continuous insulin therapy has been shown to improve metabolic control, but further studies are needed to verify that such therapy also reduces the numerous long-term complications of diabetes. An excellent review of the use of pumps in diabetes therapy is given by Pickup and Rothwell (1984).

Wearable pumps have also been used for delivering agents that were previously used only in a hospital environment (Applefield, 1983). Anti-cancer drugs can be continuously infused in ambulatory patients and recently devices have been developed that deliver pain suppressants under patient control.

Since currently used wearable pumps require that the skin be punctured for the drug-delivery catheter, implantable pumps are being developed to obviate the need for skin puncture and to alleviate the corresponding inconvenience and risk of infection. While most of those just undergoing initial testing implantable pumps are expected to be increasingly utilized for the delivery of potent pharmacological agents, chemotherapy and insulin infusion are two of the primary applications.

▻ 35.2 COMPONENTS OF DRUGS INFUSION SYSTEMS

The drug infusion systems basically consist of two components: a mechanism that delivers the drug, and a means of controlling the rate of delivery. In open loop systems the rate of delivery is set by the nurse or physician on the basis of past experience, mathematical computation, or by trial and error. The fluid is delivered at the set rate until the setting is changed. In closed-loop systems, the effects of the drugs are monitored by appropriate transducers, and the desired delivery rate is computed and set automatically.

The pump operates in such a way as to keep the physiological variable as close as possible to a desired value. An example of a closed loop system is the use of controlled infusion of the drug sodium netropruside (a vasodilator) for the control of blood pressure. A pressure transducer measures the blood pressure and this information is sent to the control algorithm which determines the rate at which the drug is infused into the patient. This is obviously a more effective method of controlling the blood pressure as compared to the manual control using the same chemical agent.

35.2.1 Delivering the Drug

The traditional and simplest intravenous infusion systems consist of a fluid container, administration set, and a clamp to control the flow from the set to the patient. The driving pressure is the difference between the hydrostatic pressure generated by the column of liquid in the administration set, and the venous pressure. Since the latter is typically 4-8 mm Hg, the driving pressure is approximated by the level of reservoir, usually adjusted to be 60-100 cm above the patient. Infusion into an artery would require raising the reservoir to some 2 m, making this method of intra-arterial infusion impractical.

The major difficulty with traditional intravenous infusion systems is that the flow rate cannot be accurately controlled (Crass and Vance, 1985). The flow rate is counted in drops rather than measured volumetrically. The rate is difficult to adjust, and even if adjusted correctly at first, it will change with time.

Intravenous 'controllers' are mechanical or electrical devices that automatically control the flow rate of fluids, even though the driving pressure is still generated by gravity. These sets contain chambers and valves that meter fixed volumes of liquid for infusion. The devices generally raise an alarm when malfunctions such as empty container, occlusion, or low batter are detected.

Infusion pumps, rather than depending on gravity to generate flow develop pressure by one of several electro-mechanical means. The two commonly used methods are discussed below:

35.2.2 Syringe Pumps

In syringe pumps a motor, through a gear-reducing mechanism and a lead screw, applies force to the plunger of a syringe containing the drug (Fig. 35.1). The device is mainly convenient for applications that require the delivery of volumes limited by the syringe size.

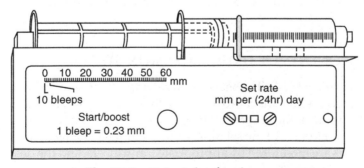

> **Fig. 35.1**   *Principle of syringe pump*

Syringe pumps are usually of a reciprocating type. A plunger or piston delivers a fixed volume of fluid on each stroke. They require valves and normally furnish a pulsating flow which can be evened out by the use of reservoirs or by polyplex arrangements in which the discharge portion of the stroking cycle is spread over more than the conventional 180°. Piston pumps can be used at high pressures. Control is achieved by varying the stroke length or the stroke rate.

35.2.3   Peristaltic Pumps

These pumps squeeze a flexible bag or tube to produce movement of the liquid inside the compressed container. Linear peristaltic pumps have a row of fingers that compress the tube in a wave-like motion, squeezing the liquid as the wave progresses. The more often used rotary peristaltic pumps use a rotor that pushes rollers against a tube along a semicircular path. Peristaltic pumps have the advantage that the fluid does not come into contact with the pump, avoiding contamination. If true volumetric pumps are required, however, special tubing must be used. Figure 35.2 shows two examples of peristaltic driving heads.

The various means for achieving peristaltic action are: cam-operated fingers pressing the tubing in succession; a rotor on an eccentric shaft which squeezes an eccentric cylindrical liner; and a row of eccentric cams which moves a cam follower, producing a clamping effect and imparting a squeezing action to plastic tubing, among others. They are self-priming, as no valves or seals are required, and the accuracy rate is high. Peristaltic pumps use speed control of the device motor, which may be a synchronous electric motor.

The foremost consideration in the design of infusion pumps must be patient safety. Apart from providing an accurate volume flow, pumps and their associated devices must ensure patient safety even in the presence of misuse or equipment malfunction. Alarms that detect air in the infusion line, depletion of the infusate, or malfunction of the pump itself

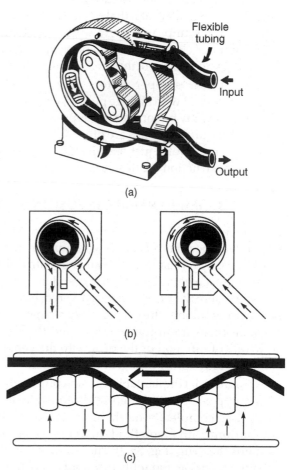

(a)

(b)

(c)

➤ **Fig. 35.2(a)** *Principle of peristaltic pump using rotor and flexible tubing*

   (b) *Rotor on eccentric shaft to squeez plastic tube*

   (c) *Pump with cam-operated fingers on flexible tubing* are usually integral parts of infusion systems. The maximum pressure that a pump can generate should be limited in order to avoid tissue infiltration or trying to pump into an occluded line.

The accuracy of volume infusion over a substantial length of time is essential when drugs with long half-lives are infused or in applications wherein the total fluid load (as in infants) must be limited. Accuracy of volume becomes the prime consideration when drugs with short half-lives are infused. Current infusion pumps are rated to give a volumetric accuracy of better than ± 5%.

Special design considerations apply to pumps that are used in an ambulatory environment. Although most of them also use peristaltic or syringe pumping, their weight and size must be considerably smaller than that of bedside devices. Especially compact (but complex) are positive displacement pumps in which a solenoid-activated piston alternatively withdraws drugs from a reservoir into a closed chamber, and then injects the contents of the chamber into the patient. Inlet and outlet valves control the direction of flow through the chamber.

Either DC or stepper motors may be used to drive the infusion pumps. DC motors are customarily used with syringe pumps, while stepper motors generally drive the peristaltic devices. The infusion may be continuous or pulsatile, but because continuous speed control at very small rates is difficult, the drugs are often delivered in very small boli. Pumps that deliver long-acting hormones may pulse just once in 120 minutes. Re-chargeable power packs and self-contained rugged construction are desirable features.

35.3 IMPLANTABLE INFUSION SYSTEMS

Implantable drug delivery pumps which are slightly larger than the cardiac pacemakers have also been developed. However, they require further miniaturization and increased reliability. They must be easily refillable and should be controllable from the outside. These pumps apply a known pressure to a reservoir of the drug, and there is a high-resistance connection between the pump and site where the drug is to be delivered, which is usually a vein. The high resistance connection is generally a long, thin capillary tube that is wound around the periphery of the pump. The constant pressure in the reservoir and the fixed resistance of the tube maintain a steady but slow rate of infusion of the drug into the venous circulation. These pumps, therefore, utilize a concentrated form of the agent to be infused.

A commercial implantable infusion pump which is available for human use is the "INFUSAID". This pump can be easily refilled but its rate of infusion cannot be regulated (Blackshear, et al., 1972). It is completely mechanical. The driving force is generated by Freon vapour (being in equilibrium with Freon fluid) in a closed compartment pushing against a bellows chamber that contains the drug (Fig. 35.3). The bellows chamber can be refilled by means of a needle that can enter the reservoir and refill it without any of the drug leaking into the surrounding tissue. The rate of infusion is determined by the temperature, fluid viscosity, and a flow restrictor. Since insulin therapy requires at least two different flow rates, an experimental version of the infused has been developed that allows bypassing the restrictor by operating a valve magnetically from the outside (Buchweld et al.,1980). Figure 35.4 shows an implantable pump from M/s. Medtronic, USA.

Controlling the Drug Infusion Rate: Drug delivery is said to be open loop if the rate of infusion (possibly a function of time) is set a-prior and it is not automatically altered by the patient's response. In closed-loop systems, the patient's response is used to automatically adjust the infusion rate.

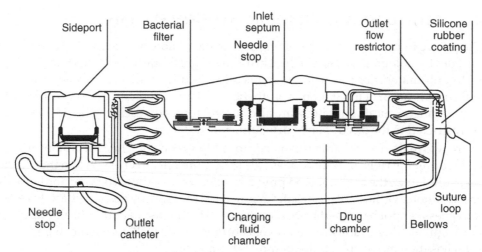

> **Fig.** 35.3   *Cross-section of an implantable pump (Redrawn after INFUSAID, USA)*

Most current systems operate in open loop. Both 'controllers' and pumps allow setting the infusion rate by dialing into them the desired rate of delivery. "Controllers" may count the drops of intravenous fluid through a photo-electric device or use special cassettes that accurately meter the flow through the device. Some controllers allow the independent setting of primary and secondary ('piggyback') infusion rates. Volumetric pumps may be controlled by the step size or frequency of stepper motor, or the speed of DC motors. Special cassettes may also be used with pumps to prevent free flow and to increase accuracy. The performance, features and cost of several commercial devices have been tabulated in 'Health Devices, 1984 and Health Devices, 1985.

Ambulatory systems for infusing insulin mostly allow considerable control by the diabetic

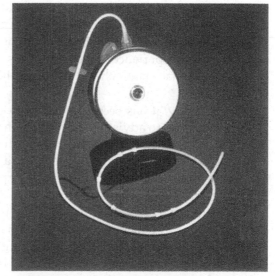

> **Fig.** 35.4   *Implantable pump (Courtesy: M/s Medtronic USA)*

patient. In addition to delivering an adjustable basal rate, these pumps can also deliver an increased prandial rate whenever commanded by the patient.

Implantable pumps are programmed remotely using a patient's control unit. The unit is used both for selecting the basal flow rate, and for commanding the prandial bolus. In one of the systems (Pacesetter/Johns Hopkins APL) the patient's control unit can also be used to provide two-way telephone communication between a programming unit in the physician's office and the microprocessor in the implanted device.

ⅢⅢ▷ 35.4 CLOSED-LOOP CONTROL IN INFUSION SYSTEMS

Closed-loop drug infusion systems are among a growing number of systems designed to automate the control of physiological variables either in a clinical or laboratory setting (Franetzki, 1984). In addition to automated drug infusion systems, closed-loop control of anaesthesia (primarily through the application of inhalation anaesthetics) and arterial blood gases (primarily through adjustments of mechanical ventilators) are also current topics of research and development. This section is restricted to the discussion of drug infusion systems.

There are two major reasons for automated closed-loop control. The first is to improve patient care by delivering the right amount of agent for maximum effectiveness. When the effectiveness can be measured (e.g., the level of blood pressure when vaso-active drugs are applied), the ideal drug concentration, within limits, is whatever is needed to cause the desired effect. In cases where the effectiveness cannot be immediately measured (as in the case of antibiotic therapy), the ideal treatment might quickly establish a steady concentration level well within the therapeutic range, a level which is then maintained until therapy is terminated.

The second reason for using automated closed-loop control is to reduce the cost of medical care. It takes almost undivided attention to manually control the infusion rate of some commonly used short-acting agents. Regardless of whether nurses or physicians act as controllers, automating the infusion would assist health professionals in performing their duties more cost-effectively.

In traditional drug therapy the rate or amount of infusion is selected by the health professional on the basis of experience and observations made about the patient (Fig. 35.5(a)). The control action is taken manually, either by adjusting the rate of infusion or by injecting the desired amount of the drug.

More recently, it has become possible to enhance the traditional mode of drug therapy by using a computer to advise the physician what the appropriate dosage should be (Fig. 35.5(b)). Information about the patient is entered into the computer which, when combined with the desired drug concentration, future injection regimen, and a model of the drug kinetics, allows an algorithm to compute a recommended drug dosage. The physician may accept or refuse the recommendation; the infusion is done manually.

In a fully automated system (Fig. 35.5(c)) the transducer senses the controlled variable, a computer algorithm determines the infusion rate on the basis of the discrepancy between actual and desired variables, and the computed infusion rate is automatically delivered by a pump. The system operates without human intervention until the therapy is completed or until a malfunction is detected.

Closed-loop control of insulin infusion was first described in 1964, but practical systems have been in existence only since 1974. All these systems are only for bedside use in hospitals because they are bulky and non-portable.

A commercially available closed-loop controlled type unit is the biostator from Miles Laboratories (Clemens et al, 1977). It withdraws blood from the patient continuously, but the blood glucose determinations are made intermittently. The peristaltic infusion pump also runs continuously, allowing the infusion of not only insulin, but dextrose and saline as well. Dextrose infusion was intended to counteract any possible excess insulin infusion, but its use has generally not been found necessary.

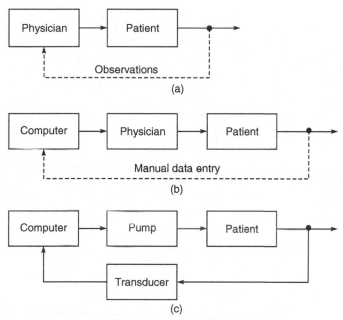

> **Fig.** 35.5 *Evolution of various levels of automation in the control of drug delivery*
>   (a) *Conventional method* (b) *computer assisted therapy* (c) *closed loop control*

⁞⁞⁞⧐ 35.5   EXAMPLES OF TYPICAL INFUSION PUMPS

35.5.1   Drop Rate Counter Type Infusion Pump

Figure 35.6 shows a block diagram of a drop rate counter type infusion pump. The heart of the system is a customized digital (LSI) device which performs all the logic functions. It operates under the control of a clock frequency of 71.68 kHz obtained from a crystal controlled oscillator which provides various timing and control signals.

The drop sensor is attached to the administration set drip chamber and closes the servo loop by providing rate feedback information. The drop sensor contains an array of light emitting diodes (LEDs) and phototransistors which generate a signal each time a drop of IV fluid falls into the drip chamber. This signal is applied to the drop detector which causes the drop indicator to flash.

Four signals are required to activate and run the motor. $\overline{\text{ALA}}$, ACM, MD2 and MD3. These signals except $\overline{\text{ALA}}$ are developed within the LSI device. $\overline{\text{ALA}}$ is a logical "1" when the unit is not in alarm. Activate motor (ACM) is the variable width signal which applies power to the motor. Motor drive (MD2 & MD3) signals are outputs from the LSI device which provide quadrature voltage to the motor. After power is applied to the motor, these two symmetrical signals, 90° apart in phase, cause the motor to step at about 360 Hz.

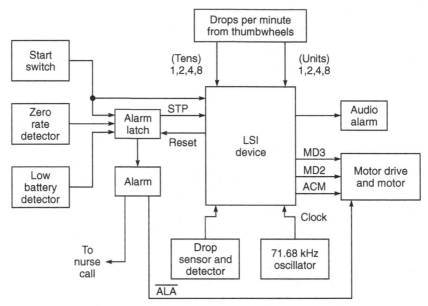

> **Fig. 35.6** *Block diagram of drop rate counter*

Each time the motor is activated by ACM, a series of MD2 and MD3 pulses is obtained. ACM increases in width each time the motor is pulsed and more MD2 and MD3 signals are applied. This causes the motor to run for longer periods. When a drop is sensed, the width of the ACM is reduced thereby decreasing the number of MD2 and MD3 pulses applied to the motor. This is the servo action.

ACM is designed to step the motor 10½ times for each drop. If 20 drops per minute were dialled in, the motor would step a minimum of 210 times per minute. The pulse width of ACM is internally limited to prevent the pump from producing a steady flow condition of IV fluid.

When an external circuit signals the unit to alarm, a stop STP signal is generated stopping the motor from stepping. Two thumbwheels are used to set the desired rate (drops per minute) in terms of 'tens' and units. A zero rate detector causes the unit to alarm if the drops per minute detector is inadvertently set to 00. When the unit goes to alarm for any reason an audible alarm signal is produced. When operating on a battery power, the low battery alarm puts the unit into alarm when the battery voltage drops to a pre-determined minimum level. This prevents inaccurate or erratic operation because of low voltage conditions.

35.5.2   Programmable Volumetric Infusion Pump

Figure 35.7 shows a block diagram of the programmable volumetric infusion pump from Smith and Nephew SIGMA, Inc. USA. Volumetric implies flow rate calibration in units of known volume, ml/hr, rather than non-volumetric units of drops /minutes. It incorporates a peristaltic (tube squeezing) pump mechanism. The flow rate is calibrated on known inner diameter tubing. Flow rate is a function of both pump speed, tubing inner diameter, and tubing elasticity. The flow rate

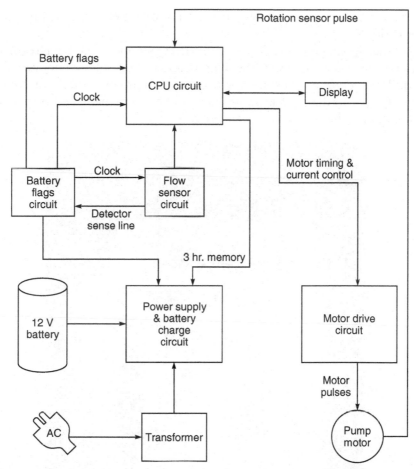

> **Fig. 35.7** *Block diagram of the programmable volumetric infusion pump (Courtesy: M/s Smith and Nephew, USA)*

tolerances are almost totally controlled by the inner diameter dimensional variations of the sets used. The speed of the pump is maintained electronically within ± 0.1%.

The heart of the system is a central processing unit. It has a 2.5 MHz external clock supplied by a crystal. The CPU outputs a stream of pulses for operating the motor drive circuit. An ultrasonic sensor is used to detect the drop which then signals to the microprocessor that a drop has been detected by the flow sensor. The ultrasonic signal is amplified and de-modulated before it is given to an 8-bit A/D converter. The A/D converter is controlled by the microprocessor.

For a feedback from the pump motor to the CPU, a hall effect sensor is used to sense the rotation. The transducer gives a single pulse per revolution and is mounted on top of the pump assembly. Similarly, the door opened/closed signals are generated by hall effect switches. A circuit is incorporated to prevent the improper operation of the pumping mechanism that could result from a loop outside of the programme or failure and a subsequent lock-up of the microprocessor. The flow rate is set using a membrane type key pad which is connected to the microprocessor through a key board decoder.

Figure 35.8 shows another type of infusion pump from AVI Inc., USA. which operates under the control of a microprocessor (Fig. 35.9). The microprocessor controls all operator programmable functions and alarm condition sensors. A synchronous stepping motor is used to operate the pumping mechanism accordingly. The pump offers user programmable control over both the dose volume and the rate at which the dose is delivered.

➤ **Fig. 35.8** *Programmable microprocessor based infusion pump (Coutesy: AVI, Inc. USA)*

The key component of the disposable administration set is a flexible, three chambered cassette. The administration sets may also be used for gravity administration independent of the pump due to their valve-less design. Precise control over delivery rate and volume infused is provided by the mechanical manipulation of the flexible cassette. The pumping system results in continuous, non-pulsatile delivery to the patient. The pump provides for patient safety by monitoring several alarm

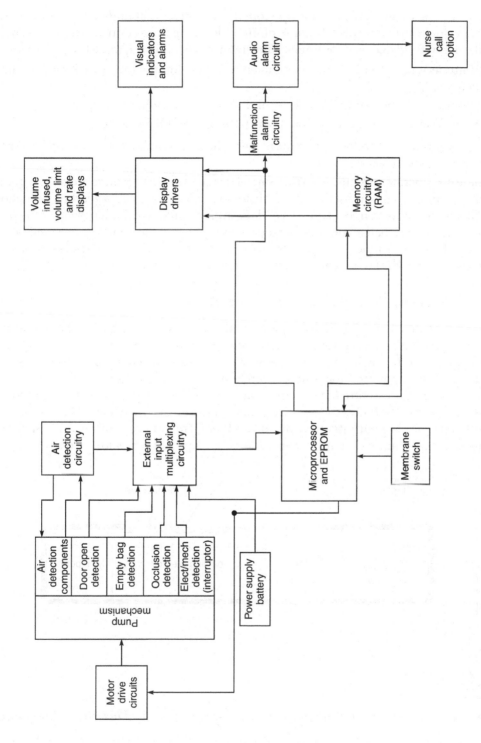

➤ **Fig.35.9** *Block diagram of the programmable microprocessor based infusion pump (Courtesy: AVI, Inc., USA)*

situations. In such situations, the pump stops fluid delivery, sounds an audible alarm and displays a visual alarm indicator. In some pumps, alert situations sound a separate audible and visual alarm but fluid delivery continues. The unit has the following three types of displays:

— Volume infused which indicates the cumulative volume (in ml) which the pump has delivered;

— Volume limit indicating the volume limit (dose), in ml, which the operator has selected.

— Rate indicating the rate (ml/hour) which the operator has selected.

The nurse call circuit monitors the inputs to the audio alarm circuit. If any of the alarm situations arise, the nurse call relay is energized.

The motor is operated according to the control programme in the microprocessor memory. A four-phase stepping motor is used, with a step angle of 7.5 degrees and a gearbox reduction ratio of 20:1. Due to this reduction, 1920 pulses yield one full revolution of the cam shaft in the pumping mechanism. Each cam shaft revolution and the resulting cycle of the pumping mechanism delivers approximately 2 ml of fluid. The delivery of fluid to the patient is non-pulsatile and nearly continuous since each ml is delivered in approximately 960 increments.

35.5.3   Programme Controlled Insulin-Dosing Device

There has been intensive research to develop an 'artificial pancreas'. By this it is meant a device which consists of a blood sugar sensor, a computer, an insulin pump (Fig. 35.10). Such a system would bypass the body's own pancreas. However, the development of a glucose sensor is probably the most difficult component of the system. Therefore, a more practical approach has been followed to develop a programme controlled insulin pump (Fig. 35.11). This pump delivers insulin through a catheter inserted in the body. The device can deliver insulin in two ways: The fixed programme units with a pre-set 24 hour rate/time profile and the demand programme units with which a basal rate is pre-selected and a supplementary dose is triggered at mealtimes (Renner, 1981).

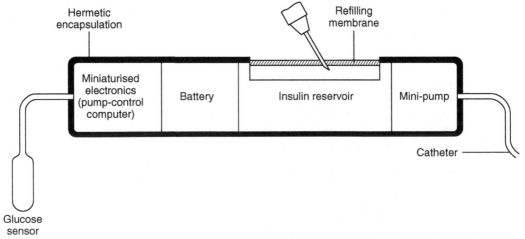

> **Fig. 35.10**   *Schematic diagram of glucose-sensor controlled insulin dosing device*

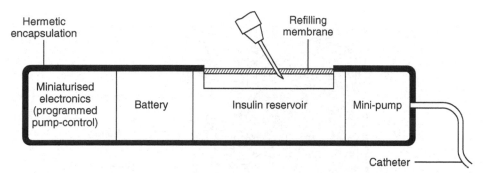

> **Fig. 35.11** *Schematic diagram of program controlled insulin dosing device*

An external programme controlled portable insulin dosing device is described by Franetzki et al (1981). The device has the following design features:

— It should be possible to wear the device in comfort for long periods inconspicuously and under all normal living conditions, inside and outside the clinic. The patient should be able to go about his normal work.

— It must be easy and safe to operate both by the physician and by the patient.

— All infusion paths should be possible and practicable including intravenous, intra-peritoneal, sub-cutaneous and intra-muscular.

— The pump must have sufficient supply pressure reserve to prevent possible catheter clogging through microthrombi.

— The highest possible level of patient safety must be assured. Malfunctioning of the device must be excluded as far as possible or must be signalled immediately to the patient and must never lead to dangerous overdosing.

In order to fulfil these requirements, a stepping motor-driven roller pump with swing-out counter jaws is employed as a pump (Fig. 35.12). The device is of the demand-rate type, i.e. on a pre-set basal rate, a supplementary dose of insulin of selectable size is called up by the patient at mealtimes. The supplementary dose is administered as a rectangular rate profile with variable time or amplitude. The insulin reservoir consists of a collapsible laminated foil bag and forms a one-trip article together with the attached silicone pump tubing. The catheter can be connected to the pump tubing with a standard Luer connector. A number of alarm and safety circuits ensure that a dangerous condition cannot occur. In the case of failure of a system component which would lead to unintentional overdosing, the pump is automatically switched off and an acoustic signal is triggered.

A very important parameter in such pumps is the accuracy of the pump delivery rate which refers to the average rate per complete revolution of the pumping head. The delivery rate is influenced by: (i) electronic component tolerances and drifts, which can be limited to $\pm 0.5\%$ by proper selection of components; (ii) manufacturing tolerances of the pumping head which can be limited to $\pm 1\%$ with proper pre-production control; and (iii) manufacturing tolerances of the pump tubing. The internal diameter, wall thickness and hardness of the pump tubing have an effect on the supply rate. The error caused by the tubing could be of the order of $\pm 4\%$. These errors

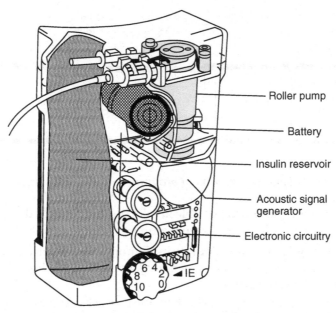

> ➤ **Fig. 35.12** *Promedos E1 external dosing device (Redrawn after Franetzki et al, 1981)*

are usually tolerable in patient treatment since the individual dose is determined with a particular dosing device and particular tubing.

References

Abdulla, U., S. Campbell, C.J. Dewhurd, D. Talbert, M. Lucas, and M. Mullarkey, 1971, Effect of diagnostic ultrasound on maternal and foetal chromosomes, *Lancet*, 2, 829.

ACR-NEMA (American College of Radiology: National Electrical Manufacturers Association), 1993, *Digital Imaging and Communications in Medicine (DICOM). Version 3.0*, Washington DC.

Ackerman, S.W. and P. Weith, 1995, Knowing your pulse oximetry monitors, *Medical Electronics*, February 1995, 82.

Agoston, M. and P. Zillich, 1971, The picoscale as one of the most important sets of the future blood diagnostic laboratory, *Asedico' News*, 1971–2, 15.

Ahn, B. K., A. O. Wist, C. C. Lia, and W. II. Ko, 1975, Development of a miniature pH glass electrode with field-effect transistor amplifier for biomedical applications, *Med. and Bio. Eng.*, 13, 450.

Anbar, M., 1998, Clinical thermal imaging today, *IEEE Eng. Med. and Biol.*, 17, 4, 25.

Andrews, A. H. and H. W. Moss, 1974, Experiences with the carbon dioxide laser in the larynx, *Annals of Otology, Rhinology and Laryngology*, 83, 462.

Alexander, J. and H.J. Krumme, 1988, SOMATOM PLUS, New perspectives in computer tomography, *Electromedia*, 56, 2, 50.

Alfonso, V.X *et al*, 1996, Comparing stress ECG enhancement algorithms, *IEEE Eng. in Med. and Biol.*, 15, 3, 37.

Allen, J. 1978, NTC thermistor microprocessor, *Measurements and Control*, April 1978, 97.

Amundsen, L. D., 1977, Sensing threshold, *Impulse* (Cardiac Pacemakers, Inc.), April 1977, 7.

Anger, H.O., 1958. Scintillation Camera, *Rev. Scientific Instruments* 29, p. 27.

Applefield, M.M., 1983, Intermittent, continuous outpatient dobutamine infusion in the management of congestive heart failure, *Am. J. Cardiol.* 51, 455.

Baggish, M. S. and J. Dorsey, 1981, CO_2 laser for the treatment of vulvular carcinoma in situ, *Obst. and Gynae.*, 57, 371.

Baker, D. W. (1970), Pulsed ultrasonic Doppler blood flow sensing, *IEEE Trans. Sonics and Ultrasonics*, SU-17, 170.

Baker, D. W., D. E. Strandness and S. L. John, 1976, Pulsed Doppler techniques: Some examples from the University of Washington, *Ultrasound in Med. and Blol.*, 2, 251.

Baker, L. E., L. A., Geddes, H. E. Hoff, and C.J. Cahput, 1966, Physiological factors underlying transthoracic impedance variations in respiration, *J. Appl. Physiol.*, 21, 1491.

Balles, M.W., C.A. Puliafito, *et al*, 1990, Semiconductor diode laser photocoagulation in retinal vascular disease, *Ophthalmology*, Vol. 97, p. 1553.

Barbaro, V. and V. Macellari, (1979), Intracranial pressure monitoring by means of a passive radiosonde, *Med. and Biol. Eng. and Comput.*, 17, 81.

Barber, F. E., D. W. Baker, A. W. C. Nation, D. E. Strandness, and J. M. Reid, 1974, Ultrasonic Duplex Echo-Doppler scanner, *IEEE Trans. Blamed. Eng.*, BME-21, 109.

Barker, A.T. *et al*, 1985, Non-invasive magnetic stimulation of human motor cortex, *Lancet*, 2, 1106.

Barlow, J.S., A. Kamp and H. B. Morton, 1974, EEG Instrumentation Standards: Report of the Committee on EEG Instrumentation Standards of the International Societies for Electroencephalography and Clinical Neurophysiology, Electroencephalogr, *Clin. Neurophysiol* 37, p. 539.

Basmajian, J. V. and J. E. Hudson, (1974), Miniature source-attached differential amplifier for electromyograph, *Am. J. Phys. Med.*, 53, 234.

Bashhur, R.L., 1995, Telemedicine effects: Cost Quality and Access, *Jr. of Medical Systems*, 19, 81.

Benchimol, A. and K. B. Desser, (1975), Advances in clinical vector-cardiograph, *Am. J. Cardlo.*, 36, 76.

Benders, D., 1976, Electrosurgery interference: Minimize its effects on ECG monitors, *Hewlett Packard Application Note AN 743*.

Bennett, P. L. and V. C. Jones, 1982, Portable defibrillator-monitor for cardiac resuscitation, *Hewlett Packard Journal*, 34, 2, 22.

Benz, P.D. 1999, Ambulatory cardiac event recorder, *Medical Electronics*, Sept. 1999, p. 38.

Bergveld, P. and N. F. de Rooji, 1979, From conventional membrane electrodes to ion-sensitive field-effect transistors, *Med. and Biol. Eng. Comput.*, 17, 647.

Becker, C.R. *et al*, 1999, First experiences with multi-slice CT SOMATOM PLUS Volume Zoom, *Electromedica*, 67,1, 47.

Beerwinkle, K. R. and J. J. Burch, 1976, A low-power combination electrocardiogram-respiration telemetry transmitter, *IEEE Trans. Biomed. Eng.*, BME-23, 484.

Bekkering, D. H., and E. Van Vollenhoven, 1967, The technical development of phonocardiography, *Digest of the 7th International Conference on Medical and Biol. Eng.*, Stockholm, 33.

Benz, P.D., 1999, Ambulatory cardiac event recorder, *Med. Electronics*, Sept.1999, p. 38.

Bernstein, A.D. *et al*, 1987, The NASPE/BPEG generic pacemaker code for anti-brady arrhythmias and adaptive rate pacing and anti-tachy arrhythmia devices, *Pace, Pacing Clin. Electrophysiolg.* 10, 794.

Bhullar, H.K, G.H. Loudou, J.C. Fothergil, and N.B. Jones, 1990, Selective non- invasive electrode to study myoelectric signals, *Med. and Biol Engg and Camp*, p. 581.

Blackshear, P.J. *et al*, 1972, The design and initial testing of an implantable infustion pump, *Surg. Gynaecol. Obslet.*, 135, 51.

Blais, M. R. and J. L. Fanton, 1979, Automated pulmonary function measurements, *Hewlett Packard Journal*, 31, 9, 20.

Blanco,C.E., W.J. Maerta Dorf and F.J. Walther, 1987, Use of combined high frequency oscillatory and intermittent mandatory ventilation in rabbits with saline lavaged lungs, *J. Intersive Care Med.*, 2, 214.

Blumenfeld, W., P. D. Wilson and S. Turney, 1974, A mathematical model for the ultrasonic measurement of respiratory flow, *Med. and Biol. Eng.*, 12, 621.

Blumenfeld, W., S. Z. Turney and R. J. Denmann, 1975, A coaxial ultrasonic pneumotachometer, *Med. and Biol. Eng.*, 13, 855.

Bobb, A.L., R.D. Popovich, G. Christopher and B.H. Scribner, 1971, The genesis of the square meter hour hypothesis, *Trans Am. Soc. Artif. Int. organs*. 17, 81.

Bom, N., C. T. Lances and F. C. Van Egmond, (1972), An ultrasonic intracardiac scanner, *Ultrasonic*, 10, 2, 72.

Bom, N., C. T., Lances, G., Van Zwieten, F. E. Kloster and J. Roelandt, 1973, Multiscan echocardiography. I. Technical description, *Circulation*, 48, 1066.

Bommer. W.J. and R.L. Haerten, (1991), The expanding role of ultrasound in cardiovascular medicine, *Electromedia*, 59,4, 115.

Boter, J., A. Den Hertog and J. Kuiper, 1966, Disturbance-free skin electrodes for persons during exercise, *Med. Electr. and Biol. Eng.*, 4, 91.

Bottomley, P. A., 1983, Nuclear magnetic resonance: Beyond phyological imaging, *IEEE Spectrum*, February 1983, 32.

Bowdle, T.A. *et al*, 1993, Cardiac Output (Part-2), *Medical Electronics*, April 1993, 53.

Bourne, P. R., 1974, Automated vector E.C.G. recording, *Med. and Biol. Eng.*, 12, 859.

Boynton, B.R. *et al*, 1984, Combined high frequency oscillatory ventilators and intermittent mandatory ventilation in critically ill neonates, *J. Pediatr*, 105, 297.

Branthwaite, M. A. and R. D. Bradley, (1968), Measurement of cardiac output by thermal dilution in man, *J. Appl. Physiol.*, 24, 434.

Brimbal, M. and J.C. Robillard, (1990), Thermal Array Recorder, *Medical Electronics*, April 1990, p. 100.

Bruner, J. M., 1967, Hazards of electrical apparatus, *Anesthesiology*, 28, 396.

Buchweld, H. *et al*, 1980, A totally implantable drug infusion device: Laboratory and clinical experience using a model with single flow rate and new design for modulated insulin infusion, *Diabetes Care*, 3, 351.

Burbank, D. P. and J.G. Webster, (1978), Reducing skin potential motion artefact by skin abrasion, *Med. and Biol. Eng. and Comput.*, 16, 31.

Cain, C. P. and A. J. Welch, 1974a, Thin film temperature sensors for biological measurements, *IEEE Trans. Biomed. Eng.*, BME-21, 421.

Cain, C. P. and A. J. Welch, 1974b, Measured and predicted laser induced temperature rise in the rabbit fundus Invest., *Ophthalmol.*, 13, 60.

Carter, L. M., 1978, The clinical role of thermography, *J. Med. Eng. and Tech.*, 2, 125.

Caspo, A., 1970, The diagnostic significance of the intra-uterine pressure, I: General considerations and techniques, *Obstet. Gynae. Surg.*, 15, 403.

Cayton, M. M., 1983, Nursing responsibilities in laser surgery, *Med. Inst.*, 17, 419.

Collier, D., 1991 Linear array recorders, *Medical Electronics*, Sept. 1991, p. 125.

Chamberlain, J.H., 1975, Cardiac output measurement by indicator dilution, *Biomed. Eng.*, 10, 92.

Chan, V., and A. Greenough. 1993, Determinants of oxygenation during high frequency oscillations, *Eur. J. Pediatr.*, 152, p. 350.

Charbonnier, F.M., 1996, External defibrillators and emergency external pacemakers, *Proceed. of the IEEE*, 84, 3, 487.

Chaussy, C., W. Bendel and E. Schmiedt, 1980, Extracorporeally induced destruction of kidney stones by shock waves, *Lancet*, 2,1265.

Cheng, E. M., W. H. Ko, R. J. Lorig, W. D. Beveridge, J.S. Brodkey and F. E. Nulsen, 1975, Intracranial pressure and temperature telemetry system using asynchronous PFM's with RF shifting for multiplexing, *28th ACEMB*, Sept. 20–24, 1975, 106.

Christensen, D. A., 1977, A new non-perturbing temperature probe using semiconductor band edge shift, *J. Bioeng.*, 1, 541.

Clark, L. C., Jr., 1956, Monitor and control of blood and tissue oxygen tensions, *Trans. Am. Soc. Artif Internal Organs,* 2, 41.

Clemens, A.H., P.H. Chang and R.W. Myers, 1977, The development of Biostator, a glucose controlled insulin infusion system, *Horm. Metab. Res. Suppl.*, 7, 23.

Clynes, M. and J. H. Milsum, 1970, *Biomedical Engineering Systems,* McGraw-Hill, New York.

Cohen, A. and R. L. Longini, 1971, Theoretical determination of the blood's relative oxygen saturation in vivo, *Med. and Biol. Eng.*, 9, 61.

Cohen, A. and N. Wadsworth, 1972, A light emitting diode skin reflectance oximeter, *Med. and Biol. Eng.*, 10, 385.

Cole, K. S. and U. Kishimoto, 1962, Platinised silver chloride electrodes, *Science,* 136, 381.

Coleman, A.J., J.E. Saunders and E.L.H. Palfrey, 1987. The destruction of renal calculi by external shockwaves: Practical operation and initial results with the Dornier lithotriptor, *Jr. of Medical Engg. and Technology* 11, 1, 4.

Collier, D., 1991, Recording trends and developments, *Medical electronics*, Sept. 1991, p. 128.

Conrad, D.P., 1990, Capacitive Electrodes, *Medical Electronics* Oct., 1990, p. 185.

Courtin, E., W. Ruchay, P. Salfield and H. Sommer, 1977, A versatile, semiautomatic fetal monitor for nontechnical users, *Hewlett Packard J.*, 28, 5, 16.

Cowell, T. and D. Bray, 1970, Measuring the heart's output, *Electronics and Power*, April 1970.

Cox, J. R. and F. M. Nolle, 1968, AZTEC, A preprocessing program for real-time ECG rhythm analysis, *IEEE Trans. Biomed. Eng.*, BME-2, 128.

Crass, R.E. and J.R. Vance, 1985, In vivo accuracy of gravit, flow IV infusion systems, *Am. J. Hosp. Pharm*, 42, 238.

Crenner, F., F. Angel, and C. Ringwald, 1989, Ag/AgCl electrode assembly for their smooth muscle electromyography, *Med and Biol. Eng. and Comput.*, p. 346.

Cronwell, J. B., 1965, The matching and linearising of thermistor probes, *World Med. Electron. Instrum.*, 3, 233.

Culshaw, B., 1982, Optical fibre transducers, *The Radio and Electronic Engineer*, Vol. 52, 6, 283.

Cunningham, L. N., C. Labrie, J.S. Socldner, R. F. Gleason and H. G. Doll, 1983, A non-invasive electromagnetic flowmeter, *Med. Inst.*, 17, 237.

Dalziel, C. F. and W. R. Iee, 1968, Re-evaluation of lethal electric currents', *IEEE Trans. Gen. and Ind. Appl.*, Vol. IGA-4, 46.

Dalziel, C. F. 1970, Transistorised ground-fault interrupter reduces shock hazard, *IEEE Spectrum*, Jan. 1970, 55.

Day *et al*, 1968, Auscultation of foetal heart rate: An assessment of the error and significance, *Brit. Med. J.*, 4, 422.

Deloskey, A. F., W. W. Nichols, C. R. Conti and C. J. Pepine, 1978, Estimation of beat-to-beat stroke volume from the pulmonary arterial pressure contour in man, *Med. and Biol. Eng. and Comput.*, 16, 707.

DiChiro, G., R. A. Brooks, L. Dubal and E. Chew, 1978, Elevated attenuation values toward the apex of the skull, *J. Comput. Assist. Tomog.*, 2, 65.

Dipietro, D. M. and J. D. Meindl, 1973, Integrated circuits for an implantable blood flowmeter, presented at the Tenth Annual Rocky Mountain, *Bioengineering Symposium*, May 7–9, 1973, Boulder, Colorado.

Dodd, F.G.B., 1993, Neural network entrances ICU patient monitoring, *Med. Electronics*, Dec. 1993, p. 54.

Doyle, F.H. and I.R. Young (1981), Imaging of the brain by nuclear magnetic resonance, *Lancet*, (8237), 53.

Easterling, R.E., O.G. Haig and J.A. Green Jr., 1969, Evaluation of disposable hemodialysers for home hemodialyses, *Trans. Am. Soc. Art. Int. Organs*, 15, 74.

Eberhard, P., W. Mindt, K. Hammacher and F. John, 1973, Oxygen monitoring of newborns by skin electrodes. Correlation between arterial and cutaneously determined pO_2, *Adv. Exp. Med. Biol.*, 37, 1697.

Eberhard, P., W. Mindt, F. John and K. Hammacher, 1975, Continuous pO_2 monitoring in the neonate by skin electrodes, *Med. and Biol. Eng.*, 13, 436.

Eberhard, P. and W. Mindt, 1976, *An Introduction to Cutaneous Oxygen Monitoring in the Neonate*, F. Hoffmann-La Roche and Co., AG. Bersle, Switzerland.

Elmquist, R. and A. Senning, 1960, An implantable cardiac pacemaker, *Medical Electronics*, Illiff and Sons, London, 253.

Engebretson, A.M., 1994, Benefits of digital hearing aids, *IEEE Engg. In Med. and Biol.* 13,2, 238.

Evans, R., 1995, Infusion delivery systems, *Medical Electronics*, April 1995, p.159.

Feder, W., 1963, Silver-Silver chloride electrode as a non-polarizable bioelectrode, *J. Appl Physiol*, 18, 397.

Fegler, G., 1954, Measurement of cardiac output in anaesthetised animals by a thermo-dilution method, *Quart. J. Exp. Physiol*, 39, 153.

Fielder, F. D., 1968, Ultrasonic foetal blood flow detector, *Biomed. Eng.*, June 1968.

Fitzsimmons, J. R., 1982, Gradient control system for nuclear-magnetic resonance imaging, *Rev. Sci. Instrum.*, 53(9), 1338.

Fish, P. J., 1975, Doppler vessel imaging for flow measurement, In Proceedings of the Second European Congress on Ultrasonics in Medicine, Munich, 153–159.

Fish, P. J., 1978, Doppler vessel imaging and its aid to flow measurement, In *Doppler Ultrasound* in the Study of the Central and Peripheral Circulation, J. P. Woodcock and R. F. Sequeria (Eds) , University of Bristol, UK, p. 50.

Flam, T., M. Beurlion, *et al*, 1994, Electroconductive lithotripsy: Principles, experimental data and first clinical results of the SONOLITH 4000, *Journal of Endourology*, 8, 4, 249.

Flax, S. W., J. G. Webster, and S. J. Updike, 1973, Pitfalls using Doppler ultrasound to transducer blood flow, *IEEE Trans. Biomed. Eng.*, BME-20, 306.

Follett, D. H. and P. Ackinson, 1976, Ultrasonic pulse-echo system design, transmitter-receiver matching', *Med. and Biol. Eng.*, 14, 362.

Foster, T.A., (1974), An easily calibrated, versatile platinum resistance thermometer, *Hewlett Packard J.*, April 1974, 13.

Fostik, M., T. Conway, R. Dwinell and J. Singer, 1980, Low power electrocardiographic data acquisition module for microprocessor system, *Med. and Biol. Eng. and Comput.*, 18, 95.

Fox, I. J. and E. H. Wood, 1957, Application of dilution curves recorded from the right side of the heart or various circulation with the aid of a new indicator dye, Proc. Staff Meeting, Mayo Clinic, 32, 541.

Franetzki, M., 1984, Drug delivery by program or sensor controlled infusion devices. *Electromedica*, No. 2, p. 75.

Franetizki, M. *et al*, 1981, State of development of program-controlled insulin-dosing devices, *Electromedica*, No. 1, 41.

Frank, R., 1993, Micromachined and integrated silicon sensors for medical instrumentation, *Medical Electronics*, Sept. 1993, 78.

Friesen *et al,* 1990, A comparison of the noise sensitivity of nine QRs detection algorithms, *IEEE Trans Biomed. Eng.,* 37, 1, 85.

Fujimasa, I., Y. Sakurai, and K. Atsumi, 1973, Digitalizing approaches to thermogram analysis, In *Medical Thermography,* K. Atsumi (Ed.), University of Tokyo Press, p.62.

Furman, S., J. W. Escher, B. Parker and N. Solomon, 1969, Clinical analysis of pacemaker function, *Am. J. Cardiol.,* 23, 112.

Fusfeld, R. D., 1978, Instrument for quantitative analysis of the electromyogram, *Med. and Blol. Eng. and Comput.,* 16, 290.

Gambino, S. R., 1967, Blood, pH, pCO_2, Oxygen saturation and pO_2, *ASCP Commission on Continuing Education.*

Gandikota, K.C., 2000, Biotelemetry : The wireless diagnosis, *Electronics for You,* March 2000, p. 35.

Ganz, W., and H.J.C. Swan, 1972, Measurement of blood flow by thermodilution, *Am. J. Cardiol.,* 29, 241.

Gaskill, D.M., 1991, Recorder resolution and print head density, *Medical Electronics,* 1991, Page 80.

Geddes, L. A., 1972, Electrodes and the Measurement of Bloelectric Events, *Wiley-Inter-science,* John Wiley and Sons, New York, pp. 10–32.

Geddes, L. A., R. Steinberg and G. Wise, (1973), Dry electrodes and holder for electro-oculography, *Med. and Biol. Eng.,* 11, 69.

Geddes, L. A. and L. E. Baker, 1968, *Principles of Applied Biomedical Instrumentation,* John Wiley and Sons, New York, 411.

Geddes, L. A. and L. E. Baker, (1975), *Principles of Applied Biomedical Instrumentation,* John Wiley and Sons, New York.

Geddes, L. A., L. E. Banker and A. G. Moore, (1969), Optimum electrolytic chloriding of silver electrodes, *Med. and Biol. Eng.,* 7, 49.

Geddes, L. A., P. Cabler, A. G. Moore, J. Rosborough and W. A. Tacker, (1973), Threshold 60 Hz current required for ventricular fibrillation in subjects of various body weights, *IEEE Trans. Biomed. Eng.,* BME-20, 465.

Gentner, D. and J. Winkler, 1973, Instrumentation requirements for indirect (external) cardiotocography, *Measuring for Medicine,* 8, 1, 6.

Getzel, W. A. and J. G. Webster, 1976, Minimizing silver-silver chloride electrode impedance, *IEEE Trans. Biomed. Eng.,* BME-23, 87.

Gilbert, B. K., S. K. Kenue, R. A. Robb, A. Chu, A. H. Lent and E. E. Swartzlander, (1981), Rapid evaluation of fan beam image reconstruction algorithms using efficient computational techniques and special-purpose processors, *IEEE Trans. Blomed. Eng.,* BME-28, 98.

Golay, M. J. E., 1969, 'Hexagonal parallel pattern transformations, *IEEE Trans. Comput.,* C-20, 551.

Golden, D. P., R. A. Wolthuis and G. W. Hoffier, 1973, A spectral analysis of the normal resting electrocardiogram', *IEEE Trang. Biomed. Eng.,* BME-20, 366.

Golden, D. P., R. A. Wolthuis, G. W. Hoffier and R.B. Gowen, 1974, Development of a Korotkov sound processor for automatic identification of ausculatory events-Part I, Specification of preprocessing bandpass filters, *IEEE Trans. Biomed. Eng.,* BME-21, 114.

Goodale, R. L., A. Okada, R. Gonzales *et al,* 1970, Rapid endoscopic control of bleeding gastric erosions by laser radiation, *Arch. Surg.,* 101, 211.

Goodman, A. H., 1969, A transistorised square wave electromagnetic flowmeter-1: The amplifier system, *Med. and Biol. Eng.,* 7, 115.

Greatbatch, W. and T. S. Bustard, 1973, A $Pu^{238}O_2$ nuclear power source for implantable cardiac pacemakers, *IEEE Trans. Biomed. Eng.,* BME-20, 332.

Griffiths, C. A. and D. W. Hill, 1969, Some applications of microelectronics to patients, *World Med. Instrum.*, 7, 8.

Grimnes, S., 1983, Impedance measurement of individual skin surface electrodes, *Med. and Biol. Eng. and Comput.*, 21, 750.

Grobstein, S. R. and R. D. Gatzke, 1977, A battery powered ECG monitor for emergency and operating room environments, *Hewlett Packard J.*, 29, 1,26.

Grolin, R. and S. G. Grolin, 1957, Hydraulic formula for calculation of the area of the stenotic mitral valve, other cardiac valves and central circulatory shunts, *Am. Heart J.*, 41, 1.

Grubbs, D. S. and D. S. Worley, 1983, New technique for reducing the impedance of silver-silver chloride electrodes, *Med. and Bio. Eng. and Comput.*, 21, 232.

Guerci, A. and S.H. Kornhousee, 1994, Electronic beam computed tomography, *Medical Electronics*, October, 1994, p. 94.

Haag, R. and A. Cuschieri, 1993, Recent advances in high frequency electrosurgery: development of automated systems, *J.R. Coll. Surg. Edin.*, 38, Dec. 1993, 354.

Haerten, R. *et al*, 1999, Ensemble tissue harmonic imaging; *The technology and clinical utility electromedica*, 67,1, 50.

Haerten, R., 1994, The role of sonography in diagnostic imaging, *Electromedica*, 62, 2, 42.

Hahn, C. E. W., 1969, The measurement of microcathode currents by means of a field effect transistor operational amplifier system with digital display, *J. of Sc. Instruments (Physics E.)*, 2, 48.

Haller, J.D. and M.H. Wholey, 1992, The current status of laser angioplasty: Coronary and peripheral results, *Proc. of the IEEE*, 80, 6, 861.

Hallett, M. and G.H. Leonardo, 1990, Magnetic TENS, *Medical Electronics*, p. 117, Feb. 1990.

Halliwell, M., 1987, Ultrasonic imaging in medical diagnosis, *IEE Proc.* 134 (Part A), 2, 179.

Hamilton, P.P., A. Onayemi and J.A. Smyth, (1983), Comparison of conventional and high frequency ventilation oxygenation and lung pathology, *J. Appl. Physiol*, 55,131.

Handelsman, H., 1990, Real-time cardiac monitor, *Med. Electronics*, Sept. 1990, p. 95.

Hanna, K. L., 1980, Firmware for a patient monitoring system, *Hewlett Packard J.*, 31, 11, 23.

Hardy, J. D., 1939, The radiating power of human skin in the infrared, *Am. J. Physiol.*, 127, 454.

Hartitzsch, B. and N.A. Hoenich, *et al*, 1973, A clinical evaluation of the dialyzers, *Kidney International*, 3, 35.

Heal, J. W., 1974, A Physiological radiotelemetry system using mark/space ratio modulation of a square wave subcarrier, *Med. and Biol. Eng.*, 12, 843.

Health Devices, 1984, Infusion Pumps, 13, 31.

Health Devices, 1985, *Infusion Controllers*, 14, 219.

Hector, M. L., 1968, Technique De I' Evergistrement, *Electroencephalographique*, Masson et Cie, Paris.

Henderson, L. W., 1976, Hemodialysis: Rationale and Physical Principles, In *The Kidney*, B. M. Brenner and F. C. Rector (Eds), W. B. Saunders, Philadelphia, 1643.

Hertz, C. H. and B. Siesjo, 1959, A rapid and sensitive electrode for continuous measurement of pCO_2 in liquids and tissues, *Acta Physiol. Scand.*, 47, 115.

Hewlett Packard, 1997, *ST Segment Monitoring, Application Note*.

Hewlett Packard, 1999a, *ST/AR Arrhythmia Algorithm, Application Note*.

Hewlett Packard, 1999b, Assessing *ST/AR Arrhythmia Performance, Application Note*.

Hill, D.R., 1979, *Principles of Diagnostic X-ray Apparatus*, Philips Technical Library, The MacMillan, London.

Hill, D.W. and A. M. Dolan, 1976, *Intensive Care Instrumentation*, Grune and Stratton, Inc., New York.

Hill, D. W. and R. S. *Khandpur*, 1969, The performance of transistor ECG amplifiers, *World Med. Electron. Instrum.*, 7, 12.

Hill, D. W. and F. D. Thompson, 1975, The importance of blood resistivity in the measurement of cardiac output by the thoracic impedance method, *Med. and Biol. Eng.*, 13,18,7.

Hill, D. W. and Tilslcy, 1973, A comparative study of the performance of five commercial blood gas and pH analysers, *Br. J. Anesth.*, 45, 467.

Hinshaw, W. S., 1976, Image formation by nuclear magnetic resonance: The sensitive point method, *J. of App. Physics*, 47, 3709.

Hinshaw, W. S. and A. H. Lent, 1983, An introduction to NMR imaging: From the Bloch equation to the imaging equation, *Proc. IEEE*, 71, 338.

Hirose, Y. *et al*, 1982, A hybrid emission CT, HEADTOME II, *IEEE Trans. Nucl. Sc.*, Vol. NS-29, 523.

Hobbes, A. F. T., 1967, A comparison of methods of calibrating the pneumotachograph, *Br. J. Anaesth.*, 39, 899.

Hoffman, E.J. *et al*, 1985, ECTA III - new PET system for heart and whole body dynamic imaging, *J. Nuc. Med.*, Vol. 26, p. 28.

Holm, H. H., J. K., Kristensen, J. F. Padersen, S. Hancks and A. Hortheved, 1975, A new mechanical real time ultrasonic contact scanner, *Ultrasound in Med. and Biol.*, 2, 19.

Homberg, R. and R. Koppel, 1994, An X-ray tube assembly with rotating anode spiral groove bearing of the 2nd generation, *Electromedica*, 66, 2, 65.

Hon, E. H. and S. T. Lee, 1963, Noise reduction in foetal electrocardiography, II: Averaging Technique, *Am.Obstet. Gyhe.*, 87, 1086.

Hsue, R. and M. Graham, 1976, Microprocessor monitor for EKG and blood pressure, *Wescon. Tech. Papers*, No 22/3.

Hubel, D. H., 1957, Tungsten microelectrode for recording from single units, *Science*, 125, 549.

Hunt, J. W., M. Arditi and F. S. Foster, 1983, Ultrasound transducers for pulse-echo medical imaging, *IEEE Trans. Biomed. Eng.*, BME-30, 453.

Huntsman, L. *et al*, 1983, Non-invasive Doppler determinations of cardiac output in man, *Circulation*, 67, 593.

Hussey, M., 1975, *Diagnostic Ultrasound*, Blackie, London.

(IEC) International Electrotechnical Commission, 1978, 'Technical Committee No.62, Sub-committee 62D: Electro-medical equipment. Draft-*High Frequency Surgical Equipment*. Particular requirements for safety and performance (IEC, Central Office, 5).

Isley, M.R. *et al*, 1998, Electromyography, Electroencephlography, *Med. Elect.*, Oct. 1999, p. 27.

Jacobson, E. D. and K. G. Swan, 1966, Hydraulic occeluder for chronic electromagnetic blood flow determinations, *J. App. Physiol.*, 21, 1400.

Janata, J., 1989, *Principles of Chemical Sensors*, Plenum, New York, 1989.

Jarlov, A. L. and P. M. Holmkjer, 1972, A dye densitometer for measuring cardiac output, *Med. and Biol. Eng.*, 10, 97.

Jaszczak, R. J., 1988, Tomographic radiopharmaceutical imaging, *Proc. IEEE*, 76, 9, 1079.

Jaszczak, R.J., L.T. Chang, *et al*, 1979, Whole body single-photon emission computed tomography using dual, large field-of-view scintillation cameras, *Phys. Med. Biol.* Vol. 24, p. 1123.

Jenkins, J. M. and S.A. Caswell, 1996, Detection algorithms in implantable cardiovascular defibrillators, *Proc. IEEE*, 84, 3, 428.

Jenkner, F. L., 1967, A new electrode material for multi-purpose biomedical application, *Electroenceph. Clin. Neurophysiol.*, 23, 370.

Jobling, D. T., J. G. Smith, and H. V. Wheal, 1981, Active microelectrode array to record from the mammalian central nervous system in vitro, *Med. and Biol. Eng. and Comput.*, 19, 553.

Johnson, S. W., P. A. Lynn, J. S.G. Miller and G. A. L. Reed, 1977, Miniature skin-mounted preamplifier for measurement of surface electromyographic potentials, *Med. and Biol. Eng. and Comput.*, 15, 710.

Johnston, K. W., B. C. Maruzzo and R. S. C. Cobbold, 1978, Doppler method for quantitative measurement and localisation of peripheral arterial occlusive disease by analysis of the blood flow velocity waveform, *Ultrasound in Med. and Biol.*, 4, 209.

Jones, J. L. and O.H. Tovar, 1996, The mechanism of defibrillation and cardioversion, *Proc. IEEE,* 84,3,392.

Jordan, J. A., 1977, The CO_2 laser in gynaecology, Presented at the *British Society for Colposcopy and Cervical Pathology and the British Society of Clinical Cytology*, Sept.1977, London.

Jordanoglou, Jr and N. B. Pride, 1968, Factors determining maximum inspiratory flow and maximum expiratory flow of the lung, *Thorax,* 23, 33.

Kahn, A., 1965, Motion artefacts and steaming potentials in reation to biochemical electrodes, Digest of the 6th Internal, *Conf on Med. and Biol. Eng.*, Tokyo, 562.

Kalender, W.A., 1993, Quo Vadis CT ? CT in the year 2000, *Electromedica*, 61,2/93, 30.

Kassal, J., W. Reeves and R.L. Donnerstein, 1994, Polymer based adherent differential output sensor for cardiac auscultation, *Med. Elect.*, Sept. 1994, p. 54.

Katinis, L. M., 1982, Nuclear magnetic resonance imaging: Methods and current status', *Med. Inst.*, 16, 213.

Ken-Itch, Ito, Ito. Masayasu, Yuta. Shin-ichi, Y. Hiromu, S., Yoshihiro, S., Hirafuku, H. Yoshihiro and K. Ueda, 1977, A real-time ultrasonic diagnostic system for dynamic and still images, Wireless Echovision, *Japan Electron. Eng.*, Dec.1977, p.20.

Keyes, W.I., 1987, Radionuclude imaging, *IEE Proc.* 134,(Part. A), 2 , 161.

Kimura, W. D., C. Gulacsik, D. C., Auth, F. B. Silverstein and R. L. Protell, 1978, Use of gas jet appositional pressurization in endoscopic laser photocoagulation, *IEEE, Trans. Biomed. Eng.*, BM~25, 218.

Kim, Y. and P.S. Schimpf, 1996, Electrical behaviour of defibrillation and pacing electrodes, *Proc. IEEE,* 84,3, 446.

Kistler, J. and A. Miller, 1982, The AAMI standard test load for electrical risk current measurements, *Med. Inst.*, 16, 224.

Klingler, D. R., H. E. Booth and A. A. Schoenberg, 1979a, Effects of dc bias currents on ECG electrodes, *Med. Instrum.*, 13, 257.

Klingler, D. R., A. A. Schoenberg, N. P. Worth, C. F. Egleston and J. A. Burkart, 1979b, A comparison of gel-to-gel and skin measurements of electrode impedance', *Med. Instrum.*, 13, 266.

Kochevar, R.E., 1992, Biological effects of excimer laser radiation, *Proc. IEEE,* 80, 6, 833.

Kolff, W. J., S. Jacobsen, R. L. Stephen and D. Rose, 1976, Towards a wearable artificial kidney, *Kidney International*, 10, 300.

Krelner, T., 1977, Heat switches the PTC thermistor, *Elect. Design.*, 25, 232.

Kuhl, D.E., 1976 The Mark IV system for radionuclude computed tomography of the brain, *Radiology*, 121, 405.

Kuiper, J., J. Bosman and J. Boter, 1966, Improvements in measuring physical load by wireless transmission of the ECG, *World Med. Elect. Instrum.*, 4,304.

Kulkarni, K. 1991, Apnoea Monitors : Past, Present and Future, *Med. Electronics*, June 1991, p. 1125.

Kumar, A. E., I. Welti and R. R. Ernst, 1975, *J. Magn. Reson.*, 18, 69.

Labuda, E.F., E.I Gordon and R.C. Miller, 1965, Continuous-duty argon ion lasers, *IEEE J. Quantum Flection.*, 1, 273.

Laβmann, M., P. Schneider and Chr. Reiners, 1998, Modern nuclear medical diagnostics with efficient gamma cameras, *Electromedica*, 66, 2, 43.

Lai, N. C., C. C. Lie, E.G. Brown, M. R. Neumann and W. H. Ko., 1975, 'development of a miniature pCO_2 electrode for biomedical applications', *Med. and Biol. Eng.*, 13, 876.

Larsen, J. L., R. F. Dilman, A. M. Nardizzi and R. N. Tverdoch, 1972, An effective ECG telemetry system, *Hewlett Packard J.*, April 1972, 2.

Laursen, H. N., M. H. Hochberg and E. D. M. George, 1976, Evaluation of the accuracy of a new ultrasonic foetal heart rate monitor, *Am. J. of Obstet. and Gyne.*, 125, 1125.

Lauterbur, P. C., 1973, Image formation by induced local interactions: Examples employing NMR, *Nature*, 242, 190.

Lawson, R. N., 1957, Thermography—A new tool in the investigations of breast lesions, *Canad. M. Ser. Med. J.*, 13, 517.

Lee, W. R., 1966, *Proceedings of IEE.*, 113, 144.

Lee, A. L., A. J. Tahmoust and R. Jennings, 1975, An LED transistor photoplethysmograph, *IEEE Trans. Biomed. Eng.*, BME-22, 248.

Lele, P. P., 1979, Safety and potential hazards in the current applications of ultrasound in obstetrics and gynecology, *Ultrasound Med. Biol*, 5, 307.

Leonard, M. A., 1980, Electrospinal Instrumentation, *Hospital Equipment and Supplies J.*, Sept.1980, 39.

L'Esperarance, F. A. 1968, An ophthalmic argon laser photocoagulation system: Design, construction and laboratory investigations, *J. Am. Ophth. Soc.*, 66, 827.

Levitan, E. and G.T. Herman, 1987, A maximum a posteriori probability expectation maximizing algorithm for image reconstruction in emission tomography, *IEEE Trans. Med. Imag.* Vol. M1–6, p. 185.

Levkov, C., G. Michov, *et al*, 1984, Subtraction of 50 Hz interference from the electrocardiogram, *Med. Bio. Eng. and Comp.*, 22, 371.

Lewes, D., 1966, Multipoint electrocardiography without skin preparation, *World Med. Electron. Instrum.*, 4, 240.

Lewis, D. and D. W. Hill, 1967, Application of Multipoint Electrodes to telemetry in Patient Monitoring and during physical exercise, *British Heart J.*, 29, 289.

Lim, C.B. *et al*, 1985, Triangular SPECT System for 3-D total organ volume imaging: Design concept and preliminary imaging results, *IEEE Trans. Nucl. Sci.* Vol. NS-32, Ch.1, p. 741.

Lin, S.C., 1999, Applying telecommunication technology to health care delivery, *IEEE Eng., in Med. and Biol*, 18,4, 28.

Lippold, O. C. J., 1952, The relationship between integrated action potentials in a human muscle and its isometric tension, *J. Physiol.*, 117, 492.

Loizou, P.C., 1999, Introduction to cochlear implant, *IEEE Eng. in Med and Biol.*, 18,1,32.

Lopez, A. and P. Richardson, 1969, Capacitive electrocardiographic and bioclectric electrodes, *IEEE Trans. Biomed. Eng.*, BME-16, 99.

Lown, B., R. Amarasingham, and J. Neuman, 1962, A new method for terminating cardiac arrhythmias; use of a synchronised capacitor discharge', *J. Am. Med. Ass.*, 182, 548.

Luca, C. J. de, R. S. le Fever and F. B. Stulen, 1979, Pasteless electrode for clinical use, *Med. and Biol. Eng. and Comput.*, 17, 387.

Macpherson, P. C., S. J. Meldrum, and P. D. S. Tunstall, 1980, A real-time spectrum analyser for ultrasonic Doppler signals using a Chirp-Z-Transform technique, *J. Med. Eng. and Tech.*, 4, 25.

Maginness, M. G., 1979, Methods and terminology for diagnostic ultrasound imaging systems, *Proceedings IEEE*, 67, 641.

Makino, H. *et al*, 1988, Implantable defibrillator with high output pacing function after defibrillation, *Proc. IEEE*, 76, 9, 1187.

Mansfield, P. and P. G. Morris, 1982, *NMR Imaging in Biomedicine*, Academic Press, New York.

Mansouri, S. and J.S., Schultz, 1984. A miniature optical glucose sensor based on affinity binding, *Biotechnology*, p. 885.

Marcus, B.L., 1992, Photodynamic therapy of human cancer, *Proc. IEEE*, 80, 6, 869.

Mark, J.R.S. *et al*, 1986, Continuous non-invasive monitoring of cardiac output with esophageal Doppler ultrasound during cardiac surgery, *Anaesthesiol.* 65, 1013.

Martin, M.J. *et al*, 1987, Fibre optics and optical sensors in Medicine, *Med. and Biol. Eng and Comp.*, Nov. 1987, p. 597.

Marx, J. L., 1980, NMR opens a new window into the body, *Science*, 210, 302.

Matsumoto, H., M. Saegusa, K. Saito and K. Mizoi, 1978, The development of a fiber optic catheter tip pressure transducer, *J. Med. Eng. and Tech.*, 2, 239.

Matthes, K. and F. Gross, 1939b, Untersuchungen, uber die Absorption von rotem and ultrarotem Licht durch kohlenoxydgesattiges sauerstoffgesattgtes und reduziertes, *Blut. Arch.f. Exper. Path. u. Pharmakol*, 191, 369.

McCann, R. and J. S. Robinson, 1963, 'Notes on the oxygen electrode, *Br. J. Anaesth.*, 35, 679.

McCann, H.A. *et al*, 1988, Multidimensional ultrasonic imaging for cardiology, *Proc. IEEE*, 76, 9, 1063.

McGann L. E., A. R. Turner and J. M. Turc, 1982, Microcomputer interface for rapid measurements of average volume using an electronic particle counter, *Med. and Biol. Eng. and Comput.*, 20, 119.

McRobbie, D.W. 1990, Rapid recovery physiological pre amplifier without AC coupling capacitors., *Med. and Bio. Eng. and Comput.*, March 1990, p. 198.

Mcshane, J. L., 1974, Ultrasonic flowmeters, In *Flow*, Vol.1, R. B. Dowdell (Ed.), Pittsburgh, P.A. USA, p. 897.

Meindl, J. D., 1976, *Acoustic Imaging*, Glen Wade (Ed.), Plenum Publishing Corpn., New York, p.175.

Mekjavic, I. B., J. B. Marrison and G. L. Brengelmann, 1984, Construction and position verification of a thermocouple esophageal temperature probe, *IEEE Trans. Biomed. Eng.*, BME-31, 486.

Mendelson, Y, *et al*, 1988, Design and evaluation of a new reflectance pulse oximeter sensor, *Medical Instrum.*, 22(4), p. 167.

Mercier, A. C., 1973, How to select a galvanometer, *Measur. Data. J.*, July-Aug., 1989.

Merrick, E. B. and T. J. Hayes, 1976, Continuous, non-invasive measurements of arterial blood oxygen levels, *Hewlett-Packard J.*, Oct. 1976, p. 2.

Miller, A., 1969, Electrode contact impedance, its measurement and its effect on the electrocardiogram, *Report No. IM-ECI*, Sanborn Co. USA.

Miller, M. N., 1976, Design and clinical results of Hematrack, an automated differential counter, *IEEE Trans. Biomed. Eng.*, BME-23, 400.

Miller, H. D. and L. E. Baker, 1973, A stable ultraminiature catheter tip pressure transducer, *Med. and Biol.Eng.*, 11, 86.

Miller, S.L., 1993, Intra-arterial blood gas monitoring : Applying optical sensing technology, *Medical Electronics*, April 1993, p. 82.

Metting Van Rijn, A.C. *et al*, 1990, High quality recording of bioelectric events, *Med. Biol. Eng. and Comput.*, Sept. 1990, p. 389.

Miyamoto. Y., K. Sakakibara, T. Tamura, T. Takahashi, T. Hiura and T. Mikami, (1981), On-line computer for assessing respiratory and metabolic function during exercise, *Med. Biol. Eng. and Comput.*, 19, 340.

Montecalvo, D.A. and D. Rolf, 1990, ECG Electrodes, *Medical Electronics Products*, Oct. 1990, p. 24.

Moore, S.C. *et al*, 1984, Improved performance from modifications to the multidetector SPECT brain scanner, *J. Nucl. Med.*, Vol. 25, p. 688

Moores, B.M., 1987, Digital X-ray imaging, *IEE Proc.* 134 (Part A), 2, 115.

Myers, G. H. and V. Parsonnet, 1969, *Engineering in the Heart and Blood Vessels*, John Wiley and Sons, New York.

Narayana Swamy, R. and F. Sevilla, 1988, Optical fibre sensors for chemical species, *J. Phys. E. Sc. Instruments*, Vol. 21, p. 10.

Neame, R. L. B., D. A. Plewis and F. J. Imms, 1977, Construction of thermistor probes suitable for the estimation of cardiac output by the thermodilution method in small animals, *Med. and Biol. Eng. and Comput.*, 15, 43.

NFPA (National Fire Protection Association), 1975, The safe use of high frequency equipment in hospitals, *NFPA 76C*, Boston.

Nichols, W. W. and W. E. Walker, 1974, Experience with the Miller PC-350 catheter tip pressure transducer, *Biomed. Eng.*, 9, 58.

Nissen, S.E. and J.C. Gurley, 1991, Application of intravascular ultrasound for detection and quantitation of coronary atherosclerosis, *Intravascular ultrasound*, Kluwer Academic Publishers, p. 165.

Nitz, W., (1996a), Magnetic resonance imaging sequences and their clinical application, (Part -1), *Electromedica*, 64, 1, 23.

Nitz, W., (1996b), Magnetic resonance imaging sequences and their clinical application, (Part -2), *Electromedica*, 64, 2, 48.

Noshiro, M. and S. Suzuki, 1978, Synchronisation of respiratory rhythm with electrical stimulation of the phrenic nerve, *IEEE Trans. Biomed. Eng.*, BME-25, 550.

Odman, S. and P. Ake Oberg, 1982, Movement induced potentials in surface electrodes', *Med. Bio. Eng. and Comp.*, 20, 159.

Oppelt, A., 1984, New applications and improved image quality: Trends in magnetic resonance tomography, *Electromedica*, 52, 57.

Parker, D., D. Delpy and M. Lewis, 1978, Catheter tip electrode for cnntinuous measurement of pO_2 and pCO_2, *Med. and Biol. Eng. and Comput.*, 16, 599.

Parsons, R., 1964, Electrode double layer, *The Encyclopaedia of Electrochemistry*, C. A. Hampel (Ed.), Reinhold Publishing Company, New York, 1206.

Pastakia, B., 1978, Biological effects of electromagnetic fields, *N. Engl. J. Med.*, 298, 1366.

Patten, C. W., F. B. Ramme and J. Roman, 1966, Dry electrodes for physiological monitoring, *NASA Tech. Note*. NASA TN D-3414, National Aeronautics and Space Administration, Washington-DC.

Pauling, L., R. Wood and C. O. Sturdevant, 1946, An instrument for determining the partial pressure of oxygen in a gas, *Science*, 103, 338.

Pfeiler, M., E. Matura, *et al*, 1989, Lithotripsy of renal and biliary calculi : Physics, technology and medical – technical application, *Electromedica*, 57, 2, 52.

Philip, J. H., M. C. Long, M.D. Quinn and R. S. Newbower, 1984, Continuous thermal measurement of cardiac output', *IEEE Trans. Biomed. Eng.*, BME-31, 393.

Pickering, S. G. and F. D. Stott, 1980, Ambulatory blood pressure—A review, Proc. of the Third International Symposium on *Ambulatory Monitoring*, F. D. Stott, E. B. Raftary and L. Goulding, (Eds), Academic Press, London, p.135.

Pickup, J.C. and D. Rothwell, 1984, Technology and the diabetic patient, *Med. Biol. Eng. and Comp* 22, 385.

Plaut, D. I. and J. G. Webster, 1980a, Ultrasonic measurement of respiratory flow, *IEEE Trans. Blomed Eng.*, BME-27, 549.

Plaut, D. I. and J. G. Webster, 1980b, Design and construction of an ultrasonic Pneumotachometer, *IEEE Trans. Biomed. Eng.*, BME-27, 590.

Polanyi, M. L. and R. M. Hehir, 1960, New reflection oximeter, *Rev. Sc. Instrum.*, 31, 401.

Polanyi, T. G. S. Stellar and H. C Bredemeier, 1970, Experimental studies with the carbon dioxide laser as a neurological instrument, *Med. and Biol. Eng.*, 8, 549.

Preston, K., M. J. B. Duff, *et al*, 1979, Basics of cellular logic with some applications in medical imaging processing, *Proc. IEEE*, 67, 827.

Racine, P. and H. Massie, 1971, An experimental internally powered cardiac pacemaker, *Med. Res. Eng.*, 3, 18.

Rader, R.D., J.P. Mechan and J.K.C. Henriksen, 1973, An implantable blood pressure and flow transmitter, *IEEE Trans. Biomed. Eng.*, BME-20, 37.

Ragheb, T. and L.A. Geddes, 1990, Electrical properties of metallic electrodes, *Med. and Biol. Eng. and Comp.* p. 182.

Rastogi, S.P., J. Dewar, T.H. Frost and D.N.S. Kerr, 1969, In vivo comparison of KIIL and Alwall Gambro dialysers, *Proc. Eur. Dial. Transp. Asson.*, 6, 363.

Reichenberger, H., 1988, Lithotriptor Systems, *Proc. IEEE* 76, 9, 1236.

Renner, R. *et al*, 1981, Clinical aspects in development of dispensing devices for continuous insulin infusion, *Electromedica*, No. 3, 159.

Reuss, J.L. 2000, Digital oximetry, *Biomed*, May 2000, p. 75.

Reyes, R.J. and J.R. Neville, 1967, An electrochemical technic for measuring carbon dioxide content of blood, USAF School Aerospace Med. Tech. Rept. SAM-TR-67-23.

Rezazadeh, M. and N.E. Evans, 1988, Remote vital-signs monitor using a dial-up telephone line, *Med. Biol. Eng. and Comp.*, Sept. 1988, p. 557.

Riederer, S.J., 1988, Recent advances in magnetic resonance imaging, *Proc. IEEE*, 76, 9, 1095.

Ring, E.E.J. 1998, Progress in measurement of human body temperature, *IEEE. Eng. In Med. and Biol.*, 17.4, 19.

Rhodine. C.N. and J.W. Steadman, 1976, *Renal dialysis membrane leak detector*, ISA BM 76323, p. 123.

Riemann, H.E. and P. Marholff, 1981, The clinical value of high-resolution X-ray television with a high number of scanning lines, *Electromedica*, 49, 18.

Rittgers, S.K., W.W. Putney and R.W. Barnes, 1980, Real time spectrum analysis and display of directional Doppler ultrasound blood velocity signal, *IEEE Trans. Biomed. Eng.*, BME-27, 723.

Roelandt, J.R.T.C., F.J., Cata, W.B., Vletter and MA. Txams, 1994, Ultrasonic dynamic three-dimensional visualization of the heart with a multiple transesophageal imaging transducer, *J. AM. Soc. Echocardiography*, 7, 217, 219.

Rogers, W.L., N.H. Clinthorne and J. Stamose, 1984, Performance evaluation of SPRINT, a single photon ring tomograph of brain imaging. *J. Nucl. Med.*, 25, 1013.

Roy, O. Z., 1980, Summary of cardiac fibrillation threshold for 60 Hz currents and voltages applied directly to the heart, *Med. and Biol. Eng. and Comput.*, 18, 657.

Ruben, S., 1969, Sealed zinc-mercuric oxide cells for implantable cardiac pacemakers, *Ann. New York Acad. Sd.*, 167, 627.

Sahambi, J. S., S.N. Tandon and R.K.P. Bhatt, 2000, An automated approach to beat-by-beat QT-interval analysis, *IEEE Engg. in Med. and Bio.*, 19, 3, 97.

Sakurai, Y., I. Fujimasa and K. Atsumi, 1973, Principles and requirements of medical thermography, *In Medical Thermography*, Ed. K. Atsumi, p. 11.

Sanders, R.S. and M.T. Lee, 1996, Implantable Pacemakers, *Proc. IEEE* 84,3, 480.

Sandler, H., T. B. Fryer, S. A. Rositano and R. D. Lee, 1973, The application of aerospace technology to patient monitoring, *IEEE Trans. Biomed. Eng.*, BME-20, 189.

Sapoff, M., 1982, thermistors, optimum Linearity Techniques *Med. Electronics*. 13(1), 87.

Sarnoff, S. J., E. Hardenburgh and J. L. Whitter-berger, 1948, Electro-phrenic respiration, *Am. J. Physiol.*, 155, 1.

Scacci, R., J. L. McMohon and W. F. Miller, 1976, Oxygen tension monitoring with cutaneous electrodes in adults, *Med. Instrum.*, 10, 192.

Schaldach, M., 1971, Implantable electrochemical energy sources, Proc. 9th Int. Conf *Med. and Biol. Eng.*, Melbourne, M-8-18.

Scheggi, A. M. and Brenci, M. *et al*, 1984, Optical-fibre thermometer for medical use, *IEE Proc.*, 131, 270.

Schlindwein, F.S., M.J. Smith and D.H. Evans, 1988, Spectral analysis of Doppler signals and computation of the normalized first moment in real time using a digital signal processor, *Med. Biol. Eng. and Comput.*, Vol. 26, p. 228.

Schittenhelm, R., 1986, Imaging systems for digital radiography, Present status and future prospects, *Electromedica*, 59, 2, 115.

Schleberger, R. and Th Senge, 1992, The non-invasive treatment of long bone pseudarthrosis by shock waves, *Arch. Orth. Trauma Surg.*, 11, 4, 224.

Schoenberg, A. A., H. E. Booth, and P. C. Lyon, 1979, Development of standard test methods for evaluating defibrillation recovery charactersitics of disposable ECG electrodes, *Med. Instrum.*, 13, 259.

Schuette, W. H., G. F. Norris and J. L. Doppman, 1976, Real time two-dimensional mechanical ultrasonic sector scanner with electronic control of sector width, *Proc. of the Soc. of Photo-optical Instrum. Engineers*, 96, 345.

Seitz, W.R., 1984, *Chemical sensors based on fiber optics*, 56, 1, 16A.

Severinghaus, J. W. and A. F. Bradley, 1958, Electrodes for blood PO_2 and pCO_2 determination', *J. App. Physiol*, 13, 515.

Severinghaus, J. W., 1962, Electrodes for blood gas pCO_2, pO_2 and blood pH, *Acta Anaesthesiol. Scand.*, 6 (Suppl. XI), 207, 18, 45.

Shackil, A.F. 1981, Microprocessor and the M.D., *IEEE spectrum*, 18, 4, 33.

Shimizu, K. (1999), Telemedicine by mobile communication, *IEEE Engg. in Med. and Biol.*, 18, 4, 33.

Show, D., 1971, *Fourier Transform NMR Spectroscopy*, Elsevier, New York.

Siggaard—Andersen, O., 1963, Blood acid base alignment nomogram, Scales for pH, pCO_2, base excess of whole blood of different hemoglobin concentration, plasma bicarbonate and plasma total CO_2, *Scand. J. Clin. Lab. Invest.*, 15, 211.

Silverstein, F. E. *et al*, 1976, High power argon laser treatment via standard endoscopes, *Gastroenterology*, 71, 558.

Slye, D.A., 1995, Customized monitoring systems, *Med. Electronics*, Feb.1995, p. 68.

Silvola, J., 1989, New non-invasive piezoelectric transducer for recording of respiration, heart rate and body movements, *Med. Biol. Eng. and Comput.*, July 1989, p. 423.

Skrzypek, J. and P. Keller, 1975, Manufacture of metal microelectrodes with the scanning electron microscope, *IEEE Trans. Biomed. Eng.*, BME-22, 435.

Smith, D.M., R.H. Propsi, and R.R. Mercer, 1979, An FM electronic system for biomedical data recording, *IEEE Trans. Biomed. Engg.*, BME–26, 170.

Soderquist, D. and J. Simmons, 1979, Temperature measurement method based on matched transistor pair requires no reference, *Precision Monolithics Inc. Catalogue,* pp. 16–24.

Soller, B.R. 1994, Design of intravascular fiber optic blood gas sensors., *IEEE Engg. in Med. and Biol.,* 13, 3, 327.

Spelman, F.A., 1999, The past, present and future of cochlear prostheses, *IEEE Eng. in Med. and Biol.,* 18, 3, 28.

Spooner, R. B., 1977, EKG amplifiers, *Hospital Instrumentation, Care and Servicing,* Instrument Society of America, p.11.

Staewen, W. S., 1982, ECG electrode dc offset potentials, *Med. Inst.,* 16, 179.

Stafi, A.D., E.J. Wilkinson and R.J. Mattingly, 1977, Laser treatment of carvical and vaginal neoplasia, *Obst. and Gynae,* 128.

Stark, A. M. and S. Way, 1974, The use of thermovision in the detection of early breast cancer, *Cancer,* 33, 1664.

Stegenga, J. V., 1980, Ultrasonic feedback recording potentiometer, *Measur. Data.,* June 1980, 131.

Stevens, W. G.S., 1963, The current voltage relationship in human skin, *Med. Elec. and Biol. Eng.,* 1, 389.

Stockret, J. and B. R. Nave, 1974, Operational amplifier circuit for linearising temperature readings from thermistors', *IEEE Trans. Biomed. Eng.,* BME-21, 164.

Stokes, K., 1996, Cardiac Pacing Electrodes, *Proceedings of the IEEE,* 84, 3, 457.

Stokes, K.B. *et al,* 1983, A steroid eluting, low-threshold, low polarizing electrode in Cardiac Pacing, K. Steinback. Ed. Darmstadt : Stein Kopff Verlag, p. 369.

Stokes, K., K. Cobian and T. Lathrop, 1979, Polyurethane Insulators: A design approach to small pacing leads, *Symp. Cardiac Pacing,* Oct. 1979.

Stow, R. W., R. F. Baer and B. F. Randall, 1957, Rapid measurement of the tension of carbon dioxide in blood, *Arch. Phys. Med. Rehabil.,* 38, 646.

Strong, M.S., G.J. Jako, T. Polanyi and R.A., Wallace, 1973, Laser surgery in the aerodigestive tract, *Am. J. Surgery,* 126, 529.

Strong, M.S., G.J. Jako, *et al,* 1975, The use of CO_2 laser in Otolaryngology, A Progress Report. Presented at the Eighteenth Annual Meeting of the Am. Acad. Ophtha and Otolaryngology. Dallas, Sept. 21–25, 1975.

Strong, Peter., 1973, *Biophysical Measurements,* Tektronix Instruments Inc., Oregon, USA.

Strotzer, M. *et al,* 1998, Experimental examinations with initial clinical experience with a flat panel detector in radiography, *Electromedia,* 66, 2, 52.

Swan, H. B. and W. Ganz, *et al,* 1970, Catheterisation of the heart in man with use of a flow directed, balloon tipped catheter, *New. Engl. J. Med.,* 283, 447.

Tacker, W.A. and L.A. Geddes, 1996, The laws of electrical stimulation of cardiac tissue, *Proc. IEEE,* 84, 3, 355.

Takatani, S. and J. Ling, 1994, Optical oximetry sensors for whole blood and tissue, *IEEE Eng. in Med. and Biol.,* 13, 3, 347.

Takeuchi, Y. and M. Hogaki, 1977, Autocorrelation method for fetal heart rate measurement from ultrasonic Doppler fetal signal, *Ultrasound in Medicine,* Vol. 3B, D. White and R. E. Brown (Eds.), Plenum Press, New York, p. 1327.

Talonen, P.P. *et al,* 1990, Neurophysiological and technical considerations for the design of an implantable phrenic nerve stimulator, *Med and Bio. Eng. and Comp.,* January, 1990 p. 31.

Tam, H. and B.O. Webster, 1977, Minimising electrode motion artefact by skin abrasion, *IEEE Trans. Biomed Eng.,* BME-24, 134.

Taylor, W. B., 1970, A versatile cell detector for cell volume measurements, *Med. and Biol. Eng.,* 8, 281.

Thakor, N. V., J. G. Webster and W. J. Tompkins, 1983, Optimal QRS detector, *Med. Bio. Eng. and Com.*, 21, 343.

Thakor, N. V., J. G. Webster and W. J. Tompkins, 1984, Design, Implementation and evaluation of microcomputer-based portable arrhythmia monitor, *Med. Bio. Eng. and Comp.*, 22, 151.

Thakor, N. V., J. G. Webster and W. J. Tompkins 1994b, Estimation of QRS complex power spectra for design of a QRS filter, *IEEE Trans. Biomed. Eng.*, BME-31, 702.

Thakor, N. V. and J.G. Webster, 1985, Electrode studies for the long term ambulatory ECG', *Med. Bio. Eng. and Comp.*, 23, 116.

Theobald. O.J. *et al*, 2000, System performance of multislice spiral computed tomography, *IEEE Eng. in Med. Biol.*, October, 2000, p. 63.

Thom, R., 1972, Vergleichende Uolersuchungen Zur Electro Nischen Zellvolumenanalyse, *AEG Telefunken Publ.* NI/EP/1698.

Thomas, A.C., 1988, Implantable defibrillation: Eight years of clinical experiences, *PACE*, Vol. II, 2053–2056, Nov. 1988.

Thompson, K.P., Q.S. Ren and J.M. Parel, 1992, Therapeutic and diagnostic application of lasers in ophthalmology, *Proc. IEEE*, 80, 6, 838.

Tompkins, W. J., 1978, A portable microcomputer based system for biomedical application, *Biomed. Sci. Instrum.*, 14, 61.

Tompkins, W. J., 1980, Modular design of microcomputer-based medical instruments, *Med. Instru.*, 14, 315.

Trimby, R., 1976, Fluid column ECG electrodes, *Hewlett Packard Application Note AN-744*.

Tuck, D. L., 1981, Improved Doppler ultrasonic monitoring of foetal heart rate, *Med. and Biol. Eng. and Comput.*, 19, 135.

Tuck, D. L., 1982, Improvement in Doppler ultrasound human foetal heart rate records by signal correlation, *Med. Biol. Eng. and Comp.*, 20, 357.

Ulrich, W. D., 1971, Ultrasound dosage for experimental use on human beings, *Report No.2 Project M43060.01-l0l0Bxx9*, Naval Medical Research Institute, National Naval Medical Centre, Bethesda, Md.

Valchanov, V., and P. Michailov, 1991, High energy shock waves in the treatment of delayed and non-union fractures, *International Orthopaedics*, 15, 181.

Van Bemmel, J. H., L. Peters and S.J. Hengeveld, 1968, Influence of the maternal ECG on the abdominal fetal ECG complex, *Am. J. Obstet. Gynec.*, 102, 556.

Van Bemmel, J. H., Veth Jelte de Haan and F. L. Ton, 1971, The function of the uterus and the foetoplacental unit, *Report No. 1.8.49-4* of the Medical Physics Institute, Utrecht, Netherlands.

Venables, P. H. and E. Sayer, 1963, On measurement of the level of the skin potential, *Brit. J. Psychol.*, 54, 251.

Vurek, G.G. *et al*, 1983, A fiber optic pCO_2 sensor, *Ann. Biomed. Eng.*, 11, p. 499.

Wagner, J.W. and L.C. Phillips, 1980, 'Reducing variations in power output measurements of electrosurgical devices, *Med. Instrum*, 14, 262.

Walt, D.R., 1992, Fibre optics sensors for continuous clinical monitoring, *Proc. IEEE* 80, 6, 903.

Warburg, E., 1899, Ueber das Verhalten sogenannter unpolarisierbarer elektroden gegen Wechselstrom, *Ann. Physik und Chemie*, 67, 493.

Warren, J.A. *et al*, 1996, Implantable cardioverter defibrillators, *Proc. IEEE* 84, 3, 468.

Watkins, D. and G. A. Holloway, 1978, An instrument to measure cutaneous blood flow using the Doppler shift of laser light, *IEEE Trans. Biomed. Eng.*, BME-25, 28.

Watmough, D. J. and R. Oliver 1968, Emissivity of human skin in vivo between 2.0 microns and 5.4 microns measured at normal incidence, *Nature*, 218, 885.

Watson, A. N., J. S. Wright and J. Longhman, 1973, Electrical thresholds for ventricular fibrillation in man, *Med. J. Anst.*, 1, 1182.

Webb, S., 1987, A review of physical aspects of x-ray transmission computer tomography, *IEE Proc. 134 (Part A)*, 2, 126.

Webstar, J., 1995, *Medical Instrumentation*, John Wiley and Sons. Inc., New York.

Wells, P.N. T., 1977, *Biomedical Ultrasonics*, Acad. Press, London, 446.

Wesseling, K. H., R. Purschke, N. T. Smith, H. B. Wust, Ban de Wit and H. A. P. Weber, 1976, A continuous module for the continuous monitoring of cardiac output in the operating theatre and the ICU, *Acta Anesth. Belg.*, 27, 327.

Whalen, R. E., C. L. Starmer, and H. D. McIntosh, 1964, Electrical hazards associated with cardiac pace-making, *Ann. New York Acad. Sci.*, 3, 922.

Whalen, R. E., and C. F. Starmer, 1967, Electric stock hazards in clinical cardiology, *Modern Concepts of Cardiovascular Disease*, American Heart Association.

White, T., N. Butler and R. Murphy, 1998, An uncooled IR sensor with digital focal plane array, *IEEE Eng. in Med and Biol.*, 17, 4, 60.

Wickham, P. J. D., 1982, Microprocessor-based signal averager for analysis of the foetal ECG', *Med. Bio. Eng. and Comp.*, 20, 253.

Winkle, R.A. *et al*, 1989, Improved low energy defibrillation efficacy in man with the use of biphasic truncated exponential waveform, *Am. Heart. Journal* 117, p. 122.

Winter B. B. and J. G Webster, 1983, Reduction of interference due to common mode voltage in biopotential amplifiers, *IEEE Trans. Biomed. Eng.*, BME-30, 58.

Wohnhas, S., 1991, Air jet ECG eElectrodes, *Medical Electronics*, p. 69.

Wolfson, R. N. and M. R. Neuman, 1969, Miniature $Si-SiO_2$ insulated electrode based on semiconductor technology, Proc. 8th mt. Conf Med. Biol. Eng. 1969, Chicago, Paper No.14-6, Carl Gorr Printing Company.

Wong, D.H. *et al*, 1990, Non-invasive cardiac output: simultaneous comparison of two different methods with thermodilution, *Anaesthesiology*, 72, 784.

Wyatt, D. G., 1984, Blood flow and blood velocity measurement in vivo by electromagnetic induction, *Med. Bio. Eng. and Camp.* 22, 193.

Yang, Wen-Jei and J. H. Wang, 1979, Shortwave and microwave diathermy for deep tissue heating, *Med. and Biol. Eng. and Comput.*, 17, 518.

Zeuthen, T., 1978, Tungsten (W) as electrode material: Electrode potential and small-signal impedance, *Med. and Biol. Eng. and Comput.*, 16, 483.

Zurinski, V. and R. Haerten, 1978, Real time sonography with the linear array scanner, Multiscan 400, *Electromedica*, 46, 141.

Index

CPSIA information can be obtained
at www.ICGtesting.com
Printed in the USA
LVOW04*1914160717
541517LV00007B/10/P